Emergent Quantum Mechanics

Emergent Quantum Mechanics: David Bohm Centennial Perspectives

Special Issue Editors

Jan Walleczek
Gerhard Grössing
Paavo Pylkkänen
Basil Hiley

MDPI • Basel • Beijing • Wuhan • Barcelona • Belgrade

Special Issue Editors
Jan Walleczek
Phenoscience Laboratories
Germany

Gerhard Grössing
Austrian Institute for Nonlinear Studies
Austria

Paavo Pylkkänen
University of Helsinki
Finland

Basil Hiley
University College of London
UK

Editorial Office
MDPI
St. Alban-Anlage 66
4052 Basel, Switzerland

This is a reprint of articles from the Special Issue published online in the open access journal *Entropy* (ISSN 1099-4300) from 2018 to 2019 (available at: https://www.mdpi.com/journal/entropy/special_issues/EmQM17)

For citation purposes, cite each article independently as indicated on the article page online and as indicated below:

LastName, A.A.; LastName, B.B.; LastName, C.C. Article Title. *Journal Name* **Year**, *Article Number*, Page Range.

ISBN 978-3-03897-616-5 (Pbk)
ISBN 978-3-03897-617-2 (PDF)

Cover image courtesy of Jan Walleczek and Kakoii (Berlin).

Contents

About the Special Issue Editors

Jan Walleczek Director of Phenoscience Laboratories, Berlin, Germany, and Director of the Fetzer Franklin Fund of the John E. Fetzer Memorial Trust, USA. Previously, he was Director of the Bioelectromagnetics Laboratory at Stanford University Medical School, USA. Jan Walleczek was a doctoral fellow at the Max-Planck-Institute for Molecular Genetics, Berlin, and post-doctoral fellow at the Research Medicine and Radiation Biophysics Division of the Lawrence Berkeley National Laboratory, University of California at Berkeley. His scientific publications cover the fields of biology, chemistry, engineering, and physics. His recent work concerns the foundations of quantum mechanics and applications to living systems of concepts such as quantum coherence, emergent dynamics, and the flow of information, a long-standing interest that he summarized as an edited volume for Cambridge University Press titled "Self-organized Biological Dynamics and Nonlinear Control". In addition to metascience and advanced methodology, his professional interests include the philosophy and foundations of science.

Gerhard Grössing Director of the Austrian Institute for Nonlinear Studies (AINS) in Vienna, Austria. He studied physics and mathematics at the University of Vienna and at Iowa State University, USA. During his post-doctoral work at Vienna's Atominstitut, he coined the term and developed, together with Anton Zeilinger, the first "Quantum Cellular Automata", and he developed an early variant of an "emergent" quantum theory named "Quantum Cybernetics" whose main results were published as a monograph with Springer Verlag, New York. His major research interests cover the foundations of quantum theory and new tools in complex systems research. He also published numerous articles and two books in the fields of philosophy and foundations of science. In recent years, the research of Gerhard Grössing and the AINS has focused on the development of an Emergent Quantum Mechanics (EmQM), and he organized the first EmQM conference in 2011.

Paavo Pylkkänen is Vice Dean of Faculty of Arts and Head of Department of Philosophy, History, and Art Studies at University of Helsinki, Finland, where he has been Senior Lecturer in Theoretical Philosophy since 2008. He is also Senior Lecturer at University of Skövde, Sweden (currently on leave). He began interacting with David Bohm while doing his Master's at University of Sussex in 1984. In his doctoral work at Helsinki, Pylkkänen explored Bohm's idea of active information at the quantum level and connected it to contemporary philosophy of mind, providing a new view of mental causation. He did his post-doctoral work at Birkbeck, University of London with Basil Hiley, with whom he has authored many papers on the foundations of quantum theory and the mind-matter relation. Pylkkänen has explored the philosophical implications of Bohm and Hiley's program in his many publications, including Mind, Matter, and the Implicate Order (Springer).

Basil Hiley, Professor, is Emeritus Professor of Physics at Birkbeck, University of London and Honorary Professor of Physics at University College London (UCL). He started his collaboration with David Bohm at Birkbeck after completing a PhD from King's College, London in 1961. His main academic interests are in the foundations of quantum theory and relativity. He also has interests in exploring the relationship between mind and matter. He has written many papers on fundamental questions in quantum mechanics and on the relationship of mind and matter. At present he is theoretical advisor to an experimental group at UCL measuring weak values of spin and momentum. With the late Professor Bohm he has co-authored "The Undivided Universe", Routledge, London, 1993. He is also co-editor with David Peat of Quantum Implications, Essays in Honour of David Bohm, Routledge, London, 1987. He was awarded the Majorana Prize EJTP Best Person in Physics 2012.

Editorial

Emergent Quantum Mechanics: David Bohm Centennial Perspectives

Jan Walleczek [1,*], Gerhard Grössing [2], Paavo Pylkkänen [3,4] and Basil Hiley [5]

1 Phenoscience Laboratories, Novalisstrasse 11, Aufgang F, 10115 Berlin, Germany
2 Austrian Institute for Nonlinear Studies, Akademiehof, Friedrichstrasse 10, 1010 Vienna, Austria; ains@chello.at
3 Department of Philosophy, History, and Art Studies, P.O. Box 24 (Unioninkatu 40 A), University of Helsinki, FI-00014 Helsinki, Finland; paavo.pylkkanen@helsinki.fi
4 Department of Cognitive Neuroscience and Philosophy, University of Skövde, P.O. Box 408, SE-54128 Skövde, Sweden
5 Department of Physics and Astronomy, University College London, Gower Street, London WC1E 6BT, UK; b.hiley@bbk.ac.uk
* Correspondence: walleczek@phenoscience.com

Received: 19 January 2019; Accepted: 24 January 2019; Published: 26 January 2019

Abstract: Emergent quantum mechanics (EmQM) explores the possibility of an ontology for quantum mechanics. The resurgence of interest in realist approaches to quantum mechanics challenges the standard textbook view, which represents an operationalist approach. The possibility of an ontological, i.e., realist, quantum mechanics was first introduced with the original de Broglie–Bohm theory, which has also been developed in another context as Bohmian mechanics. This Editorial introduces a Special Issue featuring contributions which were invited as part of the David Bohm Centennial symposium of the EmQM conference series (www.emqm17.org). Questions directing the EmQM research agenda are: Is reality intrinsically random or fundamentally interconnected? Is the universe local or nonlocal? Might a radically new conception of reality include a form of quantum causality or quantum ontology? What is the role of the experimenter agent in ontological quantum mechanics? The Special Issue also includes research examining ontological propositions that are not based on the Bohm-type nonlocality. These include, for example, local, yet time-symmetric, ontologies, such as quantum models based upon retrocausality. This Editorial provides topical overviews of thirty-one contributions which are organized into seven categories to provide orientation.

Keywords: quantum ontology; nonlocality; time-symmetry; retrocausality; quantum causality; conscious agent; emergent quantum mechanics; Bohmian mechanics; de Broglie-Bohm theory

---

"Towards Ontology of Quantum Mechanics and the Conscious Agent" was the heading of the David Bohm Centennial symposium as part of the Emergent Quantum Mechanics (EmQM) conference series (www.emqm17.org). The three-day symposium was held at the University of London, right next to Birkbeck College, the final academic home of David Bohm. The symposium offered an open forum for critically evaluating the prospects and significance—for 21st century physics—of ontological quantum mechanics, an approach which David Bohm helped pioneer. The Editorial introduces contributions featuring the original research of the EmQM symposium speakers, as well as additional researchers who are exploring the ontological implications of quantum mechanics. The contributions are thematically organized as follows: (1) Quantum Ontology and Foundational Principles, (2) The Continuing Impact of the Bohmian Theory, (3) Beyond the Bohmian Theory: New Developments, (4) Quantum Ontology and Time: Retrocausality and Irreversibility, (5) Entropy, Thermodynamics, and Emergent Quantum Gravity, (6) Alternative Quantum Models and Tools, and (7) Advanced Quantum Experimentation.

1. Quantum Ontology and Foundational Principles

Foundational principles and concepts are introduced to help guide thinking about the validity of ontological propositions for quantum mechanics. Tim Maudlin starts with the key insight that any possible ontology for quantum mechanics necessitates "a radical change in our understanding of physical ontology" [1]. Employing as an example the Aharonov-Bohm effect, he develops a method for "clarifying what the commitments of a clearly formulated physical theory are". Referring to the well-known conceptual challenges in interpreting quantum theory, Maudlin concludes by noting that if "physicists were to adopt this method ... to convey physical theories clearly and unambiguously, many conceptual problems could be avoided".

Jan Walleczek advances the concept of agent inaccessibility as a fundamental principle in quantum mechanics, based on the objective uncomputability of quantum processes as a formal limit [2]. In support of the Bohmian theory, the proposal of an agent-inaccessibility principle presents an alternative position to the standard textbook view of quantum indeterminism. Walleczek concludes that the 20th century quantum revolution need not imply "a radical shift from determinism to indeterminism" but that—based on current knowledge—it is only valid to assert that "the quantum revolution signifies the profound discovery of an agent-inaccessible regime of the physical universe".

Maurice De Gosson next introduces the mathematics of Poincare's recurrence theorem, and the associated notion of 'superrecurrence', in relation to the properties of symplectic topology, as applied to quantum mechanics [3]. De Gosson suggests that these recurrence properties "are closely related to Emergent Quantum Mechanics since they belong to the twilight zone between classical (Hamiltonian) mechanics and its quantization", and he views these properties "as *imprints of the quantum world* on classical mechanics in its Hamiltonian formulation".

William Seager provides a 21st century interpretation of the philosophy and scientific metaphysics of David Bohm [4]. Specifically, Seager examines three core features of Bohm's foundational views, namely "the holistic nature of the world, the role of a unique kind of information as the ontological basis of the world, and the integration of mentality into this basis as an essential and irreducible aspect of it". Importantly, Seager corrects the persistent, but flawed, view that Bohmian ontology "is a return to a classical picture of the world", and he explains that "Bohm's metaphysics is about as far from that of the Newtonian classical metaphysical picture of the world as one could get".

2. The Continuing Impact of the Bohmian Theory

The focus of the second category is the continuing impact, based on recent assessments and conceptual innovations, of the original de Broglie–Bohm (dBB) theory and Bohmian mechanics. The opening article by Basil Hiley and Peter Van Reeth engages the historically controversial problem of the reality of Bohmian quantum trajectories [5]. The authors argue that the previous "conclusion that the Bohm trajectories should be called 'surreal' ... is based on a false argument." Specifically, Hiley and Van Reeth show that "standard quantum mechanics produces exactly the same behavior as the Bohmian approach so it cannot be used to conclude the Bohm trajectories are 'surreal'."

Robert Flack and Basil Hiley—again addressing the problem of quantum trajectories—explore "the relationship between Dirac's ideas, Feynman paths, and the Bohm approach" [6]. After studying the relationship in detail, Flack and Hiley propose that "a Bohm 'trajectory' is the average of an ensemble of actual individual stochastic Feynman paths", and that, therefore, these paths "can be interpreted as the mean momentum flow of a set of individual quantum processes and not the path of an individual particle."

Nicolas Gisin, next, clarifies the long-standing debate between those in the mainstream of physics who argue that the Bohmian approach is "disproved by experiments", and those who insist that "Bohmian mechanics makes the same predictions as standard quantum mechanics" [7]. After performing a careful analysis, Gisin arrives at the conclusion that " ... Bohmian mechanics is deeply consistent", and he notes that "Bohmian mechanics ... could inspire brave new ideas that challenge quantum physics."

Dustin Lazarovici, Andrea Oldofredi, and Michael Esfeld continue with key arguments in support of the physical consistency of Bohmian mechanics [8]. In particular, these authors offer a critical assessment of standard no-hidden-variables theorems, which have long been used to challenge the plausibility of the Bohmian ontology. In particular, they argue that "far from challenging—or even refuting—Bohm's quantum theory, the no-hidden-variables theorems, in fact, support the Bohmian ontology for quantum mechanics."

Oliver Passon, next, tackles a common misconception regarding the dBB theory [9], namely the specific criticism that the theory "not only assigns a position to each quantum object but also contains the momenta as 'hidden variables'." In response to this perceived inconsistency, he points out that the measurement of momentum in the dBB theory is strictly contextual and does not reveal a "preexisting value", and that, therefore, the Bohmian interpretation "is not only a consistent interpretation of quantum mechanics but includes also 'quantum weirdness'—like any other interpretation of quantum theory."

Travis Norsen offers an explanation of the Born-rule statistics for the dBB pilot-wave theory [10]. In the task of finding a realist account of the Born rule expressing the probability distribution of measurement outcomes, Norsen compares the two competing approaches from the literature and he finds that "there is somewhat less conflict between the two approaches than existing polemics might suggest, and that indeed elements from both arguments may be combined to provide a unified and fully-compelling explanation, from the postulated dynamical first principles, of the Born rule."

Ángel Sanz highlights the impact of Bohmian theory—beyond mere theoretical significance, namely, as a "useful resource for computational and interpretive purposes in a wide variety of practical problems" [11]. Specifically, an analysis of "the problem of the diffraction of helium atoms from a substrate consisting of a defect with axial symmetry on top of a flat surface" is performed, and the behavior of Fermatian trajectories (optical rays), Newtonian trajectories, and Bohmian trajectories is compared, whereby, the latter are shown to "behave quite differently, due to their implicit non-classicality".

The final article in this category is contributed by Roderich Tumulka [12]. He provides an overview of Bohmian mechanics, and then continues to describe "more recent developments and extensions of Bohmian mechanics, concerning, in particular, relativistic space–time and particle creation and annihilation." Tumulka concludes by emphasizing that the described theoretical work represents "the most plausible ontological theory of quantum mechanics in relativistic space–time", and it, therefore, holds great promise "as a fully satisfactory extension of Bohmian mechanics, to relativistic space–time."

3. Beyond the Bohmian Theory: New Developments

This category features research pursuing ideas beyond the Bohmian theory and its typical interpretation. Although the researchers agree that the notion of nonlocality is essential to an ontological quantum mechanics, new developments are explored, based on the assumptions and propositions that are not normally covered by the dBB theory and by Bohmian mechanics.

Gerhard Grössing, Siegfried Fussy, Johannes Mesa Pascasio, and Herbert Schwabl present a model of quantum reality that "does not need wave functions", and one that assumes a "cosmological solution" to the problem of nonlocality [13]. That is, the researchers propose "that from the beginning of the universe, each point in space has been the location of a scalar field representing a zero-point vacuum energy that nonlocally vibrates at a vast range of different frequencies, across the whole universe." Assuming this cosmological nonlocality, the authors provide classical computer simulations of double- and n-slit interference patterns, which reveal trajectories that "are in full accordance with those obtained from the Bohmian approach."

Mohamed Hatifi, Ralph Willox, Samuel Colin, and Thomas Durt present an analysis of a quantum model inspired by "the properties of bouncing oil droplets"—as observed by the so-called 'walkers' in non-equilibrium experiments—and which "have attracted much attention because they are thought to offer a gateway to a better understanding of quantum behavior" [14]. In particular, the authors perform an analysis comparing "walker phenomenology in terms of the de Broglie–Bohm dynamics and of a stochastic version, thereof." They conclude that "the programs that aim at simulating droplet dynamics with quantum tools or at describing the emergence of quantum dynamics, based on droplet dynamics ... raise challenging fundamental questions."

Mojtaba Ghadimi, Michael Hall, and Howard Wiseman describe research findings related to the Many Interacting Worlds (MIW) proposal, which is "a new approach to quantum mechanics, inspired by Bohmian mechanics" [15]. The MIW proposal represents an entirely novel way of addressing the problem of nonlocality in Bohmian mechanics, and while "it is conceptually clear how the interaction between worlds can enable this strong nonlocality", a proof by simulation has not been possible so far. In the present contribution, the authors now "report significant progress in tackling one of the most basic difficulties that needs to be overcome: Correctly modelling wave functions with nodes."

4. Quantum Ontology and Time: Retrocausality and Irreversibility

Time-related aspects and interpretations of quantum mechanics are the focus of this category. The first three articles present work that considers, in three distinct ways, the possible relationships between the implicit time-symmetry of the quantum formalism and physical ontology. The final article discusses the concept of fundamental irreversibility in nature.

Emily Adlam starts by noticing that "the physics community has come to take seriously, the possibility that the universe might contain physical processes which are spatially nonlocal, but there has been no such revolution with regard to the possibility of temporally nonlocal processes" [16]. The author suggests that "the assumption of temporal locality is actively limiting progress in the field of quantum foundations", and then offers an investigation into "the origins of the assumption, arguing that it has arisen for historical and pragmatic reasons rather than good scientific ones." Adlam concludes with the proposal that "once we accept that the universe may be generically nonlocal, across both time and space, it becomes at least plausible that quantum theory as we know it is simply the local limit of a global theory, which applies constraints across the whole of space and time."

Kenneth Wharton introduces a new class of retrocausal models that he hopes will "guide further research into space–time-based accounts of weak values, entanglement, and other quantum phenomena" [17]. This work is inspired by the recognition that "globally-constrained classical fields provide an unexplored framework for modeling quantum phenomena, including apparent particle-like behavior." In relation to prior retrocausal models in the literature, Wharton explains that "the central novelties in the class of models discussed here are: (1) Using fields (exclusively) rather than particles; and (2) introducing uncertainty to even the initial and final boundary constraints."

Nathan Argaman reconsiders a central tenet of Bell's nonlocality theorem—the causal arrow of time. He points out that "the physical assumptions regarding causality are seldom studied in this context, and often even go unmentioned, in stark contrast with the many different possible locality conditions which have been studied and elaborated upon" [18]. Argaman envisions the future generalization of "retrocausal toy-models to a full theory—a reformulation of quantum mechanics—in which the standard causal arrow of time would be replaced by a more lenient one: An arrow of time applicable only to macroscopically-available information." He concludes by suggesting that for "such a reformulation, one finds that many of the perplexing features of quantum mechanics could arise naturally, especially in the context of stochastic theories."

Lajos Diósi compares two fundamental concepts of irreversibility which, as he emphasizes in this work, have "emerged and evolved with few or even no interactions" [19]. First, the concept of universal gravity-related irreversibility, and, second, irreversibility in "quantum state reductions, unrelated to gravity or relativity but related to measurement devices". The author first summarizes the two concepts and then highlights the significant fact that the precise relationship "between the Planckian and the Schrödinger–Newton unpredictability of our space–time" remains unknown. In conclusion, Diósi notes that "Planckian unpredictability survives non-relativistically—for massive macroscopic quantized degrees of freedom."

5. Entropy, Thermodynamics, and Emergent Quantum Gravity

Theoretical issues are addressed linking entropic and thermodynamical considerations with quantum systems and possible emergent quantum phenomena. The first article by Jen-Tsung Hsiang and Bei-Lok Hu starts out by explaining that–in view of "quantum mechanics as an emergent theory"—thermodynamical theory "is perhaps one of the most powerful theories and best understood examples of emergence in physical sciences, which can be used for understanding the characteristics and mechanisms of emergent processes" [20]. The authors stress that even for the initial goal of developing a viable "quantum thermodynamics", there are "many new issues which need be addressed and new rules formulated." For the present contribution, Hsiang and Hu offer "quantum formulations of equilibrium thermodynamic functions and their relations for Jarzynski's classical thermodynamics at strong coupling".

Osvaldo Civitarese and Manuel Gadella start with a review of "the concept of entropy in connection with the description of quantum unstable systems", whereby the goal of this work is to show "that a comprehensive scheme leading to the definition of entropy for resonances can be rigorously designed by adopting path integration techniques" [21]. Specifically, these authors advance "a proper definition of this entropy based on the use of Gamow states as state vectors for resonances." In conclusion, Civitarese and Gadella explain that the "resulting entropy is complex, with an imaginary part which gives an account for the interactions of decaying states with their surroundings."

Arno Keppens pursues a "complex systems approach, as a kind of toy model, for identifying space–time's ontological micro-constituents and their interaction, i.e., their sub-quantum dynamics" [22]. Towards that end, he combines two research strategies, whereby, the first views "gravity as an entropic phenomenon", and the second derives "a sub-quantum interaction law" from the solution of Einstein's field equations. Keppens argues that "novel views on entropic gravity theory result from this approach, which eventually provides a different view on quantum gravity and its unification with the fundamental forces."

Massimo Tessarotto and Claudio Cremaschini formulate a Bohmian trajectory-based representation for the quantum theory of the gravitational field [23]. Specifically, the researchers describe "the basic principles of a new trajectory-based approach to the manifestly-covariant quantum gravity (CQG) theory." Importantly, their work provides "new physical insight into the nature and behavior of the manifestly-covariant quantum-wave equation and corresponding equivalent set of quantum hydrodynamic equations that are realized by means of CQG-theory." Remarkably, as Tessarotto and Cremaschini emphasize, based on their approach "the existence of an emergent gravity phenomenon is proven to hold."

6. Alternative Quantum Models and Tools

A wide variety of different approaches, including those proposing the construction of alternative ontologies, are grouped together in this category. The five contributions range from quantum models that seek to explain quantum phenomena by local, yet unconventional, accounts of physical reality to a quantum model based on an observer-independent event ontology.

Tim Palmer presents a cosmological model in "which the universe evolves deterministically and causally, and from which space–time and the laws of physics in space–time are emergent" [24]. Significantly, the author counters the view that a Bohm-type nonlocality—in view of Einstein–Podolsky–Rosen (EPR)-type quantum-entanglement correlations—might exist in reality. The model "challenges the conclusion that the Bell Inequality has been shown to have been violated experimentally, even approximately", and it "postulates the primacy of a fractal-like 'invariant set' geometry". Palmer concludes by discussing the relationships between the Invariant Set Theory, which is "deterministic and locally causal", and the Bohmian theory, the cellular automaton interpretation of quantum theory and the p-adic quantum theory.

Thomas Filk continues the challenge for the need of a Bohm-type, nonlocal ontology as an explanation of the EPR-type quantum-entanglement correlations [25]. In particular, as an alternative, he describes "an interpretation of the mathematical formalism of standard quantum mechanics in terms of relations", and from this he develops "the notion of a relational space." In this description, as the author explains, "entanglement is interpreted as a relation between two entities (particles or properties of particles)." Importantly, in the proposed relational view, "the concept of 'locality' receives a completely different meaning when the positions or locations of entities (objects or events) are defined in a relational sense, as compared to an absolute space or space–time." In conclusion, Filk discusses the quantum measurement problem, from the perspective of this relational interpretation.

Dimiter Prodanov's contribution describes "the mathematical foundations of the scale relativity theory, its link to stochastic mechanics, and the theory of the Burgers equation" [26]. This work is motivated "by the premise that inherently nonlinear phenomena need development of novel mathematical tools for their description." In particular, Prodanov investigates "the potential of stochastic methods for simulations of quantum–mechanical and convection–diffusive systems", whereby, the "presented numerical approaches can be used ... for simulations of nanoparticles or quantum dots, which are mesoscopic objects and are expected to have properties intermediate between macroscopic and quantum systems".

Louis Kauffman reviews "previous results about discrete physics and non-commutative worlds" [27]. As the author points out, important "aspects of gauge theory, Hamiltonian mechanics, relativity and quantum mechanics arise naturally in the mathematics of a non-commutative framework for calculus and differential geometry." The article explores "the structure and consequences of constraints linking classical calculus and discrete calculus formulated via commutators." Specifically for the reported second-order constraint, which is "based on interlacing the commutative and non-commutative worlds"—as Kauffman reports—"leads to an equivalent tensor equation at the pole of geodesic coordinates for general relativity".

Rodolfo Gambini and Jorge Pullin provide a short review of the Montevideo interpretation of quantum mechanics [28]. Briefly, in their account of quantum phenomena, Gambini and Pullin "adopt an interpretation that provides an objective criterion for the occurrence of events", whereby, for that purpose they are constructing "an ontology of objects and events". Notably, in this alternative to the more familiar realist interpretations, the quantum events represent "actual entities" which are independent of any observers. Importantly, the Montevideo interpretation "is formulated entirely in terms of quantum concepts, without the need to invoke a classical world."

7. Advanced Quantum Experimentation

The final category is devoted to experiments, and their interpretation, targeting advanced research questions in quantum foundations, as well as practical applications. The first article is by Lukas Mairhofer, Sandra Eibenberger, Armin Shayeghi, and Markus Arndt, who present quantum-interference experiments with biomolecules, and discuss the sensitivity to weak magnetic fields of the observed fringe patterns [29]. Under suitable conditions, "the molecules can ... be prepared in superpositions of position and momentum", the authors write, "even though we can assign classical attributes such as internal temperatures, polarizabilities, dipole moments, magnetic susceptibilities and so forth to them". The researchers go on to explain that "macromolecular interferometry has very practical applications in metrology, for the measurement of electronic, optical, and even magnetic molecular properties." Specifically, the authors report data for "quantum interference of the pre-vitamin 7-dehydrocholesterol", and present the key finding that "even very small magnetic contributions can become accessible in matter-wave assisted deflectometry."

Lev Vaidman and Izumi Tsutsui offer a conceptual analysis of "the history of photons in a nested Mach–Zehnder interferometer with an inserted Dove prism" [30]. The analysis refers to previous experimental results which "became the topic of a very large controversy", as the authors explain. This contribution by Vaidman and Tsutsui serves to clarify the involved issues. Included in the article is an analysis also of the "nested interferometer in the framework of the Bohmian interpretation of quantum mechanics."

Finally, Robert Flack, Vincenzo Monachello, Basil Hiley, and Peter Barker describe a method for measuring the weak value of spin, for atoms using a variant of the original Stern–Gerlach apparatus [31]. The purpose of the methodological design is to enable the testing of "the original Bohm approach", which must use "non-relativistic atoms". Specifically, the described experiment "is designed to measure the real part of the weak value of spin for an atomic system", in this case for helium atoms. Reported in this work is a "full simulation of an experiment for observing the real part of the weak value". The obtained results suggest that a "displacement of the beam of helium atoms in the metastable 2^3S_1 state ... is within the resolution of conventional microchannel plate detectors indicating that this type of experiment is feasible."

8. Outlook

The wide range of perspectives which were contributed to this Special Issue on the occasion of David Bohm's centennial celebration, provide ample evidence for the continuing possibility of an ontological quantum mechanics. In fact, the case for realist approaches towards explaining quantum phenomena, including in the account of EPR-type quantum correlations, has only strengthened, in recent years. Pivotal to this emerging development—for which stands the project of emergent quantum mechanics or EmQM—has been the following realization: A physical ontology for the quantum level represents a measurement-dependent, contextual, or relational ontology; that is, the advancement of 'quantum ontology', as a scientific concept, marks a clear break with classical ontological propositions in the form of direct or naïve realism. Indeed, such an approach to ontology is a vital part of David Bohm's legacy. He noted that, in classical ontological theories in physics, there has been a tendency to assume that the basic concepts of the theory correspond to independently existing realities, i.e., to realities that are not dependent either on context or deeper levels of being. By contrast, in his ontological interpretation of quantum theory, the basic concepts, such as "particle" or "momentum", reflect a reality that is inherently dependent either on context, or on deeper levels, or on both. For the future, instead of denying the possibility of a 'quantum reality', the mainstream of quantum physics might embrace, and join in, the search for unconventional causal structures and non-classical ontologies, which can be fully consistent with the known record of quantum observations in the laboratory.

Acknowledgments: The guest editors J.W., G.G., P.P., and B.H., have approved the content of this Editorial. We express our gratitude to Colleen Long at MDPI for the pleasant and productive collaboration on this Special Issue. Finally, we thank Bruce Fetzer of the John E. Fetzer Memorial Trust for the generous financial support of this publication project of the Fetzer Franklin Fund.

Conflicts of Interest: The authors declare no conflict of interest.

References

1. Maudlin, T. Ontological Clarity via Canonical Presentation: Electromagnetism and the Aharonov–Bohm Effect. *Entropy* **2018**, *20*, 465. [CrossRef]
2. Walleczek, J. Agent Inaccessibility as a Fundamental Principle in Quantum Mechanics: Objective Unpredictability and Formal Uncomputability. *Entropy* **2019**, *21*, 4. [CrossRef]
3. De Gosson, M.A. The Symplectic Camel and Poincaré Superrecurrence: Open Problems. *Entropy* **2018**, *20*, 499. [CrossRef]
4. Seager, W. The Philosophical and Scientific Metaphysics of David Bohm. *Entropy* **2018**, *20*, 493. [CrossRef]
5. Hiley, B.J.; Van Reeth, P. Quantum Trajectories: Real or Surreal? *Entropy* **2018**, *20*, 353. [CrossRef]
6. Flack, R.; Hiley, B.J. Feynman Paths and Weak Values. *Entropy* **2018**, *20*, 367. [CrossRef]
7. Gisin, N. Why Bohmian Mechanics? One- and Two-time Position Measurements, Bell Inequalities, Philosophy, and Physics. *Entropy* **2018**, *20*, 105. [CrossRef]
8. Lazarovici, D.; Oldofredi, A.; Esfeld, M. Observables and Unobservables in Quantum Mechanics: How the No-hidden-variables Theorems Support the Bohmian Particle Ontology. *Entropy* **2018**, *20*, 381. [CrossRef]

9. Passon, O. On a Common Misconception Regarding the de Broglie–Bohm theory. *Entropy* **2018**, *20*, 440. [CrossRef]
10. Norsen, T. On the Explanation of Born-rule Statistics in the de Broglie-Bohm Pilot-wave Theory. *Entropy* **2018**, *20*, 422. [CrossRef]
11. Sanz, Á.S. Atom-Diffraction from Surfaces with Defects: A Fermatian, Newtonian and Bohmian Joint View. *Entropy* **2018**, *20*, 451. [CrossRef]
12. Tumulka, R. On Bohmian Mechanics, Particle Creation, and Relativistic Space–time: Happy 100th Birthday, David Bohm! *Entropy* **2018**, *20*, 462. [CrossRef]
13. Grössing, G.; Fussy, S.; Mesa Pascasio, J.; Schwabl, H. Vacuum Landscaping: Cause of Nonlocal Influences Without Signaling. *Entropy* **2018**, *20*, 458. [CrossRef]
14. Hatifi, M.; Willox, R.; Colin, S.; Durt, T. Bouncing Oil Droplets, de Broglie's Quantum Thermostat, and Convergence to Equilibrium. *Entropy* **2018**, *20*, 780. [CrossRef]
15. Ghadimi, M.; Hall, M.J.W.; Wiseman, H.M. Nonlocality in Bell's theorem, in Bohm's theory, and in Many Interacting Worlds Theorising. *Entropy* **2018**, *20*, 567. [CrossRef]
16. Adlam, E.C. Spooky Action at a Temporal Distance. *Entropy* **2018**, *20*, 41. [CrossRef]
17. Wharton, K. A New Class of Retrocausal Models. *Entropy* **2018**, *20*, 410. [CrossRef]
18. Argaman, N. A Lenient Causal Arrow of Time? *Entropy* **2018**, *20*, 294. [CrossRef]
19. Diósi, L. Fundamental Irreversibility: Planckian or Schrödinger–Newton? *Entropy* **2018**, *20*, 496. [CrossRef]
20. Hsiang, J.-T.; Hu, B.-L. Quantum Thermodynamics at Strong Coupling: Operator Thermodynamic Functions and Relations. *Entropy* **2018**, *20*, 423. [CrossRef]
21. Civitarese, O.; Gadella, M. The Definition of Entropy for Quantum Unstable Systems: A View-Point Based on the Properties of Gamow States. *Entropy* **2018**, *20*, 231. [CrossRef]
22. Keppens, A. What Constitutes Emergent Quantum Reality? A Complex System Exploration from Entropic Gravity and the Universal Constants. *Entropy* **2018**, *20*, 335. [CrossRef]
23. Tessarotto, M.; Cremaschini, C. Generalized Lagrangian Path Approach to Manifestly-Covariant Quantum Gravity Theory. *Entropy* **2018**, *20*, 205. [CrossRef]
24. Palmer, T.N. Experimental Non-Violation of the Bell Inequality. *Entropy* **2018**, *20*, 356. [CrossRef]
25. Filk, T. On Ontological Alternatives to Bohmian Mechanics. *Entropy* **2018**, *20*, 474. [CrossRef]
26. Prodanov, D. Analytical and Numerical Treatments of Conservative Diffusions and the Burgers Equation. *Entropy* **2018**, *20*, 492. [CrossRef]
27. Kauffman, L.H. Non-Commutative Worlds and Classical Constraints. *Entropy* **2018**, *20*, 483. [CrossRef]
28. Gambini, R.; Pullin, J. The Montevideo Interpretation of Quantum Mechanics: A Short Review. *Entropy* **2018**, *20*, 413. [CrossRef]
29. Mairhofer, L.; Eibenberger, S.; Shayeghi, A.; Arndt, M. A Quantum Ruler for Magnetic Deflectometry. *Entropy* **2018**, *20*, 516. [CrossRef]
30. Vaidman, L.; Tsutsui, I. When Photons Are Lying about Where They Have Been. *Entropy* **2018**, *20*, 538. [CrossRef]
31. Flack, R.; Monachello, V.; Hiley, B.; Barker, P. A Method for Measuring the Weak Value of Spin for Metastable Atoms. *Entropy* **2018**, *20*, 566. [CrossRef]

Article

Ontological Clarity via Canonical Presentation: Electromagnetism and the Aharonov–Bohm Effect

Tim Maudlin

Department of Philosophy, New York University, New York, NY 10003, USA; twm3@nyu.edu;
Tel.: +1-212-387-8172

Received: 8 May 2018; Accepted: 11 June 2018; Published: 14 June 2018

Abstract: Quantum physics demands some radical revision of our fundamental beliefs about physical reality. We know that because there are certain verified physical phenomena—two-slit interference, the disappearance of interference upon monitoring, violations of Bell's inequality—that have no classical analogs. But the exact nature of that revision has been under dispute since the foundation of quantum theory. I offer a method of clarifying what the commitments of a clearly formulated physical theory are, and apply it to a discussion of some options available to account for another non-classical phenomenon: the Aharonov–Bohm effect.

Keywords: Aharonov–Bohm effect; physical ontology; nomology; interpretation; gauge freedom; Canonical Presentation

---

1. Introduction

Metaphysicians rightly look to physics for insight into the nature of the physical world. And once upon a time, they would get clear and articulate answers. Newton, in his Scholium on space and time, for example, beautifully conveys both the exact content of his account of Absolute Space and Absolute Time and provides the bucket argument as an empirical demonstration that the relationist theory of motion cannot be correct ([1], pp. 6–12). One can investigate whether Newton's empirical considerations really confirm his particular account of space and time over all others (they don't), but you know with perfect clarity what Newton thought and why.

Unfortunately, the present day situation with respect to physics and metaphysics (ontology) is nowhere in the vicinity of that clarity. Newton did not answer every ontological question one might have been interested in—first and foremost, the source of the universal force of gravitation among particles, concerning which he did not *fingere* a *hypothesis*—but Newton was both clear about some things and clear about where there was more to be said but he didn't know what to say. Nowadays nothing is clear and sharp in the area at all. This is not news. But like the weather, everybody talks about it and nobody does anything. This paper outlines a program for the steps that might be taken.

The main aim is illustrative. It has been widely accepted that the discovery of the Aharonov–Bohm effect, in 1959, forced—or at least suggested—a radical change in our understanding of physical ontology. Briefly, the effect (once it was verified) suggested that the electromagnetic scalar and vector potentials, which were regarded as mere mathematical artifacts in the classical theory, should be regarded instead as "physically real", while the electric and magnetic fields, which were the basic ontology of Maxwell's theory, should be somehow ontologically downgraded. This common suggestion has been disputed, e.g., by regarding the fundamental physical ontology to be a connection on a fiber bundle, the vector potential as a means to represent the connection, and the electromagnetic field as the curvature of the bundle [2] (p. 226). However, none of the objections to it support or even permit the reversion back to Maxwell's ontology or the Relativistic version of it: the electromagnetic field tensor.

Drawing metaphysical conclusions (or teasing metaphysical possibilities) from physics requires a certain level of clarity about just what the physical theories at issue posit. And since the advent of quantum mechanics, the practice of physics has been to relentlessly and aggressively refuse to be clear about precisely the issue that concerns the metaphysician: what exactly physically exists? In Newton's age, this question would have been regarded as one of, if not the, central concerns of physics itself. Oddly, practicing physicists nowadays are likely to direct a student asking such questions to the philosophy department. The way this plays out in quantum theory over questions such as the status and "reality" of the wave function is well known, and I do not propose to plow that over-farmed soil yet again here. Rather, I want to pursue a similar question, as far as I can, without taking issues about the ontology of quantum theory into account. That makes the particular suggestions about ontology reached here questionable and provisional, but it should have the compensating virtue of making the methodological proposal clear and easy to follow.

What is accomplished by having a general framework in which the precise ontological commitments of a theory are made manifest and unambiguous in a disciplined way? Benefits accrue on both the conceptual and heuristic sides. Conceptually, one requires this sort of clarity to truly understand a physical theory. Without it, one can do calculations and produce predictions, but not be clear about what kind of physical world the theory presents. Quantum theory, as a whole, presents an excellent example. As for developing new theories, we will see that the method of presentation suggests almost algorithmic ways to alter the ontology of a theory.

With that as prelude, then, let's begin.

2. What Is a Physical Theory?

Let's start by zooming out all the way to the most general question we can ask: what is physics? Back in the day, physics was characterized as the theory of matter in motion. That remains a wonderful place to start, although the slogan needs some tweaks and updates. One nice thing about the slogan is that it immediately indicates the "foundational" aspect of physics as opposed to all the other empirical sciences. Every empirical science—biology, geology, psychology, economics, etc.—deals with systems that are matter in motion. A living horse, whatever else it is, is matter in motion and, as such, falls under the purview of physics. It also falls under the purview of biology, evolutionary theory, economics, cognitive science, and so on. But the converse does not hold: not every physical system provides subject matter for biology or geology or psychology or economics. A red dwarf star, for example, does not. However, the red dwarf is matter in motion and must be susceptible to physical analysis.

The phrase "matter in motion" offers two targets for conceptual and physical analysis: matter and motion. In contemporary physics, there is no such objective state as "being in motion". A particle in interstellar space, for example, can be "at rest" in the sense of not moving relative to its own inertial frame and at the same time "moving" relative to other perfectly legitimate inertial frames. Since the theory of Relativity, talk of the "motion of a system" has come to be understood as talk of the trajectory of the system through space-time. If you specify a space-time structure and the worldlines of the constituents of the system, then you have specified all there is to say about the "motion" of the system.

Of course, in order to *have* a trajectory through space-time, a worldline, the constituent has to be a *local beable* in John Bell's sense [3]. That is, the constituent has to be something that has a reasonably well-defined location in space-time. And there may exist items in the physical ontology of a theory that fail to have this characteristic. Those would be the *non-local beables* of the theory. In quantum theory, the *quantum state* (the item represented by the wave function) is such a non-local beable according to most explications of the theory. Hence there arises a trivial semantic issue: should the non-local beables of a theory count as *matter* or as something else, some *tertium quid* beside the space-time and the local beables?

Nothing hangs in this semantic decision, but for the sake of clarity I will, in this paper, refer only to local beables as "matter". According to this usage, the quantum state, if it exists at all, is an immaterial and non-local physical item. If the reader dislikes this terminology, she is invited to systematically rewrite it in whatever way is congenial.

Adopting this updated terminology, then, the fundamental ontology of any physical theory contains a space-time structure, some local beables (matter), and whatever else may have come up. Physics, the theory of matter in motion, has become the theory of the trajectories of local beables through space-time.

In the course of theorizing, we may increase the ontology of a theory by postulating some new items, local or nonlocal. Methodologically, the grounds for doing that is typically in service of accounting for the motions of a particular set of local beables, the postulated constituents of familiar observable matter. Thus, for example, Maxwell introduced the electric and magnetic fields as novel local beables in order to account for the motions of observable objects such as iron filings. By our convention, these fields are material because they are local beables. On the other hand, the postulation of a quantum state, which is a non-local beable, counts as adding an immaterial entity. It is, in any case, certainly an entity that cannot be directly observed. The only grounds we have to believe in quantum states is the influence they have on observable collections of local beables such as those that constitute pointers and radios and tables and chairs.

To sum up so far, and present this account in more detail: physics is the description of *local beables and whatever else is required to account for their trajectories through space-time*. That updates the more mellifluous "Matter in motion" in a way useful for present purposes. Another, essentially equivalent characterization of physics is *the most general account of what there is and what it does, at the fundamental level*. The "what there is" part is provided by the *ontology* of the theory, and the "what it does" part by the *dynamics*. The restriction to fundamental items excludes treating derivative entities, such as horses or species or economic systems, as the basic subject matter.

Specifying both what there is and what it does has, throughout history, involved specification of a space-time structure in which the behavior occurs. Whether the spatiotemporal structure is substantival, or relational, or something else, we do not prejudge. For the purposes of this paper, we will consider mostly classical space-times, but that is just an historical accident.

We now have four basic "categories of being" postulated by a physical theory:

(1) The local beables or matter, which exists at delimited regions of the spatio-temporal structure.
(2) The non-local beables (if any) that have no particular value at any space-time location.
(3) The spatio-temporal structure, in terms of which the distinction between local and non-local beables is drawn.
(4) The dynamical laws, which specify, either deterministically or probabilistically, how the various beables must or can behave.

There is no requirement that the elements of each of these categories be definable independently of the rest. Indeed, differentiating the local from the non-local beables obviously requires reference to the spatio-temporal structure. Nor is there any transcendental argument that these four categories of being exhaust the whole of physical reality, or that all of them must be exemplified. Einstein, for example, famously opined that the progress of physics marched inexorably toward locality, in the ontology and even in the laws:

> If one asks what, irrespective of quantum mechanics, is characteristic of the world of ideas of physics, one is first of all struck by the following: the concepts of physics relate to a real outside world, that is, ideas are established relating to things such as bodies, fields, etc., which claim "real existence" that is independent of the perceiving subject—ideas which, on the other hand, have been brought into as secure a relationship as possible with the sense data. It is further characteristic of these physical objects that they are thought of as arranged in a space-time continuum. An essential aspect of this arrangement of things in physics is that they lay claim, at a certain time, to an existence independent of one another, provided these objects "are situated in different parts of space" ... This principle

> has been carried to extremes in the field theory by localizing the elementary objects on which it is based and which exist independently of each other, as well as the elementary laws which have been postulated for it, in the infinitely small (four-dimensional) elements of space [4] (pp. 170–171)

All of our four categories of being have been subject to intense philosophical scrutiny through the ages. The nature of space and time has already been mentioned. The nature of matter has been equally treated in fundamentally different way: as that which occupies space; as the center of forces; as a bundle of properties, etc. Laws, of course, have been rendered as relations of necessitation, as members of the simplest and strongest axiomatization of the Humean mosaic, as dispositions and as primitive entities not subject to further analysis. For our purposes here, though, these controversies are inessential: one way or another every physical theory does postulate a dynamics or, as we will say, a *nomology*. Just as the ontology of a theory specifies what exists in a more "concrete" sense, the nomology specifies the laws. The spatiotemporal structure has traditionally been taken to be somehow less than material but more than nothing. And the non-local ontology is a recent innovation in physics, still a matter of much dispute (See, e.g., [5]).

Let's try a couple of simple exercises, drawn from antiquity. Democritean atomism has a material ontology of solid, shaped, indivisible, and solid bodies. The spatiotemporal structure is an infinite Euclidean space that persists through time, with the opposite directions "up" and "down" intrinsic to it. The nomology specifies that the atoms will fall at a constant rate down, save for two circumstances: collisions between atoms and the occasional random "swerve". The timing and nature of these swerves are not precisely defined by Democritus, nor are the laws of collision.

Aristotelian physics, in contrast, posits a finite, spherical spatial structure that persists through time. There are five fundamental types of matter: earth, water, air, and fire (which correspond to the modern notion of the states of matter: solid, liquid, gas and plasma), as well as the quintessence, viz. aether. The spherical spatial structure defines the "up" and "down" directions as towards and away from the center of the universe. The dynamics is given in terms of nature rather than in terms of laws: the natural motions of earth and water are to the center and the natural motions of fire and air to the periphery. The natural motion of aether is uniform circular motion about an axis through the center of the universe.

The four basic categories used in our anatomy of a physical theory are to be taken *cum grano salis*. The category of non-local beables is a recent addition, and there may be more to come. In the other direction, one could imagine categories melding: the spatio-temporal and material aspects, for example, becoming so entwined as not to always be distinct. But at least in some conditions the separate categories must emerge if the theory is to be recognizable as physics at all.

The most profound conceptual addition to this basic scheme of physical theorizing is strongly associated with the scientific revolution: the geometrization or mathematization of physics. Galileo famously declared:

> Philosophy is written in this grand book—I mean the universe—which stands continually open to our gaze, but it cannot be understood unless one first learns to comprehend the language in which it is written. It is written in the language of mathematics, and its characters are triangles, circles, and other geometric figures, without which it is humanly impossible to understand a single word of it; without these, one is wandering about in a dark labyrinth [6] (p. 65)

Modern physics cannot be separated from this mathematization. Our next task is to consider the use of mathematics in physics and how to cope with it.

3. Mathematical Physics and the Canonical Presentation

As the simple examples of Democritus and Aristotle illustrate, a physical theory need not employ any sophisticated mathematical apparatus. However, both the glory and the bane of modern physics is its highly mathematical character. This has provided both for the calculation of stunningly precise predictions and for the endemic unclarity about the physical ontology being postulated. The unclarity arises from a systematic ambiguity among terms that refer to the ontology and nomology of a theory and terms that refer to the *mathematical representations* of the ontology and nomology.

Mathematical physics uses mathematical structures to represent physical states of affairs. Unfortunately, the distinction between the representations and the entities represented is often elided in the common manner of speech. As a simple example, take the term "scalar field". It would not sound in the least odd to say: "The Higgs field is a scalar field". Further, if one inquires what a scalar field is, it would not be at all out of place to be told that a scalar field is a mapping from space-time points into a set of scalars, i.e., real or complex numbers. However, if we naively put these two innocuous statements together we appear to get the result that the Higgs field (which we took to be a physical item) is a mapping from space-time points into the field of numbers (which is an abstract mathematical object). As a purely abstract matter, infinitely many such mappings exist, but that does not mean there are infinitely many physical scalar fields. Something has gone wrong.

On the other hand, to take a familiar example, we often talk of the wave function of a system in quantum mechanics, and ask whether the wave function is "real", "physical", or "objective". However, what is a wave function? As the term implies, it is a *function*, a mathematical mapping from (e.g.,) the configuration space of a system to the complex numbers. So is the dispute about the "reality" of the wave function a dispute about the ontological status of such a mapping? Obviously not.

What has gone wrong in both of these examples is straightforward: the terms "scalar field" and "wave function" are being used ambiguously. In one sense, they refer to a specific *mathematical structure*. In the other, they are used to refer to *a postulated physical item that is supposed to be represented by the mathematical structure*. It is trivial that the mathematical structure exists in whatever sense mathematical entities do. It can be highly contentious whether any physical entity exists that can be represented by that mathematical item in the way that the physical theory requires.

The most efficient way to resolve any systematic ambiguity is by a linguistic convention. In the case of the "wave function", I have adopted the convention that the mathematical representation shall be called a "wave function" (because a *function* is plainly a mathematical item) and the non-local beable that it is posited to represent shall be called the *quantum state* of a system. So the metaphysical battles are over the existence and nature (if they exist) of quantum states.

There is no canonical way for a mathematical item to represent a physical one. No amount of staring at the mathematics per se can resolve questions like: which mathematical degrees of freedom in the representation correspond to physical degrees of freedom in the system represented? Mathematical degrees of freedom in the representation that do not correspond to physical degrees of freedom in the represented system are called *gauge degrees of freedom*, and representations that differ only in their gauge degrees of freedom are called *gauge equivalent*. Transformations between gauge equivalent mathematical representations are called *gauge transformations*.

As a seemingly trivial example, a global constant change in the phase of a wave function is standardly taken to be a gauge transformation and the resulting mathematical representation to be gauge equivalent to the original. One way to remove this redundancy from the mathematical representations is to change to another mathematical object as the vehicle of representation. In this case, one says that quantum states are not properly represented by *vectors* in a Hilbert space but by *rays* in a projective Hilbert space. In one sense of "ideal", an ideal mathematical representation for a physical system would implement a one-to-one map from the space of mathematical representations to the space of physically possible states of the system. At least, such a representation would be maximally convenient for the metaphysician, while it might be severely impractical for the physicist who actually has to compute numbers. We will see an example of this anon.

If one cannot distinguish the gauge from the non-gauge degrees of freedom in a mathematical representation by any purely mathematical analysis, how is that job to be done? The only way is by a *commentary* on the mathematical representation. We can only be sure what a piece of mathematics is supposed to represent (if anything) by being told by the expositor of a physical theory. *One and the same mathematical apparatus accompanied by a different commentary can convey different physical theories, theories with different ontologies and even with different laws*. Our examples from classical electromagnetism will illustrate this possibility.

Although this paper is not particularly about quantum theory, it is worthwhile to pause for a moment to reflect on the situation there. It is commonly said that there is this thing called "quantum theory" which works splendidly well as a physical theory but nonetheless lacks an "interpretation". The project of interpreting quantum theory is assigned to (or dumped on) people who work on the foundations of physics, and perhaps most usually to philosophers. "Interpreting" quantum theory is regarded by many physicists as pointless or frivolous or unscientific or even meaningless, except insofar as it means altering the empirical predictions of the theory. This attitude has a long history: already in 1926 the physicist Charles Galton Darwin wrote to Niels Bohr: "It is a part of my doctrine that the details of a physicist's philosophy do not matter much" [7] (p. 4).

According to the linguistic usage urged here, though, all of this talk of "interpreting" quantum theory is mistaken. If a physical theory is supposed to address the questions of what there is and what it does, the questions of physical ontology and physical nomology, then what goes by the name "quantum theory" is not a theory at all. It is rather a mathematical method of making predictions. If all one cares about are the accuracy of the predictions, then one can be completely satisfied with "quantum theory", but that is the attitude of the engineer rather than the natural philosopher. "Quantum theory" is a prediction-making recipe in need of a real physical theory, a theory that specifies what exists and how it behaves, thereby accounting for the remarkable reliability of the predictive algorithm. Rather than there being a theory in need of an interpretation, there is a calculational tool in need of a theory that accounts for it.

How, then, is one to clearly and unambiguously specify a physical theory? And what difference might it make to have our physical theories clearly and unambiguously presented? The most important elements of a mathematically formulated physical theory have already been given, and all that is required is a systematic way to exhibit them. What we need to make clear are the *physical ontology*, the *spatiotemporal structure*, and the *nomology* (i.e., the fundamental laws) of the theory, by means of presenting the *mathematical representation* of these various elements along with a *commentary* that relates the mathematical representation to the physical item it is meant to represent. The commentary should make clear which mathematical degrees of freedom in the representation are *gauge* and which rather correspond to *physical degrees of freedom* in the system represented. In addition, it can be useful to specify any *mathematical fictions* that may be employed for the purpose of simplicity of presentation or of calculation, as well as any *derivative ontology*, i.e., physically real items that are composed of and analyzable into more fundamental entities. To take an obvious example, hydrogen atoms are real physical entities, but they are not physically fundamental: they are just bound states of a proton and an electron. If protons and electrons are already in one's physical ontology and the nomology allows for them to form bound states, then the recognition of hydrogen atoms as physically real does not increase one's ontology at all.

In presenting a clear and precisely articulated physical theory, then, one ought to specify the fundamental physical ontology and how the fundamental physical ontology is to be represented mathematically. The fundamental nomology—the laws—will be represented by mathematical equations that make reference to the spatiotemporal structure, and we demand that these mathematical representations of the laws contain *only representations of the fundamental ontology*. Hydrogen atoms may be physically real, but the laws of nature do not influence them qua hydrogen atoms but qua electron and proton (or better: electron and quarks) that happen to form a bound state. Carrying this insight further, we see that there can be no fundamental physical laws that mention tables or chairs or horses or economic systems or measurements *as such*. Physics deals with these items only as derivative ontology, not as fundamental ontology. The special sciences can usefully then be regarded as operating under the fiction that the sorts of items they trade in are fundamental rather than derivative, and are governed by sui generis laws. Ultimately, physics must explain the predictive effectiveness of these special science laws, just as it must explain the predictive effectiveness of the quantum predictive algorithm.

Collecting together all of these threads, we can at last make a concrete proposal. A *Canonical Presentation* of a mathematical physical theory shall specify:

(1) The fundamental physical ontology of the theory, which may further be divided into the local beables (matter and space-time structure) and the non-local beables, if any (e.g., a quantum state represented by a wave function on configuration space).
(2) The spatio-temporal structure of the theory.
(3) The mathematical items that will be used to represent both 1 and 2, with a commentary making clear which degrees of freedom in the mathematics are gauge and which are not.
(4) The nomology of the theory, which will be represented by equations couched in terms only of the items mentioned in (3).
(5) Mathematical fictions—these are mathematically defined quantities that are not intended to directly represent any part of the physical ontology. Such fictions can play an important practical role when trying to calculate with the theory.
(6) Derivative ontology—these are items that are taken to be physically real but not fundamental. They must be definable in terms of the fundamental ontology and nomology.

The only way to argue for the utility of this sort of systematic presentation of a physical theory is to see it in action. That is our next task.

4. The Canonical Presentation of Classical Electromagnetic Theory

Let's see how this general approach to specifying a theory in mathematical physics works in a relatively uncontroversial setting: classical electromagnetism. Even there, we will confront some interesting and perhaps unexpected choices when trying to lay out the theory in such an explicit way.

Expressed in words, classical electromagnetic theory, as codified in Maxwell's equations, posits the existence of an electric field, a magnetic field, matter with mass and charge, a classical space-time, and several laws.

A naïve Canonical Presentation of the physical theory, drawn straight from a standard textbook, might look like Table 1.

Table 1. Standard E-M Theory.

Theory	Physical Ontology; Spatiotemporal Structure	Mathematical Representation of Physical Ontology	Purely Mathematical Facts	Nomology	Derivative Ontology; Mathematical Fictions
Classical E & M, Mass Density Version	Electric Field Magnetic Field Charge Density Mass Density Lorentz Force; Time 3-D Euclidean Absolute Space	$\vec{E}(x, y, z, t)$ $\vec{B}(x, y, z, t)$ $\rho(x, y, z, t)$ $\mu(x, y, z, t)$ $\vec{F}_L(x, y, z, t)$ $t \in R$ $(x, y, z) \in \mathbb{R}^3$	If Curl $\vec{C} = 0$ on a simply connected space, then $\vec{C} = \mathrm{Grad}(\xi)$ for some ξ. If Div $\vec{B} = 0$ on a simply connected space, then $\vec{B} = \mathrm{Curl}\left(\vec{A}\right)$ for some $\vec{A}$ Gauge transformations $\vec{A}' = \vec{A} + \mathrm{grad}\xi$ $\phi' = \phi - \frac{\partial\xi}{\partial t}$	$\mathrm{Div}\left(\vec{E}\right) = \rho$ $\mathrm{Div}\left(\vec{B}\right) = 0$ $\mathrm{Curl}\left(\vec{E}\right) + \frac{\partial\vec{B}}{\partial t} = 0$ $\mathrm{Curl}\left(\vec{B}\right) - \frac{\partial\vec{E}}{\partial t} = \rho\vec{v}$ $\vec{F}_L = \rho\left(\vec{E} + \left(\vec{v} \times \vec{B}\right)\right)$ $\vec{F}_{net} = \mu\frac{d\vec{v}}{dt}$	Derivative Ontology: $\vec{J} = \rho\vec{v}$ $\vec{F}_{net}$ = vector sum of all forces on a body at a point. Mathematical Fictions: Let $\vec{B} = \mathrm{Curl}\left(\vec{A}\right)$ Then $\mathrm{Curl}\left(\vec{E} + \frac{\partial\vec{A}}{\partial t}\right) = 0$ so $\vec{E} + \frac{\partial\vec{A}}{\partial t} = -\mathrm{Grad}(\phi)$ or $\vec{E} = -\mathrm{Grad}(\phi) - \frac{\partial\vec{A}}{\partial t}$

In presenting the nomology, we have adopted Metaphysician's Units: the speed of light c and 4π have both been set to unity. We have allowed ourselves the liberty because for our purposes now those quantities just clutter the Presentation up.

The Canonical Presentation makes some things immediately clear. We are considering a theory in which the electric and magnetic fields are two distinct physical items, each represented by a vector field on a Euclidean space. The Lorentz force, and hence forces in general, are taken to be real parts of the physical ontology. They, like all the rest of the physical ontology, are local beables. We could provide more detail about the mathematical structure of the mathematical representations, some of which is easy to fill in. For example, both ρ and μ are functions defined over R^3 for each value of t. t ranges over some interval of the real numbers.

Looking carefully at the nomology, we see that we have not quite managed to satisfy all of the requirements of a Canonical Presentation. In particular, the mathematical term $\vec{v}$ appears thrice in the nomology but is not the mathematical representation of any part of the ontology. Intuitively, $\vec{v}$ is the velocity of the charged matter density, and one would think that specifying μ as a function of t would serve to fix $\vec{v}$. However, because μ is supposed to be a continuous matter *density*, this is just not so. For example, suppose μ is the same for all t: a positive constant inside a sphere in E^3 and zero outside it. One might immediately assume that this is the mathematical representation of a uniform sphere of matter at rest. However, a moment's thought reveals that it could just as well represent a uniform sphere of matter rotating on an axis, or performing any other motion that an incompressible continuous fluid might. In short, the physical content expressed by the function $\vec{v}(x, y, z, t)$ outruns the physical content expressed by μ(x, y, z, t). If one really wants to make a matter density on a continuum a fundamental part of the physical ontology, then one must also accept that there is a velocity function assigned to the matter density that does not supervene on the distribution of the matter density over all of space for all of time.

The theory also has as yet no mechanism to implement conservation of charge or conservation of matter. Conservation of charge is easy: add a law to the effect that $\frac{\partial \vec{J}}{\partial t} + \mathrm{Div}(\vec{J}) = 0$. Conservation of matter would require a similar law: $\frac{\partial \mu \vec{v}}{\partial t} + \mathrm{Div}(\mu\vec{v}) = 0$.

It might also strike one as odd that the physical ontology contains both a matter density and a charge density with nothing to link them together, either in virtue of a definition or in virtue of a law. As far as our principles tell us, there could be a positive charge density where the matter density is zero, just as there could be positive matter density where the charge density is zero. The latter represents something we understand: the possibility of uncharged matter. The former, though, makes no obvious sense as it represents, as it were, free-floating charge.

Both of these problems are soluble by changing the theory from one with a matter density ontology to one with a point particle ontology. A point particle has a precise location at every moment of time that it exists. If we require that point particles neither be created nor destroyed and that they always move continuously, then the spatio-temporal career of every point particle will be represented by a world line: a continuous path through space as a function of time (or, relative to a different set of representational conventions, a continuous path through space-time). Charge conservation is secured by simply associating a quantity, the charge, with each particle, and similarly for mass. The Canonical Presentation of this new theory would be as seen in Table 2.

Table 2. E-M with a Particle Ontology.

Theory	Physical Ontology; Spatiotemporal Structure	Mathematical Representation of Physical Ontology	Purely Mathematical Facts	Nomology	Derivative Ontology; Mathematical Fictions

Classical E & M, Particle Version	Electric Field
	Magnetic Field
	Point Particles
	Particle Charge
	Particle Mass
	Lorentz Force;
	Time
	3-D Euclidean Absolute Space
		$\vec{E}(x,y,z,t)$
		$\vec{B}(x,y,z,t)$
		$\vec{x_i}(t) = (x_i, y_i, z_i)$
		$q_i \in R$
		$m_i \in R > 0$
		$\vec{F_L}(x,y,z,t)$
		$t \in \mathbb{R}$
		$(x,y,z) \in \mathbb{R}^3$
			If Curl $\vec{C} = 0$ on a simply connected space, then
			$\vec{C} = \mathrm{Grad}(\xi)$
			for some ξ.
			If Div $\vec{B} = 0$ on a simply connected space, then
			$\vec{B} = \mathrm{Curl}\left(\vec{A}\right)$
			for some $\vec{A}$
			Gauge transformations
			$\vec{A}' = \vec{A} + \mathrm{grad}\xi$
			$\phi' = \phi - \frac{\partial \xi}{\partial t}$
				$\mathrm{Div}\left(\vec{E}\right) = q_i$
				$\mathrm{Div}\left(\vec{B}\right) = 0$
				$\mathrm{Curl}\left(\vec{E}\right) + \frac{\partial \vec{B}}{\partial t} = 0$
				$\mathrm{Curl}\left(\vec{B}\right) - \frac{\partial \vec{E}}{\partial t} = q_i \vec{v_i}$
				$\vec{F_{Li}} = q_i\left(\vec{E} + \left(\vec{v_i} \times \vec{B}\right)\right)$
				$\vec{F}_{net\,i} = m_i \frac{d^2\vec{x_i}(t)}{dt^2}$
					Derivative Ontology:
					$\vec{v_i} = \frac{d\vec{x_i}(t)}{dt}$
					$\vec{J_i} = q_i \vec{v_i}$
					$\vec{F}_{net}$ = vector sum of all forces on a particle.
					Mathematical Fictions:
					Let $\vec{B} = \mathrm{Curl}\left(\vec{A}\right)$
					Then $\mathrm{Curl}\left(\vec{E} + \frac{\partial \vec{A}}{\partial t}\right) = 0$
					so $\vec{E} + \frac{\partial \vec{A}}{\partial t} = -\mathrm{Grad}(\phi)$
					or $\vec{E} = -\mathrm{Grad}(\phi) - \frac{\partial \vec{A}}{\partial t}$

$\vec{v_i}$ is now rigorously defined, and we understand why there cannot be charge where there is no matter: charges are ascribed to particles and so can only exist where particles do, and particles must also have non-zero mass. But some of our mathematical issues have just been moved around to different places: officially, $\mathrm{Div}(\vec{E})$ cannot be well-defined if the charge is only non-zero at a point. So to get our ducks in order we would have to massage the Presentation even more. But since our main concern here is with the status of the electric and magnetic fields, not with the mass density or the particles, we will not pursue that problem further. We only pause to note that getting one's ducks in a row requires first figuring out which ducks have strayed, and that the discipline we have imposed on ourselves by trying to present the theory in the form of a Canonical Presentation brings those delinquent ducklings into focus.

Let us turn our attention now to the nomology of the particle theory. There are a surprising number of "laws of nature" in this presentation: six just for classical electromagnetism with charged point particles. The ideal of a completed physics is usually envisaged as a single equation that covers everything, so compact as to fit on the proverbial T-shirt. Surely we can do better than this unruly crowd of a half-dozen. In terms of our Canonical Presentation, we would like to depopulate the nomology. There are several different sorts of moves that can accomplish this.

The easiest case to consider is the "law" $\mathrm{Div}(\vec{B}) = 0$. The content of this equation is often described as saying that there are no magnetic monopoles. But the proposition "There are no magnetic monopoles" does not really have the form of a paradigmatic law. Consider, for comparison, the proposition that water is H_2O. That is certainly true, and necessarily true, but does not have the characteristics of a physical or chemical law. We do not think that there are as many distinct chemical laws as there are chemical species: that would lead to millions of laws of chemistry. Water is H_2O is rather the answer to the fundamental philosophical question: What is it? It tells of the *metaphysical nature* of water, what water fundamentally is. Although different in some respects, "There are no magnetic monopoles" is a similar sort of claim: it specifies part of the fundamental nature of magnetic charges.

Still, to make the connection between the absence of magnetic monopoles and $\mathrm{Div}(\vec{B}) = 0$, we need a connection between the divergence of the magnetic field and magnetic charges. What would cut this Gordian knot is another statement about the fundamental metaphysical nature of electric and magnetic charges: what if magnetic charges *just are* the divergences of magnetic fields, and electric charges *just are* the divergences of electric fields? Then the ontological claim that there are no magnetic monopoles is completely equivalent ontologically to the claim that the magnetic field is divergenceless, which in turn is *represented* by the mathematical equation $\mathrm{Div}(\vec{B}) = 0$. The net result of this ontological move is to shift the equation $\mathrm{Div}(\vec{B}) = 0$ from the category of the nomology to a logical consequence of an ontological analysis of the form $\mathrm{Div}(\vec{B}) = q_m$, with q_m representing a magnetic charge, together with the negative ontological claim that magnetic charges do not exist.

A non-existence claim such as this is the sort of beast that drove Parmenides around the bend, and there are various methods we might use to incorporate it into our Canonical Presentation. But the most elegant is a simple rule of silence: in a physical theory nothing exists unless we explicitly say it does. So the elegant way to deny the existence of magnetic charges is simply not to list magnetic charges in the ontology.

Electric charges, though, do exist. But we can eliminate $\mathrm{Div}(\vec{E}) = q_i$ from the nomology by exactly the same shift. Let us propose that this equation represents not a physical law but an ontological analysis: electric charges *just are* the divergences of electric fields. In this way we reduce both the physical ontology and the nomology, and further gain an explanation of why electric charges cannot exist without electric fields.

Bundling these two changes together (one could in principle do just one or the other alone) yields a new, different physical theory (Table 3).

Table 3. E-M with Derived Charges.

Theory	Physical Ontology; Spatiotemporal Structure	Mathematical Representation of Physical Ontology	Purely Mathematical Facts	Nomology	Derivative Ontology; Mathematical Fictions

Classical E & M, Particle Version With Derived Charges	Electric Field
	Magnetic Field
	Point Particles
	Particle Mass
	Lorentz Force;
	Time
	3-D Euclidean Absolute Space
		$\vec{E}(x, y, z, t)$
		$\vec{B}(x, y, z, t)$
		$\vec{x_i}(t) = (x_i, y_i, z_i)$
		$m_i \in R > 0$
		$\vec{F_L}(x, y, z, t)$;
		$t \in \mathbb{R}$
		$(x, y, z) \in \mathbb{R}^3$
			If Curl $\vec{C} = 0$ on a simply connected space, then
			$\vec{C} = \text{Grad}(\xi)$
			for some ξ.
			If Div $\vec{B} = 0$ on a simply connected space, then
			$\vec{B} = \text{Curl}\left(\vec{A}\right)$
			for some $\vec{A}$ Gauge transformations
			$\vec{A'} = \vec{A} + \text{grad}\xi$
			$\phi' = \phi - \frac{\partial \xi}{\partial t}$
				$\text{Curl}\left(\vec{E}\right) + \frac{\partial \vec{B}}{\partial t} = 0$
				$\text{Curl}\left(\vec{B}\right) - \frac{\partial \vec{E}}{\partial t} = q_i \vec{v_i}$
				$\vec{F_{Li}} = q_i\left(\vec{E} + \left(\vec{v_i} \times \vec{B}\right)\right)$
				$\vec{F_{neti}} = m_i \frac{d^2 \vec{x_i}(t)}{dt^2}$
					Derivative Ontology:
					$\text{Div}\left(\vec{E}\right) = q_i$
					$\text{Div}\left(\vec{B}\right) = q_m$
					$\vec{v_i} = \frac{d\vec{x_i}(t)}{dt}$
					$\vec{J_i} = q_i \vec{v_i}$
					$\vec{F_{net}}$ = vector sum of all forces on a particle;
					Mathematical Fictions:
					Let $\vec{B} = \text{Curl}\left(\vec{A}\right)$
					Then $\text{Curl}\left(\vec{E} + \frac{\partial \vec{A}}{\partial t}\right) = 0$
					so $\vec{E} + \frac{\partial \vec{A}}{\partial t} = -\text{Grad}(\phi)$
					or $\vec{E} = -\text{Grad}(\phi) - \frac{\partial \vec{A}}{\partial t}$

If the method of handling $\text{Div}(\vec{B}) = 0$ strikes one as too baroque, there is another tack available. Simply stipulate that magnetic fields are to be represented mathematically by divergence-free vector fields. There is nothing preventing this sort of decision, which now appears as a further restriction on the mathematical apparatus (Table 4).

Table 4. Particle Ontology with derived charge.

Theory	Physical Ontology; Spatiotemporal Structure	Mathematical Representation of Physical Ontology	Purely Mathematical Facts	Nomology	Derivative Ontology; Mathematical Fictions

Classical E & M, Particle Version With Derived Charges	Electric Field
	Magnetic Field
	Point Particles
	Particle Mass
	Lorentz Force;
	Time
	3-D Euclidean Absolute Space
		$\vec{E}(x, y, z, t)$
		$\vec{B}(x, y, z, t)$ s.t. $\text{Div}(\vec{B}) = 0$
		$\vec{x_i}(t) = (x_i, y_i, z_i)$
		$m_i \in R > 0$
		$\vec{F_L}(x, y, z, t)$
		$t \in \mathbb{R}$
		$(x, y, z) \in \mathbb{R}^3$
			If Curl $\vec{C} = 0$ on a simply connected space, then
			$\vec{C} = \text{Grad}(\xi)$
			for some ξ.
			If Div $\vec{B} = 0$ on a simply connected space, then
			$\vec{B} = \text{Curl}\left(\vec{A}\right)$
			for some $\vec{A}$.
			Gauge transformations
			$\vec{A'} = \vec{A} + \text{grad}\xi$
			$\phi' = \phi - \frac{\partial \xi}{\partial t}$
				$\text{Curl}\left(\vec{E}\right) + \frac{\partial \vec{B}}{\partial t} = 0$
				$\text{Curl}\left(\vec{B}\right) - \frac{\partial \vec{E}}{\partial t} = q_i \vec{v_i}$
				$\vec{F_{Li}} = q_i\left(\vec{E} + \left(\vec{v_i} \times \vec{B}\right)\right)$
				$\vec{F_{neti}} = m_i \frac{d^2 \vec{x_i}(t)}{dt^2}$
					Derivative Ontology:
					$\text{Div}\left(\vec{E}\right) = q_i$
					$\vec{v_i} = \frac{d\vec{x_i}(t)}{dt}$
					$\vec{J_i} = q_i \vec{v_i}$
					$\vec{F_{net}}$ = vector sum of all forces on a particle;
					Mathematical Fictions:
					Let $\vec{B} = \text{Curl}\left(\vec{A}\right)$
					Then
					$\text{Curl}\left(\vec{E} + \frac{\partial \vec{A}}{\partial t}\right) = 0$
					so $\vec{E} + \frac{\partial \vec{A}}{\partial t} = -\text{Grad}(\phi)$
					or $\vec{E} = -\text{Grad}(\phi) - \frac{\partial \vec{A}}{\partial t}$

Note that although this Canonical Presentation is not identical to the last, a strong argument can be made that they present one and the same physical theory. The two presentations have identical physical ontologies, identical spatiotemporal structures and identical nomologies. Their methods of handling magnetic monopoles can therefore be reasonably regarded as *merely* presentational differences, to which no ontological fact corresponds.

There are a couple more mathematical worries one might have about particle ontology electromagnetic theories. If the charge of a particle is concentrated at a point, then the electric field will not be defined there. That raises problems for both the divergence in $\text{Div}(\vec{E}) = q_i$ and for the equation for the Lorentz force on the particle. There are different mathematical approaches to solving problems like this, but for illustration's sake here is a sketch of one.

$\text{Div}(\vec{E})$ is problematic at every point along the worldline of the particle—exactly the points where the charged particle exists! But we can take a page from Gauss to circumvent this problem. Gauss's theorem says that the integral of the flux of a vector field over a closed surface equals the integral of the divergence over the enclosed volume. And there is no problem defining the flux over any surface enclosing the particle. So instead of defining q_i as the divergence of the electric field at a point, we can define it as the limit of the flux over surfaces that enclose the point as the maximum distance from the point to the surface goes to zero. In a similar spirit, define the *q_i-adjusted electric field* E_{qi} at a point p as the electric field at p minus $\frac{q_i}{r^2}\hat{r}$, where r is the distance from the location of particle i to p and $\hat{r}$ is the unit vector in the direction from the location of the particle to p. Let $\overline{E_{qi,\,\Sigma}}$ be the average of E_{qi} over the surface Σ. Finally, define the electric field at a point on the worldline of particle i to be the limit of $\overline{E_{qi,\,\Sigma}}$ as the distance of the points on Σ from the point on the worldline goes to zero. It is that value of E that is used in the Lorentz force law.

What are the solutions to these dynamical equations like? I make no representation that they give the correct results. I have rather just been illustrating the sorts of conceptual, technical, and mathematical issues that have to be resolved to complete the Canonical Presentation. It forces one to answer questions about the ontology of a theory, the nomology, and the way the mathematics is being used to represent both. Those questions, in turn are illuminating and suggestive.

The last and most important thing to note in all of these Canonical Presentations is the status of the vector and scalar potentials. When one learns classical electromagnetism, the ontological status of these potentials is not in doubt: they are not "real", meaning that there is nothing in the physical ontology that they directly represent. They can, of course, indirectly represent the electric and magnetic fields, which are real, but the representation relation is indirect. It goes through the mathematical truths listed on the Presentation, which imply the well-known gauge freedom in choosing the potentials. It is that very freedom—the fact that there are mathematical degrees of freedom in the representation that *do not* correspond to physical degrees of freedom in the object represented—that makes the potentials so useful. One is free to choose a convenient gauge, different for different circumstances, that makes the mathematics easier to handle. One of the most important entries in our charts so far, then, is the characterization of the potentials as mathematical fictions. So far we have been considering moves—changes in the fundamental ontology of the theory—that can shift entries out of the nomology category and the ontology category and so reduce the basic posits of the physics. However, the theory of electromagnetism had to deal with a shocking empirical discovery that shifted in the other direction: from mathematical fiction to ontological posit. That is the next chapter of our tale.

5. The Aharonov–Bohm Effect

The physical theories presented above are not capable—even in principle—of accounting for some observable results that can be obtained in the lab. It is true that these results were not discovered at random: they were predicted by quantum theory and then confirmed. However, there is no reason in principle why they could not have been stumbled across. Since the illustrative points I want to make would become too bogged down if we tried to deal with quantum theory here, I will ignore all the theoretical details and just focus on the phenomenon and the trouble it causes.

The story is well known. There is both an electric and a magnetic Aharonov–Bohm effect described in [8], but the magnetic is more familiar and certainly striking enough. Take the usual two-slit set-up for demonstrating interference effects with electrons and embed a solenoid between the two slits. Shield the solenoid so no magnetic field inside leaks out and no electrons from outside can penetrate in. Run the experiment and note where the interference bands are formed. Then change the magnetic flux in the solenoid *without altering either the electric or magnetic fields outside the solenoid.* Run the experiment again. The interference bands will be found to have shifted. The exact amount of the shift as a function of the flux in the solenoid can be calculated and the prediction checked. They match. Figure 1 shows the experimental set-up as depicted in the original paper of Aharonov and Bohm [8] (p. 486).

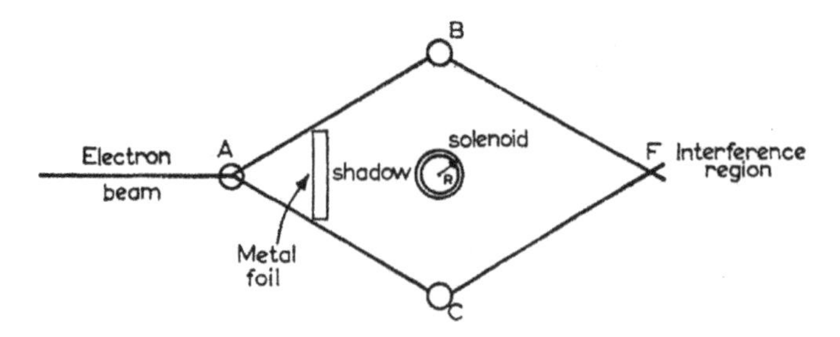

Figure 1. The Aharonov–Bohm set-up.

It is clear from our Canonical Presentations above that none of those theories are capable, even in principle, of accounting for the effect if the experimental conditions are as described. Note that everything in these theories—both the ontology and the nomology—is local in Einstein's sense. There are no non-local beables, and the equations in the nomology take the form of local differential equations. That means the if there are no changes in the experimental conditions outside of the solenoid on the two trials of the experiment, then there can be no changes in the outcome of the experiment outside of the solenoid either. But changing the flux inside the solenoid, according to the theory, does not alter either the electric or the magnetic field outside. And the electrons themselves cannot penetrate the solenoid to be affected. In principle, an action-at-a-distance theory could account for the difference in outcomes, but none of the laws in any of the nomologies are action-at-a-distance laws. So none of our ways of articulating classical electromagnetic theory is up to the task of explaining the effect.

How was the effect originally predicted? The quantum state of the electron gets coupled to the *vector potential*, not to the electric and magnetic fields. Furthermore, when the magnetic field inside the solenoid changes, the potentials *outside* the solenoid have to change as well. This follows from the Generalized Stokes Theorem: the integral of a one-form over the surface enclosing a volume must equal the integral over the enclosed volume of the exterior derivative of the one-form. In three dimensions, as stated in the more familiar vector calculus, this amounts to the claim that the integral of a vector field over the surface of a volume equals the integral over the enclosed interior of the Curl of the vector field. Now as noted in the Canonical Presentations above, since the divergence of the magnetic field $\vec{B}$ is zero, there exists a vector field $\vec{A}$ such that $\vec{B} = \mathrm{Curl}(\vec{A})$. So if we change $\vec{B}$ inside the solenoid, we change $\mathrm{Curl}(\vec{A})$ inside the solenoid, and hence we change $\mathrm{Curl}(\vec{A})$ in the interior of any closed surface that contains the solenoid. However, by the Generalized Stokes Theorem that means that $\vec{A}$ must change on the surface of the volume, even if the surface lies as far as you like from the solenoid. Changing $\vec{B}$ inside forces a change in $\vec{A}$ outside, even though $\vec{B}$ may not change a bit outside. And since the electron wavefunction couples to $\vec{A}$, this allows there to be local laws that predict the shift in the interference bands so long as those laws advert to $\vec{A}$, rather than to $\vec{B}$.

But according to our rules, $\vec{A}$ can only be used in specifying the nomology if it appears in the ontology. So the obvious way to try to account for the Aharonov–Bohm effect is to *move the scalar and vector fields from the category of Mathematical Fictions to the category of Physical Ontology.* This is the formal implementation in the setting of Canonical Presentations of the moral that Aharonov and Bohm draw in their paper. The title is straightforward: "Significance of Electromagnetic Potentials in Electromagnetic Theory". The paper opens ([8], p. 485):

> In classical electrodynamics, the vector and scalar potentials were first introduced as a convenient mathematical aid for calculating the fields. It is true that in order to obtain a classical canonical formalism, the potentials are needed. Nevertheless, the fundamental equations of motion can always be expressed directly in terms of the fields alone.

> In the quantum mechanics, however, the canonical formalism is necessary, and as a result, the potentials cannot be eliminated from the basic equations. Nevertheless, these equations, as well as the physical quantities, are all gauge invariant; so that it may seem that even in quantum mechanics, the potentials themselves have no independent significance.
>
> In this paper we shall show that the above conclusions are not correct and that a further interpretation of the potentials is needed in quantum mechanics.

Colloquially, this change in the status of the potentials is often reported by saying that in quantum theory, but not in the classical Maxwell theory, the scalar and vector potentials are physical, physically real, or real.

The Canonical Presentation offers a direct way to indicate this change: move the two potentials from the category of Mathematical Fictions to the category of Physical Ontology. There is no obvious need to make them non-local beables, so we won't.

Having added the potentials to the Physical Ontology, there are then a series of other decisions to make, which can be made in many ways. The first one is whether the addition of the potentials to the ontology should be accompanied by the elimination of the electric and magnetic fields.

Were we to leave the fields in place there would be several consequences. One is simply that the ontology becomes more bloated, presumably to Ockham's consternation. However, that is the least of it. The real problem is that the potentials and the fields are not independent degrees of physical freedom: they cannot vary independently of each other. In order to avoid such dependent behavior, we would have to add a new law to the nomology. In particular, the relation $\vec{B} = \text{Curl}(\vec{A})$, which has so far appeared only as a mathematical observation, would have to become a law relating two distinct physical magnitudes. While Ockham is known for trying to reduce the ontology to the smallest possible set, our main goal has been to reduce the *nomology* to just one compelling equation. Any sort of behavior can be accounted for by any ontology if you allow yourself enough laws, but that is not the case given a single fixed law, even if you increase the ontology.

There is one obvious new home for $\vec{B} = \text{Curl}(\vec{A})$: a constitutive definition of the magnetic field in terms of the vector potential. In this picture, magnetic fields are *real* but also *derivative*: magnetic fields are, as it were, *made out of* or *aspects of* vector potentials. As derivative, they come with no ontological cost. Since the equation is now just part of a (real) definition, they come with no nomological cost either. They are, in a sense, descriptive conveniences as the potentials were originally calculational conveniences. However, while the original potentials did not exist at all, in the new scheme the fields do exist as derivative physical entities. These derivative entities will certainly obey Humean laws: the very laws of Maxwell. However, just as the ontology is derivative and comes at no cost, so too does that Humean nomology.

The same move can be made by swapping out the scalar potential for the electric field. However, having made the potentials "real", there are knock-on consequences for the nomology. Recall that the mathematical expression of the Nomology should be couched in only in terms of the Physical Ontology and the Spatiotemporal Structure. So the reference to $\vec{E}$ and $\vec{B}$ must be expunged in favor of $\vec{A}$ and ϕ. Rewriting Maxwell's equations in terms of the potentials was the bread and butter of physicists using the classical theory because, as we have mentioned, the gauge freedom of the potentials offered opportunities to simplify calculations. If we do nothing more than rewrite the equations in this way and return from the point particle ontology to the mass density ontology, we get this Canonical Presentation (Table 5).

Table 5. Naïve Theory with Ontology of Potentials.

Theory	Physical Ontology; Spatiotemporal Structure	Mathematical Representation of Physical Ontology	Purely Mathematical Facts	Nomology	Constitutive Principles of Derivative Ontology
Vector and scalar potentials, Mass density, Newtonian Space and Time	Vector Potential Scalar Potential Charge density Mass density Lorentz Force; Time 3-D Euclidean Absolute Space	$\vec{A}(x, y, z, t)$ $\phi(x, y, z, t)$ $\rho(x, y, z, t)$ $\mu(x, y, z, t)$ $\vec{F}_L(x, y, z, t)$ $t \in \mathbb{R}$ $(x, y, z) \in \mathbb{R}^3$	The nomology does not fix the history of $\vec{A}$, or ϕ, given complete initial values. Radical indeterminism $\vec{A}' = \vec{A} + \text{Grad}\xi$ $\phi' = \phi - \frac{\partial \xi}{\partial t}$	$\nabla^2\phi - \frac{\partial\, \text{Div}\, \vec{A}}{\partial t} = -\rho$ $\nabla^2\vec{A} - \frac{\partial^2 \vec{A}}{\partial t^2} - \text{Grad}\left(\text{Div}\vec{A} + \frac{\partial \phi}{\partial t}\right) = \rho\vec{v}$ $\vec{F}_{net} = \mu\frac{d\vec{v}}{dt}$ $\vec{F}_L = \rho\left(-\text{Grad}\phi - \frac{d\vec{A}}{dt} + \text{Grad}\left(\vec{v} \cdot \vec{A}\right)\right)$	$\vec{J} = \rho\vec{v}$ $\vec{B} = \text{Curl}\ \vec{A}$ $\vec{E} = -\text{Grad}(\phi) - \frac{\partial \vec{A}}{\partial t}$

This theory has the same problem with the $\vec{v}$ term in the nomology as our original theory did. However, a much more severe issue has arisen: the new dynamics of $\vec{A}$ and ϕ is now radically indeterministic. The gauge freedom so prized by mathematicians has been converted into a real unconstrained physical freedom. Given any solution of the dynamical equations with a given set of initial conditions one can generate a physically different solution with the same initial conditions. Simply choose an arbitrary function $\zeta(x, y, z, t)$ whose initial gradient is zero and initial time derivative is zero. Plugging that into the equations for what used to be a gauge transformation yields a new solution from the same initial state.

The Canonical Presentation above is the most naïve and ham-handed way to implement the command to "regard the scalar and vector potentials as real". It illustrates the dangers of mindlessly reifying some mathematical object to serve some end. It is true that according to the theory when the physical state in the solenoid changes (by changing the $\vec{A}$ field so that it's Curl changes), the state of $\vec{A}$ out in the region available to the electron also changes. It becomes possible, then, to account for the phenomena with a theory that posits only local interactions. However, the radical indeterminism is surely too high a price to pay, especially when the phenomena themselves display no indeterminism or unpredictability (at the level of the location of the interference bands).

The problem is that we now have more physical degrees of freedom than are required for explanatory purposes. The solution is to posit some new constraint or restriction that kills off that surplus physical degree of freedom. In classical electromagnetic theory, carried out using the potentials, this would be called "fixing a gauge", but keep in mind that in that theory nothing at all physical was at stake. The most convenient gauge for each problem could be chosen, varying the choice from one situation to the next. However, exactly because of the practical utility of finding a convenient gauge, various gauge-fixing conditions were developed. Our job now is to consider what happens if we use one or another of these conditions to eliminate the indeterminism. What physical features do these various theories have?

The first, and most famous, gauge is Lorenz gauge (after Ludvig Lorenz). Lorenz gauge kills off some of the extra degrees of freedom by requiring that $\text{Div}\vec{A} = -\frac{\partial \phi}{\partial t}$. This condition clearly simplifies the nomology, in that the second equation becomes $\nabla^2\vec{A} - \frac{\partial^2 \vec{A}}{\partial t^2} = \rho\vec{v}$. That has two notable effects. First, it separates the variables, so there is one equation that mentions only the scalar potential and another than mentions only the vector potential. Second, the two equations have the same form, namely $\nabla^2 X - \frac{\partial^2 \vec{X}}{\partial t^2} = Y$ or $\Box^2 X = Y$.

Here is the Canonical Presentation (Table 6).

Table 6. Lorenz-Gauge Theory with Charge Density.

Theory	Physical Ontology; Spatiotemporal Structure	Mathematical Representation of Physical Ontology	Purely Mathematical Facts	Nomology	Constitutive Principles of Derivative Ontology
Vector and scalar potentials in Lorenz Gauge, Mass density, Charge density Newtonian Space and Time	Vector Potential Scalar Potential Charge density Mass density Lorentz Force; Time 3-D Euclidean Absolute Space	$\vec{A}(x, y, z, t)$ $\phi(x, y, z, t)$ $\rho(x, y, z, t)$ $\mu(x, y, z, t)$ $\vec{F}_L(x, y, z, t)$ $t \in R$ $(x, y, z) \in \mathbb{R}^3$	Definition of Lorenz Gauge: $\mathrm{Div}\vec{A} = -\frac{\partial\phi}{\partial t}$ ϕ is fixed by initial boundary conditions, e.g., requiring it to go to zero sufficiently fast at ∞.	$\nabla^2\phi - \frac{\partial^2\phi}{\partial t^2} = -\rho$ $\nabla^2\vec{A} - \frac{\partial^2\vec{A}}{\partial t^2} = -\rho\vec{v}$ or, using the d'Alembertian: $\Box^2\phi = -\rho$ $\Box^2\vec{A} = -\rho\vec{v}$ $\mu\frac{d\vec{v}}{dt} = \rho\left(-\mathrm{Grad}\phi - \frac{d\vec{A}}{dt} + \mathrm{Grad}\left(\vec{v}\cdot\vec{A}\right)\right)$ (if $\vec{F}_L = \vec{F}_{net}$)	$\vec{J} = \rho\vec{v}$ $\vec{B} = \mathrm{Curl}\,\vec{A}$ $\vec{E} = -\mathrm{Grad}(\phi) - \frac{\partial\vec{A}}{\partial t}$

This method of gauge-fixing does not kill off the gauge degrees of freedom completely. In addition, one must put a constraint on how the scalar field behaves as it goes to spatial infinity. That is, the Lorenz condition itself only partially fixes a gauge. The remaining freedom—such as how the potentials behave as one goes to spatial infinity—can be specified in the initial conditions, but no empirical considerstions can dictate what those initial conditions are.

Lorenz gauge is mathematically convenient because in it the dynamical equations for the vector and scalar potentials decouple, which explains why Lorenz would have employed it in 1867, long before any hint of the Theory of Relativity. However, to a more modern eye, something quite different jumps off the page. The d'Alembertian operator is obviously exactly the right form to be invariant under a Lorentz transformation (that's Hendrik Lorentz). In other words, having decided to take the potentials ontologically seriously, we find that in one gauge the nomology simplifies and shows common structure for the scalar and vector potentials. And if one had the Canonical Presentation of Maxwellian electromagnetism in Lorenz gauge, it might possibly occur to you that the fundamental dynamical equations for the potentials could suggest a change in the spatiotemporal structure. At least as far as the potentials go, all the spatiotemporal structure would need to do is to allow for the definition of the d'Alembertian. By such a route, one could have arrived at Special Relativity—as understood by Minkowski—as a new theory of space-time structure. In short, to a modern sensibility, Maxwell's theory expressed in Lorenz gauge virtually shouts "I want to live in Minkowski space-time". This gives some indication of the heuristic power that may accrue to presenting a theory in Canonical form. It becomes clear exactly what is being postulated and how the parts fit together.

We can, of course, play exactly the same game on the new theory as we did on the other theories: instead of the ontologically independent scalar potential and charge density requiring a law in the nomology to keep them correlated, one can just decide to cut the Gordian knot by making the ontological identification of ρ with $\frac{\partial^2\phi}{\partial t^2} - \nabla^2\phi$. Just as we made the electric charge density into a structural feature of the electric field above, so we can make it a structural feature—indeed, a Lorentz invariant structural feature—of the scalar potential here. That move reduces the whole dynamics of the potentials to one law, as shown below (Table 7).

Table 7. Lorenz Gauge, Derived Charge Density.

Theory	Physical Ontology; Spatiotemporal Structure	Mathematical Representation of Physical Ontology	Purely Mathematical Facts	Nomology	Constitutive Principles of Derivative Ontology
Vector and scalar potentials in Lorenz Gauge, Mass density, Derived Charge Density Newtonian Space and Time	Vector Potential Scalar Potential Mass density Lorentz Force; Time 3-D Euclidean Absolute Space	$\vec{A}(x, y, z, t)$ $\phi(x, y, z, t)$ $\mu(x, y, z, t)$ $\vec{F}_L(x, y, z, t)$ $t \in \mathbb{R}$ $(x, y, z) \in \mathbb{R}^3$	Definition: Lorenz Gauge $\mathrm{Div}\vec{A} = -\frac{\partial\phi}{\partial t}$ ϕ is also gauge-fixed if it has to go to zero sufficiently fast at ∞.	$\nabla^2\vec{A} - \frac{\partial^2\vec{A}}{\partial t^2} = -\rho\vec{v}$ or, using the d'Alembertian: $\Box^2\vec{A} = -\rho\vec{v}$ $\mu\frac{d\vec{v}}{dt} = \Box^2\phi\left(-\mathrm{Grad}\phi - \frac{d\vec{A}}{dt} + \mathrm{Grad}\left(\vec{v}\cdot\vec{A}\right)\right)$ (if $\vec{F}_L = \vec{F}_{net}$)	$\vec{J} = \rho\vec{v}$ $\vec{B} = \mathrm{Curl}\,\vec{A}$ $\vec{E} = -\mathrm{Grad}(\phi) - \frac{\partial\vec{A}}{\partial t}$ $\nabla^2\phi - \frac{\partial^2\phi}{\partial t^2} = -\rho$ or $\rho = -\Box^2\phi$

Of course, in direct parallel to the discussion above, the status of the velocity vector is obscure when using a mass density rather than a particle theory. And the same move to a particle theory is available again.

We are now in a position to make the most important observation so far. By trading off the electric and magnetic fields in the original ontology for the scalar and vector potentials, we have been able to save locality even while handling the Aharonov–Bohm effect: in fact, in the quantum mechanical one-particle theory the wave function of the electron couples to the vector potential in a completely local way. Changing the physical state of the solenoid necessarily changes the vector potential outside the solenoid, so the trick behind the effect seems to be revealed. However, one price we pay for taking the potentials seriously is taking the gauge ontologically seriously as well. What in the classical theory was a merely conventional and unphysical change of gauge becomes the means of changing out one theory in favor of a rival, distinct theory. Our next example illustrates this.

Another of the most popular gauges used in classical Electromagnetism is the Coulomb gauge. The Coulomb gauge condition is $\mathrm{Div}\vec{A} = 0$. Looking back at the form that Maxwell's laws take in terms of the potentials, the Coulomb condition again simplifies the nomology. Further, unlike the Lorenz condition, the Coulomb condition is a full gauge-fixing condition: imposing it eliminates all the original freedom in picking a gauge. Hence the resulting theory—once we take the potentials as elements of the physical ontology—is both deterministic and does not require any choice of initial conditions not motivated empirically. We take for granted that the local charge distribution, unlike the value of the scalar or vector potential, is empirically observable.

Here is a naïve attempt at a Canonical Presentation of Classical Electromagnetism formulated in terms of the potentials in Coulomb Gauge. We have eliminated the charge density by ontological definition in terms of the scalar potential (Table 8). To improve readability, we have employed $\vec{j}$ in presenting the nomology, but it can be eliminated in favor of the fundamental Physical Ontology via the Constitutive Principles at will.

Table 8. A Naïve Attempt at Coulomb-fixed Potentials.

Theory	Physical Ontology; Spatiotemporal Structure	Mathematical Representation of Physical Ontology	Purely Mathematical Facts	Nomology	Constitutive Principles of Derivative Ontology
Vector and scalar potentials in Coulomb Gauge, Mass density, Charge Density Newtonian Space and Time	Vector Potential Scalar Potential Mass Density Charge Density Lorentz Force; Time 3-D Euclidean Absolute Space	$\vec{A}(x, y, z, t)$ $\phi(x, y, z, t)$ $\mu(x, y, z, t)$ $\rho(x, y, z, t)$ $\vec{F}_L(x, y, z, t)$ $t \in \mathbb{R}$ $(x, y, z) \in \mathbb{R}^3$	Definition of Coulomb Gauge $\mathrm{Div}\vec{A} = 0$	$\frac{\partial^2 \vec{A}}{\partial t^2} - \nabla^2\vec{A} = \vec{J} - \mathrm{Grad}\left(\frac{\partial\phi}{\partial t}\right)$ or, using the d'Alembertian: $-\Box^2\vec{A} = \vec{J} - \mathrm{Grad}\left(\frac{\partial\phi}{\partial t}\right)$ $\mu\frac{d\vec{v}}{dt} = -\nabla^2\phi\left(\mathrm{Grad}\left(\phi - \left(\vec{v}\cdot\vec{A}\right)\right) + \frac{d\vec{A}}{dt}\right)$ (if $\vec{F}_L = \vec{F}_{net}$) $\phi\left(\vec{x}, t\right) = \int \frac{\rho\left(\vec{x}', t\right)}{\lvert\vec{x} - \vec{x}'\rvert} d^3\vec{x}'$	$\vec{J} = \rho\vec{v}$ $\vec{B} = \mathrm{Curl}\,\vec{A}$ $\vec{E} = -\mathrm{Grad}(\phi) - \frac{\partial\vec{A}}{\partial t}$ $-\nabla^2\phi = \rho$

A careful examination of the Canonical presentation reveals an anomaly: there are two separate equations that relate the scalar potential ϕ and the charge density ρ. One is the local equation that derives from the desire to eliminate the charge density by definition as first just the Divergence of the electric field and then as the negative of the Laplacian of the scalar potential. This is a local equation in that the scalar potential in any neighborhood of a point determines the charge at that point. The second equation is most naturally read the other way. It defines the scalar potential in a non-local way: the scalar potential is the sum of all the contributions of all the charge densities in the universe with an inverse squared-distance dependency. This definition makes the value of ϕ at any point a function of the contemporaneous charge distribution throughout the universe.

Clearly, one does not want to define the charge distribution in terms of the scalar potential and then turn around and define the scalar potential as a function of the charge distribution. So at best one of the two equations can survive: either the local or the non-local one. Einstein's choice, of course, would be the local one, but we are going to make the other decision: keep the equation relating the scalar potential to the contemporaneous charge distribution. The next question is the ontological status of this equation. Is it an ontological analysis of charge in terms of scalar potential, an ontological analysis of scalar potential in terms of charge, or an element of the nomology: a law relating the scalar potential to the contemporaneous charge distribution? Again, we will make the choice of treating it as a law, an "instantaneous production-at-a-distance" law. The picture is that the scalar potential is an *effect* of the contemporaneous global charge distribution.

One could obviously make any of these decisions differently, and so end up with a different theory. The whole universe of such theories would be interesting to map, but for reasons that will soon become apparent, this is the articulated theory I want to pursue here.

The Canonical Presentation is seen in Table 9.

Table 9. Coulomb Condition, Derived Charge Density.

Theory	Physical Ontology; Spatiotemporal Structure	Math Representation of Physical Ontology	Purely Mathematical Facts	Nomology	Constitutive Principles of Derivative Ontology
Vector and scalar potentials in Coulomb Gauge, Mass density, Derived Charge Density Newtonian Space and Time	Vector Potential Scalar Potential Mass density Lorentz Force; Time 3-D Euclidean Absolute Space	$\vec{A}(x, y, z, t)$ $\phi(x, y, z, t)$ $\mu(x, y, z, t)$ $\vec{F}_L(x, y, z, t)$ $t \in \mathbb{R}$ $(x, y, z) \in \mathbb{R}^3$	Definition of Coulomb Gauge $\mathrm{Div}\vec{A} = 0$	$\frac{\partial^2 \vec{A}}{\partial t^2} - \nabla^2 \vec{A} = \vec{J} - \mathrm{Grad}\left(\frac{\partial \phi}{\partial t}\right)$ or, using the d'Alembertian: $-\Box^2 \vec{A} = \vec{J} - \mathrm{Grad}\left(\frac{\partial \phi}{\partial t}\right)$ $\mu \frac{d\vec{v}}{dt} = -\nabla^2 \phi \left(\mathrm{Grad}\left(\phi - \left(\vec{v} \cdot \vec{A}\right)\right) + \frac{d\vec{A}}{dt}\right)$ (if $\vec{F}_L = \vec{F}_{net}$) $\phi\left(\vec{x}, t\right) = \int \frac{\rho\left(\vec{x}', t\right)}{\lvert\vec{x} - \vec{x}'\rvert} d^3\vec{x}'$	$\vec{J} = \rho\vec{v}$ $\vec{B} = \mathrm{Curl}\,\vec{A}$ $\vec{E} = -\mathrm{Grad}(\phi) - \frac{\partial \vec{A}}{\partial t}$ $-\nabla^2 \phi = \rho$

Of course, as a theory with matter densities rather than particles, the significance of the velocity that occurs explicitly in the Lorenz Force Law and implicitly in the dynamical equation for the vector potential (via the current) is somewhat obscure. So, as our final adjustment of the theory we will replace the mass density and charge density with point particles that have characteristic masses and charges. That yields Table 10.

Table 10. Potentials with Coulomb Condition and Particles.

Theory	Physical Ontology; Spatiotemporal Structure	Mathematical Representation of Physical Ontology	Purely Mathematical Facts	Nomology	Constitutive Principles of Derivative Ontology
Vector and scalar potentials in Coulomb Gauge, Particles with Mass and Charge, Newtonian Space and Time	Vector Potential Scalar Potential Point Particles Particle Charge Particle Mass Lorentz Force Time 3-D Euclidean Absolute Space	$\vec{A}(x, y, z, t)$ $\phi(x, y, z, t)$ $\vec{x}_i(t) = (x_i, y_i, z_i)$ $q_i \in R$ $m_i \in R > 0$ $\vec{F}_L(x, y, z, t)$ $t \in \mathbb{R}$ $(x, y, z) \in \mathbb{R}^3$	Definition of Coulomb Gauge $\mathrm{Div}\vec{A} = 0$	$\frac{\partial^2 \vec{A}}{\partial t^2} - \nabla^2 \vec{A} = \vec{J} - \mathrm{Grad}\left(\frac{\partial \phi}{\partial t}\right)$ or, using the d'Alembertian: $-\Box^2 \vec{A} = \vec{J} - \mathrm{Grad}\left(\frac{\partial \phi}{\partial t}\right)$ $m_i \frac{d\vec{v}_i}{dt} = q_i \left(\mathrm{Grad}\left(\phi - \left(\vec{v}_i \cdot \vec{A}\right)\right) + \frac{d\vec{A}}{dt}\right)$ (if $\vec{F}_L = \vec{F}_{net}$) $\phi\left(\vec{x}, t\right) = \sum_{n=1}^{n=N} \frac{q_i}{\lvert\vec{x} - \vec{x}'\rvert'}$ omitting points on particle worldlines	$\vec{J} = \rho\vec{v}$ $\vec{B} = \mathrm{Curl}\,\vec{A}$ $\vec{E} = -\mathrm{Grad}(\phi) - \frac{\partial \vec{A}}{\partial t}$ $\vec{v} = \vec{v}_i = \frac{d\vec{x}_i(t)}{dt}$ $\vec{J}_i = q_i\vec{v}_i$

There are various 't's to cross and 'i's to dot, but let us stop here. So far we have articulated eight separate theories—theories with distinct ontologies and nomologies—based on classical electromagnetic theory. If one uses the phrase "Maxwell's theory" or "classical electromagnetic theory" in a way that is neutral between all or some of these precisely specified theories then one is using the term "theory" too loosely for metaphysical or ontological purposes. If you want to know what a theory posits about the world then you need to have a theory that is clearly enough articulated to correspond to a single Canonical Presentation.

Note that there is nothing that would have prevented Maxwell from considering and accepting some of these had it occurred to him to "take the potentials seriously". Note also that of all the theories we have considered yield the same empirical predictions, on any clear notion of "empirical predictions".

6. Adjusting the Spatiotemporal Structure

So far, we have considered ways of changing a theory so that particular equations in the Canonical Presentation get moved around: from the Nomology to the Constitutive Principles of Derivative Ontology, from Mathematical Fictions to Mathematical Representations of Fundamental Ontology, etc. In order to account for the Aharonov–Bohm effect, either we needed something more than the electric and magnetic fields in the ontology or we needed an explicit action-at-a-distance law. Aharonov and

Bohm's original suggestion was that we have to take the potentials more seriously, an idea that can be implemented in many ways (as we have seen). However, the immediate price to pay for changing the status of the potentials is simple: radical indeterminism. One obvious way to get rid of that is to fix gauge. Both the Lorenz and Couloumb conditions fix the gauge enough to eliminate the indeterminism. A weaker reduction in the gauge degrees of freedom, could leave us with radical indeterminism.

However, one thing has remained untouched so far: the spatio-temporal structure. Every one of our theories has been constructed in the same space-time setting: Newtonian Absolute space and Absolute Time. Obviously, this is a component that not only can but must be adjusted to account for Relativistic effects in a plausible way. Our final topic is reflection on this situation.

Just as the physical ontology and the nomology are mutually constraining—only mathematical representations of items in the physical ontology should appear in the nomology and every such representation should appear somewhere in the nomology—so too are the contents of the nomology and the spatio-temporal structure. You need enough spatio-temporal structure to express the laws, and don't want more structure than is required for that purpose. Having the full Absolute Space and Time gives one a lot of structure to work with, and indeed more structure than one needs. In particular, it allows one to define the absolute velocity of a particle, which is a notion that has long been regarded as suspect. Playing the same game with Newtonian mechanics reveals that some of the structure of Absolute Space and Time is otiose for that theory, and one can set a similar theory in Galilean space-time. What considerations apply to the question of spatio-temporal structure here?

If we set r(x, y, z, t) or all the q_i to zero, all that is left of Maxwell's theory is the homogeneous Maxwell equations. In the theory that makes the potentials fundamental, choosing the Lorenz gauge condition converts the nomology to $\Box^2\phi = 0$ and $\Box^2\vec{A} = 0$. As noted above, if this where the whole story ended then the theory would be suggesting that it lives naturally in a Minkowski space. The d'Almerbertian is easily and naturally definable in Minkowski space-time, where it is Lorentz invariant. Indeed, the Lorentz invariance of the nomology in Lorenz gauge is one of the reasons so many people refer to Lorenz gauge as Lorentz gauge. Even more importantly, the manifest Lorentz invariance of the theory cast in Lorenz gauge provides an easy argument to the conclusion that Maxwellian electro-magnetic is a Lorentz invariant theory. One should then switch the spatio-temporal structure to Minkowski, as it is simpler.

There are a few flies in the ointment. One is that the homogeneous equations are not the whole story. Charged matter does have to be introduced into the theory somehow, so we need an updated version of the Lorentz force law and a dynamics to go with it. The current also needs to be reintroduced and the dynamical equation for the vector potential made sensitive to the matter. Nonetheless, the very existence of the Lorenz gauge has convinced physicists to regard Maxwell's theory as implicitly Relativistic.

Another fly, concerns the fact that the Lorenz condition makes the exact value of the vector potential empirically inaccessible. This can be settled in an island universe by imposing a demand on how the potential behaves as it goes to infinity, but what is the physical motivation for such a constraint?. So long as the potentials were regarded as mere mathematical fictions this made no difference, and the potentials in Lorenz gauge could be used to prove the Lorentz invariance of the field theory. However, if the potentials are not mere fictions, then we would prefer a deterministic theory with empirically justified attribution of values to the potentials.

These reflections suggest a different conclusion: instead of Lorenz gauge, consider a theory in which the potentials are subject to the Coulomb condition. Now the gauge is completely fixed given the charge distribution, yielding both determinism and empirical justification of the initial conditions. What sort of spatiotemporal structure would be needed to express the nomology of this theory?

The presence of the d'Alembertian in the equation for the vector field once again suggests a Minkowski space-time structure. But Coulomb's Law for the scalar potential points in a very different direction. Since the scalar potential, in Coulomb gauge, is a function of the contemporaneous charge distribution throughout all space, one requires a structure akin to Absolute Simultaneity in order to define the theory. It is natural, from this point of view, to add such a structure—a preferred foliation—to the Minkowski space-time.

By this line of argument, merely noticing the Aharonov–Bohm effect, without having a clue about where it originates, could motivate replacing Absolute Space and Absolute Time with a Lorentzian space-time plus a preferred foliation. This suggestion has arisen without any consideration of experimental violation of Bell's inequality. To review the argument: the Aharonov–Bohm phenomenon suggests that the electromagnetic situation outside of the solenoid must change when the flux inside changes. But according to the field theory, the fields outside do not change at all. The vector potential, however, does. So we reify the vector and scalar potentials. Now when the flux changes inside the solenoid the vector potential outside *must* change since the Curl of it has changed inside (via Stokes' Law). So by reifying the potentials we provide the resources needed (ultimately by quantum mechanics) to account for the effect using only local interaction of the vector potential and the electron quantum state. However, in Coulomb gauge, one posits at the very same time a non-local law relating the charge distribution on a leaf of the preferred foliation to the value of the scalar field there. This combination of Relativistic locality with foliation-dependent non-locality is at least strongly suggested by taking the Coulomb condition seriously. It is worthy of both careful consideration and astonishment that the corresponding space-time structure is precisely what one needs to adapt Bohmian mechanics to a Relativistic setting.

It is worthwhile to reflect on this astonishing fact. A little counterfactual history illustrates the point. Suppose that, before the development of quantum theory, an experimentalist just stumbled on the two-slit interference phenomenon. Then again, quite by accident, the experimentalist stuck a solenoid in the experimental design and discovered the Aharonov–Bohm effect. Armed only with classical Maxwellian electro-magnetism in a classical space-time and the familiar Maxwellian calculational techniques, what conclusions might such a physicist entertain?

Without any prompting, the experimentalist would notice that there are shifts in the interference bands even though the electric and magnetic fields outside the solenoid are unchanged. Wary of action-at-a-distance, the experimentalist would look for some physical magnitude that does change outside the solenoid. It is not far to seek in the mathematics: when the magnetic flux in the solenoid changes, the vector potential outside the solenoid must of necessity change too: Gauss's Law demands it. So the experimentalist would first be enticed by Aharonov and Bohm's conclusion: rather than merely being mathematical conveniences for solving problems, the mathematical scalar and vector potentials directly represent something physically real!

However, now the specter of gauge freedom raises its head. Our physicist has become accustomed to choosing whatever gauge equivalent vector and scalar potentials happen to be most convenient for the problem at hand, and has done so with a clear conscience because the potentials were regarded as just mathematical fictions, mere conveniences. "Making them real", whatever that precisely may come to mean, makes the conscience uneasy. If a changed vector potential in a region with an unchanged electric and magnetic field can make an observable difference, then the choice of a gauge cannot just be written off as unproblematic.

We now imagine a chain of possible reactions (not the only possible ones by any means!). First, it occurs to the physicists to cut out the former gauge degrees of freedom by gauge fixing. However, Alice chooses Lorenz gauge and Bob Coulumb gauge. Each puts the appropriate fundamental equation into the nomology, and asks: "What sort of a space-time structure do I really need to make sense of these equations?". Alice has chosen Lorenz gauge, and is now paying much more attention to it. She sees that the equations for A and phi decouple, and furthermore that the decoupled equations both have the form of a d'Alembertian acting on the potential. Noting that the d'Alembertian looks just like a Laplacian with an extra term that has flipped parity, we have already come perilously close to considering the Minkowski metric and what it could describe. So we have the familiar straight-line route from Maxwellian E & M to Special Relativity.

Meanwhile, Bob always liked to work in Coulomb gauge, so he imposes it as the condition on A and phi. Now there are two interlocked equations, one using the d'Alembertian again and the other a simple instantaneous action-at-a-distance formula for the electric potential. Having heard about Alice's work, the d'Alembertian suggests a Minkowski space-time. However, the phi dynamics require, laid over this space-time, a physical foliation. Thus we arrive at a picture of the space-time structure containing both a Relativistic metric and a privileged set of level surfaces.

The Aharonov–Bohm effect does not violate any Bell's Inequality, and can be explained in a completely local way as Alice's theory shows. Nonetheless, the extensive use of Coulomb gauge makes a natural opening—once you decide to reify the vector and scalar potentials—for a space-time structure with both a Lorentzian metric and a preferred foliation. It is exactly that space-time structure that makes it easy to explain Relativistic effects in Bohmian Mechanics, as well as to implement the non-locality that we know any empirically successful theory requires.

Our examination of various proposals for how to account for the Aharonov–Bohm effect by altering the fundamental ontology and/or nomology of the Maxwellian electrodynamics has been for illustrative rather than substantial purposes. As we have seen, even before trying to take quantum theory explicitly into account and operating in a purely classical setting, there are many, many options. We have only just touched on the changes in spatio-temporal structure that Relativity introduces. Further, the effect itself has more complex and subtle forms (see, for example, [9,10]). There is a tremendous amount of detailed work to be done in order to really come to grips with the ultimate ontology of what Maxwell thought of as the electric and magnetic fields. This essay has been concerned with what the general nature of that work is, and how it can be pellucidly displayed, not with what the ultimate outcome should be.

7. Conclusions

Having seen how many different ways the basic structure of classical electromagnetism can be used when constructing alternative, precisely defined theories (as articulated by a Canonical Presentation) it may come as no surprise that we have barely scratched the surface. Electromagnetic theory can, for example, be reformulated in the mathematical language of fiber bundles. In that setting, the fields are derivative, corresponding to the curvature of the connection on the bundle [2]. That mathematical formalism is suggestive of yet another bevy of precise physical theories that the Canonical Presentation could help keep straight.

If physicists were to adopt this method—or any other standardized method—to convey physical theories clearly and unambiguously, many conceptual problems could be avoided. First and foremost is the unfortunate tendency to portray different theories as nothing but different "interpretations" of one and the same theory. Further, the discipline that a standardized format of this sort imposes can make it easier to notice alterative theories that have not yet been considered.

Talk of physical ontology vs. nomology; of derivative ontology vs. mathematical fiction; of spatio-temporal structure; and of fundamentality may strike one as philosophical rather than physical. But these sorts of distinctions lie at the heart of physics, even if they are not often acknowledged. Aharonov and Bohm recognized this perfectly well, so it seems apt to give the last word to them:

> In classical mechanics, we recall that potentials cannot have such significance because the equation of motion involves only the field quantities themselves. For this reason, the potentials have been regarded as purely mathematical auxiliaries, while only the field quantities were thought to have a direct physical meaning.
>
> In quantum mechanics, the essential difference is that the equations of motion for a particle are replaced by the Schrödinger equation for a wave. This Schrödinger equation is obtained from a canonical formula, which cannot be expressed in terms of the fields alone, but which also requires the potentials. Indeed, the potentials play a role, in Schrödinger's equation, which is analogous to that of the index of refraction in optics. The Lorentz force

> $[e\mathbf{E} + (e/c)\mathbf{v} \times \mathbf{H}]$ does not appear anywhere in the fundamental theory, but appears only as an approximation appearing in the classical limit. It would therefore seem natural at this point to propose that, in quantum mechanics, the fundamental physical entities are the potentials, while the fields are derived from by differentiations ([1], p. 490).

Funding: This research was supported by a fellowship from the American Counsil of Learned Societies.

Conflicts of Interest: The author declares no conflict of interest.

References

1. Newton, I. *Principia Vol. 1*; University of California Press: Berkeley, CA, USA, 1934.
2. Baez, J.; Muniain, J. *Gauge Fields, Knots and Gravity*; World Scientific: Singapore, 1994.
3. Bell, J.S. The Theory of Local Beables. In *Speakable and Unspeakable in Quantum Mechanics*, 2nd ed.; Cambridge University Press: Cambridge, UK, 2004; pp. 52–62.
4. Born, M. *The Born-Einstein Letters*; Walker: New York, NY, USA, 1971.
5. Ney, A.; Albert, D.Z. (Eds.) *The Wave Function: Essays on the Metaphysics of Quantum Mechanics*; Oxford University Press: Oxford, UK, 2013.
6. Popkin, R.H. *The Philosophy of the Sixteenth and Seventeenth Centuries*; Free Press: New York, NY, USA, 1966.
7. Beller, M. *Quantum Dialogue*; University of Chicago Press: Chicago, IL, USA, 2001.
8. Aharonov, Y.; Bohm, D. Significance of Electromagnetic Potentials in the Quantum Theory. *Phys. Rev.* **1959**, *115*, 485–491. [CrossRef]
9. Greenberger, D. Reality and Significance of the Aharonov-Bohm Effect. *Phys. Rev. D* **1981**, *23*, 1460. [CrossRef]
10. Olariu, S.; Popescu, I.I. The Quantum Effects of Electromagnetic Fluxes. *Rev. Mod. Phys.* **1985**, *57*, 339. [CrossRef]

Article

Agent Inaccessibility as a Fundamental Principle in Quantum Mechanics: Objective Unpredictability and Formal Uncomputability

Jan Walleczek

Phenoscience Laboratories, Novalisstrasse 11, Aufgang F, 10115 Berlin, Germany; walleczek@phenoscience.com

Received: 25 October 2018; Accepted: 20 December 2018; Published: 21 December 2018

Abstract: The inaccessibility to the experimenter agent of the complete quantum state is well-known. However, decisive answers are still missing for the following question: What underpins and governs the physics of agent inaccessibility? Specifically, how does nature prevent the agent from accessing, predicting, and controlling, individual quantum measurement outcomes? The orthodox interpretation of quantum mechanics employs the metaphysical assumption of indeterminism—'intrinsic randomness'—as an axiomatic, in-principle limit on agent–quantum access. By contrast, ontological and deterministic interpretations of quantum mechanics typically adopt an operational, in-practice limit on agent access and knowledge—'effective ignorance'. The present work considers a third option—'objective ignorance': an in-principle limit for ontological quantum mechanics based upon self-referential dynamics, including undecidable dynamics and dynamical chaos, employing uncomputability as a formal limit. Given a typical quantum random sequence, no formal proof is available for the truth of quantum indeterminism, whereas a formal proof for the uncomputability of the quantum random sequence—as a fundamental limit on agent access ensuring objective unpredictability—is a plausible option. This forms the basis of the present proposal for an agent-inaccessibility principle in quantum mechanics.

Keywords: ontological quantum mechanics; objective non-signaling constraint; quantum inaccessibility; epistemic agent; emergent quantum state; self-referential dynamics; dynamical chaos; computational irreducibility; undecidable dynamics; Turing incomputability

---

1. Introduction

The fast rising interest in ontological quantum mechanics has brought to the fore again the problem of the fundamental limits of experimenter agency in quantum mechanics. For example, the physical consistency of de Broglie-Bohm (dBB) theory [1–4] and Bohmian mechanics [5–8], as well as recent quantum models within the ontological model framework [9–12], depends strictly on the imposition of a limit on agent access to nature. However, what governs the physics of 'agent inaccessibility'? How and why does nature prohibit the experimenter agent from having unlimited access to reality at the level of the quantum? Is the universe "fine-tuned" against agent access to the quantum state? What is the difference between 'agents' and observers' in relation to quantum inaccessibility? Finally, if agent inaccessibility is fundamental, then what is the ontological status of inaccessible quantum states?

The specific choice of an answer to these foundational questions strongly constrains the plausibility of any type of quantum-ontological formalism, whether for ψ-ontic or ψ-epistemic interpretations [9,11], including for quantum models that involve globally deterministic constraints [13–17], such as those exploring the possibility of an emergent quantum mechanics (e.g., see the Special Issue on Emergent Quantum Mechanics in the *Entropy* journal). Critically, this suggests that an *informal* principle like 'agent inaccessibility' can decide whether—or not—a *formal* quantum model, or related mathematical theorem, might be physically realistic in view of the known record of quantum observations in the laboratory.

In terms of advancing a physical account of EPR-type quantum correlations, for example, how to assess whether a proposed quantum formalism is prone to causal-paradox formation? The ineliminable dependence—apparently—of the respective answers upon an informal agent-centric notion should cause concern and motivate the development of a model or theory of the physics of agent inaccessibility.

The present work considers an agent-inaccessibility principle (AIP) as a fundamental principle in quantum mechanics. This analysis adopts the standard assumption that individual quantum detection events are *objectively* unpredictable, i.e., unpredictable by *any* experimenter agents. In search of an explanation for quantum unpredictability, three distinct physical scenarios will be compared, as captured by the concepts of (i) intrinsic randomness, (ii) effective ignorance, and finally (iii) absolute or objective ignorance (see Section 5). The latter concept introduces the possibility of an in-principle limit for agent inaccessibility based upon formal uncomputability and objective unpredictability. As a definition of objective unpredictability, and of objective non-signaling, in quantum mechanics, three types of uncomputability will be considered, all of which are based upon self-referential relations: (i) uncomputability due to the impossibility to know initial conditions with infinite precision, as in dynamical chaos, (ii) uncomputability due to 'computational irreducibility' [18,19], and (iii) uncomputability due to the halting problem as specified in the Church-Turing thesis [20,21]. Regarding the latter concept, the term 'Turing incomputability' will also be employed in this article. Next, without adopting an AIP, how could an ontological quantum theory be physically realistic?

2. Many-World and Single-World Quantum Interpretations

A well-known instance of an ontological quantum interpretation that might—possibly—do without an AIP is Everett's many-worlds (MW) interpretation [22,23]. The problem of (non-signaling) agent access is circumvented in the MW interpretation by branching—upon the agent's measurement of the quantum state—into parallel world ontologies. However, in the MW interpretation, the agent is prohibited from accessing any world ontology but the agent's own, which is, again, a notion of agent inaccessibility, and one that lacks a *physical* explanation in the MW proposal. For many-interacting-worlds interpretations, see References [24–26]. For any single-world (SW) quantum ontology, in particular, such as dBB-theory and Bohmian mechanics, but also for theories involving time-symmetric ontologies, the adoption of an AIP appears to be strictly required in view of possible violations of the non-signaling theorem of quantum mechanics (for an overview see, e.g., Reference [27]). Consequently, the question of whether an experimenter agent can access, predict, compute, and control, quantum information, e.g., as involved in EPR-type quantum correlations during tests of Bell's inequality [28], is crucial for assessing the plausibility of any proposed quantum formalism, whether the formalism posits (local) retrocausality [12–16,29–32], or nonlocality [1–8], including in the development of an emergent quantum mechanics (e.g., Reference [17]).

The target of the present analysis will be SW quantum interpretations in relation to agent inaccessibility. A defining feature of any SW interpretation, whether it is an operational or an ontological one, is that " ... from the viewpoint of an agent who carries out a measurement, this measurement has one single outcome", as was explained by Frauchiger and Renner [33] in the context of their recent argument against the self-consistency of quantum theory due to self-referential relations—in Wigner's friend paradox—between multiple experimenter agents. The significance of the phrase "from the viewpoint of an agent" concerns the additional question—in relation to the single outcome in a SW interpretation—of whether a quantum detection event, e.g., a 'spin-up' observation by an agent in the laboratory, does—or does not—constitute an "objective fact" of nature. For different criticisms of the argument by Frauchiger and Renner [33], see References [34–36].

3. Restricting Agent Access to Ontological Quantum States and Quantum Information

The physical plausibility of SW realist quantum theories, including those based upon nonlocal or retrocausal quantum ontologies, has long been recognized to depend strictly on the assumption that an ontic state (λ) exists whose exact properties are inaccessible to, and hence unobservable by, an experimenter agent. For example, in reference to a time-symmetric quantum ontology, Leifer and Pusey [12] have found that regarding the " ... exact ontic state ... we cannot actually construct an experiment that would reveal it". An example from a wholly different context is decoherence theory, where " ... definite, classical pointer states are selected in the interaction between environment and system" as Zwolak and Zurek [37] explained. There, constraints on agent access are also adopted, of course, and it was noted by these authors [37] that " ... a world where objective information is present is also a world with quantum information inaccessible to all but the most encompassing observer".

The above examples serve as reminders that agent inaccessibility is a central and unavoidable concept in quantum mechanics. That is, the existence of inaccessible quantum information is assumed in diverse quantum-theoretical contexts, and ontological quantum mechanics must typically posit the existence of an ontic state λ whose exact properties are experimentally unobservable. Is it physically feasible, however, that strictly inaccessible, i.e., unobservable, ontic states may—in fact—exist? What is the ontological status of a property or information that does *exist* but that could not be accessed and predicted either *in-practice* or *in-principle*?

3.1. On the Reality of an Indefinite Quantum Ontology: Contextuality and Relationality

An ontological regime whose exact properties are unobservable because they cannot—actually—be revealed experimentally, will be called an indefinite ontology. The term 'indefinite' was chosen as a neutral term in reference to an ontological state prior to its measurement, whether or not that state might possess relational or contextual properties. For clarification, regarding an indefinite (possibly relational or contextual) ontology, the question is not whether a property exists "when no one is looking", but whether some property, or value, exists that cannot be predicted by any amount of "looking", i.e., by any local or nonlocal tests, including computer simulations, prior to performing the actual measurement. That question is closely tied, of course, to the well-known fact that the predictions of orthodox quantum mechanics are wholly incompatible with the (naïve realist) notion of pre-existing quantum properties, i.e., with the false notion that quantum states may possess (non-contextual) definite properties or values prior to, and independent of, their measurement. Put generally, a non-contextual (non-relational) property of an ontological system is one whose outcome state or value is entirely independent of whether, or how, the property is measured by the agent. It is well-known that Einstein, Podolsky, and Rosen, first introduced a definition of definite, non-contextual ontic states in relation to the problem of "action-at-a-distance" in quantum mechanics—the concept of "the elements of reality" [38]. To be sure, quantum ontologies that could be consistent with orthodox quantum predictions must—by contrast—possess value-indefinite properties, thereby allowing consistency with the physical demands of the theorem by Kochen and Specker [39].

Again, the term 'indefinite ontology' is employed because, prior to any measuring interaction, ontic state λ exists in an indefinite state, i.e., a state whose exact value is not accessible, computable, or predictable; by contrast, again, a definite, i.e., measurement-independent, state is one whose value could—in principle—be accessed in nature by the experimenter agent. Importantly, using the present terminology, a quantum-measurement process entails the transformation of an indefinite ontic state (IOS) into a definite ontic state (DOS). Consequently, the standard measurement problem of quantum mechanics is recast as the problem of how to explain, and how to conceptualize, an IOS–DOS transition event. By that definition, a contextual, or relational, ontology is simply one that is governed by IOS-DOS transitioning during the (dynamical) process when the agent performs a measurement upon the quantum state as defined by a particular ontological quantum model.

For explanation, take a typical, experimentally generated quantum random sequence. In the orthodox interpretation of quantum mechanics, each individual random event is presumed to be objectively unpredictable as a function of quantum indeterminacy (see also Section 5.1.1). However, and this is the main proposal of the present analysis, there may be another option for explaining quantum unpredictability—an explanation that is compatible with the presence of an underlying ontology. In the ontological option, each one of the individual quantum detection events that together constitute a quantum random sequence, existed—prior to the measurement-dependent DOS transformation—in the form of an IOS, which is a state possessing value-indefinite properties (e.g., see References [40–42]); only the actual measurement interaction induces the IOS-DOS transition which results in the definite value of the measured ontic state. The present work considers the proposal that an indefinite, likely contextual, ontic state λ represents either (i) an effectively uncomputable element in the weak option of 'effective ignorance' (Section 5.2), or (ii) an objectively uncomputable element in the strong option of 'objective ignorance' (Section 5.3).

Notably, contextuality might represent the "non-information-theoretic kernel" of quantum theory, Koberinski and Müller [43] have suggested recently, and that therefore contextuality could be a genuine physical (ontological) feature of a possible quantum reality. With respect to contextuality as a "genuine physical feature", the investigators cited Fuchs [44] who had also expressed the similar hope earlier that the " ... distillate that remains—a piece of quantum theory with no information-theoretic significance—will be our first unadorned glimpse of 'quantum reality'." Related to that suggestion, the following idea is here pursued also: what Koberinski and Müller [43] have referred to as the "non-information-theoretic kernel" of quantum theory may refer directly to the uncomputable ontological features that are the topic of the present article (compare Section 6).

3.2. *The Inaccessible Universe and the Limits of Science*

In addition to an ontology being contextual or relational, further important questions are (i) whether the ontology is nonlocal [1–8], locally time-symmetric [12–16,29–32], or locally time-asymmetric [45–47], and (ii) what an experimenter agent can know exactly about a given quantum ontology. For example, in the case of the above-mentioned, time-symmetric ontological model, Leifer and Pusey [12] have noted that the exact ontic state λ—although it " ... may be unknown to the experimenter"—" ... is in principle knowable". If so, then in what specific manner? To address questions such as these, the present work proceeds by investigating this general question: What is the ontological status of an empirically inaccessible regime of physical reality?

An objective limit on access to the nature of reality is, of course, anathema to the goals of the project of modern science. Science is thought to be about an understanding of reality based upon the capacity to measure, predict, compute, and control. By contrast, the revolutionary discovery of the fundamental quantization of matter and energy has long been held to imply that—at its smallest dimensions—the universe is intrinsically random, which—from the start—prevents an agent from accessing, predicting, and controlling, individual measurement outcomes. This is, of course, the standard position known as the orthodox interpretation of quantum mechanics. With the advent of ontological quantum mechanics, however, science started to consider the possibility that (ontic) "elements of reality" might exist—at the quantum level—in a form that is both compatible with (i) determinism as well as with (ii) contextuality and single-event unpredictability. However, prior to answering the question of how a fully deterministic system may produce outcome states that are unpredictable and uncomputable as a matter of principle, three related issues will be considered first: (i) the no-hidden-variables theorems in quantum mechanics (Section 3.2.1), (ii) the concept of agent-inaccessible variables (Section 3.3), and (iii) the definition of the experimenter agent (Section 4).

3.2.1. On No-Hidden-Variables Theorems in Ontological Quantum Mechanics

As a way to begin to frame the above question of unpredictability in deterministic systems, the ontological status will be reviewed briefly of the variables called 'hidden' in the original formulation of an ontological quantum theory, namely in dBB-theory [1–3]. The present analysis argues that the introduction of the 'hidden variable' (HV) marked a turning point, not only for quantum physics, but for modern science in general. That is, if proven valid, the HV-concept necessitates the introduction of a radical limit for science: the idea that an inaccessible, or hidden, ontology of nature exists, which is beyond the scientific method to measure, predict, compute, and control (compare Section 6). Importantly, it is the very HV-concept which may ensure that a model of quantum reality could be free from causal-paradox formation, by prohibiting, for example, superluminal signaling and communication, in the typical thought experiments that envision physical inconsistencies due to unorthodox ontological propositions such as nonlocality (e.g., [27]). The opposite and orthodox view has long been defended by those who have employed the traditional "no-hidden-variables" theorems, i.e., the no-go theorems against the physical plausibility of, for example, dBB-theory and Bohmian mechanics (see also Section 5.1).

For critical views arguing against standard interpretations of no-hidden-variables theorems see, for example, Mermin [48], Maudlin [49,50], Lazarovici et al. [51], Passon [52], Tumulka [8,53], Norsen [6,54], Palmer [47], De Gosson [55], Wharton [13,32], Adlam [14,15], Ghadimi et al. [26], Khrennikov [56,57], Hiley and Van Reeth [58], Flack and Hiley [59], and Walleczek [60]. After performing a careful analysis, Gisin [61] noted recently that " ... Bohmian mechanics is deeply consistent", and he remarked that "Bohmian mechanics ... could inspire brave new ideas that challenge quantum physics."

3.3. *Hidden-Variables in Quantum Mechanics are Agent-Inaccessible Variables*

The concept of the HV in quantum mechanics was introduced by David Bohm [1,2]. In original dBB-theory, the mathematical formalism refers to hypothetical ontic elements such as the quantum potential [1–4]. Crucially, to avoid any misunderstanding, it should be mentioned that dBB-theory, which has also been developed in another context as Bohmian mechanics [5–8], is not a classical, ontological theory, but an ontological theory manifesting entirely non-classical properties, including nonlocality. The term 'hidden' usually explains this in Bohm's theory: no measurement can be performed that might reveal exact information about the ontic state in a way that allows an experimenter agent to controllably direct nonlocal information transfers. For example, Holland [4] commented that " ... the quantum potential implies that a certain kind of 'signaling' does, in fact, take place between the sites of distantly separated ... particles in an entangled state", but that this " ... transfer of information cannot, however, be extracted by any experiment which obeys the laws of quantum mechanics". More recently, Valentini [62] had also remarked that this " ... information flow is not visible at the statistical level". Walleczek and Grössing [27] have clarified the point that this nonlocal quantum information transfer must not be understood as information transfer in any communication-theoretic sense. That is, for an ontological quantum theory, such as dBB-theory, which is both contextual and nonlocal (e.g., [48]), the adoption of an AIP—as an informal non-transfer-control theorem in Reference [27]—prohibits access to, and the instrumental control of, nonlocal information transfers for the purpose of sending superluminal (Shannon-type) signals, or messages, between sender and receiver, while—at the same time—allowing the presence of non-Shannon signals [27]. Please note that the term 'hidden signaling' has also been used recently, for example by Bendersky et al. [63], in reference to the concept of non-Shannon signaling [27].

In summary, in a quantum theory such as dBB-theory, the HV indicates the presence of an indefinite ontological element in the theory (i.e., ontic state λ) whose exact value cannot be accessed, predicted, or controlled (e.g., a spin property). That is, again, the HV-concept refers to an unobservable property, not merely to one that is unobserved, and—as a consequence—it cannot be controlled by an observing agent (see Section 3.1). Therefore, John Bell [64], for example, noted that "The usual nomenclature, hidden variables, is most unfortunate", and he proposed that "Perhaps uncontrolled variable would have been better, for these variables, by hypothesis, for the time being, cannot be manipulated at will by us." The present work continues in the spirit Bell's understanding that a variable called 'hidden' represents an uncontrollable variable, i.e., a variable that "cannot be manipulated at will by us" [64]—an *agent*-inaccessible variable using the present terminology. Therefore, before proceeding any further, a definition should be given for what constitutes an 'agent'—as opposed to an 'observer'—in quantum physics and for science in general. How to define the experimenter agent to begin with?

4. Defining the Experimenter Agent

In the particular context of assessing the role of the agent in relation to the non-signaling theorem, John Bell [65] insisted that needed is at least " ... a fragment of a theory of the human being" to be able to address the question of whether or not "we can signal faster than light?". Put differently, Bell requested having a partial theory, at least, of what defines agency in the context of quantum physics. Specifically, the definition should be relevant, as Bell [65] requested, to the question of who "we think we are, we who can make measurements, we who can manipulate external fields, we who can signal at all, even if not faster than light?". In the context of Bell's theorem, a consistent understanding of the notions of agent-dependent versus agent-independent signaling—in terms of Shannon versus non-Shannon signaling—is available from the above-mentioned analysis that applied the operational framework of Shannon's mathematical theory of communication to answer Bell's questions regarding the valid interpretation of the non-signaling theorem [27].

As was described by Walleczek and Grössing [27], an experimenter agent is not merely an observer in the world but is an entity capable of acting in the world in the pursuit of goals, such as (i) in setting-up an experiment for the purpose of asking questions of nature, or (ii) in selecting specific measurement settings (for details see Section 4.3). However, the continuing lack of a model of, or of a theory for, the experimenter agent in quantum physics, and in science in general, impedes making progress towards understanding the foundations of quantum mechanics. The present work suggests that the success to counter the no-go theorems against the possibility of an ontological quantum mechanics also depends (i) on the particular model of the experimenter agent, and (ii) on an understanding of the distinctive role of an AIP in ontological and deterministic interpretations of quantum mechanics (see Section 3).

4.1. The Quantum Measurement Problem

For a long time, the observing agent was considered in the context only of the familiar quantum-measurement problem, especially vis-à-vis collapse-type interpretations such as the Copenhagen interpretation (for an introduction see, e.g., Reference [66]). In recent years, however, the distinct significance of the notions of observation *versus* agency has been recognized well beyond the issue of collapsing the wave function. It is increasingly understood that the concept of the experimenter agent is central to any plausible SW interpretation of ontological quantum theories, not only for ψ-epistemic or purely operational interpretations, such as for quantum Bayesianism [67]. The present work, therefore, seeks to establish a minimum framework, one that is capable of addressing the question of the limits of 'observer agency' in the context of new ontological perspectives for quantum physics. For example, as was described above, traditional assumptions and theorems such as nonlocality, contextuality, free choice, and non-signaling, need not necessarily contradict the existence of certain quantum ontologies. Importantly, the non-contradiction, i.e., the theoretical consistency, of permissible ontologies, such as in the measurement problem as captured by the concept of IOS-DOS transitioning described in Section 3.1, depends on the validity of an AIP in relation to a given quantum formalism. In light of an AIP, who or what is the experimenter agent?

4.2. An Early Definition of the Experimenter Agent: "Maxwell's Demon"

An early notion of the experimenter agent was introduced into physics by James Clerk Maxwell [68]. To review briefly, in Maxwell's thought experiment, an intelligent being or agent was proposed to be capable of lowering the entropy of a "closed" physical system. This being or agent became known of course as 'Maxwell's demon'—a 'demon' because of the apparent supernatural powers to observe, and act in, the world. The term 'super-natural' is used to characterize the kind of exceptional demon agency which Maxwell (falsely) presumed to be "free" from known natural constraints, such as from the Law of Energy Conservation. In short, Maxwell's agent adopts therefore an isolated and quasi-transcendent position towards the rest of the physical universe (see also Section 5.2). In this pre-quantum thought experiment, the feat of entropy reduction is achieved by micro-causal interventions of the observing demon-agent who is granted unlimited access to, and predictive control over, the relevant microphysical processes of the targeted system: first, the agent observes microscopic events, and, second—based on observational knowledge—selectively acts upon the physical system so that the system becomes increasingly ordered. That is, in Maxwell's thought experiment, knowledge-based agent interventions can predictably counter the intrinsic tendency of the closed system to spontaneously disorganize. The problem of the apparent violation of the Second Law of Thermodynamics by the "ordering agent influence" was, of course, first resolved formally by Szilard [69]. It is noteworthy in the present context that elements of Szilard's proof assisted in the development of von Neumann's mathematical foundations of quantum mechanics [70]. The point will be made next that, despite the known shortcomings, the concept of Maxwell's demon captures key features that are still relevant to recent definitions of the experimenter agent (Section 4.3).

4.3. Recent Definition of the Experimenter Agent: "Epistemic Agency"

Already in the early concept of Maxwell's demon were implicit two distinct capacities which continue to be employed in recent definitions of the experimenter agent: (i) the capacity of the agent to observe and to obtain knowledge (the epistemic dimension), and (ii) the capacity of the agent to act in the world in the pursuit of a goal (the agentic dimension). Hence, the term 'epistemic agent' can be used synonymously with the term 'observing agent'. The following informal definition for epistemic agency was introduced previously [27]:

> *"Agency is generally defined as the capacity of humans or other entities to act in the world. Put differently, an agent is defined initially by possessing the capacity to influence causal flows in nature. By prefacing "agent" with the term "epistemic", attention is drawn to the fact that a complete definition of agency represents more than the mere "capacity to influence causal flows": an agent possesses knowledge-based, i.e., epistemic, capacity for predictably directing, and redirecting, causal flows, and thus for directing, and redirecting, information flows as well. That is, an epistemic agent holds the power to (statistically) control physical activity based upon an ability to predict the outcome of specific actions on targeted processes in reference to a known standard or goal. In short, an epistemic agent thus manifests in the world a genuine source of operational control".*

Importantly, the above definition of 'operational control'—as a criterion for epistemic agency—ensures that entities other than human systems, such as artificial devices implementing goal-driven control systems, including devices and algorithms capable of computation and message communication, qualify as complete epistemic agents. Finally, in contrast to the pre-quantum conception of the agent in Maxwell's thought experiment (Section 4.2), after the quantum revolution, from the perspective of the agent as an effective actor in the world, agent inaccessibility is now characterized by the denial of operational control in relation to an inaccessible quantum regime of nature. For example, 't Hooft [71] noted recently also that what " ... distinguishes quantum systems from classical ones is our fundamental inability to control the microscopic details of the initial state ... ". Critically, in the present proposal for an AIP, the measure of 'operational control' is the computational accessibility and predictability of physical processes by the agent. This raises the all-important question of exactly how nature—after the quantum revolution—prohibits (computational) access to the experimenter agent in a way that the purely classical world view—apparently—could not.

5. How does Nature Prohibit Access to the Experimenter Agent?

No scientific consensus exists concerning the question of how nature denies unlimited access to the experimenter agent of quantum states and quantum information. Entirely different physical explanations are on offer—as part of different quantum interpretations—regarding how nature limits agent access to quantum states or information and, therefore, how nature prohibits the prediction, and operational control by epistemic agents of individual quantum measurement outcomes. As was noted already, pre-quantum, classical, physics, by contrast, knows of no fundamental limits regarding agent access to nature (compare Section 4.2).

In the textbook, SW operational interpretation, which is orthodox quantum mechanics, it is the metaphysical assumption of 'intrinsic randomness', i.e., 'quantum indeterminism', which fundamentally limits the powers of the agent to predict the value of a single measurement outcome (see Section 5.1). By contrast, an ontological quantum theory, such as dBB-theory, typically derives its constraint on quantum predictability from the technological inability of the experimenter agent to collect complete information about initial conditions (see Section 5.2). These opposing explanations are frequently discussed in terms of in-principle versus in-practice limits of agent-access to quantum systems. It is often presumed that an in-principle limit to agent-quantum access can only be posited in the case of operational quantum approaches, whereas only an apparently weaker, in-practice limit is available for ontological quantum mechanics.

The present work introduces a third option: the possibility of an in-principle limit for ontological quantum mechanics based upon self-referential dynamics which may produce outcome states whose predictability would require either (i) access to infinitely precise knowledge about initial conditions and/or (ii) the availability of infinite computational resources (see Section 5.3). In the following, the three distinct options will be compared, whereby each one, albeit based on completely different physical assumptions, seeks to explain how nature prevents the agent from computing, predicting, and controlling, individual quantum events. First, the standard position of 'universal indeterminism' will be briefly discussed and criticized in Section 5.1.

5.1. Orthodox Quantum Mechanics: "Universal Indeterminism"

In orthodox quantum mechanics, the assumption of 'intrinsic randomness' serves as an absolute barrier to agent knowledge at the quantum level. Importantly, in the orthodox interpretation, the observed randomness is viewed as an *a priori* property of nature herself, e.g., prior to any additional physical constraints involving the agent. Remarkably, in universal indeterminism, a single random event can initiate an entirely new causal chain—apparently "out of nothing" (e.g., [72]). Nevertheless, and this—again—is the remarkable feature, the detection, for example, of a single 'spin-up' event by the measuring apparatus manifests a classical (pointer) state from which may propagate new causal flows, such as those triggering the formation of new biophysical events during sensory perception in the agent who observes the 'spin-up' measurement outcome. However, the question of what the exact nature might be of that initiating event, i.e., the question of 'what is a quantum?', is not addressed—famously—in the orthodox interpretation, and therefore, Plotnitsky [73], for example, has noted that "... quantum objects are seen as indescribable and possibly even as inconceivable", in the indeterministic interpretation of textbook quantum physics.

What is problematic, however, is that the very same indeterminism, or quantum randomness, which already serves as an absolute limit on agent knowledge, is often—at the same time—held to be the source also of the free-willed agency of the experimenter as in the free-will theorem by Conway and Kochen [74,75]. This is the exact opposite of being the source of a universal constraint. How could this be? How could one and the same (quantum) randomness be the source of both (i) objective chance and (ii) free-willed agent control of physical events in the world, such as freely selecting a measurement setting? This self-contradictory view, which has previously been captured in the concept of quantum super-indeterminism (see Figure 1), has long obscured insight into the plausibility of those no-go arguments against the possibility of ontological quantum mechanics which are based upon the freedom of choice of the experimenter agent (for an overview see Walleczek [60]).

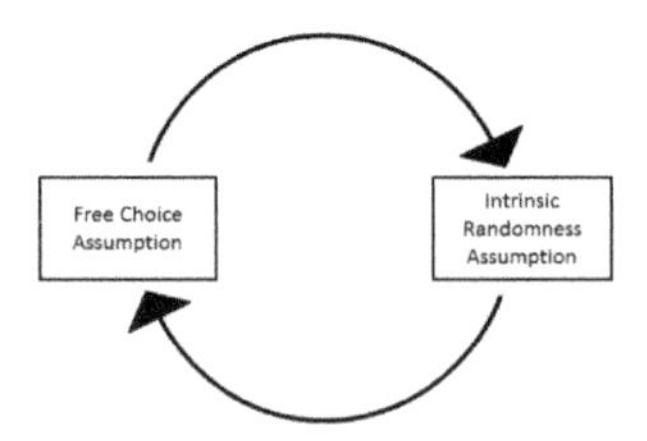

Figure 1. Quantum super-indeterminism [60]. The shortcomings of the orthodox view, which are revealed by the simple concept of super-indeterminism, in the attempt to prove, or justify, the metaphysics behind quantum indeterminacy, are recognized increasingly. The fallacy of circular reasoning is illustrated in Figure 1, which arises from the use of the intrinsic randomness assumption in support of the free choice assumption, which—in turn—rationalizes the presumably "free" selection of measurement settings. Bera et al. [76], for example, have confirmed the fact of 'super-indeterminism' by noting that there is indeed present " ... an unavoidable *circulus vitiosus*" in any tests for true randomness, because any available tests for " ... the indeterministic character of the physical reality" must presume that " ... it is, in fact, indeterministic." Similar arguments have been put forth by, and prior developments were summarized in, Landsman [77].

Standard no-go theorems, such as Bell's theorem [28] or, again, the Conway-Kochen free-will theorem [74,75] fail to account for this contradiction within the orthodox view, which is implied by super-indeterminism (see the legend to Figure 1). Therefore, such no-go theorems, i.e., the theorems claiming the impossibility of particular ontological propositions, imply conclusions of debatable value against the validity of deterministic quantum theories. For example, John Bell recognized the shortcomings himself regarding his own (no-go) theorem in view of an axiomatic interpretation of the non-signaling theorem, and he later adjusted his views [78–82]. For a detailed analysis of Bell's evolving positions—from an axiomatic to an effective non-signaling constraint—see Walleczek and Grössing [27]. Concluding, the simple concept of super-indeterminism (Figure 1) explains why the free choice assumption of the experimenter agent in selecting measurement settings does not imply the necessary rule of the standard, i.e., axiomatic, non-signaling theorem (for details see Figure 2).

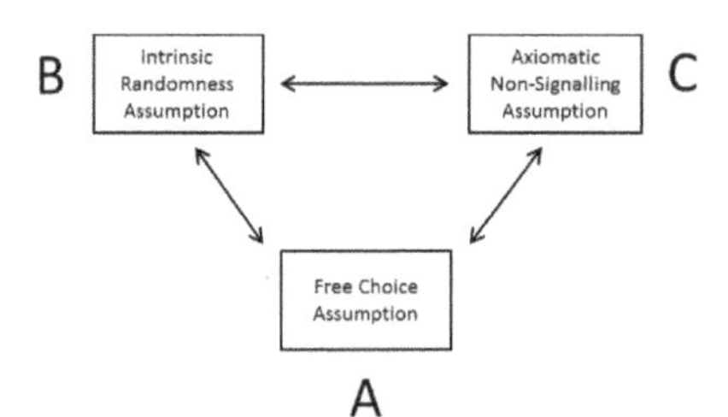

Figure 2. Illustration of the irreducible interdependency of basic assumptions that are implicit in standard interpretations of orthodox quantum mechanics (adapted from Walleczek and Grössing [27,83]). (**A**) Free choice assumption, (**B**) Intrinsic randomness assumption, and (**C**) Axiomatic non-signaling assumption. Importantly, the validity of interpreting the non-signaling theorem as a foundational theorem, or axiom, for quantum mechanics, i.e., one which would imply strict indeterminism as the only viable option for interpreting quantum theory, depends on the independent validity of assumptions (**A**,**B**). However, neither assumption (**A**) nor assumption (**B**) can be confirmed independently if the possibility of 'free choice' depends on the existence of a process that is intrinsically random and vice versa (compare Figure 1). Therefore, for example, the observation of EPR-type nonlocal correlations in the laboratory does not represent empirical proof for the indeterministic nature of the locally observed measurement outcomes, if that proof relies on the employment of an axiomatic non-signaling theorem (for more details see Walleczek and Grössing [27]).

5.1.1. On the Impossibility of Proving the Truth of Quantum Indeterminism

Long-running arguments against the possibility of deterministic, quantum-ontological approaches are increasingly criticized as falling short of their stated aims, in particular those based upon the free-will theorem and the non-signaling constraint as an axiom (see Figures 1 and 2). Importantly, it is widely accepted that quantum indeterminism in the form of actual or objective chance can neither be proven by empirical tests nor by mathematical reasoning (e.g., [84]). However, what might be provable instead is the objective unpredictability of individual quantum measurement outcomes, as defined, for example, by a formal theorem such as Turing incomputability (see Sections 5.3 and 6). Again, "indeterminism" captures a metaphysical assumption about how nature really is—prior to any formal theorizing. Furthermore, an empirical proof of indeterminism is out of reach, likely always, as a final loophole-free test seems to be a logical impossibility (compare Figure 1).

Finally, for a long time, because of the widespread belief that constraints such as absolute unpredictability, free will, nonlocality, non-signaling, or contextuality, could be compatible only with quantum indeterminism, any deterministic approaches to quantum theory have met with little interest by the mainstream of quantum physics, except often in reference to the perceived implausibility of the SW ontological quantum approaches (compare Section 2). This lack of interest has long been due to the near exclusive use—in the mainstream discourse on quantum foundations—of definite, non-contextual ontological assumptions, i.e., those that are consistent only with the classical, metaphysical assumption known as direct or naïve realism. As was mentioned before, the proposed "elements of reality" in the argument by Einstein et al. [38] represent, of course, entirely non-contextual ontic states in agreement with the classical metaphysics of naïve realism; there, the "elements" merely reveal their own "intrinsic", already given, properties at the moment of their measurement.

As was reviewed in Section 3.1, any non-contextual, measurement-independent ontology, such as naïve realism, is wholly incompatible with the measurement predictions of orthodox quantum mechanics [39]. In recent years, however, new research has been pushing the frontiers of ontological possibilities beyond naïve realism, such as in the form of relational ontologies (e.g., Esfeld [85]), time-symmetric ontologies (e.g., Leifer and Pusey [12]), including unconventional causal structures such as retrocausality (e.g., Sutherland [29], Price [30], Wharton [13,31], Price and Wharton [32]). In addition, there has been a revival of interest in the nonlocal and contextual ontologies related to dBB-theory [1–4] and Bohmian mechanics [5–8], which are ontological propositions that posit the fundamental interconnectedness, instead of the intrinsic randomness, of the physical universe (e.g., Walleczek and Grössing [86]).

The focus of the subsequent Sections 5.2 and 5.3 will be an assessment of the continuing possibility of ontology and determinism in quantum theory in relation to the experimenter agent. Specifically, what is sought is a scientifically based notion of "determinism without pre-determination" [60,86]. Next, Section 5.2 presents the traditional option for quantum mechanics in a globally deterministic universe.

5.2. *Ontological Quantum Mechanics: "Effective Ignorance in Global Determinism"*

Instead of the metaphysical assumption of intrinsic randomness (Figure 3A), an ontological quantum mechanics opts for an alternative approach to explain the origins—in a globally deterministic universe—of experimentally observed quantum randomness. That is, ontological approaches typically seek an agent-dependent explanation based upon the unpredictability of individual measurement outcomes as a function of an epistemic limit, which—in the present analysis—is introduced as 'effective ignorance' (Figure 3B).

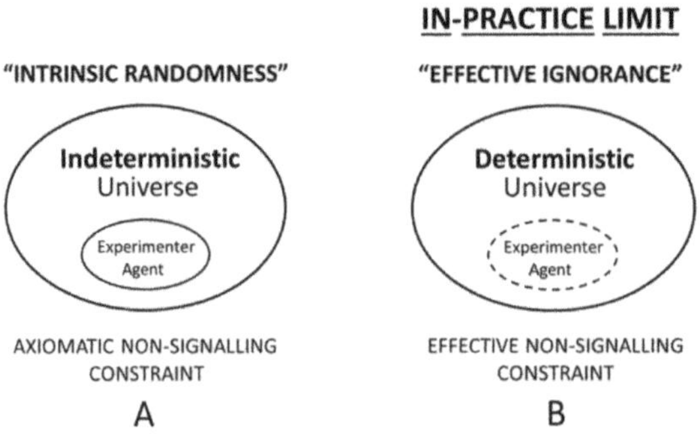

IN-PRACTICE LIMIT

"INTRINSIC RANDOMNESS"
"EFFECTIVE IGNORANCE"

Indeterministic Universe
Experimenter Agent
AXIOMATIC NON-SIGNALLING CONSTRAINT
A

Deterministic Universe
Experimenter Agent
EFFECTIVE NON-SIGNALLING CONSTRAINT
B

Figure 3. Agent inaccessibility as a function of (**A**) Intrinsic randomness versus (**B**) Effective ignorance (adapted from Walleczek [60]). Intrinsic randomness represents the orthodox interpretation of quantum mechanics, which is universal indeterminism. There, the presence of the experimenter agent introduces an apparent metaphysical dualism between agent and world (see the main text for additional explanations), which is indicated by the *closed* line that encloses the presence of the experimenter agent (Figure 3A). By contrast, in universal or global determinism, agents and the physical universe are subject to the same fundamental determinism, whereby, there, the experimenter agent is an integral element of the physical universe, i.e., agent and universe together constitute a lawful, physical continuum (e.g., Szilard [69]), as is indicated by the *open* line (see Figure 3B). In this picture, the experimenter agent constitutes an entity possessing distinct 'epistemic' as well as 'agentic' properties (for definitions see Section 4.3). For a detailed explanation of an axiomatic (Figure 3A) versus an effective (Figure 3B) non-signaling constraint—in the context of Bell's nonlocality theorem—consult Walleczek and Grössing [27]. Briefly, an axiomatic non-signaling constraint (see also Figure 2) is compatible with the violation of measurement outcome independence, which is the standard violation in the context of orthodox quantum theory; by contrast, an effective non-signaling constraint is thought to be compatible with the violation of setting or parameter independence (Shimony [87]), which is the standard violation in the context of an ontological quantum mechanics such as dBB-theory in a universally deterministic universe (Section 3.3).

Importantly, the approach towards an "effective randomness"—by way of the concept of 'effective ignorance'—is an option that can be consistently adopted if agent and universe are not metaphysically separated entities as suggested by the *open* line in Figure 3B (for details see legend to Figure 3). This is in contrast to the orthodox view shown in Figure 3A, where the agent stands in a physically isolated (quasi-transcendent) position towards the rest of the physical universe. For explanation, in the orthodox interpretation of quantum indeterminism, the agent is presumed to be capable of somehow initiating new cause-effect chains "out of nothing", e.g., in violation of Leibniz' Principle of Sufficient Reason (compare Section 5.1). This extra-physical agentic power is reminiscent of Maxwell's demon-agent who was—falsely—thought to be unconstrained by the Laws of Nature, such as by the Second Law of Thermodynamics (see Section 4.2). This isolated, or dualistic, notion of agency in the orthodox picture is indicated by the *closed* line in Figure 3A (for details see legend to Figure 3).

The essential point of 'effective ignorance' is the following (Figure 3B): If assuming that the complete initial conditions of some deterministic system could be obtained, then the exact prediction of outcome states is possible—at least in principle. An example is a computer-generated pseudorandom bit sequence that becomes fully predictable once the (random) seed, i.e., the initial condition, as well as the algorithm, which is used to generate the bit sequence from the seed, is known to the scientific agent. By analogy, having complete knowledge of initial conditions, the properties of a (deterministic) quantum state could be computed, e.g., for the purpose of prediction and control, even if possessing finite computational resources only. Significantly, in the case of effective ignorance—when discrete events are finite—while access to initial conditions (compare the "seed" above) is technologically impractical, there exists, however, no *formal* limit that fundamentally constrains access to the complete initial state. For explanation, the definition of finite resources includes the whole universe as a finite resource, which—again—imposes an in-practice, effective limit, but not an in-principle, objective limit. In summary, the notion of "effective" quantum randomness as a result of the weak epistemic option is—at least in principle—computable by a Turing machine, even if the whole universe is to be recruited as a super-computational resource to achieve quantum predictability.

5.2.1. Understanding John Bell's Concept of "Free Variables" for Quantum Mechanics

The weak epistemic option of effective ignorance is consistent with, and explains, Bell's own proposal of effectively "free variables" [79]. "I would expect a serious theory to permit ... 'pseudorandomness' for complicated subsystems (e.g., computers)," Bell [65] suggested "... which would provide variables sufficiently free for the purposes at hand." In addition, Bell provided the following explanation [79]:

> *"Consider the extreme case of a 'random' generator which is in fact perfectly deterministic in nature—and, for simplicity, perfectly isolated. In such a device the complete final state perfectly determines the complete initial state—nothing is forgotten. And yet for many purposes, such a device is precisely a 'forgetting machine'. A particular output is the result of combining so many factors, of such a lengthy and complicated dynamical chain, that it is quite extraordinarily sensitive to minute variations of any one of many initial conditions. It is the familiar paradox of classical statistical mechanics that such exquisite sensitivity to initial conditions is practically equivalent to complete forgetfulness of them."*

This in-practice limit, which Bell [65,78,79] had argued for, does not, however, deny the theoretical possibility that the evolution of a deterministic system could be (computationally) predicted—at least in principle—if it *were* possible to access and determine "the complete initial state" [79]. By contrast, under the assumption that there exists a *fundamental* limit on computability and agent knowledge about the initial state (compare Section 5.3) that theoretical possibility would be denied also. Although Bell did mention 'deterministic chaos' in the context of 'pseudorandomness' [65], he did *not* propose that chaotic dynamics may represent a limit in any *fundamental* sense. On that specific point, the present work revises the conclusions of an earlier discussion of Bell's effectively free-variables concept [27,60].

By relying on an additional principle, sometimes the powers of the weak option of effective ignorance are sought to be enhanced (e.g., Aharonov et al. [88]): the Uncertainty Principle prevents the simultaneous determination with arbitrary precision of, e.g., particle properties, thereby failing to characterize the relevant initial conditions for the same instant of time. However, the concept of 'uncertainty' is an operational, epistemic notion also, and the physical foundations of the Uncertainty Principle also remain to be identified (e.g., Rozema et al. [89]). Summarizing, the weak epistemic option represents an instance of subjective agent-inaccessibility, because that option depends upon the incomplete state of knowledge of the experimenter agent, i.e., upon an "uncertainty", about the physical universe, including about initial conditions. However, note that even if the entire universe were available as a super-computational resource, then the presence of a black-hole singularity, for example, might render impossible even the purely theoretical prospect—in the weak epistemic option—of the cosmic computability of an individual quantum measurement outcome.

5.2.2. Criticizing the Weak Option Interpretation

The weak option described above has often been criticized on the grounds that quantum randomness cannot possibly be a function of merely some in-practice limit on agent knowledge (Figure 3B). That skeptical position is echoed, for example, by Bub [35], who noted that quantum probabilities that describe the "nonlocal probabilistic correlations that violate Bell's inequality" must be "intrinsically random events", and that these probabilities "do not quantify incomplete knowledge about an ontic state (the basic idea of 'hidden variables')." For a counterpoint to Bub's skeptical position, consult, for example, Figures 1 and 2 in the present article (Section 5.1). Finally, Bub [35] also reaffirmed the popular position that this very fact in particular " ... means that quantum mechanics is quite unlike any theory we have dealt with before in the history of physics."

Indeed, the perceived uniqueness of quantum mechanics, and it is supposed 'weirdness', is often cited as an "explanation" for strange or surprising features that are encountered in quantum studies involving single-particle observations. Specifically, concepts such as superposition (e.g., Schrödinger's cat) and objective chance (i.e., intrinsic randomness)—in the form of objectively unpredictable measurement outcomes—are presumed to operate exclusively in the domain of the quantum, but never in the classical domain. However, what equally 'weird' phenomena may be produced as part of entirely classical systems? One example is the notion of 'undecidable dynamics' in classical systems as a function of self-referential systems dynamics. The present work introduces self-referential dynamics as a novel explanation that might underpin the physics of agent inaccessibility (see Section 5.3). This third and final option counters the idea that what distinguishes a quantum from a classical system is the capacity to generate objectively unpredictable outcomes.

5.3. *Ontological Quantum Mechanics: "Objective Ignorance in Global Determinism"*

The hypothesis that objective ignorance, as opposed to effective ignorance, can be the source of the unpredictability of individual quantum events in a deterministic system, represents the strong ontological option for explaining the physics of agent inaccessibility. Specifically, it had previously been proposed that agent inaccessibility in ontological quantum mechanics might be due to the limit that " ... self-referential processes may generate physical observables whose values are universally uncomputable, i.e., their computation would require an infinite amount of computational resources" (Walleczek [60]). Briefly, the key feature of a nonlinear dynamical process called 'self-referential' is that a system output becomes a new input for the system within the same system (e.g., Walleczek [90]). In *dynamical chaos*, the constant action of feedback loops (recursive processes) is responsible for the generation of the chaotically evolving dynamics. In physical systems that can be characterized by *undecidable dynamics*, self-referential, recursive processes are, again, responsible for the objective unpredictability of outcome states. Importantly, the presence of self-referential dynamics (see Table 1 below) can be identified both in concrete physical systems as well as the computational models that describe them.

Table 1. Two types of self-referential dynamics are considered as a basis for the proposed physics of agent inaccessibility. For the proposal of an AIP as a fundamental principle in quantum mechanics (objective ignorance), the objective unpredictability of an individual measurement outcome as part of a typical quantum random sequence is a function of formal uncomputability; both, dynamical chaos as well as undecidable dynamics posit "infinity"—the lack of infinite resources—as a fundamental limit on computability. Regarding the limit of infinite precision detection in relation to the concept of formal uncomputability, note that—in computational predictions of chaotic dynamics—an *arbitrarily* small difference in initial conditions may lead to a vastly different future outcome state. Note also that the concept of undecidable dynamics underpins both computational irreducibility [18,19] as well as the halting problem in the Church-Turing thesis [20,21].

Self-Referential Dynamics	Formal Uncomputability
Dynamical chaos	Infinite precision detection of initial conditions is impossible in-principle
Undecidable dynamics	Infinite computational resources are unavailable in-principle

The strong option based upon fundamental uncomputability of outcome states—as a necessary and sufficient criterion for objective ignorance—is illustrated in Figure 4B. This proposal is contrasted with the orthodox position of intrinsic randomness shown in Figure 4A. Importantly, two different types of self-referential dynamics are currently known to support the concept of formal uncomputability—dynamical chaos and undecidable dynamics; each type posits the lack of *infinite* resources as a fundamental limit on computability (see Table 1). The question of the physical plausibility of the notion of formal uncomputability in the account of the objective unpredictability of quantum processes in nature will be discussed in Section 6.

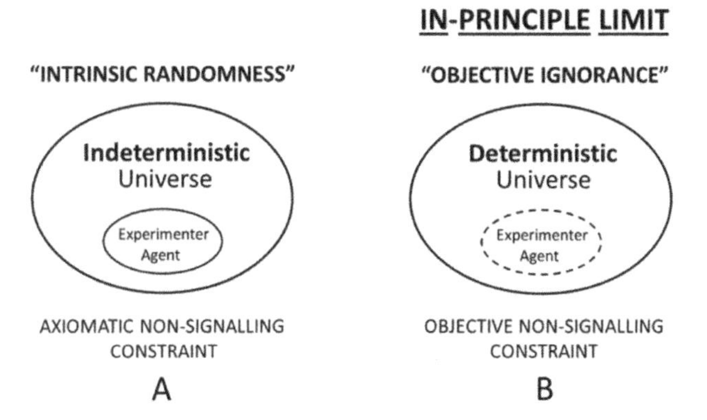

Figure 4. Agent inaccessibility as a function of (**A**) Intrinsic randomness versus (**B**) Objective ignorance (adapted from Walleczek [60]). Intrinsic randomness represents the orthodox interpretation of quantum mechanics, which is universal indeterminism (see legend to Figure 3 for an explanation of the nature of the experimenter agent). Objective ignorance, by contrast, advances the alternative proposal that quantum mechanics in a universally deterministic universe (i.e., global determinism) could account for (objective) quantum unpredictability as defined by an in-principle limit (Figure 4B). Please note that a prior report referred to a related proposal by the term 'intrinsic complexity' [60] due to the fact that such an option is available for complex systems dynamics. An objective non-signaling constraint, which is proposed here as an option that may underlie the non-signaling theorem of quantum mechanics, is equally governed by an objective, in-principle constraint; that is, the capacity for operational control by the experimenter agent (for definition see Section 4.3) of, for example, time-symmetric, or nonlocal, ontic influences, or information transfers, is formally and objectively limited by the unavailability to the agent of either (i) infinitely precise knowledge about (time-symmetric) initial conditions, or (ii) infinite computational, or generally technological, resources, or a combination of (i) and (ii). For an overview, see Table 1.

A key distinguishing feature of the concept of objective ignorance—in contrast to that of effective ignorance—is the following (Figure 4B): Even if assuming that the *complete* initial conditions of some deterministic system could be obtained, then the exact prediction of outcome states is still impossible—even in principle. That is, in the option of objective ignorance (Figure 4B), the lack of *infinite* computational resources as a criterion places an *objective* limit on the experimenter agent as a function of undecidable dynamics (see Table 1), which, as Bennett [91] put it, is dynamics that is " ... unpredictable even from total knowledge of the initial conditions". This type of objective unpredictability is exemplified also in the halting problem for Turing machines, with the essential point being that Turing machines " ... are unpredictable", as Moore [92] noted, "even if the initial conditions are known exactly".

A second key distinguishing feature which is covered by the strong option of objective ignorance, but not by effective ignorance (Section 5.2), concerns the emergence of dynamical chaos in physical systems. Importantly, due to the theoretical impossibility of gathering information with *infinite* precision about the initial state from which evolves a dynamically chaotic system, an *objective* limit is imposed on the computability of the system's outcome states. For explanation, note that *arbitrarily* small differences in initial conditions may generate strongly divergent outcome states in computational models of dynamical chaos (see Table 1).

Because the strong option is also a knowledge-constraining option, the term 'ignorance' has been retained as part of the present proposal of an AIP for quantum mechanics. However, in contrast to effective ignorance, in the concept of objective ignorance, agent knowledge is not incomplete in the sense that gathering more information about initial conditions, or amassing more computational power, might eventually lead to complete knowledge and total predictability. Instead, an in-principle limit guarantees the incompleteness of agent knowledge, and therefore the agent's inability to control and predict even a single quantum measurement outcome is ensured (see Table 1).

Therefore, the concept of objective ignorance represents an instance of *objective* agent-inaccessibility, which—obviously—is a more restrictive notion than *subjective* or *effective* agent-inaccessibility. Accordingly, the difference between the *effective* non-signaling constraint (Figure 3B) and the *objective* non-signaling constraint (Figure 4B) is that the latter constraint adopts a fundamental, and not a practical, limit on complete agent access towards an ontic state λ, and towards quantum information transfers, in ontological quantum mechanics in general. For example, this holds true for (SW) quantum ontologies that are locally time-symmetric [12–16,29–32], locally time-asymmetric [45–47], or strictly nonlocal [1–8]. Finally, the here proposed principle (AIP) is fundamental in the sense that a Turing oracle only could predict the exact value of an individual outcome state as a function of physical systems and computational model evolution. The strong option of objective ignorance (Figure 4B) might represent a fundamental principle by which nature prohibits access to the experimenter agent in the quantum regime. In the subsequent Section 6, a selection of available views and results are reviewed briefly which may support the present proposal for an AIP based upon the concepts of objective unpredictability, undecidability, and uncomputability.

6. In Search of Incomputable Nature: Quantum Reality and Quantum Randomness

The use of computational concepts and terminology in the search of the origins of the observed randomness in quantum systems, in combination with the recent "ontological turn" in quantum foundations (see Section 1), offers a new pathway towards exploring the physics of agent inaccessibility. In regard to the radical concept of incomputability in nature, one of its pioneers, S. Barry Cooper, once remarked—in reference to the puzzling features of nonlinear emergent states and chaos in nature—that " ... many of the troublesome problems can be placed in a helpful explanatory context ... " if one " ... admits the possibility that the Universe is deeply imbued with incomputability and its mathematics" [93].

How realistic is the proposal that notions such as computability and uncomputability are relevant for physical laws, i.e., for the laws that explain the behavior of concrete micro-physical systems in nature, including those that are quantum-based? For example, Lloyd [94] has recently advanced the position that " ... uncomputability is ubiquitous in physical law", and that this is a natural consequence, he argued, of the fact that many " ... physical systems are capable of universal computation". Importantly, " ... it is difficult to find an extended system with nonlinear interactions that is not capable of universal computation", he explained, " ... given proper initial conditions and inputs". Furthermore, he argued that there may be special cases when " ... quantum systems that evolve according to nonlinear interactions are capable of universal computation", which yields the path-breaking possibility that " ... the halting problem arises in the computation of basic features of many physical systems" [94].

Crucially, therefore, the concepts of uncomputability and undecidable dynamics [18–21,91–94] may have far greater significance to physics, and to the limits of science in general (compare Section 3.2), than—merely—as a concept that describes an abstract problem in recursive logic. For example, Rucker [95] has also argued that " ... we should be able to find numerous examples of undecidability in the natural world". Consequently, the formal concepts of undecidability and uncomputability may challenge the need for the (unprovable) metaphysical assumption of indeterminism as an explanation for the objective unpredictability in quantum systems. For example, Cubitt et al. [96] offered a physical model demonstrating the notion of objective unpredictability, not however as a function of quantum indeterminism, but due to self-referential, undecidable dynamics operating in the quantum regime (compare Table 1 in Section 5.3).

6.1. Computational Approaches to Quantum Theory Invoking Nonlinear Interactions

The method of conceptualizing, or even explaining, the physical universe as a (quantum) computational process has a long history, and for recent overviews, see, e.g., Cooper and Soskova [97] and Fletcher and Cuffaro [98]. For example, in relation to quantum mechanics, researchers such as 't Hooft [45,71] and Elze [46] have long promoted the idea that the probabilistic aspect of quantum physics does not necessarily have to contradict its possible algorithmic nature as demonstrated in work with quantum cellular automata. Generally, cellular automata (CA) can present models of the physical world and for the following discussion the equivalence of Turing machines and CA is assumed. "It is conceivable that the physical processes described by the laws of nature never come to an end", Franke [99], for example, remarked, and that in adopting a CA-simulation of the physical world, " ... we are simulating the behavior of a cellular automaton which runs deterministically, but is not computable." Franke [99] emphasized that in such a model—therefore—the apparent randomness in the world might be due to an " ... equivalent of chaos as understood in dynamical chaos theory, which as we know, is not based on actual chance, but on non-computability". For explanation, Franke [99], in that quote, refers to 'actual chance' as denoting the standard indeterminism of orthodox quantum theory. By contrast, the non-computability stems from the fact that the possession of knowledge about the initial conditions of a dynamically chaotic process is not possible with infinite precision, which imposes a fundamental, in-principle limit on computability (see Table 1 in Section 5.3).

Very recently, the potential power of the approach that combines the notion of universal computation with unconventional *ontological* propositions has also been noted, for example, by Koberinski and Müller [43]. They considered the kind of information-theoretic properties of quantum theory " ... which are directly linked to the possibility of having a universal computing machine, like the quantum Turing machine", which is " ... in principle able to simulate the time evolution of any physical system". These authors have proposed that the " ... notion of 'universal computation' ... is powerful enough to uniquely determine the state space, time evolution, and possible measurements (and thus also other properties like the maximal amount of non-locality) of quantum theory." Again, however, as was emphasized by Lloyd [94], any computational interpretation of quantum systems might give rise to uncomputable elements, i.e., undecidable outcome states, which—within the constraints of a universal Turing machine—may therefore yield, again, a fundamental limit on agent-quantum access and predictability regarding the calculation of exact outcome values or individual ontological properties. One specific model of undecidable dynamics operating in the quantum regime was mentioned above [96]. Besides the notion of objective or fundamental *uncomputability*, how might the notion of the indefinite, contextual, or relational, *ontology* (for details see Section 3.1) enter the picture of the information-theoretic approach towards a quantum reality?

6.2. Quantum Ontology and the Information-Theoretic Paradigm in Quantum Mechanics

As was described in Section 5.1.1, novel ontological possibilities beyond naïve realism are increasingly considered as a basis for quantum mechanics, given that indeterminacy proofs are impossible. This includes relational ontologies such as ontic structural realism (e.g., [85,100]), locally time-symmetric ontologies (e.g., [12]), including unorthodox causal structures such as retrocausality (e.g., [29–32]). In the pursuit of possible *ontological* features of quantum mechanics, Koberinski and Müller [43] have also speculated about the presence of a relational ontology as part of a future construction of quantum theory, in particular, in reference to the proposal of ontic structural realism [85,100]. They have acknowledged that while " ... the information-theoretic reconstructions ... do not typically tell us what quantum states are, or what is really going on in the world when we perform a Bell experiment, for example", the possibility might be considered of an " ... ontology of structural relations in some sense—simply of the relational structure uniquely picked out by the information-theoretic postulates ... ", which is an approach, they suggested, that " ... does not rule out the possibility of discovering a constructive successor to quantum theory, in particular since ontological stability across theory change is a characteristic of ontic structural realism." The combined computational-ontological research strategy, such as the one described above, may chart a new course also towards understanding the original HV-concept in Bohm's ontological quantum theory (Section 3.3). That is, the application of ideas such as dynamical chaos and undecidable dynamics (see Table 1), to quantum ontology in dBB-theory and Bohmian mechanics, may in the future allow a new understanding of the variables traditionally called 'hidden' as uncomputable variables (compare Section 3.3).

6.3. *Could Hidden Variables Represent Uncomputable Variables Such as Turing-Incomputable Variables?*

Following the above analysis, the HVs of original dBB-theory may not only be 'hidden', and uncontrollable, in the familiar sense of the weak option known as 'effective ignorance' (Section 5.2); instead, the HV-concept might represent a case of uncontrollability and unpredictability as a function of the strong option involving nonlinear relations as described by the concept of 'objective ignorance' (Section 5.3). That proposal suggests the presence of a fundamental limit on agent inaccessibility in dBB-theory based upon the interpretation of the HV-concept as, for one speculative possibility, a Turing-incomputable variable (TIV). At a minimum, for starters, the proposal of TIVs in an ontological quantum mechanics, such as dBB-theory [1–3], would require—in the *constructive* approach, at least—the presence of nonlinear, self-referential interactions as part of the ontology of a quantum theory, i.e., an ontology that is compatible with emergence and chaos theory (see Table 1). Where in the Bohmian approach could that be found? Could Bohm's theory manifest self-referential, chaotic behavior in a way similar to that seen in some constructions of an emergent quantum mechanics, which implements self-referential dynamics as a basic resource also?

The original writings of Bohm and Hiley [3] reveal that the nonlinear perspective on the quantum state in Bohm's theory was evident already 25 years ago: "The general behavior described", Bohm and Hiley [3] wrote, " ... is similar to that obtained in the study of non-linear equations whose solution contain what are called stable limit cycles", whereby, however, the " ... difference from the usual kind of non-linear equations is that for each stable motion we have a whole set of possible limit cycles rather than just a single cycle. Each quantum state thus corresponds to a different set of limit cycles and a transition corresponds to an orbit going from one of these to another". Importantly, quantum state transitions, as the authors further explained, happen at " ... bifurcation points dividing those orbits entering one channel from those entering another. Near these points, the motion is highly unstable and, indeed, chaotic in the sense of modern chaos theory" [3]. To mention only one new example: Work by Tzemos et al. [101] has described the origins of chaos in a mathematical model of a generalized Bohmian quantum theory. To be sure, there are additional reports that Bohmian trajectories could be chaotic and that chaotic dynamics could be the source of 'quantum relaxation' in Bohmian mechanics (e.g., References [102–104]).

Work such as the above may pave the way towards conceptualizing the HV as an (effectively) uncomputable variable, or possibly even a TIV should, e.g., evidence for undecidability emerge in a future quantum-theoretic construction (see Table 1). Next, one topic of debate has long been the potential risk of violating the non-signaling condition of quantum mechanics as a function of the intrinsic nonlocality of ontological quantum theories such as dBB-theory (compare Section 3.3). For prior work which defined an *effective* non-signaling constraint for ontological quantum mechanics based upon an analysis of the concept of free variables by John Bell (see Section 5.2), consult Walleczek and Grössing [27]. Here, the concepts of an *effective* (Section 5.2.) and of an *objective* (Section 5.3) non-signaling constraint will be discussed briefly in the context of approaches considering computational constraints towards fashioning an understanding of the non-signaling theorem.

6.4. *The Non-Signaling Theorem and Effective versus Objective Computational Constraints*

Bendersky et al. [63] have implemented a computational protocol to assess whether, or not, the nonlocal features associated with results from EPR-type quantum correlation experiments could be used to communicate messages between two space-like separated locations—in apparent violation of the non-signaling theorem. That study concluded that this is impossible because the " ... computability of results imposes a strong limitation on how nature can behave if it only had computable resources to generate outputs for the experiments." Central to that conclusion is, of course, the *standard* assumption that computational (Turing-type) processes " ... cannot generate random sequences", and that therefore, as Berendsky et al. [63] have added " ... we need to accept the existence of truly unpredictable physical processes."

Significantly, Berendsky et al. [63] concluded with the message that their findings are not in " ... conflict with the different interpretations of quantum mechanics", and they further noted that in " ... the Copenhagen interpretation, the measurement process is postulated as random, whereas, for example, in Bohmian mechanics, it is deterministic but the initial conditions are randomly distributed and fundamentally unknowable." For quantum theories operating in a universally deterministic universe (see Figures 3 and 4), such as dBB-theory and Bohmian mechanics, the quantum randomness would be generated by uncomputable processes, whether they be *effectively* uncomputable (see the effective non-signaling constraint in Section 5.2), or (ii) *objectively* uncomputable in the strong sense of dynamical chaos and/or undecidable dynamics, e.g., in the form of Turing incomputability (see the objective non-signaling constraint in Section 5.3); only the strong option of objective ignorance in deterministic systems could entail objective or true unpredictability. However, the specific topic of self-referential dynamics in formal uncomputability (see Table 1) was not addressed in the work by Berendsky et al. [63], although these workers did make the important point that "...in Bohmian mechanics ... the initial conditions are ... fundamentally unknowable." Previously, Islam and Wehner [105] had also suggested that quantum mechanics must entail the presence of (agent-inaccessible) uncomputable states as otherwise a violation of the non-signaling constraint would inevitably ensue, and these researchers noted that " ... in any theory in which the Church-Turing principle holds, certain states and/or measurements are not available to us as otherwise any (approximate) no-signaling computation could be performed." To employ the present terminology, in order (i) to prevent superluminal Shannon-type signaling in nonlocal quantum ontologies or, alternatively, (ii) to prohibit (future-to-past) retro-signaling in time-symmetric quantum ontologies, these "states and/or measurements" must be subject to an AIP as a fundamental principle in quantum mechanics.

6.5. *Quantum Randomness and Turing Incomputability*

How might the proposed link be explored further between Turing-incomputable processes and the problem of quantum randomness? On the one hand, a skeptic might argue against the notion of a successor to standard quantum theory, i.e., against the successful construction of a future quantum theory which could provide a *physical* account of quantum randomness. On the other hand, as was investigated in the present work, a new research movement is fast gaining traction which seeks to reanalyze, and explore again, the validity of ontological propositions for quantum mechanics (see Section 1). Could undecidable, Turing-incomputable processes be of significance for the research program towards an ontological quantum mechanics, including in the account of objective quantum unpredictability? Further evidence in favor of the plausibility of such a program has come forward in recent years. "Is quantum randomness Turing incomputable?", asked Calude [106], and he described " ... a procedure to generate quantum random bits that provably are not reproducible by any Turing machine". Based on work that employed an operational version of the theorem by Kochen and Specker [39], the author suggested that quantum randomness might be the best evidence, so far, for the existence of a Turing-incomputable phenomenon in the natural world [106]. For a detailed analysis of that possibility, which posits the existence of value-*indefinite* observables in nature (compare

Section 3.1), see Abbott et al. [40–42]. Given that a formal proof for quantum indeterminism is lacking in principle (see Section 5.1.1), the promise of a formal proof for the uncomputability of the observed randomness in quantum experiments both suggests and motivates the present proposal of an AIP for quantum mechanics (Section 1).

In summary, if the best available evidence for the true randomness of a sequence (that is generated by a quantum-based randomness generator) is the uncomputability of that sequence by a Turing machine, then this does not—necessarily—imply that the origins of that randomness is to be found in the metaphysics of quantum indeterminism. Consult Section 5.1 regarding arguments against the possibility of indeterminacy proofs, and Section 5.3 regarding the possibility of objective (true) unpredictability in fully deterministic systems (see Table 1). Given an AIP based upon objective ignorance, the following question remains unanswered at present: How to determine *empirically* whether the source of experimentally observed quantum randomness is either (i) 'intrinsic randomness' as in the orthodox position of Section 5.1 (Figure 3A), or (ii) 'objective ignorance' as in the strong option proposed in Section 5.3 (Figure 4B)? At present it remains unknown whether decisive experimental tests could be identified and performed. Until such tests might become available, the decision should be postponed between quantum indeterminism versus agent inaccessibility as a fundamental principle in quantum mechanics.

7. Conclusions

The question remains an open one as to whether agent inaccessibility in quantum experiments is either (i) due to metaphysical indeterminism or (ii) due to a quantum ontology of a form where the *exact* ontic state λ is either effectively or objectively uncomputable. The latter option is the basis of the present proposal for agent inaccessibility as a fundamental principle (AIP) in quantum mechanics. What is the ontological status of a fundamentally agent-inaccessible quantum state? The status is *indefinite* of the agent-inaccessible ("hidden") ontic state (IOS) because only an infinite amount of measurement information, and/or access to infinite computational resources, might enable the exact prediction of a *definite* measurement outcome (DOS). Finally, the concepts of self-referential dynamics and formal uncomputability may represent key elements in a physical theory of agent inaccessibility. Instead of framing the 20th century quantum revolution as a radical shift from determinism towards indeterminism, this work has argued that—given the available scientific evidence—it is valid only to claim the following: the quantum revolution signifies the profound discovery of an agent-inaccessible regime of the physical universe.

Funding: This research was funded by the Fetzer Franklin Fund of the John E. Fetzer Memorial Trust.

Acknowledgments: J.W. at Phenoscience Laboratories (Berlin) wishes to thank Gerhard Grössing and the research group at the Austrian Institute for Nonlinear Studies (Vienna), as well as Nikolaus von Stillfried at Phenoscience Laboratories, for invaluable insights during many excellent discussions.

Conflicts of Interest: The author declares no conflict of interest.

References

1. Bohm, D. A suggested interpretation of the quantum theory in terms of "hidden" variables. I. *Phys. Rev.* **1952**, *85*, 166–179. [CrossRef]
2. Bohm, D. A suggested interpretation of the quantum theory in terms of "hidden" variables. II. *Phys. Rev.* **1952**, *85*, 180–193. [CrossRef]
3. Bohm, D.; Hiley, B.J. *The Undivided Universe: An Ontological Interpretation of Quantum Theory*; Routledge: London, UK, 1993.
4. Holland, P.R. *The Quantum Theory of Motion*; Cambridge University Press: Cambridge, UK, 1993; ISBN 0-521-35404-8.
5. Bricmont, J. *Making Sense of Quantum Mechanics*; Springer: Heidelberg, Germany, 2016.
6. Norsen, T. *Foundations of Quantum Mechanics*; Springer: Heidelberg, Germany, 2018.
7. Dürr, D.; Teufel, S. *Bohmian Mechanics*; Springer: Heidelberg, Germany, 2009.

8. Tumulka, R. Bohmian mechanics. In *The Routledge Companion to the Philosophy of Physics*; Wilson, A., Ed.; Routledge: London, UK, 2018.
9. Harrigan, N.; Spekkens, R.W. Einstein, incompleteness, and the epistemic view of quantum states. *Found. Phys.* **2010**, *40*, 125–157. [CrossRef]
10. Pusey, M.F.; Barrett, J.; Rudolph, T. On the reality of the quantum state. *Nat. Phys.* **2012**, *8*, 475–478. [CrossRef]
11. Leifer, M.S. Is the quantum state real? An extended review of ψ-ontology theorems. *Quanta* **2014**, *3*, 67–155. [CrossRef]
12. Leifer, M.S.; Pusey, M.F. Is a time symmetric interpretation of quantum theory possible without retrocausality? *Proc. R. Soc. A* **2017**, *473*, 20160607. [CrossRef] [PubMed]
13. Wharton, K. A new class of retrocausal models. *Entropy* **2018**, *20*, 410. [CrossRef]
14. Adlam, E.C. Quantum mechanics and global determinism. *Quanta* **2018**, *7*, 40–53. [CrossRef]
15. Adlam, E.C. Spooky action at a temporal distance. *Entropy* **2018**, *20*, 41. [CrossRef]
16. Argaman, N. A lenient causal arrow of time? *Entropy* **2018**, *20*, 294. [CrossRef]
17. Grössing, G.; Fussy, S.; Mesa Pascasio, J.; Schwabl, H. Vacuum landscaping: Cause of nonlocal influences without signaling. *Entropy* **2018**, *20*, 458. [CrossRef]
18. Wolfram, S. Undecidability and intractability in theoretical physics. *Phys. Rev. Lett.* **1985**, *54*, 735–738. [CrossRef] [PubMed]
19. Wolfram, S. *A New Kind of Science*; Wolfram Media: Champaign, IL, USA, 2002; ISBN 1-57955-008-8.
20. Church, A. An unsolvable problem of elementary number theory. *Am. J. Math.* **1936**, *58*, 345–363. [CrossRef]
21. Turing, A. On Computable Numbers, with an Application to the Entscheidungsproblem. *Proc. Lond. Math. Soc.* **1936**, *42*, 230–265.
22. Everett, H. Relative state formulation of quantum mechanics. *Rev. Mod. Phys.* **1957**, *29*, 454–462. [CrossRef]
23. Vaidman, L. All is ψ. *J. Phys. Conf. Ser.* **2016**, *701*, 012020. [CrossRef]
24. Schiff, J.; Poirier, B. Quantum mechanics without wave functions. *J. Chem. Phys.* **2012**, *136*, 031102. [CrossRef] [PubMed]
25. Hall, M.J.W.; Deckert, D.-A.; Wiseman, H.M. Quantum phenomena modeled by interactions between many classical worlds. *Phys. Rev. X* **2014**, *4*, 041013. [CrossRef]
26. Ghadimi, M.; Hall, M.J.W.; Wiseman, H.M. Nonlocality in Bell's theorem, in Bohm's theory, and in many interacting worlds theorising. *Entropy* **2018**, *20*, 567. [CrossRef]
27. Walleczek, J.; Grössing, G. Nonlocal quantum information transfer without superluminal signalling and communication. *Found. Phys.* **2016**, *46*, 1208–1228. [CrossRef]
28. Bell, J.S. On the Einstein Podolsky Rosen paradox. *Physics* **1964**, *1*, 195–200. [CrossRef]
29. Sutherland, R.I. Bell's theorem and backwards-in-time causality. *Int. J. Theor. Phys.* **1983**, *22*, 377–384. [CrossRef]
30. Price, H. *Time's Arrow & Archimedes' Point: New Directions for the Physics of Time*; Oxford University Press: Oxford, UK, 1997.
31. Wharton, K. Quantum states as ordinary information. *Information* **2014**, *5*, 190–208. [CrossRef]
32. Price, H.; Wharton, K. Disentangling the quantum world. *Entropy* **2015**, *17*, 7752–7767. [CrossRef]
33. Frauchiger, D.; Renner, R. Quantum theory cannot consistently describe the use of itself. *Nat. Commun.* **2018**, *9*, 3711. [CrossRef] [PubMed]
34. Brukner, C. A no-go theorem for observer-independent facts. *Entropy* **2018**, *20*, 350. [CrossRef]
35. Bub, J. In defense of a "single-world" interpretation of quantum mechanics. *arXiv*, 2018; arXiv:quant-ph/1804.03267v1. [CrossRef]
36. Lazarovici, D.; Hubert, M. How single-world quantum mechanics is consistent: Comment on Frauchiger and Renner. *arXiv*, 2018; arXiv:quant-ph/1809.08070v1.
37. Zwolak, M.; Zurek, W. Complementarity of quantum discord and classically accessible information. *Sci. Rep.* **2013**, *3*, 1729. [CrossRef]
38. Einstein, A.; Podolsky, B.; Rosen, N. Can quantum-mechanical description of physical reality be considered complete? *Phys. Rev.* **1935**, *47*, 777–780. [CrossRef]
39. Kochen, S.; Specker, E.P. The problem of hidden variables in quantum mechanics. *J. Math. Mech.* **1967**, *17*, 59–87. [CrossRef]
40. Abbott, A.A.; Calude, C.S.; Conder, J.; Svozil, K. Strong Kochen-Specker theorem and incomputability of quantum randomness. *Phys. Rev. A* **2012**, *86*, 6. [CrossRef]

41. Abbott, A.A.; Calude, C.S.; Svozil, K. Value-indefinite observables are almost everywhere. *Phys. Rev. A* **2014**, *89*, 032109. [CrossRef]
42. Abbott, A.A.; Calude, C.S.; Svozil, K. A quantum random number generator certified by value indefiniteness. *Math. Struct. Comput. Sci.* **2014**, *24*, e240303. [CrossRef]
43. Koberinski, A.; Müller, M.P. Quantum Theory as a Principle Theory: Insights from an Information-theoretic Reconstruction. In *Physical Perspectives on Computation, Computational Perspectives on Physics*; Fletcher, S.C., Cuffaro, M.E., Eds.; Cambridge University Press: Cambridge, UK, 2018; pp. 257–280.
44. Fuchs, C. Quantum mechanics as quantum information, mostly. *J. Mod. Opt.* **2003**, *50*, 987–1023. [CrossRef]
45. 't Hooft, G. *The Cellular Automaton Interpretation of Quantum Mechanics*; Fundamental Theories of Physics; Springer: Berlin/Heidelberg, Germany, 2016; Volume 185.
46. Elze, H.-T. Quantum features of natural cellular automata. *J. Phys. Conf. Ser.* **2016**, *701*, 012017. [CrossRef]
47. Palmer, T.N. Experimental non-violation of the Bell inequality. *Entropy* **2018**, *20*, 356. [CrossRef]
48. Mermin, N.D. Hidden variables and the two theorems of John Bell. *Revs. Mod. Phys.* **1993**, *65*, 803–815. [CrossRef]
49. Maudlin, T. *Quantum Non-locality and Relativity: Metaphysical Intimations of Modern Physics*; Wiley-Blackwell: West Sussex, UK, 2011.
50. Maudlin, T. Ontological clarity via canonical presentation: Electromagnetism and the Aharonov–Bohm effect. *Entropy* **2018**, *20*, 465. [CrossRef]
51. Lazarovici, D.; Oldofredi, A.; Esfeld, M. Observables and unobservables in quantum mechanics: How the no-hidden-variables theorems support the Bohmian particle ontology. *Entropy* **2018**, *20*, 381. [CrossRef]
52. Passon, O. On a common misconception regarding the de Broglie–Bohm theory. *Entropy* **2018**, *20*, 440. [CrossRef]
53. Tumulka, R. On Bohmian mechanics, particle creation, and relativistic space-time: Happy 100th birthday, David Bohm! *Entropy* **2018**, *20*, 462. [CrossRef]
54. Norsen, T. On the explanation of Born-rule statistics in the de Broglie-Bohm pilot-wave theory. *Entropy* **2018**, *20*, 422. [CrossRef]
55. De Gosson, M.A. The symplectic camel and Poincaré superrecurrence: Open problems. *Entropy* **2018**, *20*, 499. [CrossRef]
56. Khrennikov, A. *Contextual Approach to Quantum Formalism*; Springer: Berlin/Heidelberg, Germany; New York, NY, USA, 2009.
57. Khrennikov, A. Classical probability model for Bell inequality. *J. Phys. Conf. Ser.* **2014**, *504*, 012019. [CrossRef]
58. Hiley, B.J.; Van Reeth, P. Quantum trajectories: Real or surreal? *Entropy* **2018**, *20*, 353. [CrossRef]
59. Flack, R.; Hiley, B.J. Feynman paths and weak values. *Entropy* **2018**, *20*, 367. [CrossRef]
60. Walleczek, J. The super-indeterminism in orthodox quantum mechanics does not implicate the reality of experimenter free will. *J. Phys. Conf. Ser.* **2016**, *701*, 012005. [CrossRef]
61. Gisin, N. Why Bohmian mechanics? One- and two-time position measurements, Bell inequalities, philosophy, and physics. *Entropy* **2018**, *20*, 105. [CrossRef]
62. Valentini, A. Signal-locality in hidden-variables theories. *Phys. Lett. A* **2002**, *297*, 273–278. [CrossRef]
63. Bendersky, A.; Senno, G.; de la Torre, G.; Figueira, S.; Acin, A. Non-signaling deterministic models for non-local correlations have to be uncomputable. *Phys. Rev. Lett.* **2017**, *118*, 130401. [CrossRef] [PubMed]
64. Bell, J.S. Einstein-Podolsky-Rosen experiments. In *Speakable and Unspeakable in Quantum Mechanics*; Cambridge University Press: Cambridge, UK, 1987; pp. 81–92.
65. Bell, J.S. La nouvelle cuisine. In *Speakable and Unspeakable in Quantum Mechanics*; Cambridge University Press: Cambridge, UK, 2004; pp. 232–248.
66. Bohm, D. *Quantum Theory*; Prentice-Hall: Englewood Cliffs, NJ, USA, 1951.
67. Fuchs, C.A.; Schack, R. QBism and the Greeks: Why a quantum state does not represent an element of physical reality. *Phys. Scr.* **2015**, *90*, 015104. [CrossRef]
68. Maxwell, J.C. *Theory of Heat*; Reprinted Dover: New York, NY, USA, 2001.
69. Szilard, L. Über die Entropieverminderung in einem thermodynamischen System bei Eingriffen intelligenter Wesen (On the reduction of entropy in a thermodynamic system by the intervention of intelligent beings). *Zeitschrift für Physik* **1929**, *53*, 840–856. [CrossRef]
70. Von Neumann, J. *Mathematische Grundlagen der Quantenmechanik*; Springer: Berlin, Germany, 1932.
71. 't Hooft, G. Time, the arrow of time, and quantum mechanics. *Front. Phys.* **2018**, *6*, 81. [CrossRef]

72. Colbeck, R.; Renner, R. A short note on the concept of free choice. *arXiv*, 2013; arXiv:quant-ph/1302.4446.
73. Plotnitsky, A. What is complementarity? Niels Bohr and the architecture of quantum theory. *Phys. Scr.* **2014**, *163*, 014002. [CrossRef]
74. Conway, J.; Kochen, S. The free will theorem. *Found. Phys.* **2006**, *36*, 1441–1473. [CrossRef]
75. Conway, J.; Kochen, S. The strong free will theorem. *Not. AMS* **2009**, *56*, 226–232.
76. Bera, M.N.; Acín, A.; Kuś, M.; Mitchell, M.; Lewenstein, M. Randomness in Quantum Mechanics: Philosophy, Physics, and Technology. *Rep. Prog. Phys.* **2017**, *80*, 124001. [CrossRef]
77. Landsman, K. *Foundations of Quantum Theory: From Classical Concepts to Operator Algebras*; Fundamental Theories of Physics; Springer: Berlin/Heidelberg, Germany, 2017; Volume 188, ISBN 978-3-319-51777-3.
78. Bell, J.S. The theory of local beables. *Epistemol. Lett.* **1976**, *9*, 11, reprinted in *Dialectica* **1985**, *39*, 85–96.
79. Bell, J.S. Free variables and local causality. *Epistemol. Lett.* **1977**, *15*, 15, reprinted in *Dialectica* **1985**, *39*, 103–106.
80. Bell, J.S. De Broglie-Bohm, delayed-choice, double-slit experiment, and density matrix. *Int. J. Quantum Chem.* **1980**, *14*, 155–159. [CrossRef]
81. Bell, J.S. On the impossible pilot wave. *Found. Phys.* **1982**, *12*, 989–999. [CrossRef]
82. Bell, J.S. Speakable and unspeakable in quantum mechanics. In *Speakable and Unspeakable in Quantum Mechanics*; Cambridge University Press: Cambridge, UK, 1987; pp. 169–172.
83. Walleczek, J.; Grössing, G. The non-signalling theorem in generalizations of Bell's theorem. *J. Phys. Conf. Ser.* **2014**, *504*, 012001. [CrossRef]
84. Chaitin, G.J. *Information, Randomness, and Incompleteness: Papers on Algorithmic Information Theory*; Series in Computer Science; World Scientific: Singapore, 1987; Volume 8.
85. Esfeld, M. Ontic structural realism and the interpretation of quantum mechanics. *Eur. J. Philos. Sci.* **2013**, *3*, 19–32. [CrossRef]
86. Walleczek, J.; Grössing, G. Is the world local or nonlocal? Towards an emergent quantum mechanics in the 21st century. *J. Phys. Conf. Ser.* **2016**, *701*, 012001. [CrossRef]
87. Shimony, A. Controllable and Uncontrollable Non-locality. In *Foundations of Quantum Mechanics in Light of the New Technology*; Kamefuchi, S., Butsuri Gakkai, N., Eds.; Physical Society of Japan: Tokyo, Japan, 1984; pp. 225–230.
88. Aharonov, Y.; Cohen, E.; Shushi, T. Accommodating retrocausality with free will. *Quanta* **2016**, *5*, 53–60. [CrossRef]
89. Rozema, L.A.; Darabi, A.; Mahler, D.H.; Hayat, A.; Soudagar, Y.; Steinberg, A.M. Violation of Heisenberg's measurement-disturbance relationship by weak measurements. *Phys. Rev. Lett.* **2012**, *109*, 100404. [CrossRef]
90. Walleczek, J. (Ed.) *Self-Organized Biological Dynamics and Nonlinear Control*; Cambridge University Press: Cambridge, UK, 2000.
91. Bennett, C.H. Undecidable dynamics. *Nature* **1990**, *346*, 606–607. [CrossRef]
92. Moore, C. Unpredictability and undecidability in dynamical systems. *Phys. Rev. Lett.* **1990**, *64*, 2354–2357. [CrossRef]
93. Cooper, S.B.; Odifreddi, P. Incomputability in Nature. In *Computability and Models*; Cooper, S.B., Goncharov, S.S., Eds.; Kluwer Academic, Plenum Publishers: New York, NY, USA, 2003; pp. 137–160.
94. Lloyd, S. Uncomputability and Physical Law. In *The Incomputable: Journeys Beyond the Turing Barrier*; Cooper, S.B., Soskova, M.I., Eds.; Springer: Berlin/Heidelberg, Germany, 2017; pp. 95–104.
95. Rucker, R. An Incompleteness Theorem for the Natural World. In *Irreducibility and Computational Equivalence*; Zenil, H., Ed.; Springer: Berlin/Heidelberg, Germany, 2013; pp. 185–198.
96. Cubitt, T.S.; Perez-Garcia, D.; Wolf, M.M. Undecidability of the spectral gap. *Nature* **2015**, *528*, 207–211. [CrossRef] [PubMed]
97. Cooper, S.B.; Soskova, M.I. (Eds.) *The Incomputable: Journeys Beyond the Turing Barrier*; Springer: Berlin/Heidelberg, Germany, 2017.
98. Fletcher, S.C.; Cuffaro, M.E. (Eds.) *Physical Perspectives on Computation, Computational Perspectives on Physics*; Cambridge University Press: Cambridge, UK, 2018.
99. Franke, H.W. Cellular Automata: Models of the Physical World. In *Irreducibility and Computational Equivalence*; Zenil, H., Ed.; Springer: Berlin/Heidelberg, Germany, 2013; pp. 3–10.
100. French, S. *The Structure of the World: Metaphysics and Representation*; Oxford University Press: Oxford, UK, 2014; ISBN 9780199684847.

101. Tzemos, A.C.; Efthymiopoulos, C.; Contopoulos, G. Origin of chaos near three-dimensional quantum vortices: A general Bohmian theory. *Phys. Rev. E* **2018**, *97*, 042201. [CrossRef]
102. Towler, M.D.; Russell, N.J.; Valentini, A. Time scales for dynamical relaxation to the Born rule. *Proc. R. Soc. A* **2012**, *468*, 990. [CrossRef]
103. Efthymiopoulos, C.; Contopoulos, G.; Tzemos, A.C. Chaos in de Broglie-Bohm quantum mechanics and the dynamics of quantum relaxation. *Ann. Fond. Louis Broglie* **2017**, *42*, 133–160.
104. Underwood, N.G. Extreme quantum nonequilibrium, nodes, vorticity, drift, and relaxation retarding states. *arXiv* **2018**, arXiv:quant-ph/1705.06757v2. [CrossRef]
105. Islam, T.; Wehner, S. Computability limits non-local correlations. *Phys. Rev. A* **2012**, *86*, 042109. [CrossRef]
106. Calude, C.S. Quantum Randomness: From Practice to Theory and Back. In *The Incomputable: Journeys Beyond the Turing Barrier*; Cooper, S.B., Soskova, M.I., Eds.; Springer: Berlin/Heidelberg, Germany, 2017; pp. 169–181.

 entropy

Article

The Symplectic Camel and Poincaré Superrecurrence: Open Problems

Maurice A. de Gosson

Faculty of Mathematics, NuHAG, University of Vienna, 1090 Vienna, Austria; maurice.de.gosson@univie.ac.at

Received: 26 May 2018; Accepted: 26 June 2018; Published: 28 June 2018

Abstract: Poincaré's Recurrence Theorem implies that any isolated Hamiltonian system evolving in a bounded Universe returns infinitely many times arbitrarily close to its initial phase space configuration. We discuss this and related recurrence properties from the point of view of recent advances in symplectic topology which have not yet reached the Physics community. These properties are closely related to Emergent Quantum Mechanics since they belong to a twilight zone between classical (Hamiltonian) mechanics and its quantization.

Keywords: Poincaré recurrence; symplectic camel; quantum mechanics; Hamiltonian

---

1. Introduction

In his famous prize-winning 1890 paper [1], Henri Poincaré proved that almost every phase space trajectory of an isolated three-body system must return arbitrarily close to its initial position, and this infinitely many times. Poincaré called this result *théorème de stabilité à la Poisson*, but it is nowadays universally known as *Poincaré's recurrence theorem*. Poincaré's theorem lies at the foundations of ergodic theory, and is actually true of any dynamical system moving in a bounded set under the action of measure-preserving transformations. One of the most well known applications of Poincaré's recurrence is that a bounded Hamiltonian system ("Universe") must return after some (usually extraordinarily large) time arbitrarily close to its initial configuration. Let us focus on a subsystem of that Universe (a galaxy, or more modestly, the solar system, are good examples). This subsystem will return to its initial configuration after some time—but, what time? If this subsystem does not interact with the rest of the Universe, it will have its own private return time, and it is reasonable to expect (and possible to prove) that this return time is usually shorter than the return time of the whole Universe. Things being what they are, subsystems do usually interact with the rest of the Universe, and it is then no longer reasonable to expect (or possible to prove) that the subsystem will return to its initial configuration before the whole Universe does. The aim of this paper is to briefly and tentatively discuss the possibility of such "superrecurrence" in the presence of interactions: an interacting subsystem of a Hamiltonian system will have its own return time, which is of the same order of magnitude as if there were no interaction. The difficulty lies in the fact that an interacting subsystem of a Hamiltonian system is not Hamiltonian in its own right, but has a much more complicated structure due to the interactions. We use a tool from symplectic topology, related to Gromov's symplectic non-squeezing theorem (also known as the "principle of the symplectic camel") which we have abundantly explained and discussed elsewhere [2–5] (cf. [6]). This theorem plays an essential role in quantum mechanics, and also in the study of entropy [7], aeronautics [8,9], and statistical mechanics [10]. Its importance and potential applications have however not yet been fully exploited in physics. This might be due to the mathematical difficulty of the result, which was only discovered in the mid-1980s by the mathematician Mikhail Gromov, who was awarded the Abel Prize (the equivalent of the Nobel Prize in mathematics) for his discovery. In fact, the principle of the symplectic camel can be seen as an imprint of the uncertainty principle of quantum mechanics in classical mechanics—or the other way around! Without becoming embroiled in a sterile polemic, let me just say that if one considers (as one should do!) quantum theory as the "master theory" of which classical mechanics (in its Hamiltonian formulation) is a macroscopic approximation, then one should find traces of the mathematical structure of quantum theory in the macroscopic domain. This is exactly what happens here: as we have shown in the papers cited above, uncertainty relations (in their strong Robertson–Schrödinger form) are not per se quantum mechanical, but also exist in classical mechanics, but this time for an arbitrary value of Planck's constant h, which is now viewed as a free parameter. It is then the more exact theory (quantum mechanics) which forces us to choose a lower limit for the indeterminacy by fixing a lower bound for the admissible value of the parameter h (in more physical terms, it is the quantum phenomena which force us to do so: quantum theory exists only to describe quantum phenomena, and not the other way round). However, more about this is in the Discussion ending the paper.

2. Subsystems of Hamiltonian Systems

2.1. Description of the Problem

Consider a mechanical system of N point-like particles whose motion is determined by a Hamilton function H. If the system is confined to a bounded region of phase space $\mathbb{R}^{6N}_{\mathbf{q},\mathbf{p}}$, Poincaré's recurrence theorem tells us that any initial pattern of positions and velocities (specified within a given error) will recur, independently of any permutation in the numbering of the particles of the system. The recurrence time is however generally extremely long, (see [11] for a recent analysis of recurrence time), except of course for periodic (or quasi-periodic) systems. Of course, the boundedness condition is essential: a free particle in an infinite Universe will never return to its initial position. Suppose indeed that the system, represented by a phase point $(\mathbf{q},\mathbf{p}) = (\mathbf{q}_1,...,\mathbf{q}_N,\mathbf{p}_1,...,\mathbf{p}_N)$, with $\mathbf{q}_i = (x_i, y_i, z_i)$ and $\mathbf{p}_i = (p_{x_i}, p_{y_i}, p_{z_i})$, is confined to a "universe" $\mathcal{U}$. We are not asking for an exact return of $(\mathbf{q},\mathbf{p})$ but we content ourselves with the return of some (arbitrarily) small neighbourhood Ω of that point. Then, an upper bound for the first return time of that neighbourhood has a magnitude of order $T \approx \mathrm{Vol}(\mathcal{U})/\mathrm{Vol}(\Omega)$. This number is usually very large. Let us now focus on a subsystem, identified with a point $(\mathbf{q}',\mathbf{p}') = (\mathbf{q}_1,...,\mathbf{q}_n,\mathbf{p}_1,...,\mathbf{p}_n)$ with $n < N$. Assume first that the total Hamiltonian function is of the type

$$H = H'(\mathbf{q}',\mathbf{p}') + H''(\mathbf{q}'',\mathbf{p}'') \tag{1}$$

where $(\mathbf{q}'',\mathbf{p}'') = (\mathbf{q}_{n+1},...,\mathbf{q}_N,\mathbf{p}_{n+1},...,\mathbf{p}_N)$. Due to the absence of interaction between the two subsystems $(\mathbf{q}',\mathbf{p}')$ and $(\mathbf{q}'',\mathbf{p}'')$, their motions are independent; the time-evolution of $(\mathbf{q}',\mathbf{p}')$ is thus governed solely by its own private Hamiltonian H'; the equations of motions are

$$\dot{\mathbf{q}}_j = \frac{\partial H'}{\partial \mathbf{p}_j}(\mathbf{q}',\mathbf{p}'),\ \dot{\mathbf{p}}_j = -\frac{\partial H'}{\partial \mathbf{q}_j}(\mathbf{q}',\mathbf{p}') \text{ with } 1 \leq j \leq n \tag{2}$$

and their solutions only dependent on the initial values $\mathbf{q}'(0)$ and $\mathbf{p}'(0)$. The corresponding universe $\mathcal{U}'$ consists of the set of all points $(\mathbf{q}',\mathbf{p}')$ such that $(\mathbf{q}',\mathbf{q}'',\mathbf{p}',\mathbf{p}'')$ is in $\mathcal{U}$ for some $\mathbf{q}'',\mathbf{p}''$; it is thus the projection of $\mathcal{U}$ on the reduced phase space $\mathbb{R}^{6n}_{\mathbf{q}',\mathbf{p}'}$ and, accordingly, the corresponding neighbourhood Ω' is the projection of Ω on $\mathbb{R}^{6n}_{\mathbf{q}',\mathbf{p}'}$. Let us now compare the return time $T' \approx \mathrm{Vol}(\mathcal{U}')/\mathrm{Vol}(\Omega')$ for the system $(\mathbf{q}',\mathbf{p}')$ with that of $(\mathbf{q},\mathbf{p})$. To fix the ideas, we choose for $\mathcal{U}$ a hypercube with sides of length L and for Ω a hypercube with sides of length $\varepsilon \ll L$. It follows that $T \approx (L/\varepsilon)^{6N}$ and that $T' \approx (L/\varepsilon)^{6n}$ so that the ratio T/T' between both return times is of order $(L/\varepsilon)^{6(N-n)}$. Consider next the general case, where the subsystems interact; we can no longer separate the variables that $\mathbf{q}_i$ and $\mathbf{p}_i$; this is the case if for instance,

$$H(\mathbf{q},\mathbf{p}) = \sum_{j=1}^{N} \frac{|\mathbf{p}_j|^2}{2m_j} + V(\mathbf{q}_1,...,\mathbf{q}_N) \tag{3}$$

although everything will hold for an arbitrary function of the variables $\mathbf{q}_j, \mathbf{p}_j$. We consider again the subsystem $(\mathbf{q}', \mathbf{p}')$; its motion will now depend on the global behaviour of the system $(\mathbf{q}, \mathbf{p})$, since the solutions of the corresponding Hamilton equations

$$\dot{\mathbf{q}}_j = \frac{\partial H}{\partial \mathbf{p}_j}(\mathbf{q}, \mathbf{p}),\ \dot{\mathbf{p}}_j = -\frac{\partial H}{\partial \mathbf{q}_j}(\mathbf{q}, \mathbf{p}) \text{ with } 1 \leq j \leq n \tag{4}$$

now depend on the initial values of *all* variables $\mathbf{q}_j, \mathbf{p}_j$, not only the n first. It follows that the motion of the subsystem $(\mathbf{q}', \mathbf{p}')$ is not governed by a Hamiltonian; this can be easily seen by finding the explicit solutions for simple systems. Let us illustrate this using Sharov's argument [12]. Suppose Equation (4) represents the time-evolution of a bona fide Hamiltonian system. Denoting by $\mathbf{q}_1^0, ..., \mathbf{q}_n^0; \mathbf{p}_1^0, ..., \mathbf{p}_n^0$ any set of initial conditions we have

$$\int |\mathcal{J}(t)| d\mathbf{p}_1^0 \cdots d\mathbf{p}_n^0 d\mathbf{q}_1^0 \cdots d\mathbf{q}_n^0 = \int d\mathbf{p}_1 \cdots d\mathbf{p}_n d\mathbf{q}_1 \cdots d\mathbf{q}_n$$

where $\mathcal{J}(t)$ is the Jacobian of the transformation from the initial conditions to $(\mathbf{q}', \mathbf{p}')$. If the system Equation (4) is Hamiltonian, then this transformation must be canonical, so we should have $|\mathcal{J}(t)| = 1$. However, we have, as Sharov [12] showed, $d\mathcal{J}(t)/dt \neq 0$, hence $\mathcal{J}(t) \neq \mathcal{J}(0) = 1$. In fact, the principle of the symplectic camel which we discuss below implies, without any calculation at all, that $|\mathcal{J}(t)| \geq 1$. Thus:

A subsystem of a Hamiltonian system is usually not a Hamiltonian system in its own right.

What about the return time? A first educated guess is that since the subsystem interacts (perhaps very strongly) with the rest of the system this interaction will influence the return time which will become much longer than in the interaction-free case, perhaps even of the order $T \approx (L/\varepsilon)^{6N}$, at which the total system returns.

2.2. Non-Squeezing and Packing

Liouville's theorem tells us that Hamiltonian motions are volume preserving: this is one of the best known results from elementary mechanics. However, in addition to being volume-preserving, Hamiltonian motions have unexpected "rigidity properties", which distinguish them from ordinary volume-preserving diffeomorphisms. The most famous is Gromov's *non-squeezing theorem*. Assume that we are dealing with a Hamiltonian system consisting of a large number N of particles with coordinates $\mathbf{q}_i = (x_i, y_i, z_i)$ and momenta $\mathbf{p}_i = (p_{x_i}, p_{y_i}, p_{z_i})$. If these points are sufficiently close to each other, we may, with a good approximation, identify that set with a "cloud" of phase space fluid; by phase space we mean the space $\mathbb{R}^{6N}_{\mathbf{q},\mathbf{p}}$ with $\mathbf{q} = (\mathbf{q}_1, \mathbf{q}_2, ..., \mathbf{q}_N)$ and $\mathbf{p} = (\mathbf{p}_1, \mathbf{p}_2, ..., \mathbf{p}_N)$. Suppose that this cloud contains at time $t = 0$ a ball with radius R:

$$B_R : |\mathbf{q} - \mathbf{q}_0|^2 + |\mathbf{p} - \mathbf{p}_0|^2 \leq R^2. \tag{5}$$

The orthogonal projection of the cloud of points on any plane of coordinates (x, p_x), (x, p_y), (x, p_z), etc. will thus have area at least πR^2. Let us now watch the motion of this phase-space cloud. As time evolves, it will distort and may take after a while a very different shape, while keeping constant volume in view of Liouville's theorem. However—and this is the surprising result—the projections of that deformed cloud on any of the planes of *conjugate* coordinates (x, p_x), (y, p_y) or (z, p_z) will never decrease below the value πR^2. This fact is of course strongly reminiscent of the uncertainty principle of quantum mechanics, of which it is in fact a classical version; we have discussed this analogy in detail in de Gosson [4,10]. Scheeres et al. [9] used our results to study orbit uncertainty in space craft navigation.

The phenomenon described above seems at first sight to conflict with the usual conception of Liouville's theorem: according to folk wisdom, the ball B_R can be stretched in all directions, and eventually get very thinly spread out over huge regions of phase space, so that the projections on any plane could a priori become arbitrary small after some time t; this stretching generically increases with time. In fact, one may very well envisage that the larger the number N of degrees of freedom, the more that spreading will have chances to occur since there are more and more directions in which the ball is likely to spread! This possibility has led to many quasi-philosophical speculations about the stability of Hamiltonian systems (in [4,5] we have discussed Penrose's claim ([13], p. 183, l.–3) that phase space spreading suggests that "*classical mechanics cannot actually be true of our world*"). However, the phenomena we shortly described above show that such statements (which abound in the literature) come from a deep misunderstanding of the nature of Hamiltonian mechanics. The non-squeezing theorem prevents anarchic and chaotic spreading of the ball in phase space which would be possible if it were possible to stretch it inside arbitrarily thin tubes in directions orthogonal to the conjugate planes. This possibility is perfectly consistent with Katok's lemma [14], which can be stated as follows: consider two bounded domains Ω and Ω' in $\mathbb{R}^{2n}$ which are both smooth volume preserving deformations of the ball B_R. Then, for every $\varepsilon > 0$, there exists a Hamiltonian function H and a time t such that $\mathrm{Vol}(f_t^H(\Omega)\Delta\Omega') < \varepsilon$. Here, $f_t^H(\Omega)\Delta\Omega'$ denotes the set of all points that are in $f_t^H(\Omega)$ or Ω', but not in both. Katok's lemma thus shows that up to sets of arbitrarily small measure ε any kind of phase-space spreading is a priori possible for a volume-preserving flow, because $f_t^H(\Omega)$ can become arbitrarily close to Ω'.

The properties outlined above are best understood (and proved) in terms of a new generation of theorems from symplectic topology, the first of which goes back to the mid-1980s, and is known as *Gromov's non-squeezing theorem* [15], which is often referred to as the *principle of the symplectic camel* (for various interpretations of this Biblical metaphor, see the comments to the online version of Reich's review [16] of my paper [4]. This theorem—whose implications to physics have not yet been fully explored—has allowed us to give a symplectically invariant topological version of the principle of quantum indeterminacy [3–5], and to describe in a precise classical uncertainties arising in some systems [10].

2.3. One Step Further: Subsystems

Gromov's non-squeezing theorem can be reformulated rigorously as follows: let f be a canonical transformation (often called a symplectic diffeomorphism, or symplectomorphism in the mathematical literature). This means that, if $(\mathbf{q}', \mathbf{p}') = f(\mathbf{q}, \mathbf{p})$, then the Jacobian matrix

$$Df(q,p) = \frac{\partial(q',p')}{\partial(q,p)} \tag{6}$$

is symplectic, and thus, for every point $(\mathbf{q}, \mathbf{p})$ in $\mathbb{R}^{6N}$ (Arnol'd [17]): $Df(q,p) \in \mathrm{Sp}(3N)$. Gromov's theorem says that

$$\mathrm{Area}\,\Pi_j(f(B_R)) \geq \pi R^2 \tag{7}$$

where Π_j is the orthogonal projection on the plane of conjugate variables (q_j, p_j); here the index j is any of the integers $1, ..., N$. We now address the following more general question: is there a generalization of this result to higher dimensional subspaces of $\mathbb{R}^{6N}$? More specifically, what we have in mind is the volume of the projection Π' of $f(B_R)$ on a subspace $\mathbb{R}^{6n}$ in the conjugate coordinates $(\mathbf{q}', \mathbf{p}') = (q_1, ..., q_{3n}, p_1, ..., p_{3n})$; $1 < n < N$. Such a subspace of $\mathbb{R}^{6N}$ inherits in a natural way a symplectic structure, and we ask: is the "obvious" generalization

$$\mathrm{Vol}_{6n}\,\Pi'(f(B_R)) \geq \frac{\pi^{3n}}{(3n)!} R^{6n} \tag{8}$$

of Equation (7) true? Let us call a canonical transformation satisfying the property in Equation (8) a "hereditary" canonical transformation. It has been very recently proved by Abbondandolo and Matveyev [18] that:

All linear (or affine) canonical transformations are hereditary.

An immediate consequence of this fact is that if (f_t^H) is the flow determined by the Hamilton equations for a Hamiltonian function of the type

$$H(q,p) = \sum_{j=1}^{3N} a_j(t)p_j^2 + b_j(t)q_j^2 + c_j(t)p_j + d_j(t)q_j$$

where a_j, b_j, c_j, d_j are continuous real functions of the time t, then each f_t^H is hereditary. This applies, in particular to the independent oscillator model of a heat bath where

$$H(q,p,x,p_x) = \frac{1}{2m}p_x^2 + Kx^2 + \sum_{j=1}^{3N} \frac{1}{2m_j}(p_j^2 + m\omega_j^2(x-q_j)^2)$$

is the Hamilton function of a linear oscillator in the (x, p_x) variables coupled with N isotropic oscillators with frequencies $\omega_1, ..., \omega_N$.

Let us discuss the non-linear case. In [18], the authors constructed a counterexample showing that there exist nonlinear canonical transformations which are not hereditary. However, the transformation they constructed deforms the ball B_R tremendously and seems to be very unphysical. Now, in the same paper, Abbondandolo and Matveyev discussed the validity of Equation (8) for more general canonical transformations when the radius R is small; they conjecture that this property is generically true of all Hamiltonian systems. At the time of writing, there is however no convincing proof of this conjecture. We are thus in the unusual (and unpleasant) situation where we would like to use a theorem valid for a class of Hamiltonians which has not, at the time of writing, been fully characterized! (However, see the comments in Schlenk's review paper [19]). If true, there would be large subclasses of Hamiltonian (sub)systems exhibiting superrecurrence.

2.4. A Simple Case of Superrecurrence

As was pointed out by Polterovich (see Schlenk [19]), the original motivation for Gromov to study "packing numbers" in symplectic topology was his search for recurrence properties which are stronger than those of volume preserving mappings. Consider first the following simple planar situation: we have a disk $D_R(0)$ in the plane $\mathbb{R}^2$, that is, the set of points (x, p_x) such that $x^2 + p_x^2 \leq R^2$. We have $\text{Area}(D_R(0)) = \pi R^2$. We now ask the question: For which radius r can we embed two smaller *disjoint* disks $D_r(a)$ and $D_r(b)$ inside $D_R(0)$ using a canonical transformation? The answer is easy: in the plane, canonical transformations are just the area preserving diffeomorphisms, so it suffices that

$$\text{Area}(D_r(a)) + \text{Area}(D_r(b)) \leq \text{Area}(D_R(0))$$

that is, $2r^2 \leq R^2$. We can thus embed at best two disjoint disks $D_{R/\sqrt{2}}(a)$ and $D_{R/\sqrt{2}}(b)$ inside a disk with radius R, and that disk is then completely filled by the images of deformed smaller disks. Choose now a general phase space $\mathbb{R}^{2n}$ (with for instance $n = 3N$) and consider the same problem for a ball $B_R(0)$ with radius R. We want to pack two smaller disjoint balls $B_r(a)$ and $B_r(b)$ inside $B_R(0)$ using general canonical transformations. Calculating the volumes, we have

$$\text{Vol}_{2n}(B_R) = \frac{(\pi R^2)^n}{n!} \quad , \quad \text{Vol}_{2n}(B_r) = \frac{(\pi r^2)^n}{n!}$$

and hence

$$\text{Vol}_{2n}(B_R) / \text{Vol}_{2n}(B_r) = (R/r)^{2n}. \tag{9}$$

This indicates that, at first sight, we could pack 2^n balls with radius $r = R/\sqrt{2}$ inside $B_R(0)$ and this number becomes very large when the number n of degrees of freedom increases.

The reality is, however, very different. The argument above, while true for arbitrary volume preserving diffeomorphisms, does not take into account the fact that we are dealing here with canonical transformations, and that the latter are much more "rigid" than ordinary volume preserving diffeomorphisms. In fact, when we are dealing with canonical transformations (and hence, in particular, Hamiltonian flows), the following result holds:

Gromov's Two Balls Theorem [15]: *If two disjoint phase space balls with radius r are mapped inside a ball B_R by a canonical transformation, then we must have $r \leq R/\sqrt{2}$.*

In other words, there is no quantitative difference between the packing number in two-dimensional phase plane and a general phase space $\mathbb{R}^{2n}$ with $n > 1$. This implies that any obstruction to symplectically embedding a ball into a larger ball is much stronger than the volume constraint given by Equation (9). This is a very strong result, and allows proving a simple superrecurrence theorem: assume that we have a "Universe" $\mathcal{U}$ that is the image of a very large ball B_R by some canonical transformation (it may be a symplectic ellipsoid, or more generally any compact symplectic manifold with boundary). Take a subset $\mathcal{B}$ of $\mathcal{U}$ which is the image of a ball with radius $(R+\varepsilon)/\sqrt{2}$ where ε is a small number, say $\varepsilon = (1/n)R$. Then,

$$\mathrm{Vol}_{2n}(\mathcal{U})/\,\mathrm{Vol}_{2n}(\mathcal{B}) = N \approx e^{-2}2^n. \tag{10}$$

Now, let H be a Hamiltonian function whose flow (f_t^H) preserves the universe $\mathcal{U}$, that is $f_t^H(\mathcal{U})$ (it is sufficient that the Hamiltonian vector field X_H is tangent to the boundary $\partial\mathcal{U}$). The flow f_t^H displaces the "subuniverse" $\mathcal{B}$ which becomes $f_t^H(\mathcal{B})$ after time t. If we only use the fact that f_t^H volume preserving; then Equation (10) would imply that the recurrence time could be very large: choosing $t = 1$ as a unit of time and setting $f = f_t^H$ the sets $f(\mathcal{B}), f^2(\mathcal{B}),...., f^{N-1}(\mathcal{B})$ cannot be all disjoint, and hence the return time can a priori be as large as $N-1$. However, by Gromov's Two Balls Theorem, we will have $f(\mathcal{B}) \cap \mathcal{B} \neq \emptyset$ so the first return time is $t = 1$.

Remark 1. *Gromov's Two Balls Theorem' whose classical consequences we discussed above is related to the notion of dislocation of quantum states as discussed by Polterovich [20] and Charles and Polterovich [21].*

3. Discussion

I have discussed in this contribution to EMQM17 some consequences of symplectic topology on Poincaré recurrence from a perfectly classical point of view. However, as I have explained elsewhere with Basil Hiley [6], the properties of symplectic topology described here are reminiscent of certain aspects of quantum mechanics (for instance, the uncertainty principle). I view them as *imprints of the quantum world* on classical mechanics in its Hamiltonian formulation. Of course, this point of view might be felt as controversial by some physicists, so let me explain what I have in mind (thus, partially answering some interesting remarks and objections made by a Referee). In either of its formulations, quantum mechanics is built on classical mechanics. In the Heisenberg picture, one wants to give an operator-theoretical meaning to Hamilton's equations of motion

$$\frac{dq}{dt} = \frac{\partial H}{\partial p}(q,p,t)\ ,\ \frac{dp}{dt} = -\frac{\partial H}{\partial q}(q,p,t)$$

and this is done by replacing the classical position and momentum variables q and p with operators $\widehat{q}$ and $\widehat{p}$ satisfying the Born condition $[\widehat{q},\widehat{p}] = i\hbar$; after some work, one is led to the quantum Hamilton equations

$$\frac{d\widehat{q}}{dt} = \frac{\partial H}{\partial p}(\widehat{q},\widehat{p},t)\ ,\ \frac{d\widehat{p}}{dt} = -\frac{\partial H}{\partial \widehat{q}}(\widehat{q},\widehat{p},t). \tag{11}$$

In the Schrödinger picture, which is based on de Broglie's wave mechanics, the time-evolution of the wavefunction ψ is governed by Schrödinger's equation

$$i\hbar\frac{\partial\psi}{\partial t} = \widehat{H}\psi \tag{12}$$

where $\widehat{H}$ is an operator associated with the classical Hamiltonian function H. Now, it is often claimed that both pictures lead to the same physical predictions. However, this is only true if one uses the same quantization procedure for both the Heisenberg and Schrödinger theories. In fact, the quantum Hamilton equation (Equation (11)) only make sense if one quantizes products $q^m p^n$ using the prescriptions given in 1923 by Born and Jordan, together with Heisenberg, in their famous "Dreimännerarbeit"; it follows that the operator $\widehat{H}$ in Schrödinger's Equation (12) must be derived from the Hamiltonian function H also using the Born–Jordan prescription (for otherwise both pictures would no longer be equivalent, as I have discussed in detail in [22,23]). This short digression is intended to explain that, no matter how one "defines" quantum mechanics, the classical (Hamiltonian) theory is always present as a watermark. In fact, in my opinion, this is a quite logical consequence of the fact that we, humans, are macroscopic objects and as such the only direct experience we have from our World is of a macroscopic nature, and there is no "pedagogical" way to reverse this approach, that is to make us in first place become aware of the quantum nature of our environment, and then to deduce the classical properties as an approximation thereof. Thus, all this brings me to the following observation: quantum physics is a mathematical construct. However, mathematics is an exact Science; there is no place for polemic, interpretations, or controversy. Mathematics is not as "emotional" as physics is: a mathematical statement is either *true*, or it is *false*. It turns out that I have shown in [24] that symplectic geometry is the common mathematical background of classical and quantum mechanics (also see our paper with Hiley [6]), and that both theories are mathematically *equivalent*. The proof mainly relies on the fact that a Hamiltonian isotopy automatically generates a quantum isotopy (it is a consequence of the theory of the metaplectic group) and vice versa to every quantum isotopy we can associate a Hamiltonian isotopy. This property shows that there is a *canonical* isomorphism between quantum and classical theory. A *caveat* here: the Reader is invited to observe that this is by no way a provocative or paradoxical statement: it is just a mathematical theorem, which may be perceived as counterintuitive by many physicists. Now, a mathematical theory has no a priori physical meaning unless one creates an interpretational apparatus allowing to draw real-life consequences from the mathematical objects: an equation is not a physical theory! This explains the statement I made above, namely that "...properties of symplectic topology ... can be viewed as *imprints of the quantum world* on classical mechanics...". If both the quantum and classical theory are mathematically equivalent, the sentence could indeed be reversed by saying that it is classical mechanics which leaves imprints on the quantum world. However, from the physical point of view, we are in a "Cheshire cat" scenario: since experience shows that quantum mechanics yields a better description of Nature than classical mechanics, quantum theory contains classical mechanics as an approximation, and leaves in this approximation some features reminiscent of the true theory, exactly as when the legendary cat disappears but leaves his grin.

Funding: This research was funded by Austrian Research Agency FWF grant number P20442-N13.

Acknowledgments: I wish to thank the Referees for valuable comments.

Conflicts of Interest: The author declares no conflict of interest.

References

1. Poincaré, H. Sur le problème des trois corps et les équations de la dynamique, Mémoire couronné du prix de S.M. le roi Oscar II de Suède. *Acta Math.* **1890**, *13*, A3–A270. (In French)
2. De Gosson, M. *Principles of Newtonian and Quantum Mechanics, the Need for Planck's Constant*, 2nd ed.; With a Foreword by Hiley, B.; World Scientific: Singapore, 2016.

3. De Gosson, M. *Symplectic Geometry and Quantum Mechanics*; Birkhäuser: Basel, Switzerland, 2006.
4. De Gosson, M. The Symplectic Camel and the Uncertainty Principle: The Tip of an Iceberg? *Found. Phys.* **2009**, *39*, 194–214.
5. De Gosson, M.; Luef, F. Symplectic Capacities and the Geometry of Uncertainty: The Irruption of Symplectic Topology in Classical and Quantum Mechanics. *Phys. Rep.* **2009**, *484*, 131–179.
6. De Gosson, M.A.; Hiley, B.J. Imprints of the quantum world in classical mechanics. *Found Phys.* **2011**, *41*, 1415–1436.
7. Kalogeropoulos, N. Time irreversibility from symplectic non-squeezing. *Phys. A Stat. Mech. Appl.* **2018**, *495*, 202–210.
8. Maruskin, J.M.; Scheeres, D.J.; Bloch, A.M. Dynamics of Symplectic Subvolumes. *SIAM J. Appl. Dyn. Syst.* **2009**, *8*, 180–201.
9. Scheeres, D.J.; de Gosson, M.A.; Maruskin, J.M. Applications of Symplectic Topology to orbit Uncertainty and Spacecraft Navigation. *J. Astronaut. Sci.* **2012**, *59*, 63–83.
10. De Gosson, M. On the Use of Minimum Volume Ellipsoids and Symplectic Capacities for Studying Classical Uncertainties for Joint Position—Momentum Measurements. *J. Stat. Mech.* **2010**, *2010*, P11005.
11. Altmann, E.G., Kantz, H. Recurrence time analysis, long-term correlations, and extreme events. *Phys. Rev. E.* **2005**, *71*, 056106.
12. Sharov, S.R. Basis of Local Approach in Classical Statistical Mechanics. *Entropy* **2005**, *7*, 122–133.
13. Penrose, R. *The Emperor's New Mind*; Oxford University Press: Oxford, UK, 1989.
14. Katok, A. Ergodic perturbations of degenerate integrable Hamiltonian systems. *Math. USSR Izv.* **1973**, *7*, 535–571.
15. Gromov, M. Pseudoholomorphic curves in symplectic manifolds. *Invent. Math.* **1985**, *82*, 307–347
16. Reich, E.S. How camels could explain quantum uncertainty. *New Sci.* **2009**, *12*, 2697.
17. Arnol'd, V.I. *Mathematical Methods of Classical Mechanics*, 2nd ed.; Graduate Texts in Mathematics; Springer: Berlin/Heidelberg, Germany, 1989.
18. Abbondandolo, A.; Matveyev, S. Middle-dimensional squeezing and non-squeezing behaviour of symplectomorphisms. *arXiv* **2011**, arXiv:1105.2931v1.
19. Schlenk, F. Symplectic Embedding Problems, Old and New. *Bull. Am. Math. Soc.* **2018**, *55*, 139–182.
20. Polterovich, L. Quantum Footprints of Symplectic Rigidity. *EMS Newsl.* **2016**, *12*, 16–21.
21. Charles, L.; Polterovich, L. Quantum Speed Limit Versus Classical Displacement Energy. In *Annales Henri Poincaré*; Springer: Berlin/Heidelberg, Germany, 2018; Volume 19, pp. 1215–1257.
22. De Gosson, M. Born–Jordan Quantization and the Equivalence of the Schrödinger and Heisenberg Pictures. *Found. Phys.* **2014**, *44*, 1096–1106.
23. De Gosson, M. *Introduction to Born–Jordan Quantization: Theory and Applications*; Series Fundamental Theories of Physics; Springer: Berlin/Heidelberg, Germany, 2016.
24. De Gosson, M. Paths of Canonical Transformations and their Quantization. *Rev. Math. Phys.* **2015**, *27*, 1530003.

Article

The Philosophical and Scientific Metaph) David Bohm

Entropy 2018 col

William Seager

Department of Philosophy, University of Toronto Scarborough, Scarborough, ON M1C 1A4, Canada; bill.seager@utoronto.ca; Tel.: +1-416-287-7151

Received: 22 May 2018; Accepted: 25 June 2018; Published: 26 June 2018

Abstract: Although David Bohm's interpretation of quantum mechanics is sometimes thought to be a kind of regression towards classical thinking, it is in fact an extremely radical metaphysics of nature. The view goes far beyond the familiar but perennially peculiar non-locality and entanglement of quantum systems. In this paper, a philosophical exploration, I examine three core features of Bohm's metaphysical views, which have been both supported by features of quantum mechanics and integrated into a comprehensive system. These are the holistic nature of the world, the role of a unique kind of information as the ontological basis of the world, and the integration of mentality into this basis as an essential and irreducible aspect of it.

Keywords: interpretations of quantum mechanics; David Bohm; mind–body problem; quantum holism

---

David Bohm is famous for re-invigorating and developing the "pilot wave" interpretation of quantum mechanics (QM) originally articulated in 1926 by Louis de Broglie. Bohm's theory envisages a world of particles which all have definite momenta and positions, albeit the values of which are generally inaccessible. (It is important to point out that the concept of momentum in Bohm's theory is not straightforward. The fact that we can assign a value of mv to a particle is not directly related to what we would find if we performed a QM measurement of momentum [1,2].) The particles are deterministically "steered" or "guided" by a universal field which is described by the quantum wave function. It is sometimes said that Bohm's view is a return to a classical picture of the world, embracing atomistic particularity and determinism. For example, the philosopher David Albert forthrightly claims that "the metaphysics of [Bohm's] theory is exactly the same as the metaphysics of classical mechanics" ([3], p. 174). A recent text book casually characterizes Bohm's account as one endorsing "local realism" ([4], p. 65). Christopher Fuchs once wrote that "Bohmism" represents a hopeless "return to the womb of classical physics ... yuck!" ([5], p. 417).

A core classical theory is of course that of Newton. A "Newtonian world view" is a metaphysical interpretation of a theory which can plausibly be extended to embrace the entire world instead of and speculatively beyond the systems to which it can actually be successfully applied in experiment and technology. The Newtonian viewpoint at issue is that of a world of locally interacting particles which obey well defined laws of nature and whose proclivities for combination lead to all of the complexity of form and the variety of composite systems that we so abundantly observe. As is well known, Newton himself was unable fully to subscribe to Newtonianism in this sense because his theory of gravitation postulated a non-local and instantaneously active "force" generated by every material object which permeated the universe. At the time, the notion of such a thing as "force" was dubious, carrying the taint of the occult (forces are akin to older notions of the "spirit") and the retrograde Scholastic concept of substantial forms (see [6]). Moreover, Newtonian non-locality is considerably more radical than the more recently discovered quantum variety. It permits (in principle) faster than light, indeed instantaneous, signaling via the mere rearrangement of matter. This extension of the Newtonian metaphysics of nature adds mysterious forces to the push and pull of particle

...sions which many at the time regarded as an illicit intrusion of immaterial entities into a part of the world—the material universe—that should be intelligible solely in terms of mechanical principles.

If we take a hard line on Newtonianism—as surrogate for the mechanical metaphysics—then it is hard to seriously maintain that Bohm's account of quantum mechanics is Newtonian. The quantum potential invoked by Bohm and required to duplicate the empirical success of standard quantum mechanics is irredeemably non-local, and Bohm held views entirely at odds with the mechanical view of the world as consisting of independent, causally interacting individual parts.

Even if we take a softer line, more in line with what Newton himself was willing to postulate, then Bohm's view is still at odds with a Newtonian picture of the world. There is a split worth noting here in those who work on theories that develop Bohm's original insight. Bohm himself took the radical and philosophical line I will investigate in this paper. Others, those who develop so-called Bohmian Mechanics, strongly resist any need for a new metaphysical outlook and cleave to a particle-based picture in which the world evolves via, in the words of Peter Holland, "objective processes" ([7], p. 25), albeit non-local ones; numerous papers by Sheldon Goldstein, Detlef Dürr, and their co-workers would also fall on this side of the split (for an overview, see [8]). Newton was not averse to the postulation of forces in nature. However, such forces come in at least two varieties: local forces that are properties of kinds of material bodies and non-local forces such as gravitation which are suspiciously uncaring about the nature of the bodies giving rise to it. The former he welcomed and hypothesized that they would ultimately explain chemistry: "... many things lead me to have a suspicion that all phenomena may depend on certain forces by which the particles of bodies, by causes not yet known, either are impelled toward one another and cohere in regular figures, or are repelled from one another and recede" ([9], p. 382–383). The latter were anathema to Newton, who disparaged those who might favour the idea of action at a distance: "I believe no Man who has in philosophical Matters a competent Faculty of thinking can ever fall into it" ([10], p. 102). For many at the time, even the local forces were suspicious. Pure mechanical contact interaction based upon the impenetrability of matter was the "gold standard" for explanations of the natural world. Bohm's view could hardly be more different than this vision of classical physics.

Of course, Newton was right; the "forces brigade" won the day over pure mechanism, and Newton's theory funded the development of classical physics. Still, although physicists became inured to the scandal of action at a distance and non-local instantaneous forces, there were regular calls to recast physical theory in terms of local forces smoothly transmitted through space within some kind of genuinely physical medium. This persistent attitude culminated in Maxwell's field theory of electromagnetism and, later, Einstein's revolutionary field-based account of gravitation (for a brief history of the field concept, see [11]). QM entanglement apparently introduces an entirely new kind of non-local relation, which was strongly suggestive to Bohm that a similarly new picture of reality was needed to accommodate it.

Bohm's account of QM introduces some new ideas and a radically different general outlook on nature. However, it does not make any *empirical* difference: Bohmian predictions are identical to those of "standard" QM. It's worth noting that there have been some controversial attempts to empirically distinguish the views. Bohm's account assumes that the initial conditions of a system satisfy the quantum equilibrium condition (that is, the probability distribution of the initial positions of the particles is given by $|\psi|^2$). It is conceivable that (parts of) the universe do not abide by this condition. It has also been argued that, although Bohmian theory matches QM statistically, it could vary from it in individual cases, and this divergence might not be absolutely impossible to measure. For references and discussion, see Riggs ([12], pp. 142 ff.). Still, the basic empirical equivalence is the view is often called the Bohmian or (de Broglie–Bohm) *interpretation* of QM. It thus joins the ranks of a host of alternative interpretations and turns into a *metaphysics* of nature.

This raises an unavoidable question of what is metaphysical about interpretations of QM and, in general, what distinguishes a metaphysical question from a scientific question about the structure of reality. Looming behind this issue is a more general one that questions the value of engaging in metaphysics at all. Metaphysical skepticism has a long and distinguished history and debates in this area remain vigorous (see, e.g., [13]). In this paper, I will proceed on the assumption that metaphysical speculation is both possible and valuable, at least to the extent that trying to understand the kind of world which our best science suggests we live in is worthwhile, as the lively debate about the meaning of QM suggests. The quest of metaphysics is succinctly expressed by Wilfred Sellars: "to understand how things in the broadest possible sense of the term hang together in the broadest possible sense of the term" ([14], p. 1). This quest goes beyond the purview of theoretical science. Yet delineating the distinctive nature of metaphysical questions is not a straightforward task because modern science is a more or less direct descendant of early thought about the general nature of reality. When, for example, Anaxogoras postulated that "everything is in everything," that matter is infinitely divisible, and that every portion of it is a mixture of all possible qualities in different proportions, was he doing proto-science or philosophical metaphysics? (Anaxogoras's views are in fact intriguing, complicated and far from clear; see [15].) At the time, and for long after, there was no such distinction. As the centuries accumulated, many such questions drifted from the metaphysical towards the scientific pole.

A good clue to a metaphysical question is its distance from empirical testability: the more remote from empirical consequences, the "more metaphysical" the question. I should add the usual rider of "in principle" testability. Technical difficulties in constructing appropriate experimental apparatus does not a metaphysics make. There may also be a general demand on the perceived significance of the question: metaphysics is supposed to tackle big questions. However, I don't see exactly why there can't be utterly trivial metaphysical questions. Contrary to positivists, this does not mean the question is empty or meaningless.

To give a pointedly philosophical illustration to illustrate how even the most scholastic seeming question can still link to scientific concerns, consider the nature and identity of composite objects. A persistent question in metaphysics is about the ontological status of such entities. The standard example is the contrast between a bulk lump of clay and the statue artistically formed from it. The lump of clay is the material from which the statue is made. The clay is still there after this operation as is the statue. Are they one and the same entity, about which we merely have two different ways of talking? They occupy exactly the same space and move inexorably together wherever they go. Surely they are one. And yet while we can destroy the statue with a hammer, the bulk lump of clay remains. It seems a reasonable principle that, if one can destroy x without destroying y, then $x \neq y$. The metaphysics of composite objects can get quite hairy (see [16]). One thing is pretty clear, however: it is hard to think of an empirical test which would answer the question whether the statue and lump are one or two. One might suggest a quick test with a scale. If there are two objects here, then we should sum the weight of lump and statue. But no one thinks that composite objects count for weight beyond that of their constituents (and the function relating mass of constituents to mass of the composite, which is not in general simply summation). Empirically speaking, we already know *everything* we could possibly need to know to answer this question.

Although this question has been selected as a paradigm example of a "purely philosophical" worry, the point is that it does link to scientific concerns. The general problem of understanding material composition is ancient but also has modern offshoots, obviously in studies of the chemical bond and solid state physics. A pure reductionist might dismiss the lump and statue question as merely verbal; what is "really real" is simply the atoms arranged thus and so. However, others hold that "more is different" [17] and that composition introduces new physics into the world. Most dramatically, understanding the place of the ordinary objects of everyday experience connects to the effort to show how a "classical world" can be retrieved from its more fundamental quantum mechanical description. Despite the seeming inevitability of rampant superposition of quantum states, we experience a world of stable and determinate composite objects. The modern investigation of decoherence (see [18,19]) has gone a long way towards solving this problem, which has a history of fairly radical suggestions behind it, as in the suspension of physical law required by the orthodox but infamous projection postulate, the idea that consciousness itself somehow intervenes in the measurement process or the idea that the classical realm is somehow independent of the quantum and in some way is what is fundamentally real.

In fact, although it is true to say that metaphysical questions are remote from empirical testability, we see many suggestive connections between them and scientific theorizing. As another example, there is a perennial tension between the view of the world as an unchanging unity (traditionally represented by Parmenides) versus a view that sees in nature a universal and ceaseless dynamism of change (traditionally represented by Heraclitus). At the appropriate (metaphysical) level of analysis, no empirical test will favour either side of this debate. It is true that we seem to experience change, but a feeling of change is not (necessarily) a changing feeling. It is hard not to see a dim foreshadowing of the "block universe" often associated with relativity theory in the Parmenidean view. Modern quantum cosmology deploys a master equation, the Wheeler–DeWitt equation, that seems to forbid change in the universe. Of course, temporal dynamicists (if I can call them that) push back (see, e.g., [20]). No experiment can decide this, the question seems obviously significant, and important and philosophical reflection here is greatly aided and extended by scientific development.

The grandest metaphysical question of all was posed by Leibniz in 1697: Why is there something rather than nothing at all [21]? It is hard to see how to even begin to grapple with this. Leibniz's sensible answer was that there must be an absolutely necessary ground of being (which he naturally equated with God) for there could not be a chance "eruption" of contingent reality out of nothingness. Even here, modern physics is not entirely disconnected from this problem. In a recent book, Lawrence Krauss [22] outlines how random but statistically inevitable fluctuation in quantum fields could give rise to particle states from the vacuum state. A trenchant review of Krauss's book by the philosopher (and physicist by degree) David Albert led to a testy exchange in the New York Times, which makes for an amusing read. Albert's main point (apparently revealing him to be, in Krauss's words, a "moronic philosopher") was that the QFT vacuum is not nothing: "Krauss seems to be thinking that these vacuum states amount to the relativistic-quantum-field-theoretical version of there not being any physical stuff at all," but this has "nothing whatsoever to say on ... why there should have been a world in the first place" [23]. Clearly, nothingness is incompatible with the existence of any quantum field state, vacuum or otherwise. Perhaps an analogy is this. If you shut your eyes, you see "nothing," but it appears to you as a blacked out visual field. Contrast that with your visual sense of what is behind your head. That is a nothing which is not any kind of "blackness" but simply an absence. Metaphysical nothingness is pure absence.

While Albert is obviously right about this, the relation between the vacuum state and various particle states nonetheless provides an interesting perspective in the philosophy of nothingness. In general, scientific development illuminates and, it must be admitted, usually deepens rather than answers metaphysical questions (The situation is thus reminiscent of the following anec dote: In a summary of lectures on electrodynamics delivered at Moscow University by A. A. Blasov, the following sentence was stated: "The purpose of the present course is the deepening and development of difficulties underlying contemporary theory ... " (as reported in the delightful [24], p. 88)). It would be hard to overstate the significance of the transformation in our metaphysical outlook occasioned by the scientific revolution's mechanistic metaphysics, which replaced the largely Aristotelean theological metaphysics, which dominated thought for more than a millennium.

In all these examples, to a greater or lesser extent, we see how advances in science serve not to eliminate metaphysical questions, but illuminate them and sometimes to reawaken metaphysical options that had faded from view.

Such a metaphysical question, and one that Bohm was deeply concerned with, is whether the universe is primarily a unified whole, as opposed to a collection of ultimate fundamental parts. As noted above, for a long time, the second, mechanical or part-to-whole, view received vast support from the advance of scientific understanding. In recent times, QM has with its discovery and experimental verification of entanglement, revived universal holism. Theoretical advances once again can underpin or weaken philosophical views in the absence of any decisive empirical test.

Therefore, what is philosophically spectacular about the development of quantum mechanics is the possibility that it heralds an equally momentous shift in metaphysical outlook. The still dominant mechanistic metaphysics, which pictures the world as made of independent interacting parts, will be hard to overturn. And for good reason. It has generated vast insights, leading to revolutions in technology that are now generating consequences on a planetary scale. Its philosophical impact has been no less significant. The rise of physicalism or scientific naturalism as the "default" metaphysics can in large measure be traced to its long history of success (see [25]).

The fundamental appeal of the mechanical metaphysics is its promise of maximum intelligibility and conceptual simplicity. Discover the fundamental parts of which everything is composed, and discover the laws which govern how they interact, and you have in principle the key to understanding the entire world. The idea of part–whole intelligibility also seems to be ingrained in the human psyche. When faced with something we don't understand our natural instinct is to take it apart and "see how it works." This has without question served us extremely well, probably since before we were fully human. Part–whole intelligibility arises from understanding how the properties of the parts and their interactions determine the property of the whole. Newton nicely codified this procedure, breaking it into "analysis" and "synthesis":

> By this way of Analysis we may proceed from Compounds to Ingredients, and from Motions to the Forces producing them... And the Synthesis consists in assuming the Causes discover'd, and established as Principles, and by them explaining the Phænomena proceeding from them, and proving the Explanations ([26], Query 31).

Note here that Descartes, and others, had a similarly named distinction but applied it in its standard domain of logic and mathematics. Newton's use of the notions in the context of material constitution presumably harks back to the alchemical tradition in which Newton was very well versed.

Let us call this still ongoing attempt to understand reality in terms of a construction out of independent components the *Parts Project*. Rooted in common experience of the material world, after the 17th century, science or, as it was then known, natural philosophy, and most especially what became physics and chemistry, was charged with completely vindicating the commonsense vision that the material world has a part–whole structure. The initial, seemingly crystal clear conception of pure mechanism slowly gave way to a picture which permitted interactions governed by novel forces. In the 19th century, fields were added to the ontology of interacting particles. However, the electromagnetic field had material sources and, initially, a special material substrate in which it inhered. Recall that Maxwell devoted considerable energy to developing mechanical models of the electromagnetic field (see [27], pp. 451 ff.; for discussion of Maxwell's "ontological intent" with regard to these models, see [28], pp. 55 ff.). As for the ether, Maxwell wrote that "there can be no doubt" about the existence of the "luminiferous aether" whose properties "have been found to be precisely those required to explain electromagnetic phenomena" [29]. The Parts Project assimilated these changes without difficulty.

The Parts Project is arguably humanity's most successful intellectual endeavour. Its effect on the material conditions of life is undeniable, and its associated physicalist metaphysics of an intelligible—albeit rather aloof, cold and comfortless—picture of reality is both comprehensive and possesses a still growing cultural influence.

But there is a spectre haunting this history of success. Leaving aside the instrumental and technological accomplishments, the *metaphysical* dream behind the Parts Project was exploded with the birth of QM. One can find many, often astonished, expressions of this:

> ... a particle certainly is... not a durable little thing with individuality ([30], p. 241);
>
> the historical idea... that the material world is... structured by some kind of interacting "elementary systems" is in sharp contradiction [with] quantum mechanics ([31], p. 88);
>
> quantum phenomena require us to think in a radical new way, a way in which we will have to ultimately give up both the notion of particles and fields ([32], p. 116).

David Bohm himself expressed the collapse of the Parts Project in strong terms that prefigured his favoured replacement metaphysics:

> [The] entire universe must, on a very accurate level, be regarded as a single indivisible unit in which separate parts appear as idealizations permissible only on a classical level of accuracy of description ([33], p. 167).

Given our discussion of the nature of metaphysical questions above, it is clear that the problem of interpreting QM is a metaphysical problem. This in part explains the reluctance and even distaste some physicists, notoriously, for example, Richard Feynman, have about the interpretation project. However, one cannot really avoid the metaphysical side of things since it forms a kind of backdrop or implicit viewpoint that conditions thought.

The literature on the interpretation of QM is truly vast and comes with a corresponding proliferation of interpretations (Wikipedia currently lists 18). Some interpretations do indeed posit an in principle empirically detectable change in QM (e.g., dynamic collapse theories). We have to say "in principle" because these new theories must duplicate the predictive successes of standard QM, and these are so numerous and so rigorous that alternative theories must put any empirical divergence from QM in hard to reach corners of experimental search space. However, many interpretations do not imply any distinct empirical predictions and can be regarded as providing relatively pure metaphysical pictures of the world which their proponents take to be the deep lesson of QM.

Far from an attempt to return to something like a classical mechanistic world view of independent interacting particles, Bohm's interpretation is philosophically extremely radical. Three key features of Bohm's view are especially worth emphasizing:

- holism;
- information;
- mind.

In metaphysics, the main claim of holism is that the whole is prior to, or more fundamental than, the parts (an excellent philosophical discussion and defense of holism can be found in Schaffer [34]; Ismael and Schaffer [35] explore the connection between holism and QM). Thus, it is in absolute contradiction with the mechanistic picture of the whole being determined by the system of interaction of a set of independent parts. Instead, the *parts* are the derivative entities. Bohm sometimes uses the analogy of mathematical projection from higher to lower dimensional spaces. For example, he writes

> we may regard each of the "particles" constituting a system as a projection of a "higher-dimensional" reality, rather than as a separate particle, existing together with all the others in a common three-dimensional space ([36], p. 238).

Since each entangled particle is a projection of a single encompassing higher dimensional whole, it is not surprising that particle properties are correlated. Bohm explicates the Bell correlations in these terms.

However, holism should not be identified with non-locality. All that non-locality shows is the possibility of interaction (of some sort) between spatially separated features of reality. One could imagine a world of particles that are able to "talk" to one another after they have met and established a "special bond." But once again we see some signs from QM that favour the holistic interpretation. It seems that the distant connection supported by entanglement does not permit the communication of information. This is so even in the case of non-relativistic QM. There is no a priori reason to expect that; if non-relativistic entanglement had predicted superluminal signaling, this would simply be a false prediction of a false theory. This is noted in Haroche and Raimond ([4], p. 65): "Non-relativistic quantum physics is non-local in a way subtle enough not to contradict the inherently relativistic causality principle." A holistic view makes better sense of this as a global constraint rather than some very peculiar and highly tuned property of individual particles.

Again, the wave function seems to be more fundamental than the particles. Experimenters are free to choose any measurement basis they want (e.g., position vs. momentum) but cannot as it were "mix and match." Both the lack of a privileged set of properties associated with the parts and the inability to measure "across" bases suggests the priority of the whole. We can so to speak pull out particulate features if we wish but only as permitted by the nature of the wave function.

There is no doubt that Bohm embraced a holistic interpretation on which the universe is "undivided":

> Ultimately, the entire universe (with all its "particles," including those constituting human beings, their laboratories, observing instruments, etc.) has to be understood as a single undivided whole, in which analysis into separately and independently existent parts has no fundamental status ([36], p. 221).

Instead of being the signature of ultimate reality, the world dreamt of by the mechanical philosophers, which is more or less the world of everyday experience, dissolves into a shadowy realm: non-fundamental, derivative, and merely approximate. The development of physics has not proven this, but current theories at least strongly hint if not outright suggest that the holistic metaphysics is to be favoured.

The embrace of holism leaves open the question: what is the nature of (holistic) reality? The metaphysical atomism of the mechanical world view had a simple answer to this question, based upon our intuitive familiarity with objects in the everyday world. According to this view, the world is basically material and matter itself is ultimately resolved into impenetrable, movable, independent, but capable of causal interaction, "chunks" (quite analogous to microscopic lego bricks). This "lego world" is essentially what Richard Feynman was talking about in this famous pronouncement:

> If, in some cataclysm, all of scientific knowledge were to be destroyed, and only one sentence passed on to the next generations of creatures, what statement would contain the most information in the fewest words? I believe it is the atomic hypothesis... that all things are made of atoms—little particles that move around in perpetual motion, attracting each other when they are a little distance apart, but repelling upon being squeezed into one another ([37], v. 1, p. 2).

As attractive and as useful as this picture of the world is, another deep lesson of QM is that we do not live in a lego world. The rather disturbing philosophical consequence of this is that we have lost any positive conception of the nature of matter itself. Matter, or "the physical" in general, has disappeared into an obscurity masked by our vast knowledge of how "it" structures experience. The unease this should engender is suppressed by our false impression that ordinary perception reveals, more or less directly, the nature of matter as hard, massy, and space-filling. Both the growing mathematical abstractness of physical theory and the realization that whatever lies behind our experiential contact with the material world is completely unlike the tiny "marbles" envisioned by traditional atomism leads to the insight that theory reveals only structural or relational properties of the world. These properties tell us how things interact without telling us what those things are. Mass is the "resistance" a body has to motion when a given force is applied; force is that which induces motion in mass. A certain pattern of observable effects is codified by theory, but these patterns tell us nothing about the intrinsic nature of what lies behind them. In the early to mid twentieth century, this was frequently noted. In 1927, Bertrand Russell wrote:

> Physics is mathematical not because we know so much about the physical world, but because we know so little: it is only its mathematical properties that we can discover ([38], p. 125).

and this view is echoed by Arthur Eddington:

> Physical science consists of purely structural knowledge, so that we know only the structure of the universe which it describes ([39], p. 142).

Note that the general thesis that the structural relations which science is limited to revealing require an intrinsic base often goes by the name Russellian Monism and is enjoying a current renaissance of interest (see e.g., [40]).

Now, it may not be part of the job of science to dig down into the intrinsic nature of things; maybe all it can and should deliver is this kind of structural knowledge, which is, in principle, within the realm of empirical testability. This leads to the movement in philosophy of science called structuralism (see [41] for discussion) and its radical offspring, ontological structural realism (see [13]). The metaphysical project of investigating the *what-it-is* that is being structured remains. Here, Bohm's (with Basil Hiley) notion of "active information" may be important (see [42]). Active information is a proprietary concept of Bohm (again, along with Hiley). It is supposed to explicate the relation of the quantum wave function's "guidance" role to the particles being guided. It must be admitted that the concept is not without some obscurity. As Bohm and Hiley describe it, it is unclear how exactly active information operates in the world, specifically whether it is simply another operative causal feature, and hence another aspect of the relational structure of the empirical world, or whether it is something deeper which is involved in the structuring itself. I wish to explore the latter interpretation.

Bohm and Hiley characterize active information in terms of the original etymology of the word "information": to *in-form* or to give form to something. This notion goes back at least to Aristotle's core distinction between form and matter. In our terms, "form" would refer to the structural features of the world: the pattern of interaction and system of spatial-temporal relations described in physical theory. The "matter" in this case is not *material*—the physical as scientifically characterized, but rather whatever it is that makes the structural relations investigated by science into concrete reality. We might say that this "matter" is what "breathes fire into the equations," to use Stephen Hawking's famous phrase ([43], p. 174). Bohm and Hiley hold that active information operates "actively to put form into something or to imbue something with form" ([42], p. 35). Most of their examples I regard as merely illustrative (such things as weak radio signals remotely controlling a much stronger flow of energy) for they would, if interpreted literally, just make active information into another element in the causal-structural nexus, albeit one with a distinctive role. I think the notion of active information is more radical than that.

One way to see this is, following Bohm and Hiley, to contrast active information with what they call "Shannon information." The latter is what is studied in the theory of communication and information. It is a paradigm example of how theory reveals only relational structure. Bohm and Hiley try to point this out with the claim that Shannon information is "for us," that is, the significance of information carried in some channel (information theory is in essence an analysis of such channels) is a matter of interpretation (see [44,45]). There is nothing intrinsic to a string of bits that makes it about missile guidance as opposed to, say, a Gilligan's Island rerun. However, active information is, as Hiley sometimes puts it, "for the particle" (see, e.g., [32]). Such information is intrinsically semantic as opposed to the merely syntactic or structural information of standard information theory.

Active information is not local and pervades the universe outside of or "behind" space, ready to "in-form" aspects of the world, in particular those aspects we call *particles* "to accelerate or decelerate" according to its overall content (see [42], p. 37). As Bohm and Hiley discuss this, we see again the ambiguity between positing more structural features of reality versus positing something which underlies the structure which physics investigates. For example, Bohm and Hiley conjecture that particles such as electrons (that is, those particles we take to be elementary) have "a complex and subtle inner structure" ([42], p. 37). Perhaps this structure is simply more of what physics can investigate, and will reduce to another, albeit deeper, system of causal relations holding between entities whose nature remains ultimately mysterious. Or it could be that we should interpret this as an "inner" nature that underpins the system of physical relations rather than being directly part of it. Although it cannot be disputed that Bohm and Hiley frequently encourage the former interpretation of active information, the latter avoids certain basic objections, such as that whether or how active information can involve energy transfer and the back reaction, or rather the lack of same, on this field of information by the in-formed entities (for such worries, see [46]). Regarding the wave function as embodying information also helps to solve the problem of the very high dimensionality of many body systems. Normally, configuration space is conceived of as merely a way to describe a complicated system; however, if we take the wave function ontologically seriously, then we have to grapple with the mismatch between its extremely high dimensionality and the three dimensions of space found in experience. Bohm developed the idea that the high dimensionality was an intrinsic feature of a "pool of [active] information" (though at the time he discussed this in terms of what he called the "implicate order") (see [36], pp. 236 ff.). This seems plausible insofar as information is inherently multidimensional, with no intuitive constraint to merely three dimensions. This is evident even within the realm of Shannon information, where information is quantified in terms of bits, each capable of two states. A system or pool of information of n bits is then organized in a space of n dimensions and the information state is a point in this space. The analogy with configuration space is clear, but in the case of information there is no pre-existing intuitive constraint limiting the space to three dimensions. Of course, as we have seen, Bohm did not think that Shannon information was anything like his active information. However, it is no less clear that a pool of intrinsically semantic information would also have an organizational structure of high dimensionality. Bohm thought that, if we regarded the quantum wave function as fundamentally informational, its high dimensionality might seem less mysteriously connected to the world of experience.

Seeing the universe as based upon an underlying field of information might also help with the so-called problem of "empty branches." This is the worry that, although the wave function evolves throughout its space, the particles are restricted to certain regions (there are no genuine superpositions of particles in Bohm's view). This can seem a rather arbitrary imposition of reality on a mere portion of the world as described by the wave function. As David Deutsch put it, "pilot-wave theories are parallel-universes theories in a state of chronic denial" ([47], p. 225).

But this misunderstands the nature of the particles within Bohm's metaphysics. Particles are abstractions of the holistic reality or can be regarded as projections from the higher dimensional underlying reality where we find active information. They are not to be thought of as privileged markers of what is physically real as opposed to the ghostly empty branches of the universal wave function. Of course, thinking in terms of particles can be useful, perhaps even indispensable. As Bohm puts it:

> Under the ordinary conditions of our experience, these projections will be close enough to independence so that it will be a good approximation to treat them in the way that we usually do, as a set of separately existing particles all in the same three-dimensional space. ([36], p. 239).

If the introduction of this new sort of intrinsically semantic information into the heart of a world hypothesized to be fundamentally holistic was not strange enough, the final aspect of Bohm's metaphysical interpretation of QM is more peculiar still. We can approach—gingerly—this feature of Bohm's philosophy by asking if we are familiar with any source of intrinsic semantic information? Information is everywhere, but sources of information that do not require interpretation are rare. The need for *interpretation* is of course the clue we need. Mental states are the terminus of interpretation and seem to be the only carriers of information which is intrinsically semantic. This suggests a possible connection between active information and mentality. Bohm did indeed try to forge such a connection, pointing the way to an unorthodox solution to the mind–body problem.

At a very general level, Bohm endorsed a vision that connects an underlying reality (active information) with mind:

> ...reality can be considered as in essence a set of forms in an underlying universal movement or process...Thus, the way could be opened for a world view in which consciousness and reality would not be fragmented from each other ([36], p. xiv).

More directly, Bohm suggested that "the particles of physics have certain primitive mind-like qualities" ([48], p. 272). We must always recall that for Bohm "particles" are not anything like tiny, individual entities. In the quoted philosophical article, he is writing to be understood by a wider and non-scientific, or at least non-physicist, audience. This audacious proposal integrates well with the metaphysical viewpoint we have been developing: mentality possesses the intrinsic semantics needed for active information and active information's non-local universal presence provides support for the doctrine of holism.

The idea that mental features are a fundamental and ubiquitous feature of the world is the ancient doctrine of panpsychism. It has seen a remarkable revival in recent philosophical work (see, e.g., [49–52]). The general metaphysical outlook that places the mental as the intrinsic ground of the structural relations studied by science provides a viewpoint that integrates mind and the physical world, which leaves the physical world causally complete, avoiding outside influences distorting the laws of nature, but nonetheless provides a role for mind in the world. We can see Bohm as a kind of pioneer for this rebirth (The Bohmian approach to the mind–body problem and panpsychism is explored in depth in [53]; I have tried to explore the connection to Russellian Monism in [54]).

I will conclude with one more puzzle. The most intractable aspect of the mind–body problem is the problem of understanding consciousness. Although the identification of mentality with consciousness was philosophical orthodoxy for centuries, in modern times it has been generally accepted that mentality does not automatically imply consciousness. Bohm would seem to accept this. After the above quoted endorsement of the mentality of fundamental physical entities, he quickly goes on to add that, "of course, they do not have consciousness" ([48], p. 272).

This raises a question that has almost as many answers as there are those who ask it: what is consciousness? We can to some extent cut the complexity of this question by focusing on two basic conceptions of consciousness, call them the "thick" and the "thin" conceptions of consciousness. The thick conception is one that sees consciousness as bound up with self-awareness, or a reflective appreciation of our own mental lives and a palpable sense of knowing that one has awareness. Such a conception of consciousness is not uncommon and has a distinguished pedigree going back at least to Aristotle, who arguably equated consciousness with "awareness of awareness" (see [55]) and Leibniz who defined consciousness as "reflective knowledge of this [i.e., perceptual] inner state" (see [56]). There is evidence that Bohm too subscribed to a thick conception of consciousness. For example, he characterizes "conscious awareness" in terms of "attention, sensitivity to incoherence, all sorts of subtle feeling and thoughts and creative imagination as well as much more" ([42], p. 300). It would indeed be strange to assign all these mental functions to the lowly electron!

But there is also a thin conception of consciousness. Most think that animals can feel things such as pain and pleasure, though there is much disagreement about how far bare sentience is spread throughout nature. Nonetheless, it seems clear that primitive feelings occur without such higher functions as self-reflection or creative imagination. But feeling pain is a kind of consciousness. This thin conception is what Thomas Nagel [57] was trying to get at when he pointed out that there is "something it is like" to be an experiencing creature (famously, a bat). This basic sort of consciousness is the essence of the difficulty we have integrating consciousness into a physicalist metaphysics, because how could subjective experience arise from entirely non-experiential constituents? This is in essence David Chalmers' famous "hard problem of consciousness" (see [58]). Given a world of physical entities entirely lacking any subjective aspect, how could the intrinsic subjectivity of conscious experience ever arise in the world?

The seeming intractability of the problem of consciousness suggests that it is not a phenomena amenable to direct scientific understanding. We can investigate the links between consciousness and physical processes, most especially of course those of the brain (though we do not know whether non-neural substrates, such as make up digital computers, are possible). However, these linkages will not reveal what consciousness is or how it arises. Bohm's view offers a novel explanation for both the elusiveness of consciousness when examined from the ordinary scientific standpoint and offers a place for consciousness within the natural world.

It seems fairly easy to imagine that basic sentience comes in degrees of complexity, ranging down to extremely simple forms that would be little more than the merest spark of feeling. Although intellectually challenging and, to many, intuitively implausible, it is not so hard to assign such forms to the fundamental physical entities. This will be the basic case of intrinsic semantically significant active information. Presumably then, more complex forms of consciousness will emerge via some process of increasing physical complexity of structure. Bohm and Hiley have some remarks along these line in ([42], pp. 381 ff.). How exactly this kind of "mental chemistry" would work is of course mysterious, but the idea is not incoherent. If we take on board a thin conception of consciousness, we can perhaps equate it with the primitive mind-like qualities, which Bohm assigned to the foundation of the world. We would then have the outline of a complete, and anti-reductionist, solution to the mind–body problem.

In the end, Bohm's metaphysics is about as far from that of the Newtonian classical metaphysical picture of the world as one could get. It is highly speculative and audacious. However, it appears to hold the promise of a new view of nature that integrates consciousness into the world, which science studies in a way that does not presume to dictate how science ought to proceed, nor does it suggest that mind or consciousness in any way "interferes" with natural law. At the same time, the view does not attempt to reduce or eliminate consciousness but rather offers it a place in the world as an irreducible fundamental feature of it. Overall, it is an inspiring and even exhilarating combination of philosophical and scientific metaphysics.

Funding: This research received no external funding.

Acknowledgments: The author would like to thank the anonymous referees for the journal. They made a number of important points and objections which addressing has measurably improved the paper.

Conflicts of Interest: The author declares no conflict of interest.

References

1. Myrvold, W. On some early objections to Bohm's theory. *Int. Stud. Philos. Sci.* **2003**, *17*, 7–24. [CrossRef]
2. Passon, O. On a Common Misconception Regarding the de Broglie–Bohm Theory. *Entropy* **2018**, *20*, 440. [CrossRef]
3. Albert, D. *Quantum Mechanics and Experience*; Harvard University Press: Cambridge, MA, USA, 1992.
4. Haroche, S.; Raimond, J.M. *Exploring the Quantum: Atoms, Cavities and Photons*; Oxford University Press: Oxford, UK, 2006.
5. Fuchs, C.A. *Coming of Age With Quantum Information: Notes on a Paulian Idea*; Cambridge University Press: Cambridge, UK, 2011.
6. Normore, C. The Matter of Thought. In *Representation and Objects of Thought in Medieval Philosophy (Ashgate Studies in Medieval Philosophy)*; Lagerlund, H., Ed.; Ashgate: Aldershot, UK, 2007; pp. 117–133.
7. Holland, P. *The Quantum Theory of Motion: An Account of the de Broglie-Bohm Causal Interpretation of Quantum Mechanics*; Cambridge University Press: Cambridge, UK, 1993.
8. Goldstein, S. Bohmian Mechanics. In *The Stanford Encyclopedia of Philosophy*, Summer 2017 ed.; Zalta, E.N., Ed.; Metaphysics Research Lab, Stanford University: Stanford, CA, USA, 2017.
9. Newton, I. *The Principia: Mathematical Principles of Natural Philosophy*; Cohen, I.B., Whitman, A., Eds.; University of California Press: Los Angeles, CA, USA, 1999.
10. Newton, I. *Isaac Newton: Philosophical Writings*; Janiak, A., Ed.; Cambridge University Press: Cambridge, UK, 2004.
11. McMullin, E. The Origins of the Field Concept in Physics. *Phys. Perspect.* **2002**, *4*, 13–39. [CrossRef]
12. Riggs, P. *Quantum Causality: Conceptual Isuses in the Causal Theory of Quantum Mechanics*; Springer: Dordrecht, The Netherlands, 2009.
13. Ladyman, J.; Ross, D.; Spurrett, D.; Collier, J. *Everything Must Go: Metaphysics Naturalized*; Oxford University Press: Oxford, UK, 2007.

14. Sellars, W. Philosophy and the Scientific Image of Man. In *Science, Perception and Reality*; Routledge and Kegan Paul: London, UK, 1963; pp. 1–40.
15. Marmodoro, A. *Everything in Everything: Anaxagoras's Metaphysics*; Oxford University Press: Oxford, UK, 2017.
16. Korman, D.Z. Ordinary Objects. In *The Stanford Encyclopedia of Philosophy*, Spring 2016 ed.; Zalta, E.N., Ed.; Metaphysics Research Lab, Stanford University: Stanford, CA, USA, 2016.
17. Anderson, P. More is Different. *Science* **1972**, *177*, 393–396. [CrossRef] [PubMed]
18. Joos, E.; Zeh, H.D.; Kiefer, C.; Giulini, D.; Kupsch, J.; Stamatescu, I.O. *Decoherence and the Appearance of a Classical World in Quantum Theory*, 2nd ed.; Springer: Berlin/Heidelberg, Germany, 2003.
19. Wallace, D. *The Emergent Multiverse: Quantum theory According to the Everett Interpretation*; Oxford University Press: Oxford, UK, 2012.
20. Smolin, L. *Time Reborn: From the Crisis in Physics to the Future of the Universe*; Hought Mifflin Harcourt: Boston, MA, USA, 2013.
21. Leibniz, G.W. On the Ultimate Origination of Things. In *G. W. Leibniz: Philosophical Essays*; Ariew, R., Garber, D., Eds.; Hackett: Indianapolis, IN, USA, 1989; pp. 149–154.
22. Krauss, L. *A Universe from Nothing: Why There is Something Rather than Nothing*; Free Press: New York, NY, USA, 2012.
23. Albert, D. On the Origin of Everything. *New York Times*, 15 April 2012.
24. Weber, R.; Mendoza, R. (Eds.) *A Random Walk in Science*; The Institute of Physics: London, UK, 1973.
25. Papineau, D. The Rise of Physicalism. In *The Proper Ambition of Science*; Stone, M., Wolff, J., Eds.; London Studies in the History of Science; Routledge: London, UK, 2000; pp. 174–205.
26. Newton, I. *Opticks, or, A Treatise of the Reflections, Refractions, Inflections and Colours of Light*, 4th ed.; Dover: New York, NY, USA, 1979.
27. Maxell, J.C. *The Scientific Papers of James Clerk Maxwell*; Niven, W.D., Ed.; Dover: New York, NY, USA, 1965.
28. Siegel, D.M. *Innovation in Maxwell's Electromagnetic Theory: Molecular Vortices, Displacement Current and Light*; Cambridge University Press: Cambridge, UK, 2003.
29. Maxell, J.C. Ether. In *Encyclopedia Britannica*, 9th ed.; Baynes, T., Ed.; A. & C. Black: Edinburgh, UK, 1878; Volume 8, pp. 568–572.
30. Schrödinger, E. Are There Quantum Jumps? Part II. *Br. J. Philos. Sci.* **1952**, *3*, 233–242. [CrossRef]
31. Primas, H. Emergence in Exact Natural Science. *Acta Polytech. Scand.* **1998**, *91*, 83–98.
32. Hiley, B. Active Information and Teleportation. In *Epistemological and Experimental Perspectives on Quantum Physics*; Greenberger, D., Reiter, W., Zeilinger, A., Eds.; Kluwer: Dordrecht, The Netherlands, 1999.
33. Bohm, D. *Quantum Theory*; Prentice-Hall: Englewood Cliffs, NJ, USA, 1951.
34. Schaffer, J. The Internal Relatedness of All Things. *Mind* **2010**, *119*, 341–376. [CrossRef]
35. Ismael, J.; Schaffer, J. Quantum holism: Nonseparability as common ground. *Synthese* **2016**, 1–30. [CrossRef]
36. Bohm, D. *Wholeness and the Implicate Order*; Routledge and Kegan Paul: London, UK, 1980.
37. Feynman, R.; Leighton, R.; Sands, M. *The Feynman Lectures on Physics*; Addison-Wesley: Reading, MA, USA, 1963.
38. Russell, B. *An Outline of Philosophy*; George Allen & Unwin: London, UK, 1927.
39. Eddington, A. *The Philosophy of Physical Science*; Macmillan: New York, NY, USA, 1939.
40. Alter, T.; Nagasawa, Y. (Eds.) *Consciousness in the Physical World: Perspectives on Russellian Monism*; Oxford University Press: Oxford, UK, 2015.
41. French, S. *The Structure of the World: Metaphysics and Representation*; Oxford University Press: Oxford, UK, 2014.
42. Bohm, D.; Hiley, B. *The Undivided Universe: An Ontological Interpretation of Quantum Mechanics*; Routledge: London, UK, 1993.
43. Hawking, S. *A Brief History of Time*; Bantam: New York, NY, USA, 1988.
44. Hiley, B. From the Heisenberg Picture to Bohm: A New Perspective on Active Information and its relation to Shannon Information. In *Quantum Theory: Reconsideration of Foundations*; Khrennikov, A., Ed.; Växjö University Press: Växjö, Sweden, 2002; pp. 141–162.
45. Pylkkänen, P. Can Bohmian Quantum Information Help us to Understand Consciousness? In *Quantum Interaction. Lecture Notes in Computer Science, v. 9535*; Atmanspacher, H., Filk, T., Pothos, E., Eds.; Springer: Berlin/Heidelberg, Germany, 2016; pp. 76–87.
46. Riggs, P. Reflections on the deBroglie–Bohm Quantum Potential. *Erkenntnis* **2008**, *68*, 21–39. [CrossRef]
47. Deutsch, D. Comment on Lockwood. *Br. J. Philos. Sci.* **1996**, *47*, 222–228. [CrossRef]
48. Bohm, D. A new theory of the relationship of mind and matter. *Philos. Psychol.* **1990**, *3*, 271–286. [CrossRef]

49. Skrbina, D. *Panpsychism in the West*; MIT Press: Cambridge, MA, USA, 2005.
50. Skrbina, D. (Ed.) *Mind That Abides: Panpsychism in the New Millennium*; John Benjamins: Amsterdam, The Netherlands, 2009.
51. Brüntrup, G.; Jaskolla, L. (Eds.) *Panpsychism*; Oxford University Press: Oxford, UK, 2016.
52. Seager, W. (Ed.) *The Routledge Handbook on Panpsychism*; Routledge: London, UK, 2018.
53. Pylkännen, P. *Mind, Matter and the Implicate Order*; Springer: Berlin/Heidelberg, Germany, 2007.
54. Seager, W. Classical Levels, Russellian Monism and the Implicate Order. *Found. Phys.* **2013**, *43*, 548–567. [CrossRef]
55. Caston, V. Aristotle on consciousness. *Mind* **2002**, *111*, 751–815. [CrossRef]
56. Leibniz, G.W. Principles of Nature and Grace, Based on Reason. In *G. W. Leibniz: Philosophical Essays*; Ariew, R., Garber, D., Eds.; Hackett: Indianapolis, IN, USA, 1989; pp. 206–213.
57. Nagel, T. What Is It Like to be a Bat? *Phil. Rev.* **1974**, *83*, 435–450. [CrossRef]
58. Chalmers, D. *The Conscious Mind: In Search of a Fundamental Theory*; Oxford University Press: Oxford, UK, 1996.

Article

Quantum Trajectories: Real or Surreal?

Basil J. Hiley * and Peter Van Reeth *

Department of Physics and Astronomy, University College London, Gower Street, London WC1E 6BT, UK

* Correspondence: ubap727@mail.bbk.ac.uk (B.J.H.); p.reeth@ucl.ac.uk (P.V.R.)

Received: 8 April 2018; Accepted: 2 May 2018; Published: 8 May 2018

Abstract: The claim of Kocsis et al. to have experimentally determined "photon trajectories" calls for a re-examination of the meaning of "quantum trajectories". We will review the arguments that have been assumed to have established that a trajectory has no meaning in the context of quantum mechanics. We show that the conclusion that the Bohm trajectories should be called "surreal" because they are at "variance with the actual observed track" of a particle is wrong as it is based on a false argument. We also present the results of a numerical investigation of a double Stern-Gerlach experiment which shows clearly the role of the spin within the Bohm formalism and discuss situations where the appearance of the quantum potential is open to direct experimental exploration.

Keywords: Stern-Gerlach; trajectories; spin

---

1. Introduction

The recent claims to have observed "photon trajectories" [1–3] calls for a re-examination of what we precisely mean by a "particle trajectory" in the quantum domain. Mahler et al. [2] applied the Bohm approach [4] based on the non-relativistic Schrödinger equation to interpret their results, claiming their empirical evidence supported this approach producing "trajectories" remarkably similar to those presented in Philippidis, Dewdney and Hiley [5]. However, the Schrödinger equation cannot be applied to photons because photons have zero rest mass and are relativistic "particles" which must be treated differently. In fact details of how to treat photons and the electromagnetic field in the same spirit as the non-relativistic theory have already been given in Bohm [6], Bohm, Hiley and Kaloyerou [7], Holland [8] and Kaloyrou [9], but this work seems to have been ignored. Flack and Hiley [10] have re-examined the results of the experiment of Kocsis et al. [1] in the light of this electromagnetic field approach and have reached the conclusion that these experimentally constructed flow lines can be explained in terms of the momentum components of the energy-momentum tensor of the electromagnetic field. What is being measured is the *weak value* of the Poynting vector and not the classical Poynting vector suggested in Bliokh et al. [11].

This leaves open the question of the status of the Bohm trajectories calculated from the non-relativistic Schrödinger equation [4,5] for particles with finite rest mass. The validity of the notion of a quantum particle trajectory is certainly controversial. The established view has been unambiguously defined by Landau and Lifshitz [12]:—"In quantum mechanics there is no such concept as the path of a particle". This position was not arrived at without an extensive discussion going back to the early debates of Bohr and Einstein [13], the pioneering work of Heisenberg [14] and many others [15]. We will not repeat these arguments here.

In contrast to the accepted position, Bohm showed how it was possible to define *mathematically* the notion of a local momentum, $\boldsymbol{p}(\boldsymbol{r},t) = \boldsymbol{\nabla}S(\boldsymbol{r},t)$, where $S(\boldsymbol{r},t)$ is the phase of the wavefunction. From this definition it is possible to calculate flow-lines which have been interpreted as 'particle trajectories' [5]. To support this theory, Bohm [4] showed that under polar decomposition of the wave function, the real part of the Schrödinger equation appears as a deformed Hamilton-Jacobi equation, an equation that had originally been exploited by Madelung [16] and by de Broglie [17].

Initially this simplistic approach was strongly rejected as it seemed in direct contradiction to the arguments that had established the standard interpretation, even though the approach was based on the Schrödinger equation itself with no added new mathematical structures. However, recently this approach has received considerable mathematical support from the extensive work that has been ongoing in the literature exploring the deep relation between classical mechanics and the quantum formalism which has evolved from a field called "pseudo-differential calculus". Specific relevance of this work to physics can be found in de Gosson [18] and the references found therein.

In this paper we want to examine one specific criticism that has been made against the notion of a "quantum trajectory", namely the one emanating from the work of Englert et al. [19] (ESSW). They conclude, "the Bohm trajectory is here macroscopically at variance with the actual, that is: observed track. Tersely: Bohm trajectories are not realistic, they are surreal". A similar strong criticism was voiced in Scully [20] who added that these trajectories were "at variance with common sense". However the claim of an "observed track" in the above quotation should arouse suspicion coming from authors who claim to defend the standard interpretation as outlined in Landau and Lifshitz [12].

The first part of the ESSW argument involved what they called the 'standard analysis' of a gedanken experiment consisting of several Stern-Gerlach magnets, an experiment that is discussed in Feynman [21]. It is this part of the argument that we examine in this paper. We show that they arrive at the wrong conclusion because they have not carried through the analysis correctly. Although Hiley [22] and Hiley and Callaghan [23] have presented a detailed criticism of this topic before in a different context, the point that we make in this paper is new. The standard use of quantum mechanics itself shows that what ESSW call the "macroscopically observed track" is identical to what has been called the "Bohm trajectory". We support our arguments with detailed simulations of potential experiments that are being planned at present with our group at UCL.

2. Re-Examination of the Analysis of ESSW

2.1. General Results Using Wave Packets

The ESSW paper [19] contains an error in their analysis of the Stern-Gerlach experiment as shown in Figure 1 which is similar to the set-up shown in Figure 4 appearing in ESSW [19]. It depicts the tracks of spin one-half particles entering two Stern-Gerlach (SG) magnets. The particles enter along the y-axis with their spins initially pointing along this axis. The orientation of the magnetic field in each SG magnet is as shown in the figure, the second SG magnet being twice the length of the first.

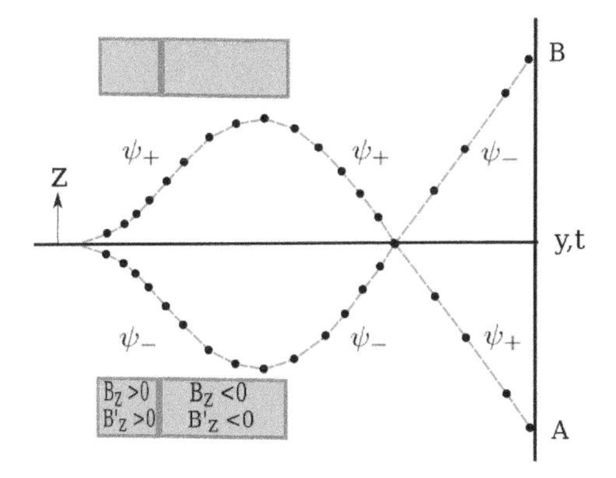

Figure 1. Sketch of Particle Tracks Presented in ESSW.

On entering the first magnet, the wave packet begins to split into two wave packets which move apart in the magnetic field. The packet, ψ_+, moves in the $+z$ direction while the other, ψ_-, moves in the $-z$ direction. Thus the ψ_+ packet follows the upper track, while the ψ_- packet follows the lower track. Note here it is the *wave packet* we are discussing, *not* the particle.

To account for the z-motion of the packets, we use standard quantum mechanics as in ESSW [19], where the spin-dependent Hamiltonian is

$$H = \frac{1}{2m}\mathbf{P}^2 + \mathcal{E}(t)\sigma_z - F(t)z\sigma_z,$$

where $\mathcal{E}(t)\sigma_z$ is the magnetic energy at $z = 0$ and $F(t)z\sigma_z$ is the energy due to the inhomogeneous field. The two components of the wave function are initially chosen to be

$$\psi_+(z,0) = \psi_-(z,0) = (2\pi)^{-1/4}(2\delta z_0)^{-1/2}\exp\left[-\left(\frac{z}{2\delta z_0}\right)^2\right],$$

where δz_0 is the initial spread in z which is assumed small compared with the eventual maximum separation of the two beams.

At a later time, the equations of motion of the two wave packets are

$$\psi_\pm(z,t) \quad = \quad A(t)\exp\left[-B(t)[z \mp \Delta z]^2 \pm \frac{i}{\hbar}[z\Delta p + \frac{\hbar}{2}\Phi(t)]\right],$$

where $A(t) = (2\pi)^{-1/4}\left[2\left(\delta z_0 + i\frac{\hbar t}{2m\delta z_0}\right)\right]^{-1/2}$ and $B(t) = \frac{1}{4\delta z_0(\delta z_0 + \frac{i\hbar t}{2m\delta z_0})}$. In arriving at this expression we have used the impulse approximation as presented in Bohm [24]. Here $\Delta p(t) = \int_0{}^t dt' F(t')$ is the momentum transferred to the "up" wave packet. The actual magnitude is not relevant to our discussion; the interested reader is referred to the original ESSW paper for these details. The magnitude of $\Phi(t) = 2/\hbar \int_0{}^t dt'\mathcal{E}(t')$ is again not relevant to our argument.

Since no measurement has been made and the two beams are still coherent, the wave function after it has traversed the magnet is written in the form

$$|\Psi\rangle = |\psi_+\rangle| + z\rangle + |\psi_-\rangle| - z\rangle. \tag{1}$$

This gives the final probability density as

$$\rho(z,t) = |\psi_+(z,t)|^2 + |\psi_-(z,t)|^2,$$

showing that there is no interference as the wave packets no longer overlap.

The z-component of the current is given by

$$\begin{aligned} j(z,t) \quad &= \quad \frac{\hbar}{2im}\left[\Psi^*\frac{\partial}{\partial z}\Psi - \Psi\frac{\partial}{\partial z}\Psi^*\right] \\ &= \quad \frac{\hbar}{m}\left[(\psi_+^*\psi_+ + \psi_-^*\psi_-)\,C(t)z + (\psi_+^*\psi_+ - \psi_-^*\psi_-)\left(C(t)\Delta z + \frac{\Delta p}{\hbar}\right)\right] \end{aligned} \tag{2}$$

where $C(t) = -\hbar t/[2m((\delta z_0)^4 + (\hbar t/2m)^2)]$. Note that the probability density is symmetric about the $z = 0$ plane, while $(\psi_+^*\psi_+ - \psi_-^*\psi_-)$ is anti-symmetric, showing that the probability current is therefore antisymmetric, therefore,

$$\rho(z,t) = \rho(-z,t) \quad \text{with} \quad j(z,t) = -j(-z,t). \tag{3}$$

Also, as $(\psi_+^*\psi_+ - \psi_-^*\psi_-) = 0$ on the $z = 0$ plane, $j(z,t) = 0$ at $z = 0$. Until this stage we agree totally with the calculations of ESSW using standard quantum mechanics based on conventional wave packet calculations, but it should be noted that this argument only holds when the incident spin is in the y-direction as in the ESSW thought experiment. Particle trajectories have not been discussed so far.

2.2. What Can Be Said about the Behaviour of Individual Particles?

Now we turn to consider what can be inferred about the behaviour of the individual particles, if anything. To answer this question let us return to Landau and Lifshitz [25] who argue that although we cannot talk about a precise particle trajectory, we can talk about the probability of finding a particle in a volume ΔV, provided the volume is large enough so that we avoid any problems associated with the uncertainty principle. Particles will flow into and out of the volume by crossing the boundary of the small volume. In this process we must ensure that probability is conserved.

To see how this works in detail, let us write the well-known conservation of probability equation in integral form. Thus

$$\frac{d}{dt}\int |\Psi|^2 dV = -\int \nabla.j dV = -\oint j d\Sigma \tag{4}$$

where at the last stage we have used Stokes' theorem. Here j is the probability current density used to ensure probability conservation. The integral of this current over the surface Σ is the probability that a particle will cross the surface in unit time. By considering a series of connected volumes we can construct what can be regarded as a "macroscopic particle track". Mott [26] has given a deeper analysis of this process.

Let us now apply this analysis to the situation shown in Figure 1. Construct a surface Σ comprising the $z = 0$ plane and a surface enclosing the upper half of the figure so as to include the upper parts of the magnet. Since the current density is zero everywhere on the $z = 0$ plane, no particles can cross this plane. Thus the particles that arrive in the upper-half of the experimental setup must remain in the upper-half and can never cross the $z = 0$ plane as long as the wave packets remain coherent. This clearly shows that the continuation of the trajectories sketched in Figure 4 of the ESSW paper (as in Figure 1 here) is *not* correct.

In Figure 5 of their paper, ESSW show more explicitly the spin directions together with a sketch of two Bohm trajectories. This shows that their spin wave packets cross the $z = 0$ axis whereas the Bohm trajectories do not. ESSW take this to mean that at first, part of the Bohm trajectories follow one of the wave packets and then, after their spin wave packets cross this axis, the trajectories follow the other wave packet. We will show in Section 4.3 the behaviour of their wave packets is not correct because they have not included spin correctly into the Bohm model.

3. The Bohm Approach When Spin Is Included

To give an account of the behaviour of a particle with spin in the non-relativistic limit, we must widen the scope of the Bohm approach. An extended model for a spin-half particle based on the Pauli equation has already been presented in Bohm, Schiller and Tiomno (BST) [27]. Full details of this model have also been discussed in a series of papers by Dewdney et al. [28–31] and by Holland [32]. This simple model has been applied to neutron diffraction and a single Stern-Gerlach magnet, the results being reported in [29,30]. It should be noted that none of this work is referred to in the ESSW paper and yet this is clearly significant as the Stern-Gerlach magnets operate on the magnetic moments of the particles.

If they had been aware of this work they would not have made the statement that in the Bohm theory a particle has a position and *nothing else*. In the BST extension, not only do we have position, but also the orientation of the spin vector. Here the Euler angles (θ, ϕ, ψ) are used to specify the spin direction. This is essentially the precursor of the flag picture of the spinor presented in Penrose and Rindler [33]. Bell [34] has a simpler model which was also based on the three components of the spin vector. A more general approach using Clifford algebras in which the Pauli spin matrices play a fundamental role has been presented in Hiley and Callaghan [35]. This approach shows how the BST model emerges as a particular representation using Euler angles.

3.1. Spin and the Use of the Pauli Equation

We start with the Pauli equation

$$i\hbar\frac{\partial\xi}{\partial t} = H\xi, \tag{5}$$

where ξ is the two-component spinor which we write in the form

$$\xi = Re^{i(\psi/2)}\begin{pmatrix}\cos(\theta/2)e^{i(\phi/2)}\\ i\sin(\theta/2)e^{-i(\phi/2)}\end{pmatrix}. \tag{6}$$

Here (θ, ϕ, ψ) are the three Euler angles.

The Hamiltonian H is then written in the form

$$H = -\frac{\hbar^2}{2m}\left(\boldsymbol{\nabla} - \frac{ie}{2m}\boldsymbol{A}\right)^2 + \mu\boldsymbol{\sigma}.\boldsymbol{B} + V, \tag{7}$$

where μ is the magnetic moment of the particle.

The original physical idea here was to assume the particle is a spinning object whose orientation is specified by the three Euler angles (θ, ϕ, ψ). The probability of the particle being at a given position, $(\boldsymbol{r}, t)$, is $\rho(\boldsymbol{r}, t) = R^2(\boldsymbol{r}, t) = |\xi(\boldsymbol{r}, t)|^2$. This means the properties of the Pauli particle are specified by four real numbers $(\rho, \theta, \phi, \psi)$ given at the point $(\boldsymbol{r}, t)$. The time evolution of these parameters is determined by the Pauli Equation (5) as we will now show.

It is more convenient to rewrite the wave function in the form

$$\xi(\boldsymbol{r}, t) = \begin{pmatrix}R_+e^{i\frac{S_+}{\hbar}}\\ R_-e^{i\frac{S_-}{\hbar}}\end{pmatrix},$$

where

$$\theta = 2\tan^{-1}\frac{R_-}{R_+}; \quad \psi = \frac{S_+ + S_-}{\hbar} - \pi/2; \quad \phi = \frac{S_+ - S_-}{\hbar} + \pi/2. \tag{8}$$

To find the velocity of the particle, let us first write the quantity $\xi^\dagger\nabla\xi$ in terms of the Euler angles,

$$\xi^\dagger\nabla\xi = R\nabla R + \tfrac{i}{2}R^2\nabla\psi + \tfrac{i}{2}\cos\theta R^2\nabla\phi.$$

Then following Hiley [36] we can define a complex local velocity

$$v = v_{Re} + iv_{Im} = \frac{-i\hbar}{m}\frac{\xi^\dagger\nabla\xi}{\xi^\dagger\xi}$$

where the probability density is given by $R^2 = \xi^\dagger\xi$.

The real part of the local velocity is

$$v_{Re} = \frac{\hbar}{2m}\hat{z}(\nabla\psi + \cos\theta\nabla\phi) \tag{9}$$

which replaces $v(r,t) = \nabla S(r,t)/m$ defined for the spin-less particle. The imaginary part, which was not discussed by Bohm in his original paper (but see Bohm and Hiley [37]) is called the "osmotic velocity" and has the form

$$v_{Im} = -\frac{i\hbar}{m}\hat{z}\left[\frac{\nabla R}{R}\right]. \tag{10}$$

We will now use Equations (9) and (10) to simulate the detailed behaviour of the particles and their spin orientations as they traverse the set-up illustrated in Figure 1.

4. Detailed Calculation of the Trajectories

4.1. One Stern-Gerlach Magnet

We begin by simulating the behaviour of the particles having passed through a single Stern-Gerlach magnet. For simplicity we use the impulse approximation given in Bohm [24] to analyse the evolution of a wave packet as it leaves the magnet (A full treatment using Feynman propagators is being prepared by Hiley and Callaghan. This allows us to calculate trajectories inside the SG. Preliminary results confirm the results presented here.).

In the Hamiltonian given in Equation (7), we replace B by the field in the SG magnet, which we write as $B \approx \mu(B_0 + zB_0')$, where B_0' is the field gradient inside the magnet and set A and V to zero.

Following Dewdney et al. [30] and Holland [32], we choose the initial wave function to be

$$\xi_0 = \xi_+ + \xi_- = f(z)(c_+u_+ + c_-u_-) = (2\pi)^{-1/2}\int g(k)(c_+u_+ + c_-u_-)e^{ikz}dk,$$

where $g(k) = (2\sigma^2/\pi)^{1/4}e^{-k^2\sigma^2}$ is a normalised Gaussian packet centred at $k = 0$ in momentum space. Here u_+ and u_- are the eigenstates of the spin operator σ_z. The solution of the Pauli equation at time t after the particle has left the SG magnetic field is

$$\xi = (2\pi)^{-1/2}\int dk g(k)\Big\{c_+u_+\exp\left[i\left(-\Delta + (k-\Delta')z - \frac{\hbar t}{2m}(k-\Delta')^2\right)\right]$$
$$+c_-u_-\exp\left[i\left(\Delta + (k+\Delta')z - \frac{\hbar t}{2m}(k+\Delta')^2\right)\right]\Big\}$$

where $\Delta = \mu B_0\Delta t/\hbar, \Delta' = \mu_0 B_0'\Delta t/\hbar$ and Δt is the time spent in the field. Carrying out the integral we find

$$\xi(z,t) = (2\pi s_t^2)^{-1/4}\Big\{c_+u_+\exp[-(z+ut)^2/4\sigma s_t]\exp\left[-i(\Delta + (z+\tfrac{1}{2}ut)\Delta')\right]$$
$$+c_-u_-\exp[-(z-ut)^2/4\sigma s_t]\exp\left[i(\Delta + (z-\tfrac{1}{2}ut)\Delta')\right]\Big\}. \tag{11}$$

Here $s_t = \sigma(1 + i\hbar t/2m\sigma^2)$, and $u = \hbar\Delta'/m$. We now write $\xi(t)$ in the form

$$\xi(z,t) = c_+R_+e^{iS_+/\hbar}u_+ + c_-R_-e^{iS_-/\hbar}u_- \tag{12}$$

where

$$R_\pm = \left(2\pi\sigma^2\right)^{-1/4}(1+\hbar^2t^2/4m^2\sigma^4)^{-1/4}\exp\left(\frac{-(z\pm ut)^2}{4\sigma^2(1+\hbar^2t^2/4m^2\sigma^4)}\right) \tag{13}$$

and

$$S_\pm/\hbar = \mp\Delta \mp (z \pm \tfrac{1}{2}ut)\Delta' - \tfrac{1}{2}\tan^{-1}(\hbar t/2m\sigma^2) + \frac{\hbar t(z \pm ut)^2}{8m\sigma^4(1 + \hbar^2 t^2/4m^2\sigma^4)}. \tag{14}$$

We are now in a position to calculate the local velocities from the specific solution given by Equation (12). Since the real part of the local velocity is given by Equation (9), namely, $\hbar(\nabla\psi + \cos\theta\nabla\phi)/2m$, we need only evaluate $\partial\psi/\partial z$ and $\partial\phi/\partial z$ since we are only considering the motion along the z-direction. In order to find these derivatives, and those required for the osmotic velocity and the quantum potential, we express the parameters $(\rho, \theta, \phi, \psi)$ in terms of (R_+, R_-, S_+, S_-) using Equations (8), (13) and (14), and obtain,

$$\frac{\partial\psi}{\partial z} = \frac{4\hbar t z}{8m\sigma^4(\hbar^2 t^2/4m^2\sigma^4 + 1)},$$

$$\frac{\partial\phi}{\partial z} = -2\Delta' + \frac{4\hbar u t^2}{8m\sigma^4(\hbar^2 t^2/4m^2\sigma^4 + 1)},$$

$$\frac{\partial\theta}{\partial z} = \sin\theta \frac{ut}{\sigma^2(\hbar^2 t^2/4m^2\sigma^4 + 1)},$$

and

$$\frac{1}{R}\frac{\partial R}{\partial z} = -\frac{z + ut\cos\theta}{2\sigma^2(\hbar^2 t^2/4m^2\sigma^4 + 1)}.$$

The Bohm velocity given by Equation (9) then becomes

$$v_{Re} = \frac{\hbar\hat{z}}{2m}\left(-2\Delta'\cos\theta + \frac{\hbar t[z + ut\cos\theta]}{2m\sigma^4(\hbar^2 t^2/4m^2\sigma^4 + 1)}\right). \tag{15}$$

Note here that the second term in the above expression corresponds to the spreading of the wave packet and contributes little to the overall behaviour. The main effect of the field comes from the first term $\Delta'\cos\theta$, which reveals clearly how the velocities and therefore the trajectories are strongly affected by the behaviour of the spin vector. This term depends implicitly on (z, t, u) and is responsible for the splitting of the beam.

The imaginary part or osmotic local velocity given in Equation (10), namely, $v_{Im} = -i\hbar[\nabla R/R]/m$, now becomes

$$v_{Im} = \frac{2\hbar}{m}\frac{\hat{z}[z + ut\cos\theta]}{\sigma^2(\hbar^2 t^2/4m^2\sigma^4 + 1)}. \tag{16}$$

Note there is no explicit dependence on the magnetic field gradient but there is an implicit dependence through u and $\cos\theta$.

These results enable us to calculate specific trajectories and spin vectors for various particle initial positions and for various values of (c_+, c_-) should that become necessary. The choice of the latter determine the initial value of the spin vector direction θ which, in our case was chosen to be along the y-direction, hence $(c_+, c_-) = 1/\sqrt{2}$. The results shown in figures below are calculated for parameters listed in Table 1.

Table 1. Parameters used in the numerical investigation.

Atom	Ag
Mass	1.8×10^{-25} Kg
Width of magnets	4 and 8×10^{-4} m
Length of magnets	1 and 2×10^{-2} m
Velocity of atoms	$v_y = y/t = 500$ m/s
Time within magnets	$\Delta t = 2$ and 4×10^{-5} s
Magnetic field strength at centre	$B_0 = 5$ Tesla
Magnetic field gradient	$B_0' = 1000$ Tesla/m
Wave packet width	$\sigma = 1 \times 10^{-4}$ m
Wave packet speed	$u = \mu_B B_0' \Delta t / m = 1$ m/s
$\Delta' = \mu_B B_0' \Delta t \hbar = mu/\hbar$	$\Delta' = 1.714 \times 10^9$ m^{-1}

4.2. Numerical Values for Single Stern-Gerlach Magnet

Integrating Equation (9) will give us the Bohm trajectories. In Figure 2 we show the ensemble of Bohm trajectories and the spin orientations as they leave the Stern-Gerlach magnet, shown in brown at the LHS of the figure. The background colours show the probability density, black being the greatest, while blue is zero.

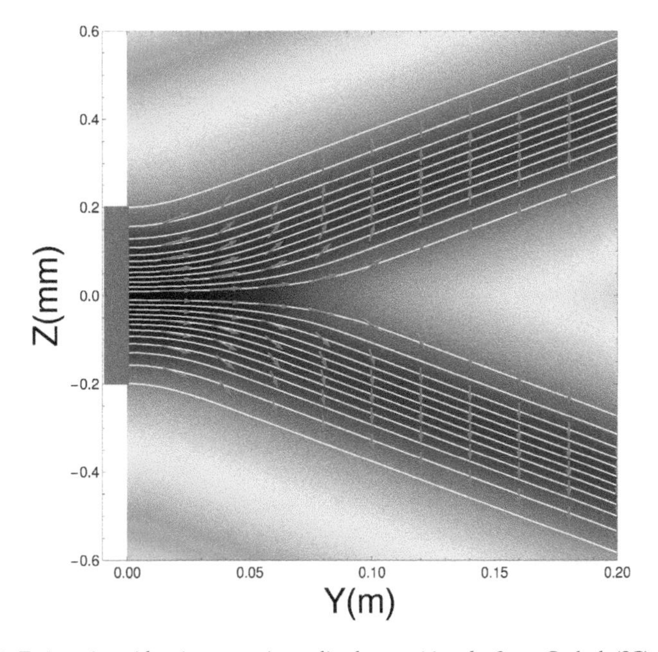

Figure 2. Trajectories with spin vectors immediately on exiting the Stern-Gerlach (SG) magnet.

The dark background shows how the wave packets diverge along straight lines, as do the trajectories. Superimposed on the trajectories are the spin orientations.

Notice that, contrary to the conventional view, the atoms do not immediately "jump" into one or other z-spin eigenstates, rather the spin vectors undergo continuous evolution until they reach their final z-spin eigenstate. This occurs once the two wave packets $\psi_+(z,t)$ and $\psi_-(z,t)$ have separated and have no significant overlap. The upper beam will contain only atoms with spin "up" in the z-direction while those in the lower beam will all be "down" in the z-direction. Notice also that the rotational changes occur in a *magnetic field-free region* . We can also see that the alignment of the spin vector at a given y value close to the magnet depends on z, with the spin associated to trajectories closer to the $z = 0$ axis rotated least. In Section 4.7 we will see that the cause of these behaviours is a torque produced by the quantum potential. These results for a single magnet confirm what was already found in Dewdney et al. [29–31].

Figure 3 shows the effect of the osmotic velocity, which we have represented by arrows. They are responsible for maintaining the wave packet profile and will be discussed further in Section 4.6.

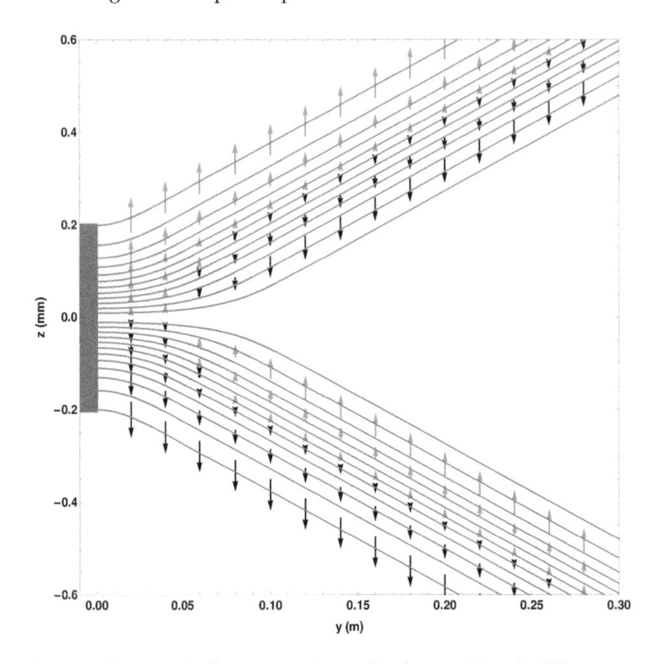

Figure 3. The osmotic flow vectors immediately on exiting the SG magnet.

4.3. *Two Stern-Gerlach Magnets*

Having seen how the atoms behave in a single SG magnet, let us now move on to consider two SG magnets with opposite field directions as shown in Figure 1. Note here the second SG magnet is double the length of the first.

The method is similar to the case of the single magnet, except now we use, as initial wave packet, the inverse Fourier transform of the wavefunction at the second magnet at time $t = t_1$. We obtain the real part of the local velocity as

$$v_{2Re} = \frac{\hbar\hat{z}}{2m}\left(\left[-2\Delta_2' - 2\frac{\Delta_1'}{\left(\frac{\hbar^2(t_1+t)^2}{4m^2\sigma^4}+1\right)}\right]\cos\theta + \frac{\hbar(t_1+t)}{2m\sigma^4\left(\frac{\hbar^2(t_1+t)^2}{4m^2\sigma^4}+1\right)}\left[z+u_2 t\cos\theta\right]\right) \tag{17}$$

and the osmotic velocity as

$$v_{2Im} = \frac{\hbar\hat{z}}{m}\frac{1}{2\sigma^2\left(\frac{\hbar^2(t_1+t)^2}{4m^2\sigma^4}+1\right)}\left[z+\left(u_1(t_1+t)+u_2 t\right)\cos\theta\right] \tag{18}$$

where $t = 0$ at the exit of the second magnet. In Figure 4 we have plotted the trajectories together with the spin orientations as the atoms pass through two SG magnets. The details of the parameters used in the calculations are again as listed in Table 1. The position of the second magnet is as indicated by the brown bar between $y = 0.1$ m and $y = 0.12$ m.

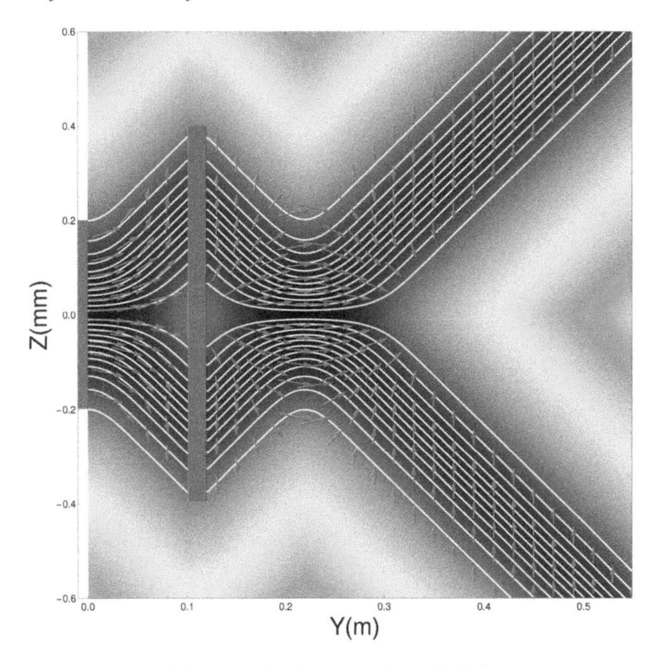

Figure 4. Spins emerging from two Stern-Gerlach magnets.

There are several features of the ensemble of trajectories that are noteworthy. Firstly, at the exit of the second magnet, the wave-packets are refocused toward the y-axis until the inner edge of the packets reaches the axis at $y \approx 0.22$ m at which point they diverge again.

Secondly, no trajectories are found to cross the $z = 0$ plane. This should, in fact, not be surprising since v_{Re} can also be obtained from $j(z,t)/\rho(z,t)$. This means that the "Bohm trajectories" are identical to the probability flow lines and, as we have seen, the probability flow lines do not cross the $z = 0$ plane. Thus there is no experimental difference between the Bohm approach and standard quantum mechanics at this stage. It could be argued that it is quantum mechanics that is "at variance with common sense"!

Thirdly, notice once again that the spins do not immediately "jump" into the eigenstates as assumed by the standard theory. Rather they take a small but finite time to reach the final eigenstate as discussed above in Section 3.1. Furthermore note that when the beams are refocused close to the $z = 0$ plane, at about $y = 0.22$, the spin vectors are rotated so that they all become aligned with the $y-$axis before being rotated again until they end up anti-parallel to the direction with which they entered the second magnet. This rotation is very surprising but is generated by the quantum torque that arises from the quantum potential as we show in the next section in Equation (21).

Furthermore this is in contradiction with Figure 5 of ESSW where they argue that the Bohm trajectories are not realistic because in order to get the observed final spin state, their particles must cross the $z = 0$ axis. Therefore the present work shows clearly the importance of coupling the spin and the centre of mass motion in order to obtain a correct and consistent analysis of the problem.

Figure 5 shows the direction of the osmotic velocity in the two SG magnets case. Its behaviour is again exactly the same as in the one SG magnet case.

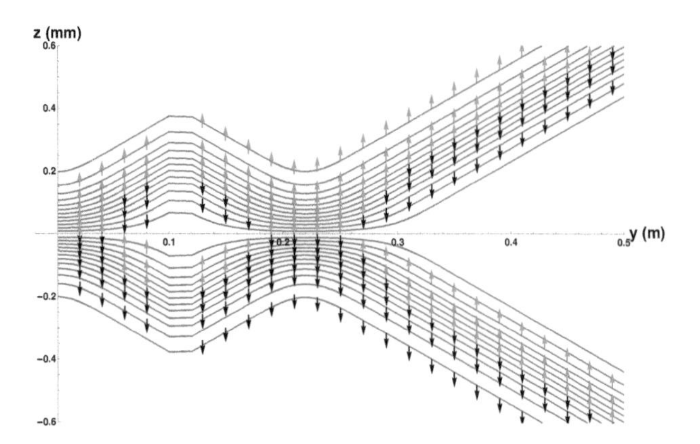

Figure 5. The osmotic velocity superimposed on the trajectories for two Stern-Gerlach magnets.

To return the packet to its original state with all the spins pointing in the y-direction, we have to add a third magnet as indicated in the original diagram in Feynman et al. [38]. Thus the Bohm approach gives a complete account of the average behaviour of the individual quantum processes.

4.4. The Appearance of the Quantum Torque

Now let us show the source of the quantum torque. We start by examining the real part of the Pauli Equation (5) under polar decomposition of the wave function, which can be written in the form

$$\tfrac{1}{2}\hbar\left(\frac{\partial\psi}{\partial t}+\cos\theta\frac{\partial\phi}{\partial t}\right)+\tfrac{1}{2}mv^2+Q_P+\frac{2\mu}{\hbar}\sigma.s+V=0. \tag{19}$$

Here once again we see, as in the case of the Schrödinger equation, an extra energy term, Q_P, the quantum potential energy, appears. In the present case Q_P takes the form

$$Q_P=-(\hbar^2\nabla^2R)/2mR-\frac{\hbar^2}{8m}[(\nabla\theta)^2+\sin^2\theta(\nabla\phi)^2]. \tag{20}$$

The first term will be recognised as the quantum potential found in the Schrödinger equation. The second term determines the evolution of the spin vector which is given by

$$s=\tfrac{1}{2}\hbar\xi^{\dagger}\sigma\xi=\tfrac{1}{2}(\sin\theta\sin\phi,\sin\theta\cos\phi,\cos\theta).$$

The equation of motion for the spin vector s, is then found to be

$$\frac{ds}{dt}=T-\frac{2\mu}{\hbar}(s\times B). \tag{21}$$

Here B is an external magnetic field and

$$T=(m\rho)^{-1}s\times\sum_i\frac{\partial}{\partial x_i}\left(\rho\frac{\partial s}{\partial x_i}\right). \tag{22}$$

It is the quantum torque, T, that acts on the individual atoms, rotating their spin vectors and the flag plane.

4.5. Detailed Calculation of the Quantum Potential

To understand better the role played by the quantum potential, let us examine in more detail its mathematical structure as shown in Equation (20). We restrict our analysis to the case of a single magnet. As the quantum Hamilton-Jacobi Equation (19) is an equation that conserves energy, the appearance of Q implies that some of the kinetic energy of the particle is transferred to the quantum potential energy Q. As we see from Equation (20), the quantum potential energy has two components

$$Q_{trans} = -\frac{\hbar^2 \nabla^2 R}{2mR} \quad \text{and} \quad Q_{spin} = \frac{\hbar^2}{8m}[(\nabla\theta)^2 + \sin^2\theta(\nabla\phi)^2].$$

We will examine the two terms independently. First consider Q_{trans}. Since the particle is moving in one-dimension

$$\nabla^2 R \to \frac{\partial^2 R}{\partial z^2} = -\frac{2}{bd}\left(-\frac{2R(z+ut\cos\theta)^2}{bd} + R\left(1 - \frac{4u^2t^2}{bd}\right)\sin^2\theta\right),$$

where we have written

$$b = \left(\frac{\hbar^2 t^2}{4m^2\sigma^4} + 1\right) \quad \text{and} \quad d = 4\sigma^2.$$

Then

$$Q_{trans} = -\frac{\hbar^2 \nabla^2 R}{2mR} = \frac{\hbar^2}{bdm}\left(-\frac{2}{bd}[(z+ut\cos\theta)^2 + 2u^2t^2\sin^2\theta] + 1\right).$$

Now we turn to evaluate the spin part of the quantum potential, Q_{spin}, where we need to evaluate

$$\nabla\phi = -2\Delta' + \frac{4\hbar ut^2}{8mb\sigma^4} \quad \text{and} \quad \nabla\theta = \sin\theta\frac{ut}{b\sigma^2}.$$

This gives

$$Q_{spin} = \frac{\hbar^2\sin^2\theta}{8bm}\left(\frac{u^2t^2}{\sigma^4} - 2\Delta'\frac{\hbar ut^2}{2m\sigma} + 4\Delta'^2\left(\frac{\hbar^2t^2}{4m^2\sigma^4} + 1\right)\right).$$

The expression for the total quantum potential, $Q = Q_{trans} + Q_{spin}$ is rather complex so it will be helpful if we can make an approximation without significantly altering the final result. This can be done by noticing the magnitude of $b = \left(\frac{\hbar^2t^2}{4m^2\sigma^4} + 1\right) \approx 1$. This means that we are assuming the wave packet does not spread significantly during the flight times considered. We arrive at the final expression for the total quantum potential:

$$Q \approx \frac{\hbar^2}{m\sigma}\left(-\frac{2}{m\sigma^4}[(z+ut\cos\theta)^2 + 2u^2t^2\sin^2\theta] + 1\right) \\ + \frac{\hbar^2\sin^2\theta}{8m}\left(\frac{u^2t^2}{\sigma^4} - 2\Delta'\frac{\hbar ut^2}{m\sigma^4} + 4\Delta'^2\right).$$

4.6. Numerical Details: Quantum Potential Single Stern-Gerlach Magnet

In Figure 6 below we plot the transverse quantum potential Q_{trans} and the spin quantum potential, Q_{spin} for the single SG magnet. The end of the SG magnet is again along the z-axis at $y = 0$, with the atoms flowing along the y-axis out of the page.

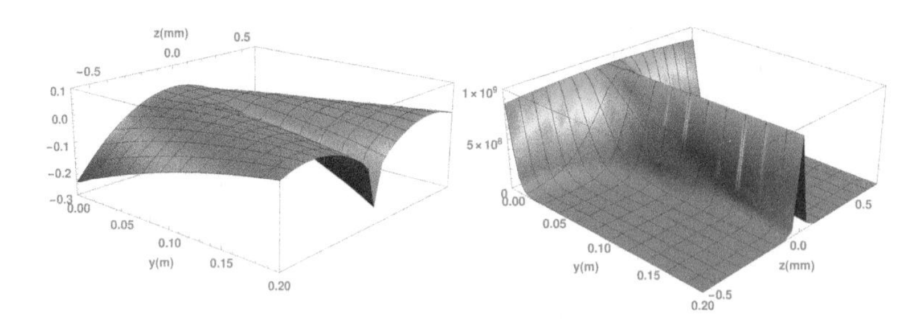

Figure 6. Transverse (**left**) and spin (**right**) quantum potential at exit of a single SG magnet.

The atoms initially experience the first part of the quantum potential where the beam begins to split into two as shown in Figure 2. Both quantum potentials split symmetrically into two parts about the y-axis. The two "domes" of Q_{trans}, shown in the left hand of the figure, cover each beam as they separate. The width of each dome characterises the spreading wave packet as it evolves in time. Also, when compared to the osmotic velocities shown in Figure 3, we can see how these velocities are related to the gradient of Q_{trans}. The trajectories are seen to follow paths of constant gradient and the osmotic velocities are constant along the trajectories in Figure 3. Furthermore, those trajectories in the wings of the wave packets experience a more steep gradient and the osmotic velocities are indeed found to be larger there. At the maximum of the packet, the osmotic velocity is zero. An interpretation of the Q_{trans} would therefore be that it gives rise to a force, which is anti-parallel to the osmotic velocity and restricts the spreading of the wave packet.

The spin part of the quantum potential Q_{spin} is shown in the right hand of Figure 6. The upward slope produces the quantum torque that rotates the spin vectors of the atoms as the two beams separate. This rotation continues until the two packets are completely separate. When this happens all the spins point "up" in the upper beam, while they all point "down" in the lower beam. At this stage the $Q_{spin} \to 0$ ensuring the atoms remain in their final spin eigenstates. Figure 7 shows the projection of the Q_{spin} of Figure 6 on the trajectories and spin orientation. Note also that the trajectories close to the y-axis do not experience the same steepness of Q_{spin} as do those which are off-axis. This explains why, as remarked earlier, the spin vectors closer to the y-axis take longer to align themselves either up or down.

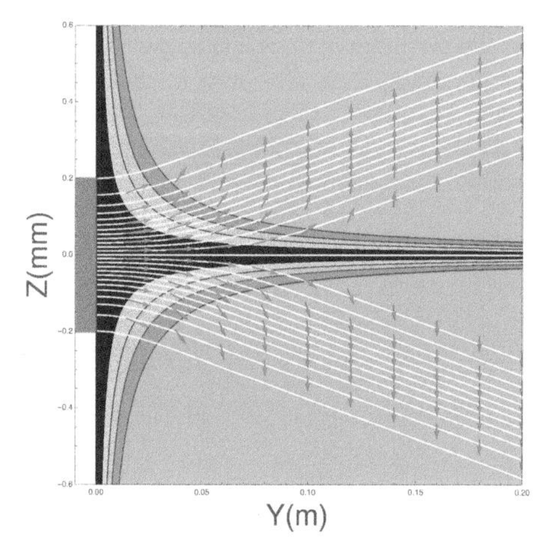

Figure 7. Trajectories with spin vectors overlaid on the spin quantum potential immediately on exiting a single SG magnet.

4.7. Numerical Details: Quantum Potential in Two SG Magnet Case

Now let us consider the case when the two Stern-Gerlach magnets are in place. The positions of the magnets are shown in brown. Recall here that the inhomogeneities in the magnetic fields oppose each other.

In Figure 8 we show both Q_{trans} and Q_{spin} for the case of two magnets. The gap in each figure corresponds to the position of the second magnet. The quantum potential after the second magnet is similar to that of the single SG magnet as shown in Figure 6. These results give a detailed picture of the expected evolution of a non-relativistic atom with spin one-half as it goes through both SG magnets.

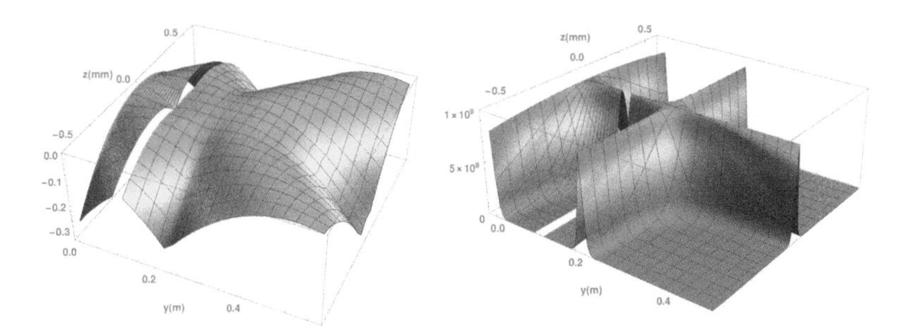

Figure 8. Q_{trans} (**left**) and Q_{spin} (**right**) quantum potential for a two SG magnets system.

Figure 9 shows the projection of the spin quantum potential superimposed on the trajectories. Notice that the quantum torque is strongest well outside the second SG magnet in the magnetic field-free region, producing a 180 degree rotation of the spin vector. It is at this point that the wave packets begin to interfere strongly. In fact the quantum torque continues to act outside the magnet until the two wave packets $\psi_+(z,t)$ and $\psi_-(z,t)$ cease to overlap. Notice once again how the spin does not immediately 'jump' into one of the two spin z-eigenstates, but undergoes a well-defined time evolution. Such a behaviour would have, perhaps, been welcomed by Schrödinger himself [39].

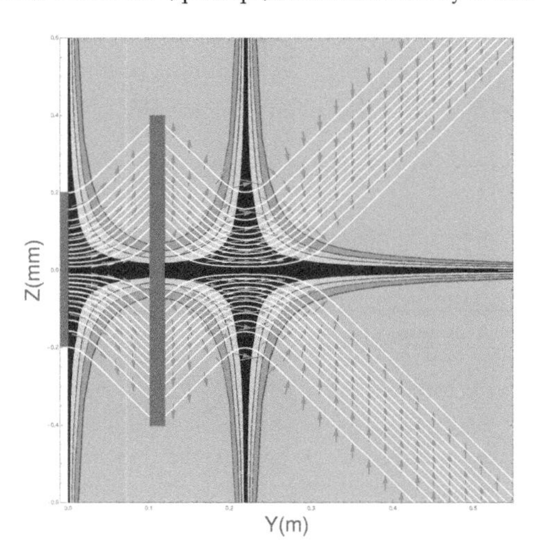

Figure 9. Trajectories with spin vectors overlaid on the spin quantum potential for a two SG magnet system.

Once they no longer overlap, each atom remains in one or the other spin eigenstates. Again, as was the case with the single SG magnet, the spin vectors along the trajectories close to the y-axis, especially at the point where the two beams are refocused, experience less of the gradient of Q_{spin}. Thus it is clear that the quantum torque arises from the interference region, implying it is an internal feature of the overall behaviour, suggesting a kind of dramatic re-structuring of the underlying process.

Bohm was intuitively well aware of this possibility and it was one of the reasons why he abandoned the view that the atom only had a local, "rock-like" property. He preferred to regard the atom as a quasi-local region of energy undergoing a new type of process that he described in more general terms as an "unfolding-enfolding" process, comparing it to a gas near its critical point, the particle itself constantly forming and dissolving, as in critical opalescence [40,41]. In other words, the quantum evolution involves an entirely new re-ordering process which should not be regarded as a particle following a well defined trajectory.

This view of the evolving quantum process becomes even more compelling since Hiley and Callaghan [42] and Takabayasi [43] have shown that the local momentum and energy are actually related to the energy-momentum tensor, $T^{\mu\nu}$, through the relations

$$\rho p^j(\boldsymbol{r},t) = T^{0j}(\boldsymbol{r},t) \quad \text{and} \quad \rho E(\boldsymbol{r},t) = T^{00}(\boldsymbol{r},t),$$

a feature of which Schwinger [44] was well aware. The question of which particular trajectory a specific atom actually takes cannot be answered because the experimenter has no way of choosing or controlling the initial position of the particle. The final result is also totally independent of the observer. A detailed discussion of the role of the experimenter in the Bohm approach can be found in Bohm and Hiley [45,46]. A more recent paper by Flack and Hiley [47] shows how the Bohm trajectories emerge from an averaging over this deeper process.

We can see from Figure 9, the above simulations predict some interesting structure in near field behaviour of the atoms after they leave the second SG magnet. This could be experimentally explored through weak measurements as suggested in [48]. At present, our group [49] is attempting to measure the weak values of momentum and spin which, if successful, would ultimately enable us to not only construct these flow lines, but also to measure the time evolution of the angle $\theta(y,t)$ of the spin vector.

We are also exploring the possibility of using the techniques we are developing to check the results shown in Figure 2. At present we are on the edge of what is technically possible and if we are successful, the experiments will show that the quantum potential energy appearing in Equation (19) has an observable experimental consequence and therefore cannot be ignored in analysing quantum phenomena.

5. Conclusions

In this paper we have shown that the differences that are claimed to exist between the standard approach to quantum mechanics and the Bohm approach do not exist when both are applied correctly. Indeed it is hard to imagine how there could be any differences in the predicted experimental results since *both* approaches use exactly the same mathematical structure. For the type of experiments considered by ESSW [19], the probability current plays a key role. In both approaches the probability current is considered as a particle flow, the conventional approach regarding it as a measure of particles flowing out of a small region, ΔV, of space, whereas the Bohm approach assumes the probability current arises from the velocities of individual particles through the relation $j(\boldsymbol{r})/\rho(\boldsymbol{r}) = \boldsymbol{\nabla} S(\boldsymbol{r})/m = \boldsymbol{p}(\boldsymbol{r})/m$. In the Bohm model this is taken as the definition of the local momentum, $\boldsymbol{p}(\boldsymbol{r})$. Clearly the behaviour of the probability currents is identical to the local momentum. This is what ESSW failed to recognise. Notice that this disagreement arises before the addition of any device to measure which path the particle actually took.

The inclusion of a which-way detector into the discussion merely confuses the issue. Traditionally it is assumed that any measurement to determine which path a particle actually takes brings about the "collapse" of the wave function. Suppose a position measurement is made after the atom has left the second SG magnet as shown in Figure 1. The wave function (1) will not then be the pure state but instead will be a mixture which must be described by a density matrix ρ with $\rho^2 \neq \rho$. This means there is no interference between the two wave packets ψ_+ and ψ_- in which case the particles actually cross the $z = 0$ plane as shown in Figure 1. Exactly the same thing happens in the Bohm model as was discussed in detail in Hiley [22] and Hiley and Callaghan [23]. We will not repeat the argument again in this paper but refer the interested reader to the original papers. Our conclusion is that the standard quantum mechanics produces exactly the same behaviour as the Bohmian approach so it cannot be used to conclude the Bohm trajectories are "surreal".

Since these earlier objections were raised, an entirely new way of experimentally constructing the "Bohm particle trajectories" has been developed by Kocsis et al. [1] as discussed in the introduction. Furthermore in the case of atoms the claim that these are "particle trajectories" has been re-examined recently by Flack and Hiley [47] who have concluded that the flow lines, as we shall now call them, are not the trajectories of single atoms but an average momentum flow, the measurements being taken over many individual particle events. In fact they have shown that they represent an average of the ensemble of actual individual stochastic Feynman paths.

The calculations we have presented in this paper provide a detailed background to the experiments of Monachello et al. [49] and Morley et al. [50]. This means that we will not have to rely on theoretical arguments alone to reach an understanding of the behaviour reported in this paper but we hope to be able to provide experimental evidence to further clarify the situation.

Author Contributions: Both authors contributed in the same manner to the research and wrote the paper together.

Acknowledgments: Special thanks to Bob Callaghan, Robert Flack and Vincenzo Monachello for their helpful discussions. Thanks also to the Franklin Fetzer Foundation for their financial support.

Conflicts of Interest: The authors declare no conflict of interest.

References

1. Kocsis, S.; Braverman, B.; Ravets, S.; Stevens, M.J.; Mirin, R.P.; Shalm, L.K.; Steinberg, A.M. Observing the Average Trajectories of Single Photons in a Two-Slit Interferometer. *Science* **2011**, *332*, 1170–1173. [CrossRef] [PubMed]
2. Mahler, D.; Rozema, L.; Fisher, K.; Vermeyden, L.; Resch, K.; Braverman, B.; Wiseman, H.; Steinberg, A.M. Measuring Bohm trajectories of entangled photons. In Proceedings of the CLEO: QELS-Fundamental Science, Optical Society of America, San Jose, CA, USA, 8–13 June 2014; p. FW1A-1.
3. Coffey, T.M.; Wyatt, R.E. Comment on "Observing the Average Trajectories of Single Photons in a Two-Slit Interferometer". *arXiv* **2011**, arXiv:1109.4436.
4. Bohm, D. A Suggested Interpretation of the Quantum Theory in Terms of Hidden Variables I. *Phys. Rev.* **1952**, *85*, 166–179. [CrossRef]
5. Philippidis, C.; Dewdney, C.; Hiley, B.J. Quantum Interference and the Quantum Potential. *Nuovo Cimento* **1979**, *52*, 15–28. [CrossRef]
6. Bohm, D. A Suggested Interpretation of the Quantum Theory in Terms of Hidden Variables II. *Phys. Rev.* **1952**, *82*, 180–193. [CrossRef]
7. Bohm, D.; Hiley, B.J.; Kaloyerou, P.N. An Ontological Basis for the Quantum Theory: II—A Causal Interpretation of Quantum Fields. *Phys. Rep.* **1987**, *144*, 349–375. [CrossRef]
8. Holland, P.R. The de Broglie-Bohm Theory of motion and Quantum Field Theory. *Phys. Rep.* **1993**, *224*, 95–150. [CrossRef]
9. Kaloyerou, P.N. The Causal Interpretation of the Electromagnetic Field. *Phys. Rep.* **1994**, *244*, 287–385. [CrossRef]
10. Flack, R.; Hiley, B.J. Weak Values of Momentum of the Electromagnetic Field: Average Momentum Flow Lines, Not Photon Trajectories. *arXiv* **2016**, arXiv:1611.06510.
11. Bliokh, K.Y.; Bekshaev, A.Y.; Kofman, A.G.; Nori, F. Photon trajectories, anomalous velocities and weak measurements: A classical interpretation. *New J. Phys.* **2013**, *15*, 073022. [CrossRef]

12. Landau, L.D.; Lifshitz, E.M. *Quantum Mechanics: Non-Relativistic Theory*; Pergamon Press: Oxford, UK, 1977; p. 2.
13. Einstein, A. *Albert Einstein: Philosopher-Scientist*; Schilpp, P.A., Ed.; Library of the Living Philosophers: Evanston, IL, USA, 1949; pp. 665–676.
14. Heisenberg, W. *Physics and Philosophy: The Revolution in Modern Science*; George Allen and Unwin: London, UK, 1958.
15. Jammer, M. *The Philosophy of Quantum Mechanics*; Wiley: New York, NY, USA, 1974.
16. Madelung, E. Quantentheorie in hydrodynamischer Form. *Z. Phys.* **1926**, *40*, 322–326. [CrossRef]
17. de Broglie, L. La mécanique ondulatoire et la structure atomique de la matière et du rayonnement. *J. Phys. Radium* **1927**, *8*, 225–241. [CrossRef]
18. de Gosson, M. *The Principles of Newtonian and Quantum Mechanics: The Need for Planck's Constant*; Imperial College Press: London, UK, 2001.
19. Englert, J.; Scully, M.O.; Süssman, G.; Walther, H. Surrealistic Bohm Trajectories. *Z. Naturforsch.* **1992**, *47*, 1175–1186. [CrossRef]
20. Scully, M. Do Bohm trajectories always provide a trustworthy physical picture of particle motion? *Phys. Scr.* **1998**, *76*, 41–46. [CrossRef]
21. Feynman, R.P.; Leighton, R.B.; Sands, M. *The Feynman Lectures on Physics III*; Addison-Wesley: Reading, MA, USA, 1965; Chapter 5.
22. Hiley, B.J. *Welcher Weg Experiments from the Bohm Perspective, Quantum Theory: Reconsiderations of Foundations-3, Växjö, Sweden 2005*; Adenier, G., Krennikov, A.Y., Nieuwenhuizen, T.M., Eds.; AIP: College Park, MD, USA, 2006; pp. 154–160.
23. Hiley, B.J.; Callaghan, R.E. Delayed Choice Experiments and the Bohm Approach. *Phys. Scr.* **2006**, *74*, 336–348. [CrossRef]
24. Bohm, D. *Quantum Theory*; Prentice-Hall: Englewood Cliffs, NJ, USA, 1951.
25. Landau, L.D.; Lifshitz, E.M. *Quantum Mechanics: Non-Relativistic Theory*; Pergamon Press: Oxford, UK, 1977; pp. 56–57.
26. Mott, N.F. The Wave Mechanics of α-Ray Tracks. *Proc. R. Soc.* **1929**, *126*, 79–84. [CrossRef]
27. Bohm, D.; Schiller, R.; Tiomno, J. A causal interpretation of the Pauli equation (A). *Nuovo Cimento* **1955**, *1*, 48–66. [CrossRef]
28. Dewdney, C. Particle Trajectories and Interference in a Time-dependent Model of Neutron Single Crystal Interferometry. *Phys. Lett.* **1985**, *109*, 377–384. [CrossRef]
29. Dewdney, C.; Holland, P.R.; Kyprianidis, A.; Vigier, J.-P. Spin and non-locality in quantum mechanics. *Nature* **1988**, *336*, 536–544. [CrossRef]
30. Dewdney, C.; Holland, P.R.; Kyprianidis, A. What happens in a spin measurement? *Phys. Lett. A* **1986**, *119*, 259–267. [CrossRef]
31. Dewdney, C.; Holland, P.R.; Kyprianidis, A. A Causal Account of Non-local Einstein-Podolsky-Rosen Spin Correlations. *J. Phys. A Math. Gen.* **1987**, *20*, 4717–4732. [CrossRef]
32. Holland, P.R. *The Quantum Theory of Motion: An Account of the de Broglie-Bohm Causal Interpretation of Quantum Mechanics*; Cambridge University Press: Cambridge, UK, 1995.
33. Penrose, R.; Rindler, W. *Spinors and Space-Time*; Cambridge University Press: Cambridge, UK, 1984; Volume 1.
34. Bell, J.S. *Speakable and Unspeakable in Quantum Mechanics*; Cambridge University Press: Cambridge, UK, 1987.
35. Hiley, B.J.; Callaghan, R.E. The Clifford Algebra approach to Quantum Mechanics A: The Schrödinger and Pauli Particles. *arXiv* **2010**, arXiv:1011.4031.
36. Hiley, B.J. Weak Values: Approach through the Clifford and Moyal Algebras. *J. Phys. Conf. Ser.* **2012**, *361*, 012014. [CrossRef]
37. Bohm, D.; Hiley, B.J. Non-locality and Locality in the Stochastic Interpretation of Quantum Mechanics. *Phys. Rep.* **1989**, *172*, 93–122. [CrossRef]
38. Feynman, R.P.; Leighton, R.B.; Sands, M. *The Feynman Lectures on Physics III*; Addison-Wesley: Reading, MA, USA, 1965; Chapter 5.2.
39. Schrödinger, E. Are There Quantum Jumps? Part I. *Br. J. Philos. Sci.* **1952**, *3*, 109–123. [CrossRef]
40. Bohm, D. *The Implicate Order: A New Approach to the Nature of Reality*; A Talk Given at Syracuse University; Syracuse University: Syracuse, NY, USA, 1982.

41. Bohm, D. A proposed Explanation of Quantum Theory in Terms of Hidden Variables at a Sub-Quantum Mechanical Level. In *Observation and Interpretation, Proceedings of the Ninth Symposium of the Colston Research Society, Bristol, UK, 1–4 April 1957*; Korner, S., Ed.; Butterworth Scientific Publications: London, UK, 1957; pp. 33–40.
42. Hiley, B.J.; Callaghan, R.E. The Clifford Algebra Approach to Quantum Mechanics B: The Dirac Particle and its relation to the Bohm Approach. *arXiv* **2010**, arXiv:1011.4033.
43. Takabayasi, T. Remarks on the Formulation of Quantum Mechanics with Classical Pictures and on Relations between Linear Scalar Fields and Hydrodynamical Fields. *Prog. Theor. Phys.* **1953**, *9*, 187–222. [CrossRef]
44. Schwinger, J. The Theory of Quantised Fields I. *Phys. Rev.* **1951**, *82*, 914–927. [CrossRef]
45. Bohm, D.J.; Hiley, B.J. Measurement Understood Through the Quantum Potential Approach. *Found. Phys.* **1984**, *14*, 255–264. [CrossRef]
46. Bohm, D.; Hiley, B.J. *The Undivided Universe: An Ontological Interpretation of Quantum Theory*; Routledge: London, UK, 1993.
47. Flack, R.; Hiley, B.J. Feynman Paths and Weak Values. *Preprints* **2018**, 2018040241. [CrossRef]
48. Flack, R.; Hiley, B.J. Weak Measurement and its Experimental Realisation. *J. Phys. Conf. Ser.* **2014**, *504*, 012016. [CrossRef]
49. Monachello, V.; Flack, R.; Hiley, B.J.; Callaghan, R.E. A method for measuring the real part of the weak value of spin using non-zero mass particles. *arXiv* **2017**, arXiv:1701.04808.
50. Morley, J.; Edmunds, P.D.; Barker, P.F. Measuring the weak value of the momentum in a double slit interferometer. *J. Phys. Conf. Ser.* **2016**, *701*, 012030. [CrossRef]

MDPI

Article

Feynman Paths and Weak Values

Robert Flack and Basil J. Hiley *

Department of Physics and Astronomy, University College London, Gower Street, London WC1E 6BT, UK; r.flack@ucl.ac.uk

* Correspondence: b.hiley@bbk.ac.uk

Received: 16 April 2018; Accepted: 9 May 2018; Published: 14 May 2018

Abstract: There has been a recent revival of interest in the notion of a 'trajectory' of a quantum particle. In this paper, we detail the relationship between Dirac's ideas, Feynman paths and the Bohm approach. The key to the relationship is the weak value of the momentum which Feynman calls a transition probability amplitude. With this identification we are able to conclude that a Bohm 'trajectory' is the average of an ensemble of actual individual stochastic Feynman paths. This implies that they can be interpreted as the mean momentum flow of a set of individual quantum processes and not the path of an individual particle. This enables us to give a clearer account of the experimental two-slit results of Kocsis et al.

Keywords: Feynman paths; weak values; Bohm theory

---

1. Introduction

One of the basic tenets of quantum mechanics is that the notion of a particle trajectory has no meaning. The established view has been unambiguously defined by Landau and Lifshitz [1]: "In quantum mechanics there is no such concept as the path of a particle". This position was not arrived at without an extensive discussion going back to the early debates of Bohr and Einstein [2], the pioneering work of Heisenberg [3] and many others [4].

Yet Kocsis et al. [5] have experimentally determined an ensemble of what they call 'photon trajectories' for individual photons traversing a two-slit interference experiment. The set of trajectories, or what we will call flow-lines, they construct is very similar in appearance to the ensemble of Bohmian trajectories calculated by Philippidis et al. [6]. Mahler et al. [7] have gone further and claimed that their new experimental results provide evidence in support of Bohmian mechanics. However such a claim cannot be correct because Bohmian mechanics is based on the Schrödinger equation which holds only for non-relativistic particles with non-zero rest mass, whereas photons are relativistic, having zero rest mass.

The flow-lines are calculated from experimentally determined weak values of the momentum operator, a notion that was introduced originally by Aharonov et al. [8] for the spin operator. When examined closely, the momentum weak value is the Feynman transition probability amplitude (TPA) [9]. In fact, Schwinger [10] explicitly writes the TPA of the momentum in exactly the same form as the weak value. Recall that the TPA involving the momentum operator plays a central role in the discussion of the path integral method, an approach that was inspired by an earlier paper of Dirac [11] who was interested in developing the notion of a 'quantum trajectory'.

Weak values are in general complex numbers, as are TPAs. The real part of the momentum weak value is the local momentum, sometimes known as the Bohm momentum. The imaginary part turns out to be the osmotic momentum introduced by Nelson [12] in his stochastic derivation of the Schrödinger equation. In this paper, we will show how the weak value of momentum, Feynman paths and the Bohm trajectories are related enabling us to give a different meaning to the flow-lines constructed in experiments of the type carried out by Kocsis et al. [5] and Mahler et al. [7].

Feynman [9] also shows that in his approach the usual expression for the kinetic energy becomes infinite unless one introduces a small fluctuation in the mass of the particle. We will show that this is equivalent to introducing the quantum potential, a new quality of energy that appears in the real part of the Schrödinger equation under polar decomposition of the wave function [13].

2. Dirac's Notion of a Quantum Trajectory

2.1. Dirac Trajectories

To make the context of our discussion clear, we will begin by drawing attention to an early paper by Dirac [11] who attempted to generalise the Heisenberg algebraic approach through his unique bra-ket notation, not as elements in a Hilbert space, but as elements of a non-commutative algebra. In this approach the operators of the algebra are functions of time. Dirac argued that to get round the difficulties presented by a non-commutative quantum algebra, strict attention must be paid to the time-order of the appearance of elements in a sequence of operators.

In the non-relativistic limit, operators at different times always commute. (In this paper, we will, for simplicity, only consider the non-relativistic domain. Dirac himself shows how the ideas can be extended to the relativistic domain.) This means that a time ordered sequence of position operators can be written in the form,

$$\langle x_t|x_{t_0}\rangle = \int \cdots \int \langle x_t|x_{t_j}\rangle dx_j \langle x_{t_j}|x_{t_{j-1}}\rangle \ldots \langle x_{t_2}|x_{t_1}\rangle dx_1 \langle x_{t_1}|x_{t_0}\rangle. \quad (1)$$

This breaks the TPA, $\langle x_t|x_{t_0}\rangle$, into a sequence of adjacent points, each pair connected by an infinitesimal TPA. Dirac writes "...one can regard this as a trajectory ... and thus makes quantum mechanics more closely resemble classical mechanics".

In order to analyse the sequence in Equation (1) further, Dirac assumed that for a small time interval $\Delta t = \epsilon$, we can write

$$\langle x|x'\rangle_\epsilon = \exp[iS_\epsilon(x,x')/\hbar] \quad (2)$$

where we will take $S_\epsilon(x,x')$ to be a real function in the first instance. Then Dirac [14] shows that

$$p'_\epsilon(x,x') = \langle x|\hat{P}'|x'\rangle_\epsilon = i\hbar\nabla_{x'}\langle x|x'\rangle_\epsilon = -\nabla_{x'}S_\epsilon(x,x')\langle x|x'\rangle_\epsilon \quad (3)$$

and

$$p_\epsilon(x,x') = \langle x|\hat{P}|x'\rangle_\epsilon = -i\hbar\nabla_x\langle x|x'\rangle_\epsilon = \nabla_x S_\epsilon(x,x')\langle x|x'\rangle_\epsilon. \quad (4)$$

Here $\hat{P}$ is the momentum operator. The remarkable similarity of these objects to the canonical momentum appearing in the classical Hamilton-Jacobi theory should be noted, a fact of which Dirac was well aware. They are also the canonical momenta appearing in the real part of the Schrödinger equation under polar decomposition of the wave function exploited by Bohm [13] who identified the momentum with the gradient of the phase of the wave function.

In an earlier paper, Dirac [11] did not specify how $S_\epsilon(x,x')$ could be determined. It was Feynman [9] who later identified its relation to the classical Lagrangian $L(\dot{x},x,t)$ through the relation

$$S_{tt'}(x,x') = \mathrm{Min}\int_{t'}^{t} L(\dot{x},x,t)dt. \quad (5)$$

However, this Lagrangian determines the classical path, so using the exponent of the classical action seems puzzling. Is there a mathematical explanation for such a choice? The answer is 'yes' and is discussed in Guillemin and Sternberg [15]. The essential reason for this lies in the relation between the symmetry group, in this case the symplectic group, and its covering group. Exploiting this structure, de Gosson and Hiley [16] have shown in detail how it is possible to mathematically 'lift' classical trajectories onto this covering space. It is from this structure that the wave properties emerge. The lift is achieved by exponentiating the classical action, namely using $\exp[iS_\epsilon(x,x')]$. It is the existence of this structure that the close relation between the Dirac quantum 'trajectories' and the de Broglie-Bohm 'trajectories' first calculated by Philippidis et al. [6] emerges. We will bring out this relationship in the rest of this paper.

2.2. The Feynman Propagator

Equation (5) allows us to write the propagator in the well known form

$$K(x,x') = \int_{x'}^{x} e^{iS(x,x')}\mathcal{D}x'$$

where the integral is taken over all paths connecting x' to x. We have written $\mathcal{D}x'$ for $\frac{dx_0}{A},\ldots,\frac{dx_{j-1}}{A}$ where $(x_0, x_1, \ldots, x_{j-1})$ are points on the path and A is the normalising factor introduced by Feynman. Clearly here $S(x,x')$ is real.

For a free particle with mass m, we have $L = m\dot{x}^2/2$ and one can show that

$$K_{tt'}(x,x') = \frac{1}{A}\exp\left[\frac{im(x-x')^2}{2\hbar(t-t')}\right] \tag{6}$$

where $A = \left(\frac{2\pi i(t-t')}{m}\right)^{1/2}$. With this propagator, Feynman was able to derive the Schrödinger equation by assuming the underlying paths were continuous and differentiable.

However if we examine the terms $\langle x|x'\rangle_\epsilon$ for $\epsilon \to 0$, we find the curves, although continuous, are non-differentiable. To show this let us introduce the TPA of a function $F(x,t)$ defined by

$$\langle\phi_t|F|\psi_{t'}\rangle_S = \mathrm{Lim}_{\epsilon\to 0}\int\cdots\int \phi^*(x,t)F(x_0,x_1,\ldots,x_j)$$
$$\times\exp\left[\frac{i}{\hbar}\sum_{k=0}^{j-1}S(x_{k+1},x_k)\right]\psi(x',t')\mathcal{D}x(t).$$

Here $\mathcal{D}$ is now written as $\mathcal{D}x(t) = \frac{dx_0}{A}\ldots\frac{dx_{j-1}}{A}dx_j$.

These TPAs can be evaluated by using functional derivatives. In fact, the average of the functional derivative of a function $F(x,t)$ is given by

$$\left\langle\frac{\delta F}{\delta x(s)}\right\rangle_S = -\frac{i}{\hbar}\left\langle F\frac{\delta S}{\delta x(s)}\right\rangle_S \tag{7}$$

at the point $x(s)$ on the path $x(t)$. In the case of the specific integral

$$\int\frac{\partial F}{\partial x_k}\exp[(i/\hbar)S(x(t))]\mathcal{D}x(t),$$

Equation (7) can be written in the form

$$\left\langle\frac{\partial F}{\partial x_k}\right\rangle_S = -\frac{i}{\hbar}\left\langle F\frac{\partial S}{\partial x_k}\right\rangle_S.$$

Feynman notes that the quantities in this expression need not be observables, nevertheless the equivalence is true [17].

Let us now consider three adjacent points x_{k-1}, x_k, x_{k+1}, each separated by a small time difference ϵ, we have

$$-\frac{\hbar}{i}\left\langle\frac{\partial F}{\partial x_k}\right\rangle_S=\left\langle F\left[\frac{\partial S(x_{k+1},x_k)}{\partial x_k}+\frac{\partial S(x_k,x_{k-1})}{\partial x_k}\right]\right\rangle_S.$$

This equation is correct to zero and first order in ϵ. If we choose the action for a particle moving in a potential V, we have

$$S(x,x')=\left[\frac{m(x-x')^2}{2\epsilon}\right]-\epsilon V(x,x').$$

Then at the point x_k this gives us

$$-\frac{\hbar}{i}\left\langle\frac{\partial F}{\partial x_k}\right\rangle_S=\left\langle F\left[-m\left(\frac{x_{k+1}-x_k}{\epsilon}-\frac{x_k-x_{k-1}}{\epsilon}\right)-\epsilon\frac{\partial V}{\partial x_k}(x_k)\right]\right\rangle.$$

If F is unity and we divide by ϵ we get

$$0=\left\langle\frac{1}{\epsilon}\left[-m\left(\frac{x_{k+1}-x_k}{\epsilon}-\frac{x_k-x_{k-1}}{\epsilon}\right)-\frac{\partial V}{\partial x_k}(x_k)\right]\right\rangle. \tag{8}$$

If we follow Feynman and call $(x_{k+1}-x_k)/\epsilon$ a 'velocity', then this equation gives the 'average' over an ensemble of individual velocities. It is the *quantum equivalent* of Newton's second law of motion; the potential V at x_k gives rise to a force which changes the incoming momentum $m(x_k-x_{k-1})/\epsilon$ to the outgoing momentum $m(x_{k+1}-x_k)/\epsilon$. Notice to order ϵ, no extra term corresponding to the quantum potential appears. de Gosson and Hiley [18] have shown in a detailed analysis that this is to be expected.

These paths are reminiscent of Brownian motion, a characteristic feature of which is the appearance of two 'derivatives' at x_k, a 'forward' and a 'backward' derivative, illustrating the non-differentiable nature of the path. In this paper, we need not discuss the precise nature of these paths to arrive at our conclusion. It is sufficient for us to note that the substructure of a quantum process is *certainly not classical*. In passing we should also note that the 'velocities', being of order $(\hbar/m\epsilon)^{1/2}$, diverge as $\epsilon\to 0$ and therefore, in Feynman's terms, are not observables.

2.3. TPAs Involving the Momentum

In 1974 Hirschfelder [19,20] introduced a quantity $\psi(x,t)^{-1}\hat{p}\psi(x,t)$, which he called a 'sub-observable' as he could see no way of measuring it directly, although integrating it over the whole of configuration space gave the measurable expectation value. Using the polar form of the wave function, $\psi(x,t)=R(x,t)\exp[iS(x,t)/\hbar]$, this 'sub-observable' is the weak value of the momentum operator which can be written in the form

$$\psi(x,t)^{-1}\hat{p}\psi(x,t)=\frac{\langle x|\hat{p}|\psi(t)\rangle}{\langle x|\psi(t)\rangle}=m[v_B(x,t)-iv_O(x,t)], \tag{9}$$

where explicitly $v_B(x,t)=\nabla S(x,t)/m$ is the local Bohm velocity and $v_O(x,t)=\nabla R(x,t)/mR(x,t)$ is the localising osmotic velocity, originally introduced by Nelson [12] in a stochastic theory. The meaning of these velocities is discussed in more detail in Bohm and Hiley [21]. Much later Hiley [22] showed exactly how these expressions emerged directly from the weak value of the momentum operator. It should be noted that weak values are essentially TPAs of the type considered by Feynman [9] and Schwinger [23].

In the spirit of Schwinger [10], where he argues that "the quantum dynamical laws will find their proper expression in terms of the transformation functions" that is TPAs, we can introduce *two* momentum TPAs, $\langle x|\overrightarrow{P}|\psi(t)\rangle$ and $\langle\psi(t)|\overleftarrow{P}|x\rangle$ where $\overrightarrow{P} = -i\hbar\overrightarrow{\nabla}$ and $\overleftarrow{P} = i\hbar\overleftarrow{\nabla}$. Notice by placing the arrows over the momentum operators, we are emphasising the distinction between left and right multiplication and it is this distinction that is equivalent to the forward and backward derivatives. In fact we may identify

$$\langle X|\overrightarrow{P}|\psi(t')\rangle = \langle X|\overrightarrow{P}|x'\rangle\psi(x',t') = -i\lim_{(x'\to X)}\frac{\psi(X)-\psi(x')}{(X-x')}$$

with the forward derivative at X, a point that lies between x' and x.

$$\langle\psi(t)|\overleftarrow{P}|X\rangle = \psi^*(x,t)\langle x|\overleftarrow{P}|X\rangle = i\lim_{(X\to x)}\frac{\psi^*(x)-\psi^*(X)}{(x-X)}$$

corresponds to the backward derivative. Note that the words 'forward' and 'backward' here have nothing to do with time order.

If we again evaluate these TPAs using $\psi = R\exp(iS/\hbar)$, we find

$$\frac{1}{2}\left[\frac{\langle x|\overrightarrow{\hat{P}}|\psi(t)\rangle}{\langle x|\psi(t)\rangle} + \frac{\langle\psi(t)|\overleftarrow{\hat{P}}|x\rangle}{\langle\psi(t)|x\rangle}\right] = \nabla S(x,t) = P_B(x,t), \tag{10}$$

and

$$\frac{1}{2i}\left[\frac{\langle x|\overrightarrow{\hat{P}}|\psi(t)\rangle}{\langle x|\psi(t)\rangle} - \frac{\langle\psi(t)|\overleftarrow{\hat{P}}|x\rangle}{\langle\psi(t)|x\rangle}\right] = \frac{\nabla R(x,t)}{R(x,t)} = P_O(x,t). \tag{11}$$

Notice how the sums and differences of the left/right operators produce real values.

We can immediately connect these results with those of Dirac [11] if, in Equations (3) and (4), we replace the real value of $S_\epsilon(x,x')$ by a complex value which we will write as $S'_\epsilon(x,x') = S_\epsilon(x,x') - i\ln R_\epsilon(x,x')$. In this case, we find

$$p'_\epsilon(x,x') = -\nabla_{x'}S_\epsilon(x,x') - i\frac{\nabla_{x'}R_\epsilon(x,x')}{R_\epsilon(x,x')} \tag{12}$$

and

$$p_\epsilon(x,x') = \nabla_x S_\epsilon(x,x') - i\frac{\nabla_x R_\epsilon(x,x')}{R_\epsilon(x,x')}. \tag{13}$$

Notice also the connection with the classical relations obtained in Equations (3) and (4).

2.4. *The Relation between Weak Values and TPAs*

In the previous two sections, we have shown how TPAs of the form $\langle\phi_t|\hat{F}|\psi_{t'}\rangle$ arise from some underlying non-differentiable process. The original assumption was that these quantities could not be investigated experimentally. However starting from a different perspective, the notion of a weak value, introduced by Aharonov, Albert and Vaidman [8], allows us to experimentally measure these quantities.

A weak value of an operator $\hat{F}$ is defined by

$$\langle\hat{F}\rangle_w = \frac{\langle\phi_t|\hat{F}|\psi_{t'}\rangle}{\langle\phi_t|\psi_{t'}\rangle}.$$

Clearly these weak values are Feynman TPAs. Using the suggestions of Leavens [24] and Wiseman [25], Kocsis et al. [5] have actually measured the weak value of the transverse momentum in an optical two-slit experiment and as a result have constructed what they called *photon 'trajectories'*. We refer to their paper to explain the details of how this is done.

Unfortunately photons cannot be treated as particles that satisfy the Schrödinger equation. They have zero rest mass and are excitations of the electromagnetic field. Nevertheless this does not invalidate the notion of a momentum flow line; the question remains "How are we to understand these flow lines?" Flack and Hiley [26] have shown that if we generalise the Bohm approach to include the electromagnetic field [27], each flow line emerges as the locus of a weak Poynting vector.

To connect with the non-relativistic approach we are discussing in this paper, we need to use atoms. Indeed experiments are being developed at UCL to measure weak values of spin and momentum, $\langle \hat{p} \rangle_w$, for helium atoms [28] and argon atoms [29] respectively. The experimental details can be found in these references. In this paper, we will clarify further the relation between the Feynman paths and weak values.

3. Weak Values Are Weighted TPAs

3.1. Flow Lines Constructed from Weak Values

In quantum mechanics, the uncertainty principle does not allow us to give meaning to the 'trajectory' of a single particle so we are left with the question: "How does a particle get from A to B?". Rather than taking two points, consider two small volumes, $\Delta V'(x')$ surrounding the point $A = x'$ and $\Delta V(x)$ surrounding $B = x$. We assume these volumes are initially large enough to avoid problems with the uncertainty principle.

Now imagine a sequence of particles emanating from $\Delta V'(x')$, each with a different momentum. Over time we will have a spray of possible momenta emerging from the volume $\Delta V'(x')$, the nature of this spray depending on the size of $\Delta V'(x')$. Similarly there will be a spray of momenta over time arriving at the small volume $\Delta V(x)$ surrounding the point x.

Better still let us consider a small volume surrounding the midpoint X. At this point there is a spray arriving and a spray leaving a volume $\Delta V(X)$ as shown in Figure 1. To see how the local momenta behave at the midpoint X, we will use the real part of $S'_\epsilon(x, x')$ defined by

$$S_\epsilon(x, x') = \frac{m}{2} \frac{(x - x')^2}{\epsilon}. \tag{14}$$

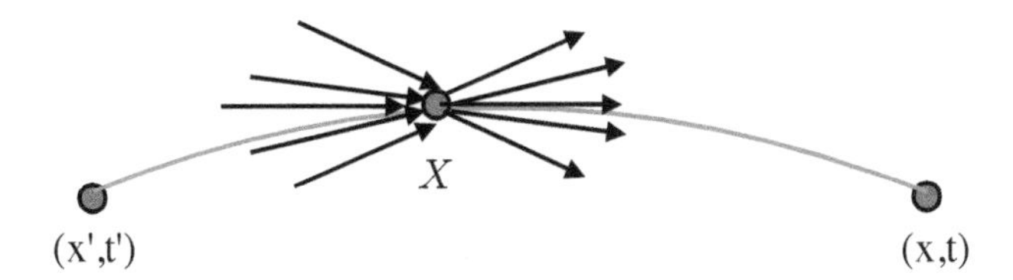

Figure 1. Behaviour of the momenta sprays at the midpoint of $\langle x, t | x', t' \rangle_\epsilon$.

Let us define a quantity

$$P_X(x, x') = \frac{\partial S_\epsilon(x, x')}{\partial X} = \frac{\partial S_\epsilon(X, x')}{\partial X} + \frac{\partial S_\epsilon(x, X)}{\partial X}, \tag{15}$$

then using the action (Equation (14)), we find

$$P_X(x, x') = m \left[\frac{(X - x')}{\epsilon} - \frac{(x - X)}{\epsilon} \right] = p'_X(x, x') + p_X(x, x'). \tag{16}$$

Not surprisingly, this is exactly what Feynman [9] is averaging over at the point X, agreeing with the term between the brace $[\ldots]$.

What is more important is the relation of Equation (16) to Equation (10) which is the real part of the weak value of the momentum operator. Thus, the mean momentum of a set of Feynman paths at X is clearly the real part of this weak value. However, this weak value is just the Bohm momentum. Thus the Bohm 'trajectories' are simply an ensemble of the average of the ensemble of individual Feynman paths.

To see how this unexpected result also emerges from a different perspective, let us consider the process in Figure 1 which we regard as an image of an ensemble of actual individual quantum processes. We are interested in finding the average behaviour of the momentum, P_X, at the point X. However, we have two contributions to consider, one coming from the point x' and one leaving for the point x. We must determine the distribution of momenta in each spray to produce a result that is consistent with the wave function $\psi(X)$ at X. Feynman suggests [9] that we can think of $\psi(X)$ as 'information coming from the past' and $\psi^*(X)$ as 'potential information appearing in the future'. This suggests that we can write

$$\lim_{x'\to X}\psi(x') = \int\phi(p')e^{ip'X}dp' \quad \text{and} \quad \lim_{X\to x}\psi^*(x) = \int\phi^*(p)e^{-ipX}dp.$$

The $\phi(p')$ contains information regarding the probability distribution of the incoming momentum spray, while $\phi^*(p)$ contains information about the probability distribution in the outgoing momentum spray. These wave functions must be such that in the limit $\epsilon \to 0$ they are consistent with the wave function $\psi(X)$.

Thus, we can define the mean momentum, $\overline{\overline{P}}(X)$, at the point X as

$$\rho(X)\overline{\overline{P}}(X) = \int\int P\phi^*(p)e^{-ipX}\phi(p')e^{ip'X}\delta(P-(p'+p)/2)dPdpdp' \tag{17}$$

where $\rho(X)$ is the probability density at X. We have added the restriction $\delta(P-(p'+p)/2)$ since momentum is conserved at X. We can rewrite Equation (17) and form

$$\rho(X)\overline{\overline{P}}(X) = \frac{1}{2\pi}\int\int P\phi^*(p+\theta/2)e^{-iX\theta}\phi(p-\theta/2)d\theta dP$$

or equivalently taking Fourier transforms

$$\rho(X)\overline{\overline{P}}(X) = \frac{1}{2\pi}\int\int P\psi^*(X-\sigma/2)e^{-iP\sigma}\psi(X+\sigma/2)d\sigma dP$$

which means that $\overline{\overline{P}}(X)$ is the conditional expectation value of the momentum weighted by the Wigner function. Equation (17) can be put in the form

$$\rho(X)\overline{\overline{P}}(X) = \left(\frac{1}{2i}\right)[(\partial_{x_1}-\partial_{x_2})\psi(x_1)\psi(x_2)]_{x_1=x_2=X} \tag{18}$$

an equation that appears in the Moyal approach [30], which is based on a different non-commutative algebra. If we evaluate this expression for the wave function written in polar form $\psi(x) = R(x)\exp[iS(x)]$, we find $\overline{\overline{P}}(X) = \nabla S(X)$ which is identical to the expression for the local (Bohm) momentum used in the Bohm interpretation.

This then confirms the conclusion we reached above, namely, that the set of Bohm 'trajectories' is an ensemble of the average ensemble of individual paths. Notice, once again, that this gives a very different picture of the Bohm momentum from the usual one used in Bohmian mechanics [31]. It is not the momentum of a single 'particle' passing the point X, but the mean *momentum flow* at the point in question.

This conclusion is supported by the experiments of Kocsis et al. [5]. They construct the flow lines from an average made over many individual input photons. Thus, the so-called 'photon' flow-lines are constructed *statistically* from an ensemble of individual events. As was shown in Flack and Hiley [26], these flow lines are an average of the momentum flow as described by the *weak* value of the Poynting vector. This agrees with what one would expect from standard quantum electrodynamics, where the notion of a 'photon trajectory' has no meaning, but the notion of a 'momentum flow' does have meaning.

Bliokh et al. [32] have presented a beautiful illustration showing the results of a two-slit interference experiment. Figure 2a shows the real part of the momentum flow lines in the electromagnetic field, while the imaginary component (osmotic) momentum flow lines are shown in Figure 2b. It is then clear that we can regard $v_B(x,t) = p_B(x,t)/m$ as a *local* velocity, while the osmotic velocity $v_O(x,t) = p_O(x,t)/m$ can be regarded as a *localising* velocity as discussed in Bohm and Hiley [33]. The osmotic velocity behaves in such a way as to maintain the form of the probability distribution.

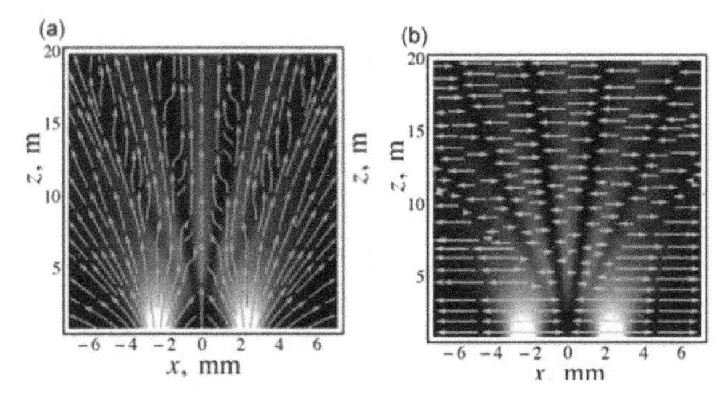

Figure 2. (**a**) Local field momentum; (**b**) Localising field momentum.

3.2. Where Is the Quantum Potential?

One of the features that many find 'mysterious' [34] is the appearance of the 'quantum potential' in the Bohm approach. Is there any trace of it in the Feynman paper [9]? To answer this question, we must first refer to de Gosson and Hiley [18] where it is shown that this energy term is absent in quantum processes when taken only to $O(\Delta t = \epsilon)$ so we must consider terms to $O(\Delta t = \epsilon^2)$.

Feynman shows that the kinetic energy is of $O(\epsilon^2)$ when written in the form K.E. $= [(x_{k+1} - x_k)/\epsilon]^2$, and diverges as $\epsilon \to 0$. Feynman points out that this quantity is not an observable functional. However, let us now define the kinetic energy to be

$$\text{K.E.}' = \frac{m}{2}\left(\frac{x_{k+1}-x_k}{\epsilon}\right)\left(\frac{x_k - x_{k-1}}{\epsilon}\right).$$

This function is finite to $O(\epsilon)$ and therefore is an observable functional. Feynman then shows that if we allow "the mass to change by a small amount to $m(1+\delta)$ for a short time, say ϵ around t_k" we can obtain the relation

$$\frac{m}{2}\left(\frac{x_{k+1}-x_k}{\epsilon}\right)\left(\frac{x_k - x_{k-1}}{\epsilon}\right) = \frac{m}{2}\left(\frac{x_{k+1}-x_k}{\epsilon}\right)^2 + \frac{\hbar}{2i\epsilon}, \tag{19}$$

the extra term arising from the normalising function A. Thus, we must add a 'correction' term to the K.E. in order for the total energy to be finite to $O(\epsilon^2)$.

This is the forerunner of mass renormalisation used in quantum electrodynamics. In that case the charged particle is subjected to electromagnetic vacuum fluctuations. The particle we are considering here is not charged and so the fluctuation must arise from a different source, but however it arises, it changes the TPA by δ.

Later in the same paper, Feynman shows that any random fluctuation in the phase function will produce the same effect. A random fluctuation at the point x_k implies we must replace $S(x_{k+1}, t_{k+1}; x_k, t_k)$ by $S_\delta(x_{k+1}, t_{k+1}; x_k, t_k - \delta)$. Thus, to the first order in δ we have

$$\langle \xi|1|\psi\rangle_S - \langle \xi|1|\psi\rangle_{S_\delta} = \frac{i\delta}{\hbar}\langle \xi|H_k|\psi\rangle_S$$

where H_k is the Hamiltonian functional

$$H_k = -\frac{\partial S(x_{k+1}, t_{k+1}; x_k, t_k)}{\partial t_{k+1}} + \frac{\hbar}{2i(t_{k+1} - t_k)}. \tag{20}$$

Apart from the minus sign, the last term is identical to the last term in Equation (19). Thus Feynman required extra energy to appear from somewhere. A more detailed discussion of this feature appears in Feynman and Hibbs [35]. The Bohm approach indicates that some 'extra' energy appears in the form of the quantum potential energy at the expense of the kinetic energy. Could it be that the source of the energy is the same?

To explore this possibility, let us use the method explained in Section 2.3 to obtain a more general result for the K.E. The real part of the weak value of the momentum operator squared is obtained from $(\langle\psi(t)|\hat{p}^2|x\rangle + \langle x|\hat{p}^2|\psi(t)\rangle)/2$. Under polar decomposition of the wave function, we find the real part of the weak value of the kinetic energy is

$$\frac{1}{2m}\langle \hat{p}^2\rangle_w = \frac{1}{2m}\left[(\nabla S)^2 - \frac{\nabla^2 R}{R}\right]. \tag{21}$$

With the identification $\nabla S \leftrightarrow m(x_{k+1} - x_k)/\epsilon$, we see that the quantum potential is playing a similar role as the mass/energy fluctuation in Feynman's approach. In fact, de Broglie's original suggestion was that the quantum potential could be associated with a change of the rest mass [36].

Notice that the quantum potential appears essentially as a derivative of the osmotic velocity, which in turn is obtained from the imaginary part of $S'(x, x')$. Any fluctuating term added to the real part of $S_\epsilon(x, x')$ should also be added to the imaginary part. This would also introduce some change in the energy relation shown in Equation (20). This interplay between the real components of the complex $S_\epsilon(x, x')$ is clearly presented as an average over fluctuations arising from some background. Here we can recall Bohr insisting that quantum phenomena must include a description of the whole experimental arrangement. More details will be found in Smolin [37] and in Hiley [38].

4. Conclusions

Our explorations of the weak values of the momentum operator [22] have led us to reconsider the basis on which Feynman [9] developed his path integral approach. We have shown that there is an unexpected close connection between the Feynman propagator, the weak values of the momentum and the original Bohm approach [13].

Feynman had already noticed that to prevent the kinetic energy tending to infinity as the time interval between steps tends to zero, it was necessary to introduce a 'fluctuation' in the mass/energy of the particle. This extra energy can be thought of as arising in a way similar to the way the quantum potential energy appears as an effect of some background field. Indeed, as we have remarked above, de Broglie [36] had already proposed that the quantum potential could be included in the mass term $M = \sqrt{[m^2 + (\hbar^2/c^2)\Box R/R]}$, R being the amplitude of the wave function. Hiley [38] has shown a similar conclusion arises for the Dirac equation.

The approach outlined in this paper shows that the basic assumption made in Bohmian mechanics, namely, that each particle follows one of the ensemble of 'trajectories' calculated by Philippidis et al. [6] from $P_B(x, t)$ cannot be maintained. Rather the trajectories should be interpreted as a statistical average of the momentum flow of a basic underlying stochastic process.

It is now possible to experimentally explore weak values, perhaps clarifying the nature of this stochastic process. In the case of the electromagnetic field this has already been done by Kocsis et al. [5], but as we have seen the notion of a 'photon trajectory' has no meaning. However, the average momentum flow does have meaning [26]. As mentioned above, new experiments using argon and helium atoms are now being carried out at UCL by Morley et al. [29] and by Monachello, Flack, and Hiley [28]. It is hoped that these future experiments will throw more light on the nature of individual quantum processes.

Author Contributions: Both authors contributed equally to this manuscript.

Funding This research was funded in part by the Fetzer Franklin Memorial Trust.

Acknowledgments: Special thanks to Bob Callaghan, Glen Dennis and Lindon Neil for their helpful discussions. Thanks also to the Franklin Fetzer Foundation for their financial support.

Conflicts of Interest: The founding sponsors had no role in the design of the study; in the collection, analyses, or interpretation of data; in the writing of the manuscript, and in the decision to publish the results.

References

1. Landau, L.D.; Lifshitz, E.M. *Quantum Mechanics: Non-Relativistic Theory*; Pergamon Press: Oxford, UK, 1977.
2. Einstein, A. *Albert Einstein: Philosopher-Scientist*; Schilpp, A.P., Ed.; Library of the Living Philosophers: Evanston, IL, USA, 1949; pp. 665–676.
3. Heisenberg, W. *Physics and Philosophy: The Revolution in Modern Science*; George Allen and Unwin: London, UK, 1958.
4. Jammer, M. *The Philosophy of Quantum Mechanics*; Wiley: New York, NY, USA, 1974.
5. Kocsis, S.; Braverman, B.; Ravets, S.; Stevens, M.J.; Mirin, R.P.; Shalm, L.K.; Steinberg, A.M. Observing the average trajectories of single photons in a two-slit interferometer. *Science* **2011**, *332*, 1170–1173. [CrossRef] [PubMed]
6. Philippidis, C.; Dewdney, C.; Hiley, B.J. Quantum interference and the quantum potential. *Il Nuovo Cimento B* **1979**, *52*, 15–28. [CrossRef]
7. Mahler, D.H.; Rozema, L.A.; Fisher, K.; Vermeyden, L.; Resch, K.J.; Braverman, B.; Wiseman, H.M.; Steinberg, A.M. Measuring bohm trajectories of entangled photons. In Proceedings of the 2014 Conference on Lasers and Electro-Optics (CLEO)—Laser Science to Photonic Applications, San Jose, CA, USA, 8–13 June 2014.
8. Aharonov, Y.; Albert, D.Z.; Vaidman, L. How the result of a measurement of a component of the spin of a spin-1/2 particle can turn out to be 100. *Phys. Rev. Lett.* **1988**, *60*, 1351–1354. [CrossRef] [PubMed]
9. Feynman, R.P. Space-time approach to non-relativistic quantum mechanics. *Rev. Mod. Phys.* **1948**, *20*, 367–387. [CrossRef]
10. Schwinger, J. The theory of quantised fields I. *Phys. Rev.* **1951**, *82*, 914–927. [CrossRef]
11. Dirac, P.A.M. On the analogy between Classical and Quantum Mechanics. *Rev. Mod. Phys.* **1945**, *17*, 195–199. [CrossRef]
12. Nelson, E. Derivation of schrödinger's equation from newtonian mechanics. *Phys. Rev.* **1966**, *150*, 1079–1085. [CrossRef]
13. Bohm, D. A suggested interpretation of the quantum theory in terms of hidden variables, I. *Phys. Rev.* **1952**, *85*, 180–193. [CrossRef]
14. Dirac, P.A.M. *The Principles of Quantum Mechanics*; Oxford University Press: Oxford, UK, 1947.
15. Guillemin, V.W.; Sternberg, S. *Symplectic Techniques in Physics*; Cambridge University Press: Cambridge, UK, 1984.
16. De Gosson, M.; Hiley, B.J. Imprints of the quantum world in classical mechanics. *Found. Phys.* **2011**, *41*, 1415–1436. [CrossRef]
17. Brown, L. *Feynman's Thesis: A New Approach to Quantum Mechanics*; World Scientific Press: Singapore, 2005.
18. De Gosson, M.; Hiley, B.J. Short-time quantum propagator and bohmian trajectories. *Phys. Lett.* **2013**, *377*, 3005–3008. [CrossRef] [PubMed]
19. Hirschfelder, J.O.; Christoph, A.C.; Palke, W.E. Quantum mechanical streamlines I. square potential barrier. *J. Chem. Phys.* **1974**, *61*, 5435–5455. [CrossRef]

20. Hirschfelder, J.O. Quantum mechanical equations of change. I. *J.Chem. Phys.* **1978**, *68*, 5151–5162. [CrossRef]
21. Bohm, D.; Hiley, B.J. Non-locality and locality in the stochastic interpretation of quantum mechanics. *Phys. Rep.* **1989**, *172*, 93–122. [CrossRef]
22. Hiley, B.J. Weak values: Approach through the Clifford and Moyal algebras. *J. Phys. Conf. Ser.* **2012**, *361*, 012014. [CrossRef]
23. Schwinger, J. The theory of quantum fields III. *Phys. Rev.* **1953**, *91*, 728–740. [CrossRef]
24. Leavens, C.R. Weak measurements from the point of view of bohmian mechanics. *Found. Phys.* **2005**, *35*, 469–491. [CrossRef]
25. Wiseman, H.M. Grounding bohmian mechanics in weak values and bayesianism. *New J. Phys.* **2007**, *9*, 165–177. [CrossRef]
26. Flack, R.; Hiley, B.J. Weak values of momentum of the electromagnetic field: Average momentum flow lines, not photon trajectories. *arXiv* **2016**, arXiv:1611.06510.
27. Bohm, D.; Hiley, B.J.; Kaloyerou, P.N. An ontological basis for the quantum theory: II—A causal interpretation of quantum fields. *Phys. Rep.* **1987**, *144*, 349–375. [CrossRef]
28. Monachello, V.; Flack, R.; Hiley, B.J. A method for measuring the real part of the weak value of spin using non-zero mass particles. *arXiv* **2017**, arXiv:1701.04808.
29. Morley, J.; Edmunds, P.D.; Barker, P.F. Measuring the weak value of the momentum in a double slit interferometer. *J. Phys. Conf. Ser.* **2016**, *701*, 012030. [CrossRef]
30. Moyal, J.E. Quantum mechanics as a statistical theory. *Proc. Camb. Phil. Soc.* **1949**, *45*, 99–123. [CrossRef]
31. Dürr, D.; Teufel, S. *Bohmian Mechanics*; Springer: Berlin/Heidelberg, Germany, 2009.
32. Bliokh, K.Y.; Bekshaev, A.Y.; Kofman, A.G.; Nori, F. Photon trajectories, anomalous velocities and weak measurements: A classical interpretation. *New J. Phys.* **2013**, *15*, 073022. [CrossRef]
33. Bohm, D.; Hiley, B.J. *The Undivided Universe: An Ontological Interpretation of Quantum Theory*; Routledge: London, UK, 1993.
34. Feynman, R.P.; Leighton, R.B.; Sands, M. *The Feynman Lectures on Physics*; III, Sec 21-8; Addison-Wesley: Boston, MA, USA, 1965.
35. Feynman, R.P.; Hibbs, A.R. *Quantum Mechanics and Path Integrals*; McGraw-Hill: New York, NY, USA, 1965.
36. De Broglie, L. *Non-Linear Wave Mechanics: A Causal Interpretation*; Elsevier: Amsterdam, The Netherlands, 1960; p. 116.
37. Smolin, L. Quantum mechanics and the principle of maximal variety. *Found. Phys.* **2016**, *46*, 736–758. [CrossRef]
38. Hiley, B.J. On the Nature of a Quantum Particle. 2018, in press.

MDPI

Article

Why Bohmian Mechanics? One- and Two-Time Position Measurements, Bell Inequalities, Philosophy, and Physics

Nicolas Gisin

Group of Applied Physics, University of Geneva, 1211 Geneva 4, Switzerland; nicolas.gisin@unige.ch; Tel.: +41-79-776-2317

Received: 21 December 2017; Accepted: 31 January 2018; Published: 2 February 2018

Abstract: In Bohmian mechanics, particles follow continuous trajectories, so two-time position correlations have been well defined. However, Bohmian mechanics predicts the violation of Bell inequalities. Motivated by this fact, we investigate position measurements in Bohmian mechanics by coupling the particles to macroscopic pointers. This explains the violation of Bell inequalities despite two-time position correlations. We relate this fact to so-called surrealistic trajectories that, in our model, correspond to slowly moving pointers. Next, we emphasize that Bohmian mechanics, which does not distinguish between microscopic and macroscopic systems, implies that the quantum weirdness of quantum physics also shows up at the macro-scale. Finally, we discuss the fact that Bohmian mechanics is attractive to philosophers but not so much to physicists and argue that the Bohmian community is responsible for the latter.

Keywords: Bohmian mechanics; quantum theory; surrealistic trajectories; Bell inequality

---

1. Introduction

Bohmian mechanics differs deeply from standard quantum mechanics. In particular, in Bohmian mechanics, particles, here called Bohmian particles, follow continuous trajectories; hence, in Bohmian mechanics, there is a natural concept of time-correlation for particles' positions. This led M. Correggi and G. Morchio [1] and more recently Kiukas and Werner [2] to conclude that Bohmian mechanics "cannot violate any Bell inequality" and hence is disproved by experiments. However, the Bohmian community maintains its claim that Bohmian mechanics makes the same predictions as standard quantum mechanics (at least as long as only position measurements are considered, arguing that, at the end of the day, all measurements result in position measurement, e.g., a pointer's positions).

Here, we clarify this debate. First, we recall why two-time position correlation is at a tension with Bell inequality violation. Next, we show that this is actually not at odds with standard quantum mechanics because of certain subtleties. For this purpose, we do not go for full generality but illustrate our point with an explicit and rather simple example based on a two-particle interferometers, partly already experimentally demonstrated and certainly entirely experimentally feasible (with photons, but also feasible at the cost of additional technical complications with massive particles). The subtleties are illustrated by explicitly coupling the particles to macroscopic systems, called pointers, that measure the particles' positions. Finally, we raise questions about Bohmian positions, about macroscopic systems, and about the large differences in appreciation of Bohmian mechanics between philosophers and physicists.

2. Bohmian Positions

Bohmian particles have, at all times, well defined positions in our three-dimensional space. However, for the purpose of my analysis, I need only to specify in which mode the Bohmian particle is.

Here I use "mode" as is usually done in optics, including atomic optics. For example, if a particle in Mode 1 encounters a beam splitter (BS) with Output Modes 1 and 2, then the Bohmian particle exits the beam splitter either in Mode 1 or in Mode 2, see Figure 1.

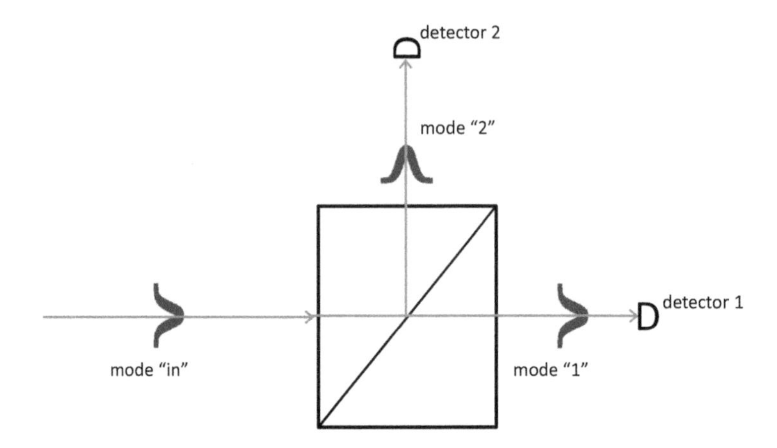

Figure 1. A Bohmian particle and its pilot wave arrive on a beam splitter (BS) from the left in Mode "in". The pilot wave emerges both in Modes 1 and 2, as per the quantum state in standard quantum theory. However, the Bohmian particle emerges either in Mode 1 or in Mode 2, depending on its precise initial position. As Bohmian trajectories cannot cross each other (in configuration space), if the initial position is in the lower half of Mode "in", then the Bohmian particle has the BS in Mode 1 or, if not, in Mode 2.

Part of the attraction of Bohmian mechanics lies then in the following assumption:

- Assumption **H**:
  Position measurements merely reveal in which (spatially separated and non-overlapping) mode the Bohmian particle actually is.

Accordingly, if Modes 1 and 2 after the beam splitter are connected to two single-particle detectors, then, if the Bohmian particle is in Mode 1, the corresponding detector clicks, and the case of Mode 2 is similar, see Figure 1.

3. Two-Time Position Correlation in a Bell Test

Let's consider a two-particle experiment with 4 modes, labeled 1, 2, 3, and 4, as illustrated in Figure 2. The source produces the quantum state:

$$\psi_0 = (|1001\rangle + |0110\rangle)/\sqrt{2} \tag{1}$$

where, e.g., $|1001\rangle$ means that there is one particle in Mode 1 and one in Mode 4, and Modes 2 and 3 are empty. This is an entangled state that can be used in a Bell inequality test. For this, Alice (who controls Modes 1 and 2) and Bob (who controls Modes 3 and 4) apply phases x and y to Modes 1 and 4, respectively, and combine their modes on a beam splitter, see Figure 2. Taking into account that a reflection on a BS induces a phase $e^{i\pi/2} = i$, the quantum state after the two beam splitters reads

$$\begin{aligned} &\frac{e^{i(x+y)}}{2^{3/2}}\left(|1001\rangle + i|0101\rangle + i|1010\rangle - |0110\rangle\right) \\ &+ \frac{1}{2^{3/2}}\left(|0110\rangle + i|0101\rangle + i|1010\rangle - |1001\rangle\right). \end{aligned} \tag{2}$$

If Modes 1, 2, 3, and 4, after the beam splitter, encounter four single-particle detectors, also labeled 1, 2, 3, and 4, then the probabilities for coincidence detection are

$$P_{14} = P_{23} = \frac{1}{8}|e^{i(x+y)} - 1|^2 = \frac{1 - \cos(x+y)}{4} \tag{3}$$

$$P_{13} = P_{24} = \frac{1}{8}|e^{i(x+y)} + 1|^2 = \frac{1 + \cos(x+y)}{4} \tag{4}$$

from which a maximal violation of the CHSH-Bell inequality of $2\sqrt{2}$ can be obtained with appropriate choices of the phase inputs.

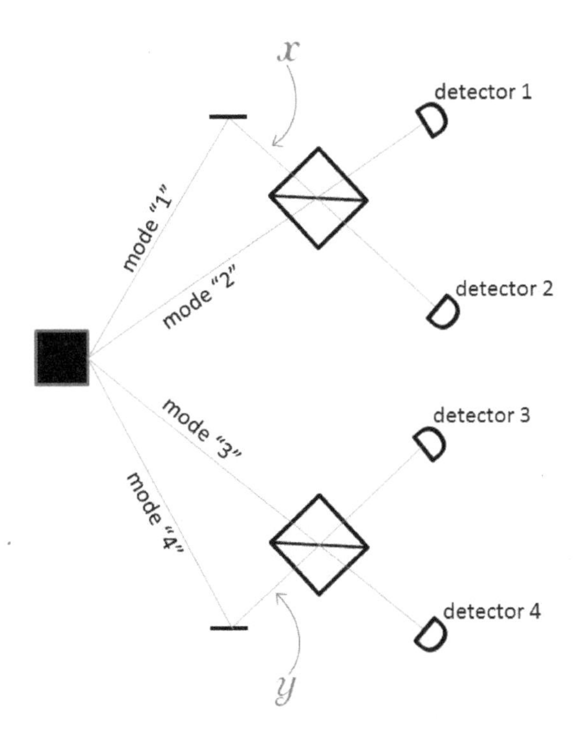

Figure 2. Two Bohmian particles spread over four modes. The quantum state is entangled, see Equation (1), so the two particle are either in Modes 1 and 4 or in Modes 2 and 3. Alice applies a phase x on Mode 1 and Bob a phase y on Mode 4. Accordingly, after the two beam splitters, the correlations between the detectors allow Alice and Bob to violate Bell inequality. The convention regarding mode numbering is that modes do not cross, i.e., the nth mode before the beam splitter goes to detector n.

In Bohmian mechanics, this experiment is easily described. Denote the two particles' positions r_A and r_B. In the initial state (Equation (1)), the particles are either in Modes 1 and 4, a situation we denote $r_A \in$ "1" and $r_B \in$ "4," or in Modes 2 and 3, i.e., $r_A \in$ "2" and $r_B \in$ "3." According to Bohmian mechanics, the particles have more precise positions, but for our argument this suffices.

Now, according to Bohmian mechanics and Assumption **H**, one does not need to actually measure the positions of the particles; it suffices to know that each is in one specific mode. Hence, one can undo Alice's measurement as illustrated in Figure 3. After the phase shift $-x$, the quantum state is precisely back to the initial state ψ_0, see Equation (1). Alice can thus perform a second measurement with a freshly chosen phase x' and a third beam splitter, see Figure 3. Moreover, as Bohmian trajectories cannot cross each other (in configuration space), if r_A is in Mode 1 before the first BS, then r_A is also in Mode 1 before the last BS.

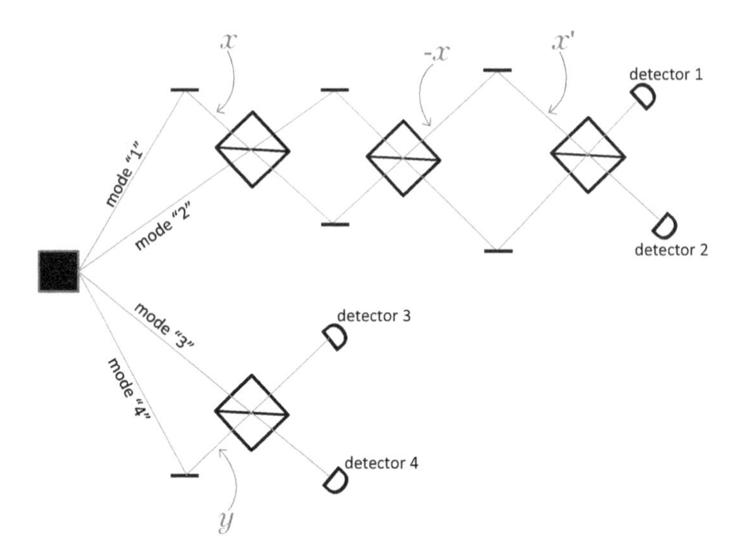

Figure 3. Alice's first "measurement", with phase x, can be undone because in Bohmian mechanics there is no collapse of the wavefunction. Hence, after having applied the phase $-x$ after her second beam splitter, Alice can perform a second "measurement" with phase x'. Mode number convention implies, e.g., that Mode 1 is always the upper mode, i.e., the mode on which all phases x, $-x$ and x', are applied.

There is no doubt that, according to Bohmian mechanics, there is a well-defined joint probability distribution for Alice's particle at two times and Bob's particle: $P(r_A, r'_A, r_B|x, x', y)$, where r_A denotes Alice's particle after the first beam splitter and r'_A after the third beam splitter of Figure 3. But here comes the puzzle. According to Assumption **H**, if $r_A \in$ "1", then any position measurement performed by Alice between the first and second beam splitter necessarily results in $a = 1$. Similarly, $r_A \in$ "2" implies $a = 2$. Thus, Alice's position measurement after the third beam splitter is determined by r'_A, and Bob's measurement is determined by r_B. Hence, it seems that one obtains a joint probability distribution for Alice's measurements results and Bob's: $P(a, a', b|x, x', y)$. However, such a joint probability distribution implies that Alice does not have to make any choice (she merely makes both choices, one after the other), and in such a situation there cannot be any Bell inequality violation. Hence, as claimed in [2], it seems that the existence of two-time position correlations in Bohmian mechanics prevents the possibility of a CHSH-Bell inequality violation, in contradiction with quantum theory predictions and experimental demonstrations [3].

Let's have a closer look at the probability distribution that lies at the bottom of our puzzle: $P(r_A, r'_A, r_B|x, x', y)$. More precisely, it suffices to consider in which modes the Bohmian particles are. That is, it suffices to consider the following joint probability distribution:

$$P(r_A \in \text{"}a\text{"}, r'_A \in \text{"}a'\text{"}, r_B \in \text{"}b\text{"}|x, x', y) \tag{5}$$

where $a, a' = 1, 2$ and $b = 3, 4$ number modes. This can be computed explicitly:

$$P(r_A \in \text{“}a\text{”}, r'_A \in \text{“}a'\text{”}, r_B \in \text{“}b\text{”}|x, x', y) = \frac{1 + (-1)^{a+b}\cos(x+y)}{4} \cdot \frac{1 + (-1)^{a'+b}\cos(x+y)}{2}. \quad (6)$$

Note that, if one sums over a', i.e., traces out Alice's second measurement, then one recovers the quantum prediction equations (Equations (3) and (4)):

$$\begin{aligned} P(r_A \in \text{“}a\text{”}, r_B \in \text{“}b\text{”}|x, y) &= \\ \sum_{a'} P(r_A \in \text{“}a\text{”}, r'_A \in \text{“}a'\text{”}, r_B \in \text{“}b\text{”}|x, x', y) &= \\ \frac{1 + (-1)^{a+b}\cos(x+y)}{4}&. \end{aligned} \quad (7)$$

It is important is to notice that $P(r_A \in \text{“}a\text{”}, r_B \in \text{“}b\text{”}|x, y)$ does not depend on Alice's second measurement setting x', as one would expect. Similarly, if one traces out Alice's first measurement,

$$P(r_A \in \text{“}a'\text{”}, r_B \in \text{“}b\text{”}|x', y) = \frac{1 + (-1)^{a'+b}\cos(x'+y)}{4} \quad (8)$$

one recovers Equations (3) and (4). Again, the probability that Equation (8) does not depend on Alice's first measurement setting.

So far so good, but now comes the catch. If one traces out Bob's measurement, one obtains a probability distribution for Alice's particle's position that depends on Bob's setting y:

$$\begin{aligned} P(r_A \in \text{“}a\text{”}, r_A \in \text{“}a'\text{”}|x, x', y) &= \\ \sum_{b} P(r_A \in \text{“}a\text{”}, r'_A \in \text{“}a'\text{”}, r_B \in \text{“}b\text{”}|x, x', y) &= \\ \frac{1 + (-1)^{a+a'}\cos(x+y)\cos(x'+y)}{4}&. \end{aligned} \quad (9)$$

Hence, the joint probability distribution (Equation (6)) is signaling from Bob to Alice! Is this a problem for Bohmian mechanics? Probably not, as the Bohmian particles' positions are assumed to be "hidden". Actually, it is already well-known that they have to be hidden in order to avoid signaling in Bohmian mechanics. Some may find this feature unpleasant, as it implies that Bohmian particles are postulated to exist "only" to immediately add that they are ultimately not fully accessible, but this is not new.

Consequently, defining a joint probability for the measurement outcomes a, a', and b in the natural way,

$$\begin{aligned} P &\quad (a, a', b|x, x', y) \equiv \\ P &\quad (r_A \in \text{“}a\text{”}, r_A \in \text{“}a'\text{”}, r_B \in \text{“}b\text{”}|x, x', y) \end{aligned} \quad (10)$$

can be done mathematically but cannot have a physical meaning, as $P(a, a', b|x, x', y)$ would be signaling.

4. What Is Going on? Let's Add a Position Measurement

In summary, it is the identification in Equation (10) that confused the authors of [1,2] and led them to wrongly conclude that Bohmian mechanics cannot predict violations of Bell inequalities in experiments involving only position measurements. Note that the identification of Equation (10) follows from Assumption **H**, so Assumption **H** is wrong. Every introduction to Bohmian mechanics should emphasize this. Indeed, Assumption **H** is very intuitive and appealing, but wrong and confusing.

To elaborate on this, let's add an explicit position measurement after the first beam splitter on the Alice side. The fact is that, according to both standard quantum theory and Bohmian mechanics, this position measurement perturbs the quantum state (hence the pilot wave) in such a way that the second measurement, labeled x' on Figure 4, no longer shares the correlation (Equation (9)) with the first measurement, see [4–6].

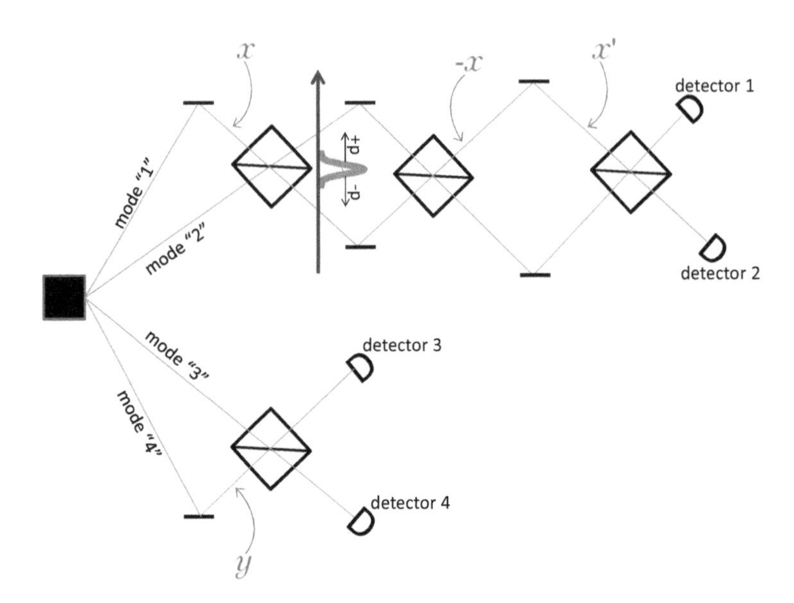

Figure 4. We add a pointer that measures through which path Alice's particle propagates between her first and second beam splitter. The pointer moves up if Alice's particle goes through the upper path, i.e., $r_A \in$ "1", and down if it goes through the lower path, i.e., $r_A \in$ "2". Hence, by finding out the pointer's position, one learns through which path Alice's particle goes, i.e., one finds out Alice's first measurement result, though it all depends how fast the pointer moves. See text for explanation.

Let's model Alice's first position measurement, labeled x (i.e., corresponding to the input phase x), by an extra system, called here the pointer, initially at rest in a Gaussian state, see Figure 4. One should think of the pointer as a large and massive system; note that it suffices to consider the state of the center of mass of the pointer. If Alice's particle passes through the upper part of the interferometer ($r_A \in$ "1"), then the pointer gets a kick in the upward direction and is left with a momentum $+p$; however, if Alice's particle passes through the lower part of her interferometer ($r_A \in$ "2"), then the pointer gets a kick $-p$. We made p large enough so that the two quantum states of the pointer $|\pm p\rangle$ are orthogonal, i.e., according to quantum theory, we consider a strong (projective) which-path measurement. Note, however, that, immediately after the pointer has interacted with Alice's particle, the two Gaussians corresponding to $|\pm p\rangle$ overlap in space, so no position measurement can distinguish them. It is only after some time that the two Gaussians separate in space and that position measurements can distinguish them. Since in Bohmian mechanics there are only position measurements, this implies that, in Bohmian mechanics, it takes some time for the pointer to measure Alice's particle.

Accordingly, if p is large enough for the pointer to have moved by more than its spread by the time Alice's particle hits the second BS, then the pointer acts like a standard measurement, and the second position measurement x' of Alice's particle is perturbed by measurement x, as discussed in the previous paragraph. However, if p is small enough, then, by the time the second measurement x' takes place, the pointer barely moves. In this case, the second position measurement is not affected [4–6], see also the appendix. However, it is now this second measurement x' that perturbs the "first" one, i.e., perturbs measurement x. Indeed, because of the entanglement between Alice's particle and the pointer, if one waits long enough for the pointer to move by more than its spread and then reads the result of the "first" measurement out of this pointer, then one will not find the expected result: the second measurement perturbed the "first" one. I put "first" in quotes because, in such a slow measurement, the result is actually read out of the pointer after the "second" measurement took place.

This is very similar to the so-called surrealistic trajectories, see [4–6]. In the appendix, I recall this counter-intuitive aspect of Bohmian mechanics.

5. What about Large Systems?

So far so good. But let's now consider, not single particles, but elephants. One of the advantages of Bohmian mechanics is that whether systems are microscopic and macroscopic makes no difference: all systems are treated alike. The price to pay, as we illustrate below, is that all the strangeness of quantum physics at the microscopic level has to show up also at the macroscopic level.

Let's consider two elephants in the state of Equation (1) corresponding to entangled elephants in Modes 1 & 4 superposed with elephants in Modes 2 & 3. Note that, instead of elephants, one may consider classical light pulses and replace, in Equation (1), the one-photon state $|1\rangle$ with a coherent state $|\alpha\rangle$ with mean photon number $|\alpha|^2$ as large as desired: $\big(|\alpha,0,0,\alpha\rangle + |0,\alpha,\alpha,0\rangle\big)/\sqrt{2}$. The beam splitters have to be replaced by EBSs—Elephant Beam Splitters—which split elephants: an incoming elephant emerges from an EBS in a superposition of elephant-transmitted and elephant-reflected. In the case of coherent states, the transformation reads:

$$|\alpha,0\rangle \rightarrow \big(|\alpha,0\rangle + i|0,\alpha\rangle\big)/\sqrt{2} \tag{11}$$

$$|0,\alpha\rangle \rightarrow \big(i|\alpha,0\rangle + |0,\alpha\rangle\big)/\sqrt{2}. \tag{12}$$

Note that the above deeply differs from the standard BS, which corresponds to $|\alpha,0\rangle \rightarrow |\alpha/\sqrt{2}, i\alpha/\sqrt{2}\rangle$.

The story of the single particles described above remains the same. In Bohmian mechanics, the elephants' positions are also hidden, or at least not fully accessible. However, this is puzzling, as it means that, when one "looks slowly" (as the pointer in Section 4, see also the Appendix) at an elephant, one may see it where it is not. Indeed, according to Bohmian mechanics, an elephant is where all the Bohmian positions of all the particles that make up the elephant are, but what does this mean if it does not correspond to where one sees the elephant? Bohmians may reply that one does not "look slowly" at elephants and that EBSs do not exist. This is certainly true of today's technology, but there will soon be beam splitters for quantum systems large enough to be seen by the naked eye. In addition, to avoid signaling, it has to be impossible to "see" or find out in any way two-time position correlations of such quantum systems, even when they are large.

Admittedly, it is an advantage that, in Bohmian mechanics, the difference between micro- and macro-worlds is immaterial. But, accordingly and unavoidably, quantum weirdness shows up at the macro-scale.

6. Assumption H Revisited

Assumption **H** is wrong. How should one reformulate it? Clearly, a position measurement does not *merely* reveal the Bohmian particle because of the following:

1. A position measurement necessarily involves the coupling to a large system, some sort of pointer, and this coupling implies some perturbation. Hence the "merely" in assumption **H** is wrong [7].
2. Whether a position measurement reveals information about the Bohmian particle or not depends on how the coupling to a large system is done and on how that large system (the pointer) evolves. Hence, not all measurements that, according to quantum theory, are position measurements,

are also Bohmian-position measurements: some quantum-position measurements do not reveal where the Bohmian particle is.

The first point above is very familiar to quantum physicists. However, it may take away some of the appeal of Bohmian mechanics. Indeed, the naive picture of particles with always well-defined positions is obscured by the fact that these positions cannot be "seen"—in fact, one can not "merely see" in which mode a Bohmian particle is. At the end of the day, Bohmian mechanics is not simpler than quantum theory. The promise of a continuously well-defined position and the associated intuition is deceptive.

The second point listed above is interesting: One should distinguish between *quantum-position* and *Bohmian-position* measurements. The latter refers to measurements that provide information about the position of Bohmian particles. It would be interesting to figure out how to characterize such Bohmian-position measurements without the need to fully compute all the Bohmian trajectories.

7. Why Bohmian Mechanics

From all we have seen so far, one should, first of all, recognize that Bohmian mechanics is deeply consistent and provides a nice and explicit existence proof of a deterministic non-local hidden variables model. Moreover, the ontology of Bohmian mechanics is pretty straightforward: the set of Bohmian positions is the real stuff. This is especially attractive to philosophers. Understandably so. But what about physicists mostly interested in research? What new physics did Bohmian mechanics teach us in the last 60 years? Here, I believe it is fair to answer: Not enough! Understandably disappointing.

It is deeply disappointing that an alternative theory to quantum mechanics, a theory that John Bell thought should be taught in parallel to standard textbook quantum mechanics [8], did not produce new physics, nor even inspirations for new ideas to be tested in the lab (though see [9–12]). How could this be? Some may conclude that not enough people worked on Bohmian mechanics. But tens or hundreds of passionate researchers worked on it for decades. Some may conclude that this lack of new ideas proves that Bohmian mechanics is a dead end. But how could a consistent theory, empirically equivalent to quantum theory, have no future?

Let me suggest some possible, albeit only partial, answers to the above puzzle. I am afraid that almost all the research on Bohmian mechanics over the last several decades remained trapped within an exceedingly narrow viewpoint and worked only on problems of interest that were highly specific to their Bohmian community. I believe this is especially disappointing, as there were several interesting open problems that Bohmian-inspired ideas could have addressed. The positive side is there are likely still interesting open problems that open-minded researchers can explore.

Let me illustrate some of the ideas I believe Bohmian mechanics should have triggered. This list is obviously subjective—it is only important that it is not empty. Bohmian mechanics, like quantum theory, is in deep tension with relativity theory. I know of Bohmians who claim that it is obvious that any non-local theory, Bohmian or not, requires a privileged universal reference frame. I also know of Bohmians who claim that it is obvious that Bohmian mechanics can be generalized to a relativistic theory (though, admittedly, I never understood their model). However, I know of no Bohmians who are inspired by their theory and its tension with relativity to try to go beyond Bohmian mechanics, as illustrated in the next two paragraphs.

According to Bohmian mechanics, particles "make decisions" at beam splitters in the sense that, after a beam splitter, the particle is definitively in one of the output modes. Admittedly, this is not a real decision as everything is determined by the initial state of the particle and of all other systems entangled with the particle. However, let me continue using this inspiring terminology. Accordingly, and following Suarez, we call such beam splitters *choice-devices* [13]. Such choice-devices take into account everything in their past. Now, a natural assumption inspired by the sketched description is that the past is not merely the past light cone, but all of the past in the inertial reference frame of the choice device. This idea led Suarze and Scarani to suggest that one should test situations in which several choice-devices, e.g., several beam splitters, are in relative motion such that what is the past for one choice-device may differ from the past of another choice-device [14]. This has the advantage (at least for researchers in physics) that it leads to experimental predictions that differ from standard quantum predictions and that can be experimentally tested. Hence, this brings Bohmian-inspired ideas to physics. This has been tested in my lab, and the result have shown that the idea, in spite of its appeal, is wrong [15].

Another Bohmian-inspired idea follows directly from an observation by Hiley and Bohm [16]: "it is quite possible that quantum nonlocal connections might be propagated, not at infinite speeds (as in standard Bohmian mechanics), but at speeds very much greater than that of light. In this case, we could expect observable deviations from the predictions of current quantum theory (e.g., by means of a kind of extension of the Aspect-type experiment)." Again, this can be experimentally tested [17–20]. The results put lower bounds on this hypothetical faster-than-light-but-finite speed influence, something like 10,000 to 100,000 times the speed of light. Aspect-type experiments between two sites can only either find that hypothetical speed or set lower bounds on it. However, recently we have been able to demonstrate that, by going to more parties, one can prove that either there is no such finite-but-superluminal speed or that one can use it for faster than light communication using only classical inputs and output (i.e., measurement settings and results) [21,22].

I am confident that Bohmian mechanics and other alternative views on quantum mechanics will inspire further ideas that will lead to experiments that might work to extend quantum theory. The real question is whether the Bohmian community will pursue such ideas.

8. Conclusions

Naive Bohmian mechanics that assumes Assumption **H** is wrong. Still, Bohmian mechanics is deeply consistent. Position measurements perturb the system, even in Bohmian mechanics. Hence, the existence of two-time position correlations is not in contradiction with possible violations of Bell inequalities.

Generally, position measurements sometimes reveal information about Bohmian positions, but never full information and sometimes none at all. Simple and handy criteria for determining when the Bohmian position measurements of a particle under test highly correlate with the position of the center of mass of some large pointer are still missing.

Bohmian mechanics is attractive to philosophers because it provides a clear ontology. However, it is not as attractive to researchers in physics. This is unfortunate because it could inspire brave new ideas that challenge quantum physics.

Acknowledgments: A preliminary version of this note was presented at a workshop at the ETh- Zurich in October 2014 organized by Gilles Brassard and Reneto Renner. There, participants drew my attention to reference [4]. The present version profited from comments by Sandu Popescu who drew my attention to [5] and by Michael Esfeld, Lucian Hardy, Franck Laloe, Tim Maudlin, and Howard Wiseman. Work was partially supported by the COST Action *Fundamental Problems in Quantum Physics* and my ERC-AG MEC.

Conflicts of Interest: The author declares no conflict of interest.

Appendix A. Slow Position-Measurements in Bohmian Mechanics

Appendix A.1. Bohmian Trajectories in a Semi-Interferometer

Let us consider a "half Mach-Zehnder" interferometer, which is a Mach-Zehnder interferometer in which the second beam splitter is removed, see Figure A1. We call such a circuit a semi-interferometer. After the beam splitter, any particle entering from the input mode is in a superposition of Modes 1 and 2: $|1\rangle \rightarrow \big(|1,0\rangle + |0,1\rangle\big)/\sqrt{2}$, with possibly a relative phase irrelevant in semi-interferometers.

According to quantum theory, if Detector 1 clicks, then the particle went through Mode 1. This should be interpreted as "if one adds position measurements in Modes 1 and 2, then there is a 100% correlation between Detector 1 and the position measurement in Mode 1 (and similarly for Detector 2 and the position measurement in Mode 2)".

According to Bohmian mechanics, things are different. If Detector 1 clicks, then the particle went through Mode 2, in sharp contrast to the quantum retro-diction. However, the interpretation is also totally different. According to Bohmian mechanics, particles follow continuous trajectories and the interpretation here is that, if Detector 1 clicks, then the Bohmian particle followed Mode 2.

In order to reconcile both views, let's add position measurements in Modes 1 and 2.

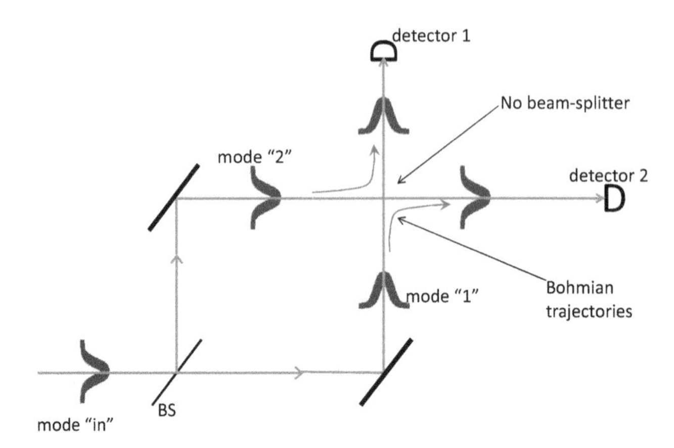

Figure A1. A Bohmian particle and its pilot wave arrive on a beam splitter (BS) from the left in Mode "in". The pilot wave emerges both in Modes 1 and 2, as the quantum state in standard quantum theory. Modes 1 and 2 meet again, but there is no beam splitter at this meeting point. Nevertheless, the Bohmian trajectories bounce at this point as indicated by the red arrows. Intuitively, this can be understood because the evolution equation of the Bohmian position is a first order differential equation, so Bohmian trajectories never cross each other. This intuition is confirmed by numerical simulations.

Appendix A.2. Position Measurements in Modes 1 and 2

In order to describe position measurements in both modes, we add two pointers, each initially at rest, denoted $|p_j = 0\rangle$, that we locally couple to Modes 1 and 2 in such a way that, if the particle is in Mode j, then the corresponding pointer gets a momentum kick k, resulting in state $|p_j = k\rangle$, while the other pointer is left unaffected, see Figure A2. The joint particle-pointers state after the two local interactions thus read $(|1,0\rangle|p_1 = k\rangle|p_2 = 0\rangle + |0,1\rangle|p_1 = 0\rangle|p_2 = k\rangle)/\sqrt{2}$. Note that the pointers are in some localized (e.g., Gaussian) states, the kets only indicate the mean momenta.

Let us emphasize that the pointer should be thought of as large and massive and consisting of many internal degrees of freedom; in short, it is a "macroscopic" object, and the result of the position measurement can merely be read of the pointer's position: if the pointer moves, then it has detected the presence of the particle; if the pointer hasn't moved, then the particle went the other way. Such a formalization of position measurements applies both to quantum and Bohmian theories.

Note that, in order for the pointer to indicate an unambiguous result, one has to wait long enough for the pointer to have moved by much more than it's spread Δx and the kick has to be large enough, $k >> \hbar/\Delta x$.

According to quantum theory, if Detector 1 clicks, then Pointer 1 got a kick and thus moves, while Pointer 2 rests in state $|p = 0\rangle$. However, the situation as described by Bohmian mehcanics is more interesting.

First, consider the case that the kick k is so large that the pointer, if kicked, moves by more than its spread before the two Modes 1 and 2 cross at the place of the "missing beam splitter". In this case, the particle and the kicked pointer become entangled, and this modifies the Bohmian trajectory of the particle. According to this modified trajectory and in full agreement with quantum predictions, if Detector 1 clicks, then Bohmian mechanics predicts that it is Pointer 1 that moved (including the

Bohmian position of pointer 1). Note that, in this situation, Bohmian trajectories can apparently cross each other, because the trajectory actually happens in a higher dimensional (configuration) space and it is only its shadow in our space that crosses.

Next, consider the case that the kick k is not that large and that, by the time Modes 1 and 2 cross, the pointer has barely moved. Bohmian mechanics predicts that, if Detector 1 clicks, the particle went through Mode 2 (as if there were no pointer); however, if one waits long enough for the pointer to eventually move by more than its spread, then one finds that it is Pointer 1 that moves. Accordingly, in the case of "slow pointers", the pointer indicates where the Bohmian particle was not. This is surprising, at least to physicists that are used to quantum theory. However, this is how Bohmian mechanics describes the situation, and one should add that there is nothing wrong with this description in the sense that all observable predictions are in agreement with quantum predictions [23,24].

Finally, we investigated numerically intermediate cases in which, at the time Modes 1 and 2 cross, the pointer moved but not much. We find that, in such cases, some trajectories of the particle bounce in the region of mode crossing, while other trajectories go through the crossing region more or less in straight lines. Accordingly, conditioned on Detector 1 clicking, there is a chance that the particle went through Mode 1 and a complementary chance that it went through Mode 2, depending on the precise value of the kick k and the exact initial position of the Bohmian particle.

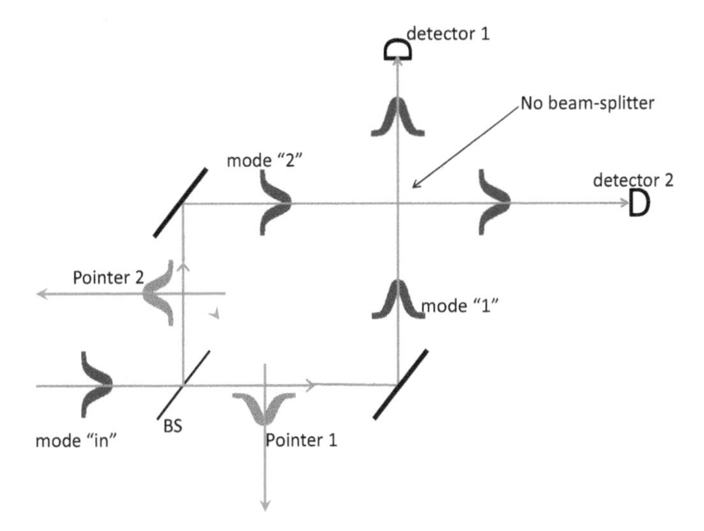

Figure A2. Semi-interferometer with two macroscopic pointers locally coupled to Modes 1 and 2. The pointers are initially at rest, $|p_j = 0\rangle$, but when detecting a particle they get a kick and end in a quantum state with momentum k: $|p_j = k\rangle$.

Note that one of the two detectors moves fast to prevent the Bohmian trajectories from bouncing in the crossing region. In fact, there is only one detector; if the kick is received by this single detector, the Bohmian trajectories bounce.

Finally, note that, for pointers composed of many internal degrees of freedom, one of the pointer's particle motion depends on whether the test particle is present or not for the counter-intuitive surreal Bohmian trajectories to disappear. However, if only the center of mass is coupled to the test particle, then there is no reason for any of the particles making up the pointer to move differently, regardless of whether or not the test particle is present.

Appendix A.3. Conclusions

What is a position measurement? Quantum theory has a clear answer to this question. However, in Bohmian mechanics, there are two possible definitions. First, the natural one: a *Bohmian position* measurement is any interaction between the particle under test and a macroscopic device (e.g., a pointer) that fully correlates the Bohmian position of the particle immediately before the interaction took place with the final state of the device (e.g., the final position of the pointer). Next, a quantum-inspired one: Anything that is a position measurement according to quantum theory (i.e., represented by the position operator q or a function of it) is also a *quantum-position* measurement in Bohmian mechanics.

Hence, in Bohmian mechanics, one should distinguish between Bohmian-position measurements and quantum-position measurements. In most situations, both types of position measurements coincide. However, there are cases, such as the slow pointer described in this note, where the two starkly differ.

When one says that Bohmian mechanics makes the same predictions as quantum theory as long as all measurements, at the end of the day, reduce to position measurements, one refers to quantum-position measurements. This may differ from Bohmian-position measurements, so Bohmian trajectories may differ from quantum expectations. This is surprising to quantum physicists, but one should emphasize that there is nothing wrong with that: different theories lead to different pictures of reality.

References

1. Correggi, M.; Morchio, G. Quantum mechanics and stochastic mechanics for compatible observables at different times. *Ann. Phys.* **2002**, *296*, 371–389.
2. Kiukas, J.; Werner, R.F. Maximal violation of Bell inequalities by position measurements. *arXiv* **2010**, arXiv:quant-ph/0912.3740.
3. Brunner, N.; Cavalcanti, D.; Pironio, S.; Scarani, V.; Wehner, S. Bell nonlocality. *Rev. Mod. Phys.* **2014**, *86*, 419.
4. Englert, B.-G.; Scully, M.O.; Süssmann, G.; Walther, H. Surrealistic Bohm Trajectories. *Z. Naturforschung A* **1992**, *47*, 1175–1186.
5. Vaidman, L. The reality in Bohmian quantum mechanics or can you kill with an empty wave bullet? *Found. Phys.* **2005**, *35*, 299–312.
6. Dewdney, C.; Hardy, L.; Squires, E.J. How late measurements of quantum trajectories can fool a detector. *Phys. Lett. A* **1993**, *184*, 6–11.
7. Maudlin, T. New York University, New York, NY, USA. Personal communication, 2016.
8. Bell, J.S. On the impossible pilot wave. *Found. Phys.* **1982**, *12*, 989–999.
9. Valentini, A. Hidden Variables and the Large-Scale Structure of Spacetime. *arXiv* **2005**, arXiv:quant-ph/0504011.
10. Mahler, D.H.; Rozema, L.; Fisher, K.; Vermeyden, L.; Resch, K.J.; Wiseman, H.M.; Steinberg, A. Experimental nonlocal and surreal Bohmian trajectories. *Sci. Adv.* **2016**, *2*, e1501466.
11. Dürr, D.; Goldstein, S.; Zanghí, N. Quantum equilibrium and the origin of absolute uncertainty. *J. Stat. Phys.* **1992**, *67*, 843–907.
12. Goldstein, S.; Zanghi, N. Reality and the Role of the Wave Function in Quantum Theory. In *The Wave Function: Essays in the Metaphysics of Quantum Mechanics*; Albert, D., Ney, A., Eds.; Oxford University Press: Oxford, UK, 2012.
13. Suarez, A. Relativistic nonlocality in an experiment with 2 *non-before* impacts. *Phys. Lett. A* **1997**, *236*, 383–390.
14. Suarez, A.; Scarani, V. Does entanglement depend on the timing of the impacts at the beam splitters? *Phys. Lett. A* **1997**, *232*, 9–14.
15. Stefanov, A.; Zbinden, H.; Gisin, N.; Suarez, A. Quantum correlations with spacelike separated beam splitters in motion: Experimental test of multisimultaneity. *Phys. Rev. Lett.* **2002**, *88*, 120404.
16. Bohm, D.; Hiley, B.J. *The Undivided Universe*; paperback ed.; Routledge: London, UK, 1993; p. 347.
17. Scarani, V.; Tittel, W.; Zbinden, H.; Gisin, N. The speed of quantum information and the preferred frame: Analysis of experimental data. *Phys. Lett. A* **2000**, *276*, 1–7.

18. Salart, D.; Baas, A.; Branciard, C.; Gisin, N.; Zbinden, H. Testing spooky action at a distance. *Nature* **2008**, *454*, 861–864.
19. Cocciaro, B.; Faetti, S.; Fronzoni, L. A lower bound for the velocity of quantum communications in the preferred frame. *Phys. Lett. A* **2011**, *375*, 379–384.
20. Yin, J.; Cao, Y.; Yong, H.-L.; Ren, J.-G.; Liang, H.; Liao, S.-K.; Zhou, F.; Liu, C.; Wu, Y.-P.; Pan, G.-S.; et al. Lower Bound on the Speed of Nonlocal Correlations without Locality and Measurement Choice Loopholes. *Phys. Rev. Lett.* **2013**, *110*, 260407.
21. Bancal, J.-D.; Pironio, S.; Acin, A.; Liang, Y.-C.; Scarani, V.; Gisin, N. Quantum non-locality based on finite-speed causal influences leads to superluminal signaling. *Nat. Phys.* **2012**, *8*, 867–870.
22. Barnea, T.; Bancal, J.D.; Liang, Y.C.; Gisin, N. Tripartite quantum state violating the hidden-influence constraints. *Phys. Rev. A* **2013**, *88*, 022123.
23. Lam, M.M.; Dewdney, C. Locality and nonlocality in correlated two-particle interferometry. *Phys. Lett. A* **1990**, *150*, 127–135.
24. Guay, E.; Marchildon, L. Two-particle interference in standard and Bohmian quantum mechanics. *J. Phys. A* **2003**, *36*, 5671–5624.

Article

Observables and Unobservables in Quantum Mechanics: How the No-Hidden-Variables Theorems Support the Bohmian Particle Ontology

Dustin Lazarovici *, Andrea Oldofredi * and Michael Esfeld *

Section de Philosophie, Université de Lausanne, 1015 Lausanne, Switzerland
* Correspondence: dustin.lazarovici@unil.ch (D.L.); andrea.oldofredi@unil.ch (A.O.); michael-andreas.esfeld@unil.ch (M.E.)

Received: 23 April 2018; Accepted: 17 May 2018; Published: 18 May 2018

Abstract: The paper argues that far from challenging—or even refuting—Bohm's quantum theory, the no-hidden-variables theorems in fact support the Bohmian ontology for quantum mechanics. The reason is that (i) all measurements come down to position measurements; and (ii) Bohm's theory provides a clear and coherent explanation of the measurement outcome statistics based on an ontology of particle positions, a law for their evolution and a probability measure linked with that law. What the no-hidden-variables theorems teach us is that (i) one cannot infer the properties that the physical systems possess from observables; and that (ii) measurements, being an interaction like other interactions, change the state of the measured system.

Keywords: no-hidden-variables theorems; observables; measurement problem; Bohmian mechanics; primitive ontology

---

1. Introduction

The famous no-hidden-variables theorems have played a crucial, though often questionable role in the history of quantum mechanics. For decades, they have been employed to defend the quantum orthodoxy and to argue, nay prove, that any attempt to go beyond the statistical formalism of standard quantum mechanics in providing a "complete" description of the microcosm is bound to fail. Even after David Bohm [1] got "the impossible done" (as Bell [2] (p. 160) later put it) and showed how the statistical predictions of quantum mechanics can be derived from an ontology of point particles and a deterministic law of motion, many scientists and philosophers refused to pay attention to this theory on the basis that the no-hidden-variables theorems had established that it couldn't be correct (one striking example of such a misunderstanding is Wigner [3] (pp. 53–55)).

Of course, Bohm's theory is not a counterexample to these theorems qua mathematical theorems. It is rather the most striking demonstration of the fact that these mathematical results do not support the ideological conclusions in defense of which they have been generally cited. That notwithstanding, it would be premature to dismiss the "no-go theorems" as physically and philosophically irrelevant. They capture something not only about the nature of measurements and the statistical predictions of quantum mechanics that strikes us as remarkable and contrary to classical intuitions, but also about the nature of physical objects. The aim of this paper is to work out what exactly these theorems show and how they support in fact Bohm's quantum theory, instead of being an argument against it.

In the next section, we briefly recall the quantum orthodoxy and Bohm's quantum theory. Section 3 outlines three of the most important theorems useful for our discussion. Section 4 rebuts the conclusions that are commonly drawn from them. Section 5 provides an account of the Bohmian theory of measurements. Section 6 shows how it supports an ontology of point particles that are characterized by their positions only. Section 7 draws a general conclusion.

2. Quantum Orthodoxy and Bohmian Mechanics

In the words of David Mermin [4] (p. 803), the scope of the no-hidden-variables theorems is to defend "a fundamental quantum doctrine", namely that

> (Q) A measurement does not, in general, reveal a preexisting value of the measured property.

However, accepting this doctrine leads to at least two urgent questions:

1. *How do the quantum observables acquire definite values upon measurement?*
   It is now generally acknowledged that measurements are not a new type of interaction—let alone a primitive metaphysical concept—that requires a special treatment, but come under the common types of physical interactions (electromagnetism, gravitation, etc.). Hence, our physical theories should be able, at least in principle, to describe them. This, in turn, entails that the notion of measurement must not be part of the axioms of a physical theory. Thus, if quantum theory implies that the observable values are not merely revealed but produced by the measurement process—that is, by the interaction between the measurement device and the measured system, the theory should tell us how they are produced.
2. *What characterizes a physical system prior to—or better: independent of—measurement?*
   After all, there must be some sort of ontological underpinning to the measurement process and the empirical data that it yields. That is, there must be *something* in the world on which the measurement is actually performed—something with which the observer or measurement device interacts, and there must be something definite about the physical state of the observer or measurement device that does not, in turn, require a measurement of the measurement (and so on, *ad infinitum*).

According to Mermin [4] (p. 803), the orthodox response to question 1 is that "Precisely how the particular result of an individual measurement is brought into being—Heisenberg's 'transition from the possible to the actual'—is inherently unknowable". The response to 2 seems to be some sort of radical idealism, expressed in his now famous assertion (and belated response to Einstein) according to which the moon is demonstrably not there when nobody looks (Mermin [5]). Bohm's theory entirely rejects this way of talking. Its presentation as a "hidden variables theory" suggests that it denies the doctrine Q. However, most contemporary Bohmians actually endorse this doctrine, and quite emphatically so. Let us briefly recall why this is the case.

For present purposes, we use the formulation of Bohm's theory that is today known as Bohmian mechanics (see Dürr et al. [6]; for a discussion of the different contemporary formulations of Bohm's theory, see Belousek [7]; Bohm and Hiley [8] is the latest elaborate treatment by Bohm himself). Bohmian mechanics can be defined in terms of the following four principles:

1. *Particle configuration*: There always is a configuration of N permanent point particles in the universe that are characterized only by their positions $X_1, \ldots, X_N$ in three-dimensional, physical space at any time t.
2. *Guiding equation*: A wave function Ψ is attributed to the particle configuration, being the central dynamical parameter for its evolution. On the fundamental level, Ψ is the *universal* wave function attributed to all the particles in the universe together. The wave function has the task to determine a velocity field along which the particles move, given their positions. It accomplishes this task by figuring in the law of motion of the particles, which is known as the guiding equation:

$$\frac{\mathrm{d}X_k}{\mathrm{d}t} = \frac{\hbar}{m_k} \mathrm{Im} \frac{\nabla_k \Psi}{\Psi}(X_1, \ldots, X_N). \tag{1}$$

   This equation yields the evolution of the k-th particle at a time t as depending on, via the wave function, the position of all the other particles at that time.

3. *Schrödinger equation*: The wave function always evolves according to the Schrödinger equation:

$$i\hbar\frac{\partial\Psi}{\partial t} = -\sum_{k=1}^{N}\frac{\hbar^2}{2m_k}\Delta_k\Psi + V\Psi. \tag{2}$$

4. *Typicality measure*: On the basis of the universal wave function Ψ, a unique stationary (more precisely: equivariant) typicality measure can be defined in terms of the $|\Psi|^2$–density (see Goldstein and Struyve [9] for a proof and precise statement of the uniqueness result). Given that typicality measure, it can then be shown that for nearly all initial conditions, the distribution of particle configurations in an ensemble of sub-systems of the universe that admit of a wave function ψ of their own (known as effective wave function) is a $|\psi|^2$–distribution. A universe in which this distribution of the particles in sub-configurations obtains is considered to be in quantum equilibrium.

Assuming that the actual universe is a typical Bohmian universe in that it is in quantum equilibrium, one can hence deduce Born's rule for the calculation of measurement outcome statistics on sub-systems of the universe in Bohmian mechanics (instead of simply stipulating that rule). In a nutshell, the axiom of $|\Psi|^2$ providing a typicality measure with Ψ being the universal wave function justifies applying the $|\psi|^2$–rule for the calculation of the probabilities of measurement outcomes on particular sub-systems within the universe, with ψ being the effective wave function of the particular systems in question (see Dürr et al. [6] (Chapter 2); cf. Section 5 for the notion of effective wave functions). Thus, the quantum probabilities have in Bohmian mechanics exactly the same status as the probabilities in classical statistical mechanics: they are derived from a deterministic law of motion via an appropriate probability measure that is linked with the law. Moreover, if a sub-system admits for an autonomous description in terms of an effective wave ψ, its complete physical and dynamical state at any time t is given by the pair (X_t, ψ_t), where $X_t = (X_1(t), \ldots, X_M(t))$ describes the actual spatial configuration of the system.

Consequently, measurements of observables such as energy, angular momentum, spin, etc. do not reveal predetermined properties of the particles, because Bohmian mechanics does not admit them as intrinsic properties of the particles to begin with. Similarly, a simple analysis of the theory shows that a measurement of the momentum observable does not, except under special circumstances, measure the instantaneous velocity of a particle. It is a crucial feature of the theory that the only property of the particles is their position in space. The particles have a velocity, of course, but velocity is nothing else than the change of position in time. The Bohmian velocity is not an observable (see Dürr et al. [6] (Chapter 3.7.2) for a simple proof, but also Wiseman [10] for the possibility of weak measurements; see [6] (Chapter 7) for a good discussion of both results). And velocity is not—in contrast to the Newtonian case—a dynamical degree of freedom that can be specified independently of the position, because the guiding law (1) is a first order differential equation, requiring only positions as initial data.

The first and foremost role of the wave function is a dynamical one, namely to yield the motion of the particles as output, given their positions as input. This explains the name "pilot-wave theory" historically given to Bohm's theory, as if the particles were literally guided or piloted by a wave in physical space. This way of speaking, however, cannot be taken literally, since the wave function is defined on configuration space; it is not a wave propagating in physical space (for the debate about the status of the wave function in Bohmian mechanics, see [11]). Even in the special case when the wave function of a subsystem happens to be an eigenstate ψ_α of a certain observable $\hat{A}$ with eigenvalue α—for instance after an ideal measurement—and it would be safe to say that "the particle possesses a definite value of A", this way of speaking is unwarranted. It should be replaced by the statement that a (repeated) experiment, whose statistics is encoded in the operator $\hat{A}$, would yield the outcome α with certainty; or simply by the statement that the effective wave function, guiding the motion of the system, is ψ_α. In summary, the validity of doctrine Q is not denied, but substantiated by Bohmian mechanics on the basis of this theory recognizing only a position as a property of the physical system.

3. No-Hidden-Variables Theorems

The basic question that the no-hidden-variables theorems set out to address is whether the probabilistic nature of the quantum formalism allows for an ignorance interpretation in the sense that the measured values of quantum observables are in fact predetermined by additional parameters, whose actual values, in individual runs of an experiment, are unknown to us, but whose statistical distributions over a series of measurements reproduce the observed outcome statistics. In more formal terms, the question is whether for any relevant family of quantum observables $\hat{A}, \hat{B}, \hat{C}, \ldots$ there exists a corresponding family of random variables $Z_A, Z_B, Z_C, \ldots$ on a common probability space Ω such that the values of these random variables correspond to the possible measurement outcomes—that is, the eigenvalues of the observable operators. Any $\omega \in \Omega$ would then be a value of the hypothetical hidden variable(s), determining the measurement values $Z_A(\omega), Z_B(\omega), Z_C(\omega), \ldots$, and the quantum predictions, for some quantum state ψ, would be reproduced by a probability distribution μ_ψ over this hidden variable, such as $\langle\psi|\hat{A}|\psi\rangle = \int_\Omega Z_A(\omega) \mathrm{d}\mu_\psi(\omega)$, etc.

A no-hidden-variables theorem is thus, in general, a result of the following form (cf. Dürr et al. [6] (Chapter 3)):

> There is no "good" map $\hat{A} \mapsto Z_A$ from the set of self-adjoint operators on a Hilbert space $\mathcal{H}$ to random variables on a common probability space Ω such that the possible values of Z_A correspond to the eigenvalues of $\hat{A}$ (that is, the possible measurement values).

The term "good map" is not quite precise, but deliberately so, for it is essentially on this point—the requirements on the assignment $\hat{A} \mapsto Z_A$—that the various no-hidden-variables theorems differ.

3.1. Von Neumann

The first no-hidden-variables theorem was proven by von Neumann in his seminal 1932 book *Mathematische Grundlagen der Quantenmechanik* ([12], English translation [13]). In this theorem, a "good" map from observables to random variables was supposed to be linear, that is, in particular:

$$\hat{A} + \hat{B} \mapsto Z_{A+B} = Z_A + Z_B. \tag{3}$$

It is easy to see that such a map cannot exist, since, for non-commuting operators, the eigenvalues of their sum are in general not sums of their eigenvalues. Von Neumann's linearity assumption was arguably motivated by the additivity of quantum mechanical expectations values ($\langle\psi|\hat{A} + \hat{B}|\psi\rangle = \langle\psi|\hat{A}|\psi\rangle + \langle\psi|\hat{B}|\psi\rangle$ holds for all observables $\hat{A}, \hat{B}$ and any state ψ), but is nowadays considered as rather naive (Mermin [4] (pp. 805–806) calls it "silly"). As Mermin [4] (p. 806) points out, requiring Equation (3) "is to ensure that a relation holds in the mean by imposing it case by case—a sufficient, but hardly a necessary condition". In addition, the physical significance of this assumption—in particular for non-commuting observables that cannot even simultaneously measured—is rather obscure. If, let us say, $\hat{X}$ is the position and $\hat{P}$ the momentum observable, what is a "measurement of $\hat{X} + \hat{P}$" even supposed to mean? For decades, von Neumann's impossibility proof was a key element in the defense of the quantum orthodoxy, but it started to fall apart rather quickly, once people began to study it more systematically.

3.2. Kochen–Specker

The theorem of Kochen and Specker [14] was a considerable improvement because it makes a requirement for the "goodness" of the assignment $\hat{A} \mapsto Z_A$ that seems a priori much more plausible:

(NC) Whenever the quantum mechanical joint distribution of a set of self-adjoint operators $(A_1, \ldots, A_m)$ exists, that is, when they form a commuting family, the joint distribution of the corresponding set of random variables, that is, of $(Z_{A_1}, \ldots, Z_{A_m})$, must agree with the quantum mechanical joint distribution.

This assumption actually implies that all algebraic identities that hold between the observable operators must also hold between the random variables (e.g., if $\hat{A} \times \hat{B} = \hat{B} \times \hat{A} = \hat{C}$, it means that the joint distribution is zero on the value set $\{(c \neq ab) \mid a, b, c \text{ eigenvalues of } \hat{A}, \hat{B}, \hat{C}\}$ and hence $Z_A \times Z_B = Z_C$ almost surely), but the condition is now only imposed on families of commuting observables that can be jointly measured.

Families of commuting observables always have a common probability distribution (as random variables on a classical probability space). So what could possibly go wrong? One can consider an observable $\hat{A}$ once as part of a commuting family $(\hat{A}, \hat{B}, \hat{C}, \ldots)$ and once as part of a commuting family $(\hat{A}, \hat{L}, \hat{M}, \ldots)$ such that $\hat{B}, \hat{C}, \ldots$ and $\hat{L}, \hat{M}, \ldots$ are incompatible—that is, non-commuting—with each other. Assumption (NC) would be trivial if the observable $\hat{A}$ could be associated with a random variable Z_A, as part of the family $(Z_A, Z_B, Z_C, \ldots)$, and another random variable $\tilde{Z}_A$ as part of the family $(\tilde{Z}_A, Z_L, Z_M, \ldots)$. The considered hidden-variables-schemes presuppose, however, a rigid assignment $\hat{A} \mapsto Z_A$, independent of the measurement context. In other words, Z_A must be the same, whether $\hat{A}$ is measured together with $\hat{B}, \hat{C}, \ldots$ or together with $\hat{L}, \hat{M}, \ldots$. The crucial assumption underlying the no-go theorem of Kochen and Specker has thus been named *non-contextuality*. The upshot is that non-contextual hidden variables are incompatible with the predictions of quantum mechanics.

A particularly nice and simple proof is due to Mermin [4] (p. 810). It consists in the following arrangement of 3×3 observables on a four-dimensional Hilbert space:

$$\begin{array}{ccc} \sigma_x^1 & \sigma_x^2 & \sigma_x^1\sigma_x^2, \\ \sigma_y^2 & \sigma_y^1 & \sigma_y^1\sigma_y^2, \\ \sigma_x^1\sigma_y^2 & \sigma_x^2\sigma_y^1 & \sigma_z^1\sigma_z^2. \end{array}$$

Using the standard commutation relations of the Pauli-matrices ("spin observables") and the fact that the possible eigenvalues are ± 1, it is easy to verify that:

(a) The observables in each of the three rows and each of the three columns are mutually commuting.
(b) The product of the three observables in each of the three rows is 1.
(c) The product of the three observables in first two columns is 1, while the product of the right column is -1.

Thus, no consistent assignment of predetermined values to the nine observables is possible, since identity (b) would require the product of all nine values to be $+1$, while (c) would require it to be -1. This proves the Kochen–Specker theorem.

3.3. Bell

One of the more tragic chapters in the history of quantum mechanics is that, for many defenders of the supposed orthodoxy, Bell's theorem (reprinted in Bell [2] (Chapter 2)) has replaced von Neumann's as the mathematical result that finally spells the dead for any "completion" of the quantum formalism. Certainly, the physical significance of Bell's theorem can hardly be overstated, but to understand it as just another no-hidden-variables argument is to miss the point entirely. Bell himself has addressed the misunderstanding on various occasions, for instance:

> My own first paper on this subject (*Physics* 1, 195 (1965)) starts with a summary of the EPR argument *from locality to* deterministic hidden variables. However the commentators have almost universally reported that it begins with deterministic hidden variables. (Bell [2] (p. 157))

The point of Bell's theorem is not hidden variables but *nonlocality* (see Maudlin [15] for an excellent discussion). Bell's analysis starts from the EPR argument that assumes locality and concludes that the quantum formalism must be incomplete. EPR did indeed attack the quantum doctrine that observables do not have predetermined values prior to measurement. In brief, they did so by noticing that, when considering two entangled systems A and B, certain observable values of A can be determined by measurements on the distant system B (and vice versa). However, this would presuppose some sort of nonlocal influence *unless* these values were actually predetermined, prior to the measurement on the distant system, by hidden variables (and thus only *revealed* rather than *determined* by our interaction with the distant system).

Three decades later, Bell proved that even by introducing additional variables, the statistical predictions of quantum mechanics cannot be reproduced without nonlocal influences. The conclusion is thus that quantum mechanics is nonlocal, no matter what. In addition, since a substantial amount of experimental evidence confirms the predictions of quantum mechanics, the conclusion is that any correct theory of nature is nonlocal, no matter what. Nonlocality, in other words, is not the price that we pay for introducing hidden variables. Hidden variables were Einstein's hope for avoiding the nonlocality of standard quantum mechanics, and Bell proved that this hope cannot be realized because nonlocality is a fact of nature. Hence, using nonlocality as an argument against Bohmian mechanics, or so-called "hidden variables theories" in general, gets the issue completely wrong. Quantum mechanics is nonlocal, and any extension of—or alternative to—quantum mechanics better be nonlocal as well; otherwise, it is demonstrably wrong.

4. The Message of the Quantum

Thus, what is the upshot of the no-hidden-variables theorems? In this section, we consider some common responses and briefly indicate why they are wrong-headed.

4.1. *Completeness of Quantum Mechanics*

The no-hidden-variables theorems are usually cited in support of the claim that standard quantum mechanics is "complete", that is, in particular, that the wave function or quantum state —with its role in determining the probabilities of measurement outcomes—represents the complete physical state of a quantum system. However, when used in this context, the traditional hidden variables program seems to commit the following mistake that Einstein warned the young Heisenberg about:

> I suspect that you will run into problems at exactly that part of your theory that we just talked about ... You pretend that you could leave everything as it is on the side of observations, that is, that you could just talk in the former language about what physicists observe. (Quoted after Heisenberg [16] (p. 89); translation by the authors.)

Indeed, the idea that physical observations must be reported in "classical language" (while the same language is unable to provide an objective description of the microcosm) became one of the core tenants of the so-called Copenhagen interpretation. This included the (at least tacit) assumption that the relevant observables of quantum physics are just the familiar properties known from Newtonian mechanics, or at least that the physical and ontological status of the properties, once measured, is the same as had been generally assumed in classical physics, namely that the observables refer to intrinsic properties of the physical systems. The no-hidden-variables theorems then show that the intrinsic properties of physical systems, insofar as they are captured by observables, cannot have predetermined values (unless one buys into undesirable consequences such as "contextuality" that seem to defeat the purpose of assuming predetermined properties).

However, this orthodox way of reading the no-hidden-variables theorems directly runs into the two questions raised in Section 2: How do the quantum observables acquire definite values upon measurement? What characterizes a physical system prior to—or better: independent of—measurement? The "industry of no-go theorems" (Laudisa [17]) drives us towards the negative conclusion of no predetermined values, but it does not provide an answer to these questions. Instead of this reading of no predetermined values of intrinsic properties of physical systems, there also is another, arguably more radical reading of the no-hidden-variables theorems possible: they tell us that observables do not correspond to properties of physical systems at all, so that the question of predetermined values of such properties does not even arise. This is the Bohmian reading, which then does provide an answer to these questions.

4.2. *Metaphysical Indeterminacy*

Following the lead of mainstream physics, the philosophical literature has recently developed a renewed interest in the concept of metaphysical indeterminacy, which is intended in this context to capture the idea that the values of quantum observables, prior to measurement, are not merely unknown but, in a metaphysically robust sense, unspecified. According to Calosi and Wilson [18], properties of quantum systems are metaphysically indeterminate in the sense that they have a determinable property without a unique corresponding determinate. Thus, an electron, for instance, possesses a determinable property "spin", but its value is indeterminate until we actually measure it.

In contrast, and arguing against the concept of metaphysical indeterminacy, Glick [19] (p. 207) proposes what he calls a "sparse view" of standard quantum mechanics:

> Sparse view: when the quantum state of A is not in an eigenstate of $\hat{O}$, it lacks *both* the determinate and determinable properties associated with $\hat{O}$.

Obviously, none of these views does anything to address the measurement problem, that is, to clarify how a measurement turns an indeterminate—or non-existent—property of a physical system into a determinate one. In addition, while this is not the issue of this particular philosophical discussion, it certainly is dubious to base metaphysics on imprecise or even inconsistent physics.

Bohmian mechanics, by contrast, shows that there is no work to do for a concept of metaphysical indeterminacy: the state of a physical system is completely and precisely determined, at any moment in time, by the actual particle positions and the wave function, fixing how the positions change in time. Furthermore, this theory supports a metaphysical view that is even sparser than the one advocated by Glick: neither the determinate nor the determinable property associated with an observable $\hat{O}$ is part of the ontology, independent of whether or not the quantum state of a system is an eigenstate or not. The only property that particles have—and need—is a position in physical space (see Esfeld and Deckert [20] for an elaboration on a sparse ontology in that sense).

4.3. *Quantum Logic*

One of the more audacious claims in support of which the no-hidden-variables theorems are employed is that quantum mechanics compels us to give up classical logic in favor of a new quantum logic. It is easy to see where this idea comes from. If we consider the simple example of spin (for a spin-1/2-particle, to be discussed in detail in Section 5), it is tempting to assign to the proposition

$q \vee \neg q$: The particle has z-spin up or z-spin down

the truth-value *true*. However, according to the doctrine Q, neither

q: The particle has z-spin up

nor

$\neg q$: The particle has z-spin down

can be considered *true* prior to a measurement or unless the particle happens to be in a z-spin eigenstate.

Since Quine's seminal paper "Two dogmas of empiricism" [21], it is widely accepted in philosophy that not even a revision of the rules of logic is out of bounds when adjusting a theoretical system to new empirical evidence, though they are the last knob to turn. In that vein, the first and most important objection to quantum logic (as a proposal for the "true" logic of the physical world) is not that it is a priori absurd, but that it is hardly justified by theoretical or explanatory merits. Giving up on classical logic does nothing to address the two crucial questions formulated in Section 2. The various proposed systems of "quantum logic" are merely modeled on the standard theory and thus inherit all its problems—including the measurement problem. In particular, changing a logical formalism does not elucidate the ontology of quantum mechanics, nor does it provide for a physical account of when and why propositions involving quantum observables acquire definitive truth values. Conversely, the example of Bohmian mechanics shows that once we have a clear ontology, and take the measurement process seriously as part of the theory, no departure from classical logic is called for.

5. Measurements in Bohmian Mechanics: Spin

In this section, we explain how Bohmian mechanics treats measurement experiments, how this treatment supports doctrine Q and what the consequences for the status of observables are.

5.1. The Bohmian Treatment of the Measurement Process

The solution to the measurement problem offered by Bohmian mechanics comes from a simple idea: to describe quantum mechanically also the experimental devices, since macroscopic objects are composed of microscopic objects. Thus, to describe experimental situations in Bohmian mechanics, we split the total configuration (of, in the last resort, the entire universe) into $(X,Y) \in \mathbb{R}^{3M} \times \mathbb{R}^{3(N-M)}$ where the former variable refers to the particle configuration of the investigated M-particle sub-system and the latter to the configuration of the environment, which includes the particles of the measurement device registering the outcomes in "pointer positions". Fundamentally, in the Bohmian theory, there is only one wave function, the universal wave function $\Psi = \Psi(x,y)$, guiding all the particles together (the lower case variables refer to the possible configurations—Ψ is a function on the entire configuration space—in contrast to the actual configurations denoted by upper case letters). However, by inserting the actual configuration of the environment at time t, we get a *conditional wave function*, which is a function of the degrees of freedom of the sub-system only:

$$\psi_t(x) := \Psi_t(x, Y_t). \tag{4}$$

This conditional wave function is always well-defined but not very useful in practice, since it has a non-trivial dependence on the exact configuration of the environment. However, in some situations, when the universal wave function takes the form

$$\Psi(x,y) = \psi(x)\Phi(y) + \Psi^{\perp}(x,y), \tag{5}$$

where $\Phi(y)$ and $\Psi^{\perp}(x,y)$ have macroscopically disjoint support in the y-variables and $Y_t \in \operatorname{supp}\Phi$, i.e., $\Psi^{\perp}(x,Y_t) = 0 \ \forall x$, we can for all practical purposes forget about the "empty" wave $\Psi^{\perp}$ and provide an autonomous description of the subsystem in terms of the *effective wave function* ψ, which is the Bohmian analog to the usual wave function used in textbook quantum mechanics. Now, let us consider an ideal measurement associated with an "observable" with eigenvalues $\alpha_1, \dots, \alpha_n$ and corresponding eigenstates $\varphi_1, \dots, \varphi_n$. In general, ψ will be a superposition $\psi = \sum_{i=1}^{n} c_i\varphi_i$, $c_i \in \mathbb{C}$. Under the Schrödinger evolution—after the subsystem has coupled to the measurement device in the course of the measurement process—the state of system + environment (ignoring again the empty part of the wave function $\Psi^{\perp}$) will thus have the form

$$\Psi(x,y) = \sum_{i=1}^{n} c_i \varphi_i(x) \Phi_i(y), \tag{6}$$

where the environment states Φ_i are concentrated, in particular, on different pointer configurations, indicating the measurement outcomes α_i, and have therefore pairwise disjoint supports in configuration space. Note that it is only for simplicity that we do not distinguish between the degrees of freedom of the measurement device and the rest of the universe, subsuming both in the "environment" (y-system). However, the *actual* configuration Y of the universe (pointer) will lie inside only one of the branches, let us say $Y \in \operatorname{supp} \Phi_k$. Hence, the actual pointer configuration will indicate the measurement outcome α_k and the new effective (=conditional) wave function of the subsystem becomes $\psi^Y(x) = c_k \varphi_k(x) \Phi_k(Y) \triangleq \varphi_k$ after normalization. Hence, while the universal wave function always evolves according to the linear Schrödinger equation, the effective wave function automatically collapses into the eigenstate corresponding to the registered measurement result (for a detailed exposition see Dürr and Teufel [22] (Chapter 9)).

This account notably has the following five features:

1. There never are superpositions of anything in physical space. All there is in physical space are particle configurations with always definite positions. Thus, Schrödinger's cat always is in a configuration of either a live cat or a dead cat. Superpositions concern only the wave function in physical space in its role to determine the trajectories on which the particles move.
2. Consequently, quantum logic is irrelevant when it comes to an account of measurement: the particle configuration belongs unambiguously to one of the possible supports of the wave function, which in turn correspond to macroscopically different components of the experimental device, determining in this way the final outcome of the observation at hand.
3. Nevertheless, there is entanglement in physical space: the motion of any particle depends on, strictly speaking, the positions of all the other particles in the universe via the wave function. Thus, for instance, in the double slit experiment, the motion of any particle after having passed one slit depends on the position of all the particles making up the experimental set-up, in particular on whether or not the other slit is open. This is the way in which Bohmian mechanics implements the quantum nonlocality proven by Bell's theorem. The consequence is that the trajectories of the particles often are highly non-classical.
4. A measurement is an interaction that will in general change the wave function of the measured system. "Incompatible measurements"—corresponding to non-commuting observables—are simply experiments in which the first measurement interaction changes the wave function in a way that influences the statistics of the second, etc.
5. The fact that we cannot go beyond Born's rule in making predictions is explained not by any indeterminacy of the properties of the particles, or any indeterminism of the dynamics, but by the fact that we cannot have more precise knowledge of the initial particle configuration. As mentioned in Section 2, in Bohmian mechanics, Born's rule is derived from the laws of motion plus a probability (more precisely: typicality) measure linked with these laws.

Once "measurements" and "observations" are no longer treated as primitive but as physical processes, to be analyzed on the basis of a precise microscopic theory, it turns out that the quantum orthodoxy was right about the fact that measurements do not reveal preexisting values of observables, but wrong about the idea that these observables correspond to properties of physical systems. The important contrast between classical and quantum mechanics that the no-hidden-variables theorems reveal is thus not that quantum phenomena are irreducibly random, but rather that quantum phenomena are at odds with a metaphysics of intrinsic properties that classical mechanics did not necessitate but indulge.

5.2. What Is Measured in a Spin Measurement?

Let us now discuss a Stern–Gerlach spin measurement, as the simplest but maybe most instructive example of a measurement process in Bohmian mechanics. In this famous experiment, a spin-1/2-particle (originally a silver atom) is sent through an inhomogeneous magnetic field (Stern–Gerlach magnet) and then registered on a detector screen, where one observes a deviation perpendicular to the flight direction and parallel or anti-parallel to the gradient of the magnetic field.

To describe the experiment theoretically, we consider the propagation of a concentrated wave packet

$$\Phi_0 = \varphi_0(z)\left(\alpha\begin{pmatrix}1\\0\end{pmatrix} + \beta\begin{pmatrix}0\\1\end{pmatrix}\right) \tag{7}$$

through an inhomogeneous magnetic field with the gradient in the z-direction. We ignore the components of the wave-function in the x, y-directions and the spatial spreading of the wave function, assuming that the flight time is reasonably short. A straightforward computation using the Pauli equation (which is the non-relativistic limit of the Dirac equation, describing the time evolution of a spinor-valued wave function in an external electromagnetic field) then shows that the equations for the two spin-components decouple and that each acquires a phase

$$\Phi^{(n)}(\tau) = \exp[\mathrm{i}(-1)^{n+1}\frac{\mu b\tau}{\hbar}z]\,\Phi_0^{(n)},$$

where τ is the time spent in the magnetic field, corresponding to a group velocity of

$$v_z = (-1)^{n+1}\frac{\mu b\tau}{m}.$$

The inhomogeneous magnetic field thus leads to a spatial separation of the wave packets, corresponding to the spin-components: The wave packet $\Phi^{(1)}(t) = \alpha\varphi_1(t,z)\binom{1}{0}$ propagates in the positive z-direction (in the direction of the gradient of the magnetic field) and the wave packet $\Phi^{(2)}(t) = \beta\varphi_2(t,z)\binom{0}{1}$ in the negative z-direction. Assuming that the two wave packets remain reasonably well localized, they will have approximately disjoint supports after a little while, that is, $\Phi^{(1)}$ is concentrated above the symmetry axis and $\Phi^{(2)}$ below. It is important to emphasize that this is purely a result of the Schrödinger (respectively Pauli) time evolution, which is part of every quantum theory, independent of interpretative issues.

However, in Bohmian mechanics (and only there), it now makes sense to ask whether the particle moves upwards—guided by the wave packet $\Phi^{(1)}$—or downwards, guided by the wave packet $\Phi^{(2)}$. In the first case, it would hit a detector screen above the symmetry axis and one says that "the particle has z-spin up"; in the second case, it would hit a detector screen below the symmetry axis and one says that "the particle has z-spin down". However, this is a rather unfortunate way of speaking. Spin is not a property that the particle possesses over and above its position. To "have" spin up or spin down means nothing more and nothing less than to be guided by the part of the wave function that corresponds to the upper or lower spinor-component (in the z-spin basis)—that is, to *move*, in the pertinent measurement context, in the respective way. In other words: spin is a degree of freedom of the wave function (related to its transformation under rotations) that manifests itself, under certain circumstances, in a particular kind of particle motion. As such, it belongs to the dynamical structure of the theory, not to the ontology of objects in physical space (see also Bell [2] (Chapter 4) and Norsen [23]).

According to Born's rule for the particle positions, we can compute the probability of finding the particle with "spin up", that is, in the support of $\Phi^{(1)}$, or "spin down", that is, in the support of $\Phi^{(2)}$ as:

$$\begin{aligned}\mathbb{P}(\text{"z-spin up"}) &= \mathbb{P}(X \in \operatorname{supp}\Phi^{(1)}) = \int_{\operatorname{supp}\Phi^{(1)}} |\Phi^{(1)}(t,z)|^2 \mathrm{d}z = |\alpha|^2,\\ \mathbb{P}(\text{"z-spin down"}) &= \mathbb{P}(X \in \operatorname{supp}\Phi^{(2)}) = \int_{\operatorname{supp}\Phi^{(2)}} |\Phi^{(2)}(t,z)|^2 \mathrm{d}z = |\beta|^2.\end{aligned} \tag{8}$$

Obviously, these probabilities can already be computed from the initial state, using the projections on the respective spin-components:

$$\mathbb{P}(\text{"z-spin up"}) = \langle \Phi_0 | \uparrow \rangle \langle \uparrow | \Phi_0 \rangle = |\alpha|^2,$$
$$\mathbb{P}(\text{"z-spin down"}) = \langle \Phi_0 | \downarrow \rangle \langle \downarrow | \Phi_0 \rangle = |\beta|^2. \tag{9}$$

Finally, assigning to "spin up" and "spin down" the numerical values $\pm\frac{\hbar}{2}$, the expectation value is computed as

$$\frac{\hbar}{2}\langle \Phi_0 | \Big(|\uparrow\rangle\langle\uparrow| - |\downarrow\rangle\langle\downarrow| \Big) |\Phi_0\rangle = \frac{\hbar}{2}\langle \Phi_0 | \sigma_z | \Phi_0 \rangle. \tag{10}$$

In standard quantum mechanics, the operator $\frac{\hbar}{2}\sigma_z$ has developed a certain life of its own as the "spin observable". The Bohmian analysis reveals it to be nothing more and nothing less than a convenient book-keeper of the measurement statistics (for a general discussion of observables and operators in Bohmian mechanics, see Dürr et al. [6]) (Chapter 3). We should note that the example of spin is particular in Bohmian mechanics in that the statistical analysis does not require the coupling to a measurement device. It makes sense to ask whether the particle moves upwards or downwards after passing the Stern–Gerlach magnet, without considering a screen or detector in which its position is finally recorded. In many cases, though, the "observable values" have meaning only insofar as their are registered in some sort of "pointer" configuration.

It is interesting to observe that all precise formulations of quantum mechanics, which solve the measurement problem, agree on this basic point that the measured values are *produced* rather than *revealed* by the interaction between system and measurement device. According to spontaneous collapse theories (such as the Ghirardi–Rimini–Weber (GRW) theory), it is the Stern–Gerlach magnet that causes the wave packets to separate and the subsequent coupling to a detector (screen) that (very very likely) causes a collapse and forces the system to go into one of the possible outcomes. According to the more sophisticated versions of many-worlds, it is the splitting of the wave packets in the Stern–Gerlach magnet and the subsequent interaction with a detector that leads to decoherence and a branching into "worlds", in which the detector has registered "spin up" and "spin down", respectively.

Only in Bohmian mechanics, however, is a unique measurement outcome determined by the initial position of the particle and the deterministic law of motion (Collapse theories are fundamentally stochastic, while, in many-worlds theories, measurements do not have unique outcomes). That notwithstanding, it would be misleading to say that the particle possesses a predetermined spin, irrespective of the measurement context. In particular, what we end up calling the "spin value" is a number that encodes the result of the measurement interaction—how the particle moves after passing the magnetic field—by contrast to an additional physical quantity that determines it.

5.3. Is Bohmian Mechanics "Contextual"?

In fact, this confusion between "predetermined outcomes" and "predetermined properties" is all there is to the discussion of contextuality in Bohmian mechanics. What this theory rejects is the "naive realism about operators" or observables (Daumer et al. [24])—these unholy and categorically confused amalgams of self-adjoint operators, physical properties, and observed data points. As mentioned before, observables play no fundamental role in the theory; they merely arise, in a statistical analysis, as book-keepers of outcome statistics. Consequently, they are not properties of anything. It is simply wrong, and giving rise to further confusion, to call them "contextual properties" of physical systems. In fact, different experimental setups associated with the same "observable" may have nothing in common besides the fact that they are associated with the same statistical book-keeping operator.

To illustrate this point, let us return to Mermin's proof of the Kochen–Specker theorem (see Section 3.2) and focus, for instance, on the observable $\sigma_x^1\sigma_x^2$ in the upper right corner of his scheme. This observable can be trivially measured together with σ_x^1 and σ_x^2: Take two spin-1/2-particles and measure their x-spin separately in the way described above. Assign the value $+1$ if the particle moves in positive x-direction and -1 if the particle moves in negative x-direction and compute the product of the outcome values to obtain "the value of $\sigma_x^1\sigma_x^2$". However how to measure $\sigma_x^1\sigma_x^2$ together with $\sigma_y^1\sigma_y^2$ and $\sigma_z^1\sigma_z^2$? We have no idea, actually. In any case, one cannot simply measure the x-spin of particle 1 and 2 separately, as before, since this would preclude the simultaneous measurement of $\sigma_y^1\sigma_y^2$ and $\sigma_z^1\sigma_z^2$. Hence, whatever an experimentalist would have to do to perform a joint measurement of $(\sigma_x^1\sigma_x^2, \sigma_y^1\sigma_y^2, \sigma_z^1\sigma_z^2)$—and whatever the physical significance of this measurement might be—it certainly requires a completely different experiment than the measurement of $(\sigma_x^1, \sigma_x^2, \sigma_x^1\sigma_x^2)$.

In Bohmian mechanics, the initial state (wave-function + positions) of the particles (possibly together with the initial state of the experimental setup) would determine the outcome of "the $\sigma_x^1\sigma_x^2$-measurement" in both experiments, but there is simply no reason why these outcomes must in every case agree. A disagreement would be troubling only if one assumed that the particles actually have a preexisting $\sigma_x^1\sigma_x^2$-property that both experiments are supposed to reveal by different methods. However this is just not the case in Bohmian mechanics. Furthermore, taking the physical situation seriously, there is no reason why it should be the case in any reasonable theory. As Goldstein [25] notes: "If we avoid naive realism about operators, contextuality amounts to little more than the rather unremarkable observation that results of experiments should depend on how they are performed ...".

5.4. Why Measurements?

Nonetheless, since, according to Bohmian mechanics, the outcome of any measurement is determined by the initial state of the system (or at least of system + apparatus), the measurement outcome does reveal a certain amount of information about the state of the system prior to measurement. In fact, in some cases, the Bohmian theory allows us to infer significantly more information about the measured system than standard quantum mechanics does. If we consider, for instance, a z-spin measurement on a particle in the spin state $\frac{1}{\sqrt{2}}(|\uparrow_z\rangle + |\downarrow_z\rangle)$ and assume that the setup is reasonably symmetric about the incident axis, we can infer from the "no-crossing property" of Bohmian trajectories that if a particle hits the screen above/below the symmetry axis (corresponding to z-spin up or z-spin down, respectively), its initial position must have been above/below the symmetry axis as well.

In general, though, a quantum experiment provides more information about the state of the system *after* the measurement process. In particular, if we perform an ideal (projective) measurement and find a non-degenerate eigenvalue α of some observable $\hat{A}$, we know that the effective quantum state of the system after the measurement is the corresponding eigenstate ψ_α. According to Bohmian mechanics, this quantum state is an objective physical degree of freedom of the system (in accordance with the Pusey–Barrett–Rudolph (PBR) theorem [26]), providing statistical information about the particle configuration and determining its state of motion. It is thus highly informative about the future behavior of the system. Note, however, that it would be wrongheaded to interpret the effective quantum state as an additional intrinsic property of the particles, a) because one can, in general, assign a wave function only to the subsystem as a whole but not to each particle individually (non-separability) and b) because the effective wave function depends—implicitly—on the universal wave function and the configuration of all the other particles in the universe (cf. Equations (4) and (5)).

Orthodox quantum mechanics agrees that a measurement provides, in general, more information about the post-measurement state of the system, but would, strictly speaking, disagree on what the information is actually about. The disagreement can be summarized as follows: According to Bohmian mechanics, the "observable values" are best understood as encoding information about the quantum state (i.e. the dynamical state) of the system, while according to standard quantum mechanics (or at least most versions thereof), the quantum state is understood as encoding information about the observable values. What makes the Bohmian view more coherent is the fact that the observable values per se—in contrast to the quantum state—have no causal role within the theory (except maybe for conserved quantities, but even those get their physical significance mostly in the "classical limit").

To appreciate this point, it might be helpful to engage in a little thought exercise: Suppose we write down some abstract self-adjoint operator $\hat{A}$ on a Hilbert space and tell you that a certain physical system (an electron, let us say) has the value α of this "observable". What information have we actually given you about the world? How would you (or any other physical system) have to interact with the electron to "notice" that it has the $\hat{A}$-value α rather than α'? Try to answer these questions by taking the physical theory seriously, whatever you consider quantum theory to be.

6. Are Observables Observable?

The suggestive but misleading terminology of "measuring an observable" has not only lead to a naive realism, but also to a naive empiricism about observables in quantum mechanics. It is usually taken for granted that all empirical data underlying quantum physics consist in measured values of observables, represented by—or corresponding to—self-adjoint operators. Against this backdrop, our previous analysis seems to lead to a certain dilemma. Since the measured values of quantum observables are emergent in a measurement process, they must emerge from an underlying ontology that is not itself characterized in terms of definite values of quantum observables. This seems to leave us with two possible options:

1. The physical properties are not observable.
2. The physical properties are a small subset of the observables (small enough to avoid the no-hidden-variables results).

Both options invite criticism. In the first case, the underlying ontology would have no direct empirical basis. The second option is open to the charge of arbitrariness, as it seems to reify some observable properties but not others that have the same empirical status. In fact, both lines of attack are occasionally used against Bohmian mechanics, the first in form of the claim that "the Bohmian trajectories cannot be observed", the second in form of the question "why take the position as your 'hidden variable' and not something else?". While some interesting remarks could be made in response to these objections, we want to take a step back and question the basic assumption that the "observables" are somehow a priori given as fundamental objects of empirical observation.

Consider the following image (Figure 1) from an original Stern–Gerlach experiment, reported as the first experimental observation of a "quantized direction (Richtungsquantelung)" of the angular momentum/magnetic moment of atoms in an external magnetic field (Gerlach and Stern [27]). Should we say that what was actually observed in this experiment—what the empirical data consists in—is the particles' spin?

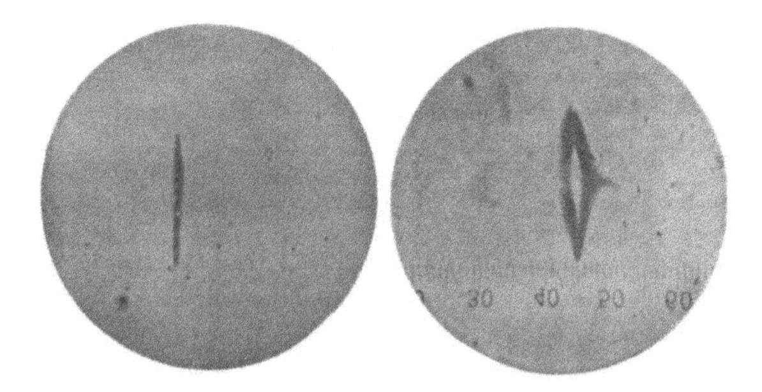

Figure 1. Pattern created by a ray of silver atoms in the original Stern–Gerlach experiment: **left**: without, **right**: with magnetic field.

Evidently, our more immediate observation is that of dark marks on a screen, the "non-classical two-valuedness" being manifested in the distinct separation of the arcs on both sides of the symmetry axis, when the magnetic field is turned on. In addition, evidently, the statistics of "spin up" and "spin down" (deviation to the right/left) alone are too coarse-grained to capture all observable details of the pattern.

However, this now puts the orthodox view in a predicament. Either quantum mechanics could describe the experiment, in more detail, as series of position measurements (the points of impact of the atoms building up the pattern on the screen); then, the spin observable is redundant or, at least, derivative upon the observable "position"—or standard quantum mechanics somehow compels us to describe this experiment as a measurement of "spin". Then, the theory is empirically incomplete, since it cannot—even statistically and in principle—account for all observable details of the experimental outcome.

In general, all that we observe are the positions of discrete objects and the change of these positions. Of course, there is more to these discrete objects than their mere positions, that is, spatial relationships and change of these relationships. They notably have different colors, which makes it possible to discern them in perception. However, color perception is not an observable that figures in any physical theory, and the quantum observables do not help us to come up with an account of color perception. In electromagnetism, "colors" are identified with certain wavelengths in the electromagnetic field. However, the electromagnetic field should be first and foremost understood in terms of its role for the motion of particles (and be it particles in our visual receptors). In other words, we do not observe fields, but only certain patterns of motion that we explain and calculate in terms of fields (cf. Lazarovici [28]). For classical electrodynamics, even a field free formulation is available, namely the one of Wheeler and Feynman [29], which may have a number of drawbacks, but certainly does not fail for the reason that it denies alleged field observations. By the same token, even in the case of the gravitational waves detected by LIGO (Laser Interferometer Gravitational-Wave Observatory) in 2016, all the evidence is evidence of change in the relative positions of particles, which is then mathematically described in terms of a wave rippling through the gravitational field.

Bell [2] (p. 166) considered it to be the first and foremost lesson of Bohmian mechanics that

> in physics the only observations we must consider are position observations, if only the positions of instrument pointers. It is a great merit of the de Broglie–Bohm picture to force us to consider this fact. If you make axioms, rather than definitions and theorems, about the "measurement" of anything else, then you commit redundancy and risk inconsistency.

This crucial point applies to the whole of physics. Also in classical mechanics, we do not observe mass when we observe gravitational attraction, and we do not literally see angular momentum when we notice the regular motion of the moon around the earth. What we observe is just that: certain regularities in the motion of matter, which are captured by the dynamical structure of the theory.

Hence, even in classical physics, quantities like energy, momentum, angular momentum, etc. get their meaning and relevance from what they tell us about the way matter moves. The same applies also to the classical parameters of mass and charge. Ernst Mach [30] (p. 241) highlighted this issue when he emphasized in his comment on Newton's *Principia* that "The true definition of mass can be deduced only from the dynamical relations of bodies". In Bohmian mechanics, then, the way matter moves is encoded in the wave function, making all additional properties unnecessary or redundant (mass and charge, as well, are best understood as situated on the level of the wave function, instead of being intrinsic properties of the particles, see most recently Pylkkänen [31] and Esfeld et al. [32]). This is the basic reason why Bohmian mechanics endorses doctrine Q. In that respect, the lesson of the no-hidden-variables theorems is that in quantum mechanics, one cannot treat the observables as properties of the physical systems, whereas in classical mechanics, one does not run into a problem with the physics if one regards quantities like energy, momentum, angular momentum, etc. as properties of the physical systems (although there is no cogent reason to do so in classical physics either).

Any quantum theory that admits what is known as a primitive ontology of matter in physical space privileges position—be it the position of permanent particles as in the Bohm theory, be it the value of the density of matter at the points of physical space as in the GRWm theory, or be it single events (flashes) occurring at some points of space as in the GRWf theory (see Allori et al. [33]). In all these theories, the quantum observables are construed on the basis of the positions of objects, namely in terms of how these positions behave in certain experimental contexts. Also in the many worlds theory, which does not recognize a primitive ontology of matter in physical space, but proposes an ontology in terms of the universal wave function, position is privileged: it is the position basis in which the wave function decoheres, splitting into different branches, which constitute "many worlds" on this view.

7. Conclusions

What we perceive with the naked eye are the positions of macroscopic objects. However, we know from scientific experience that the macroscopic objects are composed of discrete microscopic objects. If the macroscopic objects have precise positions when we observe them, so do the microscopic objects. There is no coherent theory of a magic power of the mind to change macroscopic objects in such a way that they acquire positions only when a being with a mind perceives them. Thus, the macroscopic objects better have positions independently of someone observing them. If not the moon, so surely the desk in my office is there also when I do not observe it. However then it follows that also the microscopic objects that compose these macroscopic objects do have positions independently of them being observed. Again, there is no coherent theory according to which there is something special about the microscopic objects that compose my desk and the like. Thus, the conclusion is that the microscopic objects *tout court* have a position independently of them being observed.

Bohmian mechanics shows how to build a quantum theory on this simple and obvious reasoning. Superpositions then concern only the parameter that encodes the dynamics of the particles, namely the wave function, but not the particles themselves. This insight is the key to answering the two questions raised at the beginning of this paper and to avoid all the puzzles of standard quantum mechanics, such as notably the measurement problem. However, as it is trivial that physical objects have positions, so it is trivial that in order to access these positions, we have to interact with these objects and thereby change their positions. Generally speaking, for one particle configuration, say a macroscopic object, to contain information about the positions of other particles, there must be a correlation between them, which is, furthermore, reliable in the sense of being reproducible. This applies in particular to correlations between particle configurations in human brains and particles outside the brains, assuming that all the perceptual knowledge that persons acquire passes through their brains.

Hence, for reasons stemming from the very way in which we acquire knowledge about the natural world, a limited accessibility of physical objects is to be expected. In that sense, classical mechanics is an idealization, and quantum mechanics brings out that limitation on our knowledge. In Bohmian mechanics, this is done in the theorem of "absolute uncertainty" (Dürr et al. [6] (Chapter 2)), stating that we cannot have more information about the actual particle configuration of a sub-system than what is provided by the $|\psi|^2$-distribution in terms of its effective wave function. That notwithstanding, there is, of course, no question of an a priori deduction of this theorem—or the Heisenberg uncertainty relations—from general conditions of our knowledge. It is just that some principled limit on our knowledge of particular matters of fact—such as initial conditions of physical systems—is to be expected.

If the evolution of the physical systems is highly sensitive to slight variations in their initial conditions, as is the case with quantum systems, it then follows that in general we can only make statistical predictions about the behavior of ensembles of physical systems prepared under the same conditions, but not predictions about the evolution of an individual system, although the laws of motion that govern the evolution of these systems can be fully deterministic (cf. Oldofredi et al. [34]). Again, classical mechanics is deceptively generous in this respect, and quantum mechanics brings out a fact that turns out to be trivial upon reflection (and actually comes out already in classical statistical mechanics): deterministic laws require a probability measure to yield predictions, which then are statistical. However, all these are facts about epistemology, the theory of knowledge—as the word "uncertainty relations" clearly brings out, and not about ontology, that is, about what there is in the world.

Nonetheless, the no-hidden-variables theorems have a great merit: they tell us that a sparse ontology of positions is not just good metaphysics, but strongly suggested by our best theory of physics. In classical mechanics, one can attribute dynamical parameters and all sorts of "observables", which are functions of the particle positions and momenta, as intrinsic properties to the particles. This does not lead to conflict with the phenomena because the active role of the measurement process—both in producing the data and in changing the state of the measured system—can be usually neglected in the classical regime. In quantum mechanics, as we have seen, the situation is markedly different. The moral then is of course not that there is nothing if one cannot go from observables to ontology, but that one has to start with conceiving a—provisional, hypothetical—ontology for whose evolution the dynamical parameters then are formulated. The guideline for this is the experimental evidence together with the coherence and explanatory fruitfulness of the proposed ontology. Bohmian mechanics shows how the simplest suggestion in that respect—the evidence of discrete objects and their composition by discrete micro-objects suggesting to try out a particle ontology—can go through also in the quantum case and yield all the explanations that one can reasonably demand.

In a nutshell, the lesson of the no-hidden-variables theorems is that it is *position only* when it comes to the ontology of the physical world, and Bohmian mechanics teaches us how to do physics on that basis (see Esfeld and Deckert [20] for a general treatment of that insight from classical mechanics to quantum field theory). Note that this is not about classical vs. quantum. The ontology neither is classical nor quantum. The dynamics may be classical (as in local field theories) or quantum. What a quantum dynamics has to be subsequent to Bell's theorem is clearly brought out by the nonlocality implemented in the Bohm theory. There is no a priori explanation of why the dynamics of the world is nonlocal. However, this nonlocality fundamentally deviates from the ideas that drive classical field theory, showing a profound interconnectedness (holism) of the things in the universe.

Author Contributions: Conceptualization, D.L. and M.E.; Writing—Original Draft Preparation, D.L.; Writing—Review & Editing, A.O. and M.E.; Visualization, A.O.; Supervision, M.E.

Funding: Andrea Oldofredi acknowledges funding from the Swiss National Science Foundation (SNSF), grant number 105212_175971.

Acknowledgments: We are grateful to Detlef Dürr and three anonymous referees for helpful suggestions to improve the manuscript.

Conflicts of Interest: The authors declare no conflict of interest.

References

1. Bohm, D. A suggested interpretation of the quantum theory in terms of "hidden" variables, I and II. *Phys. Rev.* **1952**, *85*, 166–193. [CrossRef]
2. Bell, J.S. *Speakable and Unspeakable in Quantum Mechanics*, 2nd ed.; Cambridge University Press: Cambridge, UK, 2004.
3. Wigner, E. Review of Quantum Mechanical Measurement Problem. In *Quantum Optics, Experimental Gravity and Measurement Theory*; Meystre, P., Scully, O., Eds.; Plenum Press: New York, NY, USA, 1983.
4. Mermin, N.D. Hidden variables and the two theorems of John Bell. *Rev. Mod. Phys.* **1993**, *65*, 803–815. [CrossRef]
5. Mermin, N.D. Is the moon there when nobody looks? Reality and the quantum theory. *Phys. Today* **1985**, *38*, 38–47. [CrossRef]
6. Dürr, D.; Goldstein, S.; Zanghì, N. *Quantum Physics without Quantum Philosophy*; Springer: Berlin/Heidelberg, Germany, 2013.
7. Belousek, D.W. Formalism, ontology and methodology in Bohmian mechanics. *Found. Sci.* **2003**, *8*, 109–172. [CrossRef]
8. Bohm, D.; Hiley, B.J. *The Undivided Universe. An Ontological Interpretation of Quantum Theory*; Routledge: London, UK, 1993.

9. Goldstein, S.; Struyve, W. On the uniqueness of quantum equilibrium in Bohmian mechanics. *J. Stat. Phys.* **2007**, *128*, 1197–1209. [CrossRef]
10. Wiseman, H.M. Grounding Bohmian Mechanics in Weak Values and Bayesianism. *New J. Phys.* **2007**, *9*, 165. [CrossRef]
11. Esfeld, M.; Lazarovici, D.; Hubert, M.; Dürr, D. The ontology of Bohmian mechanics. *Br. J. Philos. Sci.* **2014**, *65*, 773–796. [CrossRef]
12. Von Neumann, J. *Mathematische Grundlagen der Quantenmechanik*; Springer: Berlin/Heidelberg, Germany, 1932.
13. Von Neumann, J. *Mathematical Foundations of Quantum Mechanics*; Princeton University Press: Princeton, NJ, USA, 1955.
14. Kochen, S.; Specker, E. The problem of hidden variables in quantum mechanics. *J. Math. Mech.* **1967**, *17*, 59–87. [CrossRef]
15. Maudlin, T. What Bell Did. *J. Phys. A Math. Theor.* **2014**, *47*, 424010. [CrossRef]
16. Heisenberg, W. *Der Teil und das Ganze: Gespräche im Umkreis der Atomphysik*, 9th ed.; Piper Verlag: Munich, Germany, 2012. (In German)
17. Laudisa, F. Against the 'no-go' philosophy of quantum mechanics. *Eur. J. Philos. Sci.* **2014**, *4*, 1–17. [CrossRef]
18. Calosi, C.; Wilson, J. Quantum Metaphysical Indeterminacy. 2017. Available online: https://philpapers.org/rec/CALQMI (accessed on 5 May 2018).
19. Glick, D. Against quantum indeterminacy. *Thought J. Philos.* **2017**, *6*, 204–213. [CrossRef]
20. Esfeld, M.; Deckert, D.A. *A Minimalist Ontology of the Natural World*; Routledge: New York, NY, USA, 2017.
21. Van Orman Quine, W. Two dogmas of empiricism. *Philos. Rev.* **1951**, *60*, 20–43. [CrossRef]
22. Dürr, D.; Teufel, S. *Bohmian Mechanics: The Physics and Mathematics of Quantum Theory*; Springer: Berlin/Heidelberg, Germany, 2009.
23. Norsen, T. The pilot-wave perspective on spin. *Am. J. Phys.* **2014**, *82*, 337–348. [CrossRef]
24. Daumer, M.; Dürr, D.; Goldstein, S.; Zanghì, N. Naive realism about operators. *Erkenntnis* **1996**, *45*, 379–397.
25. Goldstein, S. *Bohmian Mechanics*; Stanford Encyclopedia of Philosophy: Stanford, CA, USA, 2017.
26. Pusey, M.F.; Barrett, J.; Rudolph, T. On the Reality of the Quantum State. *Nat. Phys.* **2012**, *8*, 476–479. [CrossRef]
27. Gerlach, W.; Stern, O. Der experimentelle Nachweis der Richtungsquantelung im Magnetfeld. *Z. Phys.* **1922**, *9*, 349–352. (In German)
28. Lazarovici, D. Against Fields. *Eur. J. Philos. Sci.* **2018**, *8*, 145–170. [CrossRef]
29. Wheeler, J.A.; Feynman, R.P. Interaction with the absorber as the mechanism of radiation. *Rev. Mod. Phys.* **1945**, *17*, 157–181. [CrossRef]
30. Mach, E. *The Science of Mechanics: A Critical and Historical Account of Its Development*, 4th ed.; Open Court: Chicago, IL, USA, 1919.
31. Pylkkänen, P.; Hiley, B.J.; Pättiniemi, I. Bohm's approach and individuality. In *Individuals across the Sciences*; Guay, A., Pradeu, T., Eds.; Oxford University Press: Oxford, UK, 2015; Chapter 12, pp. 226–246.
32. Esfeld, M.; Lazarovici, D.; Lam, V.; Hubert, M. The physics and metaphysics of primitive stuff. *Br. J. Philos. Sci.* **2017**, *68*, 133–161. [CrossRef]
33. Allori, V.; Goldstein, S.; Tumulka, R.; Zanghì, N. On the common structure of Bohmian mechanics and the Ghirardi-Rimini-Weber theory. *Br. J. Philos. Sci.* **2008**, *59*, 353–389. [CrossRef]
34. Oldofredi, A.; Lazarovici, D.; Deckert, D.A.; Esfeld, M. From the universe to subsystems: Why quantum mechanics appears more stochastic than classical mechanics. *Fluct. Noise Lett.* **2016**, *15*, 164002. [CrossRef]

Commentary

On a Common Misconception Regarding the de Broglie–Bohm Theory

Oliver Passon

School for Mathematics and Natural Sciences, University of Wuppertal, Gaußstr. 20, 42119 Wuppertal, Germany; passon@uni-wuppertal.de; Tel.: +49-202-439-2490

Received: 26 April 2018; Accepted: 5 June 2018; Published: 5 June 2018

Abstract: We discuss a common misconception regarding the de Broglie–Bohm (dBB) theory; namely, that it not only assigns a position to each quantum object but also contains the momenta as "hidden variables". Sometimes this alleged property of the theory is even used to argue that the dBB theory is inconsistent with quantum theory. We explain why this claim is unfounded and show in particular how this misconception veils the true novelty of the dBB theory.

Keywords: quantum theory; de Broglie–Bohm theory; contextuality

---

1. Introduction

Bohm's interpretation of quantum theory [1] is an example for a (non-local) hidden variable theory which avoids the notorious measurement problem of quantum mechanics. It is usually assumed that this theory reproduces all predictions of ordinary quantum theory. Some times this claim is qualified by the remark "as long as the latter are unambiguous" (see e.g., [2]). This refers e.g., to the fact that the quantum mechanical predictions regarding "time" (e.g., the "arrival time" or the "tunneling time") are problematic.

However, there is a common misconception with regard to the question which quantities exactly are promoted to the "hidden variable" status. For example Mario Bunge writes ([3], p. 453):

> "In particular, Bohm enriched standard non-relativistic quantum mechanics with a classical position coordinate and the corresponding momentum [...]."

While it is true that Bohmian mechanics assigns a well defined position to each quantum object at any moment, this quote suggests, that each particle possesses also a well defined momentum. Given that a well defined position apparently translates into a velocity which may be multiplied with the mass of the corresponding object (say, the electron) this claim is seemingly very natural. However, it turns out to be wrong nevertheless. At the same time this alleged property of Bohmian mechanics has been used to argue that this theory is inconsistent with quantum theory. An argument along these lines was recently given by Michael Nauenberg [4]. We think that the explanation why this charge is unfounded provides a good opportunity to combat certain long-standing prejudices surrounding the Bohm interpretation. In addition it provides a opportunity to unveil what we take to be the true novelty of the Bohm theory.

In Section 2 we will briefly outline the basics of the Bohm theory. Section 3 contains the most recent example for the criticism outlined above, namely the interesting contribution of Michael Nauenberg [4]. Section 4 explains the underlying misconception on which these claims are based. Finally, we will summarize our discussion in Section 5.

2. A (Very) Brief Introduction to the dBB Theory

In Bohm's interpretation a system of N particles is described by the wave function (i.e., the solution of the corresponding Schrödinger equation) and the configuration q_k, i.e., the actual positions of the quantum objects. Thus, Bohm has to add an "equation of motion" or "guiding equation" for the positions to the formalism. Assuming a wave function, $\psi = R\exp(iS/\hbar)$, the guiding equation for the position q of a spin-less particle takes (in the 1-particle case) the form:

$$\frac{dq}{dt} = \frac{\vec{\nabla} S}{m}. \quad (1)$$

Note, that this equation is of first order in time, i.e., the initial configuration alone fixes the motion uniquely. The generalization to the many particle case and including spin is straightforward [5]. Choosing initial conditions for the position according to Born's rule (i.e., $\rho = |\psi|^2$) the continuity equation

$$\frac{d\rho}{dt} + \nabla j = 0 \quad (2)$$

with the usual quantum mechanical probability current

$$j = \frac{\hbar}{2mi}[\psi^*(\nabla\psi) - (\nabla\psi^*)\psi] \quad (3)$$

$$= \rho\frac{\nabla S}{m} \quad (4)$$

ensures that the positions remain $|\psi|^2$ distributed. Hence, in terms of position any measurement yields exactly the result of the standard formalism. Thus, the guidance equation is consistent with the requirements of quantum mechanics. An other way to bring out the difference between ordinary QM and the dBB theory is the following: In QM the probability current refers to the probability to measure a certain position. Within the dBB theory it can be viewed as the probability of the particle to be at a certain position—independent of any measurement. The different strategies to explain why the $|\psi|^2$ distribution holds initially are critically examined by Norsen [6].

We may add a remark with respect to further references which provide a full exposition. Bohm's work was essentially an independent re-discovery of work that was done by Louis de Broglie already in the 1920s [7]. Thus the term "de Broglie–Bohm" (dBB) theory is more appropriate. Further more there are different schools of the dBB theory. While [5] presents a version called "Bohmian mechanics", the books by Bohm and Hiley or Holland [8,9] stick closer to the original presentation of Bohm from 1952 [1]; sometimes called "ontological" or "causal" interpretation of quantum mechanics. However, these distinctions play a minor role in what follows.

3. The Asymmetry between Position and Momentum

We now turn to the criticism mentioned above, which arises if the special role of the position observable within the de Broglie–Bohm theory is not taken into account. Reference [4] claims that the equivalence holds only in the "coordinate representation" while moving into the "momentum representation" leads to conflicting results. To reach this conclusion it introduces the "velocity operator" (i.e., the momentum operator divided by the mass):

$$\vec{v} = -\frac{i\hbar}{m}\vec{\nabla}_q. \quad (5)$$

The author investigates the expectation value of this operator (in particular its second moment) and finds a result which differs from the second moment of the expression $v = \frac{\nabla S}{m}$, i.e., the velocity of the Bohmian particles (called $\vec{v}_B$ by this author). In [4] it is further claimed that this feature has gone unnoticed by the recent literature on the Bohm interpretation and we will come to this point shortly.

Reference [4] discusses as a specific example the velocity of a Bohmian particle for stationary solutions with vanishing phase, i.e., described by real wave functions. Such wave functions describe for example the electron in the ground state of the hydrogen or the energy eigenstates of the quantum mechanical harmonic oscillator. In all these cases the velocity vanishes ($v = \frac{\nabla S}{m} = 0$). Examples like this are among the oldest objections raised against the Bohm interpretation [10]. Already in 1953 Einstein discussed a particle-in-a-box example with vanishing velocity. Clearly, that these examples are old calls into question the originality of this argument but as such not its soundness. Interestingly, Einstein did not conclude that the Bohm interpretation is in contradiction with quantum mechanics but finds fault with it since [11]:

> "The vanishing of the velocity contradicts the well-founded requirement, that in the case of a macro-system the motion should agree approximately with the motion following from classical mechanics."

Now, Reference [4] goes a step further and claims the refutation of Bohm's interpretation:

> "But this result contradicts the fact that in quantum mechanics the velocity or momentum distribution for stationary solutions, given by the absolute square of the Fourier transform of ψ in coordinate space, is not a delta function at $\vec{v} = 0$, as is implied by Bohm's interpretation." (p. 44)

Before we turn to the question why this statement is unfounded we may note that along similar lines the argument could have been developed further. Comparing the expectation values for the kinetic energy with the expression $\langle \frac{mv_B^2}{2} \rangle$ or the angular momentum with the alleged Bohmian prediction "$\vec{L} = \vec{r} \times m\vec{v}_B$" would have yielded a host of "predictions" which differ from the corresponding quantum mechanical result. In addition, perhaps most devastating: given a momentum $\vec{p} = m\vec{v}_B$ would in general contradict with Heisenberg's uncertainty principle.

4. Contextuality of All Observables Other than Position

Now, we finally should explain why and where all this reasoning goes astray. In Reference [4] it is apparently presumed that in Bohm's theory the expression "$\vec{p} = m\vec{v}_B$" should describe a momentum and should relate to the momentum operator of standard QM. After having shown that this does not work it is claimed that:

> "[...] this interpretation is not only inconsistent with the standard formulation of quantum mechanics, but also with classical mechanics, where momentum is defined by the relation $\vec{p} = m\vec{v}$." (p. 45)

Let us discuss these points in reversed order. Bohm's interpretation is certainly inconsistent with classical mechanics since it is a quantum theory. In the case of Bohmian mechanics there can be something added to this general argument. Given that the guiding equation is of first order one would not even expect that concepts of (second order) Newtonian mechanics (like momentum and work) play any role on the level of individual particles whatsoever.

However, the actual mistake in Nauenberg's argument is the assumption, that via $\vec{p} = m\vec{v}_B$ the momentum of the particles should be constrained, i.e., that Bohm's theory has not only a "hidden variable" for the particle position, but also for momentum (and why stop here and not add energy, spin, angular momentum etc.—as indicated above). That this cannot be the case is acknowledged by all scientists working on the field. Technically speaking this follows from the Kochen–Specker no-go theorem which implies that such a scheme contradicts quantum theory [12]. Now, Bohm's theory avoids this problem by not introducing such additional variables for momentum, spin and the like. Reference [4] re-introduces them in the case of momentum (or rather velocity) and demonstrates (correctly) the inconsistency of this *modified* theory.

However, all this still leaves open how one should actually think about these quantities within Bohm's theory. The starting point is the observation that any measurement of, say, momentum or spin involves a position measurement. The momentum of a charged particle is usually inferred from the bending inside a homogeneous magnetic field or the spin from the position measurement after the particle passing a Stern–Gerlach magnet. The outcome of these experiments is—according to Bohm—determined by the wave function and the initial *position(s)* and not by the value of any other "hidden variable". This is expressed by saying that all quantities but position are "contextualized" (compare the discussion in [13]).

To see better what this means take the above example of a Stern–Gerlach experiment to determine the spin of a silver atom. Suppose that the north-pole of the magnet is situated above the south-pole and that a single silver atom gets deflected up (call this deflection towards the north pole "spin up"). The reason for this specific outcome lies—according to Bohm's interpretation—in the fact that the initial position of the coordinate was above the symmetry plane of the system. Thus, a reversed orientation of north- and south-pole would have led to an up deflection still while this time ("deflection towards the south pole") the opposite spin ("down") would have been assigned to the atom. Note, that this example illustrates also *in nuce* how the Bohm interpretation explains definite outcomes for each single measurement, hence, solves the infamous measurement problem.

In other words, according to this view the specific spin value is not an intrinsic property of the particle but depends on the wave function, ψ, the configuration, q_i, and the experimental arrangement (viz. the "context") as well. The same holds for the momentum, energy etc.

All this illustrates that the "Bohmian particle" should not be confused with, say, an electron. The latter is a fermion with specific mass and charge. In addition the "measurement" of momentum, energy and the like gives certain results which can be predicted by the theory (the term "measurement" has been put into scare quotes since it does not reveal a preexisting value of these quantities). The former just has the properties "position". To describe the "electron" the de Broglie–Bohm interpretation needs both, the configuration *and* the wave function. In Daumer et al. this contextuality is even put into a wider context and the authors argue against what they call "naive realism about operators". They conclude [14]:

> "We thus believe that contextuality reflects little more than the rather obvious observation that the result of an experiment should depend upon how it is performed!"

5. Summary

We have dealt with the common misconception, that the de Broglie–Bohm theory does not only assign position to each quantum object but also a well defined momentum (or the value of any other observable). This idea is apparently very reasonable since the velocity of the Bohm-particles can certainly be multiplied with its mass. Michael Nauenberg has turned this misconception into an ingenious argument against the dBB theory—or rather into an argument against this modified version of it.

However, within the de Broglie–Bohm theory the product of velocity and mass has no physical meaning. In general its ambition is not to restore a classical world-view, but to solve the conceptual problems of quantum theory. In doing so any system is described by the pair of wave function and configuration. This implies an interesting reinterpretation of the property-concept. Properties like mass, charge, spin or momentum cannot be assigned to the object moving along the well-defined trajectories. Instead, they "belong" to the wave function, or rather: the result of a measurement of these quantities is determined by the wave function, the configuration and the specific experimental arrangement and does not reveal a previously existing ("hidden") value. On the level of the individual trajectories concepts like momentum, energy and the like lose their meaning. This should not confuse anybody, since this motion is ruled by Bohmian mechanics and not by Newtonian mechanics.

All this may appear odd and nobody has to like or even to support the Bohm interpretation. However, it has to be acknowledged that it is not only a consistent interpretation of quantum mechanics but includes also "quantum weirdness"—like any other interpretation of quantum theory.

Funding: This research received no external funding.

Acknowledgments: The author is grateful for the remarks of two anonymous referees.

Conflicts of Interest: The author declares no conflict of interest.

References

1. Bohm, D. A Suggested Interpretation of the Quantum Theory in Terms of "Hidden" Variables. *Phys. Rev.* **1952**, *85*, 166–193. [CrossRef]
2. Bell, J. De Broglie–Bohm, Delayed-Choice Double-Slit Experiment, and Density Matrix. *Int. J. Quantum Chem.* **1980**, *14*, 155–159. [CrossRef]
3. Bunge, M. Twenty-Five Centuries of Quantum Physics: From Pythagoras to Us, and from Subjectivism to Realism. *Sci. Educ.* **2003**, *12*, 445–466.:1025336332476. [CrossRef]
4. Nauenberg, M. Is Bohm's Interpretation Consistent with Quantum Mechanics? *Quanta* **2014**, *3*, 43–46. [CrossRef]
5. Dürr, D.; Teufel, S. *Bohmian Mechanics: The Physics and Mathematics of Quantum Theory*; Springer: Berlin/Heidelberg, Germany, 2009.
6. Norsen, T. On the Explanation of Born-Rule Statistics in the de Broglie-Bohm Pilot-Wave Theory. *Entropy* **2018**, *20*, 422. [CrossRef]
7. Bacciagaluppi, G.; Valentini, A. *Quantum Theory at the Crossroads Reconsidering the 1927 Solvay Conference*; Cambridge University Press: Cambridge, UK, 2009.
8. Bohm, D.; Hiley, B.J. *The Undivided Universe: An Ontological Interpretation of Quantum Theory*; Routledge & Kegan Paul: London, UK, 1993.
9. Holland, P.R. *The Quantum Theory of Motion*; Cambridge University Press: Cambridge, UK, 1993.
10. Myrvold, W.C. On some early objections to Bohm's theory. *Int. Stud. Phil. Sci.* **2003**, *17*, 7–24. [CrossRef]
11. Einstein, A. Elementare Überlegungen zur Interpretation der Grundlagen der Quanten-Mechanik. In *Scientific Papers Presented to Max Born*; Oliver and Boyd: Edinburgh, UK, 1953; pp. 33–40. (In German)
12. Mermin, N.D. Simple unified form for the major no-hidden variables theorems. *Phys. Rev. Lett.* **1990**, *65*, 3373. [CrossRef] [PubMed]
13. Pagonis, C.; Clifton, R. Unremarkable contextualism: Dispositions in the Bohm theory. *Found. Phys.* **1995**, 25, 281–296. [CrossRef]
14. Daumer, M.; Dürr, D.; Goldstein, S.; Zanghí, N. Naive Realism About Operators. *Erkenntnis* **1997**, *45*, 379–397. [CrossRef]

Article

On the Explanation of Born-Rule Statistics in the de Broglie-Bohm Pilot-Wave Theory

Travis Norsen

Department of Physics, Smith College, Northampton, MA 01063, USA; tnorsen@smith.edu

Received: 8 May 2018; Accepted: 28 May 2018; Published: 31 May 2018

Abstract: The de Broglie-Bohm pilot-wave theory promises not only a realistic description of the microscopic world (in particular, a description in which observers and observation play no fundamental role) but also the ability to derive and explain aspects of the quantum formalism that are, instead, (awkwardly and problematically) postulated in orthodox versions of quantum theory. Chief among these are the various "measurement axioms" and in particular the Born rule expressing the probability distribution of measurement outcomes. Compared to other candidate non-orthodox quantum theories, the pilot-wave theory suffers from something of an embarrassment of riches in regard to explaining the Born rule statistics, in the sense that there exist, in the literature, not just one but two rather compelling proposed explanations. This paper is an attempt to critically review and clarify these two competing arguments. We summarize both arguments and also survey some objections that have been given against them. In the end, we suggest that there is somewhat less conflict between the two approaches than existing polemics might suggest, and that indeed elements from both arguments may be combined to provide a unified and fully-compelling explanation, from the postulated dynamical first principles, of the Born rule.

Keywords: pilot-wave theory; Bohmian mechanics; Born rule statistics; measurement problem

---

1. Introduction

In standard textbook formulations of quantum mechanics, microscopic systems are described by wave functions which, under normal circumstances, obey Schrödinger's equation. However, there are also abnormal circumstances in which the normal rules cease to apply. When a measurement occurs, for example, the system's wave function momentarily ceases to evolve in accordance with Schrödinger's equation and instead "collapses" to one of the eigenstates of the operator corresponding to the type of measurement being performed. Simultaneously, the outcome of the measurement comes to be registered in some directly-observeable (and separately-posited) classical object which we can think of as the pointer on the measuring device.

Which particular outcome is realized (i.e., to which particular eigenstate the quantum system's wave function collapses and to which particular position the pointer ends up pointing) is supposed to be irreducibly random, with probability given by the Born rule: the probability for a particular eigenstate is given by the absolute square of the coefficient of that eigenstate in the linear expansion of the pre-measurement wave function.

For example, consider a position measurement on a quantum system with wave function $\psi(x)$. The apparatus is initially in its ready state (including, say, a pointer with position $Y = Y_0$, resting at the left end of its scale next to an illuminated green light indicating that the device is able to make a measurement). The position operator $\hat{x}$ satisfies

$$\hat{x}\,\delta(x - X) = X\,\delta(x - X), \tag{1}$$

i.e., the eigenfunctions of the position operator are the delta functions $\delta(x - X)$ corresponding to particles which are definitely located at $x = X$. The initial wave function of the quantum system can be written as a linear combination of these eigenfunctions as follows:

$$\psi(x) = \int \psi(X)\delta(x - X)\, dX. \tag{2}$$

Then, according to the quantum measurement postulates, during the measurement, the wave function of the quantum system collapses to $\delta(x - X)$ for some particular value of X, with probability (density) $P(X) = |\psi(X)|^2$—and, simultaneously, the apparatus pointer jumps from $Y = Y_0$ to $Y = \lambda X$ (where λ is some proportionality constant capturing the calibration of the device) with, say, the light changing from green to blue to indicate that the measurement has been successfully completed.

More generally, if a measurement of the property corresponding to operator $\hat{A}$ is performed on a system with quantum state

$$\psi = \sum_m c_m \psi_m \tag{3}$$

where $\hat{A}\psi_m = a_m \psi_m$, the probability that the measurement outcome is a_m (and that the system is left in quantum state ψ_m) is

$$P(a_m) = |c_m|^2. \tag{4}$$

This is the Born rule.

Followers of orthodox (and/or, here equivalently, Copenhagen) quantum theory have tended to read allegedly deep truths about nature (such as the failure of determinism and the ineliminable dynamical role of observation in creating the observed reality) off from this account. More sober critics have instead regarded the orthodox formulation as obviously not providing the final truth, with the cluster of related implausible aspects of the orthodox formulation (for example, its division of the world into separate quantum and classical realms, with different kinds of ontologies and different dynamical laws, but also ad hoc exceptions to the usual dynamics when the two realms interact) being described as the "measurement problem" [1,2]. Recognition of the measurement problem has motivated the search for theories which cure it—or better, avoid it from the outset—by providing a description of the world (in both its ontological and dynamical aspects) that is *uniform* (i.e., consistent, coherent) and in which, in particular, "measurement" can be understood as simply another physical interaction to be treated just like any other, and with the orthodox measurement postulates (including especially the Born rule) becoming *theorems* derived from the more basic dynamical postulates.

There is thus a high degree of overlap between the quest to find a believable "quantum theory without observers" [3,4] and the quest to *explain* (rather than simply postulate) the Born rule. Perhaps the simplest such alternative theory to understand is the spontaneous collapse theories, in which Schrödinger's equation is replaced by a stochastic differential equation that incorporates occasional, random, partial collapses which occur at a certain fixed rate per particle [5]. Such theories (approximately) reproduce the Born-rule randomness by, in effect, modifying the Schrödinger equation to include it [6]. So it is relatively easy to understand how the Born rule statistics arise in this kind of theory. By contrast, in the Everettian (a.k.a. "many worlds") formulation of quantum theory [7], the collapse posulates of orthodox QM are eliminated entirely, with the wave function (of the universe) obeying Schrödinger's linear equation all of the time. This cleanly removes the schizophrenia that gave rise to the measurement problem in the context of ordinary QM, but makes it rather difficult to understand how the Born rule might be derived—or even what the probabilities the Born rule is *about* could possibly *mean* in the context of a (deterministic!) theory in which every "possible" outcome is guaranteed to actually occur. This situation has given rise to a lively debate [8–16]. In addition, there have been many other attempts to similarly derive and/or explain Born's rule, in the context of other alternative approaches to quantum theory, as well [17–24].

Here we will focus on the de Broglie-Bohm pilot-wave formulation of quantum theory [25]—a so-called "hidden variable theory" in which the usual quantum mechanical wave function (obeying Schrödinger's equation) is supplemented with actual particle positions (A note on terminology: The phrase "hidden variable theory" is used here simply to denote a theory in which, contra orthodox QM, the complete description of a quantum system is given by a wave function *plus some additional variables*. However, as applied to the pilot-wave theory, the word "hidden" is rather inappropriate and misleading—Bell called it a piece of historical silliness [26]—since, if anything, it is the wave function and not the particle positions that are "hidden". According to the theory, the world we literally see around us—including tables, chairs, other humans, planets, etc.—is composed of these particles, with the wave function playing a rather mysterious, invisible, background role of choreographing the particles' motion). In terms of the simple example from a few paragraphs back, the pilot-wave theory can be understood as extending the quantum world all the way up into the macroscopic classical realm in the sense that the entire universe (including both the "quantum sub-system" and the "measuring apparatus") will now be (partially) described by a wave function $\Psi(x, y)$. The theory also extends the classical realm down into the microscopic quantum realm in the sense that now not only the particles composing the apparatus, but also the "quantum sub-system" particle, will be assigned definite positions (Y and X, respectively).

The introduction of additional variables (in particular the definite position X of the "quantum sub-system") has often been regarded as pointless (or worse) in so far as the pilot-wave theory, at the end of the day, makes the same statistical predictions as ordinary QM. This concern is however wrong-headed for two (overlapping) reasons. First, to whatever extent one accepts that orthodox QM suffers from a measurement problem, one should be willing to (and should expect to) pay some price for its solution. It is ridiculous, that is, to dismiss the positing of new things, simply on the grounds of wanting as few things as possible, without considering what problems the new things are posited to solve. For example, should Pauli's postulation of the existence of the neutrino have been dismissed, simply on the grounds that its introduction would make particle theory more complicated? However—second—it is not even clear that, overall, the pilot-wave theory is more complicated than orthodox QM. It is true that, as mentioned above, it extends the realm to be described in terms of wave functions from the microscopic to the entire universe, and extends the realm to be described in terms of particles with definite positions from the macroscopic to the entire universe. However, in so far as this supplemented ontology allows one to *derive*—rather than awkwardly postulate—the Born rule (and other rules about measurements), a strong case can be made that the pilot-wave theory is actually, overall, not more complicated but rather *simpler* than the orthodox formulation.

In the rest of the paper we attempt to explain and clarify this derivation of the Born rule in the context of the pilot-wave theory. We focus on the two recent, prominent, and fully-developed programs, ignoring for the most part their historical roots; [27] gives a helpful overview of some of the territory we will cover, including more historical references. Note also that virtually all of the issues to be discussed have close parallels in the context of (classical) statistical mechanics. We again want to focus on the explanation of Born rule statistics in the pilot-wave theory and avoid getting into generalized questions about the nature of statistical explanation, the relative merits of competing approaches to formulating and understanding the second law of thermodynamics, etc. However, readers who already know something about such controversial questions in the foundations of statistical mechanics will recognize parallel controversies playing out here. Readers who want to learn more about the foundations of statistical mechanics might see, for example, [28].

In the following section we give a more technical overview of the pilot-wave theory. Section 3 then provides an overview of one of the extant approaches to deriving the Born rule—the dynamical relaxation program of Antony Valentini. Some objections to Valentini's program are reviewed in Section 4. Then, in Sections 5 and 6 we review and then discuss some objections to the other extant program for deriving the Born rule—the "typicality" approach of Dürr, Goldstein, and Zanghí. The points discussed in these sections are illustrated with the results of numerical simulations for a simple example system. Finally, in Section 7, we take stock of the situation and suggest that elements from both the dynamical relaxation and the typicality programs are needed for—and make possible—a very clear and satisfying explanation of the origin of the Born rule in the context of the pilot-wave theory.

2. Pilot-Wave Theory and the Born Rule

The de Broglie-Bohm pilot-wave theory is best thought of as a candidate *explanation* of the measurement formalism of ordinary quantum mechanics, much as the kinetic/atomic theory (coupled with ideas from statistical mechanics) can be understood as the explanation of phenomenological thermodynamics. We begin here by reviewing the basic dynamical and ontological postulates of the theory as well as the "quantum equilibrium hypothesis" (QEH), the use of which allows a straightforward derivation of the standard quantum phenomenology. This overview is intended to establish a foundational context for the sketches of the two leading proposed analyses of the QEH which will be presented in the following sections.

For an N-particle system of, for simplicity, spinless, non-relativistic particles, the pilot-wave theory posits a wave function $\Psi(q,t) = \Psi(\vec{x}_1, \vec{x}_2, ..., \vec{x}_N, t)$ obeying the usual Schrödinger equation

$$i\hbar \frac{\partial \Psi}{\partial t} = -\sum_{i=1}^{N} \frac{\hbar^2}{2m_i} \vec{\nabla}_i^2 \Psi + V(\vec{x}_1, \vec{x}_2, ..., \vec{x}_N, t)\Psi \tag{5}$$

as well as N (literal, point) particles with configuration $Q(t) = \{\vec{X}_1(t), \vec{X}_2(t), ..., \vec{X}_N(t)\}$ evolving according to

$$\frac{dQ}{dt} = v(q,t)\big|_{q=Q(t)}. \tag{6}$$

The configuration-space velocity field v is given by $v(q,t) = \{\vec{v}_1(q,t), \vec{v}_2(q,t), ..., \vec{v}_N(q,t)\}$ where

$$\vec{v}_i(q,t) = \frac{\hbar}{m_i} \mathrm{Im}\left[\frac{\vec{\nabla}_i \Psi}{\Psi}\right] = \frac{\vec{j}_i(q,t)}{\rho(q,t)}. \tag{7}$$

Here $\vec{j}_i = \frac{-i\hbar}{2m_i}(\Psi^* \vec{\nabla}_i \Psi - \Psi \vec{\nabla}_i \Psi^*)$ is the ith particle component of the usual quantum "probability current"

$$j(q,t) = \{\vec{j}_1, \vec{j}_2, ..., \vec{j}_N\} \tag{8}$$

and

$$\rho = |\Psi|^2 \tag{9}$$

is the usual quantum "probability density". Note, though, that despite these traditional names ρ and j have not yet been invested with any probabilistic significance, and should instead be thought of simply as properties of the pilot-wave field Ψ. (It is perhaps helpful here to regard ρ and j as analogous, respectively, to the "field energy density" and "Poynting vector" in classical electromagnetism. However, of course the fact that Ψ, and hence ρ and j, are functions on the $3N$-dimensional configuration space, rather than 3-dimensional physical space, strains the analogy to some extent).

It is a purely mathematical consequence of Equation (5) and the definitions (8) and (9) that ρ and j jointly obey the continuity equation

$$\frac{\partial \rho}{\partial t} + \nabla \cdot j = 0 \tag{10}$$

where $\nabla = \{\vec{\nabla}_1, \vec{\nabla}_2, ..., \vec{\nabla}_N\}$. In light of Equation (7), this can be re-written as

$$\frac{\partial \rho}{\partial t} + \nabla \cdot (\rho v) = 0. \tag{11}$$

On the other hand, if at some initial time (say, $t = 0$), the particle configuration Q is *random* with distribution $P(q,0)$, then the distribution will evolve, under Equation (7), according to

$$\frac{\partial P}{\partial t} + \nabla \cdot (Pv) = 0. \tag{12}$$

The wave function "intensity" ρ and the particle probability distribution P thus evolve the same way in time and will hence remain identical for all times if they are equal at any one time. As this important fact is sometimes expressed, the particular distribution $P_B = \rho = |\Psi|^2$ is "equivariant".

One should think here of the more familiar case of the Liouville distribution (that is, the Lebesque measure restricted to the energy surface in phase space) being "invariant" in classical statistical mechanics: The Hamiltonian flow guarantees that if the phase space point is Liouville-distributed at $t = 0$ it will remain Liouville-distributed for all times. So the Liouville distribution has a dynamically privileged, equilibrium status, that is usually thought to play an important role in, for example, justifying the consideration of micro-canonical ensembles.

In the case of the pilot-wave theory, the analogous—dynamically privileged, equilibrium—distribution is not *in*variant, i.e., is not constant in time: P_B will evolve non-trivially in time whenever Ψ does. However, the time-evolutions of Ψ and P_B "track" one another, i.e., P_B retains a constant relationship to Ψ even as both evolve: $P_B(q,t) = |\Psi(q,t)|^2$. Hence "equivariance."

The subscript "B" on P_B is meant to stand for "Born". This is because the assumption—that the initial ($t = 0$) configuration of a quantum system was random with distribution P_B, i.e., the so-called "quantum equilibrium hypothesis" (QEH)—allows a straightforward proof that the pilot-wave theory reproduces the phenomenological predictions of ordinary quantum theory, i.e., the Born rule statistics. To see this, consider a typical measurement scenario involving a setup with generic configuration $q = \{x, y\}$ which we decompose into two parts representing the system-to-be-measured (x) and the measuring device (y). The measuring device should include a "pointer" whose final configuration indicates the outcome of the measurement. Suppose it is the property of the system corresponding to Hermitian operator $\hat{A}$ that is to be measured. Then (in order to justify calling the interaction a "measurement") it must be that an initial state

$$\Psi(x,y,0) = \psi_m(x)\phi_0(y) \tag{13}$$

(where $\psi_m(x)$ is an eigenfunction of $\hat{A}$ with eigenvalue a_m and $\phi_0(y)$ represents the "ready" state for the measuring apparatus) evolves, under the Schrödinger equation, into

$$\Psi(x,y,T) = \psi_m(x)\phi_m(y) \tag{14}$$

where $\phi_m(y)$ is "narrowly peaked" around configurations in which the apparatus pointer is pointing to the value "a_m". (By "narrowly peaked" here we mean that $\phi_m^*(y)\phi_n(y) \approx 0$ whenever $m \neq n$. That is, the functions $\phi_m(y)$ for distinct values of m have approximately non-overlapping support.)

Now suppose that the initial state of the system-to-be-measured is instead an arbitrary superposition of eigenfunctions, i.e.,

$$\Psi(x,y,0) = \left(\sum_m c_m \psi_m(x)\right)\phi_0(y). \tag{15}$$

(Note, we assume here that things are normalized in the standard ways so that, for example, $\sum_m |c_m|^2 = 1$). It then follows, from the linearity of Schrödinger's equation, that the post-measurement wave function will be

$$\Psi(x,y,T) = \sum_m c_m \psi_m(x)\phi_m(y). \tag{16}$$

The post-interaction system-apparatus wave function is an entangled superposition which picks out no one particular measurement outcome. In ordinary quantum mechanics, where there is supposed to be nothing but the wave function, Equation (16) thus tends to lead to apoplexy, evasion, or the hand-waving introduction of new "measurement axioms". However, in the pilot-wave theory, the actual measurement outcome is to be found, not in the wave function, but in the actual post-interaction configuration, Y, of the particles composing the apparatus pointer.

In addition, it is easy to see that if we adopt the QEH and thus regard the configuration $Q(0) = \{X(0), Y(0)\}$ as random and P_B-distributed, then, because of the equivariance property discussed above, the post-interaction configuration $Q(T)$ will have distribution $P(q,T) = P_B(q,T) = |\Psi(x,y,T)|^2$. Because of the property mentioned just after Equation (14), the cross terms will vanish and the marginal distribution for the pointer configuration Y becomes

$$P_T[Y = y] = \int P_B(q,T)dx = \sum_m |c_m|^2 |\phi_m(y)|^2. \tag{17}$$

Or equivalently, calling "Y_m" the set of (macroscopically indistinguishable) configurations in the support of $\phi_m(y)$, we can express this as follows:

$$P_T[Y = Y_m] = |c_m|^2. \tag{18}$$

That is: the probability is $|c_m|^2$ that, at the end of the experiment, the pointer indicates that the measurement had outcome a_m. This is of course just exactly the Born rule, Equation (4).

Please note that it may—and probably should—seem that there is hardly any distinction at all between the quantum equilibrium hypothesis (QEH) that we put in, and the Born rule that we get out. In effect, we put in the Born rule (for particle positions) at $t = 0$, and get out the Born rule (for particle positions and so, in particular, for the positions of the particles composing the apparatus pointer) at $t = T$. Of course, it is noteworthy that the Born rule for particle positions is preserved in time ("equivariance") and it is noteworthy that the Born rule for particle positions is sufficient to guarantee the generalized Born rule for arbitrary measurements. Still, though, to the extent that we simply assert the QEH as an additional postulate of the pilot-wave theory, the claim that the theory *predicts* Born rule statistics has a somewhat embarrassingly circular feel to it. This is why a genuinely convincing *derivation* of Born-rule statistics (in the context of the pilot-wave theory) needs to go a little deeper, by providing some kind of *justification* for the QEH. Let us then turn to exploring the two extant candidate justifications of this type.

3. The Dynamical Relaxation Justification

Following up an idea that was suggested by Bohm already in the 1950s [25,29], Antony Valentini proved, in his 1992 Ph.D. thesis [30] and associated papers [31–33], a "sub-quantum H-theorem" purporting to establish that non-equilibrium probability distributions will undergo a kind of dynamical relaxation toward coarse-grained equilibrium. This can be understood as a justification of the QEH, the idea being that even if, say, back at the big bang, particle positions were *not* P_B-distributed, they would inevitably become P_B-distributed (at least in a good-enough-for-all-practical-purposes, FAPP, coarse-grained sense) during the subsequent dynamical evolution of the universe, thereby justifying the application of the QEH to the "initial" conditions relevant to contemporary experimental investigations.

Let us review Valentini's proof (Actually we will rehearse just one of several related arguments that he presents). Define $g(q)$ as the *ratio* of $\rho = |\Psi|^2$ and the particle probability distribution P:

$$g(q,t) \equiv \frac{\rho(q,t)}{P(q,t)} \tag{19}$$

and define the sub-quantum entropy "S" (which is really a kind of relative entropy between ρ and P) as

$$S \equiv -\int dq\, P\, g\, \ln(g) = -\int dq\, \rho\, \ln[\rho/P]. \tag{20}$$

To begin with, we note that in "quantum equilibrium" $\rho = P$, so $g = 1$ and hence $S = 0$. Also, since $x \ln(x/y) \geq x - y$ for all x and y, we have that

$$S \leq -\int dq\,(\rho - P) = 0. \tag{21}$$

Thus quantum equilibrium is a maximum (sub-quantum) entropy condition.

It follows from the fact that P and ρ obey the same continuity equations—(11) and (12)—that their ratio g is constant along particle trajectories. That is:

$$\frac{dg}{dt} = \frac{\partial g}{\partial t} + \frac{dQ}{dt} \cdot \nabla g = 0. \tag{22}$$

It is then a closely related fact that the exact (sub-quantum) entropy remains constant in time:

$$\frac{dS}{dt} = 0. \tag{23}$$

What Valentini showed, however, (following a long history of parallel demonstrations in the case of classical mechanics) is that an appropriately defined *coarse-grained* (sub-quantum) entropy can be shown to relax toward equilibrium.

Let us thus divide the configuration space up into cells of finite volume δV and define coarse-grained versions of the probability measure P and wave function intensity ρ as follows:

$$\bar{P}(q) = \frac{1}{\delta V} \int_{\delta V} dq'\, P(q') \tag{24}$$

and

$$\bar{\rho}(q) = \frac{1}{\delta V} \int_{\delta V} dq'\, \rho(q') \tag{25}$$

where, in both cases, the integration is over the cell containing the point q. We may also define a coarse-grained version of g as follows:

$$\bar{g}(q) = \frac{\bar{\rho}(q)}{\bar{P}(q)}. \tag{26}$$

In terms of these quantities we may then also define the coarse-grained (sub-quantum) entropy

$$\bar{S} = -\int dq\, \bar{\rho}\, \ln(\bar{g}). \tag{27}$$

Let us then consider the change in $\bar{S}$ from some initial time (0) to some later time (t):

$$\bar{S}(t) - \bar{S}(0) = -\int dq\, \bar{\rho}(t)\, \ln[\bar{g}(t)] + \int dq\, \bar{\rho}(0)\, \ln[\bar{g}(0)] \tag{28}$$

Now comes an important assumption on the initial conditions: we assume what Valentini describes as "no 'micro-structure' for the initial state", which basically means that the functions $\rho(0)$ and $g(0)$ are "smooth" and are therefore equal to their coarse-grained counterparts. Thus

$$\bar{S}(t) - \bar{S}(0) = -\int dq\, \bar{\rho}(t)\, \ln[\bar{g}(t)] + \int dq\, \rho(0)\, \ln[g(0)]. \tag{29}$$

However, then, the constancy of the exact S implies

$$\bar{S}(t) - \bar{S}(0) = -\int dq\, \bar{\rho}(t)\, \ln[\bar{g}(t)] + \int dq\, \rho(t)\, \ln[g(t)]. \tag{30}$$

The constancy of $\bar{g}$ and hence $\ln[\bar{g}]$ over the cells implies that $\int dq\, \bar{\rho}(t)\, \ln[\bar{g}(t)] = \int dq \rho(t) \ln[\bar{g}(t)]$ so that

$$\bar{S}(t) - \bar{S}(0) = \int dq\, \rho(t)\, \ln\left(\frac{g(t)}{\bar{g}(t)}\right) = \int dq\, P(t)\, g(t)\, \ln\left(\frac{g(t)}{\bar{g}(t)}\right). \tag{31}$$

Finally, it can be shown that $\int dq\, P\, (\bar{g} - g) = 0$ so we may add zero on the right hand side of the previous equation, yielding

$$\bar{S}(t) - \bar{S}(0) = \int dq\, P(t) \left[g(t) \ln\left(\frac{g(t)}{\bar{g}(t)}\right) + \bar{g}(t) - g(t)\right]. \tag{32}$$

The term in square brackets is (again using $x \ln(x/y) \geq x - y$) non-negative, so

$$\Delta\bar{S} = \bar{S}(t) - \bar{S}(0) \geq 0. \tag{33}$$

That is, the coarse-grained (sub-quantum) entropy S *increases* from its initial value.

Let us illustrate the implied dynamical approach to equilibrium with a numerical simulation. We follow [34,35] in considering a particle of mass m moving in a two-dimensional square box potential: $V(x,y) = 0$ if $0 < x < L$ and $0 < y < L$, and $V = \infty$ otherwise. The energy eigenstates can be labeled with positive integers m, n with the energy eigenvalues given by

$$E_{m,n} = \frac{\hbar^2 \pi^2 (m^2 + n^2)}{2mL^2}. \tag{34}$$

We take the initial quantum state to be a superposition (with equal amplitudes but randomly chosen phases) of the 6 lowest-lying energy eigenstates. The state $\Psi(x,y,t)$ then evolves periodically with period $T = 4mL^2/\hbar\pi$. At $t = 0$ the wave function intensity $\rho(x,y) = |\Psi(x,y)|^2$ is as shown in Figure 1.

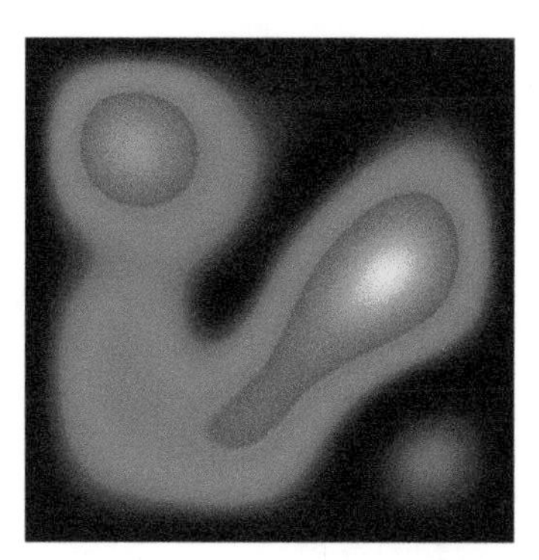

Figure 1. Density plot of $\rho(0) = |\Psi(0)|^2$. Please note that ρ evolves periodically with period T. Note also that, for the particular non-equilibrium initial distribution $P(0) =$ constant, $g(0) = \rho(0)/P(0) = \rho(0)$. So this same figure can be taken also as illustrating $g(0)$ for this particular non-equilibrium distribution. The corresponding $g(t)$ for $t = T$, $t = 2T$, and $t = -T$ are shown in subsequent figures.

As a concrete example of a non-equilibrium distribution, we will take the *uniform distribution* in which $P(x,y)$ is constant (inside the "box", and zero outside):

$$P(x,y) = \begin{cases} \frac{1}{L^2} & \text{if } 0 < x < L \text{ and } 0 < y < L \\ 0 & \text{otherwise} \end{cases} \tag{35}$$

Please note that the quantity $g = \rho/P$ discussed above is simply proportional to ρ for P = constant, so Figure 1 may also be understood as a graph of $g(x,y,0)$ for this particular non-equilibrium distribution.

Valentini's dynamical relaxation is perhaps best illustrated by considering the time-evolution of an ensemble of particles whose initial positions are chosen randomly in accordance with the non-equilibrium distribution P. The results of such a simulation, with an ensemble of 5000 particles, are shown in Figure 2. One can literally see that the initial $t = 0$ distribution evolves steadily, over the course of several periods of the background wave function dictating the particle trajectories via Equation (6), from one that "looks uniform" to one that instead "looks $|\Psi|^2$". Remember here that the wave function evolves in a periodic way, so ρ at each of the moments pictured is just the ρ shown in Figure 1.

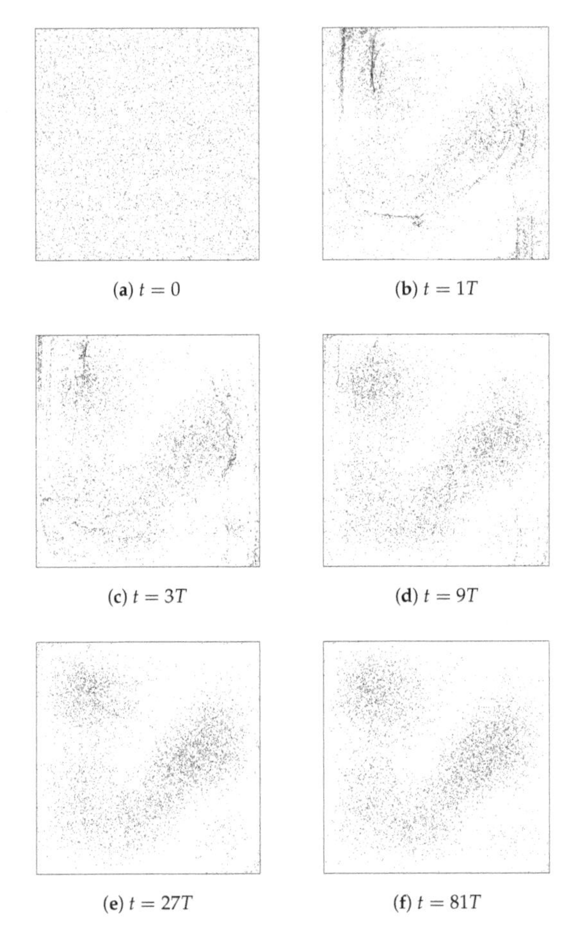

(**a**) $t = 0$ (**b**) $t = 1T$ (**c**) $t = 3T$ (**d**) $t = 9T$ (**e**) $t = 27T$ (**f**) $t = 81T$

Figure 2. Relaxation of a uniform distribution toward coarse-grained equilibrium

In this simulation, these 5000 particles *appear* $|\Psi|^2$-distributed by (something of order) $t = 10T$ or maybe $t = 100T$. However, the truth, of course, is that really the distribution is only approaching the $P = \rho$ equilibrium in a coarse-grained sense; it is just that, for this particular system, the grain-size at which the dis-equilibrium can be seen happens to be smaller than the typical inter-particle spacing by $t \approx 10T$.

To illustrate this point, we can run another identical simulation, but with a much greater number or particles. Figure 3 thus shows, in the left panel, the same thing as Figure 2's frame (d)—that is, an ensemble of initially-uniformly-distributed particles at $t = 9T$—but with 50,000 instead of 5000 particles in the ensemble. The right panel shows a zoomed-in portion of the box for an even larger ensemble, making the fine-grained (dis-equilibrium) structure easily noticeable.

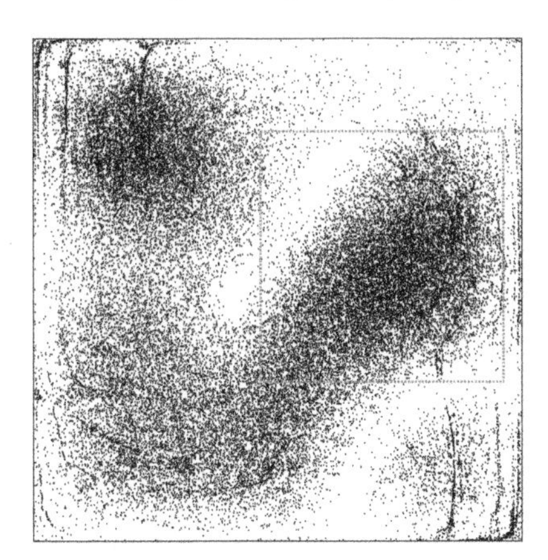
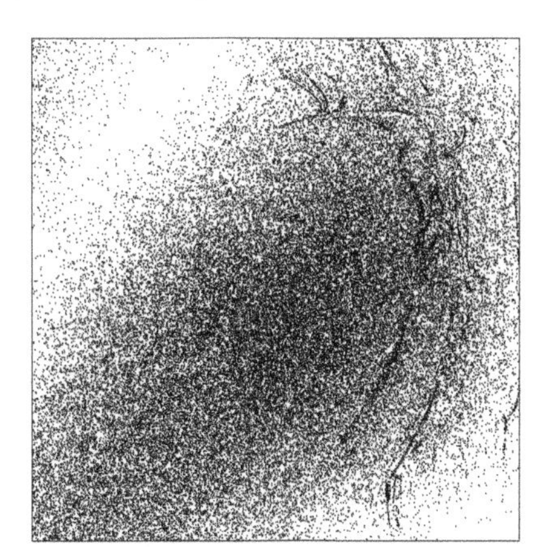

Figure 3. The left panel is the same as panel (d) of the previous figure, but for an ensemble of 50,000 initially-uniformly-distributed particles, allowing one to see some of the fine-grained structure in the distribution. The right panel zooms in on the portion of the box highlighted by the gray box on the left, for an ensemble of 200,000 particles, allowing even more fine-grained structure to be visible.

It is of course also possible to numerically track the evolution of $P(x,y)$ through time. Compared to tracking an ensemble of particle positions, doing this in detail is somewhat computationally intensive. However, it gives a very clear sense of what is going on to see the evolution of an initially-constant P over just a few periods. Actually, it is somewhat more convenient to consider $g = \rho/P$ since, as mentioned previously, the value of g is constant along particle trajectories. We can thus follow the numerical strategy introduced by Valentini and Westman to create a detailed map of $P(x,y,t)$ as follows: consider a position X,Y where we want to know $P(t)$; solve the equation of motion for the trajectory, backwards in time, to find the $t = 0$ location X_0, Y_0 of a particle which arrives at X,Y at time t; use the fact that g is constant along trajectories to assign

$$g(X,Y,t) = g(X_0,Y_0,0) = \frac{\rho(X_0,Y_0,0)}{P(X_0,Y_0,0)} \sim \rho(X_0,Y_0,0) \tag{36}$$

since, for the particular non-equilibrium distribution we consider, $P(X_0,Y_0,0)$ is a constant.

To really see the fine-grained-stucture in $P(t)$, one already needs to consider an $N \times N$ grid of points X,Y, with N of order several thousand, after just a few periods (That is, already after just a few periods, the grain-size of the structure in P is of order 10^{-3} L). So computationally, making a beautifully detailed map of g is equivalent to running a simulation with something like 10 million particles. This is all by way of explaining why it would be computationally prohibitive to continue making detailed maps of g all the way out to, say, $t = 81T$. However, as I remarked above, one nevertheless gets a very clear idea of what's happening from looking at the evolution over just a few periods. This is illustrated in Figure 4.

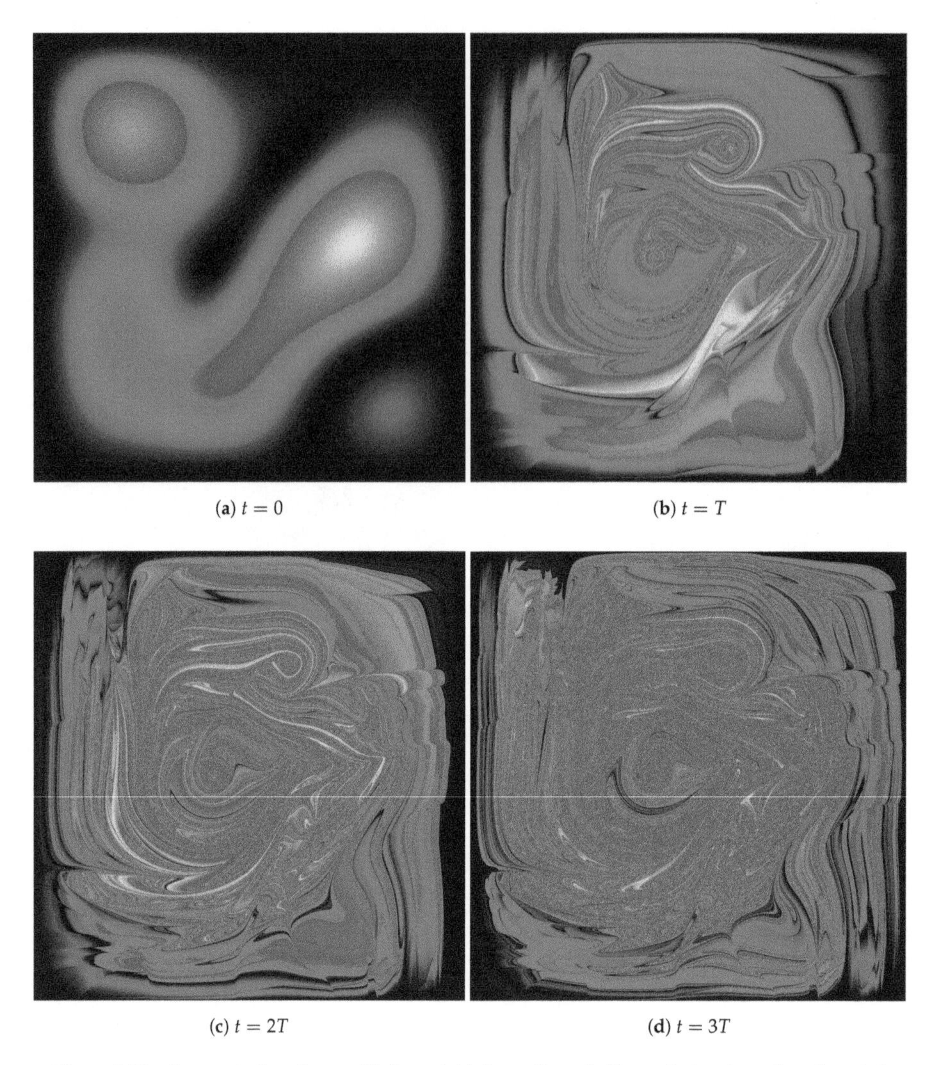

(**a**) $t = 0$ (**b**) $t = T$

(**c**) $t = 2T$ (**d**) $t = 3T$

Figure 4. The time-evolution of $g = \rho/P$, for an initially-uniform P. The evolution through each period is like a "kneading" operation which results in the non-uniformities in g being systematically mixed down to smaller and smaller length scales. Further time-evolution would eventually result in a map whose $\bar{g}$ was uniform.

Valentini characterizes the relaxation of $\bar{P}$ toward ρ—i.e., the relaxation of $\bar{g}$ to uniformity—as follows:

> "The exact (fine-grained) density is given by [$P = \rho/g$]. Now, starting from an arbitrary [$g(q, 0)$], the initial values of [g] are carried along the system trajectories in configuration space. If the system is sufficiently complicated, the chaotic wandering of the trajectories [$Q(t)$] will distribute the [g] values in an effectively random manner over the accessible region of configuration space. On a coarse-grained level, P will then be indistinguishable from [$\rho = |\Psi|^2$]. Another (equivalent) picture sees the increase of subquantum entropy as associated with the effectively random mixing of two 'fluids', with densities P and [ρ], each of which obeys the same continuity equation, and is 'stirred' by the same

> velocity field [so that] the two 'fluid' densities... will be thoroughly mixed, making them indistinguishable on a coarse-grained level". [31]

To summarize, the illustrations in this section clearly demonstrate that Valentini's notion of dynamical relaxation to sub-quantum coarse-grained equilibrium *happens* in at least some systems. In particular we have seen how an initially *smooth* g (recall Valentini's "no initial micro-structure" assumption) evolves toward a g with a very detailed, fine-grained structure which approaches perfect uniformity on a coarse-grained level. This is exactly in accordance with Valentini's demonstration (and previous numerical simulations) and seems to add something quite relevant and substantial to our quest to understand the Born rule in the context of the pilot-wave theory.

In particular, as we discussed earlier, the invocation of the Quantum Equilibrium Hypothesis (QEH) at an initial time gives the explanation of Born rule statistics a somewhat circular appearance: you in effect put the Born rule in at $t = 0$ only to get it back out again at a later time (corresponding, say, to the moment when some measurement process is completed). What Valentini's sub-quantum H-theorem purports to show is that you do not need to postulate the QEH: a broader class of initial distributions, with $P \neq \rho$, will relax towards $P = \rho$ in a coarse-grained sense that is sufficient to account for the appearance of Born rule statistics in actual, finite-resolution experiments.

4. Objections to Dynamical Relaxation

If it could be shown that *any* possible non-equilibrium initial distribution $P(q, 0)$ would relax toward coarse-grained equilibrium for any possible initial wave function $\Psi(q, 0)$, that would of course constitute a rigorous derivation of the Born rule for the pilot-wave theory. However, it should be immediately clear that this is not possible.

For one thing, there certainly exist wave functions which generate particle trajectories that are insufficiently complicated/chaotic to produce relaxation—e.g., trivially, stationary states for which the velocity field is identically zero forever. For such a Ψ the distribution will stay constant, $P(q, t) = P(q, 0)$, and there can be no relaxation toward sub-quantum equilibrium. So at best the claim can be that we might expect relaxation toward coarse-grained equilibrium for the kinds of wave functions that might plausibly obtain in our world. There is, for example, some reason to think that the wave function of our universe is not a stationary state.

However, even for such appropriately non-trivial wave functions, it cannot be the case that *all* initial particle distributions relax monotonically toward equilibrium. This is clear from the time-reversal symmetry of the theory. We illustrate here with simulations paralleling those shown in the previous section, but now letting time run backwards from the initial $t = 0$.

Figure 5 shows the evolution of an ensemble of particles (with $t = 0$ positions randomly chosen with probability distribution $P(x, y, 0) =$ constant) backwards in time. It is clear that the ensemble approaches coarse-grained equilibrium moving away from $t = 0$ in the negative temporal direction, just as it does—recall Figure 1—in the positive temporal direction. Or, considering the evolution from negative times to $t = 0$, as the images are arranged in the figure, one would say that an ensemble of particles that *appears* to have an equilibrium distribution (and is certainly in equilibrium in a coarse-grained sense) at one time may in fact evolve, at least for some considerable period of time, toward a dramatically non-equilibrium (here, uniform!) distribution.

This possibility is of course familiar from classical statistical mechanics as well (recall Loschmidt) and is not strictly speaking in conflict with Valentini's sub-quantum H-theorem: the early-time ensemble depicted, for example, in frame (a) of Figure 5 is in fact not a typical ensemble associated with the equilibrium distribution $P = \rho$. Instead, it is a typical ensemble for a very different P—one which is equivalent to ρ in a coarse-grained sense, but which in fact contains incredible "detailed microstructure". The flavor of this is conveyed by the image in Figure 6, which shows the distribution $g = \rho/P$, not at $t = -81T$ but at $t = -T$, which forward-evolves into $g = \rho$ (i.e., into $P =$ constant) at $t = 0$ (The distribution g that would be required at even earlier times is qualitatively similar, but of course with a considerably more fine-grained micro-structure). In any case, one can see that the initial distribution P for which the ensemble depicted in frame (a) of Figure 5 would be a typical example, will fail to respect Valentini's "no initial fine-grained micro-structure" assumption, $P(q,0) = \bar{P}(q,0)$.

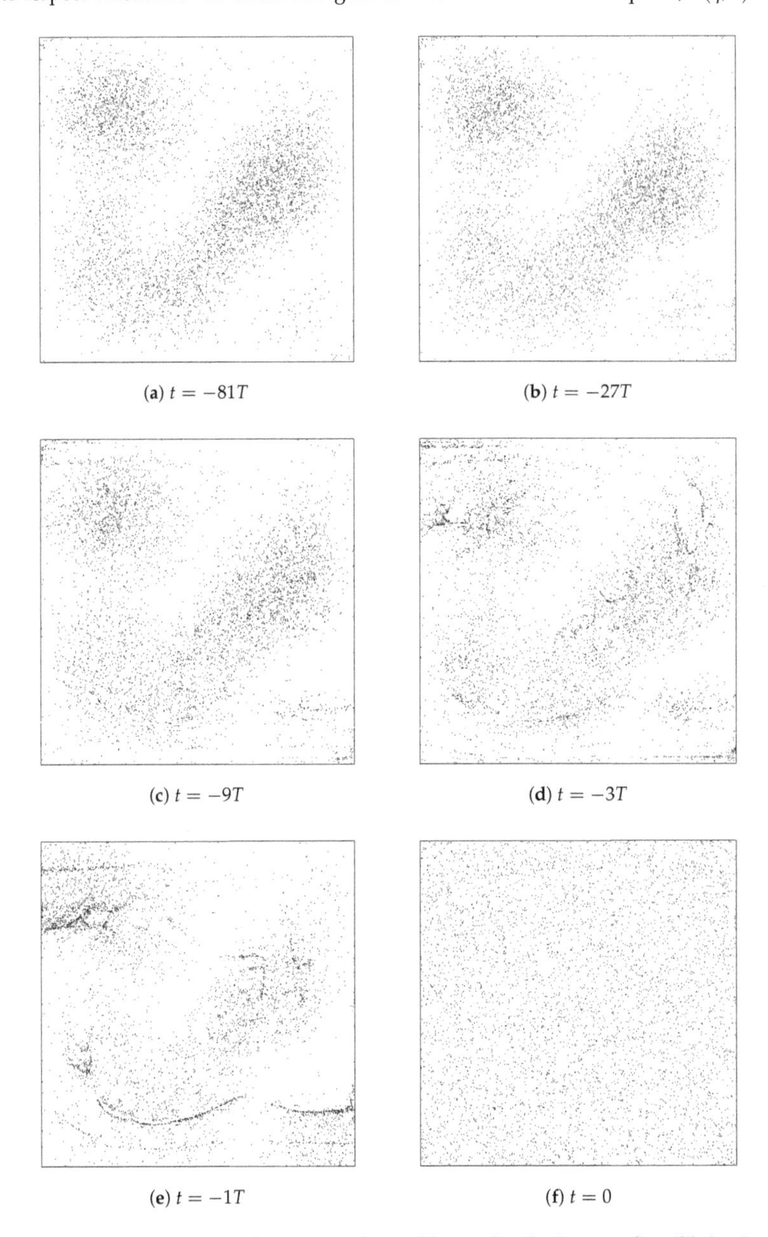

(**a**) $t = -81T$ (**b**) $t = -27T$

(**c**) $t = -9T$ (**d**) $t = -3T$

(**e**) $t = -1T$ (**f**) $t = 0$

Figure 5. Anti-relaxation of an apparently-equilibrium distribution out of equilibrium!

Callender has questioned whether this assumption of Valentini's proof, that $P = \bar{P}$ initially, is any improvement over just assuming the quantum equilibrium hypothesis, namely, that $P = \rho$ initially: "In both cases we are assuming that the early configuration distribution had a rather special profile" [27]. I think it would indeed be an improvement, at very least in the sense that it is a much weaker assumption: there are many possible "smooth" initial distributions satisfying $P = \bar{P}$ which do not correspond to coarse-grained equilibrium ($\bar{P} = \rho$).

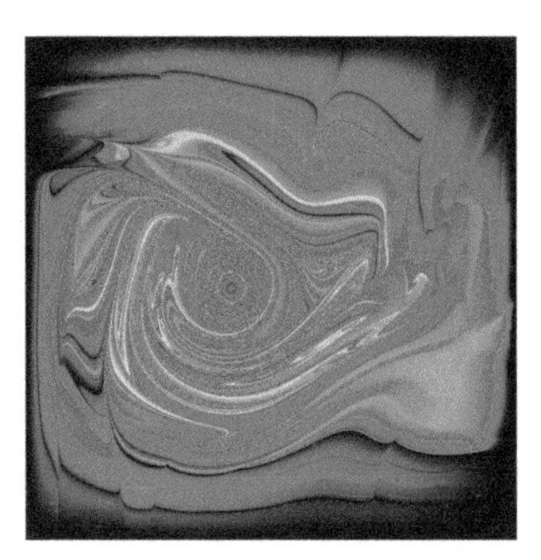

Figure 6. This is the distribution for $g = \rho/P$ that is required at $t = -T$ in order to make $g = \rho$, i.e., to make $P =$ constant, at $t = 0$. A very specific fine-grained micro-structure in P is required, at the earlier time, to generate a smooth, easily-describable non-equilibrium distribution at a later time.

Another set of possible concerns about the dynamical relaxation approach involves the distinction between continuous distributions P—whose relation, over time, to ρ is illustrated, for example, in Figure 4—and ensembles of particles with definite positions, as illustrated, for example, in Figure 2. It should be clear that if the quantum system we are studying is the universe as a whole—and the pilot-wave theory is fundamentally a theory about this ultimate system—then $\rho = |\Psi|^2$ (where Ψ is the universal wave function) is in some sense a really-existing thing, but P, the probability distribution over particle configurations Q of the universe as a whole, is not: what the pilot-wave theory posits as existing in addition to the universal wave function $\Psi(q,t)$ is just some definite configuration $Q(t) = \left\{\vec{X}_1(t), \vec{X}_2(t), ..., \vec{X}_N(t)\right\}$. There is no such additional thing as the probability distribution P over such configurations.

Callender criticizes Valentini on something like these grounds, namely for failing to clearly distinguish between the (effective) wave function ψ of an individual sub-system (within the universe) and the wave function Ψ of the universe. In particular, Callender stresses that "what is really needed is a justfication of" what, in our terminology, would be expressed as $p = |\psi|^2$, i.e., a justfication for taking the probability distribution p associated with an ensemble of sub-systems with identical effective wave functions ψ to be $|\psi|^2$. Callender states that universal quantum equilibrium, $P = |\Psi|^2$, is not sufficient for establishing what is really needed, namely, $p = |\psi|^2$, and adds:

> "Even if one proves that the universe as a whole is in quantum equilibrium, we really want to prove that patterns inherent in subsystems of the universe are in quantum equilibrium. Whether any [of Valentini's results] survive the move to the proper understanding of [$p = \rho$] is not clear". [27]

I think Callender is on the right track about what is needed and is right to question whether Valentini's argument provides it. That said, I do not think Callender quite puts his finger on the problem. If one could prove that "the universe as a whole is in quantum equilibrium"—$P = |\Psi|^2$—I think it would indeed follow that ensembles of sub-systems with identical effective wave functions would also be in quantum equilibrium—$p = |\psi|^2$. Indeed, Valentini demonstrates this very thing in Section 4 of [31].

The problem, though, is that it cannot be possible to prove that "the universe as a whole is in quantum equilibrium" because the "P" in "$P = |\Psi|^2$" does not and cannot refer to anything real. That is, there is simply no such thing as the probability distribution P for particle configurations of the universe as a whole, because there is just one universe. This point, raised as a criticism of Valentini's dynamical relaxation program, was made sharply by Dürr, Goldstein, and Zanghí: given the assumption that the configuration Q of the universe is random with probability distribution $|\Psi|^2$,

> "you might well imagine that it follows that any variable of interest, e.g., X, has the 'right' distribution. However, even if this were so (and it is), it would be devoid of physical significance! As Einstein has emphasized, 'Nature as a whole can only be viewed as an individual system, existing only once, and not as a collection of systems'." [36]

Coming fully to grips with this does not immediately render Valentini's analysis irrelevant, for we can still attempt to understand the analysis as a description of ensembles of systems with identical effective wave functions. However, it does lead to several further points which I think are problematic for the relaxation approach to understanding Born rule statistics in the context of the pilot-wave theory.

Let's start by straightening out the terminology a bit. Suppose the universe includes n sub-systems, with coordinates $x_1, x_2, ..., x_n$, and that the universal wave function has the form

$$\Psi = \psi(x_1)\psi(x_2)\cdots\psi(x_n)\Phi(y) + \Psi^{\perp} \tag{37}$$

where y denotes the coordinates of the environment of our collection of sub-systems, and $\Psi^{\perp}$ is macroscopically disjoint from $\Phi(y)$. In the pilot-wave theory we of course also have definite particle positions $X_1(t), X_2(t), ..., X_n(t)$ and $Y(t)$. If Y is in the support of $\Phi(y)$, then $\Psi^{\perp}$ will be irrelevant to the future motion of the particles and we may assign the effective wave function ψ to each of the n sub-systems.

We can define a (really-existing, meaningful) empirical distribution for the ensemble by taking

$$p(x,t) = \frac{1}{n}\sum_{i=1}^{n}\delta\left(x - X_i(t)\right). \tag{38}$$

Valentini pursues this line in [30] (p. 18) and notes that "In the limit of large n, $[p(x,t)]$ may for our purposes be replaced by a purely smooth function, again denoted just $[p(x,t)]$, which in practical circumstances will behave like a probability distribution." The relaxation analysis is then intended to show that the coarse-grained $\bar{p}$ approaches the Born-rule distribution $\rho = |\psi|^2$ for the sub-system.

Valentini acknowledges that

> "[t]he very concept of a smooth distribution $[p]$ is limited, being strictly valid *only in the purely theoretical limit of an infinite ensemble* ($n \to \infty$). This implies for example that in a laboratory consisting of a finite number of atoms, the actual distribution (say of electron positions) has the discrete form [of our (38), just above] so that one *necessarily* has some disequilibrium $[p \neq |\psi|^2]$ on a fine-grained level". [30]

He acknowledges, that is, that for realistic finite ensembles the approach to equilibrium can in some sense only be approximate. This raises, for me, two concerns.

The first is that there would seem to be an additional, but unacknowledged, approximation that is relevant: the sub-system (effective) wave functions ψ will only evolve according to their own sub-system Schrödinger equations if the sub-systems are *perfectly* isolated from their environments. However, again, for more realistic ensembles of similarly-prepared systems, the isolation will be imperfect and hence the sub-system wave functions ψ for the individual ensemble members will not even evolve the same way in time. On the one hand, this renders certain key aspects of Valentini's analysis as either questionable or downright inapplicable. For example, if, after some evolution time, the initially-identical effective wave functions for the individual ensemble members are no longer the same, there will not even exist any particular $\rho = |\psi|^2$ to which one might meaningfully compare the distribution $\bar{p}$! On the other hand, though, random variations in the evolution of the individual ensemble members' effective wave functions (arising from random variations in the precise way in which each member is only imperfectly isolated from its environment) may constitute an additional source of noise that could accelerate the dynamical relaxation toward Born-rule distributions, in some appropriate sense. This issue deserves further study and attention.

However, there is a second and somewhat more fundamental worry associated with the fact that real ensembles will always be finite so that the distribution function p will be given by Equation (38) and hence not be a continuous function. The worry is not quite what Valentini acknowledges, namely that the fine-grained structure of the distribution p (for any realistic finite ensemble of identically-prepared sub-systems) implies "some disequilibrium ... on a fine-grained level". Rather, the worry is that the fine-grained structure in p will mean that p cannot possibly equal its coarse-grained partner $\bar{p}$. However, that $p = \bar{p}$ at the initial time was one of the crucial assumptions of the theorem. So the worry is that the theorem is thus rendered simply inapplicable to realistic finite ensembles, i.e., irrelevant to goings-on in our universe.

In his 1991 paper, Valentini characterizes the "no fine-grained initial micro-structure" assumption as follows:

> "It is this assumption which introduces a distinction between past and future: Essentially, it is assumed that there is no special 'conspiracy' in the initial conditions, which would lead to 'unlikely' entropy-decreasing behaviour". [31]

The basic problem that this leads us to seems to be the following: any *specific* initial condition for the configuration of the particles in our ensemble will possess "fine-grained initial micro-structure", i.e., will be extremely 'unlikely'—one might even say 'conspiratorial'—just in the sense that, no matter which smooth probability distribution p one regards the realized configuration as having been drawn from, it will be one out of a continuously infinite set of possible such configurations, and hence highly "unlikely".

The real question is: what *portion* of the continuously infinite set of possible configurations (each of them, considered alone, being extremely "unlikely"!) would lead to the troublesome "entropy-decreasing behaviour"? Do these "bad" initial configurations constitute rare exceptions to the typical behavior, or are the overwhelming majority of possible initial configurations of this "bad" sort, or what?

It is true, for example, that many possible initial configurations (for the ensemble of 2-D particle-in-a-box systems), drawn randomly from the probability distribution $p = |\psi|^2$, will evolve into different configurations that still nevertheless have the property of being typical exemplars of $p = |\psi|^2$, i.e., of looking like equilibrium ensembles. However, it is also true that there are initial configurations, drawn randomly from the probability distribution $p = |\psi|^2$, which "lead to... entropy-decreasing behaviour", i.e., which evolve into configurations that are extremely unusual, atypical exemplars of $p = |\psi|^2$, i.e., which do not look like equilibrium distributions at all. Figure 5 of course provides a concrete example here.

Similarly, we have shown (by numerical simulation, see Figure 2) that many possible configurations, drawn randomly from a "non-equilibrium" distribution such as p = constant, evolve into configurations that appear to be in equilibrium, i.e., which have $\bar{p} = |\psi|^2$. However, of course there will also exist initial configurations for the ensemble, that could be drawn randomly from p = constant, which do *not* evolve into something that looks equilibrated, but instead evolve into something that is even further from quantum equilibrium than p = constant.

All of these behaviours are possible and are realized for some of the continuously infinite set of possible configurations for the ensemble. To be genuinely convincing, then, an argument for "relaxation toward sub-quantum equilibrium" would need to establish that, in some appropriate sense, *most* of these possible non-equilibrium configurations will relax toward coarse-grained equilibrium. It would need to establish, in a word, that relaxation toward equilibrium is *typical*. I believe this is actually true: establishing it is very closely-related to what Valentini has already done, and anyway it is strongly reinforced by the numerical simulations we've used for illustrative purposes here and those more serious versions done by Valentini and Westman and Towler et al. in [34,35].

However, reframing what is needed from a relaxation argument in this way—in terms of typicality—also severely undercuts the *need* for a relaxation argument in the first place, in order to understand the origin of the Born rule statistics in the pilot-wave theory. Understanding that point is the subject of the following section.

5. The Argument from Typicality

J.S. Bell, in his 1981 essay "Quantum Mechanics for Cosmologists", discusses the pilot-wave theory and the explanation of Born rule statistics therein. He notes that "it is easy to construct in the pilot-wave theory an ensemble of worlds which gives the [Born rule] exactly". Given what we've called equivariance, "it suffices to specify... that the initial configuration $[Q(0)]$ is chosen at random from an ensemble of configurations in which the distribution is" $P = |\Psi(q,0)|^2$ [37].

Bell then continues, echoing some of the concerns raised in the previous section:

> "However, this question arises: what is the good of ... giving distributions over a hypothetical ensemble (of worlds!) when we have only one world. The answer [... is that...] a single configuration of the world will show statistical distributions over its different parts. Suppose, for example, this world contains an actual ensemble of similar experimental set-ups. [...I]t follows from the theory that the 'typical' world will approximately realize quantum mechanical distributions over such approximately independent components. The role of the hypothetical ensemble is precisely to permit definition of the word 'typical'." [37]

The idea, then, is that there is no need to try to explain how quantum equilibrium statistics could arise from some earlier out-of-equilibrium distribution; if the overwhelming majority of possible initial configurations of the universe will exhibit Born rule statistics—if Born-rule statistics are *typical*—then we should expect to see them and there is no further mystery to explain if we do.

We will give a brief rehearsal of the detailed argument, presented by Dürr, Goldstein, and Zanghí in 1992 [38], below. However, first, to help set the idea (which is often mis-understood), we discuss a couple of simple warm-up examples.

As a first example, suppose you pull a coin from your pocket and flip it 100 times. In addition, suppose that the particular sequence that you see happens to include roughly equal numbers of Heads and Tails (for example, 47 Heads and 53 Tails). How can you explain this fact? It would normally be regarded as a sufficient explanation to simply note that, of all the 2^{100} possible sequences you might conceivably have seen, *almost all* of them have "roughly equal numbers of Heads and Tails" (Some concrete numbers: there are approximately 10^{29} 100-flip sequences which have precisely 50 Heads, and about that same number again with 49, 48, and 47 Heads... whereas there is only a single sequence which has zero Heads. More than 99% of the 2^{100} possible sequences have between 35 and 65 Heads). The property of having "roughly equal numbers of Heads and Tails", that is, is *typical* of 100-flip sequences. In addition, this means that, unless we have some specific relevant information about the way the flips were conducted (e.g., the coin is unevenly weighted and therefore highly biased towards Heads, or the "flips" were not really independent flips at all but were in some way highly correlated) we should not be surprised by the observed results. We saw, in short, just the kind of behavior that we should have expected to see, so there is nothing further to explain.

This same kind of reasoning is of course common and crucial in statistical mechanics. For example, consider the distribution of velocities for the N individual molecules in a certain macroscopic sample of (say, monatomic ideal) gas. Every student of thermodynamics learns that we should expect to observe the Maxwell velocity distribution

$$P(\vec{v})\, d^3v \sim e^{-\alpha v^2} d^3v \tag{39}$$

(where $\alpha = m/2k_BT$), but what is the explanation for this? Again, the usual explanation is simply that the overwhelming majority of states that the gas might conceivably be in, will exhibit this distribution. Here is Boltzmann:

> "... by far the largest number of possible velocity distributions have the characteristic properties of the Maxwell distribution, and compared to these there are only a relatively small number of possible distributions that deviate significantly from Maxwell's". [39]

Of course, unlike the case of the sequence of coin flips in which the total number of sequences is finite, the number of distinct points in phase (or just velocity) space, consistent say with some constraint on the total energy of the gas, is continuously infinite. Any statement about properties exhibited by the "overwhelming majority" of the states thus presupposes a measure μ over the states.

The usual thing in the context of classical statistical mechanics is to take μ to be the restriction to the energy surface of the Lebesque measure on phase space; this measure is "natural" in the sense that it is invariant under the flow generated by the Hamiltonian equations of motion (Liouville's theorem). It is then straightforward to prove that the overwhelming number of points on the energy surface exhibit the Maxwell velocity distribution. More formally, in the $N \to \infty$ limit, the μ-measure of the set of points for which the velocity distribution differs significantly from Equation (39), approaches zero. See [40] for a more detailed and very clear discussion.

Please note that it would be very strange, having shown that the Maxwell distribution is *typical* with respect to the measure μ, to worry that the specific choice of measure made any difference to the velocity distribution one regards as typical. Any other measure μ' that is absolutely continuous with μ will, by definition, agree about the size of measure-0 and measure-1 sets. Of course, by hand-picking a measure that is concentrated on special points in phase space, one could diagnose any distribution one wants as "typical". For example, consider the measure which is zero everywhere except at the point in velocity space where all N particles have the same velocity, $\vec{v}_0$:

$$\mu' = \delta^3(\vec{v}_1 - \vec{v}_0)\delta^3(\vec{v}_2 - \vec{v}_0) \cdots \delta(\vec{v}_N - \vec{v}_0). \tag{40}$$

It would then follow that μ'-most of the accessible phase space points exhibit the highly non-Maxwellian velocity distribution

$$P(\vec{v})\, d^3v \sim \delta(\vec{v} - \vec{v}_0) \tag{41}$$

in which all particles have identical velocities $\vec{v}_0$. However, such games are as transparently ridiculous as they are possible. The idea is that any "reasonable" measure—any measure which is not specifically hand-tailored to give special weight to phase space points not exhibiting the Maxwell distribution—will agree with the "natural" measure about the Maxwell distribution being typical.

In the case of the pilot-wave theory, in which the state of the universe at a given moment t is given by the universal wave function $\Psi(q,t)$ and the particle configuration $Q(t)$, the natural measure of typicality is the one given by $\mu = |\Psi|^2$. The equivariance property discussed above means that, although μ itself will be time-dependent (because Ψ is time-dependent), the form "$\mu = |\Psi|^2$" will be timelessly true: $\mu(q,0) = |\Psi(q,0)|^2$ implies $\mu(q,t) = |\Psi(q,0)|^2$ for all t. (The uniqueness of this equivariant measure was established in [41].)

We may divide the universe into a sub-system of interest (with degrees of freedom x) and its environment (with degrees of freedom y), so $q = (x,y)$ and $Q = (X,Y)$. It is then trivial to derive what Dürr, Goldstein, and Zanghí call the "fundamental conditional probability formula"

$$\mu(X_t \in dx|Y_t) = |\psi(x,t)|^2\, dx \tag{42}$$

where

$$\psi(x,t) = \frac{\Psi(x,Y(t),t)}{\int |\Psi(x,Y(t),t)|^2\,dx} \tag{43}$$

is the (normalized) conditional wave function (CWF) for the sub-system.

It is reasonable to think of the CWF as the wave function of the sub-system, in the context of the pilot-wave theory, for at least two reasons: first, the guidance formula for the sub-system configuration X can be written directly in terms of $\psi(x)$; and second, the CWF will obey the obvious sub-system Schrödinger equation when the sub-system is suitably decoupled from its environment. (Note, however, that in general the CWF does *not* obey a simple sub-system Schrödinger equation, but instead evolves in a more complicated way; this is a feature, not a bug, since the complicated non-linear evolution in fact reproduces, in a precise and continuous way, the complicated non-linear evolution that would be predicted, in the context of ordinary QM, by the *ad hoc* combination of Schrödinger evolution and intermittent applications of the collapse postulate). In the simple case that the full system wave function is a product,

$$\Psi(x,y) = \psi(x)\phi(y), \tag{44}$$

the CWF for the x sub-system coincides with $\psi(x)$. One should thus appreciate that the CWF is a generalization of the ordinary quantum mechanical wave function: the Bohmian CWF agrees with the wave function that would be attributed to a system in ordinary QM *in those situations where ordinary QM would attribute any definite wave function at all to the sub-system*; but the Bohmian CWF always exists and provides a rigorous interpolation (consistent with the overall Bohmian dynamics for the particles) between such times, even through preparations and measurements. Note also that, in the situations where the *effective* wave function for a sub-system exists, it is given by the CWF, so we use those interchangeably here.

To understand Dürr, Goldstein, and Zanghí's main statistical result, let us consider again the situation described in Section 3 in which the universal wave function has the structure given in Equation (37). Suppose that $\Psi^{\perp}$ and Φ have macroscopically disjoint y-supports and that $Y_t \in supp(\Phi)$ so that each of the n sub-systems has identical CWF ψ. It then follows from the fundamental conditional probability formula that

$$\mu(X_1 \in dx_1, X_2 \in dx_2, ..., X_n \in dx_M|Y) = |\psi(x_1)|^2 \cdots |\psi(x_n)|^2\,dx_1 \cdots dx_n. \tag{45}$$

Thus, the configurations X_i of the particles composing the members of our ensemble of identically-prepared sub-systems are independent, identically distributed random variables, with common distribution $|\psi(x)|^2$.

It is then essentially an immediate and standard application of the law of large numbers to infer that, in the $n \to \infty$ limit, the empirical distribution

$$p_{emp}(x) = \frac{1}{n}\sum_{i=1}^{n}\delta(x - X_i) \tag{46}$$

is very close to $\rho = |\psi|^2$ for μ-most initial configurations $Q(0)$ (A slightly more precise statement is that the measure of the "agreement set" for which $||p_{emp} - \rho|| \leq \epsilon$, for a suitable notion of $||\cdot||$, approaches unity as $n \to \infty$. Note also that, in addition to the "equal time analysis" we have sketched here, Dürr, Goldstein, and Zanghí provide in addition an analysis of the statistics of measurements performed on sub-systems across time; they show that, in this case as well, Born-rule statistics are typical. See [38] for elaboration and details).

The argument is thus completely parallel to the standard statistical-mechanical explanation of the Maxwell velocity distribution sketched above: using an appropriate natural measure over the space of possible states, it is possible to infer that the overwhelming majority of those states will realize a certain statistical property—namely, that ensembles of sub-systems, each member of which is prepared with (effective) wave function ψ, should exhibit Born-rule statistics $p = |\psi|^2$. One should therefore expect the positions of particles in identically-prepared quantum mechanical sub-systems to be Born-rule distributed, according to the pilot-wave theory, for exactly the same reason that one should expect, in the context of classical mechanics, to see a Maxwell velocity distribution: "... by far the largest number of possible ... distributions have [this characteristic] distribution, and compared to these there are only a relatively small number of possible distributions that deviate significantly from [it]."

There is thus no need, according to the proponents of the typicality program, to explain how "quantum equilibrium" might have arisen, via relaxation, from some earlier non-equilibrium distribution (and so whether, in particular, relaxation toward equilibrium is typical for non-equilibrium distributions). If "by far the largest number of possible ... distributions" exhibit Born rule statistics, then there is precisely the same motivation for expecting non-Born-rule distributions in, say, the early universe as there is for expecting them today—namely, none.

6. Objections to the Typicality Argument

Proponents of the typicality argument insist that the measure μ be thought of (merely) as a measure of typicality and not as a fully detailed probability distribution. They stress, that is, that in the argument the *only* use to which μ is put is in assessing certain sets as having μ-measure of approximately unity (or approximately zero) (See [42] for further discussion).

It must be admitted, though, that the distinction between regarding $\mu = |\Psi|^2$ as a *typicality* measure, and regarding it instead as a full-fledged *probability* measure, is rather subtle and perhaps has the air of a distinction without much of a difference (Valentini, for example, writes in a footnote of [43]: "Note that if the word 'typicality' is replaced by 'probability', the result of Dürr et al. becomes equivalent to the 'nesting' property proved by Valentini, which states that an equilibrium probability for a many-body system implies equilibrium probabilities for extracted subsystems...") (Indeed, such skepticism about the importance, or meaningfulness, of the difference is perhaps supported by the name Dürr, Goldstein, and Zanghí give to Equation (42)). The most common and important objection to the claim that Born-rule statistics are *typical* (and hence not in need of some special explanation involving, for example, dynamical relaxation toward equilibrium) is thus that the argument purporting to establish this is *circular*: we only get Born-rule statistics out for subsystems, according to this objection, because we assume Born-rule statistics apply to the universe as a whole.

In [43], for example, Valentini writes that the approach of Dürr et al.

> "may be illustrated by the case of a universe consisting of an ensemble of n independent subsystems (which could be complicated many-body systems, or perhaps just single particles), each with wavefunction $\psi_0(x)$. Writing $\Psi_0^{univ} = \psi_0(x_1)\psi_0(x_2)\cdots\psi_0(x_n)$ and $X_0^{univ} = (x_1, x_2, x_3, ..., x_n)$, a choice of X_0^{univ} determines – for large n – a distribution $\rho_0(x)$ which may or may not equal $|\psi_0(x)|^2$.
>
> "Now it is true that, with respect to the measure $|\Psi_0^{univ}|^2$, as $n \to \infty$ almost all configurations X_0^{univ} yield equilibrium $\rho_0 = |\psi_0|^2$ for the subsystems. It might then be argued that, as $n \to \infty$, disequilibrium configurations occupy a vanishingly small volume of configuration space and are therefore intrinsically unlikely. However, for the above case, with respect to the measure $|\Psi_0^{univ}|^4$ almost all configurations X_0^{univ} correspond to the *dis*equilibrium distribution $\rho_0 = |\psi_0|^4$. This has led to charges of circularity: that an equilibrium probability density $|\Psi_0^{univ}|^2$ is in effect being assumed for X_0^{univ}; that the approach amounts to inserting quantum noise into the initial conditions themselves..."

In [44], Jean Bricmont rehearses a closely related objection. He considers an ensemble of particles in the ground state of a one-dimensional length-L "box" potential so that

$$\Psi(x_1, x_2, ..., x_N, t) = \phi(x_1, t)\phi(x_2, t) \cdots \phi(x_N, t) \tag{47}$$

where $\phi(x,t) = \sqrt{\frac{2}{L}} \sin(\pi x/L) e^{-iE_0 t/\hbar}$. Bricmont explains that

> "because of the law of large numbers, the set of typical points [for which the statistical distribution of positions in the ensemble approximately matches $|\phi|^2$)] will have a $|\Psi|^2$-measure close to 1, for N large. So, if one picks a microscopic configuration of the universe Q that is typical relative to $|\Psi|^2$, it will give rise... to an empirical distribution satisfying Born's statistical law...
>
> "[Dürr, Goldstein, and Zanghí] claim that quantum equilibrium and therefore Born's law is actually very natural. However, for that last claim to be right, one needs to argue that the measure with respect to which the configurations under discussion are typical is itself 'natural' (every configuration is typical with respect to at least one measure—the delta measure concentrated on itself)". [44]

Bricmont goes on to state that, for example, on Bayesian grounds, one might reasonably expect a uniform distribution rather than the $|\phi|^2$-distribution which vanishes at the edges of the "box". The implication is that one has not really *explained* the $|\phi|^2$ distribution, but only showed that it follows from a typicality measure that has been specifically selected to produce this very result:

> "In fact the only 'explanation' of the fact that we obtain a $|\phi|^2$ distribution rather than a uniform distribution is probably that God likes quantum equilibrium and Born's law and so put it there at the beginning of times.
>
> "The upshot of this discussion is that quantum equilibrium, in Bohmian mechanics, should, in my view, be presented as a postulate, independent of the other ones, rather than as somehow being the only natural or reasonable choice. It is not a particularly unnatural choice and it is true that quantum equilibrium is still far less mysterious than classical non equilibrium at the origin of the universe... But one should not present it as more natural than it is". [44]

The objections here share the following common structure: when the universal wave function has the structure of Equation (37), where there is an ensemble of n sub-systems with identical effective wave functions ψ, it is true that μ-most configurations will display empirical statistics consistent with the Born rule, $p = |\psi|^2$, if the typicality measure is given by $\mu = |\Psi|^2$. But it is also true, for example, that μ-most configurations will display (non-Born rule!) $p = |\psi|^4$ statistics, if, instead, the typicality measure is given by $\mu = |\Psi|^4$... and that μ-most configurations will display (differently non-Born rule!) $p =$ constant statistics if, instead, the typicality measure is given by $\mu =$ constant... and so on.

The basis for the feeling of circularity is clear: it seems that one simply gets, as the "typical" statistics for ensembles, whatever one wants, and in particular the sub-system equivalent of whatever one uses for the fundamental typicality measure μ: $\mu = |\Psi|^2$ gives $p = |\psi|^2$, but $\mu = |\Psi|^4$ gives $p = |\psi|^4$, etc. Of course, the equivariance of the specific measure $\mu = |\Psi|^2$ may suggest that this particular measure is dynamically privileged and hence in some sense "natural". But, like other "naturalness" arguments in physics, this may seem rather subjective and insufficiently substantial as a foundation for a genuine *explanation* of Born rule statistics.

7. Discussion

The debate, between those who think that Born-rule statistics in the pilot-wave theory should be explained by some kind of dynamical relaxation and those who think the Born-rule should be derived by a typicality analysis, has raged for several decades in a rather sectarian way. My view is that both perspectives contain valuable insights and both sides offer critiques of their opponents' arguments which can sharpen our understanding of the truth. We should look to combine the insights emerging from both sides of the debate, and thereby aim to construct a single unified explanation of the Born-rule, instead of feeling constrained to make a binary, either-or choice between them.

For example, I think it must be recognized that the dynamical relaxation perspective at least tacitly relies on appeals to typicality. It is emphatically *not* the case that *every* initial configuration, consistent with some particular out-of-equilibrium, non-Born-rule distribution, will evolve monotonically toward coarse-grained equilibrium. See, e.g., Figure 5, which shows an ensemble that is initially in coarse-grained equilibrium evolving through a sequence of increasingly *out-of*(-coarse-grained)-equilibrium distributions. In addition, as discussed in Section 4, there is not much comfort in the fact that dynamical relaxation has only been shown to occur for initial distributions respecting the "no fine-grained micro-structure condition" which, strictly speaking, the probability distributions for which the ensembles pictured in frames (a)–(e) of Figure 5 are typical exemplars violate. There *is* no actually-existing universal "probability distribution" $P(q)$. All that exists (in addition to the wave function) is the actual configuration Q and its subsets, including *finite* ensembles of similarly-prepared subsystems whose empirical distributions, given by Equation (38), necessarily contain fine-grained micro-structure.

Moreover, such finite-ensemble empirical distributions simply do not allow any meaningful statements about a hypothetical associated continuous probability distribution $p(x)$. For example, what does it even mean to say that the probability distribution $p(x)$—associated with the ensemble depicted in frame (f) of Figure 2—has *no* fine-grained micro-structure, whereas the $p(x)$ associated with the ensemble depicted in frame (a) of Figure 5 *does* have fine-grained micro-structure? I think such a claim is utterly empty in the final analysis. All one can say is that the two ensembles share the same equilibrium, Born-rule coarse-grained distribution $\bar{p}(x) = |\psi(x)|^2$, but one of them has the property that it *stays* Born-rule distributed (in the coarse-grained sense) whereas the other one (at least for a while!) does not.

As discussed in Section 4, I think these considerations imply that, at most, the claim defended by the dynamical relaxation program must be that relaxation toward coarse-grained equilibrium (i.e., toward FAPP Born-rule statistics) is *typical*: not all configurations (that do not already exhibit Born-rule statistics) will evolve into configurations that do exhibit Born-rule statistics, but the overwhelming majority of them (in some appropriate sense) will.

The sorts of numerical simulations pioneered by Valentini and Westman (and which I have reproduced here for illustrative purposes) strongly suggest that this modified dynamical relaxation claim is *true*: evolution toward coarse-grained sub-quantum equilibrium really is typical, such that, if one believes that the universe may have started in a (globally atypical) configuration for which Born-rule statistics did not obtain, we should expect Born-rule statistics to emerge over time and hence be observed today (As a concrete example in support of this claim, I ran the simulation depicted in Figure 2, several times—with different initial positions randomly drawn from the distribution $p =$ constant—and it looks qualitatively the same every time). But the explicit recognition of the role of typicality in that explanation should also remove most of the motivation for thinking the universe may have started in some "out-of-equilibrium" initial state in the first place, such states representing, after all, a vanishingly small fraction (...at least, if one uses the supposedly "natural" measure of typicality, $\mu = |\Psi|^2$...) of those which could evolve into configurations consistent with what is observed today.

Note here the dramatic contrast with the case of thermodynamic equilibrium, for which there is compelling observational evidence that the out-of-thermodynamic-equilibrium state we see today did arise from an earlier state that was even further out-of-equilibrium. In the sub-quantum case that we have been discussing in this paper, all observational evidence to date is consistent with the (sub-) Quantum Equilibrium Hypothesis, and—contrary to claims made sometimes by Valentini (for example, Valentini and Westman write in [34]: "a relaxation process from an earlier non-equilibrium state.... leads naturally to the suggestion that quantum non-equilibrium may have existed in the early Universe...")—there is no compelling reason to think that today's equilibrium arose from an earlier state of disequilibrium. The idea that *most* non-equilibrium configurations evolve toward coarse-grained equilibrium (which is how we are suggesting Valentini's claims should be understood, and which idea we are willing to grant is probably true) does *not* imply that equilibrium is most likely to have arisen from earlier disequilibrium. Just the opposite is true! Concretely, if one randomly chooses a large number of equilibrium ensembles that look like frame (f) of Figure 2 and runs them all backwards in time, very few will look like frame (a) of Figure 2 (or any other disequilibrium distribution) at $t = 0$; the overwhelming majority will still look just like frame (f), i.e., will exhibit equilibrium, Born-rule statistics at the initial time as well.

In any case, the melding of the typicality and dynamical relaxation perspectives seems to cast a calming and clarifying light on the subject. Invoking typicality helps one understand exactly what the dynamical relaxation argument purports to prove, and helps one more fairly assess how promising it might be, for example, to search for violations of Born-rule statistics in relic particles from the early universe.

In addition, conversely, the time-evolution that is the focus of the dynamical relaxation approach—and which is so vividly portrayed in the numerical simulations of [34,35]—also helps clarify the typicality argument in the face of the sorts of objections reviewed in Section 6. It is definitely true that, as long as one simply considers a specific moment in time, the explanation of $p = |\psi|^2$ by appeal to $\mu = |\Psi|^2$ has an unconvincing and circular character. But that single-moment-in-time snapshot is really a thin caricature of the actual argument.

The real argument is that Born rule $p(x,t) = |\psi(x,t)|^2$ statistics—across a range of nonzero times t—arise for typical initial configurations $X(0)$ if we measure typicality using $\mu = |\Psi(q,0)|^2$. Note in particular here that it is *not* the case that if one instead measured typicality using, say, the $\mu =$ constant measure, one would get $p(x,t) =$ constant! That is, once we move beyond the single-moment-in-time caricature, it is simply not the case that the typical $p(x)$ is the same function of ψ that the typicality measure μ is as a function of Ψ.

Dürr, Goldstein, and Zanghí explain, for example, that one could infer the typicality of the non-Born-rule statistics $p(x,t) = |\psi(x,t)|^4$

> "*provided* the sense [μ] of typicality were given, not by $|\Psi|^4$ (which is not equivariant), but by the density to which $|\Psi_t|^4$ would backwards evolve as the time decreases from t to THE INITIAL TIME 0. This distribution, this sense of typicality, would presumably be extravagantly complicated and exceedingly artificial.
>
> "More important, it would depend upon the time t under consideration, while equivariance provides a notion of typicality that works for all t". [38]

This is an extremely important point. With (non-trivial) time-evolution between the initial time and the time at which one is interested in considering the statistical distribution, one does not simply get out for $p(x)$ what one puts in for μ. Note as well that what DGZ describe here as an "extravagantly complicated and exceedingly artificial" typicality measure μ could also perhaps be described by saying that μ has an implausibly fine-grained micro-structure.

I also find it illuminating to turn Dürr, Goldstein, and Zanghí's point around and consider the sorts of statistics one would diagnose as typical if one began with a more plausibly smooth (but non-$|\Psi|^2$) typicality measure, such as $\mu =$ constant or $\mu = |\Psi|^4$, over initial configurations. Assuming the wave function of the universe has a non-trivial dynamics, such a μ would forward-evolve into something "extravagantly complicated and exceedingly artificial" at later times t. Indeed, this is precisely the sort of evolution we have illustrated in, for example, Figure 4: even for an extremely simple (two-dimensional!) system, an initially uniform distribution evolves into something with an incredible degree of fine-grained micro-structure in a very short period of time. This is what Valentini's relaxation argument predicts, and one thus expects that something qualitatively similar will happen with the (evidently much more complicated!) evolution of the wave function of the entire universe, such that (say) an initially-uniform μ will forward-evolve into something with an incredibly filimented, fine-grained, non-uniform structure which will diagnose, as typical for ensembles of similarly-prepared sub-systems, a $p(x,t)$ function with its own extravagantly complicated fine-grained structure... but whose coarse-grained partner $\bar{p}(x,t)$ matches the Born rule distribution $|\psi(x,t)|^2$.

Dürr, Goldstein, and Zanghí have shown that Born-rule statistics are to be expected from the pilot-wave theory, if one assesses typicality using the "natural" equivariant measure, $\mu = |\Psi|^2$. What I mean to be recommending here is that Valentini's analysis, and the associated numerical simulations, can be taken as suggesting that other, non-equivariant typicality measures (which are, at $t = 0$, in some sense "reasonably smooth", i.e., which contain no fine-grained micro-structure, i.e., which are not extravagantly complicated and exceedingly artificial) will *also* end up diagnosing, as typical, Born-rule statistics at later times. So to whatever extent one finds the typicality argument presented by DGZ to be circular, or to rely too heavily on unconvincing "naturalness" type assumptions, one can rest assured that Born-rule statistics are typical—i.e., that one should expect to see Born-rule statistics in practice—not *only* according to that one special equivariant measure of typicality, but according to virtually *any* "reasonable" measure of typicality.

That, of course, does not mean that Born rule statistics are absolutely guaranteed by the pilot-wave theory's dynamics. They aren't. There are possible initial configurations of the universe—indeed, there are an infinite number of them—that will, for example, give rise to perfect $p = |\psi|^4$ statistics today. But these are like the possible states for a box of gas molecules in which every molecule has exactly the same velocity: the existence of such possible states should not undermine one's expectation to see a Maxwell velocity distribution. Similarly here: I think what the typicality analysis together with Valentini's relaxation argument and the associated numerical solutions show is that any "reasonable"—smooth, simply-expressable, non-artificial—measure over initial configurations will imply that *we should expect to see Born-rule statistics.* If that does not constitute a genuine statistical *explanation* of the Born rule, from the dynamical first principles of the pilot-wave theory, I truly do not know what would or could.

Funding: This research received no external funding.

Acknowledgments: Thanks to two anonymous referees for helpful comments on the paper's first draft.

Conflicts of Interest: The author declares no conflict of interest.

References

1. Bell, J.S. Against Measurement. In *Speakable and Unspeakable in Quantum Mechanics*, 2nd ed.; Cambridge University Press: Cambridge, UK, 2004.
2. Norsen, T. *Foundations of Quantum Mechanics*; Springer: Berlin/Heidelberg, Germany, 2017.
3. Goldstein, S. Quantum Theory without Observers. Part I. *Phys. Today* **1998**, *51*, 42–46. [CrossRef]
4. Goldstein, S. Quantum Theory without Observers. Part II. *Phys. Today* **1998**, *51*, 38–42. [CrossRef]
5. Ghirardi, G.C.; Rimini, A.; Weber, T. Unified dynamics for microscopic and macroscopic systems. *Phys. Rev. D* **1986**, *34*, 470, doi:10.1103/PhysRevD.34.470. [CrossRef]
6. Goldstein, S.; Tumulka, R.; Zanghí, N. The Quantum Formalism and the GRW Formalism. *J. Stat. Phys.* **2012**, *149*, 142–201. [CrossRef]
7. Everett, H. 'Relative State' formulation of quantum mechanics. *Rev. Mod. Phys.* **1957**, *29*, 454–462. [CrossRef]
8. Deutsch, D. Quantum theory of probability and decisions. *Proc. R. Soc. Lond. A* **1999**, *455*, 3129, doi:10.1098/rspa.1999.0443. [CrossRef]
9. Wallace, D. *The Emergent Multiverse*; Oxford University Press: Oxford, UK, 2012.
10. Saunders, S. Chance in the Everett Interpretation. In *Many Worlds?* Saunders, S., Barrett, J., Kent, A., Wallace, D., Eds.; Oxford University Press: Oxford, UK, 2010.
11. Wallace, D. How to Prove the Born Rule. In *Many Worlds?* Saunders, S., Barrett, J., Kent, A., Wallace, D., Eds.; Oxford University Press: Oxford, UK, 2010.
12. Greaves, H.; Myrvold, W. Everett and Evidence. In *Many Worlds?* Saunders, S., Barrett, J., Kent, A., Wallace, D., Eds.; Oxford University Press: Oxford, UK, 2010.

13. Kent, A. One World Versus Many: The Inadequacy of Everettian Accounts of Evolution, Probability, and Scientific Confirmation. In *Many Worlds?* Saunders, S., Barrett, J., Kent, A., Wallace, D., Eds.; Oxford University Press: Oxford, UK, 2010.
14. Albert, D. Probability in the Everett Picture. In *Many Worlds?* Saunders, S., Barrett, J., Kent, A., Wallace, D., Eds.; Oxford University Press: Oxford, UK, 2010.
15. Price, H. Decisions, Decisions, Decisions: Can Savage Salvage Everettian Probability? In *Many Worlds?* Saunders, S., Barrett, J., Kent, A., Wallace, D., Eds.; Oxford University Press: Oxford, UK, 2010.
16. Rae, A.I.M. Everett and the Born rule. *Stud. Hist. Philos. Sci. B* **2009**, *40*, 243–250. [CrossRef]
17. Gleason, A.M. Measures on the closed subspaces of a Hilbert space. *J. Math. Mech.* **1957**, *6*, 885, doi:10.1512/iumj.1957.6.56050. [CrossRef]
18. Barnum, H.; Caves, C.M.; Finkelstein, J.; Fuchs, C.; Schack, R. Quantum probability from decision theory? *Proc. R. Soc. Lond.* **2000**, *456*, 1175, doi:10.1098/rspa.2000.0557. [CrossRef]
19. Caves, C.M.; Fuchs, C.A.; Schack, R. Quantum probabilities as Bayesian probabilities. *Phys. Rev. A* **2002**, *65*, 022305, doi:10.1103/PhysRevA.65.022305. [CrossRef]
20. Farhi, E.; Goldstone, J.; Gutman, S. How probability arises in quantum mechanics. *Ann. Phys.* **1989**, *192*, 368, doi:10.1016/0003-4916(89)90141-3. [CrossRef]
21. Clifton, R.; Bub, J.; Halvorson, H. Characterizing quantum theory in terms of information-theoretic constraints. *Found. Phys.* **2003**, *33*, 1561, doi:10.1023/A:1026056716397. [CrossRef]
22. Zurek, W.H. Environment-assisted invariance, entanglement, and probabilities in quantum physics. *Phys. Rev. Lett.* **2003**, *90*, 120404, doi:10.1103/PhysRevLett.90.120404. [CrossRef] [PubMed]
23. Squires, E.J. On an alleged 'proof' of the quantum probability law. *Phys. Lett. A* **1990**, *145*, 67–68. [CrossRef]
24. Schlosshauer, M.; Fine, A. On Zurek's derivation of the Born rule. *Found. Phys.* **2005**, *35*, 197–213. [CrossRef]
25. Bohm, D. A suggested Interpretation of the Quantum Theory in Terms of Hidden Variables, I and II. *Phys. Rev.* **1952**, *85*, 166–193. [CrossRef]
26. Bell, J.S. On the impossible pilot wave. In *Speakable and Unspeakable in Quantum Mechanics*, 2nd ed.; Cambridge University Press: Cambridge, UK, 2004.
27. Callender, C. The emergence and interpretation of probability in Bohmian mechanics. *Stud. Hist. Philos. Sci. B* **2007**, *38*, 351–370. [CrossRef]
28. Sklar, L. *Physics and Chance*; Cambridge University Press: Cambridge, UK, 1993.
29. Bohm, D. Proof that Probability Density Approaches $|\psi|^2$ in Causal Interpretation of the Quantum Theory. *Phys. Rev.* **1953**, *89*, 15, doi:10.1103/PhysRev.89.458. [CrossRef]
30. Valentini, A. On the Pilot-Wave Theory of Classical, Quantum, and Sub-Quantum Physics. Ph.D. Thesis, International School for Advanced Studies, Trieste, Italy, 1992.
31. Valentini, A. Signal-locality, uncertainty, and the subquantum H-theorem. I. *Phys. Lett. A* **1991**, *156*, 5–11. [CrossRef]
32. Valentini, A. Signal-locality, uncertainty, and the subquantum H-theorem. II. *Phys. Lett. A* **1991**, *158*, 1–8. [CrossRef]
33. Valentini, A. Pilot-wave theory of fields, gravitation, and cosmology. In *Bohmian Mechanics and Quantum Theory: An Appraisal*; Cushing, J.T., Fine, A., Goldstein, S., Eds.; Kluwer Academic Publishers: Dordrecht, The Netherlands, 1996.
34. Valentini, A.; Westman, H. Dynamical Origin of Quantum Probabilities. *Proc. R. Soc. A* **2005**, *8*, doi:10.1098/rspa.2004.1394. [CrossRef]
35. Towler, M.D.; Russell, N.J.; Valentini, A. Time scales for dynamical relaxation to the Born rule. *Proc. R. Soc. A* **2012**, *468*, 990–1013. [CrossRef]
36. Dürr, D.; Goldstein, S.; Zanghí, N. Bohmian Mechanics as the Foundation of Quantum Mechanics. In *Bohmian Mechanics and Quantum Theory: An Appraisal*; Cushing, J.T., Fine, A., Goldstein, S., Eds.; Kluwer Academic Publishers: Dordrecht, The Netherlands, 1996.
37. Bell, J.S. Quantum mechanics for cosmologists. In *Speakable and Unspeakable in Quantum Mechanics*; Cambridge University Press: Cambridge, UK, 2004.
38. Dürr, D.; Goldstein, S.; Zanghì, N. Quantum equilibrium and the origin of absolute uncertainty. *J. Stat. Phys.* **1992**, *67*, 843.
39. Boltzmann, L. Annalen der Physik. In *Kinetic Theory*; Pergamon: Oxford, UK, 1966; Chapter 8.

40. Oldofredi, A.; Lazarovici, D.; Deckert, D.A.; Esfeld, M. From the universe to subsystems: Why quantum mechanics appears more stochastic than classical mechanics. *Fluct. Noise Lett.* **2016**, *15*, 1640002, doi:10.1142/S0219477516400022. [CrossRef]
41. Goldstein, S.; Struyve, W. On the Uniqueness of Quantum Equilibrium in Bohmian Mechanics. *J. Stat. Phys.* **2007**, *128*, 1197–1209. [CrossRef]
42. Goldstein, S. Typicality and Notions of Probability in Physics. In *Probability in Physics*; Ben-Menahem, Y., Hemmo, M., Eds.; Springer: Berlin/Heidelberg, Germany, 2012; pp. 59–71.
43. Valentini, A. Hidden Variables, Statistical Mechanics and the Early Universe. In *Chance in Physics*; Bricmont, J., Dürr, D., Galavotti, M.C., Ghirardi, G., Petruccione, F., Zanghi, N., Eds.; Springer: Berlin, Germany, 2001.
44. Bricmont, J. Bayes, boltzmann, and Bohm: Probabilities in Physics. In *Chance in Physics*; Springer: Berlin, Germany, 2001.

Article

Atom-Diffraction from Surfaces with Defects: A Fermatian, Newtonian and Bohmian Joint View

Ángel S. Sanz

Department of Optics, Faculty of Physical Sciences, Universidad Complutense de Madrid, Pza. Ciencias 1, Ciudad Universitaria, 28040 Madrid, Spain; a.s.sanz@fis.ucm.es

Received: 1 May 2018; Accepted: 7 June 2018; Published: 9 June 2018

Abstract: Bohmian mechanics, widely known within the field of the quantum foundations, has been a quite useful resource for computational and interpretive purposes in a wide variety of practical problems. Here, it is used to establish a comparative analysis at different levels of approximation in the problem of the diffraction of helium atoms from a substrate consisting of a defect with axial symmetry on top of a flat surface. The motivation behind this work is to determine which aspects of one level survive in the next level of refinement and, therefore, to get a better idea of what we usually denote as quantum-classical correspondence. To this end, first a quantum treatment of the problem is performed with both an approximated hard-wall model and then with a realistic interaction potential model. The interpretation and explanation of the features displayed by the corresponding diffraction intensity patterns is then revisited with a series of trajectory-based approaches: Fermatian trajectories (optical rays), Newtonian trajectories and Bohmian trajectories. As it is seen, while Fermatian and Newtonian trajectories show some similarities, Bohmian trajectories behave quite differently due to their implicit non-classicality.

Keywords: atom-surface scattering; bohmian mechanics; matter-wave optics; diffraction; vortical dynamics

---

1. Introduction

In the last several decades, there has been a fruitful and beneficial transfer of the ideas involved in David Bohm's formulation of quantum mechanics [1–4] from the domain of the quantum foundations to the arena of the applications [5–9]. The conceptual and mathematical background provided by Bohmian mechanics [3,4,10–12] has become a resourceful tool to investigate quantum problems from an alternative viewpoint regardless of the always ongoing hidden-variable debate. This includes both fresh interpretations to (known and also new) quantum phenomena and novel implementations of alternative numerical algorithms to tackle them [13]. The essential ingredients of Bohmian mechanics have also inspired methodologies and descriptions aimed at providing effective trajectory-like explanations of wave phenomena beyond the quantum realm [14]. For instance, Bohmian-like trajectories have been synthesized from experimental data with light [15], while Bohmian-like behaviors have been recreated in classical fluid-dynamics experiments [16–22].

Getting back to quantum mechanics, one of the advantages of Bohmian mechanics is, perhaps, its capability to put on the same level quantum and classical analyses or descriptions of the same physical phenomenon by virtue of the concept of trajectory, well defined in both contexts. Now, because Bohmian trajectories are in compliance with quantum mechanics, they can be considered to be at a descriptive level above classical trajectories. Thus, an interesting question that naturally arises is how much or what kind of information is kept when passing from a descriptive level to the next one. This is precisely the question addressed here. To this end, a realistic working system is considered, although it is still simple enough to provide a clear answer to the question. Specifically, the phenomenon analyzed is the helium-atom diffraction from a carbon monoxide molecule (CO) adsorbed on a platinum (111) single-crystal surface. This is a system that has been extensively studied in the literature both experimentally and theoretically [23–31], even from a Bohmian viewpoint [32,33]. An appropriate description and knowledge of the CO-Pt(111) interaction is important to the understanding of the role of Pt as a catalyst of the electrochemical oxidation of the CO, with industrial and technological applications (of course, an extensive literature on other analogous systems is also available). In order to determine such an interaction, one of the experimental techniques employed is the He-atom diffraction (or scattering) at low energies (typically energies are between 10 meV and 200 meV, about three orders of magnitude below the range for low-energy electron diffraction). This is a rather convenient tool to investigate and characterize surfaces at relatively low energies with neutral probes, which provides valuable information about the surface electronic distribution without a damage of the crystal—the atoms remain a few Ångstroms above the surface, in contrast to low-energy electron diffraction, where electrons penetrate a few crystal layers, strongly interacting with the crystal atoms.

As is well-known, when a matter wave is diffracted by a crystal lattice, either by reflection (the case of He atoms or low-energy electrons) or by transmission (the case of neutron diffraction or high-energy electrons), the resulting spatial intensity distribution is characterized by a series of maxima along the so-called Bragg directions. The lattice structure can be determined from these characteristic patterns by means of a convenient modeling. Sometimes, however, these patterns have distortions due to a break of the translational symmetry typical of a perfect lattice. This symmetry-breaking can be induced by intrinsic thermal (lattice) atom vibrations (phonons) or by the presence of different types of defects randomly distributed across the lattice [34], for instance, adsorbed particles (adsorbates). In the case of a periodic surface, the presence of an adsorbate on top of it translates into a blurring of the well-defined Bragg features and the appearance of broad intensity features. This diffuse scattering effect [24] is analogous to the image distortion produced by a rough mirror—the larger the number of CO adsorbates on the clean Pt surface, the larger the distortion with respect to neat Bragg-like diffraction intensity peaks.

Instead of considering a large number of randomly distributed CO adsorbates, we are going to focus on the effects produced by a single isolated CO adsorbate. Moreover, we shall not focus on the quantitative description of the diffraction process itself, but on the analysis of how the features associated with a given descriptive level manifest on the next level, which is assumed to be more refined and, therefore, accurate. Accordingly, we will see that although such features are transferred from the model characterizing one level to the model corresponding to the next level, it not always easy to establish an unambiguous one-to-one correspondence. In simple terms, appealing to a biological metaphor, it is like considering a body and its skeleton. The skeleton is the structure upon which the body rests. However, although the body reveals some features of the skeleton (we can perceive the position of some bones under our flesh), it is a much more complex super-structure. In particular, here the problem considered is approached at three descriptive levels:

- The **Fermatian level**, which refers to the analysis of the problem assuming a bare hard-wall-like (fully repulsive) model to describe the He-CO/Pt(111) interaction. Because the trajectories here are of the type of sudden impact (free propagation except at the impact point on the substrate wall, where the trajectory is deflected according to the usual law of reflection), they are going to be straight-like rays, as in optics (this is why it is referred to as Fermatian).
- The **Newtonian level**, where the He-CO/Pt(111) interaction is modeled in terms of a potential energy surface that smoothly changes from point to point. This model has a repulsive wall that avoids He atoms to approach the substrate beyond a certain distance (for a given incidence energy), and an attractive tail that accounts for van der Waals long-range attraction. The existence of these two regions, repulsive and attractive, gives rise to an attractive channel around the CO adsorbate and that continuous along the flat Pt surface, inducing the possibility of temporary trapping for the He atoms.

- The **Bohmian level** is the upper one and, to some extent, makes an important difference with the previous models because here the trajectories are not only dependent on the interaction potential model, but also on the particular shape displayed by the wave function at each point of the configuration space at a given time (the "guiding" or "pilot" wave).

From a physical view point, we are going to focus on three different aspects or phenomena, namely *reflection symmetry interference*, *surface rainbows* and ısurface trapping. As it will be seen, all these aspects have a manifestation in the corresponding diffraction intensity patterns. *Reflection symmetry interference* is the mechanism considered to explain the oscillations displayed by the intensity pattern at large diffraction angles [24,26], and is based on the hypothesis that the diffracted wave can be assumed to be constituted by two interfering waves. In terms of trajectories, as we shall see, this means that there are pairs of homologous paths (either Fermatian or Newtonian), which, when using semiclassical arguments in terms of the optical concept of paths difference, explain the appearance of such type of interference [24,26]. *Rainbow* features arise as a consequence of the local changes in the curvature of the interaction potential model [27,35], which give rise to accumulations of classical trajectories along some privileged directions (rainbow deflection angles), but that quantum-mechanically leave some uncertainties when we look at the corresponding diffraction patterns [29,32]. Finally, *surface trapping* along the clean Pt surface arises for some deflections from the adsorbate at grazing angles. This phenomenon takes place when, by virtue of the interaction with the adsorbate, the He atoms lose too much energy along their perpendicular direction, transferring it to their parallel degree of freedom (perpendicular and parallel are defined with respect to the clean Pt surface). The energy associated with their perpendicular degree of freedom becomes negative, while the parallel energy gets larger than the incident one (by conservation), which ends up with the atom moving in the form of jumps along the surface until it encounters another adsorbate that, by means of the reverse process, can be used to release the atom from the surface [36].

According to the above discussion, this work has been organized as follows. Details about the interaction potential models considered to determine the different trajectory dynamics are provided and discussed in Section 2. In this section, a brief discussion is also given concerning the numerical approaches used to determine the calculations shown here. The diffraction intensity patterns for both the hard-wall model and the realistic interaction potential are shown and discussed in Section 3. Section 4 is devoted to the description and analysis of the different types of trajectories (Fermatian, Newtonian and Bohmian), emphasizing the features that are both common to all approaches and also their main differences. To conclude, a series of final remarks are summarized in Section 5.

2. Potential Model and Computational Details

Interaction potential models for systems like the one we are dealing with here are determined from information extracted from experimental diffraction patterns. Thus, let us consider that, as it is typically done, the intensity distribution in these patterns is specified in terms of the transfer of He-atom momentum from the incidence direction to the direction parallel to the surface or, in brief, parallel momentum transfer [24], i.e.,

$$\Delta K = k_{d,x} - k_{i,x} = k_i(\sin\theta_d - \sin\theta_i), \quad (1)$$

where θ_d and θ_i are the diffraction and the incident angles, respectively, and $k_i = \sqrt{2mE_i/\hbar}$ is the incident wavenumber. Then, diffraction features in the large-angle region of the pattern (large parallel momentum transfers) typically convey information about the repulsive part of the interaction, while the attractive part has a more prominent influence on the small-angle region (low parallel momentum transfers). From these potential models, it is possible to estimate the effective size of the adsorbed particles [23,24] as well as the local curvature of the surface electron density, which additionally may induce the presence and contribution of rainbow-like features [35,37] whenever the interaction potential consists of a short range repulsive region followed by a longer range attractive one (accounting for the van der Waals interaction between the adsorbate-surface system and the impinging neutral atom).

The model considered here, proposed by Yinnon et al. [27], gathers the above mentioned features and nicely fits the experimental data [27–29]. This potential model consists of two terms:

$$V(\mathbf{r}) = V_{\mathrm{Pt}(111)}(z) + V_{\mathrm{CO}}(\mathbf{r}), \tag{2}$$

where $V_{\mathrm{Pt}(111)}(z)$ and $V_{\mathrm{CO}}(\mathbf{r})$ describe, respectively, the separate interaction of the He atom with the clean Pt surface and the adsorbate. The functional form of these two potential functions are a Morse potential for the He-Pt(111),

$$V_{\mathrm{Pt}(111)}(z) = D\left[1 - \mathrm{e}^{-\alpha(z-z_m)}\right]^2 - D, \tag{3}$$

and a Lennard–Jones potential for the He-CO,

$$V_{\mathrm{CO}}(\mathbf{r}) = 4\epsilon\left[\left(\frac{\sigma}{r}\right)^{12} - \left(\frac{\sigma}{r}\right)^{6}\right], \tag{4}$$

with $r = \sqrt{x^2 + y^2 + z^2}$. The reference system for the potential (2) is centered at the CO center of mass, with $D = 4.0$ meV, $\alpha = 1.13$ Å^{-1}, $z_m = 1.22$ Å and $\epsilon = 2.37$ meV [28,29]. With this, z accounts for the position of the He atom along the perpendicular direction with respect to the clean Pt surface, while x and y describe its parallel motion. Because of the rotation symmetry around the axis $x = y = 0$ exhibited by the interaction potential (2), for an illustration, in the contour plot displayed in Figure 1a only one half of the transverse section along the plane $y = 0$ is shown. In it, negative and positive energy contours are denoted, respectively, with blue solid line and red dashed line (due to symmetry, only a half of the potential is shown). On the right-hand side, panels (b)–(d) show different transverse sections of the potential to better appreciate the effect of the local curvature along the z direction at three different distances from $x = 0$ (for $x = 0$ Å, $x = 3.31$ Å and $x = 6.35$ Å, respectively), which give an idea of the well-depth on top of the adsorbate (about 2.96 meV), at the intersection of the adsorbate with the flat Pt surface (6.37 meV), and on top of the flat surface far from adsorbate (4 meV), respectively.

The existence of an attractive well around the adsorbate and also along the surface is going to induce temporary trapping both classically and quantum-mechanically—only the presence of another neighboring adsorbate may help to remove such trapping. Since far from the influence region of the adsorbate the He-Pt(111) interaction only depends on the z coordinate, the trapped motion will be ruled by the well of the Morse function, being free along the x direction. The resulting motion is thus a combination of jumps along the z direction with a uniform motion parallel to the Pt surface, with an average speed larger than $\sqrt{2E_i/m}$, with E_i being the incident energy (notice that the energy along the x direction has to be larger than along the z direction in order to achieve negative values along the latter and, hence, trapping). Classically, if the energy along the z direction is given by

$$E_z = \frac{p_z^2}{2m} + V_{Pt(111)}(z), \tag{5}$$

where $E_z = E_i - E_x < 0$ and m is the He-atom mass, it is easy to show that the turning points will be located at

$$z_{\pm} = z_m - \frac{1}{\alpha}\ln\left[1 \pm \sqrt{1 - \frac{|E_z|}{D}}\right]. \tag{6}$$

This motion is anharmonic, with frequency

$$\omega = \sqrt{\frac{2\alpha^2|E_z|}{m}}. \tag{7}$$

The length of each jump along the x direction can be easily estimated from Equation (7) according to the relation

$$\Delta x = \frac{p_x}{m}\tau = \frac{2\pi}{\alpha}\sqrt{\frac{E_x}{(E_x - E_i)}} = \frac{2\pi}{\alpha}\sqrt{\frac{(E_i - E_z)}{|E_z|}}, \tag{8}$$

where $p_x = mv_x$ and $\tau = 2\pi/\omega$. As can be noticed, at the threshold $E_x = E_i$, the jump length becomes infinity, i.e., the trajectories leave the adsorbate being and remaining parallel to the clean Pt surface.

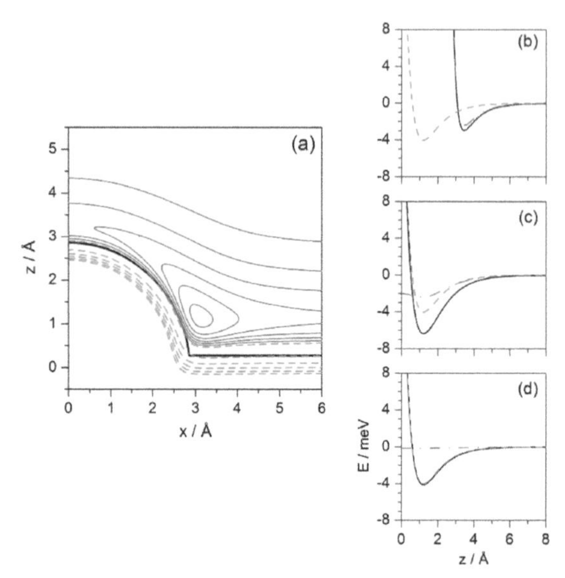

Figure 1. (**a**) contour plot of the He-CO/Pt(111) interaction potential model (2) (see text for details). The energy difference between consecutive repulsive/attractive contour levels (red dashed lines/blue solid lines) is 10 meV/1 meV. The thick black solid line denotes the repulsive boundary for an approximate hard-wall model set for an incidence energy of 10 meV (see text for details). In the right-hand side panels, energy profiles along the z direction for: (**b**) $x = 0$ Å; (**c**) $x = 3.31$ Å and (**d**) $x = 6.35$ Å. In these panels, the total interaction potential is denoted with black solid line, while red dashed and blue dash-dotted lines refer to the Morse and Lennard–Jones contributions, respectively.

Quantum-mechanically, the trapped portions of the wave packet will somehow contain information about the bound states of the Morse function, with eigenenergies given by [38]

$$E_n = \hbar\Omega\left(n + \frac{1}{2}\right)\left[1 - \frac{\hbar\Omega}{4D}\left(n + \frac{1}{2}\right)\right], \tag{9}$$

with

$$\Omega = \sqrt{\frac{2\alpha^2 D}{m}} \tag{10}$$

being the harmonic frequency resulting from approximating the Morse potential to a harmonic oscillator. From the condition to determine the total number of bound states, i.e., $\Delta E_{n'=n+1,n} = E_{n'} - E_n \geq 0$, it is found that, for the parameters here considered, there are only three bound states, namely the ground state plus two excited ones: $E_0 = -2.53$ meV, $E_1 = -0.60$ meV and $E_2 = -3 \times 10^{-3}$ meV. If we assume $E_z = E_n$, then we obtain $\Delta_x^{(0)} \approx 12$ Å, $\Delta_x^{(1)} \approx 23$ Å and $\Delta_x^{(2)} \approx 320$ Å, respectively, according to Equation (8).

The potential model (2) will be used to investigate the behavior of Newtonian and Bohmian trajectories when the He atoms are influenced by it. For the Fermatian approach, we shall consider a rather crude approximation to this potential, which consists of substituting it by a purely repulsive infinite barrier, which will be referred to as the (repulsive) hard-wall model. This barrier is an approximation to the equipotential $V(\mathbf{r}) = 10$ meV (black thick solid line in Figure 1a), so that $V_{HW}(\mathbf{r}) = \infty$ for any position $\mathbf{r}$ such that $V(\mathbf{r}) \geq 10$ meV and zero everywhere else (i.e, for all $\mathbf{r}$ such that $V(\mathbf{r}) < 10$ meV). An optimal fit to this equipotential renders a radius $a = 2.86$ Å for the adsorbate and a position $z_r = 0.28$ Å of the clean Pt surface above the CO center of mass. Since the Bohmian treatment of this model is not included here, it is worth mentioning a recent analysis of an analogous system by Dubertrand et al. [39], where the authors consider the Bohmian description of quantum diffraction by a half-flat surface (simplified by a half-line barrier), earlier considered by Prosser [40] in the context of the solution of Maxwell's equations to the problem of diffraction by a semi-infinite conducting sheet [41].

One of the advantages of the model introduced in this section is that it can describe the presence of a localized adsorbed particle on a surface (which can be regarded as a point-like defect [24]), but also a row of adsorbates aligned, for instance, along the y-axis (a linear-like defect [23]). In the first case, we have radial symmetry with respect to the z-axis, while the latter is characterized by axial symmetry, along the y-axis. From the viewpoint of a trajectory-based description, there is no difference between one case and the other, since what happens in one half of a transverse section also happens in any other section (which can be reconstructed either by rotation symmetry around the z-axis or by translational symmetry along the y-axis and/or mirror symmetry with respect to the yz plane). The difference between both models relies on the way how the trajectories distribute spatially and therefore how many of them lay within a certain solid angle, independently of whether the trajectories are classical (Fermatian or Newtonian) or quantum-mechanical (Bohmian). Having said this, since we are interested in comparing the behavior of different types of trajectories, from now on, the discussion will turn around the two-dimensional description, which is analogous to consider an axial-symmetric system.

Regarding the conventions used here [42], the incidence angle, θ_i, for Fermatian and Newtonian trajectories is defined as the angle subtended between the incident direction of a given trajectory and the normal to the clean Pt surface. Dynamically speaking, this translates into an effective way of how much momentum is provided initially to each direction, i.e.,

$$\begin{aligned} p_{i,x} &= p_i \sin\theta_i = \sqrt{2mE_i}\sin\theta_i, \\ p_{i,z} &= -p_i \cos\theta_i = -\sqrt{2mE_i}\cos\theta_i, \end{aligned} \tag{11}$$

with $p_i = \hbar k_i = \sqrt{2mE_i}$. In particular, in the calculations presented and discussed below, we have considered two values of the incident energy, namely $E_i = 10$ meV and 40 meV. In terms of the de Broglie wavelength, $\lambda_{\rm dB} = h/\sqrt{2mE_i}$, with h being Planck's constant, these energies correspond to $\lambda_{\rm dB} = 1.43$ Å and 0.72 Å, respectively. The relations (11) are also used for the quantum analysis, where the incident wave function is launched with a momentum in compliance with these expressions. The deflection (or outgoing) angle, θ_d, on the other hand, is defined as the angle subtended by the normal and the deflection direction for the corresponding trajectory—in the case of wave-function descriptions, this angle is going to be denoted as the diffraction angle —, although an analogous definition in terms of the momentum with which the particle is deflected can also be used. Once the incidence angle is established, depending on the initial position assigned to the trajectories, they will behave in a way or another. To characterize the trajectories according to their initial positions, it is common to refer them to the impact parameter, which in the present context is defined as the impact position on the clean surface in absence of interaction. In periodic surfaces, the range for impact parameters (b) is typically established in terms of the lattice parameter (the unit cell length) [42,43]. Here, because the presence of the adsorbate breaks the periodicity of the clean surface, we need to redefine this range in an alternative and slightly different way. Specifically, this range is taken as a portion of surface that covers the extension of the adsorbate and goes well inside the region where the flat surface potential is already not influenced by the adsorbate attractive tail. With the potential function used here, this means is satisfied by impact parameters taken within the range $[-10.6, 10.6]$ Å. Accordingly, the initial positions are specified as

$$\begin{aligned} x_i &= b - z_i \tan\theta_i, \\ z_i &= z_0, \end{aligned} \tag{12}$$

with $b \in [-10.6, 10.6]$ Å, and where $z_0 = 10.27$ Å has been chosen far from the surface, such that $V(\mathbf{r}) \approx 0$ (the same holds for an incident wave function, whose probability density has to be far from the influence region of the adsorbate).

With the above definitions, the computation of Fermatian trajectories is trivial, since we only need to know their incidence direction and, from it, the point on the substrate (adsorbate or flat surface) where they will feel the impact. At such a point, applying the law of reflection, we readily obtain the deflection angle and therefore the deflected part of the trajectory. Fermatian trajectories are just a pure geometric issue and, as it will be seen in next section, the corresponding quantum calculations are just analytical, so they do not imply high computational demands. For Newtonian trajectories, however, the computational task is more refined, since the action of the interaction potential introduces important changes in the curvature of the trajectories when they are approaching the substrate. Nevertheless, the computational demand is still low, since such trajectories can be readily obtained by integrating Newton's equations (actually, Hamilton's equations) with a simple fourth-order Runge–Kutta algorithm using the above momentum and position values, Equations (11) and (12), respectively, as initial conditions.

The numerical computation of the wave-function evolution and the Bohmian trajectories is, however, more subtle, since it implies the solution of a partial differential equation. In this case, integration has been carried out by means of the second-order finite-difference algorithm [44], making use of the fast Fourier transform to compute the kinetic part of the operator [45]. For the initial wave function, we would like to simulate an incident nearly plane wave, which mimics a highly collimated He-atom beam. Numerically, we can recreate this situation by considering a quasi-monochromatic wave function or wave packet that covers the substrate well beyond the effective size of the adsorbate. This can be done by linearly superimposing a large number of Gaussian wave packets, which in our cases amounts to considering 250 Gaussian wave packets [46], where the spreading of each wave packet is 0.84 Å along the x direction and 2.65 Å along the z direction. With these conditions, the wave function reaches the surface with almost no increase of its size, which is launched from an average position along the z direction $\langle z \rangle_0 = z_0 = 10.27$ Å and normal incidence conditions (again, for visual clarity). In order to ensure an optimal overlapping along the x direction, the centers of the wave packets are separated a distance of 0.21 Å. For simplicity both here and also with the hard-wall model, the quantum calculations have been performed at normal incidence conditions ($\theta_i = 0°$), although this does not diminish the generality of the results presented.

Regarding the computation of the Bohmian trajectories, they are synthesized on the fly from the wave function. That is, the wave function is made to evolve for a small time interval dt, and then the trajectories are propagated from their actual position to the new one with the phase information provided by the updated wave function. The equation of motion that rules this behavior is the guidance Equation (24) (see Section 4.3 for further details on this equation of motion). Since the value of the wave function and its derivatives is known only on the knots of the numerical grid, the guidance equation has to be solved with the aid of numerical interpolators, which render the values required at any other point other than a grid knot with a reliable accuracy. With these values, the equation of motion is solved by means of a Runge–Kutta algorithm, as in the case of Newtonian trajectories, although the degree of accuracy required is higher, particularly due to the appearance of nodal structures. As for the initial conditions, they have been chosen along lines parallel to the flat surface, at different constant distances from the latter and taking the value $\langle z \rangle_0 = z_0$ as a reference. Specifically, three of these lines have been taken above this value $[z(0) > z_0]$ and another three have been taken below $[z(0) < z_0]$, i.e., closer to the substrate (see Section 4.3 for further details).

3. Wave-Function Approach

3.1. Diffraction from a Repulsive Hard-Wall Potential

In the case of the two-dimensional (axial-symmetric) version of the hard-wall model described in Section 2, the diffracted wave far from the adsorbate can be obtained from the exact (analytical) asymptotic solution to the problem of the diffraction from a cylinder [47],

$$\Psi = e^{i\mathbf{k}_i \cdot \mathbf{r}} + \frac{f(\mathbf{k}_d)}{\sqrt{r}} e^{ik_d r} = e^{i\mathbf{k}_{i,x}x - k_{i,z}z} + \frac{f(k_{d,x}, k_{d,z})}{\sqrt{r}} e^{ik_d r}, \quad (13)$$

with $\mathbf{r} \to \infty$, and where $\mathbf{k}_i = (k_{i,x}, k_{i,z})$ is the incidence wave vector (momentum) and $\mathbf{k}_d = (k_{d,x}, k_{d,z})$ is an outgoing wave vector pointing along an arbitrary diffraction direction—the two wave vectors are expressed in terms of their parallel and perpendicular components ($k_{i,x}$ and $k_{d,x}$, and $k_{i,z}$ and $k_{d,z}$, respectively) on purpose. In this expression, the first term is the contribution from the direct wave and the second term accounts for the diffraction caused by the defect itself (the minus sign in the first contribution arises from the fact that the incidence direction is considered to be negative).

If instead of a cylinder we have a half of it on top of a flat surface, the solution (13) has to include the effect of the reflection from the flat surface, i.e.,

$$\Psi = e^{i(k_{i,x}x - k_{i,z}z)} - e^{i(k_{i,x}x + k_{i,z}z)} + \frac{f(k_{d,x}, k_{d,z})}{\sqrt{r}} e^{ik_d r}, \quad (14)$$

which has to satisfy the hard-wall boundary conditions

$$\Psi(x, z = z_r) = \Psi(r = a) = 0.$$

The first two terms in (14) satisfy this condition when $z = 0$, as expected on the flat surface. On the other hand, for the third term to satisfy these boundary conditions it is necessary that the diffraction (or scattering) amplitude, f, is given by two contributions,

$$f(\theta_i, \theta_d) = f_a(|\theta_d - \theta_i|) - f_b(\pi - |\theta_d + \theta_i|), \quad (15)$$

where the first term describes direct reflection from the adsorbate and the second, a double reflection from the adsorbate and then the flat surface. Both amplitudes can be recast in terms of the difference between the diffraction and incidence angles. In the first case, this can be readily seen; in the latter, a similar result is obtained after assuming collisions with a full cylinder, because then the diffraction angle is $\theta'_d = \pi - \theta_d$. In addition, note that the symmetry displayed by Equation (15) for the specific case $\theta_i = 0$ is analogous to the antisymmetry condition arising in fermion-fermion collisions [48], where the symmetrized amplitude for two fermions with 1/2-spin in a triplet state has the functional form

$$f_-(\theta) = \frac{1}{\sqrt{2}} [f(\theta) - f(\pi - \theta)] . \quad (16)$$

Analytically, in the short-wavelength limit ($ka \to \infty$) and for a cylindrical defect, the diffraction amplitudes in Equation (15) are of the form [47],

$$f(\theta) = -\left[\frac{a \operatorname{sen} \theta/2}{2}\right]^{1/2} \mathrm{e}^{-2ika \operatorname{sen} \theta/2} + \frac{\mathrm{e}^{-i\pi/4}}{\sqrt{2\pi k}} \frac{(1 + \cos\theta)}{\operatorname{sen}\theta} \operatorname{sen}(ka \operatorname{sen}\theta). \quad (17)$$

where the first term, known as the illuminated face term, accounts for the backward scattering of the wave and the second one describes the Fraunhofer diffraction. The final analytical expression for the diffraction amplitude f is obtained by considering the symmetrization condition (15) in these results.

The diffraction intensity for the axial-symmetric, fully repulsive hard-wall model is shown in Figure 2a as a function of the parallel momentum transfer (1), for $E_i = 10$ meV and normal incidence ($\theta_i = 0°$). In the same figure, the intensities associated with the illuminated face term and the Fraunhofer diffraction are also displayed separately (red dotted line and blue dashed line, respectively) in order to get a better idea at a quantitative level of their respective contributions to the total pattern. As it can be seen, for small momentum transfers (small diffraction angles), the leading term is the Fraunhofer one, which decreases fast as the momentum transfer increases (as $(\Delta K)^{-2}$). On the contrary, the illuminated-face term, together with its mirror image, becomes the leading contribution for large momentum transfers (large diffraction angles). The type of oscillations generated by these two terms, the illuminated-face one and its mirror image, give rise to the reflection symmetry phenomenon [24], which explains why the diffraction pattern does not decay for large momentum transfers, as happens, for instance, in simpler cases of diffraction by a wire or a slit. This behavior is observed regardless of the incident energy, as can be noticed from the intensities displayed in Figure 2b for $E_i = 40$ meV (and also normal incidence).

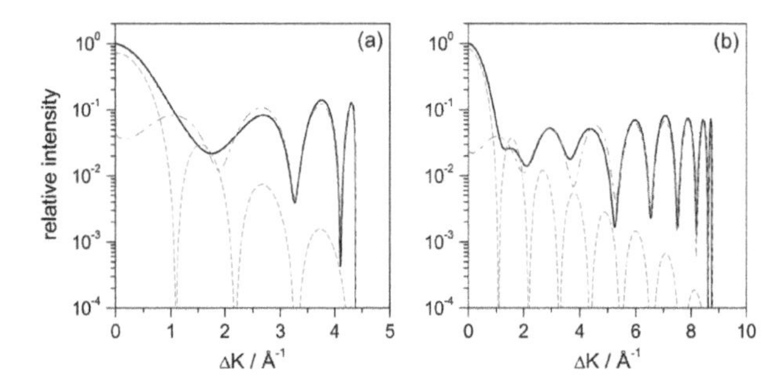

Figure 2. (**a**) Relative diffraction intensity (black solid line) produced by a radial hard-wall model for incidence conditions $\theta_i = 0°$ and $E_i = 10$ meV. To compare with, the Fraunhofer and illuminated-face intensities are also shown, which are denoted with red dashed line and blue dash-dotted line, respectively; (**b**) As in panel (**a**), but for $E_i = 40$ meV.

3.2. *Diffraction from the Potential Model (2)*

The quantum treatment for the potential model (2) does not admit analytical solutions, as it is the case of the hard-wall model of Section 3.1, which provides us with an asymptotic analytical solution of the diffraction far from the interaction region. A way to tackle the problem is by using a numerical wave-packet propagation method, as described in Section 2, which renders a description of the diffraction phenomenon in real time, i.e., providing us with direct information on the time-evolution of the He-atom wave function, of particular interest in the region where the interaction between the He atom and the substrate is stronger. Hence, although we lose analyticity, we gain insight on the dynamical process in the interaction region.

Accordingly, the evolution of the probability density as it approaches the adsorbate and then gets diffracted is shown in Figure 3 at three different instants of its evolution for $E_i = 10$ meV and normal incidence (taking advantage of the mirror symmetry with respect to $x = 0$ due to the normal incidence, only a half is plotted for simplicity). In panel (a), we observe the appearance of circular wavefronts (ripples) around the adsorbate due to the interference produced by the overlapping of the part of the wave function that is still approaching the adsorbate with the part that is already being diffracted. In panel (b), the whole of the wave function is interacting with the substrate (at about 1.5 ps). In this case, there are circular wavefronts around the adsorbate, as before, but also additional plane wavefronts produced by an analogous interference process associated with the portion of the flat Pt surface reached by the incident wave function. The superposition of the circular wavefronts with the planar ones generates around the adsorbate a web of maxima (see inset for a more detailed picture), which gets weaker and blurred as we move far from the adsorbate. The periodicity of the web of maxima is related to the incoming He-atom de Broglie wavelength, $\lambda_{dB} = 1.43$ Å, although with some distortions due to the presence of the attractive region of the interaction. As the wave function further evolves, we can observe a superposition of two contributions, as seen in panel (c). One of them is the reflection of a nearly square function, which starts displaying the typical Fresnel or near-field features of such functions when they arise from single slit diffraction. This contribution is precisely the illuminated face term that we saw in Section 3.1. The other contribution, which distributes around the adsorbate, is going to be related to the Fraunhofer diffraction and also to other features, such as trapping of the wave along the attractive well near the Pt surface. Nonetheless, notice that the final diffraction pattern, as in the case of the hard-wall model (see Equation (17)), is a superposition of the two contributions (diffraction amplitudes), and not only the direct sum of their probabilities.

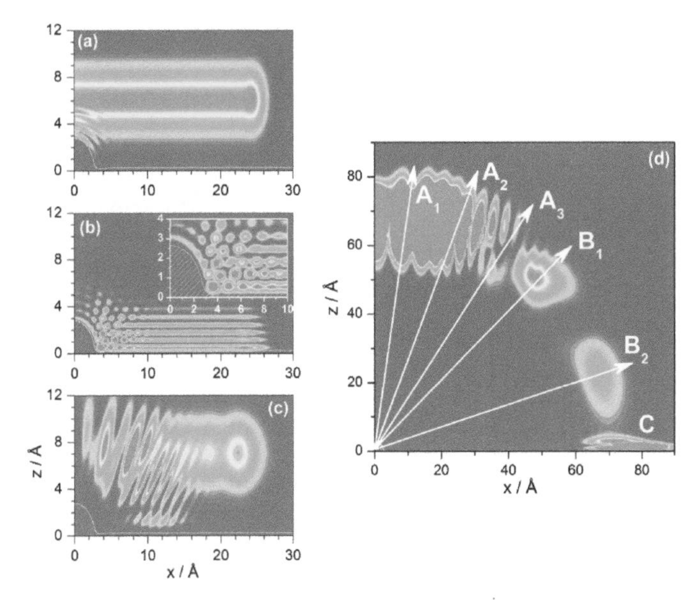

Figure 3. On the left-hand side, contour plots illustrating three different instants of the evolution of the probability density near the surface for incidence conditions $\theta_i = 0°$ and $E_i = 10$ meV: (**a**) when the density starts being influenced by the adsorbate; (**b**) when the density is totally interacting with the substrate (i.e., with both the adsorbate and the flat Pt surface) and (**c**) when the density starts leaving the substrate. In panel (**d**), on the right-hand side, plot of the probability density far from the influence of the adsorbate ($t = 11$ ps). Arrows and capital letters denote different diffraction directions to be identified in the intensity plot displayed in Figure 4a: A_i: directions identifying interference features associated with the superposition of the circular and planar wavefronts, contributing to the central maxima of the intensity pattern; B_i: associated with features arising from the reflection symmetry interference phenomenon; C: surface trapping.

Asymptotically, far from the influence of the classical interaction, as shown in Figure 3d, it is possible to observe traits related to the evolution of the three parts of the scattered wave, which can be eventually associated with the peaks characterizing the corresponding intensity pattern (see Figure 4a). Thus, interferences arising from the superposition of the circular and planar wavefronts give rise to the central features of the intensity pattern (denoted with directions labeled with A_i in the plot). However, there are also other types of peaks, which propagate along the directions denoted as B_i and, as will be seen below, can be associated with the reflection symmetry interference phenomenon. Both types of peaks, A_i and B_i, implicitly carry information about the classical rainbow (see Section 4.2) in a sort of global fashion, since this phenomenon does not manifest with a particular well-defined peak, in general. On the other hand, it is also possible to observe the presence of trapping (C), which is related to the lower part of the wave function that keeps moving close and parallel to the clean Pt surface.

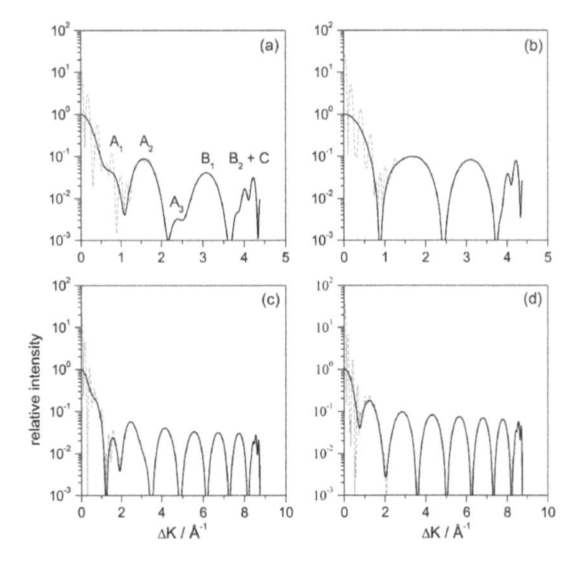

Figure 4. (**a**) Relative diffraction intensity (black solid line) produced by an axial-symmetric potential model based on (2) for incidence conditions $\theta_i = 0°$ and $E_i = 10$ meV. For comparison, the intensities before (red dashed line) and after (black solid line) removing the plane-wave contribution (see text for details) are both plotted; (**b**) the same as in panel (**a**), but for a fully repulsive model of the CO adsorbate, obtained after removal of the attractive part of the Lennard–Jones function (4). To compare, in panels (**c**,**d**), the same as in panels (**a**,**b**), respectively, but for $E_i = 40$ meV.

In order to quantify the effects produced by the diffraction process in the incoming wave, we represent the diffracted wave function as a linear combination of plane waves,

$$\Psi(\mathbf{r}, t) \sim \int \frac{S(k_{i,x}, k_{d,x})}{\sqrt{k_{d,z}}} e^{i(k_{d,z}z + k_{d,x}x)} dk_{d,x}, \tag{18}$$

where the elements $S(k_{i,x}, k_{d,x})$ provide the probability amplitudes associated with the change of parallel momentum $\Delta K = k_{d,x} - k_{i,x}$. These elements are determined by projecting the numerically computed diffracted wave onto this expression,

$$S(k_{i,x}, k_{d,x}) \sim \int \sqrt{k_{d,z}} e^{-i\mathbf{k}_d \cdot \mathbf{r}} \Psi(\mathbf{r}, t) d\mathbf{r}, \tag{19}$$

from which the probability $|S(k_{i,x}, k_{d,x})|^2$ is obtained to detect atoms that have exchanged a given amount ΔK of parallel momentum is obtained (i.e., the reflection coefficient). With periodic lattices, this calculation is typically performed within a single unit cell; here, because of the lack of periodicity, the calculation involves an artificial cell of about 53 Å, which covers a region large enough as to include both the adsorbate and a good portion of the flat surface that is not influenced by artifacts related to the adsorbate curvature—this is appropriate to capture isolated signatures of trapping (otherwise, there could be some contamination from the wave scattered in other directions). This procedure, however, has an inconvenience: the intensity pattern includes a rather high contribution from mirror reflection from the flat Pt surface. In order to remove it, the incident plane wave, ranging from $-\infty$ to

∞, is decomposed as a linear superposition of two contributions, one contained within the integration range $\mathcal{R}$ (ϕ) and another outside it (χ), i.e.,

$$\Psi_0(\mathbf{r}) = \sum_{x_i \in \mathcal{R}} \phi^i(\mathbf{r}) + \sum_{x_i \notin \mathcal{R}} \chi^i(\mathbf{r}) = \sum_{x_i \in \mathcal{R}} \left[\phi^i(\mathbf{r}) - \chi^i(\mathbf{r})\right] + \sum_{\forall x_i} \chi^i(\mathbf{r}), \tag{20}$$

where the subscript i labels functions from a given basis set established on purpose to construct the wave function. Because the χ waves essentially describe diffraction from the clean Pt surface, the last term in this expression is going to be a contribution with the form of a δ-function along the incident direction. Thus neglecting this contribution, if the diffraction process from the surface is altered, we make a projection of the wave on the plane-wave basis set, and the S-matrix elements can be recast as

$$S(k_i, k_d) = S^{\mathcal{R}}(k_i, k_d) - S_0^{\mathcal{R}}(k_i, k_d), \tag{21}$$

where $S_0^{\mathcal{R}}(k_i, k_d)$ is the matrix element corresponding to the scattering of the wave by a flat unit cell. The intensity to be compared with the experiment is thus obtained from the differential cross section or differential reflection coefficient,

$$\frac{dR}{d\theta_d} \propto k_{d,z} |S(k_i, k_d)|^2. \tag{22}$$

This intensity is displayed in Figure 4a as a function of the parallel momentum transferred (instead of the deflection angle, θ_d) after the wave function has evolved for 5.5 ps after the maximum approach to the surface. By this time, the action of the adsorbate interaction potential on the wave function is already negligible. Comparing the solid line with the dashed line, we notice the effect of removing the contribution from the plane wave. Accordingly, using (21) is analogous to "smoothing" the diffraction pattern, where the probability distribution does not show well defined maxima because of the presence of the plane-wave contribution. The clean oscillations that we observe (solid line) arise from the interference between the circular wave fronts coming from the adsorbate and the plane wavefronts coming from reflection from the flat Pt surface. This "interaction" is more prominent for small values of ΔK, this being the reason why the pattern, before removing the plane-wave contribution, displays fast oscillations. Such oscillations, however, get weaker and even meaningless as ΔK increases, because the circular wavefronts become less affected by their overlapping with the flat outgoing wavefronts. This is analogous to the behavior already observed with the hard-wall model, in Figure 2a: for small values of ΔK, the dominant contribution was the Fraunhofer one, while, for larger values of ΔK, the leading one was the illuminated-face contribution.

The intensity maxima in Figure 4a, though, do not totally correspond with those in Figure 2a, even if their number is the same. In particular, notice that some of these maxima (A_1 and A_3) display a kind of "wings". In order to elucidate their origin, the same calculation has been repeated using a repulsive model for the adsorbate, which consists of removing the attractive part of the Lennard–Jones function (4). The results for this model are displayed in panel (b). Comparing both models, we find that everything is essentially the same, except precisely for the presence of such "wings". This result is indeed close to the one displayed in Figure 2a for the hard-wall model, although the maxima are wider, which can be associated with the presence of an attractive well around the substrate. In order to determine whether it is an effect or not linked to the incidence energy, the same analysis was repeated for $E_i = 40$ meV and normal incidence. From the classical calculations, we conclude that such "wings" have to be associated with the presence of rainbows [35] (see Section 4.2), since this phenomenon is linked to the local curvature of the interaction potential around the adsorbate. In this regard, the attractive well around the adsorbate plays a key role, since its removal makes rainbow features to disappear even though there is still an attractive region along the clean Pt surface. Actually, on a more quantitative level, notice that, for instance, for $E_i = 10$ meV, the rainbow appears for $\Delta K \approx 1.89$ Å^{-1}. In Figure 4b, this value of ΔK is close to the maximum of the second lobe, which would lead to a distortion of this maximum and the adjacent ones (where the traits A_1 and A_3 appear). In the case of $E_i = 40$ meV, there is a classical rainbow at $\Delta K \approx 1.13$ Å^{-1}, which corresponds to the maximum of the second lobe and, in consequence, this lobe essentially disappears in the attractive model. Thus, the quantum manifestation of classical rainbow features does not necessarily mean a contribution in terms of given maximum in the intensity pattern [28,29], but it can also be in terms of "global" phenomenon that affects the whole pattern [35,49]. This has actually been a controversial point in the literature when assigning the origin of the different diffraction maxima [27–29].

Finally, regardless of whether the interaction model considers an attractive or repulsive adsorbate, and also independently of the value of the incident energy, we observe that the last lobe of the intensity plots in Figure 4 gathers information of both grazing deflection and trapping. In panel (a), for instance, this maximum has been label as $B_2 + C$. According to Figure 3d, the maximum for B_2 should appear at $\Delta K \approx 4.16$ Å^{-1}. On the other hand, in the same figure, for C, we should have a maximum parallel transfer. However, the trapped probability is indeed oscillating inside the well while it moves along the x direction, which makes the corresponding ΔK value to fluctuate. Thus, the probability amplitudes related to B_2 and C will display some interference, which is precisely what we observe in the last lobe of all calculations presented in Figure 4.

4. Trajectory-Based Description

4.1. Fermatian Level

Although it is a rather crude approximation, the hard-wall model is quite insightful because it allows for explaining and understanding on simple terms the reflection symmetry interference phenomenon [23,24,27,47,50,51] as well as the conditions leading to trapping. As in geometric optics, the key element is the interpretation of wave phenomena in terms of the phase difference arising from two different but equivalent paths, where by "equivalent" we mean that both leave the surface with the same deflection (outgoing) angle, even if their journeys close to the surface are quite different (actually, it is this difference that generates the phase difference). These geometric rays are what we call here Fermatian trajectories with the purpose to highlight such optical connotation and, as seen below, in the particular problem we are dealing with here, there are always homologous pairs of such trajectories. These pairs are formed by one trajectory that undergoes a single bounce from the substrate before getting deflected and another trajectory that undergoes two bounces (one with the adsorbate and another with the flat surface).

In Figure 5, there is a set of Fermatian trajectories, $\mathcal{F}$, of particular interest: they are separatrices that determine the boundaries for ensembles of Fermatian trajectories that display a particular behavior, that is, all the Fermatian trajectories confined within two adjacent separatrices are going to exhibit an analogous behavior. In general, it can be noticed that, for a given incidence angle (here, $\theta_i = 20°$), trajectories may display either a single collision (regions denoted with light gray) or double collisions (blue and purple regions, denoted with A, B and C) depending on their impact parameter. The deflection can then be forward (trajectories represented with solid line) or backward (dashed line). Moreover, there are also regions of geometric shadow (red region, denoted with S) that cannot be reached by any trajectory, except for normal incidence ($\theta_i = 0°$). This region covers an area of length

$$\ell = \frac{(1 - \cos\theta_i)}{\cos\theta_i} \left[a - z_0 \tan\theta_i/2 \right], \tag{23}$$

which depends on θ_i, vanishing ($\ell = 0$) for $\theta_i = 0°$ and reaching its maximum extension ($\ell = \infty$) for parallel incidence ($\theta_i = \pi/2$).

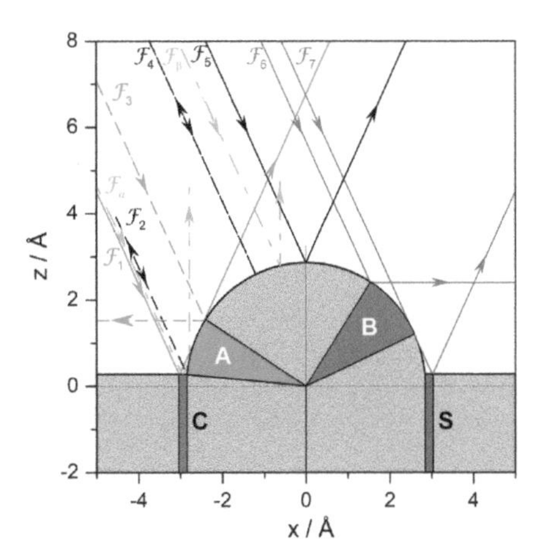

Figure 5. Separatrix Fermatian trajectories for the repulsive hard-wall model set for an incidence energy $E_i = 10$ meV. For a better illustration of all possible cases, an incidence angle $\theta_i = 20°$ has been chosen. Separatrices displaying forward/backward deflection are denoted with solid/dashed lines. Separatrices delimiting regions leading to double scattering in the rear/front part of the adsorbate (regions C-A/B) are denoted with red/blue color. Separatrices deflected perpendicularly to the flat surface are denoted with green color. The shadow region S, which depends on the incidence angle, cannot be reached by any trajectory (except for $\theta_i = 0°$, where this area goes to zero).

Each trajectory in Figure 5 carries a label, which helps to identify behavioral domains. We have that any trajectory impinging on the substrate either to the left of $\mathcal{F}_\alpha$ or to the right of $\mathcal{F}_\beta$ will undergo forward deflection; any other trajectory confined in between will be deflected backwards. Notice that $\mathcal{F}_\alpha$ and $\mathcal{F}_\beta$ are the only two trajectories that are deflected along the normal, constituting themselves a pair of homologous trajectories, with $\mathcal{F}_\alpha$ displaying double collision (first with the flat surface and then with the adsorbate) and $\mathcal{F}_\beta$ displaying a single collision (with the adsorbate). As for the other separatrices:

- Trajectories to the left of $\mathcal{F}_1$ or to the right of $\mathcal{F}_7$ only interact with the clean Pt surface and hence their deflection and incidence angles are equal. These trajectories, plus $\mathcal{F}_5$ only contribute to mirror reflection from the flat surface, only contributing the intensity for $\Delta K = 0$, since $\theta_d = \theta_i$—hence, this contribution will be more prominent as the range of impact parameters increases.
- Any trajectory between $\mathcal{F}_1$ and $\mathcal{F}_\alpha$ is deflected in an angle that goes from θ_i to $0°$ as the impact parameter increases. The same deflection angles are found for trajectories between $\mathcal{F}_\beta$ and $\mathcal{F}_5$, although here the trend is that the angle increases from $0°$ to θ_i as b increases. Here, we have two sets of pairs of homologous trajectories: trajectories from the first set undergo double collisions (first with the flat surface and then with the adsorbate) and trajectories from the latter only have a single collision (with the flat surface). For any of these pairs, the angular distance between their impact points on the adsorbate surface is $\pi/2 - \theta_i$, as can be seen in Figure 6a.
- There are also pairs of homologous trajectories with deflection angles between θ_i and $\pi/2$. These are the trajectories confined between $\mathcal{F}_5$ and $\mathcal{F}_6$, with single collisions (with the adsorbate), and between $\mathcal{F}_6$ and $\mathcal{F}_7$, with double collisions (first with the adsorbate and then with the flat surface). This second set corresponds to trajectories impinging on the adsorbate within the sector B. In this case, the angular distance between impact points is not a constant, but depends on the deflection angle as $\pi/2 - \theta_d$. This distance gradually vanishes as both trajectories approach $\mathcal{R}_6$ and is maximum when the trajectories coincide with the separatrices $\mathcal{R}_5$ and $\mathcal{R}_7$. A representative set is depicted in Figure 6b.
- Trajectories $\mathcal{F}_2$ and $\mathcal{F}_4$ are both deflected backwards along the incidence direction, i.e., $\theta_d = -\theta_i$. Accordingly, trajectories between $\mathcal{F}_\alpha$ and $\mathcal{F}_2$ are deflected between $0°$ and $-\theta_i$ after undergoing

double collisions (first with the flat surface and then with the adsorbate), while trajectories between $\mathcal{F}_4$ and $\mathcal{F}_\beta$ (to the right of $\mathcal{F}_4$) undergo single collisions. The angular distance between impact points of homologous pairs of trajectories is now $\pi/2 - \theta_d$, although not all trajectories between $\mathcal{F}_4$ and $\mathcal{F}_\beta$ have a correspondent between $\mathcal{F}_\alpha$ and $\mathcal{F}_2$. This is because the flat surface intersects the adsorbate surface at a distance $z = z_r$ above its center of mass instead of at $z = 0$. Thus, instead of reaching a maximum deflection of $-\theta_i$, we have $\theta_d^{\max} = -\theta_i + (\sin)^{-1}(z_r/a)$, which is the deflection for the trajectory $\mathcal{F}_2'$. An illustrative pair of homologous trajectories of this kind is displayed in Figure 6c.

- The trajectory $\mathcal{F}_3$ separates the sets of homologous trajectories that are backward deflected, with the second collision taking place from the flat surface. One set is confined within trajectories $\mathcal{F}_2$ and $\mathcal{F}_3$, with double collisions (first with the adsorbate and then with the flat surface), and the other set, with single collisions, is delimited by $\mathcal{F}_3$ and $\mathcal{F}_4$ (trajectories to the left of $\mathcal{F}_4$). Unlike the previous set of backward-scattered homologous pairs, here all trajectories are paired, with the angular distance between their impact points being $\pi/2 - \theta_d$, as before. A representative pair is displayed in Figure 6d.

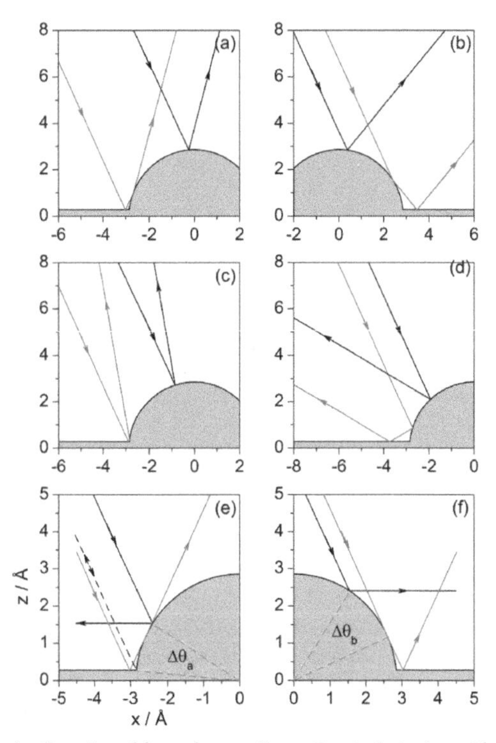

Figure 6. In panels (**a–d**), pairs of homologous Fermatian trajectories with different deflection angle θ_d: (**a**) forward deflection, with $\theta_i \geq \theta_d \geq 0$; (**b**) forward deflection, with $\pi/2 \geq \theta_d \geq \theta_i$; (**c**) backward deflection, with $-\theta_i + \delta \leq \theta_d \leq 0$, where $\delta = (\sin)^{-1}(z_r/a)$ and (**d**) backward deflection, $-\pi/2 \leq \theta_d \leq -\theta_i$. In panels (**e**,**f**), boundaries of the regions where any trajectory impinging on them will display double collisions (see text for further details).

In general, independently of whether the above pairs of homologous Fermatian trajectories describe situations of forward or backward scattering, and also regardless of the incidence angle, double collisions arise whenever the impact on the adsorbate surface takes place between the point where this surface intersects the flat Pt surface and the point at which any impinging trajectory is deflected parallel to such a flat surface (at a given incidence). These conditions determine the regions labeled with A, B and C in Figure 5. This is seen with more detail in Figure 6e, where the neighboring regions A and C are determined by the separatrices $\mathcal{F}_1$ and $\mathcal{F}_3$, and in Figure 6f for region B, delimited by the separatrices $\mathcal{F}_6$ and $\mathcal{F}_7$. Although regions A and C are close to each other, the double collision process is reversed when the impact parameter passes from the domain of one of them to the other. Thus, we find that while in regions A and B the atom first collides with the adsorbate and then with the flat surface, in the case of region C it is the opposite. The length of this latter region is nearly the same as the one for the shadow region S, given by Equation (23), because of the symmetry between the trajectories $\mathcal{F}_1$ and $\mathcal{F}_7$ On the other hand, regions A and B are defined, respectively, by the angular sectors $\Delta\theta_A = \pi/4 - \theta_i/2 - (\sin)^{-1}(z_0/a)$ and $\Delta\theta_B = \pi/4 - \theta_i/2$, which are both dependent on the incidence angle.

Regarding trapping, although this simple potential function cannot lead to this phenomenon, it is interesting to note that, to some extent, $\mathcal{F}_3$ and $\mathcal{F}_7$ can be considered as permanently trapped trajectories, since they will keep evolving parallel to the surface ($\theta_d = \pm\pi/2$). This is, of course, a rather weak case associated with a purely repulsive model, but still it is useful to understand in simple terms the types of dynamics that can be expected from a more refined classical model, such as the one specified by (2), concerning the presence of homologous pairs of trajectories (associated with single and double collisions) as well as the appearance of trapping.

4.2. Newtonian Level

The hard-wall model constitutes a sort of crude approach to the system studied, appropriate to explain some general features. However, a more realistic or refined model (closer to the experimental system) is going to include additional features, such as the rainbow phenomenon, which have also received much attention both experimentally [24,26] and theoretically [27–29], or the defect-mediated diffraction resonance [27,36], a kind of trapping induced by the presence of adsorbates on surfaces. These phenomena are associated to the particularities displayed by the potential model, such as the depth of potential wells, the stiffness of the repulsive region or the range of its attractive region. All these features are controlled by means of parameters that are found from best fit to the cross sections (intensities) experimentally measured; the optimal values eventually provide us with valuable information on the system analyze. In the particular case of the system discussed in this work, such information has to do with the way how the CO is attached to Pt(111) surface, which can be later used to better understand reactivity and diffusion properties.

The dynamics under the influence of the potential model (2) are illustrated in Figure 7a by means of a set of trajectories with initial conditions uniformly covering a wide range of impact parameters. Specifically, in this simulations $b \in [-10.6, 10.6]$ Å, for $E_i = 10$ meV and $\theta_i = 20°$ (again, as in the previous section, incidence out of the diagonal has also been chosen for these trajectories to stress some particular aspects). The attractive long-range term from Lennard–Jones contribution plays an important role here, since it is responsible for the permanent trapping (in this model) of He atoms along the Pt surface. This term accounts for the van der Waals interaction mediating between the neutral He atoms and the CO adsorbate, producing an effective transfer of energy from the perpendicular direction to the parallel one, such that the energy along this latter direction becomes larger than the incident energy to the expense of making negative the energy along the z direction. Of course, surface trapping is not totally permanent, since the presence of other adsorbates (not considered here, where we are working under single-adsorbate conditions) produces the opposite effect, that is, a trapped atom, after colliding with another neighboring adsorbates, may acquire enough energy along the z-direction (loosing it along the x direction) to escape from the surface.

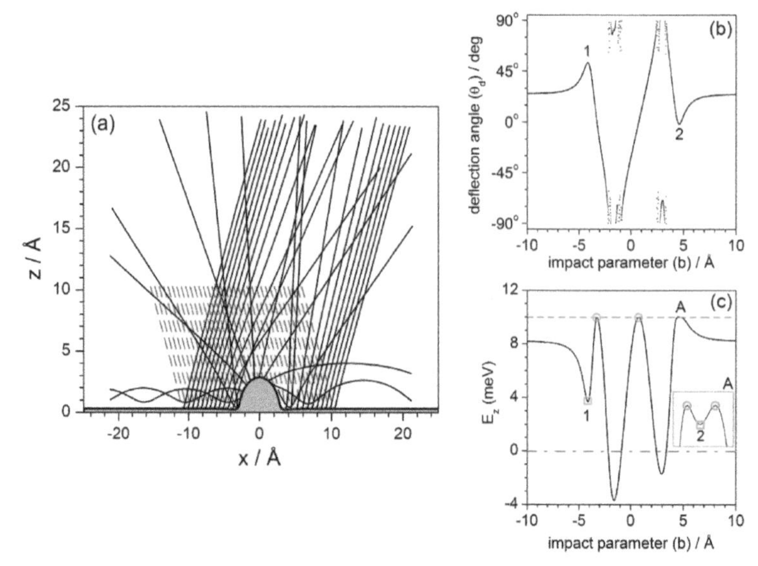

Figure 7. (**a**) Set of classical Newtonian trajectories for the interaction potential model (2) for an incidence energy $E_i = 10$ meV. As in Figure 5, for a better illustration of all possible cases, also an incidence angle $\theta_i = 20^\circ$ has been chosen. Moreover, the incident part of the trajectories has been represented with dashed line; (**b**) classical deflection function. While surface trapping gives rise to two kind of discontinuous regions, rainbow features manifest as local maxima (1) and minima (2); (**c**) Asymptotic-energy diagram. Here, trapping is detected through the two regions of the curve below the threshold $E_z = 0$ meV, while rainbows manifest with two local minima (green squares; for rainbow 2, see enlargement of region A in the inset). Orange circles denote conditions leading to perpendicular deflection with respect to the flat surface.

A useful tool to systematize and analyze the different dynamical behaviors exhibited by the trajectories of Figure 7a is the deflection function, i.e., the representation of the deflection angle with which the He atoms asymptotically leave the substrate as a function of their impact parameter. This function is represented in Figure 7b and clearly shows that it is characterized by two types of regions. One of them is smooth and continuous, which means that the deflection angle increases or decreases gradually. The local maxima and minima within this type of region denote the presence of rainbows, i.e., deflection directions characterized by an extremely high intensity (leaving aside mirror reflection from the flat surface). In the figure, we observe the presence of two of these rainbows (denoted with the numbers 1 and 2), one of them nearly along the normal to the flat Pt surface. The other type of region is seemingly random, which is a signature of trapping—this makes the function in these regions to be time-dependent, since deflection will depend on the time at which it is computed. Nonetheless, this behavior is not to be misinterpreted with presence of chaos that we observe in analogous representations for He diffraction from corrugated surfaces [42]. The difference is that, while, in those cases, this random-like region has a fractal structure [42]; here, it is very regular. This can easily be seen by plotting more values of the impact parameter within this region. Then, as this region of the deflection function becomes more dense, we will be able to better appreciate its regularity and, therefore, the lack of an underlying chaotic dynamics.

To complement as well as to disambiguate the information provided the by deflection function, particularly in the trapping regions, an alternative representation can be obtained if we focus far from the adsorbate, where the system energy is separable, as mentioned above. Accordingly, unless there is an extra energy exchange because of the presence of other defects, the He atom will keep constant its energies along the x and the z directions (there is no coupling term between both degrees of freedom in the potential describing the flat Pt surface). This thus allows to consider energy diagrams, where the energy left along the z degree of freedom is compared to the total available energy, which coincides with the incident energy (E_i), and to the dissociation threshold (0 meV), which determines the minimum energy for unbound motion. This plot is shown in Figure 7c for $E_i = 10$ meV. In this representation, we first note that a large portion of impact parameters gives rise to unbound motion, i.e., all the trajectories leave the Pt surface. Among these trajectories, those leaving the surface at an angle equal to a rainbow angle produce a local minimum in the energy diagram (green squares)—the absolute value of the rainbow angle can then be readily determined by means of the simple relation $\cos\theta_R = \sqrt{E_z^R/E_i}$ —while maxima (orange circles) correspond to trajectories that leave the surface perpendicularly (no energy along the x degree of freedom, although initially they all started with some energy in it). On the other hand, we also find some impact parameter regions for which the energy is negative. These regions correspond to conditions leading to trapping (oscillatory bound motion parallel to the flat surface), which are in correspondence with the discontinuous regions observed in the deflection function. Actually, from the energy diagram, we can determine with high accuracy the precise impact parameters that, at a certain incidence, will give determine the ranges of trapping. From the figure, we see that these limits are given by the intersection points of the energy curve with the zero-energy condition.

As we have seen, the energy diagram and the deflection function, in general terms, provide the same kind of information. Now, the energy diagram can be smartly used to determine pairs of homologous Newtonian trajectories as follows. Consider a given value for E_z. All the impact parameters that are obtained from the intersection of a horizontal line at the selected value of E_z with the energy diagram will provide sets of pairs of homologous trajectories. Some illustrative pairs are shown in Figure 8. In panel (a), for a selected energy such that $E_1 > E_z > 0$, where E_1 is the energy (along the z direction) for rainbow 1, we find two pairs of trajectories, one back-scattered and another forward-scattered. Although distinguishing between single and double collisions is not as simple as with the hard-wall model, we still can perceive that within these two pairs of homologous trajectories one of them undergoes a single collision (denoted with black color), while the other shows something that could be related with a double collision (with blue color). If E_z is below zero, we can still have two pairs of homologous trajectories, as seen in panel (b), although these trajectories exhibit permanent trapping. For energies above E_1 we can observe up to three or (for $E_z > E_2$) four pairs of homologous trajectories, with analogous behaviors. Actually, unlike the hard-wall model, here we notice that there can be several pairs contributing to reflection symmetry interference, as happens in panels (c) and (d).

Finally, Figure 9 offers a comparison between the behavior displayed by a purely repulsive adsorbate (closer to the hard-wall model) and the attractive adsorbate in terms of the corresponding deflection functions for $E_i = 10$ meV and normal incidence. The main difference between both models relies on a removal of the attractive term of the Lennard–Jones model. Notice that the repulsive model lacks the two rainbow features for $\Delta K_R = \pm 1.95$ Å^{-1} ($\theta_R = \pm 26.50°$), although both models keep nearly the same trapping rates, which has to do with the effective energy transfer from the perpendicular to the parallel directions due to the adsorbate curvature, and not that much with the presence of attractive basins around the adsorbate itself. Although not shown here, the same holds if the energy is increased. For instance, for $E_i = 40$ meV and normal incidence, rainbows are observed for $\Delta K_R = \pm 1.21$ Å^{-1} ($\theta_R = \pm 7.96°$) with the full potential model (2), but there is a complete absence of them when the repulsive model is used instead.

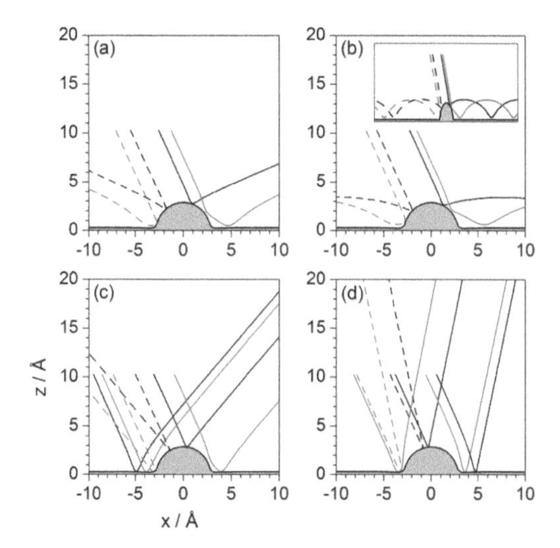

Figure 8. Pairs of homologous Newtonian trajectories with the same asymptotic value for their energy along the z, obtained from the energy diagram of Figure 7c: (**a**) $E_1 > E_z > 0$; (**b**) $0 > E_z$; (**c**) $E_i \cos^2 \theta_i > E_z > E_1$ and (**d**) $E_2 > E_z > E_i \cos^2 \theta_i$, where E_1 and E_2 are the energies (along the z direction) corresponding to rainbows 1 and 2 (see Figure 7b,c), respectively, and $E_i \cos^2 \theta_i$ is the incidence energy (also along the z direction), with $E_i = 10$ meV. Forward/backward deflected trajectories are denoted with solid/dashed line. Trajectories undergoing single/double collision/s are denoted with black/blue colors. All trajectories are started from a distance $z_i = 10.27$ Å above the flat Pt surface (beyond $z \approx 6.35$ Å, the interaction potential model (2) is negligible; see Figure 1a).

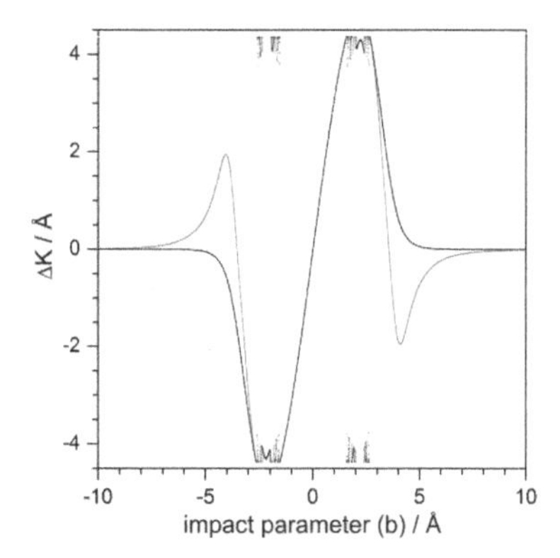

Figure 9. Deflection function for a purely repulsive model of the CO adsorbate (black solid line), obtained after removal of the attractive part of the Lennard–Jones function (4). Here, for simplicity, the incidence conditions are $\theta_i = 0°$ and $E_i = 10$ meV. To compare with, the deflection function corresponding to the full interaction potential model (2) is also represented (blue solid line), obtained for the same incidence conditions.

4.3. Bohmian Level

In the literature there has always been a controversy concerning how the different diffraction features observed in intensity patterns should be assigned in the case of the atom diffraction by impurities on surfaces. In 1988, for instance, Yinnon et al. [27] suggested the use of the quantum flux, **J**, as an interpretational tool to understand trapping processes in this type of systems. A vector representation of this field provides a reliable representation of how the Fraunhofer, rainbow and trapping features arise, although is not unambiguous at all. This procedure was employed earlier on in reactive scattering by Wyatt [52–54]. Although some dynamical information can be extracted about the distribution and change of the flux, it is difficult to understand how the system eventually evolves. The last step in our journey is precisely the use of Bohmian trajectories to get an idea of what is going on this type of systems from at a fully quantum level and, beyond quantum flux based analyses, to understand the dynamics by causally connecting a point on the initial state with another point of the final state, following a well-defined trajectory in real time. This trajectory is obtained by integration of the equation of motion [12,14]

$$\dot{\mathbf{r}} = \frac{\mathbf{J}}{\rho} = \frac{\hbar}{2mi}\left(\frac{\Psi^*\nabla\Psi - \Psi\nabla\Psi^*}{\Psi\Psi^*}\right) = \frac{\nabla S}{m}, \tag{24}$$

often regarded as guidance equation. In this equation, formerly introduced by Bohm as a postulate [1,4], ρ and $\mathbf{J}$ are respectively the probability density and the quantum flux [55]; the velocity field $\mathbf{v} = \dot{\mathbf{r}}$ is just the way how the probability density spreads through the configuration space in the form of a flux or current density. Regarding S, it is the phase field, which describes the local variations of the quantum phase and is typically obtained from the polar transformation of the wave function,

$$\Psi(\mathbf{r},t) = \sqrt{\rho(\mathbf{r},t)}e^{iS(\mathbf{r},t)/\hbar}. \tag{25}$$

The Bohmian trajectories shown below are obtained taking into account the value of the wave function at a given time (obtained by means of the propagator mentioned in Section 2), according to the second expression for the velocity field in Equation (24).

The analysis in this section is performed by studying the behavior displayed by sets of Bohmian trajectories with initial positions taken at a series of distances from the surface and uniformly distributed along the parallel direction for each one of those distances. Unlike the procedure followed in preceding sections, now the loss of translational symmetry caused by the presence of the isolated adsorbate on the Pt surface does not allow for studying the dynamics considering impact parameters only along the parallel direction for a given z value. This is the reason why different values of z are considered. This enables a better way to assign the different parts of the incident wave with final outgoing intensity peaks without ambiguity. These sets are displayed in Figures 10–12, with each set being labeled with the corresponding initial condition along the z direction referred to the center of the wave packet $\langle z \rangle_0 = 10.27$ Å. According to these figures, quantum-mechanically the dynamics cannot be understood in the same local terms as in classical mechanics, where trajectories did not display a different behavior depending on their starting distance along the vertical direction with respect to the adsorbate. In order to obtain a complete knowledge of the diffraction process, trajectories have to be chosen from across the whole region covered by the initial wave function. Different regions will give rise to different diffraction features, but also different sets of trajectories will be able to probe the surface at a different level, being able to approach it very closely or, on the contrary, will bounce backwards far away, without even having touched it physically. In this regard, one wonders whether, contrary to what is commonly stated within the "Bohmian community", the (Bohmian) trajectories can be considered to reveal the "true" motion followed by the particles they are associated with—in the present case, the motion of individual He atoms. This is a rather challenging question (as well as metaphysical), which, to the author's best knowledge, has not still been unambiguously answered, that is, with a solid, irrefutable experimental proof. Therefore, taking on a pragmatic view, here Bohmian trajectories are considered as hydrodynamic streamlines that allow us to investigate the flow dynamics of the probability flux in configuration space and, therefore, to understand towards which directions atoms are more likely redirected (deflected) after being scattered from the substrate (without providing any particular information on how each individual atom really moves). Actually, to some extent, this random view of the atomic motion is the idea behind the work that Bohm developed in 1954 in collaboration with Vigier [56] and, later on, in 1989 with Hiley [57], or the approach developed in 1966 by Nelson [58] (although this latter approach has nothing to do with Bohmian mechanics, it introduces analogous stochastic concepts). In these examples, the quantum particle follows a random-like path as a consequence of the action of a sub-quantum random medium; when motions are averaged, particle statistics reproduce the results described by Schrödinger's equation and Bohmian trajectories arise as the averaged flow-lines associated with the solutions to this equation.

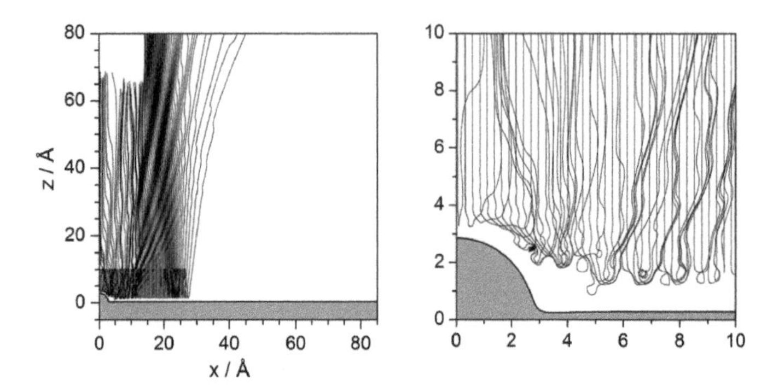

Figure 10. Set of Bohmian trajectories with initial positions uniformly distributed along the x direction and fixed value along the z direction: $z_i = \langle z \rangle_0 = 10.27$ Å, which corresponds to the center of the incident wave packet ($t = 0$) above the clean Pt surface (beyond $z \approx 6.35$ Å, the interaction potential model (2) is negligible; see Figure 1a). On the right-hand side, enlargement of the left plot near the adsorbate to illustrate the dynamical behavior displayed by the trajectories close to the substrate surface.

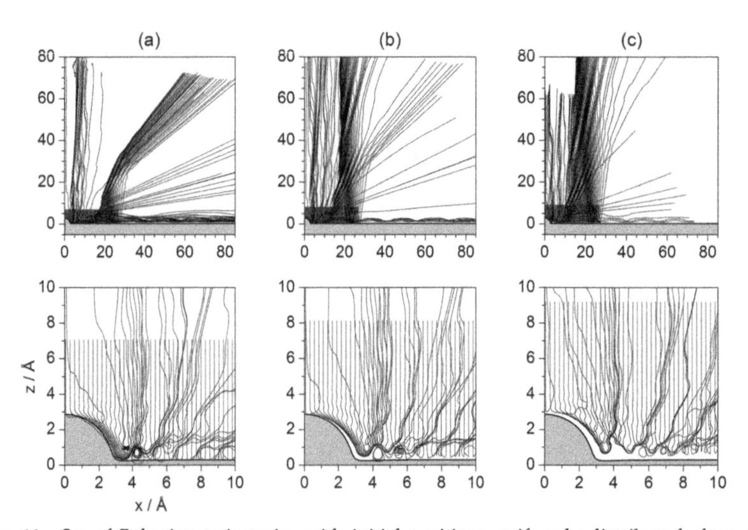

Figure 11. Set of Bohmian trajectories with initial positions uniformly distributed along the x direction and fixed value along the z direction: (**a**) $z_0 = \langle z \rangle_0 - 3.18$ Å; (**b**) $z_0 = \langle z \rangle_0 - 2.12$ Å and (**c**) $z_0 = \langle z \rangle_0 - 1.06$ Å, where $\langle z \rangle_0 = 10.27$ Å, which corresponds to the center of the incident wave packet ($t = 0$) above the clean Pt surface (beyond $z \approx 6.35$ Å, the interaction potential model (2) is negligible; see Figure 1a). In the corresponding lower panels, enlargement of the upper plots near the adsorbate to illustrate the dynamical behavior displayed by the trajectories close to the substrate surface.

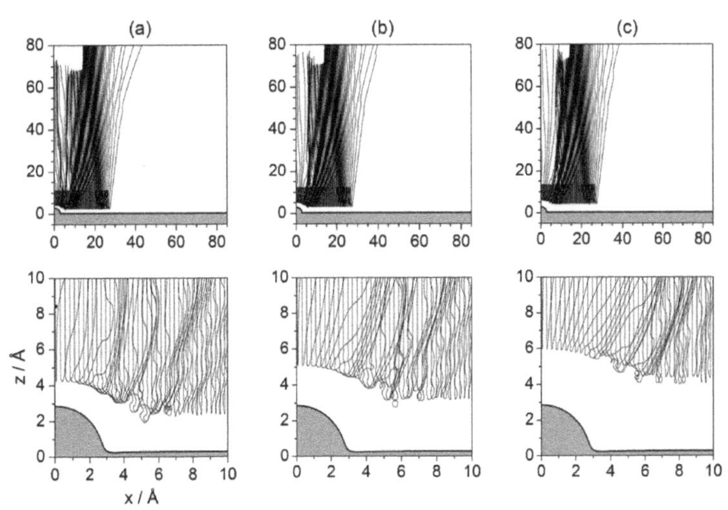

Figure 12. As in Figure 11, but considering: (**a**) $z_0 = \langle z \rangle_0 + 1.06$ Å; (**b**) $z_0 = \langle z \rangle_0 + 2.12$ Å and (**c**) $z_0 = \langle z \rangle_0 + 3.18$ Å.

In Figures 11 and 12, sets of trajectories taken from below and from above the z-value selected in Figure 10 are shown. In the three cases displayed in Figure 11, we notice that those trajectories that start closer to the surface are going to contribute more importantly to the marginal portions of the outgoing wave (B_i and C, as defined in Figure 3d). In sharp contrast, trajectories started in regions far from the adsorbate (see Figure 12) will contribute to the peaks related to small ΔK (A_i). This can be understood establishing a nice analogy with classical fluid dynamics and then introducing the notion of quantum pressure introduced by Takabayashi [59]. Accordingly, when the wave function is on the surface, we find that, while its lowest part is already bouncing backwards, the upper one is still moving downwards (towards the surface). In this situation, in the analogy of the wave function being associated with an ideal non-viscid and incompressible fluid, its upper part would be pushing the lowest one and then generating a remarkable pressure on it (this effect would increase with the incidence energy, although its duration would be shorter). This is something that cannot be seen directly on a wave-function representation, but that has an interesting counterpart in the case of the trajectories because we can see that those associated with the lowest part of the incident wave function are then pushed (or make evident the push) against the surface, remaining there for a rather long time. On the contrary, as the initial conditions are taken from upper regions of the initial wave function, the pressure on them will be lower and hence will bounce backwards further away from the physical surface. Thus, the evolution of these trajectories resembles in a closer way the behavior of classical trajectories, while the trajectories below are somehow forced to propagate parallel to the surface and escape by the borders of the wave.

The turbulent dynamics manifested by the trajectories close to the surface has to do with the web of nodal lines characterizing the wave function when it is on the surface. In particular, it is interesting that how sometimes the trajectories get trapped and whirl around some of these nodes, undergoing a sort of transient vortical trapping [32,33], different from the temporary trapping of the trajectories that will remain confined within the surface attractive well far from the adsorbate. The whirlpool motion displayed by such trajectories has actually an interesting property, namely the associated action is quantized, being an integer multiple (equal to the number of full loops) of a certain value. This type of motion is a confirmation of the former results found by Yinnon et al. [27]. Actually, recently, Efthymiopoulos et al. [60] formally determined the conditions under which such temporary nodal regions in scattering problems will appear, which happens to be in the regions where the amplitudes (modulus) of the incoming and outgoing waves become equal. This boundary, where both values are the same and that have important dynamical consequences, is what they called "separator".

Regarding the reflection symmetry interference phenomenon, this time it is not as simple as in the two previous cases because the trajectories obey the guiding rules imposed by the wave function and not the direct action of the potential. In the case of the rainbow, however, there is a more evident manifestation. Although Bohmian trajectories cannot cross one another at the same time, collectively they show a sort of rainbow-like precession, which is more prominent in the case of the trajectories shown in Figure 11: after the collision with the adsorbate, they start showing larger and larger deflections until getting trapped, then the start precessing backwards until reaching a maximum angle, and finally precess backwards again to smaller angles. In the case of the trajectories displayed in Figure 12, the behavior is analogous, even if in this case there is no trapping and the effect is more subtle. Thus, all the ensembles show a similar behavior, supporting the idea of the rainbow as being a sort of "global" effect, which translates into the "wings" observed in the intensity pattern shown in Figure 4.

5. Conclusions

In the last several years, Bohmian mechanics has been gaining ground as an appealing tool to deal with very different problems out of the area of the quantum foundations, its traditional environment. In such cases, the interest and relevance of Bohmian mechanics is emphasized by directly tackling a given problem with it. In the current work, however, the motivation has been a bit different. There are different trajectory-based approaches that have been or can be used to describe, analyze, understand and explain quantum systems and phenomena, even if they have different degrees of accuracy. The purpose here has been to establish an appropriate context to better understand the role of Bohmian trajectories within all those formulations as well as the kind of information conveyed by each one at its level of accuracy. To this end, we have considered a problem of interest out of the field of the quantum foundations, specifically the diffraction of helium atoms by a nearly flat platinum surface on top of which there is a carbon monoxide adsorbate. This is a problem that has been considered in the literature due to its intrinsic practical applications, although here it has been chosen due to its suitability to the purpose, since it is sufficiently simple to allow us its treatment at different levels.

Accordingly, the system has first been analyzed with a usual wave-function-based framework, investigating the effects associated with two different He-CO/Pt(111) interaction potential models:

- **Hard-wall model.** This model is in the form of an impenetrable (fully repulsive) wall, where the interaction is reduced to a sudden impact on the He atoms on the such a wall. The first model allows an exact asymptotic analytical treatment, convenient to elucidate the main mechanism observed in the diffraction pattern produced by single adsorbed particles on nearly flat surfaces, namely reflection symmetry interference.
- **Potential energy function.** This interaction model is determined from fitting to the experimental data and constitutes a refinement of the previous one in the sense that there is detailed information on the intensity of the interaction between the incoming atom and the substrate at each point (in this regard, the hard-wall model is just a crude approximation). Thus, in spite of its lack of analyticity, unlike the hard-wall model, it provides us with a more realistic description of the diffraction process in real time, rendering information on additional physics, such as rainbow features or surface trapping.

Although at a different degree of accuracy, these two models provide us with explanation of the features observed in the experimental diffraction patterns. The question is how to interpret or understand these diffraction features or, in other words, to elucidate the physical mechanism responsible for each of such features (reflection symmetry interference, rainbows, or surface trapping). Typically, this is done by setting protocols based on quantum-classical correspondence, e.g., analyzing the system by means of classical trajectories and then comparing the results rendered by both the classical (trajectory) model and the quantum-mechanical (wave function) one.

Bearing that in mind, as has been seen in the preceding sections, here we have tackled the issue at three different levels:

- **Fermatian level.** This first level is the simplest one, based on computing what has been here denoted as Fermatian trajectories, which are just the direct analog to optical rays reflected on a hard wall in a medium with constant refractive index. According to this trajectory model:
    - These trajectories have revealed that there are pairs of homologous trajectories, such that one of the peers undergoes single scattering off the interaction potential, while the other undergoes double scattering. The fact that a trajectory collides with the CO/Pt system at one point (single collision) or at two different points (double collision) is a function of the impact parameter. Accordingly, a simple mapping can be establish, which helps to easily localize regions of impact parameters that are going to produce homologous pairs of trajectories.
    - The mechanism of reflection symmetry interference is associated with these paired trajectories, which is explained in the same way that we explain interference from two coherent sources: interference maxima and minima arise depending on whether the path difference between the two paths (or virtual rays) joining each source with a given observation point on a distant screen is equal to an integer number of wavelengths or to half an integer, respectively. Although these paths are nonphysical (they are just a mathematical construct), they allow us to understand in simple terms the appearance of the alternating structure of bright and dark interference fringes. In the present case, the path length arises from the extra path length of the trajectory affected by the double collision with respect to the homologous pair with single collision.
    - In addition, it has also been seen that two specific trajectories are deflected parallel to the surface, which can be interpreted as a mechanism precursor of the surface trapping mechanism that appears in more refined models, such as the Newtonian and the Bohmian ones.
- **Newtonian level.** On the next level, the Newtonian one, classical trajectories are obtained for the realistic potential energy surface describing the interaction between the He atoms and the substrate. In this case, it is not so simple to distinguish between single and double collisions, because the deflection of the trajectories near the surface, where the interaction between the He atoms and the CO/Pt surface is stronger, changes gradually very smoothly. However, we have been able to extract a series of interesting conclusions:
    - By means of an energy diagram (asymptotic energy along the z direction as a function of the impact parameter), we been able to devise a method that allows to determine in a simple fashion pairs of homologous (Newtonian) trajectories. This diagram is thus a suitable method to determine a behavioral mapping of initial conditions (impact parameters) for a given incidence direction (incident energy).
    - Accordingly, also at this level, it is possible to find an underlying mechanism responsible for the reflection symmetry interference found in the corresponding quantum intensity patterns. Actually, interference patterns could be reconstructed in the same way as with the Fermatian model, although in this case we would be dealing with a space-dependent refractive index (the potential function) and the Newtonian trajectories would play the role of Feynman's paths. Nonetheless, although such a reconstruction is possible and the techniques are well known, this does not mean that trajectories, Fermatian or Newtonian, contain any information on the interference process; in both cases, they are only a tool to determine the interference pattern.
    - Regarding the trapping phenomenon, it has been found to be more prominent, with an important amount of trajectories remaining trapped permanently along the surface. This is, however, only a temporary feature, since it may disappear as son as the trapped atoms find

another adsorbate. In such a case, the collisions with this adsorbate may provoke an effective transfer of energy from the parallel to the normal direction, such that the will be able to eventually leave the surface.

- Finally, due to the attractive well surrounding the adsorbate, we have also observed the appearance of rainbow features, i.e., high accumulations of trajectories along particular deflection directions. However, rather than contributing with a specific, localized feature in the corresponding quantum intensity pattern, rainbows seem to manifest affecting them globally, i.e., giving rise to features that appear at different places. This has been noticed by computing exactly the same with an alternative repulsive adsorbate model, which lacks the surrounding attractive well and therefore does not give rise to the formation of rainbows.

- **Bohmian level.** The upper level here considered is the Bohmian one, where things change substantially if we note that the transition from the Fermatian level to the Newtonian one can be seen as a refinement associated with having a more accurate description of the interaction potential model, changing a hard wall by a "soft" wall. These are the main findings at this level:

  - First of all, since Bohmian trajectories are associated with a particular wave function, there is no freedom to choose a given set of initial conditions because depending on the positions selected relative to the region covered by the initial wave function, the trajectories are going to exhibit a different behavior. Thus, we have seen that while some of them are deflected quite far from the physical surface (more intense interaction region), other trajectories move just on top of it, displaying signatures of vorticality.
  - To better understand that point, notice that Fermatian trajectories are only ruled by the law of reflection, while Newtonian trajectories are ruled by correlations between the two degrees of freedom, x and y, that can be locally established within the interaction region (i.e., the region where the interaction potential is stronger, near the substrate). In the case of Bohmian trajectories, the dynamics is not directly ruled by the interaction potential, but by a wave field that is able to (non-classically) convey information from everywhere in the configuration space (through its phase). This makes a substantial difference between classical (Fermatian or Newtonian) and Bohmian trajectories, which may lead us to think that direct comparisons or analogies must be taken with care. That is, nothing of what has been seen at the previous levels remains at the upper one, since it is not possible to form pairs of homologous trajectories.
  - In this case, and contrary to the two previous models, the trajectories contain information about the interference process and, therefore, can be used to determine the fringe structure of the pattern by simply making statistics over them. If they are properly distributed across the region of the configuration space covered by the initial probability density, they will eventually distribute according to the final probability distribution by virtue of the continuity equation that they satisfy.
  - Regarding rainbow features, present in the Newtonian model and also, with a weak precursor, in the Fermatian one, the only a similar behavior is observed, although it is difficult to establish a unique correspondence with the phenomenon of the two previous models. In the Bohmian case, taken the trajectories that start with the same value z_0, it is seen that their final positions show, for some range of x_0 values, a certain "precession" as x_0 increases. However, it has not been possible to uniquely identify this phenomenon with the classical rainbow. In the case of surface trapping, on the contrary, there same effect has been observed in the three models (again, in the Fermatian model it is only a weak precursor).
  - Finally, it has also been observed that, depending on how close or far a Bohmian trajectory is started from the physical substrate surface, it will be able to reach this surface or just bounce backwards quite far from it (from what we could call an effective nonphysical surface). Actually, if the trajectories start close to the surface, they are influenced by the web of

maxima developed (and sustained for some time) around the adsorbate, displaying a rich vortical dynamics.

To conclude, we can say that, although there is no one-to-one correspondence between classical and Bohmian trajectories, it is still possible to understand these two alternative descriptions in a complementary way, with one being the skeleton upon which the other rests, at least at a formal level. Classical trajectories have been and are used to understand in relatively simple terms why quantum distributions are as they are, in a way analogous to how an optical path allows us to understand and explain the appearance of wave phenomena, such as diffraction or interference. Bohmian trajectories are synthesized from wave amplitudes (wave functions), so the same underlying scheme should also be valid (i.e., using classical trajectories to understand why Bohmian trajectories evolve in the way they do). At the same time, Bohmian trajectories offer a clear picture of the evolution of the quantum system by monitoring the local evolution of the quantum flux, which provides some clues on dynamical aspects that otherwise would remain hidden (e.g., the development and effects of vortical dynamics, or the appearance of effective barriers).

Acknowledgments: The author would like to express his gratitude to the Frankling-Fetzer Foundation for support to attend the Emergent Quantum Mechanics (EmQM17) conference, held in London on the occasion of the David Bohm Centennial, and the organizers for their kind invitation to participate in the meeting. Support from the Spanish MINECO is also acknowledged (Grant No. FIS2016-76110-P).

Conflicts of Interest: The author declares no conflict of interest.

References

1. Bohm, D. A suggested interpretation of the quantum theory in terms of "hidden" variables. I. *Phys. Rev.* **1952**, *85*, 166–179. [CrossRef]
2. Bohm, D. A suggested interpretation of the quantum theory in terms of "hidden" variables. II. *Phys. Rev.* **1952**, *85*, 180–193. [CrossRef]
3. Bohm, D.; Hiley, B.J. *The Undivided Universe*; Routledge: New York, NY, USA, 1993.
4. Holland, P.R. *The Quantum Theory of Motion*; Cambridge University Press: Cambridge, UK, 1993.
5. Chattaraj, P.K. (Ed.) *Quantum Trajectories*; CRC Taylor and Francis: New York, NY, USA, 2010.
6. Hughes, K.H.; Parlant, G. (Eds.) *Quantum Trajectories*; CCP6: Daresbury, UK, 2011.
7. Oriols, X.; Mompart, J. (Eds.) *Applied Bohmian Mechanics: From Nanoscale Systems to Cosmology*; Pan Standford Publishing: Singapore, 2012.
8. Sanz, A.S.; Miret-Artés, S. *A Trajectory Description of Quantum Processes. II. Applications*; Lecture Notes in Physics; Springer: Heidelberg/Berlin, Germany, 2014.
9. Benseny, A.; Albareda, G.; Sanz, A.S.; Mompart, J.; Oriols, X. Applied Bohmian mechanics. *Eur. Phys. J. D* **2014**, *68*, 286. [CrossRef]
10. Dürr, D.; Teufel, S. *Bohmian Mechanics: The Physics and Mathematics of Quantum Theory*; Springer: Heidelberg/Berlin, Germany, 2009.
11. Dürr, D.; Goldstein, S.; Zanghì, N. *Quantum Physics without Quantum Philosophy*; Springer: Heidelberg/Berlin, Germany, 2013.
12. Sanz, A.S.; Miret-Artés, S. *A Trajectory Description of Quantum Processes. I. Fundamentals*; Lecture Notes in Physics; Springer: Heidelberg/Berlin, Germany, 2012.
13. Wyatt, R.E. *Quantum Dynamics with Trajectories*; Springer: New York, NY, USA, 2005.
14. Sanz, A.S. Bohm's approach to quantum mechanics: Alternative theory or practical picture? *arXiv* **2017**, arXiv:1707.00609v1.
15. Kocsis, S.; Braverman, B.; Ravets, S.; Stevens, M.J.; Mirin, R.P.; Shalm, L.K.; Steinberg, A.M. Observing the average trajectories of single photons in a two-slit interferometer. *Science* **2011**, *332*, 1170–1173. [CrossRef] [PubMed]
16. Couder, Y.; Protière, S.; Fort, E.; Boudaoud, A. Dynamical phenomena: Walking and orbiting droplets. *Nature* **2005**, *437*, 208. [CrossRef] [PubMed]
17. Couder, Y.; Fort, E. Single-particle diffraction and interference at a macroscopic scale. *Phys. Rev. Lett.* **2006**, *97*, 154101. [CrossRef] [PubMed]

18. Protière, S.; Boudaoud, A.; Couder, Y. Particle-wave association on a fluid interface. *J. Fluid. Mech.* **2006**, *554*, 85–108. [CrossRef]
19. Fort, E.; Eddi, A.; Boudaoud, A.; Moukhtar, J.; Couder, Y. Path-memory induced quantization of classical orbits. *Proc. Natl. Acad. Sci. USA* **2010**, *108*, 17515–17520. [CrossRef]
20. Bush, J.W.M. Quantum mechanics writ large. *Proc. Natl. Acad. Sci. USA* **2010**, *107*, 17455–17456. [CrossRef]
21. Harris, D.M.; Moukhtar, J.; Fort, E.; Couder, Y.; Bush, J.W.M. Wavelike statistics from pilot-wave dynamics in a circular corral. *Phys. Rev. E* **2013**, *88*, 011001. [CrossRef] [PubMed]
22. Bush, J.W.M. Pilot-wave hydrodynamics. *Annu. Rev. Fluid Mech.* **2015**, *47*, 269–292. [CrossRef]
23. Lahee, A.M.; Manson, J.R.; Toennies, J.P.; Wöll, C. Observation of interference oscillations in helium scattering from single surface defects. *Phys. Rev. Lett.* **1986**, *57*, 471–474. [CrossRef] [PubMed]
24. Lahee, A.M.; Manson, J.R.; Toennies, J.P.; Wöll, C. Helium atom differential cross sections for scattering from single adsorbed CO molecules on a Pt(111) surface. *J. Chem. Phys.* **1987**, *86*, 7194–7203. [CrossRef]
25. Drolshagen, G.; Vollmer, R. Atom scattering from surfaces with isolated impurities: Calculations for hard wall and soft potentials. *J. Chem. Phys.* **1987**, *87*, 4948–4957. [CrossRef]
26. Graham, A.P.; Hofman, F.; Toennies, J.P.; Manson, J.R. Helium atom scattering from isolated CO molecules on copper (001). *J. Chem. Phys.* **1996**, *105*, 2093–2098. [CrossRef]
27. Yinnon, A.T.; Kosloff, R.; Gerber, R.B. Atom scattering from isolated adsorbates on surfaces: Rainbows, diffraction interferences, and trapping resonances. *J. Chem. Phys.* **1988**, *88*, 7209–7220. [CrossRef]
28. Carré, M.N.; Lemoine, D. Fully quantum study of the 3D diffractive scattering of He from isolated CO adsorbates on Pt(111). *J. Chem. Phys.* **1994**, *101*, 5305–5312. [CrossRef]
29. Lemoine, D. Atomic scattering from single adsorbates: What can we learn from the gas phase? *Phys. Rev. Lett.* **1998**, *81*, 461–464. [CrossRef]
30. Choi, B.H.; Tang, K.T.; Toennies, J.P. Interpretation of helium atom scattering from isolated CO molecules on copper (001) based on an exact quantum mechanical model. *J. Chem. Phys.* **1997**, *107*, 1631–1633. [CrossRef]
31. Choi, B.H.; Tang, K.T.; Toennies, J.P. Quantum-mechanical scattering of an atom from a rigid hemisphere on a flat surface. *J. Chem. Phys.* **1997**, *107*, 9437–9446. [CrossRef]
32. Sanz, A.S.; Borondo, F.; Miret-Artés, S. Quantum trajectories in atom-surface scattering with single adsorbates: The role of quantum vortices. *J. Chem. Phys.* **2004**, *120*, 8794–8806. [CrossRef] [PubMed]
33. Sanz, A.S.; Borondo, F.; Miret-Artés, S. Role of quantum vortices in atomic scattering from single adsorbates. *Phys. Rev. B* **2004**, *69*, 115413. [CrossRef]
34. Hofmann, F.; Toennies, J.P. High-resolution helium atom time-of-flight spectroscopy of low-frequency vibrations of adsorbates. *Chem. Rev.* **1996**, *96*, 1307–1326. [CrossRef] [PubMed]
35. Kleyn, A.W.; Horn, T.C.M. Rainbow scattering from solid surfaces. *Phys. Rep.* **1991**, *199*, 191–230. [CrossRef]
36. Glebov, A.; Manson, J.R.; Skofronick, J.G.; Toennies, J.P. Defect-mediated diffraction resonances in surface scattering. *Phys. Rev. Lett.* **1997**, *78*, 1508–1511. [CrossRef]
37. Child, M.S. *Molecular Collision Theory*; Academic Press: London, UK, 1974.
38. Morse, P.M. Diatomic molecules according to the wave mechanics. II. Vibrational levels. *Phys. Rev.* **1929**, *34*, 57–64. [CrossRef]
39. Dubertrand, R.; Shim, J.B.; Struyve, W. Bohmian trajectories for the half-line barrier. *J. Phys. A* **2018**, *51*, 085302. [CrossRef]
40. Prosser, R.D. The interpretation of diffraction and interference in terms of energy flow. *Int. J. Theor. Phys.* **1976**, *15*, 169–180. [CrossRef]
41. Born, M.; Wolf, E. *Principles of Optics: Electromagnetic Theory of Propagation, Interference and Diffraction of Light*, 7th ed.; Cambridge University Press: Cambridge, UK, 1999.
42. Guantes, R.; Sanz, A.S.; Margalef-Roig, J.; Miret-Artés, S. Atom-surface diffraction: A trajectory description. *Surf. Sci. Rep.* **2004**, *53*, 199–330. [CrossRef]
43. Sanz, A.S.; Miret-Artés, S. Selective adsorption resonances: Quantum and stochastic approaches. *Phys. Rep.* **2007**, *451*, 37–154. [CrossRef]
44. Leforestier, C.; Bisseling, R.H.; Cerjan, C.; Feit, M.D.; Friesner, R.; Guldberg, A.; Hammerich, A.; Jolicard, G.; Karrlein, W.; Meyer, H.D.; et al. A comparison of different propagation schemes for the time dependent schrödinger equation. *J. Comput. Phys.* **1991**, *94*, 59–80. [CrossRef]
45. Press, W.H.; Teukolsky, S.A.; Vetterling, W.T.; Flannery, B.P. *Numerical Recipes in Fortran 77: The Art of Scientific Computing*, 2nd ed.; Cambridge University Press: Cambridge, UK, 1992.

46. Sanz, A.S.; Borondo, F.; Miret-Artés, S. Particle diffraction studied using quantum trajectories. *J. Phys. Condens. Matter* **2002**, *14*, 6109–6145. [CrossRef]
47. Morse, P.M.; Feshbach, H. *Methods of Theoretical Physics*; McGraw-Hill: New York, NY, USA, 1953.
48. Joachain, C.J. *Quantum Collision Theory*; North-Holland: Amsterdam, The Netherlands, 1975.
49. Farías, D.; Rieder, K.H. Atomic beam diffraction from solid surfaces. *Rep. Prog. Phys.* **1998**, *61*, 1575–1664. [CrossRef]
50. Gerber, R.B.; Yinnon, A.T.; Kosloff, R. Effects of isolated impurities on atom scattering from crystalline surfaces: Exact quantum-mechanical calculations. *Chem. Phys. Lett.* **1984**, *105*, 523–526. [CrossRef]
51. Choi, B.H.; Graham, A.P.; Tang, K.T.; Toennies, J.P. Helium atom scattering from isolated CO molecules on a Pt(111) surface: Experiment versus close-coupling calculations for a realistic He-CO potential. *J. Chem. Phys.* **2000**, *112*, 10538–10547. [CrossRef]
52. McCullough, E.A.; Wyatt, R.E. Quantum dynamics of the collinear (H,H_2) reaction. *J. Chem. Phys.* **1969**, *51*, 1253–1254. [CrossRef]
53. McCullough, E.A.; Wyatt, R.E. Dynamics of the collinear $H+H_2$ reaction. I. Probability density and flux. *J. Chem. Phys.* **1971**, *54*, 3578–3591. [CrossRef]
54. McCullough, E.A.; Wyatt, R.E. Dynamics of the collinear $H+H_2$ reaction. II. Energy analysis. *J. Chem. Phys.* **1971**, *54*, 3592–3600. [CrossRef]
55. Schiff, L.I. *Quantum Mechanics*, 3rd ed.; McGraw-Hill: Singapore, 1968.
56. Bohm, D.; Vigier, J.P. Model of the causal interpretation of quantum theory in terms of a fluid with irregular fluctuations. *Phys. Rev.* **1954**, *96*, 208–216. [CrossRef]
57. Bohm, D.; Hiley, B.J. Non-Locality and Locality in the Stochastic Interpretation of Quantum Mechanics. *Phys. Rep.* **1989**, *172*, 93–122. [CrossRef]
58. Nelson, E. Derivation of the Schrödinger equation from Newtonian mechanics. *Phys. Rev.* **1966**, *150*, 1079–1085. [CrossRef]
59. Takabayasi, T. On the formulation of quantum mechanics associated with classical pictures. *Prog. Theor. Phys.* **1952**, *8*, 143–182. [CrossRef]
60. Efthymiopoulos, C.; Delis, N.; Contopoulos, G. Wavepacket approach to particle diffraction by thin targets: Quantum trajectories and arrival times. *Ann. Phys.* **2012**, *327*, 438–460. [CrossRef]

MDPI

Article

On Bohmian Mechanics, Particle Creation, and Relativistic Space-Time: Happy 100th Birthday, David Bohm!

Roderich Tumulka

Mathematisches Institut, Eberhard-Karls-Universität, Auf der Morgenstelle 10, 72076 Tübingen, Germany; roderich.tumulka@uni-tuebingen.de

Received: 24 April 2018; Accepted: 6 June 2018; Published: 14 June 2018

Abstract: The biggest and most lasting among David Bohm's (1917–1992) many achievements is to have proposed a picture of reality that explains the empirical rules of quantum mechanics. This picture, known as pilot wave theory or Bohmian mechanics among other names, is still the simplest and most convincing explanation available. According to this theory, electrons are point particles in the literal sense and move along trajectories governed by Bohm's equation of motion. In this paper, I describe some more recent developments and extensions of Bohmian mechanics, concerning in particular relativistic space-time and particle creation and annihilation.

Keywords: de Broglie–Bohm interpretation of quantum mechanics; pilot wave; interior-boundary condition; ultraviolet divergence; quantum field theory

---

1. Introduction

In 1952, David Bohm [1] solved the biggest of all problems in quantum mechanics, which is to provide an explanation of quantum mechanics. (For discussion of this problem see e.g., [2–5].) His theory is known as Bohmian mechanics, pilot-wave theory, de Broglie–Bohm theory, or the ontological interpretation. This theory makes a proposal for how our world might work that agrees with all empirical observations of quantum mechanics. Unfortunately, it is widely under-appreciated. It achieves something that was often (before and even after 1952) claimed impossible: To explain the rules of quantum mechanics through a coherent picture of microscopic reality.

In the following, I briefly review Bohmian mechanics and then discuss some extensions of it that were developed in recent years. For textbook-length introductions to Bohmian mechanics, see [4–7]; for a recent overview article, see [8].

1.1. *Significance of Bohmian Mechanics*

Bohmian mechanics is remarkably simple and elegant. In my humble opinion, some extension of it is probably the true theory of quantum reality. Compared to Bohmian mechanics, orthodox quantum mechanics appears rather incoherent. In fact, orthodox quantum mechanics appears like the narrative of a dream whose logic does not make sense any more once you are awake although it seemed completely natural while you were dreaming (e.g., [2,4]).

According to Bohmian mechanics, electrons and other elementary particles are particles in the literal sense, i.e., they have a well-defined position $Q_j(t) \in \mathbb{R}^3$ at all times t. They have trajectories. These trajectories are governed by Bohm's equation of motion (see below). In view of the widespread claim that it was impossible to explain quantum mechanics, it seems remarkable that something as simple as particle trajectories does the job. Thus, what went wrong in orthodox QM? Some variables were left out of consideration: the particle positions!

1.2. Laws of Bohmian Mechanics

According to non-relativistic Bohmian mechanics of N particles, the position $Q_j(t)$ of particle j in Euclidean three-space moves according to Bohm's equation of motion,

$$\frac{dQ_j}{dt} = \frac{\hbar}{m_j}\mathrm{Im}\,\frac{\psi^*\nabla_j\psi}{\psi^*\psi}(Q_1,\ldots,Q_N) \tag{1}$$

for every $j = 1,\ldots,N$. If some particles have spin, then $\psi^*\phi$ means the inner product in spin space. The wave function ψ of the universe evolves according to the Schrödinger equation,

$$i\hbar\frac{\partial\psi}{\partial t} = -\sum_j \frac{\hbar^2}{2m_j}\nabla_j^2\psi + V\psi\,. \tag{2}$$

The initial configuration $Q(0) = (Q_1(0),\ldots,Q_N(0))$ of the universe is random with probability density

$$\rho = |\psi_0|^2\,. \tag{3}$$

(Actually, the point $Q(0)$ need not be truly random; it suffices that $Q(0)$ "looks typical" with respect to the statistical properties of the ensuing history $t \mapsto Q(t)$ [9], much like the number π is not truly random but its decimal expansion looks like a typical sequence of digits.)

1.3. Properties of Bohmian Mechanics

It follows from Equations (1)–(3) that at any time $t \in \mathbb{R}$, $Q(t)$ is random with density $\rho_t = |\psi_t|^2$ ("equivariance theorem" or "preservation of $|\psi|^2$"). It follows further, by a theorem akin to the law of large numbers, that subsystems of the universe with wave function φ will always have configurations that look random with $|\varphi|^2$ distribution [9]. This fact, known as "quantum equilibrium", is the root of the agreement between the empirical predictions of Bohmian mechanics and the rules of the quantum formalism.

For an example of equivariance and quantum equilibrium, Figure 1 shows a selection of trajectories for the double-slit experiment with roughly a $|\varphi|^2$ distribution, where φ is a 1-particle wave function. The equivariance theorem implies that the arrival places on the right (where one may put a screen) are $|\varphi|^2$ distributed; thus, more particles arrive where $|\varphi|^2$ is larger. John Bell commented [10]:

> "This idea seems to me so natural and simple, to resolve the wave-particle dilemma in such a clear and ordinary way, that it is a great mystery to me that it was so generally ignored."

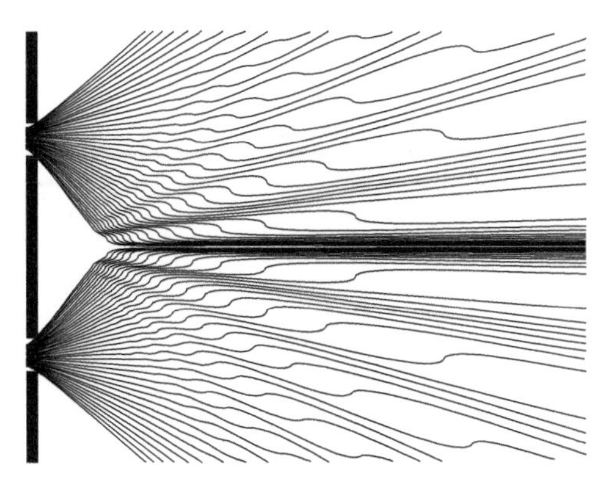

Figure 1. Several possible trajectories for a Bohmian particle in a double-slit setup, coming from the left. (Reprinted from [6], based on a figure in [11]).

Bohmian mechanics is clearly non-local (i.e., involves faster-than-light influences) because, according to Equation (1), the velocity of particle j depends on the simultaneous positions of all other particles $Q_1, \ldots, Q_N$. Of course, Bell's theorem [12] shows that every theory in agreement with the empirical facts of quantum mechanics must be non-local.

Bohmian mechanics avoids the problematical idea that the world consists only of wave function. It provides precision, clarity, and a clear ontology in space-time. It allows for an analysis of quantum measurements, thus replacing the postulates of orthodox quantum mechanics by theorems.

2. Extension of Bohmian Mechanics to Particle Creation

Bohmian mechanics has been successfully extended to incorporate particle creation. In theories with particle trajectories, particle creation and annihilation mean that trajectories can begin and end (Figure 2). Perhaps the most plausible picture would have them begin and end on the trajectories of other particles.

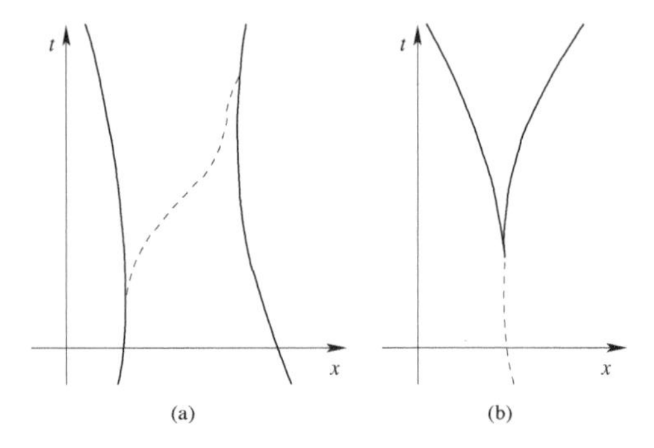

Figure 2. Possible patterns of particle world lines in theories with particle creation and annihilation: (**a**) a boson (dashed world line) is emitted by a fermion and absorbed by another; and (**b**) a boson (dashed world line) decays into two fermions. (Reprinted from [13]).

Particle creation and annihilation come up particularly in quantum field theory (QFT); since we want to connect them with particle trajectories, we make use of the particle-position representation of QFTs, a representation used also independently of the Bohmian approach, for example in [14–16]. The state vector then is a vector in Fock space $\mathscr{F}$,

$$\psi \in \mathscr{F} = \bigoplus_{n=0}^{\infty} \mathscr{H}_n \,, \tag{4}$$

or perhaps in the tensor product of several Fock spaces. Here, the n-particle Hilbert space $\mathscr{H}_n$ (also called the n-particle sector or simply n-sector of $\mathscr{F}$) is the symmetrized or anti-symmetrized n-th tensor power of the 1-particle Hilbert space $\mathscr{H}_1$. The position representation of $\psi \in \mathscr{F}$ is a function on the configuration space of a variable number of particles,

$$\mathcal{Q} = \bigcup_{n=0}^{\infty} \mathbb{R}^{3n} \,, \tag{5}$$

and $|\psi|^2$ defines a probability distribution on $\mathcal{Q}$. Here, $\mathbb{R}^{3n}$ is called the n-sector of $\mathcal{Q}$. (In fact, it is often desirable to use *unordered* configurations $\{x_1, \ldots, x_N\}$ because, in nature, configurations are not ordered. In Equation (5) and in the following, we use *ordered* configurations $(x_1, \ldots, x_N)$ because that allows for easier notation.)

2.1. Bell's Jump Process (In Its Continuum Version)

Here is the natural extension of Bohmian mechanics to particle creation [13,17–21]; Bell [17] considered this on a lattice, but it can be set up as well in the continuum [13,18,19], and we directly consider this case. The configuration curve $Q(t)$ will jump one sector up (respectively, down) whenever a particle is created (respectively, annihilated) (see Figure 3).

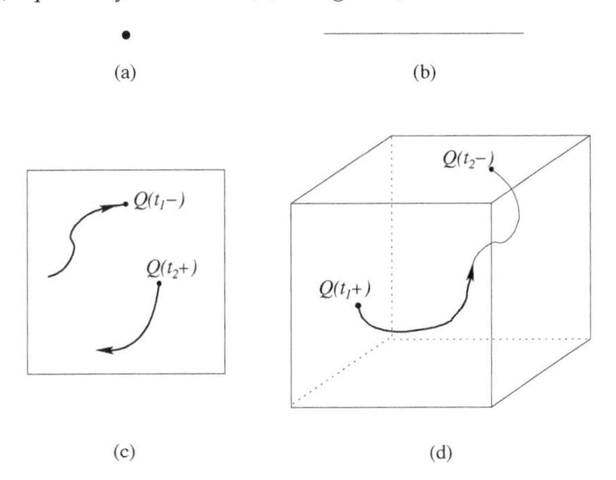

Figure 3. The configuration space in Equation (5) of a variable number of particles; drawn are, for space dimension $d = 1$, the first four sectors: (**a**) the zero-particle sector has a single element, the empty configuration; (**b**) the one-particle sector is a copy of physical space; (**c**) the two-particle sector; and (**d**) the three-particle sector. In addition, the configuration curve corresponding to Figure 2a is drawn; it jumps at time t_1 from the two-particle sector to the three-particle sector and at time t_2 back. (Reprinted from [13]).

According to (the continuum version of) Bell's proposal, jumps (e.g., from the n-sector to the $n+1$-sector) occur in a *stochastic* way, with rates governed by a further law of the theory. This means that, according to this theory, jumps occur spontaneously as an element of irreducible randomness in nature; they are not pre-determined by any further variables ("hidden" or not). It was not the point of Bohmian mechanics to restore determinism but to hypothesize what actually happens in the microscopic reality; if the most convincing hypothesis turns out to be deterministic (as it does for fixed particle number), then that is fine, if not, that is fine, too. Here, the randomness in the jumps is relevant to ensuring that, after particle creation, the configuration is still $|\psi|^2$ distributed.

Mathematically, $(Q(t))_{t\in\mathbb{R}}$ forms a stochastic process, in fact a Markov jump process. Between jumps, Bohm's equation of motion applies. The law governing the jumps reads as follows: Given that the present configuration $Q(t)$ is $q' \in \mathcal{Q}$, the rate (i.e., probability per time) of jumping to a volume element dq around $q \in \mathcal{Q}$ is

$$\sigma^{\psi}(q' \to dq) = \frac{\max\{0, \frac{2}{\hbar}\,\mathrm{Im}\,\langle\psi|q\rangle\langle q|H_I|q'\rangle\langle q'|\psi\rangle\}}{\langle\psi|q'\rangle\langle q'|\psi\rangle}dq\,. \tag{6}$$

Here, H_I is the interaction Hamiltonian as in $H = H_0 + H_I$ with H_0 the free Hamiltonian. More generally, $|q\rangle\langle q|\,dq$ could be replaced by a PVM (projection-valued measure) or a POVM (positive-operator-valued measure) $P(dq)$ on $\mathcal{Q}$ (and $|q'\rangle\langle q'|$ by $P(dq')$, as factors of dq' would cancel out). Since H_I usually links only to the next higher and lower sector, only jumps to the next higher or lower sector are allowed by Equation (6).

The jump rate in Equation (6) is so designed as to entail an equivariance theorem [19]: that is, if $Q(0)$ is $|\psi_0|^2$ distributed (that is, abstractly speaking, $\langle\psi_0|P(\cdot)|\psi_0\rangle$ distributed), then at every $t \in \mathbb{R}$, $Q(t)$ is $|\psi_t|^2$ distributed (that is, $\langle\psi_t|P(\cdot)|\psi_t\rangle$ distributed).

The jump rate Equation (6) can be thought of as an analog of Bohm's equation of motion in Equation (1) for jumps: for example, it involves quadratic expressions in ψ in both the numerator and the denominator and leads to the equivariance of $|\psi|^2$. The point of the jump law is to set up a process $Q(t)$ once a Hilbert space $\mathscr{H}$, a state vector $\Psi \in \mathscr{H}$, a (reasonable) Hamiltonian H, a configuration space $\mathcal{Q}$, and configuration operators $P(dq)$ are given. Together with Bohm's equation of motion in Equation (1), the rate Equation (6) achieves this for Hamiltonians with ultraviolet cutoff, which brings us to the problem of ultraviolet divergence.

2.2. An Ultraviolet Divergence Problem

For the sake of concreteness of our discussion, consider a simplified, non-relativistic model QFT, in which x-particles can emit and absorb bosonic y-particles. Let us suppose that there is only 1 x-particle, and it is fixed at the origin, so $\mathscr{H}$ is the bosonic Fock space of the y-particles, and the configuration space is given by Equation (5).

The naive, original expression for the Hamiltonian in the particle-position representation with creation and annihilation of y-particles at the origin $\mathbf{0}$ reads

$$\begin{aligned}(H_{\text{orig}}\psi)(\boldsymbol{y}_1,\dots,\boldsymbol{y}_n) = &-\frac{\hbar^2}{2m_y}\sum_{j=1}^{n}\nabla^2_{\boldsymbol{y}_j}\psi(\boldsymbol{y}_1,\dots,\boldsymbol{y}_n) \\ &+ g\sqrt{n+1}\,\psi(\boldsymbol{y}_1,\dots,\boldsymbol{y}_n,\mathbf{0}) \\ &+ \frac{g}{\sqrt{n}}\sum_{j=1}^{n}\delta^3(\boldsymbol{y}_j)\,\psi(\boldsymbol{y}_1,\dots,\widehat{\boldsymbol{y}}_j,\dots,\boldsymbol{y}_n)\,, \end{aligned} \tag{7}$$

where g is a real coupling constant (the charge of the x-particle), and $\widehat{\boldsymbol{y}}_j$ means that $\boldsymbol{y}_j$ is omitted. Recall that ψ is a function on $\cup_{n=0}^{\infty}\mathbb{R}^{3n}$, so $\psi(\boldsymbol{y}_1,\dots,\boldsymbol{y}_n)$ makes sense for any number n; note that $\psi(\boldsymbol{y}_1,\dots,\boldsymbol{y}_n,\mathbf{0})$ refers to the $n+1$-sector of $\psi \in \mathscr{H}$ and $\psi(\boldsymbol{y}_1,\dots,\widehat{\boldsymbol{y}}_j,\dots,\boldsymbol{y}_n)$ to the $n-1$-sector. Roughly speaking, the middle line of Equation (7) represents the annihilation of the $n+1$-st y-particle at the origin, while the last line represents the creation of a new y-particle at the origin, viz., with wave function δ^3.

Unfortunately, the Hamiltonian in Equation (7) is ultraviolet (UV) divergent and thus mathematically ill defined. This means that the creation and annihilation terms in H_{orig}, when expressed in the momentum representation, involve an integral over $\boldsymbol{k}$ that diverges for large values of $|\boldsymbol{k}|$. The root of the problem is that, according to the last line of Equation (7), the wave function of a newly created y-particle is a Dirac δ function, which has infinite energy and, what is worse, does not even lie in the Hilbert space (which contains only square-integrable functions). Many QFTs suffer from similar UV problems.

The UV problem can be circumvented by introducing an UV cut-off, i.e., by replacing the δ function by a square-integrable approximation φ as in Figure 4. The cutoff corresponds to "smearing out" the x-particle with "charge distribution" $\varphi(\cdot)$, and it leads to a well-defined Hamiltonian, given explicitly by

$$\begin{aligned}(H_{\text{cutoff}}\psi)(\boldsymbol{y}_1,\dots,\boldsymbol{y}_n) = &-\frac{\hbar^2}{2m_y}\sum_{j=1}^{n}\nabla^2_{\boldsymbol{y}_j}\psi(\boldsymbol{y}_1,\dots,\boldsymbol{y}_n) \\ &+ g\sqrt{n+1}\int_{\mathbb{R}^3} d^3\boldsymbol{y}\,\varphi^*(\boldsymbol{y})\,\psi(\boldsymbol{y}_1,\dots,\boldsymbol{y}_n,\boldsymbol{y}) \\ &+ \frac{g}{\sqrt{n}}\sum_{j=1}^{n}\varphi(\boldsymbol{y}_j)\,\psi(\boldsymbol{y}_1,\dots,\widehat{\boldsymbol{y}}_j,\dots,\boldsymbol{y}_n)\,. \end{aligned} \tag{8}$$

However, there is no empirical evidence that electrons have a nonzero radius; it is therefore unknown which size or shape φ should have; a cutoff tends to break Lorentz invariance; and, as another implausible consequence of the cutoff, emission and absorption occur anywhere in the support of φ around the x-particle, as depicted in Figure 5.

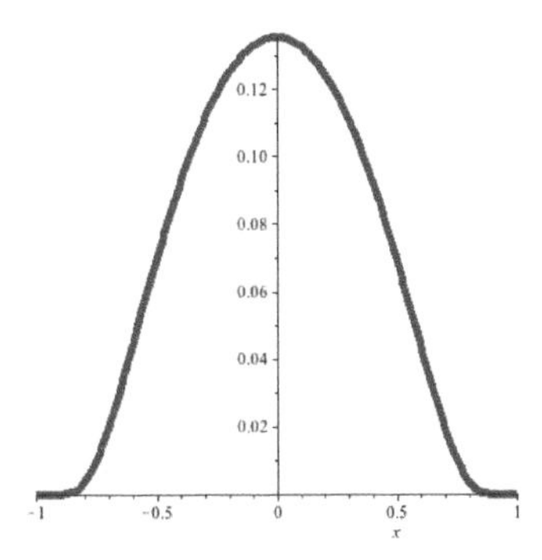

Figure 4. An example of a natural candidate for the cut-off function $\varphi(\cdot)$: a bump-shaped function that is a smooth and square-integrable approximation to a Dirac δ function and vanishes outside a small ball around the origin.

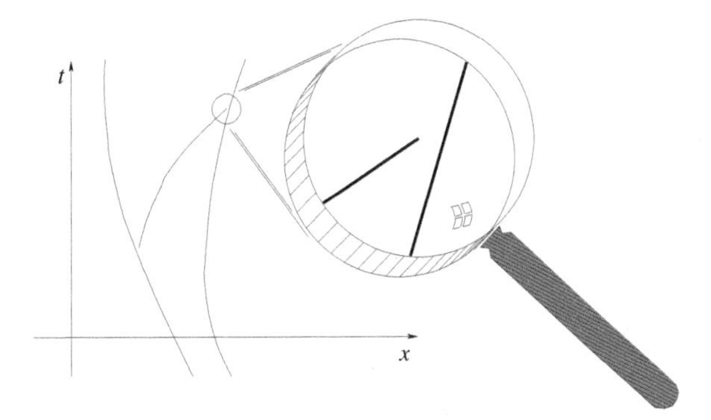

Figure 5. When using H_{cutoff}, the emission and absorption of a y-particle happens, according to Equation (6), not exactly at the location of an x-particle, but at a separation that can be as large as the radius of the support of φ. This does not happen with the alternative Hamiltonian defined by means of interior-boundary conditions.

2.3. UV Problem Solved!

Recent work [22–25] has shown that this UV problem can be solved, at least in the non-relativistic case, by means of interior-boundary conditions (IBCs): they allow the rigorous definition of a Hamiltonian H_{IBC}. In fact, for the specific Hamiltonian in Equation (7) with the x-particle fixed at the origin, it was known before [26] that, for any sequence $\varphi_n \to \delta^3$, there exist constants $E_n \in \mathbb{R}$ such that $H_{\text{cutoff}} - E_n$ possesses a limit H_∞ as $n \to \infty$, called the renormalized Hamiltonian and independent of the choice of the sequence φ_n. It has been shown [23] that H_∞ coincides with H_{IBC} up to addition of a constant (i.e., of a multiple of the identity). However, for the case of moving x-particles in three space dimensions, it is not known how to obtain a renormalized Hamiltonian, and the IBC approach has provided for the first time a mathematically well defined Hamiltonian [25].

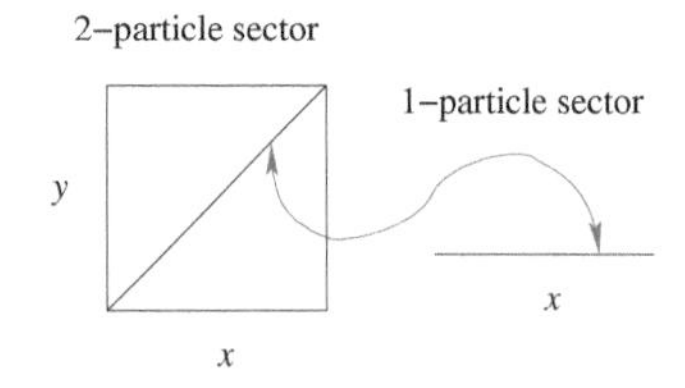

Figure 6. An interior-boundary condition is a relation between the values of ψ at two points: a point q on the boundary (that is, where two particles collide, such as (x, x) in the two-particle sector) and a point q' in the interior of a lower sector (such as x).

Here is how this approach works [22,27–29]. An interior-boundary condition is a condition that links two configurations connected by the creation or annihilation of a particle (see Figure 6). Abstractly speaking, an IBC on a function ψ on a domain $\mathcal{Q}$ with boundary $\partial\mathcal{Q}$ is a condition of the form

$$\psi(q') = (\text{const.})\, \psi(q)\,, \tag{9}$$

where q' is a boundary point and q an interior point. In our case, the boundary configurations are those in which a y-particle meets an x-particle. In the case of moving x-particles, such configurations lie on diagonal surfaces in configuration space, as depicted in Figure 6; in the case of a fixed x-particle at $\mathbf{0}$, they lie on the surfaces $\mathbf{y}_k = \mathbf{0}$ for any $k = 1, 2, \ldots$. The corresponding interior configuration q is the one with this y-particle removed, so q lies one sector lower than q'. For example, with an x-particle at $\mathbf{0}$, the IBC is roughly of the form

$$\psi(\mathbf{y}_1, \ldots, \mathbf{y}_n, \mathbf{0}) = \frac{g\, m_y}{2\pi\hbar^2\sqrt{n+1}}\, \psi(\mathbf{y}_1, \ldots, \mathbf{y}_n)\,. \tag{10}$$

In fact, the precise formula is yet a little different. That is because $|\psi|^2$ must diverge similar to $1/r^2$ as $r = |\mathbf{y}| \to 0$ to guarantee a non-vanishing flux of probability into the origin; in fact, the relevant ψs can be expanded in the form

$$\psi(\mathbf{y}_1, \ldots, \mathbf{y}_n, \mathbf{y}) = \alpha(\mathbf{y}_1, \ldots, \mathbf{y}_n)\, r^{-1} + \beta(\mathbf{y}_1, \ldots, \mathbf{y}_n)\, r^0 + o(r^0) \tag{11}$$

($r = |\mathbf{y}|$), and it is the leading coefficient α in this expansion that should appear on the left-hand side of Equation (10). Thus, the IBC reads

$$\lim_{r \searrow 0} r\psi(\mathbf{y}_1, \ldots, \mathbf{y}_n, r\boldsymbol{\omega}) = \frac{g\, m_y}{2\pi\hbar^2\sqrt{n+1}}\, \psi(\mathbf{y}_1, \ldots, \mathbf{y}_n) \tag{12}$$

for all unit vectors $\boldsymbol{\omega} \in \mathbb{R}^3$, $|\boldsymbol{\omega}| = 1$. (The limit $r \searrow 0$ means $r \to 0$ with $r > 0$.)

The expression for the corresponding Hamiltonian H_{IBC} then reads, with $\mathbb{S}^2 = \{\boldsymbol{\omega} \in \mathbb{R}^3 : |\boldsymbol{\omega}| = 1\}$ the unit sphere,

$$\begin{aligned}(H_{\text{IBC}}\psi)(\mathbf{y}_1, \ldots, \mathbf{y}_n) = &-\frac{\hbar^2}{2m_y}\sum_{j=1}^{n} \nabla^2_{\mathbf{y}_j}\psi(\mathbf{y}_1, \ldots, \mathbf{y}_n) \\ &+ \frac{g\sqrt{n+1}}{4\pi}\int_{\mathbb{S}^2} d^2\boldsymbol{\omega}\, \lim_{r \searrow 0}\frac{\partial}{\partial r}\Big(r\psi(\mathbf{y}_1, \ldots, \mathbf{y}_n, r\boldsymbol{\omega})\Big) \\ &+ \frac{g}{\sqrt{n}}\sum_{j=1}^{n} \delta^3(\mathbf{y}_j)\, \psi(\mathbf{y}_1, \ldots, \widehat{\mathbf{y}}_j, \ldots, \mathbf{y}_n)\,.\end{aligned} \tag{13}$$

The term in the last line, involving the problematical δ function, actually gets canceled by the term created when the Laplacian gets applied to the αr^{-1} term in Equation (11), which contributes a δ function; the constant prefactor in the IBC in Equation (10) or (12) is dictated by the goal of this cancellation. The middle line extracts the next-to-leading coefficient β of Equation (11) from ψ in the last variable $\boldsymbol{y}_{n+1}$. (As a consequence of the expansion of Equation (11), which is valid for ψ in the domain of H_{IBC}, the integrand is independent of $\boldsymbol{\omega}$, so that it is actually unnecessary to average over $\boldsymbol{\omega}$.)

Here is the rigorous result about H_{IBC}:

Theorem 1 ([23]). *On a suitable dense domain $\mathscr{D}_{\mathrm{IBC}}$ of ψs in $\mathscr{H}$ of the form of Equation (11) satisfying the IBC (12), H_{IBC} is well defined, self-adjoint, and positive. In particular, there is no UV divergence.*

Historically, IBCs were invented several times for various purposes [30–33], but only recently considered for the UV problem [22,27]. Rigorous results about existence and self-adjointness of the Hamiltonian were proved in [25] for moving x-particles in three dimensions, in [24] for moving x-particles in two dimensions, and also in [24] for the Nelson model [16] in three dimensions.

2.4. Particle Trajectories

This is also a jump process associated to H_{IBC} in $\mathcal{Q}$ analogous to Bell's that is $|\psi_t|^2$ distributed at every time t [34]. In this process, the world lines of y-particles begin and end on those of the x-particles (like in Figure 2a and unlike in Figure 5). We conjecture that this process is the limit of the continuum Bell process governed by Equation (6) as $\varphi \to \delta^3$.

Since the Hamiltonian is no longer of the form $H_0 + H_I$ (particularly as the functions in the domain of H_0 do not satisfy the boundary condition), the jump rate Equation (6) does not immediately apply. Nevertheless, the process can be defined as follows [34]. Between the jumps, the configuration follows Bohm's equation of motion in $\mathcal{Q}^{(n)} = \mathbb{R}^{3n}$. Every jump is either an absorption (to the next lower sector) or an emission (to the next higher sector). The absorption events are deterministic and occur when $Q(t) \in \mathcal{Q}^{(n)}$ reaches $\boldsymbol{y}_j = \mathbf{0}$ for any $j = 1, ..., n$; in that moment, the configuration jumps to $(\boldsymbol{y}_1, ..., \widehat{\boldsymbol{y}}_j, ..., \boldsymbol{y}_n) \in \mathcal{Q}^{(n-1)}$. The emission of a new y-particle at $\mathbf{0} \in \mathbb{R}^3$ occurs at a random time t in a random direction $\boldsymbol{\omega}$ (there is one trajectory starting there in each direction $\boldsymbol{\omega}$) with a rate dictated by time reversal invariance, the Markov property, and the wish for equivariance [28,34]: If $Q(t) = y = (\boldsymbol{y}_1, ..., \boldsymbol{y}_n) \in \mathcal{Q}^{(n)}$, then with jump rate

$$\sigma^{\psi}(y \to y \times 0 d^2\boldsymbol{\omega}) = \lim_{r \searrow 0} \frac{\max\{0, \frac{\hbar}{m} \operatorname{Im}[r^2 \psi(y, r\boldsymbol{\omega})^* \partial_r \psi(y, r\boldsymbol{\omega})]\}}{|\psi(y)|^2} d^2\boldsymbol{\omega} \tag{14}$$

it jumps to the solution of Bohm's equation of motion in $\mathcal{Q}^{(n+1)}$ beginning at

$$(\boldsymbol{y}_1, \ldots, \boldsymbol{y}_{j-1}, 0\boldsymbol{\omega}, \boldsymbol{y}_j, \ldots, \boldsymbol{y}_n) \tag{15}$$

with $1 \leq j \leq n+1$. That is, the newly created y-particle at the origin gets inserted at the j-th position, where j is chosen uniformly random (ψ is symmetric against permutation), and starts moving in direction $\boldsymbol{\omega}$. By virtue of Equation (11), the right-hand side of Equation (14) is actually independent of $\boldsymbol{\omega}$, so $\boldsymbol{\omega}$ is random with uniform distribution.

3. Extension of Bohmian Mechanics to Relativistic Space-Time

3.1. The Time Foliation

A *foliation* is a slicing of space-time into hypersurfaces, that is, a family of non-intersecting hypersurfaces whose union is space-time, as depicted in Figure 7. We will consider the possibility that there is a *preferred* foliation of space-time into spacelike hypersurfaces ("time foliation" $\mathscr{F}$), that is, that one foliation $\mathscr{F}$ plays a special dynamical role in nature, essentially defining a kind of simultaneity at a distance. If the existence of a time foliation is granted, then there is a simple, convincing analog of Bohmian mechanics, $BM_{\mathscr{F}}$. For a single particle, a time foliation is unnecessary, as Bohm found already in 1953 [35]. Bohm and Hiley [7] introduced the equation of motion of $BM_{\mathscr{F}}$ for flat foliations (i.e., parallel hyperplanes, i.e., Lorentz frames), Dürr et al. [36] for curved foliations, and I contributed [37] a proof of equivariance for curved space-time. The surfaces belonging to $\mathscr{F}$ will be called the *time leaves*.

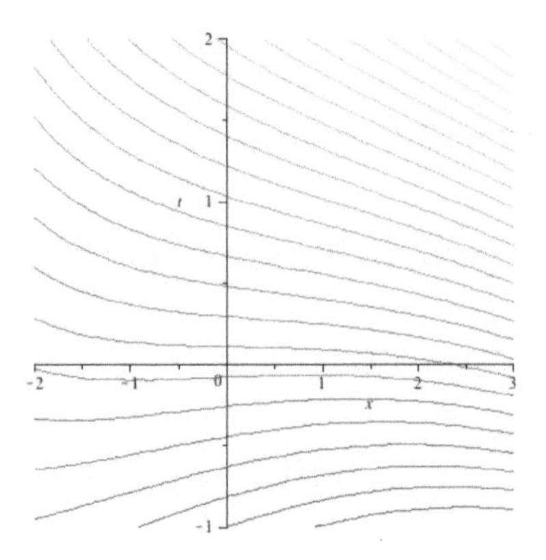

Figure 7. Example of a spacelike foliation (i.e., slicing into spacelike hypersurfaces) of Minkowski space-time in $1 + 1$ dimensions.

Without a time foliation (i.e., a preferred foliation), no version of Bohmian mechanics is known that would make predictions anywhere near quantum mechanics, and I have no hope that such a version can be found in the future.

Sutherland [38,39] has made an attempt towards such a version; he has proposed a Bohm-like equation of motion without a time foliation but involving retrocausation. While one may have reservations about retrocausation, it would be of interest to know whether such a theory can be made to work. At the present stage, Sutherland has formulated a proposal for trajectories of non-interacting particles between measurements at times t_i and t_f; for an assessment, one would need to formulate a proposal that can be applied to the universe as a whole and that can also treat measurements as just particular instances of motion and interaction of particles. I have considered a natural extension of Sutherland's equations to a universe with interaction and concluded that measurement outcomes, if their records get erased before the final time of the universe, may have a probability distribution that deviates very much from the one predicted by quantum mechanics and $BM_{\mathscr{F}}$. Thus, one would have to come up with a better proposal for an interacting version.

Let me return to $BM_{\mathscr{F}}$. To grant a time foliation seems against the spirit of relativity. However, it is a real possibility that our world is like that. It does not mean relativity would be irrelevant: After all, there is still a metric $g_{\mu\nu}$; the free Hamiltonian is still the Dirac operator (or whichever relativistic operator is appropriate); formulas are still expressed with 4-vector indices (j^μ, etc.); the statistics of experimental outcomes are independent of $\mathscr{F}$ (see below); and superluminal signaling is impossible in $BM_{\mathscr{F}}$. On the other hand, there exists also the vector n_μ normal to the time foliation, and the hypothesis of a time foliation provides a simple and straightforward explanation of the non-locality required by Bell's theorem.

A preferred foliation may be provided anyhow by the metric: If we take space-time to be curved and have a big bang singularity (which seems realistic), then the simplest choice of $\mathscr{F}$ consists of the level sets of the real-valued function T on space-time such that $T(x)$ is the timelike distance of x from the big bang; e.g., T(here–now) = 13.7 billion years (if what we call the big bang did involve a singularity).

Alternatively, $\mathscr{F}$ might be defined in terms of the quantum state vector ψ, $\mathscr{F} = \mathscr{F}(\psi)$ [40], or $\mathscr{F}$ might be determined by an evolution law (possibly involving ψ) from an initial time leaf.

Let us turn to the definition of the trajectories.

3.2. *The Single-Particle Case*

I begin with the simplest case, that of a single particle [35], which does not involve the time foliation $\mathscr{F}$. Let $\psi : \mathbb{R}^4 \to \mathbb{C}^4$ be a solution of the Dirac equation

$$i\hbar\gamma^\mu\partial_\mu\psi = m\psi\,. \tag{16}$$

The vector field

$$j^\mu = \overline{\psi}\gamma^\mu\psi \tag{17}$$

is called the probability current 4-vector field. It is formed in a covariant way (since $\psi \mapsto \overline{\psi} = \psi^\dagger\gamma^0$ is a covariant operation, whereas $\psi \mapsto \psi^\dagger$ is not); j^μ is real, future timelike-or-lightlike, and divergence free, $\partial_\mu j^\mu = 0$.

The Bohmian trajectories are the integral curves of the vector field j^μ; put differently, the equation of motion reads

$$\frac{dQ^\mu}{d\tau} \propto j^\mu(Q^\nu(\tau))\,, \tag{18}$$

where τ can be proper time or, in fact, any curve parameter, and $\propto$ means "is proportional to". In fact, it suffices to prescribe $dQ^\mu/d\tau$ only up to scalar factors (and to allow any curve parameter) because that fixes the tangent (i.e., the direction) of the world line in space-time.

It then follows that the possible world lines are timelike-or-lightlike curves. On any spacelike (Cauchy) hypersurface Σ_0, we can choose an initial condition $Q^\mu(\tau = 0) \in \Sigma_0$, and a unique solution curve $Q^\mu(\tau)$ exists for all times (except, technically speaking, for a set of measure zero of initial conditions) [41]. Equivariance holds in the following sense: On a spacelike (Cauchy) hypersurface Σ, the appropriate interpretation of "$|\psi|^2$ distribution" is the distribution whose density relative to the 3-volume d^3x defined by the 3-metric on Σ is $j^\mu n_\mu = \overline{\psi}\not{n}\psi$ with $n_\mu(x)$ the future unit normal vector to Σ at $x \in \Sigma$ and $\not{n} = n_\mu\gamma^\mu$. If the initial condition $Q^\mu(\tau = 0)$ is random with distribution $|\psi_{\Sigma_0}|^2$ then on every other Σ, the intersection point of the world line with Σ is random with distribution $|\psi_\Sigma|^2$. The evolution of ψ from Σ_0 to Σ is unitary.

All I said remains true when an external electromagnetic field is added to the Dirac equation, or when we consider a curved space-time.

3.3. *Law of Motion for Many Particles*

Here is the definition of $\mathrm{BM}_{\mathscr{F}}$ [36]. Consider N particles. Suppose that, for every $\Sigma \in \mathscr{F}$, we have a wave function ψ_Σ on Σ^N. (The next section discusses how to obtain ψ_Σ from multi-time wave functions.) For N timelike-or-lightlike world lines $Q_1, \ldots, Q_N$, the configuration on Σ consists of the intersection point of each world line with Σ,

$$Q(\Sigma) = (Q_1 \cap \Sigma, \ldots, Q_N \cap \Sigma) \tag{19}$$

The equation of motion is of the form (see Figure 8)

$$\frac{dQ_k^\mu}{d\tau} \propto \text{expression}\Big[\psi(Q(\Sigma))\Big]\,. \tag{20}$$

Specifically, for N Dirac particles, the wave function is of the form $\psi_\Sigma : \Sigma^N \to (\mathbb{C}^4)^{\otimes N}$ for every $\Sigma \in \mathscr{F}$, and the equation of motion reads

$$\frac{dQ_k^\mu}{d\tau} \propto j_k^\mu(Q(\Sigma)), \tag{21}$$

where

$$j^{\mu_1 \cdots \mu_N}(x_1, ..., x_N) = \overline{\psi}(x_1, ..., x_N)[\gamma^{\mu_1} \otimes \cdots \otimes \gamma^{\mu_N}]\psi(x_1, ..., x_N), \tag{22}$$

$$j_k^{\mu_k}(x_1, ..., x_N) = j^{\mu_1, ..., \mu_N}(x_1, ..., x_N)\, n_{\mu_1}(x_1) \cdots (k\text{-th omitted}) \cdots n_{\mu_N}(x_N), \tag{23}$$

and $n_\mu(x)$ is the future unit normal vector to Σ at $x \in \Sigma$.

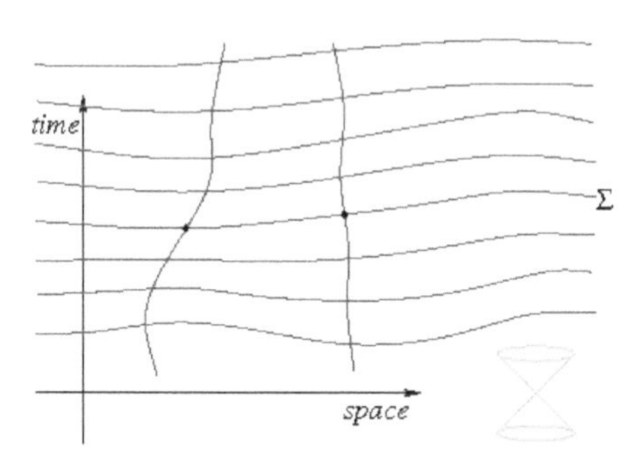

Figure 8. The equation of motion of $BM_{\mathscr{F}}$ specifies the tangent direction of a world line by means of the wave function evaluated at the configuration where all world lines intersect the same time leaf Σ.

The appropriate version of the $|\psi|^2$ distribution (which we simply call $|\psi|^2$) is the one with density

$$\rho(x_1, ..., x_N) = j_k^\mu(x_1, ..., x_N)\, n_\mu(x_k) = \overline{\psi}[\not{n}(x_1) \otimes \cdots \otimes \not{n}(x_N)]\psi \tag{24}$$

relative to the volume $d^3x_1, ..., d^3x_N$ defined by the metric $g_{\mu\nu}$ on Σ. (Actually, ρ is literally $|\psi|^2$ if for each x_j we use the Lorentz frame tangent to Σ at x_j.) It can be shown [36,37] that the $|\psi|^2$ distribution is equivariant, more precisely: If the initial configuration is $|\psi|^2$-distributed, then the configuration $Q(\Sigma)$ is $|\psi_\Sigma|^2$-distributed *on every* $\Sigma \in \mathscr{F}$. Moreover:

Theorem 2 ([42]). *If detectors are placed along any spacelike surface Σ (and if some reasonable assumptions about the evolution of ψ_Σ are satisfied), then the joint distribution of detection events is $|\psi_\Sigma|^2$.*

That is, while undetected configurations $Q(\Sigma')$ may fail to be $|\psi_{\Sigma'}|^2$ distributed if Σ' is not a time leaf, the detected configuration is $|\psi_\Sigma|^2$-distributed on *every* spacelike Σ. Consequently, $\mathscr{F}$ is invisible, i.e., experimental results reveal no information about $\mathscr{F}$. In fact, all empirical predictions of $BM_{\mathscr{F}}$ agree with the standard quantum formalism (and the empirical facts).

$BM_{\mathscr{F}}$ is a very robust theory, as it works for arbitrary foliation $\mathscr{F}$; it works even if the time leaves have kinks [43] (a case in which $\mathscr{F}$ violates a condition in the mathematicians' definition of "foliation"); it works even if the leaves of $\mathscr{F}$ overlap [44]; it can be combined with the stochastic jumps for particle creation; it works also in curved space-time [37]; and it still works if space-time has singularities [45].

3.4. Multi-Time Wave Functions

A multi-time wave function $\phi(t_1, \boldsymbol{x}_1, \ldots, t_N, \boldsymbol{x}_N)$ [46–49] is a natural relativistic generalization of the N-particle wave function $\psi(t, \boldsymbol{x}_1, \ldots, \boldsymbol{x}_N)$ of non-relativistic quantum mechanics: It is a function of N space-time points, and thus of N time variables. It is usually defined only on the set $\mathscr{S}$ of spacelike configurations, i.e., of those N-tuples $(x_1, \ldots, x_N) \in \mathbb{R}^{4N}$ of space-time points $x_j = (t_j, \boldsymbol{x}_j) \in \mathbb{R}^4$

for which any two x_j, x_k are spacelike separated or identical. ϕ is the covariant particle-position representation of the state vector. The usual (single-time) wave function ψ is contained in ϕ by setting all time variables equal,

$$\psi(t, \boldsymbol{x}_1, \ldots, \boldsymbol{x}_N) = \phi(t, \boldsymbol{x}_1, \ldots, t, \boldsymbol{x}_N)\,. \tag{25}$$

More generally, we can obtain for every spacelike hypersurface Σ a wave function ψ_Σ on Σ^N by simply setting

$$\psi_\Sigma(x_1, \ldots, x_N) = \phi(x_1, \ldots, x_N) \tag{26}$$

for all $x_1, \ldots, x_N \in \Sigma$. This is the ψ_Σ that goes into Equations (20) and (21), and the theorem from [42] reported in the previous subsection. Thus, the theorem is really a theorem about multi-time wave functions. Since ψ_Σ is closely related to the Tomonaga-Schwinger [50,51] wave function, so is ϕ; at the same time, ϕ is a simpler kind of mathematical object, as it is a function of only finitely many variables (at least locally, when we consider Fock space).

The obvious choice (though not the only possible one [52]) of time evolution equations for ϕ is to introduce an equation for each time variable,

$$i\hbar \frac{\partial \phi}{\partial t_j} = H_j \phi \quad \forall j = 1, \ldots, N. \tag{27}$$

It follows that the single-time wave function ψ as in Equation (25) will evolve according to the usual kind of Schrödinger equation

$$i\hbar \frac{\partial \psi}{\partial t} = H\psi \tag{28}$$

if and only if

$$\sum_{j=1}^{N} H_j = H \tag{29}$$

at equal times, a relation relevant to guessing suitable multi-time Schrödinger equations in Equation (27).

A big difference between multi-time and single-time Schrödinger equations is that for Equation (27) to possess solutions for all initial conditions at $0 = t_1 = t_2 =, \ldots, = t_N$, the partial Hamiltonians H_j must satisfy a *consistency condition* [47,48,53]

$$\left[i\hbar \frac{\partial}{\partial t_j} - H_j, i\hbar \frac{\partial}{\partial t_k} - H_k \right] = 0 \quad \forall j \neq k\,. \tag{30}$$

If the H_j are time-independent, then the condition reduces to $[H_j, H_k] = 0$. These conditions are trivially satisfied for non-interacting particles [15], but to implement interaction is a challenge; for example, interaction potentials violate consistency [53,54]. However, it has been shown that interaction can be consistently implemented [55], in particular in the form of zero-range interactions ("δ potentials") [56,57] and of interaction through emission and absorption of bosons [58,59].

The upshot is that the evolution of the wave function can be defined in a covariant way without using the time foliation $\mathscr{F}$, which then needs to be introduced for the trajectories. The evolution of the wave function can directly be formulated in the particle-position representation, in fact with rather simple equations [52,58].

4. Outlook and Concluding Remarks

Those who regard a theory with a preferred foliation as unacceptable may want to consider relativistic collapse theories instead [60,61], which do not need a preferred foliation. I believe, however, that we should take the possibility of a preferred foliation (depending perhaps on the space-time metric and/or the wave function) seriously. Then, $\text{BM}_{\mathscr{F}}$ seems to be the most plausible ontological theory of quantum mechanics in relativistic space-time, and I regard it as a fully satisfactory extension of Bohmian mechanics to relativistic space-time. Particle creation and annihilation can be incorporated into it in the same way as described in Section 2 for the non-relativistic case.

A goal for the future would be to formulate a version of quantum electrodynamics (QED) with particle trajectories. The particle-position representation of the quantum state in QED was formulated already by Landau and Peierls [14] in 1930, and it lends itself nicely to a multi-time formulation. Thus, what are the obstacles? The main obstacle is that defining Bohmian trajectories for a photon requires defining the probability current j^μ, so we would need a formula for photons analogous to $j^\mu = \overline{\psi}\gamma^\mu\psi$ for Dirac wave functions, but such a formula is not known to date except for plane waves (for which it is $j^\mu = |c|^2 k^\mu/\hbar$ whenever the energy-momentum tensor is $T^{\mu\nu} = |c|^2 k^\mu k^\nu$). Of course, this problem concerns not only the Bohmian approach but every approach to QED, but it is of particular importance in the Bohmian framework. Oppenheimer [62] argued in 1931 that j^μ does not exist for photons; while his argument is not completely compelling, it is by itself quite reasonable. However, since we can measure probability distributions of photons in photon counters and interference experiments, I have trouble imagining how j^μ could fail to exist for photons. Thus, I tend to suspect that there is a formula for j^μ which we have not found yet.

Another problem for future research is whether the technique of interior-boundary conditions can be applied to relativistic Hamiltonians. A further problem is how to deal in the Bohmian framework with positrons, the Dirac sea, and states of negative energy. Some authors [63,64] have proposed to take the Dirac sea literally as an infinity (or at least a very large number) of Bohmian particles. I am inclined to take positrons literally as Bohmian particles, but various questions about this approach remain open.

Let me conclude. While standard quantum mechanics is often unclear, standard quantum field theory is often even less clear. However, the developments I have described provide reasons for optimism that a clear version of serious QFTs (such as QED) can be obtained, and the Bohmian approach of using particle trajectories is in my opinion the most promising candidate for getting there. A fully satisfactory formulation of non-relativistic quantum mechanics is provided by Bohmian mechanics, and I believe that we should try hard to reach a clear formulation of QED as well. Some of the difficulties of QED are of a mathematical nature (such as the precise definition of the time evolution of the quantum state), others of an ontological nature (what is actually there), and yet others of an operational nature (such as how to compute the position probability distribution of photons for arbitrary states). Some of the difficulties can often be circumvented or ignored, while the Bohmian approach forces us to face them. I think that is ultimately an advantage.

Funding: This research received no external funding.

Conflicts of Interest: The author declares no conflict of interest.

References

1. Bohm, D. A suggested interpretation of the quantum teory in terms of "hidden" variables, I and II. *Phys. Rev.* **1952**, *85*, 166–193. [CrossRef]
2. Einstein, A. Reply to criticisms. In *Albert Einstein, Philosopher-Scientist*; Schilpp, P.A., Ed.; Open Court: La Salle, IL, USA, 1949; p. 664.
3. Bell, J.S. Against "measurement". In *Sixty-Two Years of Uncertainty*; Miller, A.I., Ed.; Plenum Press: New York, NY, USA, 1990. Reprinted as chapter 23 of *Speakable and Unspeakable in Quantum Mechanics*, 2nd ed.; Bell, J.S.; Cambridge University Press: Cambridge, UK, 2004. Also reprinted in *Phys. World* **1990**, *3*, 33–40.
4. Bricmont, J. *Making Sense of Quantum Mechanics*; Springer: Heidelberg, Germany, 2016.
5. Norsen, T. *Foundations of Quantum Mechanics*; Springer: Heidelberg, Germany, 2018.
6. Dürr, D.; Teufel, S. *Bohmian Mechanics*; Springer: Heidelberg, Germany, 2009.
7. Bohm, D.; Hiley, B.J. *The Undivided Universe: An Ontological Interpretation of Quantum Theory*; Routledge: London, UK, 1993.

8. Tumulka, R. Bohmian mechanics. In *The Routledge Companion to the Philosophy of Physics*; Wilson, A., Ed.; Routledge: London, UK, 2018.
9. Dürr, D.; Goldstein, S.; Zanghì, N. Quantum equilibrium and the origin of absolute uncertainty. *J. Stat. Phys.* **1992**, *67*, 843–907. Reprinted in *Quantum Physics without Quantum Philosophy*; Dürr, D.; Goldstein, S.; Zanghì, N.; Springer: Heidelberg, Germany, 2013. [CrossRef]
10. Bell, J.S. Six possible worlds of quantum mechanics. In Proceedings of the Nobel Symposium 65: Possible Worlds in Arts and Sciences, Stockholm, Sweden, 11–15 August 1986. Reprinted as chapter 20 of *Speakable and Unspeakable in Quantum Mechanics*, 2nd ed.; Bell, J.S.; Cambridge University Press: Cambridge, UK, 2004.
11. Philippidis, C.; Dewdney, C.; Hiley, B.J. Quantum interference and the quantum potential. *Il Nuovo Cimento* **1979**, *52B*, 15–28. [CrossRef]
12. Bell, J.S. *Speakable and Unspeakable in Quantum Mechanics*, 2nd ed.; Cambridge University Press: Cambridge, UK, 2004.
13. Dürr, D.; Goldstein, S.; Tumulka, R.; Zanghì, N. Bohmian mechanics and quantum field theory. *Phys. Rev. Lett.* **2004**, *93*, 090402. Reprinted in *Quantum Physics without Quantum Philosophy*; Dürr, D.; Goldstein, S.; Zanghì, N.; Springer: Heidelberg, Germany, 2013. [CrossRef] [PubMed]
14. Landau, L.; Peierls, R. Quantenelektrodynamik im Konfigurationsraum. *Z. Phys.* **1930**, *62*, 188–200; English translation: Quantum electrodynamics in configuration space. In *Selected Scientific Papers of Sir Rudolf Peierls with Commentary*; Dalitz, R.H., Peierls, R., Eds.; World Scientific: Singapore, 1997; pp. 71–82. [CrossRef]
15. Schweber, S. *An Introduction to Relativistic Quantum Field Theory*; Harper: New York, NY, USA, 1961.
16. Nelson, E. Interaction of nonrelativistic particles with a quantized scalar field. *J. Math. Phys.* **1964**, *5*, 1190–1197. [CrossRef]
17. Bell, J.S. Beables for quantum field theory. *Phys. Rep.* **1986**, *137*, 49–54. Reprinted in *Quantum Implications: Essays in Honour of David Bohm*; Peat, F.D., Hiley, B.J., Eds.; Routledge: London, UK, 1987; p. 227. Also reprinted as chapter 19 of *Speakable and Unspeakable in Quantum Mechanics*, 2nd ed.; Bell, J.S.; Cambridge University Press: Cambridge, UK, 2004. [CrossRef]
18. Dürr, D.; Goldstein, S.; Tumulka, R.; Zanghì, N. Trajectories and particle creation and annihilation in quantum field theory. *J. Phys. A Math. Gen.* **2003**, *36*, 4143–4149. [CrossRef]
19. Dürr, D.; Goldstein, S.; Tumulka, R.; Zanghì, N. Bell-type quantum field theories. *J. Phys. A Math. Gen.* **2005**, *38*, R1–R43. [CrossRef]
20. Vink, J.C. Quantum mechanics in terms of discrete beables. *Phys. Rev. A.* **1993**, *48*, 1808–1818. [CrossRef] [PubMed]
21. Vink, J.C. Particle trajectories for quantum field theory. *Found. Phys.* **2018**, *48*, 209–236. [CrossRef]
22. Teufel, S.; Tumulka, R. New type of Hamiltonians without ultraviolet divergence for quantum field theories. *arXiv* **2015**, arXiv:1505.04847.
23. Lampart, J.; Schmidt, J.; Teufel, S.; Tumulka, R. Particle creation at a point source by means of interior-boundary conditions. *Math. Phys. Anal. Geom.* **2018**, *21*, 12. [CrossRef]
24. Lampart, J.; Schmidt, J. On the domain of Nelson-type Hamiltonians and abstract boundary conditions. *arXiv* **2018**, arXiv:1803.00872.
25. Lampart, J. A nonrelativistic quantum field theory with point interactions in three dimensions. *arXiv* **2018**, arXiv:1804.08295.
26. Dereziński, J. Van Hove Hamiltonians—Exactly solvable models of the infrared and ultraviolet problem. *Ann. Henri Poincaré* **2003**, *4*, 713–738. [CrossRef]
27. Teufel, S.; Tumulka, R. Avoiding ultraviolet divergence by means of interior-boundary conditions. In *Quantum Mathematical Physics—A Bridge between Mathematics and Physics*; Finster, F., Kleiner, J., Röken, C., Tolksdorf, J., Eds.; Birkhäuser: Basel, Switzerland, 2016; pp. 293–311.
28. Georgii, H.-O.; Tumulka, R. Some jump processes in quantum field theory. In *Interacting Stochastic Systems*; Deuschel, J.-D., Greven, A., Eds.; Springer: Berlin, Germany, 2004; pp. 55–73.
29. Keppeler, S.; Sieber, M. Particle creation and annihilation at interior boundaries: One-dimensional models. *J. Phys. A Math. Theor.* **2016**, *49*, 125204. [CrossRef]
30. Moshinsky, M. Boundary conditions for the description of nuclear reactions. *Phys. Rev.* **1951**, *81*, 347–352. [CrossRef]
31. Moshinsky, M. Boundary conditions and time-dependent states. *Phys. Rev.* **1951**, *84*, 525–532. [CrossRef]

32. Thomas, L.E. Multiparticle Schrödinger Hamiltonians with point interactions. *Phys. Rev. D* **1984**, *30*, 1233–1237. [CrossRef]
33. Yafaev, D.R. On a zero-range interaction of a quantum particle with the vacuum. *J. Phys. A Math. Gen.* **1992**, *25*, 963–978. [CrossRef]
34. Dürr, D.; Goldstein, S.; Teufel, S.; Tumulka, R.; Zanghì, N. Bohmian trajectories for Hamiltonians with interior–boundary conditions. In preparation, 2018.
35. Bohm, D. Comments on an article of Takabayasi concerning the formulation of quantum mechanics with classical pictures. *Prog. Theor. Phys.* **1953**, *9*, 273–287. [CrossRef]
36. Dürr, D.; Goldstein, S.; Münch-Berndl, K.; Zanghì, N. Hypersurface Bohm–Dirac models. *Phys. Rev. A* **1999**, *60*, 2729–2736. Reprinted in *Quantum Physics without Quantum Philosophy*; Dürr, D.; Goldstein, S.; Zanghì, N.; Springer: Heidelberg, Germany, 2013. [CrossRef]
37. Tumulka, R. Closed 3-Forms and Random World Lines. Ph.D. Thesis, Mathematics Institute, Ludwig-Maximilians-Universität, München, Germany, 2001. Available online: http://edoc.ub.uni-muenchen.de/7/ (accessed on 11 June 2018).
38. Sutherland, R. Causally symmetric Bohm model. *Stud. Hist. Philos. Mod. Phys.* **2008**, *39*, 782–805. [CrossRef]
39. Sutherland, R. Lagrangian description for particle interpretations of quantum mechanics-entangled many-particle case. *Found. Phys.* **2017**, *47*, 174–207. [CrossRef]
40. Dürr, D.; Goldstein, S.; Norsen, T.; Struyve, W.; Zanghì, N. Can Bohmian mechanics be made relativistic? *Proc. R. Soc. A* **2014**, *470*, 20130699. [CrossRef] [PubMed]
41. Teufel, S.; Tumulka, R. Simple proof for global existence of Bohmian trajectories. *Commun. Math. Phys.* **2005**, *258*, 349–365. [CrossRef]
42. Lienert, M.; Tumulka, R. Born's rule for arbitrary Cauchy surfaces. *arXiv* **2017**, arXiv:1706.07074.
43. Struyve, W.; Tumulka, R. Bohmian trajectories for a time foliation with kinks. *J. Geom. Phys.* **2014**, *82*, 75–83. [CrossRef]
44. Struyve, W.; Tumulka, R. Bohmian mechanics for a degenerate time foliation. *Quantum Stud. Math. Found.* **2015**, *2*, 349–358. [CrossRef]
45. Tumulka, R. Bohmian mechanics at space-time singularities. II. Spacelike singularities. *Gen. Relat. Gravit.* **2010**, *42*, 303–346. [CrossRef]
46. Dirac, P.A.M. Relativistic quantum mechanics. *Proc. R. Soc. Lond. A* **1932**, *136*, 453–464. [CrossRef]
47. Dirac, P.A.M.; Fock, V.A.; Podolsky, B. On quantum electrodynamics. *Phys. Z. Sowjetunion* **1932**, *2*, 468–479; Reprinted in *Selected Papers on Quantum Electrodynamics*; Schwinger, J., Ed.; Dover: New York, USA, 1958.
48. Bloch, F. Die physikalische Bedeutung mehrerer Zeiten in der Quantenelektrodynamik. *Phys. Z. Sowjetunion* **1934**, *5*, 301–305.
49. Lienert, M.; Petrat, S.; Tumulka, R. Multi-time wave functions. *J. Phys. Conf. Ser.* **2017**, *880*, 012006. [CrossRef]
50. Tomonaga, S. On a relativistically invariant formulation of the quantum theory of wave fields. *Prog. Theor. Phys.* **1946**, *1*, 27–42. [CrossRef]
51. Schwinger, J. Quantum electrodynamics. I. A covariant formulation. *Phys. Rev.* **1948**, *74*, 1439–1461. [CrossRef]
52. Lienert, M. Direct interaction along light cones at the quantum level. *arXiv* **2017**, arXiv:1801.00060.
53. Petrat, S.; Tumulka, R. Multi-time Schrödinger equations cannot contain interaction potentials. *J. Math. Phys.* **2014**, *55*, 032302. [CrossRef]
54. Nickel, L.; Deckert, D.-A. Consistency of multi-time Dirac equations with general interaction potentials. *J. Math. Phys.* **2016**, *57*, 072301.
55. Droz-Vincent, P. Relativistic quantum mechanics with non conserved number of particles. *J. Geom. Phys.* **1985**, *2*, 101–119. [CrossRef]
56. Lienert, M. A relativistically interacting exactly solvable multi-time model for two mass-less Dirac particles in 1+1 dimensions. *J. Math. Phys.* **2015**, *56*, 042301. [CrossRef]
57. Lienert, M.; Nickel, L. A simple explicitly solvable interacting relativistic *N*-particle model. *J. Phys. A Math. Theor.* **2015**, *48*, 325301. [CrossRef]
58. Petrat, S.; Tumulka, R. Multi-time wave functions for quantum field theory. *Ann. Phys.* **2014**, *345*, 17–54. [CrossRef]
59. Petrat, S.; Tumulka, R. Multi-time formulation of pair creation. *J. Phys. A Math. Theor.* **2014**, *47*, 112001. [CrossRef]

60. Tumulka, R. A relativistic version of the Ghirardi–Rimini–Weber model. *J. Stat. Phys.* **2006**, *125*, 821–840. [CrossRef]
61. Bedingham, D.; Dürr, D.; Ghirardi, G.C.; Goldstein, S.; Tumulka, R.; Zanghì, N. Matter density and relativistic models of wave function collapse. *J. Stat. Phys.* **2014**, *154*, 623–631. [CrossRef]
62. Oppenheimer, J.R. Note on light quanta and the electromagnetic field. *Phys. Rev.* **1931**, *38*, 725–748. [CrossRef]
63. Colin, S.; Struyve, W. A Dirac sea pilot-wave model for quantum field theory. *J. Phys. A Math. Theor.* **2007**, *40*, 7309–7341. [CrossRef]
64. Deckert, D.-A.; Esfeld, M.; Oldofredi, A. A persistent particle ontology for QFT in terms of the Dirac sea. *arXiv* **2016**, arXiv:1608.06141.

Article

Vacuum Landscaping: Cause of Nonlocal Influences without Signaling

Gerhard Grössing *, Siegfried Fussy, Johannes Mesa Pascasio and Herbert Schwabl

Austrian Institute for Nonlinear Studies, Akademiehof, Friedrichstrasse 10, 1010 Vienna, Austria; siegfried.fussy@gmail.com (S.F.); johannes.mesa@gmx.at (J.M.P.); h.schwabl@padma.ch (H.S.)

* Correspondence: ains@chello.at

Received: 30 April 2018; Accepted: 11 June 2018; Published: 13 June 2018

Abstract: In the quest for an understanding of nonlocality with respect to an appropriate ontology, we propose a "cosmological solution". We assume that from the beginning of the universe each point in space has been the location of a scalar field representing a zero-point vacuum energy that nonlocally vibrates at a vast range of different frequencies across the whole universe. A quantum, then, is a nonequilibrium steady state in the form of a "bouncer" coupled resonantly to one of those (particle type dependent) frequencies, in remote analogy to the bouncing oil drops on an oscillating oil bath as in Couder's experiments. A major difference to the latter analogy is given by the nonlocal nature of the vacuum oscillations. We show with the examples of double- and n-slit interference that the assumed nonlocality of the distribution functions alone suffices to derive the de Broglie–Bohm guiding equation for N particles with otherwise purely classical means. In our model, no influences from configuration space are required, as everything can be described in 3-space. Importantly, the setting up of an experimental arrangement limits and shapes the forward and osmotic contributions and is described as vacuum landscaping.

Keywords: Schrödinger equation; de Broglie–Bohm theory; nonequilibrium thermodynamics; zero-point field

PACS: 03.65.-w, 03.65.Ta, 05.40.-a, 05.70.Ln

---

1. Introduction: Quantum Mechanics without Wavefunctions

"Emergent Quantum Mechanics" stands for the idea that quantum mechanics is based on a more encompassing deeper level theory. This counters the traditional belief, usually expressed in the context of orthodox Copenhagen-type quantum mechanics, that quantum theory is an "ultimate" theory whose main features will prevail for all time and will be applicable to all questions of physics. Note, for example, that, even in more recent approaches to spacetime, the concept of an "emergent spacetime" is introduced as a description even of space and time emerging from basic quantum mechanical entities. This, of course, need not be so, considering the fact that there is "plenty of room at the bottom," i.e., as Feynman implied, between present-day resolutions and minimally possible times and distances, which could in principle be way below resolutions reasonably argued about in present times (i.e., on Planck scales).

One of the main attractive features of the de Broglie–Bohm interpretation of the quantum mechanical formalism, and of Bohmian mechanics as well, lies in the possibility to extend its domain into space and/or time resolutions where modified behaviors different from quantum mechanical ones may be expected. In other words, there may be new physics involved that would require an explicitly more encompassing theory than quantum mechanics, i.e., a deeper level theory. Our group's approach, which we pursued throughout the last 10 years, is characterized by the search for such a theory under the premise that even for nonrelativistic quantum mechanics, the Schrödinger equation cannot be an appropriate starting point, since the wavefunction is still lacking a firm theoretical basis and its meaning is generally not agreed upon.

For a similar reason, also the de Broglie–Bohm theory cannot be our starting point, as it is based on the Schrödinger equation and the use of the wavefunction to begin with. Rather, we aim at an explicit ansatz for a deeper level theory without wavefunctions, from which the Schrödinger equation, or the de Broglie–Bohm guiding equation, can be derived. We firmly believe that we have accomplished this and we can now proceed to study consequences of the approach beyond orthodox expectations.

Throughout recent years, apart from our own model, several approaches to a quantum mechanics without wavefunctions have been proposed [1–5]. These refer to "many classical worlds" that provide Bohm-type trajectories with certain repulsion effects. From our realistic point of view, the true ontologies of these models, however, do not become apparent. So let us turn to our model. As every physical theory is based on metaphysical assumptions, we must make clear what our assumptions are. They are as follows.

We propose a "cosmological solution" in that the Big Bang, or any other model explaining the apparent expansion of the universe, is essentially related to the vacuum energy (The latter may constitute what is called the dark energy, but we do not need to specify this here). We assume that from the beginning of the universe each point in space has been the location of a scalar field representing a zero-point vacuum energy that vibrates at a vast range of different frequencies across the whole universe. More specifically, we consider the universe as an energetically open system where the vacuum energy not only drives expansion, but also each individual "particle" oscillation $\omega = E/\hbar$ in the universe. In order to maintain a particular frequency, any such oscillator must be characterized by a throughput of energy external to it. In this regard, we have time and again employed the analogy of Couder's experiments with bouncing oil drops on a vibrating bath [6–11]: The bouncer/particle is always in resonant interaction with a relevant environment.

Our model, though also largely classical, has a very different ontology from the "many classical worlds" one. We consider *one* "superclassical" world instead: a purely classical world plus "cosmological nonlocality," i.e., a nonlocal bath for every oscillator/particle due to the all-pervading vacuum energy, which—mostly in the context of quantum mechanics—is called the zero-point energy. Thus, it is the one classical world together with the fluctuating environment related to the vacuum energy that enters our definition of a quantum as an emergent system. The latter consists of a bouncer and an undulatory/wave-like nonlocal environment defined by proper boundary conditions (As an aside we note that this is not related to de Broglie's "nonlinear wave mechanics" [12], as there the nonlinear wave, with the particle as soliton-like singularity, is considered as one ontic entity. In our case, however, we speak of two separate, though synchronous elements: local oscillators and generally nonlocal oscillating fields).

In previous work, we have shown how the Schrödinger equation can be derived from a nonequilibrium sub-quantum dynamics [13–16], where in accordance with the model sketched above the particle is considered as a steady state with a constant throughput of energy. This, then, leads to the two-momentum approach to emergent quantum mechanics which shall be outlined in the next section.

2. The Two-Momenta Approach to Emergent Quantum Mechanics

We consider the empirical fact that each particle of nature is attributed an energy $E = \hbar\omega$ as one of the essential features of quantum systems (We have also presented a classical explanation for this relation from our sub-quantum model [17], but do not need to use the details for our present purposes). Oscillations, characterized by some typical angular frequency ω, are described as properties of off-equilibrium steady-state systems. "Particles" can then be assumed to be dissipative systems maintained in a nonequilibrium steady-state by a permanent throughput of energy, or heat flow, respectively. The heat flow must be described by an external kinetic energy term. Then the energy of the total system, i.e., of the particle and it's thermal context, becomes

$$E_{\text{tot}} = \hbar\omega + \frac{(\delta \mathbf{p})^2}{2m} , \tag{1}$$

where $\delta\mathbf{p}$ is an additional, fluctuating momentum component of the particle of mass m.

We assume that an effect of said thermal context is given by detection probability distributions, which are wave-like in the particle's surroundings. Thus, the detection probability density $P(\mathbf{x}, t)$ is considered to coincide with a classical wave's intensity $I(\mathbf{x}, t) = R^2(\mathbf{x}, t)$, with $R(\mathbf{x}, t)$ being the wave's real-valued amplitude

$$P(\mathbf{x}, t) = R^2(\mathbf{x}, t) , \quad \text{with normalization} \int P \, \mathrm{d}^n x = 1 . \tag{2}$$

In [13], we combine some results of nonequilibrium thermodynamics with classical wave mechanics. We propose that the many microscopic degrees of freedom associated with the hypothesized sub-quantum medium can be recast into the emergent macroscopic properties of the wave-like behavior on the quantum level. Thus, for the relevant description of the total system, one no longer needs the full phase space information of all microscopic entities, but only the emergent particle coordinates.

For implementation, we model a particle as being surrounded by a heat bath, i.e., a reservoir that is very large compared to the small dissipative system, such that that the momentum distribution in this region is given by the usual Maxwell–Boltzmann distribution. This corresponds to a "thermostatic" regulation of the reservoir's temperature, which is equivalent to the statement that the energy lost to the thermostat can be regarded as heat. Thus, one can formulate a *proposition of emergence* [13] providing the equilibrium-type probability (density) ratio

$$\frac{P(\mathbf{x}, t)}{P(\mathbf{x}, 0)} = \mathrm{e}^{-\frac{\Delta Q(t)}{kT}} , \tag{3}$$

with k being Boltzmann's constant, T the reservoir temperature, and $\Delta Q(t)$ the heat that is exchanged between the particle and its environment.

Equations (1)–(3) are the only assumptions necessary to derive the Schrödinger equation from (modern) classical mechanics. We need to employ only two additional well-known results. The first is given by Boltzmann's formula for the slow transformation of a periodic motion (with period $\tau = 2\pi/\omega$) upon application of a heat transfer ΔQ. This is needed as we deal with an oscillator of angular frequency ω in a heat bath Q, and a change in the vacuum surroundings of the oscillator will come as a heat transfer ΔQ. The latter is responsible for a change δS of the action function S representing the effect of the vacuum's "zero-point" fluctuations. With the action function $S = \int (E_{\text{kin}} - V) \, \mathrm{d}t$, the relation between heat and action was first given by Boltzmann [18],

$$\Delta Q(t) = 2\omega[\delta S(t) - \delta S(0)] . \tag{4}$$

Finally, the requirement that the average kinetic energy of the thermostat equals the average kinetic energy of the oscillator is given, for each degree of freedom, by

$$\frac{kT}{2} = \frac{\hbar\omega}{2} . \tag{5}$$

Combining these two results, Equations (4) and (5), with Equation (3), one obtains

$$P(\mathbf{x}, t) = P(\mathbf{x}, 0)\mathrm{e}^{-\frac{2}{\hbar}[\delta S(\mathbf{x},t) - \delta S(\mathbf{x},0)]} , \tag{6}$$

from which follows the expression for the momentum fluctuation $\delta\mathbf{p}$ of Equation (1) as

$$\delta\mathbf{p}(\mathbf{x},t) = \nabla(\delta S(\mathbf{x},t)) = -\frac{\hbar}{2}\frac{\nabla P(\mathbf{x},t)}{P(\mathbf{x},t)} \,. \tag{7}$$

This, then, provides the additional kinetic energy term for one particle as

$$\delta E_{\mathrm{kin}} = \frac{1}{2m}\nabla(\delta S)\cdot\nabla(\delta S) = \frac{1}{2m}\left(\frac{\hbar}{2}\frac{\nabla P}{P}\right)^2 \,. \tag{8}$$

Thus, writing down a classical action integral for $j = N$ particles in m-dimensional space, including this new term for each of them, yields (with external potential V)

$$A = \int L\,\mathrm{d}^m x\,\mathrm{d}t = \int P\left[\frac{\partial S}{\partial t} + \sum_{j=1}^{N}\frac{1}{2m_j}\nabla_j S\cdot\nabla_j S + \sum_{j=1}^{N}\frac{1}{2m_j}\left(\frac{\hbar}{2}\frac{\nabla_j P}{P}\right)^2 + V\right]\mathrm{d}^m x\,\mathrm{d}t \tag{9}$$

where the probability density $P = P(\mathbf{x}_1, \mathbf{x}_2, \ldots, \mathbf{x}_N, t)$.

With the definition of forward and osmotic velocities, respectively,

$$\mathbf{v}_j := \frac{\mathbf{p}_j}{m_j} = \frac{\nabla_j S}{m_j} \qquad \text{and} \qquad \mathbf{u}_j := \frac{\delta\mathbf{p}_j}{m_j} = -\frac{\hbar}{2m_j}\frac{\nabla_j P}{P}, \tag{10}$$

one can rewrite Equation (9) as

$$A = \int L\,\mathrm{d}^m x\,\mathrm{d}t = \int P\left[\frac{\partial S}{\partial t} + V + \sum_{j=1}^{N}\frac{m_j}{2}\mathbf{v}_j^2 + \sum_{j=1}^{N}\frac{m_j}{2}\mathbf{u}_j^2\right]\mathrm{d}^m x\,\mathrm{d}t \,. \tag{11}$$

This can be considered as the basis for our approach with two momenta, i.e., the forward momentum $m\mathbf{v}$ and the osmotic momentum $m\mathbf{u}$, respectively. At first glance, the Lagrangian in Equation (11) looks completely classical, with two kinetic energy terms per particle instead of one. However, due to the particular nature of the osmotic momentum as given in Equation (10), nonlocal influences are introduced: even at long distances away from the particle location, where the particle's contribution to P is practically negligibly small, the expression of the form $\frac{\nabla_j P}{P}$ may be large and affects immediately the whole fluctuating environment. This is why the osmotic variant of the kinetic energy makes all the difference to the usual classical mechanics, or, in other words, is the basis for quantum mechanics.

Introducing now the Madelung transformation

$$\psi = R\,\mathrm{e}^{\frac{i}{\hbar}S} \tag{12}$$

where $R = \sqrt{P}$ as in Equation (2), one has, with bars denoting averages,

$$\overline{\left|\frac{\nabla_j\psi}{\psi}\right|^2} := \int \mathrm{d}^m x\,\mathrm{d}t\left|\frac{\nabla_j\psi}{\psi}\right|^2 = \overline{\left(\frac{1}{2}\frac{\nabla_j P}{P}\right)^2} + \overline{\left(\frac{\nabla_j S}{\hbar}\right)^2}, \tag{13}$$

and one can rewrite Equation (9) as

$$A = \int L\,\mathrm{d}^m x\,\mathrm{d}t = \int \mathrm{d}^m x\,\mathrm{d}t\left[|\psi|^2\left(\frac{\partial S}{\partial t} + V\right) + \sum_{j=1}^{N}\frac{\hbar^2}{2m_j}|\nabla_j\psi|^2\right] \,. \tag{14}$$

Thus, with the identity $|\psi|^2 \frac{\partial S}{\partial t} = -\frac{i\hbar}{2}(\psi^* \dot{\psi} - \dot{\psi}^* \psi)$, one obtains the familiar Lagrange density

$$L = -\frac{i\hbar}{2}(\psi^* \dot{\psi} - \dot{\psi}^* \psi) + \sum_{j=1}^{N} \frac{\hbar^2}{2m_j} \nabla_j \psi \cdot \nabla_j \psi^* + V \psi^* \psi \,, \tag{15}$$

from which by the usual procedures one arrives at the N-particle Schrödinger equation

$$i\hbar \frac{\partial \psi}{\partial t} = \left(-\sum_{j=1}^{N} \frac{\hbar^2}{2m_j} \nabla_j^2 + V \right) \psi \,. \tag{16}$$

Note also that from Equation (9) one obtains upon variation in P the modified Hamilton–Jacobi equation familiar from the de Broglie–Bohm interpretation, i.e.,

$$\frac{\partial S}{\partial t} + \sum_{j=1}^{N} \frac{(\nabla_j S)^2}{2m_j} + V(\mathbf{x}_1, \mathbf{x}_2, \ldots, \mathbf{x}_N, t) + U(\mathbf{x}_1, \mathbf{x}_2, \ldots, \mathbf{x}_N, t) = 0 \tag{17}$$

where U is known as the "quantum potential"

$$U(\mathbf{x}_1, \mathbf{x}_2, \ldots, \mathbf{x}_N, t) = \sum_{j=1}^{N} \frac{\hbar^2}{4m_j} \left[\frac{1}{2} \left(\frac{\nabla_j P}{P} \right)^2 - \frac{\nabla_j^2 P}{P} \right] = -\sum_{j=1}^{N} \frac{\hbar^2}{2m_j} \frac{\nabla_j^2 R}{R} \,. \tag{18}$$

Moreover, with the definitions of $\mathbf{u}_j$ in Equation (10) one can rewrite U as

$$U = \sum_{j=1}^{N} \left[\frac{m_j \mathbf{u}_j^2}{2} - \frac{\hbar}{2} (\nabla_j \cdot \mathbf{u}_j) \right] \,. \tag{19}$$

However, as was already pointed out in [13], with the aid of Equations (4) and (6), $\mathbf{u}_j$ can also be written as

$$\mathbf{u}_j = \frac{1}{2\omega_j m_j} \nabla_j Q \,, \tag{20}$$

which thus explicitly shows its dependence on the spatial behavior of the heat flow δQ. Insertion of Equation (20) into Equation (19) then provides the thermodynamic formulation of the quantum potential as

$$U = \sum_{j=1}^{N} \frac{\hbar^2}{4m_j} \left[\frac{1}{2} \left(\frac{\nabla_j Q}{\hbar \omega_j} \right)^2 - \frac{\nabla_j^2 Q}{\hbar \omega_j} \right] \,. \tag{21}$$

As in our model particles and fields are dynamically interlocked, it would be highly misleading to picture the quantum potential in a manner similar to the classical scenario of particle plus field, where the latter can be switched on and off like an ordinary potential. Contrariwise, in our case the particle velocities/momenta must be considered as *emergent*. One can illustrate this with the situation in double-slit interference (Figure 1). Considering an incoming beam of, say, electrons with wave number $\mathbf{k}$ impinging on a wall with two slits, two beams with wave numbers $\mathbf{k}_A$ and $\mathbf{k}_B$, respectively, are created, which one may denote as "pre-determined" quantities, resulting also in pre-determined velocities $\mathbf{v}_\alpha = \frac{1}{m} \hbar \mathbf{k}_\alpha$, $\alpha = A$ or B.

However, if one considers that the electrons are not moving in empty space, but in an undulatory environment created by the ubiquitous zero-point field "filling" the whole experimental setup. One has to combine all the velocities/momenta at a given point in space and time in order to compute the resulting, or emergent, velocity/momentum field $\mathbf{v}_i = \frac{1}{m} \hbar \boldsymbol{\kappa}_i$, $i = 1 \text{ or } 2$ (Figure 1), where i is a bookkeeping index not necessarily related to the particle coming from a particular slit [19]. The relevant contributions other than the particle's forward momentum $m\mathbf{v}$ originate from the osmotic momentum $m\mathbf{u}$. The latter is well known from Nelson's stochastic theory [20], but its identical form has been derived by one of us from an assumed sub-quantum nonequilibrium thermodynamics [13,21] as it was described above. As shall be shown in the next section, our model also provides an understanding and deeper-level explanation of the microphysical, causal processes involved, i.e., of the guiding law [22] of the de Broglie–Bohm theory.

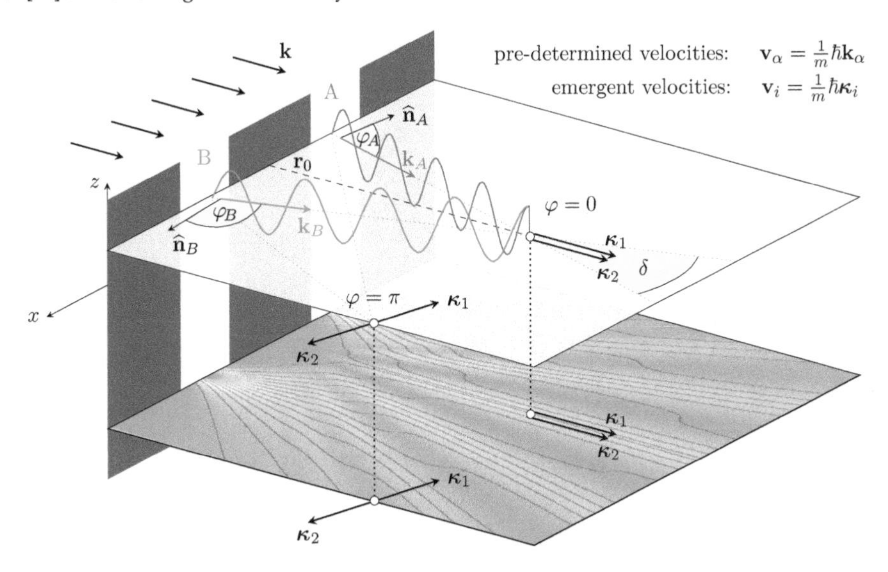

Figure 1. Scheme of interference at a double-slit. Considering an incoming beam of electrons with wave number $\mathbf{k}$ impinging on a wall with two slits, two beams with wave numbers $\mathbf{k}_A$ and $\mathbf{k}_B$, respectively, are created, which one may denote as "pre-determined" velocities $\mathbf{v}_\alpha = \frac{1}{m}\hbar\mathbf{k}_\alpha$, α=A or B. Taking into account the influences of the osmotic momentum field $m\mathbf{u}$, one has to combine all the velocities/momenta at a given point in space and time in order to compute the resulting, or emergent, velocity/momentum field $\mathbf{v}_i = \frac{1}{m}\hbar\boldsymbol{\kappa}_i$, $i = 1$ or 2. This, then, provides the correct intensity distributions and average trajectories (lower plane).

3. Derivation of the De Broglie–Bohm Guiding Equation for N Particles

Consider at first one particle in an n-slit system. In quantum mechanics, as well as in our emergent quantum mechanics approach, one can write down a formula for the total intensity distribution P which is very similar to the classical formula. For the general case of n slits, it holds with phase differences $\varphi_{ii'} = \varphi_i - \varphi_{i'}$ between the slits i, i' that

$$P = \sum_{i=1}^{n} \left(P_i + \sum_{i'=i+1}^{n} 2R_i R_{i'} \cos \varphi_{ii'} \right) \tag{22}$$

where the phase differences are defined over the whole domain of the experimental setup. As in our model, the "particle" is actually a bouncer in a fluctuating wave-like environment, i.e., analogously to the bouncers of the Couder experiments, one does have some (e.g., Gaussian) distribution, with its center following the Ehrenfest trajectory in the free case, but one also has a diffusion to the right and to the left of the mean path, which is only due to that stochastic bouncing. Thus, the total velocity field of our bouncer in its fluctuating environment is given by the sum of the forward velocity $\mathbf{v}$ and the respective osmotic velocities $\mathbf{u}_\mathrm{L}$ and $\mathbf{u}_\mathrm{R}$ to the left and the right. As for any direction α the osmotic velocity $\mathbf{u}_\alpha = \frac{\hbar}{2m}\frac{\nabla P}{P}$ does not necessarily fall off with the distance, one has long effective tails of the distributions which contribute to the nonlocal nature of the interference phenomena [23]. In sum, one has three distinct velocity (or current) channels per slit in an n-slit system.

We have previously shown [19,24] how one can derive the Bohmian guidance formula from our two-momentum approach. Introducing classical wave amplitudes $R(\mathbf{w}_i)$ and generalized velocity field vectors $\mathbf{w}_i$, which represent either a forward velocity $\mathbf{v}$ or an osmotic velocity $\mathbf{u}$ in the direction transversal to $\mathbf{v}$, we calculate the phase-dependent amplitude contributions of the total system's wave field projected on one channel's amplitude $R(\mathbf{w}_i)$ at the point $(\mathbf{x}, t)$ in the following way. We define a *relational intensity* $P(\mathbf{w}_i)$ as the local wave intensity $P(\mathbf{w}_i)$ in each channel (i.e., $\mathbf{w}_i$), recalling that there are 3 velocity channels per slit: $\mathbf{u}_{\mathrm{L}}$, $\mathbf{u}_{\mathrm{R}}$, and $\mathbf{v}$. The sum of all relational intensities, then, is the total intensity, i.e., the total probability density. In an n-slit system, we thus obtain for the relational intensities and the corresponding currents, respectively, i.e., for each channel component i,

$$P(\mathbf{w}_i) = R(\mathbf{w}_i)\hat{\mathbf{w}}_i \cdot \sum_{i'=1}^{3n} \hat{\mathbf{w}}_{i'} R(\mathbf{w}_{i'}) \tag{23}$$

$$\mathbf{J}(\mathbf{w}_i) = \mathbf{w}_i P(\mathbf{w}_i), \qquad i = 1, \ldots, 3n \tag{24}$$

with unit vectors $\hat{\mathbf{w}}_i$ and

$$\cos \varphi_{ii'} := \hat{\mathbf{w}}_i \cdot \hat{\mathbf{w}}_{i'} \,. \tag{25}$$

Consequently, the total intensity and current of our field read as

$$P_{\mathrm{tot}} = \sum_{i=1}^{3n} P(\mathbf{w}_i) = \left(\sum_{i=1}^{3n} \hat{\mathbf{w}}_i R(\mathbf{w}_i) \right)^2 \tag{26}$$

$$\mathbf{J}_{\mathrm{tot}} = \sum_{i=1}^{3n} \mathbf{J}(\mathbf{w}_i) = \sum_{i=1}^{3n} \mathbf{w}_i P(\mathbf{w}_i), \tag{27}$$

leading to the *emergent total velocity*

$$\mathbf{v}_{\mathrm{tot}} = \frac{\mathbf{J}_{\mathrm{tot}}}{P_{\mathrm{tot}}} = \frac{\sum_{i=1}^{3n} \mathbf{w}_i P(\mathbf{w}_i)}{\sum_{i=1}^{3n} P(\mathbf{w}_i)}, \tag{28}$$

which represents the *probability flux lines.*

In [16,19], we have shown with the example of $n = 2$, i.e., a double-slit system, that Equation (28) can equivalently be written in the form

$$\mathbf{v}_{\mathrm{tot}} = \frac{R_1^2 \mathbf{v}_1 + R_2^2 \mathbf{v}_2 + R_1 R_2 \left(\mathbf{v}_1 + \mathbf{v}_2\right) \cos \varphi + R_1 R_2 \left(\mathbf{u}_1 - \mathbf{u}_2\right) \sin \varphi}{R_1^2 + R_2^2 + 2 R_1 R_2 \cos \varphi} \,. \tag{29}$$

The trajectories or streamlines, respectively, are obtained according to $\dot{\mathbf{x}} = \mathbf{v}_{\mathrm{tot}}$ in the usual way by integration. As we have first shown in [16], by re-inserting the expressions for forward and osmotic velocities, respectively, i.e.,

$$\mathbf{v}_i = \frac{\nabla S_i}{m}, \qquad \mathbf{u}_i = -\frac{\hbar}{m} \frac{\nabla R_i}{R_i}, \tag{30}$$

one immediately identifies Equation (29) with the Bohmian guidance formula. Naturally, employing the Madelung transformation for each slit α ($\alpha = 1$ or 2),

$$\psi_\alpha = R_\alpha \mathrm{e}^{\mathrm{i} S_\alpha / \hbar}, \tag{31}$$

so $P_\alpha = R_\alpha^2 = |\psi_\alpha|^2 = \psi_\alpha^* \psi_\alpha$, with $\varphi = (S_1 - S_2)/\hbar$. Recalling the usual trigonometric identities such as $\cos\varphi = \frac{1}{2}\left(e^{i\varphi} + e^{-i\varphi}\right)$, one can rewrite the total average current immediately in the usual quantum mechanical form as

$$\begin{aligned} \mathbf{J}_{\text{tot}} &= P_{\text{tot}} \mathbf{v}_{\text{tot}} \\ &= (\psi_1 + \psi_2)^* (\psi_1 + \psi_2) \frac{1}{2} \left[\frac{1}{m} \left(-i\hbar \frac{\nabla(\psi_1 + \psi_2)}{(\psi_1 + \psi_2)} \right) + \frac{1}{m} \left(i\hbar \frac{\nabla(\psi_1 + \psi_2)^*}{(\psi_1 + \psi_2)^*} \right) \right] \\ &= -\frac{i\hbar}{2m} \left[\Psi^* \nabla \Psi - \Psi \nabla \Psi^* \right] = \frac{1}{m} \mathrm{Re} \left\{ \Psi^* (-i\hbar\nabla) \Psi \right\} \end{aligned} \tag{32}$$

where $P_{\text{tot}} = |\psi_1 + \psi_2|^2 =: |\Psi|^2$.

Equation (28) has been derived for one particle in an n-slit system. However, for the spinless particles obeying the Schrödinger equation, it is easy to extend this derivation to the many-particle case (As we do not yet have a relativistic model involving spin, our results for the many-particle case cannot account for the difference in particle statistics, i.e., for fermions or bosons. This will be a task for future work). Due to the purely additive terms in the expressions for the total current and total probability density, respectively, also for N particles, Equations (26) and (27) become

$$P_{\text{tot},N} = \sum_{j=1}^{N} \left[\sum_{i=1}^{3n} P(\mathbf{w}_i) \right]_j = \sum_{j=1}^{N} \left[\left(\sum_{i=1}^{3n} \hat{\mathbf{w}}_i R(\mathbf{w}_i) \right)^2 \right]_j \tag{33}$$

$$\mathbf{J}_{\text{tot},N} = \sum_{j=1}^{N} \left[\sum_{i=1}^{3n} \mathbf{J}(\mathbf{w}_i) \right]_j = \sum_{j=1}^{N} \left[\sum_{i=1}^{3n} \mathbf{w}_i P(\mathbf{w}_i) \right]_j . \tag{34}$$

Analogously, Equation (28) becomes

$$\mathbf{v}_{\text{tot},N} = \frac{\mathbf{J}_{\text{tot}}}{P_{\text{tot}}} = \frac{\sum_{j=1}^{N} \left[\sum_{i=1}^{3n} \mathbf{w}_i P(\mathbf{w}_i) \right]_j}{\sum_{j=1}^{N} \left[\sum_{i=1}^{3n} P(\mathbf{w}_i) \right]_j}, \tag{35}$$

where $\mathbf{w}_i$ is dependent on the velocities (30) with different S_i and R_i for every j. In quantum mechanical terms the only difference now is that the currents' nabla operators have to be applied at all of the locations of the respective N particles, thus providing

$$\mathbf{J}_{\text{tot}}(N) = \sum_{j=1}^{N} \frac{1}{m_j} \mathrm{Re} \left\{ \Psi^*(t) (-i\hbar\nabla_j) \Psi(t) \right\} \tag{36}$$

where $\Psi(t)$ now is the total N-particle wave function, whereas the flux lines are given by

$$\mathbf{v}_j(t) = \frac{\hbar}{m_j} \mathrm{Im} \frac{\nabla_j \Psi(t)}{\Psi(t)} \qquad \forall j = 1, ..., N. \tag{37}$$

In sum, with our introduction of a relational intensity $P(\mathbf{w}_i)$ for channels $\mathbf{w}_i$, which include sub-quantum velocity fields, we obtain the guidance formula also for N-particle systems in real 3-dimensional space. The central ingredient for this to be possible is to consider the emergence of the velocity field from the interplay of the totality of all of the system's velocity channels.

In Figures 2 and 3, trajectories (flux lines) for two Gaussian slits are shown (from [16]). These trajectories are in full accordance with those obtained from the Bohmian approach, as can be seen by comparison with [25–27], for example.

Figure 2. Classical computer simulation of the interference pattern: intensity distribution with increasing intensity from white through yellow and orange, with trajectories (red) for two Gaussian slits, and with *large dispersion* (evolution from bottom to top; $v_{x,1} = v_{x,2} = 0$). From [16].

Figure 3. Classical computer simulation of the interference pattern: intensity distribution with increasing intensity from white through yellow and orange, with trajectories (red) for two Gaussian slits, and with *small dispersion* (evolution from bottom to top; $v_{x,1} = -v_{x,2}$). From [16].

4. Vacuum Landscaping: Cause of Nonlocal Influences without Signaling

In the foregoing sections, we pointed out how nonlocality appears in our model. Particularly in discussing Equations (9)–(11), it was shown that the form of the osmotic momentum

$$m\mathbf{u} = -\frac{\hbar}{2}\frac{\nabla P}{P} \tag{38}$$

may be responsible for relevant influences. Moreover, if one assumes a particle at some position $\mathbf{x}$ in space, and with a probability distribution P, the latter is a distribution around $\mathbf{x}$ with long tails across the whole experimental setup, which may be very thin but still non-zero. Then, even at locations $\mathbf{y}$ very remote from $\mathbf{x}$, and although the probability distribution P pertaining to the far-away particle might be minuscule, it still may become effective immediately through the zero-point field.

The physical reason for bringing in nonlocality is the assumed resonant coupling of the particle(s) with fluctuations of the zero-point vacuum filling the whole experimental setup. Take, for example, a typical "Gaussian slit." We effectively describe P by a Gaussian with long non-zero tails throughout the whole apparatus. As we have seen, in order to calculate on-screen distributions (i.e., total intensities) of particles that went through an n-slit device one at a time, one only needs a two-momentum description and a calculation that uses the totality of all relational intensities involving the relative phases determined across the whole apparatus.

In general, we propose a resonant interaction of the bouncing "particle" with a *relevant environment* (In a similar vein, Bohm [28] speaks of a "relatively independent subtotality" of the universe, to account for the possible neglect of the "rest of the universe" in practical calculations). For idealized, non-interacting particles, this relevant environment would be the whole universe and thus the idealized prototype of the "cosmological solution" referred to in the introduction.

For any particle in any experimental setup, however, the relevant environment is defined by the boundary conditions of the apparatus. Whereas the idealized one-particle scenario would constitute an indefinite order of vibrations with respect to the particle oscillations potentially locking in, the very building up of an experiment may represent a dynamical transition from this indefinite order to the establishment of a definite order. The latter is characterized by the emergence of standing waves between the boundaries of the apparatus (e.g., source and detector), to which the particle oscillations lock in. Moreover, if an experimenter decides to change the boundary conditions (e.g., by altering the probability landscape between source and detector), such a "switching" would establish yet another definite order. The introduction or change of boundary conditions, which immediately affects the probability landscape, and the forward and the osmotic fields, we term "vacuum landscaping."

In other words, the change of boundary conditions of an experimental arrangement constitutes the immediate transition from one cosmological solution in the relevant environment (i.e., within the old boundary conditions) to another (i.e., the new ones). The "surfing" bouncer/particle simply locally jumps from the old to the new standing wave solutions, respectively. This is a process that happens locally for the particle, practically instantaneously (i.e., within a time span $\propto 1/\omega$), and nonlocally for the standing waves, due to the very definition of the cosmological solutions. The vacuum landscape is thus nonlocally changed without the propagation of "signals" in a communication theoretical sense (It is *exclusively* the latter that must be prohibited in order to avoid causal loops leading to paradoxes. See Walleczek and Grössing [29,30] for an extensive clarification of this issue).

We have, for example, discussed in some detail what happens in a double-slit experiment if one starts with one slit only, and when the particle might pass it, one opens the second slit [23,31]. In accordance with Tollaksen et al. [32], we found that the opening of the second slit (i.e., a change in boundary conditions) results in an uncontrollable shift in momentum on the particle passing the first slit. Due to its uncontrollability (or, the "complete uncertainty" in [32]), this momentum shift cannot be used for signaling. Still, it is necessary to *a posteriori* understand the final distributions on the screen, which would be incorrect without acknowledging said momentum kick.

Similarly, Aspect-type experiments of two-particle interferometry can be understood as alterations of vacuum landscapes. Consider, for example, the case in two-particle interferometry, where Alice and Bob each are equipped with an interfering device and receive one of the counter-propagating particles from their common source. If Alice during the time-of-flight of the particles changes her device by making with suitable mirrors one of the interferometer arms longer than the other, this constitutes an immediate switching from one vacuum landscape to another, with the standing waves of the zero-point field now reflecting the new experimental arrangement. In other words, the P-field has been changed nonlocally throughout the experimental setup and therefore all relational intensities

$$P(\mathbf{w}_i) = R(\mathbf{w}_i)\hat{\mathbf{w}}_i \cdot \sum_{i'} \hat{\mathbf{w}}_{i'} R(\mathbf{w}_{i'}) \tag{39}$$

involved. The latter represent the relative phase shifts $\delta\varphi_{i,i'} = \delta \arccos \hat{\mathbf{w}}_i \cdot \hat{\mathbf{w}}_{i'}$ occurring due to the switching, and this change is becoming manifest also in the total probability density

$$P_{\text{tot}} = \sum_i P(\mathbf{w}_i) = \left(\sum_i \hat{\mathbf{w}}_i R(\mathbf{w}_i) \right)^2, \tag{40}$$

with i running through all channels of both Alice and Bob. The quantum mechanical nonlocal correlations thus appear without any propagation (e.g., from Alice to Bob), superluminal or other. As implied by Gisin's group [33], this violates a "principle of continuity" of propagating influences from A to B, but its non-signaling character is still in accordance with relativity and the nonlocal correlations of quantum mechanics. Practically instantaneous vacuum landscaping by Alice and/or Bob thus ensures the full agreement with the quantum mechanical predictions without the need to invoke (superluminal or other) signaling. Our model is, therefore, an example of nonlocal influencing without signaling, which was recently shown to provide a viable option for realistic modeling of nonlocal correlations [29,30].

5. Conclusions and Outlook

With our two-momentum approach to an emergent quantum mechanics we have shown that one can in principle base the foundations of quantum mechanics on a deeper level that does not need wavefunctions. Still, one can derive from this new starting point, which is largely rooted in classical nonequilibrium thermodynamics, the usual nonrelativistic quantum mechanical formalism involving wavefunctions, like the Schrödinger equation or the de Broglie–Bohm guiding law. With regard to the latter, the big advantage of our approach is given by the fact that we avoid the troublesome influence from configuration space on particles in real space, which Bohm himself has called "indigestible." Instead, in our model, the guiding equation is completely understandable in real coordinate space, and actually a rather typical consequence of the fact that the total current is the sum of all particular currents, and the total intensity, or probability density, respectively, is the sum of all relational intensities. As we are working with Schrödinger (i.e., spinless) particles, accounting for differences in particle statistics is still an open problem.

As shown, we can replicate quantum mechanical features exactly by subjecting classical particle trajectories to diffusive processes caused by the presence of the zero point field, with the important property that the probability densities involved extend, however feebly, over the whole setup of an experiment. The model employs a two-momentum approach to the particle propagation, i.e., forward and osmotic momenta. The form of the latter has been derived without any recurrence to other approaches such as Nelson's.

The one thing that *is* to be digested from our model is the fact that the relational intensities are nonlocally defined, over the whole experimental arrangement (i.e., the "relevant environment"). This lies at the bottom of our deeper-level ansatz, and it is the *only* difference to an otherwise completely classical approach. We believe that this price is not too high, for we obtain a logical, realistic picture of quantum processes which is rather simple to arrive at. Nevertheless, in order to accept it, one needs to radically reconsider what an "object" is. We believe that it is very much in the spirit of David Bohm's thinking to direct one's attention away from a particle-centered view and consider an alternative option: that the universe is to be taken as a totality, which, only under very specific and delicate experimental arrangements, can be broken down to a laboratory-sized relevant environment, even if that laboratory might stretch along interplanetary distances. In our approach, the setting up of an experimental arrangement limits and shapes the forward and osmotic contributions and is described as vacuum landscaping. Accordingly, any change of the boundary conditions can be the cause of nonlocal influences throughout the whole setup, thus explaining, e.g., Aspect-type experiments. We argue that these influences can in no way be used for signaling purposes in the communication theoretic sense, and are therefore fully compatible with special relativity.

Accepting that the vacuum fluctuations throughout the universe, or at least within such a laboratory, are a defining part of a quantum, amounts to seeing any object like an "elementary particle" as nonlocally extended and, eventually, as exerting nonlocal influences on other particles. For anyone who can digest this, quantum mechanics is no more mysterious than classical mechanics or any other branch of physics.

Author Contributions: "Conceptualization, G.G.; Software, J.M.P.; Validation, S.F. and H.S.; Formal Analysis, all; Investigation, all; Resources, all; Data Curation, S.F. and J.M.P.; Writing—Original Draft Preparation, G.G.; Writing—Review & Editing, all; Visualization, H.S. and J.M.P.; Project Administration, G.G.

Acknowledgments: We thank Jan Walleczek for many enlightening discussions, and the Fetzer Franklin Fund of the John E. Fetzer Memorial Trust for partial support of the current work. We also wish to thank Lev Vaidman and several other colleagues for stimulating discussions. We thank the latter for the exchange of viewpoints sometimes closely related to our approach, as they also become apparent in their respective works: Herman Batelaan [34], Ana María Cetto and Luis de la Peña [35], Hans-Thomas Elze [36], Basil Hiley [37], Tim Maudlin [38], Travis Norsen [39], Garnet Ord [40], and Louis Vervoort [41].

Conflicts of Interest: The authors declare no conflict of interest.

References

1. Deckert, D.A.; Dürr, D.; Pickl, P. Quantum Dynamics with Bohmian Trajectories. *J. Phys. Chem. A* **2007**, *111*, 10325–10330, doi:10.1021/jp0711996.
2. Poirier, B. Bohmian mechanics without pilot waves. *Chem. Phys.* **2010**, *370*, 4–14, doi:10.1016/j.chemphys.2009.12.024.
3. Poirier, B. Trajectory-based Theory of Relativistic Quantum Particles. *arXiv* **2012**, arXiv:1208.6260.
4. Schiff, J.; Poirier, B. Communication: Quantum mechanics without wavefunctions. *J. Chem. Phys.* **2012**, *136*, 031102, doi:10.1063/1.3680558.
5. Hall, M.J.W.; Deckert, D.A.; Wiseman, H.M. Quantum Phenomena Modeled by Interactions between Many Classical Worlds. *Phys. Rev. X* **2014**, *4*, 041013, doi:10.1103/PhysRevX.4.041013.
6. Couder, Y.; Protière, S.; Fort, E.; Boudaoud, A. Dynamical phenomena: Walking and orbiting droplets. *Nature* **2005**, *437*, 208, doi:10.1038/437208a.
7. Couder, Y.; Fort, E. Single-particle Diffraction and Interference at a macroscopic scale. *Phys. Rev. Lett.* **2006**, *97*, 154101, doi:10.1103/PhysRevLett.97.154101.
8. Couder, Y.; Fort, E. Probabilities and trajectories in a classical wave-particle duality. *J. Phys. Conf. Ser.* **2012**, *361*, 012001, doi:10.1088/1742-6596/361/1/012001.
9. Bush, J.W.M. Quantum mechanics writ large. *Proc. Natl. Acad. Sci. USA* **2010**, *107*, 17455–17456, doi:10.1073/pnas.1012399107.
10. Bush, J.W.M. The new wave of pilot-wave theory. *Phys. Today* **2015**, *68*, 47–53, doi:10.1063/PT.3.2882.
11. Bush, J.W.M. Pilot-Wave Hydrodynamics. *Annu. Rev. Fluid Mech.* **2015**, *47*, 269–292, doi:10.1146/annurev-fluid-010814-014506.
12. De Broglie, L.V.P.R. *Non-Linear Wave Mechanics: A Causal Interpretation*; Elsevier: Amsterdam, The Netherland, 1960.
13. Grössing, G. The Vacuum Fluctuation Theorem: Exact Schrödinger Equation via Nonequilibrium Thermodynamics. *Phys. Lett. A* **2008**, *372*, 4556–4563, doi:10.1016/j.physleta.2008.05.007.

14. Grössing, G.; Fussy, S.; Mesa Pascasio, J.; Schwabl, H. Emergence and collapse of quantum mechanical superposition: Orthogonality of reversible dynamics and irreversible diffusion. *Physica A* **2010**, *389*, 4473–4484, doi:10.1016/j.physa.2010.07.017.
15. Grössing, G.; Fussy, S.; Mesa Pascasio, J.; Schwabl, H. Elements of sub-quantum thermodynamics: Quantum motion as ballistic diffusion. *J. Phys. Conf. Ser.* **2011**, *306*, 012046, doi:10.1088/1742-6596/306/1/012046.
16. Grössing, G.; Fussy, S.; Mesa Pascasio, J.; Schwabl, H. An explanation of interference effects in the double slit experiment: Classical trajectories plus ballistic diffusion caused by zero-point fluctuations. *Ann. Phys.* **2012**, *327*, 421–437, doi:10.1016/j.aop.2011.11.010.
17. Grössing, G.; Mesa Pascasio, J.; Schwabl, H. A Classical Explanation of Quantization. *Found. Phys.* **2011**, *41*, 1437–1453, doi:10.1007/s10701-011-9556-1.
18. Boltzmann, L. Über die mechanische Bedeutung des zweiten Hauptsatzes der Wärmetheorie. *Wien. Ber.* **1866**, *53*, 195–200.
19. Fussy, S.; Mesa Pascasio, J.; Schwabl, H.; Grössing, G. Born's Rule as Signature of a Superclassical Current Algebra. *Ann. Phys.* **2014**, *343*, 200–214, doi:10.1016/j.aop.2014.02.002.
20. Nelson, E. Derivation of the Schrödinger Equation from Newtonian Mechanics. *Phys. Rev.* **1966**, *150*, 1079–1085, doi:10.1103/PhysRev.150.1079.
21. Grössing, G. On the thermodynamic origin of the quantum potential. *Physica A* **2009**, *388*, 811–823, doi:10.1016/j.physa.2008.11.033.
22. Grössing, G.; Fussy, S.; Mesa Pascasio, J.; Schwabl, H. Implications of a deeper level explanation of the deBroglie–Bohm version of quantum mechanics. *Quantum Stud. Math. Found.* **2015**, *2*, 133–140, doi:10.1007/s40509-015-0031-0.
23. Grössing, G.; Fussy, S.; Mesa Pascasio, J.; Schwabl, H. 'Systemic nonlocality' from changing constraints on sub-quantum kinematics. *J. Phys. Conf. Ser.* **2013**, *442*, 012012, doi:10.1088/1742-6596/442/1/012012.
24. Grössing, G.; Fussy, S.; Mesa Pascasio, J.; Schwabl, H. Relational causality and classical probability: Grounding quantum phenomenology in a superclassical theory. *J. Phys. Conf. Ser.* **2014**, *504*, 012006, doi:10.1088/1742-6596/504/1/012006.
25. Holland, P.R. *The Quantum Theory of Motion*; Cambridge University Press: Cambridge, UK, 1993.
26. Bohm, D.; Hiley, B.J. *The Undivided Universe: An Ontological Interpretation of Quantum Theory*; Routledge: London, UK, 1993.
27. Sanz, Á.S.; Borondo, F. Contextuality, decoherence and quantum trajectories. *Chem. Phys. Lett.* **2009**, *478*, 301–306, doi:10.1016/j.cplett.2009.07.061.
28. Bohm, D. *Wholeness and the Implicate Order*; Routledge: London, UK, 1980.
29. Walleczek, J.; Grössing, G. The Non-Signalling theorem in generalizations of Bell's theorem. *J. Phys. Conf. Ser.* **2014**, *504*, 012001, doi:10.1088/1742-6596/504/1/012001.
30. Walleczek, J.; Grössing, G. Nonlocal Quantum Information Transfer Without Superluminal Signalling and Communication. *Found. Phys.* **2016**, *46*, 1208–1228, doi:10.1007/s10701-016-9987-9.
31. Grössing, G. Emergence of quantum mechanics from a sub-quantum statistical mechanics. *Int. J. Mod. Phys. B* **2014**, 145–179, doi:10.1142/S0217979214501793.
32. Tollaksen, J.; Aharonov, Y.; Casher, A.; Kaufherr, T.; Nussinov, S. Quantum interference experiments, modular variables and weak measurements. *New J. Phys.* **2010**, *12*, 013023, doi:10.1088/1367-2630/12/1/013023.
33. Bancal, J.D.; Pironio, S.; Acin, A.; Liang, Y.C.; Scarani, V.; Gisin, N. Quantum nonlocality based on finite-speed causal influences leads to superluminal signaling. *Nat. Phys.* **2012**, *8*, 867–870, doi:10.1038/nphys2460.
34. Batelaan, H.; Jones, E.; Huang, W.C.W.; Bach, R. Momentum exchange in the electron double-slit experiment. *J. Phys. Conf. Ser.* **2016**, *701*, 012007, doi:10.1088/1742-6596/701/1/012007.
35. De la Peña, L.; Cetto, A.M.; Valdés-Hernándes, A. *The Emerging Quantum: The Physics behind Quantum Mechanics*; Springer: Berlin/Heidelberg, Germany, 2014.
36. Elze, H.T. On configuration space, Born's rule and ontological states. *arXiv* **2018**, arXiv:1802.07189.
37. Hiley, B.J. Structure Process, Weak Values and Local Momentum. *J. Phys. Conf. Ser.* **2016**, *701*, 012010, doi:10.1088/1742-6596/701/1/012010.
38. Maudlin, T. *Quantum Non-Locality and Relativity: Metaphysical Intimations of Modern Physics*, 3rd ed.; Wiley-Blackwell: West Sussex, UK, 2011.
39. Norsen, T. Bohmian Conditional Wave Functions (and the status of the quantum state). *J. Phys. Conf. Ser.* **2016**, *701*, 012003. doi:10.1088/1742-6596/701/1/012003.

40. Ord, G. Quantum mechanics in a two-dimensional spacetime: What is a wavefunction? *Ann. Phys.* **2009**, *324*, 1211–1218. doi:10.1016/j.aop.2009.03.007.
41. Vervoort, L. No-Go Theorems Face Background-Based Theories for Quantum Mechanics. *Found. Phys.* **2016**, *46*, 458–472, doi:10.1007/s10701-015-9973-7.

Article

Bouncing Oil Droplets, de Broglie's Quantum Thermostat, and Convergence to Equilibrium

Mohamed Hatifi [1,*], Ralph Willox [2], Samuel Colin [3] and Thomas Durt [1]

[1] Aix Marseille Université, CNRS, Centrale Marseille, Institut Fresnel UMR 7249, 13013 Marseille, France; thomas.durt@centrale-marseille.fr
[2] Graduate School of Mathematical Sciences, the University of Tokyo, 3-8-1 Komaba, Meguro-ku, Tokyo 153-8914, Japan; willox@ms.u-tokyo.ac.jp
[3] Centro Brasileiro de Pesquisas Físicas, Rua Dr. Xavier Sigaud 150, 22290-180 Rio de Janeiro, RJ, Brazil; scolin@cbpf.br
* Correspondence: hatifi.mohamed@gmail.com

Received: 24 August 2018; Accepted: 3 October 2018; Published: 11 October 2018

Abstract: Recently, the properties of bouncing oil droplets, also known as "walkers," have attracted much attention because they are thought to offer a gateway to a better understanding of quantum behavior. They indeed constitute a macroscopic realization of wave-particle duality, in the sense that their trajectories are guided by a self-generated surrounding wave. The aim of this paper is to try to describe walker phenomenology in terms of de Broglie–Bohm dynamics and of a stochastic version thereof. In particular, we first study how a stochastic modification of the de Broglie pilot-wave theory, *à la* Nelson, affects the process of relaxation to quantum equilibrium, and we prove an H-theorem for the relaxation to quantum equilibrium under Nelson-type dynamics. We then compare the onset of equilibrium in the stochastic and the de Broglie–Bohm approaches and we propose some simple experiments by which one can test the applicability of our theory to the context of bouncing oil droplets. Finally, we compare our theory to actual observations of walker behavior in a 2D harmonic potential well.

Keywords: bouncing oil droplets; stochastic quantum dynamics; de Broglie–Bohm theory; quantum non-equilibrium; H-theorem; ergodicity

---

1. Introduction

"Walkers" are realized as oil droplets generated at the surface of a vibrating oil bath. As shown by Couder and Fort [1–3], the vibration of the bath prevents the coalescence of the droplets with the surface, allowing them to remain stable for very long times. Moreover, the trajectories of the walkers are guided by an external wave [4,5] that they themselves generate at the surface of the oil bath. From this point of view, walkers are reminiscent of wave-particle duality [2,6], and they seem to offer deep analogies with de Broglie–Bohm particles [7]. Up until now, different aspects of walker dynamics have been studied in a purely classical framework, typically in a hydrodynamical approach [3,5]. For instance, certain models address their deformations due to their bouncing off the surface of the bath, in function of the density and viscosity of the oil and other parameters [5]. Other studies describe the dynamics of the surface waves that the walkers generate during the bouncing process, and how those waves in turn guide their trajectories. In these models, this complex behavior is characterized by a memory time which relates the dynamics of the walker bouncing at time t to its successive bouncing positions in the past [8,9]. The presence of such a memory effect establishes a first difference with quantum mechanics. Normally, in quantum mechanics, it is assumed that all results of any possible future measurements to be performed on a quantum system are encapsulated in its present quantum state [10]: its wave function at the present time t.

Droplets also transcend the most common interpretations of quantum theory which prohibit any description of the system in terms of instantaneous, classical-like trajectories. Droplets and their trajectories are visible with the naked eye at any time and standard interpretations of quantum mechanics do not apply. This is why we believe that it is necessary and worthwhile to adapt realist (causal) formalisms such as de Broglie–Bohm (dBB) dynamics [11,12] or a stochastic version thereof à la Nelson [13] to explore the analogy with quantum systems. This is the main motivation of the present paper.

Another difference between walker trajectories and quantum trajectories is that the quantum description is intrinsically probabilistic and non-classical, while there exist regimes in which the trajectory of the walkers is deterministic and classical (for example, when they bounce exactly in phase with the bath, they can be shown to follow straight lines at constant velocity [14–17]). However, there also exist regimes in which a Brownian motion is superimposed on their flow lines (e.g., above the Faraday threshold), and other regimes where the trajectories appear to be chaotic [5]. In fact, in several regimes, droplets appear to exhibit ergodic behavior. In practice, ergodicity has been established on the basis of the following observations: if we prepare a walker at the surface of the liquid bath (a corral, for instance), it will progressively explore each part of the surface, following an apparently random motion [4]. If one then visualizes the statistics of the sojourn time of the walker in each of these regions, a striking pattern emerges, bearing more than a simple resemblance to an interference pattern [4,7]. It is this, again remarkable, manifestation of wave-particle duality that first attracted our attention and which lies at the origin of this paper. The onset of quantum equilibrium in the framework of dBB dynamics and in stochastic versions thereof is an important foundational issue in itself, which has motivated numerous studies (see, e.g., [13,18–24] as well as [25] and references therein). Several authors in the past have indeed tried to explain how the Born rule emerges from individual trajectories, which is a highly non-trivial problem. In the case of dBB dynamics, it is easy to show that in simple situations the relaxation to the Born statistical distribution does not occur at all, but recent studies [26–31] show that in sufficiently complex situations (several modes of different energies for instance) the system might exhibit mixing, which explains the onset of quantum equilibrium in such cases. As we shall show in the present paper, in the case of Nelson-type dynamics, the quantum Brownian motion imposed in such a model accelerates the relaxation to Born's distribution, and in fact ensures that relaxation to the Born rule will almost always occur (as we shall also show). In our view, for the above reasons, de Broglie–Bohm and Nelson-type dynamics are good candidates for explaining how wavelike statistics emerge after averaging a set of apparently chaotic and/or stochastic trajectories.

Briefly summarized, our main goal is to explain the emergence of aforementioned interference patterns in the framework of the dynamical models of de Broglie–Bohm and of a stochastic version thereof which is based on the models of Bohm-Vigier [18] and Bohm-Hiley [19] but which is formally close to Nelson [13]. Both models are introduced in Section 2. Here, it is worth noting that thus far there is no experimental evidence that droplets indeed follow de Broglie–Bohm and/or Nelson trajectories. Our approach therefore differs radically from previous studies on droplets, in the sense that we impose a quantum dynamics by brute force, whereas, until now, the attempt to illustrate how chaos may underlie quantum stochasticity has been a pillar of the research on walkers/droplets. In fact, Nelson's original goal, in proposing his dynamics, was to derive an effective wave equation from the properties of an underlying Brownian motion, as in classical statistical mechanics where a diffusion equation is derived from microscopic properties of the atoms. There actually exists an impressive number of attempts in that direction, as, e.g., stochastic electro-dynamics [5,32,33]. However, there exists (as far as we know) no way to derive an effective Schrödinger equation from hydrodynamical models of droplets.

By choosing exactly the opposite approach, i.e., by imposing quantum-like dynamics on the droplets, we pursue three goals. The first one is to describe the onset of quantum equilibrium (and ergodicity). A second objective is to formulate precise quantitative predictions regarding this relaxation process, which can possibly be validated by future experiments. A third objective is to show, for the first time, that certain dBB trajectories present a deep structural resemblance with certain trajectories that have been reported in the literature for droplets trapped in a harmonic potential.

A short discussion of the onset of equilibrium in de Broglie–Bohm dynamics and the importance of coarse-graining is given in Section 3. In the case of our stochastic, Nelson-type dynamics, we derive in Section 4 a new H-theorem showing the relaxation to quantum equilibrium, which does not rely on coarse-graining and is valid at all scales. We pay particular attention to the ergodicity of trajectories in the case of our stochastic dynamics (which mix properties of the de Broglie–Bohm dynamics with Brownian motion). We apply these ideas to discuss ergodicity in the case of the stochastic treatment of a particle trapped in a harmonic potential (Section 5) and to describe the dynamics of a droplet trapped in a harmonic potential (Section 6). In this latter section (in Section 6.1), we also propose some simple experiments by which one can test the applicability of a Nelson-type dynamics to the context of bouncing oil droplets, and we briefly discuss the problems caused by the presence of zeros in the interference pattern that is encoded in the statistics of the trajectories. In Section 7, we study a situation during which the attractor of the probability distribution is no longer a static eigenstate of the (static) Hamiltonian, and we compare the onset of equilibrium in the dBB and stochastic formalisms in that special framework. In Section 8, we tackle the dynamics of droplets in a 2D harmonic potential through a simple model where the pilot wave is treated as a dynamical object. This constitutes a preliminary attempt, ultimately aimed at establishing a dynamics that would combine stochastic and/or dBB dynamics with a feedback of the trajectory on the wave, which is a fundamental feature of droplet phenomenology that has never been addressed in the framework of dBB or Nelson dynamics. The last section is devoted to conclusions and open questions. A short overview of the numerical methods used in the paper is given in the Appendix A.

2. dBB and Nelson Dynamics

2.1. The dBB Theory

In the following quick overview of the dBB theory we shall limit ourselves to the case of a single particle. In the dBB theory, particle positions exist at all times and they are merely revealed by position measurements, instead of "originating" with the measurement as the standard interpretation of quantum mechanics would have it. The dynamics is described by a wave function which obeys the Schrödinger equation:

$$i\hbar\frac{\partial\Psi(\mathbf{x},t)}{\partial t} = -\frac{\hbar^2}{2m}\Delta\Psi(\mathbf{x},t) + V(\mathbf{x},t)\Psi(\mathbf{x},t) \tag{1}$$

where $V(\mathbf{x},t)$ is an external potential and m the mass of the particle, *as well as* by a position $\mathbf{x}$. In order to reproduce the predictions of standard quantum mechanics, one must have that the positions are distributed according to

$$\mathcal{P}(\mathbf{x},t) = |\Psi(\mathbf{x},t)|^2 \tag{2}$$

where $\mathcal{P}(\mathbf{x},t)$ is the distribution of particle positions over an ensemble of trajectories. An ensemble satisfying condition (2) is said to be in *quantum equilibrium*.

It is also commonly assumed that (2) is satisfied at some initial time. Therefore, in order to be at (quantum) equilibrium for all t, the condition to enforce is

$$\frac{\partial\mathcal{P}(\mathbf{x},t)}{\partial t} = \frac{\partial|\Psi(\mathbf{x},t)|^2}{\partial t}. \tag{3}$$

As is well known, the probability density $|\Psi(\mathbf{x},t)|^2$ satisfies the continuity equation

$$\frac{\partial|\Psi(\mathbf{x},t)|^2}{\partial t} + \boldsymbol{\nabla}\cdot j(\mathbf{x},t) = 0 \tag{4}$$

where
$$j = \frac{\hbar}{m} \mathfrak{Im} (\Psi^* \nabla \Psi), \tag{5}$$
is the (probability) current describing the flow of the probability due to (1).

The probability density $\mathcal{P}$, on the other hand, will satisfy a continuity equation

$$\frac{\partial \mathcal{P}}{\partial t} + \nabla \cdot (\mathcal{P}\, \mathbf{v}) = 0 \tag{6}$$

where $\mathbf{v}$ is the velocity field for the particle. Therefore, Equation (3) will be satisfied if

$$\mathbf{v}(\mathbf{x}, t) = \frac{\mathbf{j}(\mathbf{x}, t)}{|\Psi(\mathbf{x}, t)|^2}. \tag{7}$$

The expression (7) for the velocity field is of course not the only possible one: any solution of the form
$$\mathbf{v}'(\mathbf{x}, t) = \mathbf{v}(\mathbf{x}, t) + \frac{\nabla \times f(\mathbf{x}, t)}{|\Psi(\mathbf{x}, t)|^2}, \tag{8}$$

where f is a scalar function, will also give rise to Equation (3) (see [34] for more details).

Secondly, if one expresses the wave function in terms of its phase $S(\mathbf{x}, t)$ and modulus $R(\mathbf{x}, t) = \sqrt{|\Psi(\mathbf{x}, t)|^2}$,
$$\Psi(\mathbf{x}, t) = R(\mathbf{x}, t) e^{i\, S(\mathbf{x}, t)/\hbar}, \tag{9}$$

one finds that
$$j = \frac{|\Psi(\mathbf{x}, t)|^2}{m} \nabla S \tag{10}$$

and that the velocity of the particle at time t is given by

$$\frac{d\mathbf{x}(t)}{dt} = \mathbf{v}(\mathbf{x}, t) = \frac{1}{m} \nabla S(\mathbf{x}, t) \bigg|_{\mathbf{x}=\mathbf{x}(t)}. \tag{11}$$

Integrating the system given by Equation (11), we recover the dBB trajectory. From the above, it should be clear that the dBB theory is deterministic. Any stochastic element only comes from our lack of knowledge of the initial positions.

In the context of bouncing droplets, we shall view the external wave generated by the droplet as being in one-to-one correspondence with the "pilot wave" Ψ, which guides the position of the dBB particle.

2.2. A Simple Realization of de Broglie's Quantum Thermostat—Nelson Dynamics

As mentioned in the introduction, the trajectories of walkers are often characterized by a non-negligible stochastic (Brownian) component which sets them apart from the smooth dBB trajectories. From this point of view, it seems worthwhile to try to model walkers dynamics in terms of stochastic generalisations of dBB dynamics.

de Broglie himself, in fact, considered such generalizations of the deterministic dBB dynamics (which he called the "quantum thermostat hypothesis") to be highly welcome because they might provide a physically sound picture of the hidden dynamics of static quantum states. For instance, if we consider the position of an electron prepared in the ground state of a hydrogen atom, the dBB dynamics predicts that its position will remain frozen at the same place throughout time, which is counterintuitive to say the least. Adding a stochastic component to its velocity could, in principle, explain why averaging the position of the electron over time is characterized by an exponentially decreasing probability density function, in agreement with the Born rule (provided, of course, that ergodicity is present in the problem in exactly the right proportion). A first proposal in this sense was formulated by Bohm and Vigier in 1954 [18] which, later on, was made more precise by Bohm and

Hiley [19], but stochastic derivations of Schrödinger's equation by Nelson [13] (and others [32,33] in the framework of stochastic electrodynamics) can also be considered to provide models of the quantum thermostat. Quoting de Broglie: "*...Finally, the particle's motion is the combination of a regular motion defined by the guidance formula, with a random motion of Brownian character... any particle, even isolated, has to be imagined as in continuous "energetic contact" with a hidden medium, which constitutes a concealed thermostat. This hypothesis was brought forward some fifteen years ago by Bohm and Vigier [18], who named this invisible thermostat the "subquantum medium"... If a hidden sub-quantum medium is assumed, knowledge of its nature would seem desirable...*" (In [35] Ch.XI: On the necessary introduction of a random element in the double solution theory. The hidden thermostat and the Brownian motion of the particle in its wave.)

In this paper, we shall consider a particular model of the quantum thermostat in which, as in the Bohm–Vigier model, a single spinless particle suspended in a Madelung fluid moves with the local velocity of the resulting field, given by Equation (11), and is subjected to fluctuations coming from the latter (cf. Figure 1). However, following Nelson, we shall model these fluctuations by means of a particular stochastic process. To be precise: our model is formally the same as Nelson's in that it relies on the same stochastic process. However, in spirit, it is closer to the Bohm–Hiley model [19] in that we do not assume to be at quantum equilibrium (an assumption which is fundamental to Nelson's theory, as was already pointed out by Bohm and Hiley [19]; see also [20] for a detailed presentation and a comparison of both approaches).

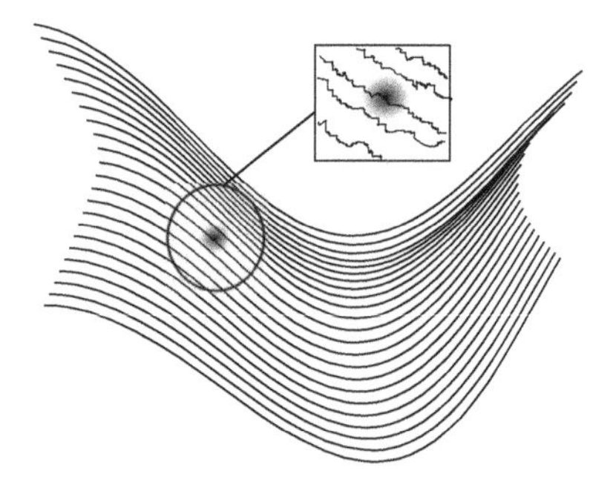

Figure 1. A particle suspended in a Madelung fluid and subject to local fluctuations.

This stochastic process is defined on a probabilistic space Ω, characterized by a probability distribution $P(\mathbf{x}, t)$ and obeying an Ito stochastic differential equation of the general form:

$$dx = \left[\frac{1}{m}\nabla S + \gamma\right]\Big|_{x=x(t)} dt + \sqrt{\alpha}\, d\mathbf{W}(t) \tag{12}$$

where α is the (constant) diffusion coefficient that characterizes the strength of the random part, and $d\mathbf{W}(t)$ is a Wiener process in three dimensions. The function $\gamma(\mathbf{x}, t)$ in (12) is a systematic drift, the so-called *osmotic velocity*, which we shall fix in the following way.

The conservation equation of the probability distribution (which we denote by P, in order to stress the difference with the probability in the dBB dynamics which is denoted by $\mathcal{P}$), can be written as the Fokker–Planck equation:

$$\frac{\partial P}{\partial t} = -\nabla \cdot \left(\frac{P}{m}\nabla S + \gamma P\right) + \frac{\alpha}{2}\Delta P. \tag{13}$$

If we now require that the quantum equilibrium $P(q,t) = |\Psi(q,t)|^2$ be a solution of this Fokker–Planck equation, we obtain from Equations (4), (10), and (13) that

$$\nabla \cdot \left(\gamma |\Psi|^2 - \frac{\alpha}{2} \nabla |\Psi|^2\right) = 0, \tag{14}$$

which is a constraint on the osmotic velocity. The simplest solution of this constraint is [36]

$$\gamma(\mathbf{x},t) = \frac{\alpha}{2} \frac{\nabla |\Psi|^2}{|\Psi|^2}. \tag{15}$$

In the rest of the paper, we choose the osmotic drift velocity to be Equation (15), with α an a priori free parameter, Nelson's choice for α ($\alpha = \hbar/m$) being irrelevant if we apply this formalism to droplets.

In summary, our Nelson dynamics is fully defined by the following Ito equation

$$d\mathbf{x}(t) = \left[\frac{1}{m}\nabla S + \frac{\alpha}{2}\frac{\nabla|\Psi|^2}{|\Psi|^2}\right]\Bigg|_{x=x(t)} dt + \sqrt{\alpha} d\mathbf{W}(t) \tag{16}$$

where $dW_i(t)$ represents a Wiener process with

$$< dW_i(t) > \;=\; 0 \qquad \text{and} \qquad < dW_i(t) dW_j(t') > \;=\; \frac{1}{2}\delta_{ij}\,\delta(t-t'), \tag{17}$$

and by the Fokker–Planck equation for the associated probability distribution $P(\mathbf{x},t)$

$$\frac{\partial P}{\partial t} = \frac{\alpha}{2}\Delta P - \nabla \cdot \left(\frac{P}{m}\nabla S + \frac{\alpha}{2}\frac{P}{|\Psi|^2}\nabla|\Psi|^2\right), \tag{18}$$

where $\Psi(\mathbf{x},t)$ satisfies the Schrodinger equation:

$$i\hbar\frac{\partial \Psi}{\partial t} = -\frac{\hbar^2}{2m}\Delta\Psi + V\Psi. \tag{19}$$

At quantum equilibrium, i.e., when $P(\mathbf{x},t) = |\Psi(\mathbf{x},t)|^2$, the diffusion velocity is balanced by the osmotic term and the Bohm velocity is recovered, on average.

We shall now discuss the details of the relaxation towards quantum equilibrium, in the dBB and stochastic formalisms.

3. Relaxation to Quantum Equilibrium in the de Broglie–Bohm Theory

In our presentation of the dBB theory for a single particle, in Section 2.1, we assumed that the particle positions are initially distributed according to Born's law

$$\mathcal{P}(\mathbf{x},t_i) = |\Psi(\mathbf{x},t_i)|^2 \tag{20}$$

over an ensemble. The dynamics then ensure that the same relation will hold for any later time. This is the assumption de Broglie and Bohm made in their original papers [11,12,37]. Although Bohm tried, already in the 1950s (first on his own—see, e.g., [11] (Section 9)—and later with Vigier [18]), to relax this assumption by modifying the dynamics, to many authors working today on the dBB theory it is still an assumption which has to be made (the final objective of de Broglie, Bohm, Vigier, and Nelson–and many other contributors to various realistic hidden variable interpretations in which quantum systems are assumed to be localized in space at any time–was to rationalize wave-like statistics in terms of individual trajectories; the same problem occurs in droplets phenomenology and, according to us, admits no fully satisfying solution yet).

According to Valentini [21,38,39], however, there is no need to assume that the particle positions are initially distributed according to Born's law or to modify the dynamics. His claim is that an ensemble in which Born's law is not satisfied (so-called quantum non-equilibrium) will evolve naturally towards quantum equilibrium, provided that the wave function leads to sufficiently complex dynamics. This relaxation process has to take place on a coarse-grained level and can only occur if the initial distributions do not display any fine-grained micro structure.

Let us first explain the need for coarse-graining. Let us introduce the function $f = \mathcal{P}/|\Psi|^2$, as in [26]. An important implication of (6) is that the function f is conserved along the dBB trajectories:

$$\frac{df(\mathbf{x},t)}{dt} \equiv \frac{\partial f(\mathbf{x},t)}{\partial t} + \dot{\mathbf{x}} \cdot \boldsymbol{\nabla} f(\mathbf{x},t) = 0. \tag{21}$$

Hence we have that

$$\mathcal{P}(\mathbf{x},t) = \frac{\mathcal{P}(\mathbf{x}_i,t_i)}{|\Psi(\mathbf{x}_i,t_i)|^2}|\Psi(\mathbf{x},t)|^2 \tag{22}$$

where $\mathbf{x}_i$ is the initial position of the particle which leads to $\mathbf{x}$, when evolving from t_i to t according to the dBB dynamics. If one assumes that $\mathcal{P}(\mathbf{x}_i,t_i)/|\Psi(\mathbf{x}_i,t_i)|^2 \neq 1$, relaxation to quantum equilibrium is clearly impossible, at least at the microscopic level. However, as argued by Valentini [21], relaxation *is* possible at the coarse-grained level, provided the initial distribution does not display any fine-grained microstructure.

The operational definition of the coarse-graining is as follows. We divide the domain of interest $A \subset \Omega$ into small cubes of equal edge length ϵ (we call them coarse-graining cells, or CG cells for short). These CG cells do not overlap and their union is equal to A. The coarse-grained densities, which we denote by $\overline{\mathcal{P}}(\mathbf{x},t)$ and $\overline{|\Psi(\mathbf{x},t)|^2}$, are then defined as

$$\overline{\mathcal{P}}(\mathbf{x},t) = \frac{1}{\epsilon^3}\int_{\text{CG cell}\ni\mathbf{x}} d^3x\mathcal{P}(\mathbf{x},t) \tag{23}$$

$$\overline{|\Psi(\mathbf{x},t)|^2} = \frac{1}{\epsilon^3}\int_{\text{CG cell}\ni\mathbf{x}} d^3x|\Psi(\mathbf{x},t)|^2 \tag{24}$$

where the domain of integration is the CG cell containing $\mathbf{x}$.

We can now discuss the second assumption: the lack of a fine-grained microstructure in the initial distribution. Let us assume we have a non-equilibrium distribution $\mathcal{P}(\mathbf{x},t_i)$ which relaxes to quantum equilibrium at the coarse-grained level, under the dynamics generated by the wave function $\Psi(\mathbf{x},t)$. As the dBB theory is time-reversal invariant, in the time-reversed situation, under the dynamics generated by $\Psi^*(\mathbf{x},-t)$, we would have a distribution that moves away from quantum equilibrium. Thus, it would seem that time-reversal invariance contradicts the possibility of relaxation to quantum equilibrium. This conclusion is unwarranted, however: as the initial distribution $\mathcal{P}(\mathbf{x},t_i)$ relaxes to quantum equilibrium, it retains information on the original values of f (which are constant in time) and thereby acquires a fine-grained microstructure, which means that at the final time t_f, $\mathcal{P}(\mathbf{x},t_f)$ will differ significantly from $\overline{\mathcal{P}}(\mathbf{x},t_f)$. Therefore, in the time-reversed situation, the initial distribution would exhibit a fine-grained microstructure, which is prohibited under our assumption, thereby breaking the time-reversal invariance.

In order to quantify the difference between the distribution $\mathcal{P}(\mathbf{x},t)$ and the quantum equilibrium condition $|\Psi(\mathbf{x},t)|^2$ at the coarse-grained level, Valentini [21,38,39] introduced the entropy-like function

$$\overline{H}(t) = \int_\Omega d^3x\,\overline{\mathcal{P}}\ln\left(\overline{\mathcal{P}}/\overline{|\Psi|^2}\right) \tag{25}$$

where $\overline{\mathcal{P}}$ and $\overline{|\Psi|^2}$ as in Equations (23) and (24), for which he has shown the (quantum) H-theorem,

$$\overline{H}(t) \leq \overline{H}(t_i), \tag{26}$$

under the assumption of no fine-grained microstructure. It should be stressed, however, that this is not necessarily a monotonic decay and therefore does not prove that quantum equilibrium will always be reached. It merely indicates a tendency towards relaxation. The strongest support for the idea of relaxation to quantum equilibrium comes from numerical simulations of the evolution of non-equilibrium distributions for various quantum systems [26–31] (see [25] and references therein for a review). The first numerical simulations were performed by Valentini and Westman [26] who showed, in the case of a 2D box, that relaxation quickly takes place for a wave function which is a superposition of the first 16 modes of energy (the superposition being equally weighted). It was also hinted that the nodes of the wave function, with their associated vorticity, play a crucial role in the relaxation process, as purveyors of chaos (or mixing) in the dynamics. This later claim was properly understood in [40]. The dependence of the relaxation timescale on the coarse-graining length ϵ and on the number of energy modes was studied in [28]. In [31], it was shown that quantum systems with a low number of modes are likely to never fully relax, in which case $\overline{H}$ reaches a non-zero residue value. However, such a scenario becomes unlikely as the number of modes increases.

According to the quantum non-equilibrium hypothesis, standard quantum mechanics is only one facet of the pilot-wave theory, that of quantum equilibrium, leaving the possibility for new physics: that of quantum non-equilibrium. One should assume of course that during our time we have only had (or can only have) access to systems for which quantum equilibrium has already been reached. But that does not mean that quantum non-equilibrium never existed in the early universe (which could be inferred from the observation of the remnants of the early fractions of seconds of the universe, just after the Big Bang [41]), or that some, yet undetetected, exotic quantum systems cannot still be in quantum non-equilibrium today [42]. This is why droplets are appealing, because if their dynamics do present analogies with dBB dynamics, their study will allow us to observe relaxation to (quantum) equilibrium "in real time" in the lab with the naked eye, which is not possible with quantum systems for which we have no direct access to individual trajectories.

4. An H-Theorem for Nelson Dynamics

Let us start by introducing an analog of Valentini's entropy, Equation (25), for the probability distribution $P(\mathbf{x}, t)$ associated with our Nelson dynamics, as defined by Equations (16–19):

$$H_V(t) = \int_\Omega d^3x \, P \ln\left(\frac{P}{|\Psi|^2}\right), \tag{27}$$

which is a special instance of a relative entropy known as the Kullback–Leibler divergence [43].

We also define a second non-negative functional

$$L_f(t) = \int_\Omega d^3x \, f(P - |\Psi|^2) \tag{28}$$

where

$$f(\mathbf{x}, t) = \frac{P(\mathbf{x}, t)}{|\Psi(\mathbf{x}, t)|^2}. \tag{29}$$

Note that we always impose the boundary conditions $|\Psi|^2|_{\partial\Omega} = P|_{\partial\Omega} = 0$ and $f|_{\partial\Omega} = 1$ so as to avoid divergence of these integrals on the boundary of Ω.

It should be noted that the entropy of Equation (27) or the functional of Equation (28) we shall use to quantify the relaxation to quantum equilibrium are very different from the entropies usually considered in the context of classical H-theorems (like e.g., the Boltzmann entropy). One should bear in mind, however, that *quantum equilibrium* is radically different from classical equilibrium [44] and has no connection whatsoever with relaxation to quantum *thermal* equilibrium, for the simple reason that the Born distribution of positions reached by an ensemble of trajectories *à la* Nelson or dBB is not a thermal distribution.

To understand why the functionals in Equations (27) and (28) are non-negative and why they are zero if and only if (quantum) equilibrium is reached (that is to say when $f = 1$ everywhere in space), it is important to note that the integrands of H_V and L_f satisfy the inequalities

$$(P - |\Psi|^2) \leq P \ln \frac{P}{|\Psi|^2} \leq \frac{P}{|\Psi|^2}(P - |\Psi|^2), \tag{30}$$

for which any of the possible equalities only hold when $P = |\Psi|^2$. (This is immediate from the trivial inequality: $\forall x > 0, \quad (1 - 1/x) \leq \ln x \leq x - 1$.)

Now, since both $P(\mathbf{x}, t)$ and $|\Psi(\mathbf{x}, t)|^2$ are probability distributions, i.e., since we have $\int_\Omega P dx = \int_\Omega |\Psi|^2 dx = 1$, it follows from (30) that whenever $H_V(t)$ and $L_f(t)$ are well-defined, they satisfy the following inequalities:

$$0 \leq H_V(t) \leq L_f(t). \tag{31}$$

Moreover, for the same reason, L_f can be re-expressed as $\int_\Omega d^3x \left[f(P - |\Psi|^2) - (P - |\Psi|^2)\right]$, the integrand in which is non-negative due to Equation (30). Therefore, L_f can only be zero if its integrand is zero, i.e., if $P = |\Psi|^2$ (if P, $|\Psi|^2$, and f are sufficiently smooth, which is something we shall always assume unless otherwise stated). Similarly [21], one also has that H_V can only be zero when $P = |\Psi|^2$ everywhere in Ω.

Let us now prove the relaxation to quantum equilibrium. Substituting $P = f|\Psi|^2$ in the Fokker–Planck Equation (18), and using Equations (4) and (10), it is easily verified that

$$|\Psi|^2 \frac{\partial f}{\partial t} = \frac{\alpha}{2} \nabla \cdot (|\Psi|^2 \nabla f) - \frac{|\Psi|^2}{m} (\nabla f)(\nabla S). \tag{32}$$

Rewriting L_f as

$$L_f = \int_\Omega d^3x\, f(f-1)|\Psi|^2, \tag{33}$$

its behavior in time can be calculated using Equations (32), (4), and (10):

$$\frac{dL_f}{dt} = \int_\Omega d^3x \left[-\nabla \cdot \left(\frac{|\Psi|^2}{m}(f^2 - f)\nabla S\right) + \frac{\alpha}{2}(2f-1)\nabla \cdot (|\Psi|^2 \nabla f)\right] \tag{34}$$

$$= \frac{\alpha}{2} \int_\Omega d^3x \left[\nabla \cdot \left[(2f-1)|\Psi|^2 \nabla f\right] - 2(\nabla f)^2 |\Psi|^2\right] \tag{35}$$

$$= -\alpha \int_\Omega d^3x\, (\nabla f)^2 |\Psi|^2, \tag{36}$$

which is of course strictly negative, for all t, as long as ∇f and $|\Psi|^2$ are not identically zero. Hence, if $|\Psi|^2$ is not zero throughout Ω, L_f will decrease monotonically for as long as f is not (identically) equal to 1 on Ω, and therefore necessarily converges to 0, a value it can only attain when $f \equiv 1$ or, equivalently, when $P \equiv |\Psi|^2$. We have thus established a strong H-theorem showing that, in the case of Nelson dynamics, any probability distribution P necessarily converges to $|\Psi|^2$, if the latter does not become zero identically. Note that this excludes the case of a free particle for which $\lim_{t \to +\infty} |\Psi(x, t)|^2 = 0$, for all x, which means that $\frac{dL_f}{dt}$ tends to zero even when f does not converge to 1.

A result, similar to the above, is also easily established for H_V since L_f dominates the latter, or alternatively from the formula

$$\frac{dH_V}{dt} = -\frac{\alpha}{2} \int_\Omega d^3x\, (\nabla f)^2 \frac{|\Psi|^2}{f}. \tag{37}$$

The above results show that (excluding the case of the free particle) Nelson dynamics, naturally, exhibits relaxation towards quantum equilibrium and that it does so for general initial probability distributions (at least, as long as the initial distribution is smooth enough). In this stochastic setting, there is therefore no need for any assumptions on the microstructure of the initial distributions, nor is there any need for the coarse-grained hypothesis when deriving an H-theorem.

Note that these results also show that we have, in fact, convergence of the distribution P to the quantum equilibrium distribution $|\Psi|^2$ in the L^1 norm. This is a consequence of the so-called Csiszár-Kullback-Pinsker inequality [43]:

$$L_1 \leq \sqrt{2H_V} \tag{38}$$

where

$$L_1 = \int_\Omega d^3x \, \big|P - |\Psi|^2\big|. \tag{39}$$

This generalizes the results by Petroni and Guerra [22,23] obtained in their study of the relaxation towards quantum equilibrium in the framework of the Nelson dynamics of a single particle in a harmonic potential. The L^1 norm is also used by Efthymiopoulos et al. [25] in the context of the dBB theory.

We shall illustrate these results by means of numerical simulations for the case of a ground state for the 1D-harmonic oscillator in Section 5.4, for the case of the 2D-harmonic oscillator in Section 6.2, and in the case of a coherent state in Section 7.1.

A last important remark concerns the influence of possible zeros in the equilibrium distribution $\Psi(\mathbf{x},t)$, which would give rise to singularities in the osmotic velocity terms in the Ito equation, Equation (16), or in the Fokker–Planck equation, Equation (18) (or equivalently in Equation (32)), and might make the functions H_V and L_f ill-defined. In Section 6.2, we discuss the case of the first excited state of the 1D-harmonic oscillator, for which $\Psi(\mathbf{x},t)$ has a node at $x = 0$. One could in fact imagine studying higher excited states for which one has a finite number of nodes. In that case, the osmotic velocity of Equation (15) will have simple poles at a finite number of positions in x. At the level of the Ito equation, one would not expect a finite set of poles to cause any particular problems, not only because the probability of hitting a pole exactly in the stochastic evolution is zero but also because the osmotic term tends to move the particle away from the pole very quickly. Similarly, a finite number of simple poles in the convection-diffusion equation, Equation (32), for f only influence the velocity field in the convection term in a finite number of distinct places, and it is to be expected that this would have the effect of actually enhancing the mixing of information in the system.

Moreover, it is also clear that simple nodes in $\Psi(\mathbf{x},t)$ only give rise to (a finite number of) logarithmic singularities in the integrand of H_V and that the integral in Equation (27) therefore still converges. The H-theorem for H_V derived above is thus still valid, and an arbitrary distribution P (sufficiently smooth) will still converge to quantum equilibrium, even in the presence of nodes for $\Psi(\mathbf{x},t)$. The same cannot be said, however, of the function L_f, as simple zeros in $\Psi(\mathbf{x},t)$ give rise to double poles in the integrand and a possible divergence of the integral of Equation (28). Hence, at the beginning of the evolution, for an arbitrary P, the function L_f might take an infinitely large value (the integrand only diverges when $|\Psi|^2 \ll P$, i.e., when it is positive) but as soon as convergence sets in (which is guaranteed by the H-theorem for H_V), the divergent parts in its integrand will be smoothed out and the function L_f will take finite values that converge to zero as time goes on. Of course, when calculating these quantities for the results of numerical simulations, there is always some amount of coarse-graining going on and genuine infinities never occur.

5. Relaxation to Quantum Equilibrium and Nelson Dynamics: The Static Case

In this section, in order to simplify the discussion, we will only consider the case of stationary states $\Psi_{st}(x)$ for the one dimensional Schrödinger equation, i.e., energy levels for which $S = -E\,t$ and which therefore have zero Bohm velocity (11): $\boldsymbol{\nabla} S \equiv S_x = 0$.

5.1. Fokker–Planck Operator and a Formal Connection to a Schrödinger Equation

There exists a wide literature [45,46] concerning a particular method for studying the convergence of solutions of the Fokker–Planck equation to a stationary one, which is only sporadically mentioned in the literature devoted to Nelson dynamics [47]. This approach makes it possible to quantify very precisely the speed of convergence to equilibrium, in terms of (negative) eigenvalues of the Fokker–Planck operator. In order to show this, let us rewrite the Fokker–Planck Equation (18) in terms of the Fokker–Planck operator $\widehat{\mathcal{L}}$:

$$\frac{\partial P}{\partial t} = \widehat{\mathcal{L}} P = \left[-\frac{\partial \gamma}{\partial x} - \gamma(x)\frac{\partial}{\partial x} + \frac{\alpha}{2}\frac{\partial^2}{\partial x^2} \right] P \tag{40}$$

where (15):

$$\gamma(\mathbf{x}) = \alpha \frac{\left(|\Psi_{st}|\right)_x}{|\Psi_{st}|}. \tag{41}$$

Note that, due to the presence of the first derivative $\frac{\partial}{\partial x}$, the $\widehat{\mathcal{L}}$ operator is not Hermitian.

Now, in order to establish the H-theorem, we must prove that in the long-time limit this equation tends to a stationary solution $P_{st} = |\Psi_{st}|^2$. The key idea here is to transform the Fokker–Planck equation to a simple diffusion equation through the transformation

$$P(x,t) = \sqrt{P_{st}(x)}\; g(x,t), \tag{42}$$

under which the r.h.s. of Equation (40) reduces to

$$\widehat{\mathcal{L}}\, P = \sqrt{P_{st}(x)}\; \widehat{\mathcal{H}}_{st}\; g(x,t) \tag{43}$$

where $\widehat{\mathcal{H}}_{st}$ is now a Hermitian operator:

$$\widehat{\mathcal{H}}_{st} = \frac{\alpha}{2}\frac{\partial^2}{\partial x^2} - \frac{1}{2}\left(\frac{\partial \gamma}{\partial x} + \frac{\gamma^2}{\alpha} \right). \tag{44}$$

The function $g(x,t)$ thus obeys a "Schrödinger-like" equation (though with imaginary time) with an effective potential ($\widehat{\mathcal{H}}_{st}$) that depends on $\gamma(x)$:

$$\frac{\partial g(x,t)}{\partial t} = \widehat{\mathcal{H}}_{st}\; g(x,t). \tag{45}$$

Note that the effective potential is exactly the Bohm-quantum potential defined by

$$Q_\Psi = -\frac{\hbar^2}{2m}\frac{1}{|\Psi_{st}|}\frac{\partial^2 |\Psi_{st}|}{\partial x^2}, \tag{46}$$

which can be expressed in terms of the osmotic velocity (41) as

$$\frac{Q_\Psi}{m\,\alpha} = -\frac{1}{2}\left(\frac{\partial \gamma}{\partial x} + \frac{\gamma^2}{\alpha} \right). \tag{47}$$

5.2. Superposition Ansatz

We can now represent the solution of Equation (45) as a superposition of discrete eigenvectors (all orthogonal, as the operator $\widehat{\mathcal{H}}_{st}$ is Hermitian) and impose the superposition ansatz [48]:

$$g(x,t) = \sum_{k=0}^{\infty} a_k(t)\, g_k(x). \tag{48}$$

Equation (45) is separable and gives rise to the eigenvalue problem:

$$\frac{1}{a_k(t)}\frac{da_k(t)}{dt} = \frac{1}{g_k(x)}\widehat{\mathcal{H}}_{st}\, g_k(x) = -\lambda_k. \tag{49}$$

As a result, we have

$$g(x,t) = \sum_{k=0}^{\infty} a_k e^{-\lambda_k t}\, g_k(x) \tag{50}$$

for a set of constants a_k and where all the λ_k are real (as $\mathcal{H}$ is Hermitian), for eigenfunctions $g_k(x)$ that satisfy the orthonormality conditions:

$$\int_{-\infty}^{\infty} dx\, g_k(x) g_l(x) = \delta_{k,l}. \tag{51}$$

Thus, we have the expression

$$P(x,t) = \sum_{k=0}^{\infty} a_k e^{-\lambda_k t} \sqrt{P_{st}(x)}\, g_k(x). \tag{52}$$

By construction, the function $\sqrt{P_{st}(x)}$ is an eigenstate of the effective Hamiltonian with energy 0. We shall associate the label λ_0 with this energy level.

In order to have a well defined probability distribution and to avoid any divergence in time, it is clear that all eigenvalues $-\lambda_k$ have to be negative, which requires Ψ_{st} to be the ground state of the effective Hamiltonian $\mathcal{H}_{st}$: just as in the case of the usual Schrödinger equation, the eigenvalues $-\lambda_k$ in Equation (49) are all negative only if $\Psi_{st}(x)$ has no zeros (see also [49], Appendix 2, for an elementary proof that all λ_k are indeed positive if $\Psi_{st}(x)$ h as no zeros).

If $\Psi_{st}(x)$ does have zeros, the osmotic velocity will have singularities. In [49] (Appendix 2), we consider what happens in the case when $\Psi_{st}(x)$ is an excited state of the harmonic oscillator and we derive a formal solution in terms of the eigenvalues $-\lambda_k$, which are now not all negative, thus revealing the appearance of instabilities for cases where the above formalism would still be valid.

5.3. One-Dimensional Oscillator and the Evolution of Gaussian Distributions for the Ground State

In [49] (Appendix 2), we discuss the application of the method of the effective Hamiltonian outlined in Section 5.1 to this particular problem, and we derive a Green function for the associated Fokker–Planck equation when Ψ_{st} is the ground state of the one-dimensional oscillator. This Green function $K_P(x, x', t)$ is defined through

$$P(x,t) = \int_{-\infty}^{\infty} dx'\, P(x',0)\, K_P(x,x',t) \tag{53}$$

where the kernel K_P is given by

$$K_P(x,x',t) = \left(\frac{a}{\pi\, sinh(\omega\, t)}\right)^{\frac{1}{2}} e^{\omega\left(n+\frac{1}{2}\right)t} \times e^{\frac{-a}{sinh(\omega\, t)}\left[(x^2+x'^2)\, cosh(\omega\, t)+(x^2-x'^2)\, sinh(\omega\, t)-2\, xx'\right]}. \tag{54}$$

An important property of the Green function for this case is that, if $|\Psi(x)|^2$ and $P(x,0)$ are Gaussian, then $P(x,t)$ will still be Gaussian (53). Let us define the ground state as

$$|\Psi_{st}|^2 \equiv |\Psi(x)|^2 = \sqrt{\frac{2a}{\pi}} e^{-2\,a\,x^2}, \tag{55}$$

for which we can then write

$$P(x,t) = \sqrt{\frac{2b(t)}{\pi}} e^{-2\,b(t)\,(x-\langle x(t)\rangle\,)^2}. \tag{56}$$

Injecting Equation (56) in the Fokker–Planck equation, Equation (40), gives a differential equation for $\langle x(t)\rangle$,

$$\frac{d\,\langle x(t)\rangle}{dt} = -2a\alpha\,\langle x(t)\rangle\,, \tag{57}$$

which is readily solved as

$$\langle x(t)\rangle = \langle x_0\rangle\, e^{-2a\alpha\, t}, \tag{58}$$

as well as an equation for $b(t)$,

$$\frac{1}{2b(t)}\frac{db(t)}{dt} + 2\alpha\,(b(t)-a) = 0 \tag{59}$$

with solution

$$b(t) = \frac{a}{1-\left(1-\frac{a}{b_0}\right)e^{-4a\alpha\, t}}. \tag{60}$$

From Equations (56) and (60), we can then calculate the width of the non-equilibrium Gaussian as

$$\begin{aligned} \sigma_x^2(t) \equiv \frac{1}{4b(t)} &= \frac{1}{4a}\left[\left(1-e^{-4a\alpha\, t}\right)+\frac{a}{b_0}e^{-4a\alpha\, t}\right] \\ &= \sigma_{eq}^2\left(1-e^{-4a\alpha\, t}\right)+\sigma_x^2(0)\,e^{-4a\alpha\, t} \end{aligned} \tag{61}$$

where σ_{eq}^2 represents the width $1/(4a)$ of the equilibrium distribution of Equation (55).

Clearly, $\langle x\rangle \overset{t\to\infty}{=} \langle x\rangle_{eq} = 0$ with a characteristic relaxation time inversely proportional to the diffusion coefficient α. Moreover,

$$\frac{d\sigma_x(t)}{dt} \propto 4a\alpha\,(\sigma_{eq}^2-\sigma_x^2(0))\,e^{-4a\alpha\, t}, \tag{62}$$

which has the same sign as that of the difference $(\sigma_{eq}-\sigma_x(0))$. Hence, $\sigma_x(t)$ converges monotonically to the equilibrium value σ_{eq}, with a characteristic time inversely proportional to the diffusion coefficient α, as can be seen in Figure 2.

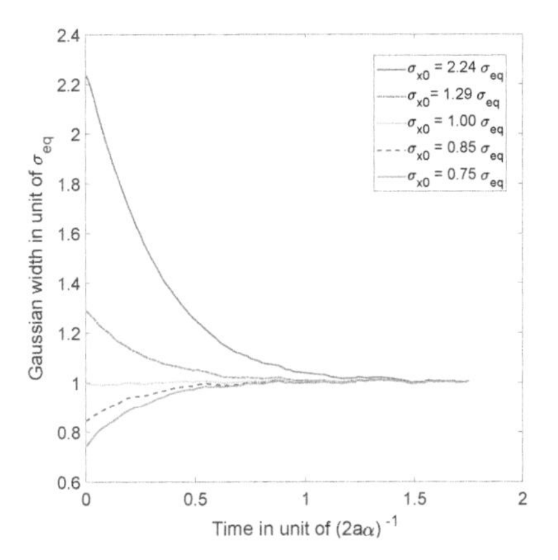

Figure 2. Simulations of 10,000 trajectories (calculated from the Ito equation, Equation (16), for the ground state (Equation (55)) of the 1D harmonic oscillator), whose initial positions are normally distributed, for 5 different choices of distribution width (for $a = 0.5$ and $\alpha = 1$). We observe, in each case, convergence to the equilibrium of Equation (55) as predicted by the theory.

5.4. *Ergodicity in the Relaxation to Quantum Equilibrium for the Ground State of the Harmonic Oscillator*

We have just shown how Gaussian initial distributions converge towards quantum equilibrium, but one could also ask the same question for non-Gaussian initial distributions. Convergence is guaranteed by the H-theorem of Section 4, but contrary to the Gaussian case, we have no clear measure for the rate of convergence, except for the entropy-like functions H_V (27) and L_f (28), or the L_1 norm (39), defined in Section 4. The evolution, in time, of these three quantities is shown in Figure 3, for the stochastic trajectories obtained from 20,000 uniformly distributed initial conditions. The relaxation towards quantum equilibrium is clearly visible in all three quantities. As expected, the convergence of H_V is extremely fast. Note that, although initially very large, L_f quickly matches L_1, up to numerical fluctuations.

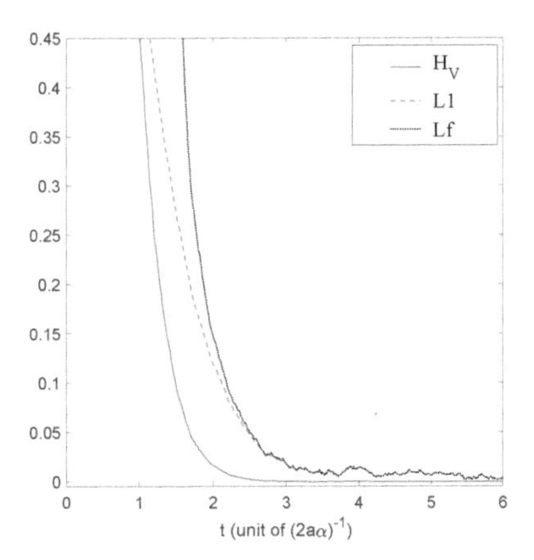

Figure 3. Time evolution of H_V, L_f and the L_1 norm, for a uniform initial probability distribution, calculated from the Ito equation, Equation (16), for the ground state of the 1D harmonic oscillator. Relaxation towards the distribution of the ground state $|\Psi_{st}|^2$ of Equation (55) is clearly visible. The simulation was performed for $\alpha = 1$, $a = 0.5$, and $\Delta t = 0.01$, for 20,000 uniformly distributed initial conditions.

One important question concerning this relaxation process is of course that of possible ergodicity. Since we want to study the ergodic properties of Nelson dynamics in a numerical way, we choose the definition of ergodicity that is, in our approach, the easiest to test. Defining the time average $\widehat{h}$ of a function h on Ω by the limit (if it exists),

$$\widehat{h} = \lim_{t\to+\infty} \frac{1}{t} \int_0^t h(\mathbf{x}_{t'})\, dt' \tag{63}$$

(where $\mathbf{x}_{t'}$ represents the position of a particle at time t', as obtained form the Ito stochastic differential equation, Equation (16), for an initial condition $\mathbf{x}$) we say [50] that the corresponding stochastic process is ergodic if the time average of any bounded function h on Ω is always independent of $\mathbf{x}$. Since for bounded h the time average is also invariant under shifts in time, we can say that we have ergodicity if all time averages of such functions are in fact constants. The main reason for choosing this particular definition is that it is well-suited to empirical testing, since it is of course sufficient to establish constancy of the time averages for all indicator functions χ_A of arbitrary (measurable) sets $A \subset \Omega$, for the analogous property to ensue automatically for all bounded functions on Ω. Another reason for choosing this particular definition is that it is also applicable to non-stationary stochastic processes, as in the case of the coherent state of Section 7.

More precisely, we need to verify that

$$\widehat{\chi}_A = \lim_{t\to+\infty} \frac{1}{t} \int_0^t \chi_A(\mathbf{x}_{t'})\, dt' \tag{64}$$

is independent of both t and $\mathbf{x}$, for any measurable $A \subset \Omega$. Remember that one has of course that $\chi_A(\mathbf{x}_t) = \chi_{\phi_t^{-1}A}(\mathbf{x})$, where $\phi_t^{-1}A = \{\mathbf{x} \in \Omega \,|\, \mathbf{x}_t \in A\}$.

In the present case, i.e., that of the Nelson dynamics defined by the stationary (ground) state of the 1D harmonic oscillator, it is clear that the distribution $|\Psi_{st}|^2$ obtained from the ground state eigenfunction Ψ_{st} is a stationary solution to the associated Fokker–Planck equation, Equation (18). This distribution provides a natural invariant measure μ on Ω: $d\mu = |\Psi_{st}|^2 dx$, for which $\int_\Omega d\mu = 1$ and

$$\mu(A) = \int_A |\Psi_{st}|^2 dx = \mu(\phi_t^{-1}A), \quad {}^\forall t > 0, {}^\forall A \in \Omega. \tag{65}$$

If a stationary stochastic process is ergodic, i.e., if for such a process all $\widehat{\chi}_A$ are indeed constants, the values of these constants are simply the measures of the subsets A [51]. Therefore, when one needs to decide whether or not a stationary stochastic process is ergodic, it suffices to verify that $\widehat{\chi}_A = \mu(A)$, for any $A \in \Omega$.

The usual way to check this condition is to consider sampling time averages for a sufficiently refined "binning" of Ω. Starting from a particular initial particle position $\mathbf{x}$, we calculate the trajectory $\mathbf{x}_t$ that follows from the Ito stochastic equation, Equation (16), for a sufficiently long time t. As was explained for the coarse-graining in Section 3, the configuration space Ω is subdivided into a large number of non-overlapping cells or "bins" A_k ($k = 1, \ldots, N_b$), each with the same volume $\Delta\mathbf{x}$. The trajectory $\mathbf{x}_{t'}$ ($t' \in [0, t]$) is then sampled at regular intervals Δt, yielding $N+1$ sample positions $\mathbf{x}_{n\Delta t}$ ($n = 0, \ldots, N$), for $N = t/\Delta t$. We then define the sampling function $\varphi_{N,k}$

$$\varphi_{N,k} = \frac{1}{N} \sum_{n=0}^{N} \chi_{A_k}(\mathbf{x}_{n\Delta t}), \tag{66}$$

which is a discretization of $\frac{1}{t}\int_0^t \chi_A(\mathbf{x}_{t'})\, dt'$ in Equation (64) and which gives the frequency with which the (sample of the) orbit visited the bin A_k. Hence, if in the limit $N \to +\infty$, for diminishing bin sizes $\Delta\mathbf{x}$ and sampling steps Δt, the normalized distribution obtained from $\varphi_{N,k}/\Delta\mathbf{x}$ tends to a constant distribution (and, in particular, does not depend on the initial positions $\mathbf{x}$), then the stochastic process is ergodic according to the above definition.

Moreover, since in that case $\widehat{\chi}_{A_k} = \mu(A_k)$, this normalized distribution must in fact coincide with that for the invariant measure for the stationary process. For example, in the case at hand, if the normalized distribution we obtain is indeed independent of the initial positions, then since $\mu(A_k) = |\Psi_{st}(x)|^2\big|_{x=\xi} \Delta x$ for some point $\xi \in A_k$, we must have that for sufficiently large N

$$\frac{\varphi_{N,k}}{\Delta x} \approx \frac{\mu(A_k)}{\Delta x} = |\Psi_{st}(x)|^2\big|_{x=\xi}, \tag{67}$$

i.e., the empirical distribution obtained from this sampling time average must coincide with the stationary quantum probability $|\Psi_{st}|^2$. This is exactly what we obtain from our numerical simulations, as can be seen from the histograms depicted in Figure 4. After a certain amount of time, the histograms we obtain indeed converge to the equilibrium distribution, and this for arbitrary initial positions. The convergence clearly improves if we increase the integration time, or if we diminish the spatial size of the bins (while diminishing the sampling time step in order to keep the occupancy rate of each bin high enough). Although purely numerical, we believe this offers conclusive proof for the ergodicity of the Nelson dynamics associated with the ground state of the harmonic oscillator in one dimension.

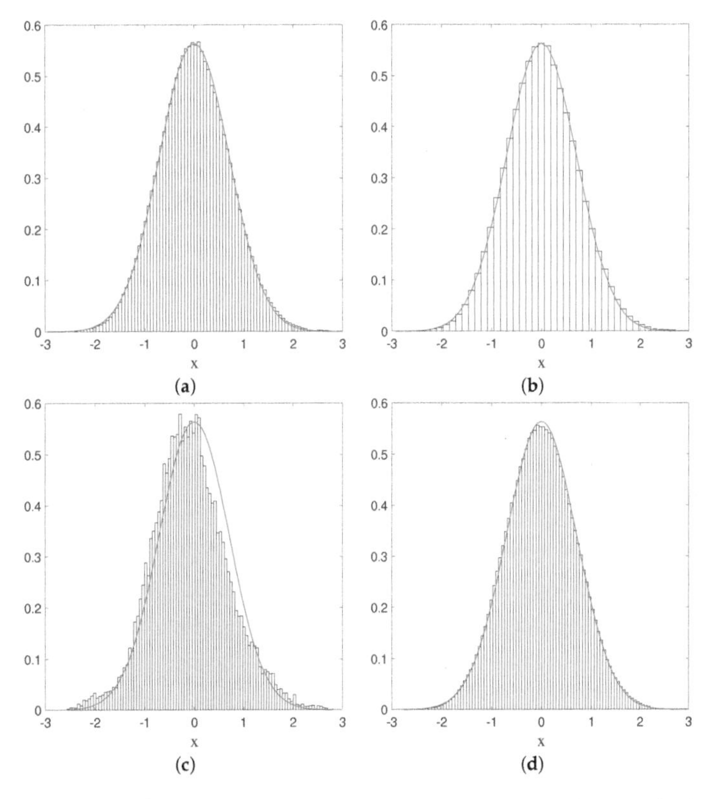

Figure 4. Histograms of the positions of a single particle, subject to Nelson dynamics for the ground state of the 1D harmonic oscillator. The full (red) curve corresponds to the quantum probability $|\Psi_{st}|^2$. Here, $a = 0.5$, $\alpha = 1$, and the total simulation time (t = 10,000) is sampled with $\Delta t = 0.01$. (**a**) The initial particle position is $x_0 = 2.5$, and the number of bins $N_b = 100$ (each with spatial size $\Delta x = 0.0635$); (**b**) Same as (**a**) but with $N_b = 50$ and $\Delta x = 0.1270$; (**c**) Same as (**a**) but with $t = 200$; (**d**) Same as (**a**) but for $x_0 = -0.85$.

The same can be said, in fact, for the 2-dimensional oscillator which will be the main topic of Section 6 (and even for the 2D corral as can be seen in [49]). Some results of a simulation of a single trajectory under the Nelson dynamics for the ground state of this system are shown in Figure 5, in which the red dot in the plot on the left-hand side indicates the (final) position of the particle at time t. The probability distribution obtained by sampling the trajectory clearly decreases with the distance to the origin along concentric circles.

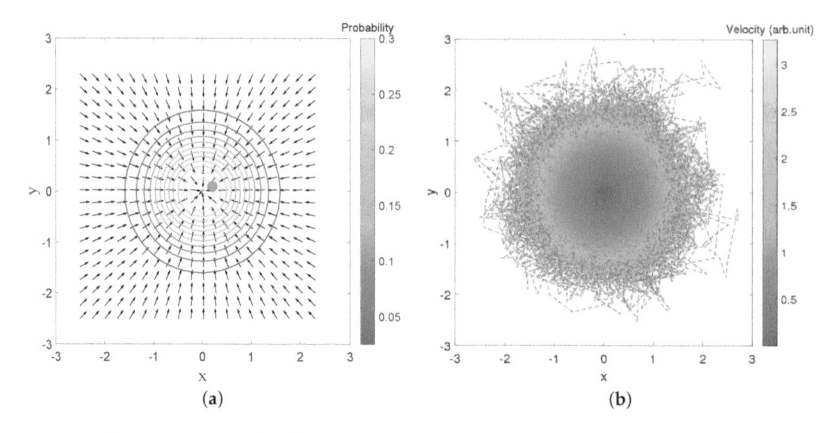

Figure 5. (**a**) A point-particle (the dot near the center) subject to the osmotic velocity field $-2a\alpha(x(t), y(t))$, due to the ground state of the 2D harmonic oscillator at time t; (**b**) Color plot of the velocities along a trajectory for the evolution under Nelson dynamics, for the ground state of the 2D harmonic oscillator. The simulation (for $a = 0.5$ and $\alpha = 1$) started from the initial position $(-2, 1)$ and was sampled up to $t = 1000$ with step $\Delta t = 0.01$.

6. Nelson Dynamics: A Phenomenological Dynamical Model for Walkers?

6.1. 2D Harmonic Oscillator

Experimentally, it has proven possible to study the dynamics of bouncing droplets under the influence of an effective harmonic potential in two dimensions, thanks to a well-chosen electro-magnetic configuration and magnetic droplets [52]. It is therefore interesting to compare predictions that we, on our side, can make in the framework of Nelson dynamics, with actual experimental observations of droplets dynamics (see [53] for a pioneering work very similar to ours in the case of the double slit experiment). We think that an important comparison to make concerns the convergence to equilibrium.

For example, if the initial distribution of positions projected along a reference axis, say X, fits a mixture of the ground state and the nth Fock state ($n = 1, 2 \cdots$) (Appendix 2, [49]) for the 2D harmonic oscillator (conveniently weighted in order to respect the ineluctable constraint of positivity), our Nelson-like model predicts that the typical time of convergence to equilibrium will scale like the inverse of the eigenvalue of the nth Fock state, i.e., as $1/n$, which constitutes a very precise quantitative prediction. This follows from Equation (52), when $\sqrt{P_{st}(x)}$ is the Gaussian ground state of the 1D harmonic oscillator and where the eigenfunctions g_k are the energy eigenstates (Fock states) of the harmonic oscillator (this, of course, because of the separability of the Schrödinger equation and of our Nelson dynamics along X and Y in the case of an isotropic 2D oscillator).

A possible way to measure this characteristic time would be to record the projections along X of trajectories that correspond to an equally spaced grid of initial positions, weighted so as to fit a mixture of the ground state with the nth Fock state ($n = 1, 2 \cdots$), and to compare the histogram constructed in this way at different times with theoretical predictions derived from Equation (52).

Another precise quantitative (theoretical) prediction, which is even simpler to verify, is that if we prepare a droplet many times at exactly the same initial position, the position obtained after averaging over all trajectories will (1) decrease exponentially in time and (2) be characterized by a decay time which scales like $1/a\alpha$, by virtue of the discussion in Section 5.3 and particularly Equation (58). It has been suggested that droplet trajectories might be characterized by a quantum-like Zitterbewegung, which can be seen in relativistic quantum dynamics as an intrinsic Brownian motion at the Compton scale [54,55], and various proposals have been formulated in order to express the amplitude and frequency of this Zitterbewegung [5,56] in terms of the parameters characterizing droplet dynamics (these are, e.g., the viscosity of the fluid, the mass of the droplets, the ratio between the amplitude of the vibrations imposed on the bath and the Faraday threshold, the oil temperature [5,57], and so on). Exploring these analogies in depth lies beyond the scope of this paper, but the aforementioned attempts ([56] in particular) pave the way for introducing a Brownian component in the description of droplet trajectories.

6.2. Presence of Zeros in the Interference Pattern

One of our first motivations, when we decided to incorporate a Brownian component in the dBB theory in order to simulate the dynamics of droplets, was the pioneering paper [4] reporting on observations of a walker trapped in a spherical 2D cavity (corral), for which the histogram of positions occupied over time by a single droplet trajectory faithfully reproduces the Bessel function J_0 (this is also related to the Green function of the Helmholtz equation, with a typical length equal to the Faraday wave length of the vibrating bath over which droplets propagate [16]). These observations reveal, in a telling way, the presence of a pilot-wave that guides the dynamics of the particles, and raise the question of ergodicity.

If we try the approach we used for the 2D harmonic oscillator in the case of the corral (effectively replacing the Gaussian ground state of the 2D harmonic oscillator by the zero order Bessel function), we are immediately confronted with problems caused by the presence of zeros in the Bessel function. In particular, the eigenvalues $-\lambda_k$ of the Fokker–Planck operator in Equation (49) are not always negative when zeros are present, which of course would menace the stability of the relaxation process. However, as we already indicated in Section 4, although the effect of zeros of the pilot wave in our Nelson dynamics is by no means trivial, there are several observations that indicate that this problem is not crucial.

First of all, as mentioned in Section 4, the Wiener process makes it in principle possible to "jump" over the zeros of the equilibrium distribution. This has actually been confirmed in numerical simulations for the case of the 1D harmonic oscillator, where we imposed that the equilibrium distribution P_{st} is the square modulus of the first excited (Fock) state ([49], Appendix 2A), with the following amplitude:

$$P_{st} = |\Psi_{st}|^2 = |\Psi_1(x,t)|^2 = \left(\frac{2a}{\pi}\right)^{\frac{1}{2}} \left(a\,x^2\right) e^{-2ax^2}. \tag{68}$$

Indeed, as can be clearly seen from Figure 6, the particle will, from time to time, jump over the zero in the middle (with the same probability from left to right as in the opposite direction), in such a way that finally the trajectory covers the full real axis, while the histogram of positions faithfully reproduces the quantum prediction $P_{st} = |\Psi_{st}|^2 = |\Psi_1(x,t)|^2$. This indicates that, even in the presence of a zero in the equilibrium distribution, the relaxation process is still ergodic.

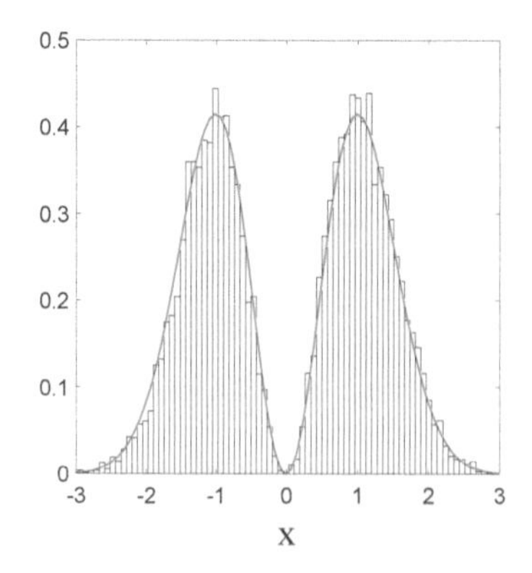

Figure 6. Histogram of the positions in x of a single particle, in the case of the first Fock state given by Equation (68). The full curve (red) corresponds to the quantum probability $|\Psi_1|^2$. Here, $a = 0.5$ and $\alpha = 1$. The total simulation time t is $t = 1000$, and the sampling time step is $\Delta t = 0.01$. The initial position is $x_i = 1$, and the number of bins $N_b = 75$, each with width $\Delta x = 0.08$.

The relaxation of a uniform initial distribution to this quantum equilibrium is shown in Figure 7, for the quantities H_V, L_f, and L_1.

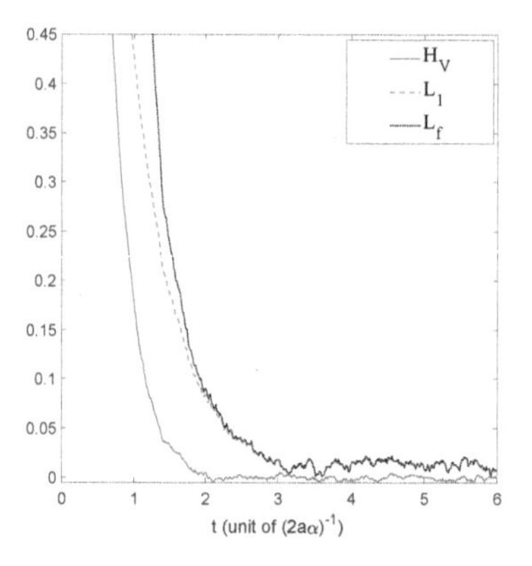

Figure 7. Evolution in time of H_V (27), L_f (28), and of the L_1 (39) norm, for a uniform initial probability distribution, showing the relaxation towards the distribution of the first excited state $|\Psi_1|^2$ (68). The simulation was performed for $\alpha = 1$, $a = 0.5$, and $\Delta t = 0.01$ and from 20,000 uniformly distributed initial conditions.

A second indication that the problem posed by the presence of zeros is not so serious stems in fact from the experimental observations. Indeed, if we study the observations reported in [4] for the case of a corral, it is clear that the minima of the histogram expressing the distribution of positions of the droplet are in fact not zeros. This, undoubtedly, is due to the presence of a non-negligible residual background. Without this background, the droplet would never pass between regions separated by zeros: due to the rotational symmetry of the corral, the zeros form circles centered at the origin and the position histogram obtained from a trajectory would remain confined to a torus comprising the initial position. This, however, is clearly not the case, which thus suggests the following strategy: to simulate Nelson dynamics with a static distribution $P_{st} = |\Psi_{st}|^2$ given by the Bessel function J_0 but supplemented with a constant positive background ϵ:

$$d\mathbf{x}(t) = \frac{\alpha}{2}\frac{\nabla J_0(r)^2}{J_0(r)^2+\epsilon}\,dt + \sqrt{\alpha}d\mathbf{W}(t). \tag{69}$$

In this case, the singularities of the Fokker–Planck equation automatically disappear and, despite the fact that we have no analytic expression for the solutions as in the case of the ground state of the harmonic oscillator, we are able to numerically simulate Nelson dynamics without difficulty. The results of these simulations are shown in Figure 8. The osmotic velocity in the Nelson dynamics clearly tends to bring back the particle to regions where $|\Psi|^2$ has extrema and the resemblance with the plot on the left is striking. The fact that this result again does not depend on the choice of initial condition strongly suggests that the relaxation process to quantum equilibrium is also ergodic in this case.

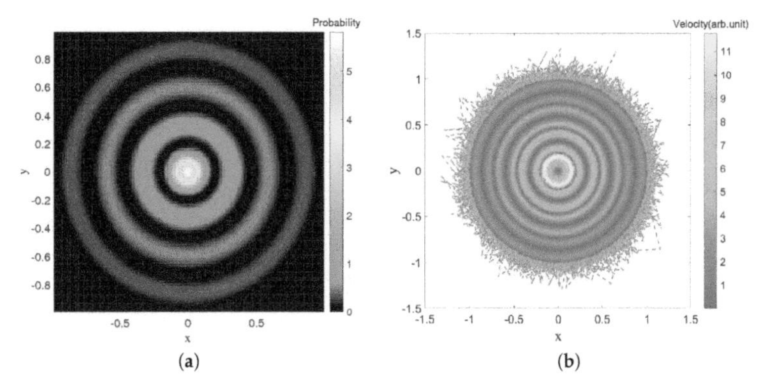

Figure 8. (**a**) The quantum probability associated to the Bessel function of the first kind J_0; (**b**) Color plot of the velocities reached along the trajectory for an evolution corresponding to (69). The initial position was $(1,1)$, the simulation time $t = 5000$, and the sampling time step $\Delta t = 0.005$. We chose $\alpha = 0.1$ and $\epsilon = 0.2$, and the size of the domain is $L = 2$. On the boundary we impose a harmonic field force of the form $-2a\alpha\,\mathbf{r}$.

7. Relaxation to Quantum Equilibrium with dBB and Nelson Dynamics: The Non-Static Case

7.1. Nelson Dynamics and Asymptotic Coherent States

Up to now, we have developed analytic and numerical tools aimed at studying the onset of equilibrium when the asymptotic equilibrium distribution is static. As the H-theorem of Section 4 is also valid for non-stationary processes, one of course expects relaxation to take place even if the asymptotic state is not static, for instance, when it is a Gaussian distribution, the center of which periodically oscillates at the classical frequency ω of the oscillator without deformation (typical for coherent states). In fact, our numerical simulations show not only that equilibrium is reached even in this case, but also that this relaxation is ergodic.

More precisely, we considered a wave function in the coherent state

$$\Psi(x,t) = \left(\frac{2a}{\pi}\right)^{\frac{1}{4}} e^{-a\,(x-\bar{x}_t)^2 + i\frac{\bar{p}_t x}{\hbar} + i\varphi(t)} \tag{70}$$

where φ is a global phase containing the energy, and $\bar{x}_t$ ($\bar{p}_t$) is the mean position (momentum) of a classical oscillator at time t:

$$\bar{x}_t = \bar{x}_0 \cos(\omega t) \quad \text{and} \quad \bar{p}_t = -m\bar{x}_0 \sin(\omega t), \tag{71}$$

with $\omega = 2a\alpha$ ($\alpha = \hbar/m$). For this ansatz we solved the Ito equation, Equation (16), numerically for a collection of initial conditions.

As can be seen in Figure 9, the trajectories are affected by the stochastic evolution but keep oscillating at the same period because of the deterministic part of the Ito process. Notice, however, that the trajectories seem to be approaching classical trajectories that only differ from each other by a simple shift. This can be explained as follows: at equilibrium (cf. Figure 10), the Brownian motion is balanced by the osmotic velocity and the dBB velocity is recovered "on average." Now, the center of the Gaussian distribution moves at a classical velocity by virtue of Ehrenfest's theorem; moreover, in the present case, the dBB velocities can only depend on time and not on space as the envelope of a coherent state moves without deformation. Hence, the dBB trajectories obtained at equilibrium are, in fact, classical trajectories that only differ by a mere shift in space (the magnitude of which, however, may change over time).

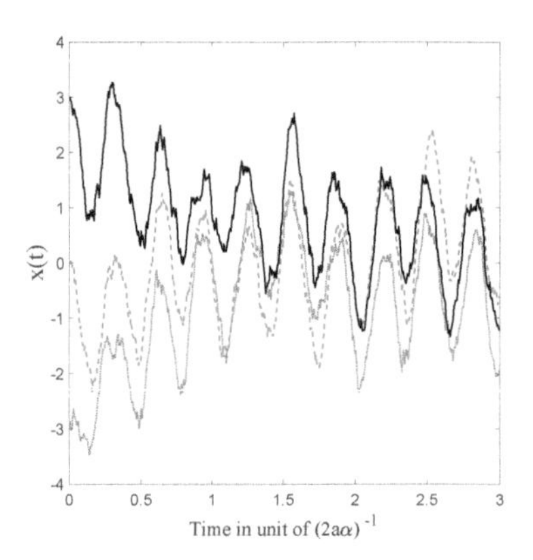

Figure 9. Numerical solutions of the Ito stochastic differential equation, Equation (16), corresponding to the coherent state of Equation (70), for three different initial conditions. We used $\bar{x}_0 = 1$, $a = 0.5$, and $\alpha = 1$ and expressed the results in natural units.

Secondly, as can be clearly seen on Figure 10, even for a uniform initial probability distribution, the convergence to the quantum equilibrium is remarkably fast and the converged distribution faithfully follows the oscillating motion of the non-stationary equilibrium distribution. The remarkable speed of the convergence to quantum equilibrium is corroborated by the decay of the functions H_V and L_f and of the L_1 norm shown in Figure 11.

Moreover, Figure 12 depicts the sampling time average (as defined in Section 5.4) of a single trajectory for this non-stationary stochastic process. The convergence of the sampling distribution to a static distribution $\Phi(x)$, described by the integral of $|\Psi(x,t)|^2$ as given by Equation (70), over a period of the oscillation

$$\Phi(x) = \frac{\omega}{2\pi}\int_0^{2\pi/\omega} |\Psi(x,t)|^2 dt \tag{72}$$

is striking. As the asymptotic distribution $\Phi(x)$ does not depend on the choice of initial condition, we conclude that the relaxation to equilibrium for the non-stationary stochastic process associated with Nelson dynamics for the coherent state (70) is ergodic (in the sense explained in Section 5.4).

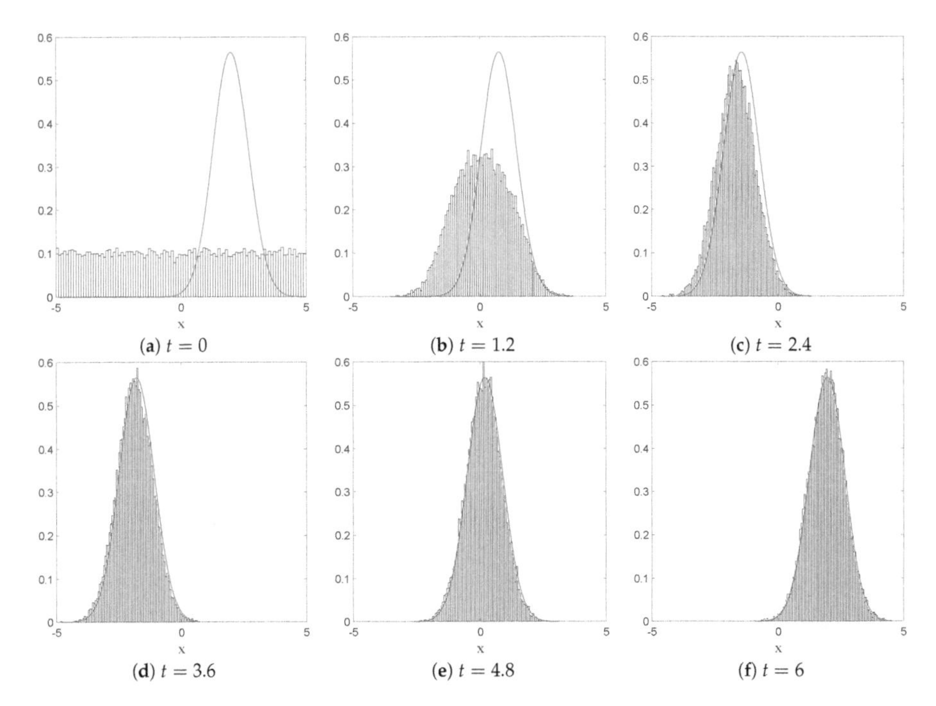

Figure 10. The time evolution of a non-equilibrium ensemble, illustrated with position histograms at six different times ((**a**): $t = 0$, (**b**): $t = 1.2$, (**c**): $t = 2.4$, (**d**): $t = 3.6$, (**e**): $t = 4.8$, (**f**): $t = 6$). The continuous curve is the squared modulus $|\Psi|^2$ for the coherent state of Equation (70). As can be seen in (**d–f**), once equilibrium is reached, the distribution clings to the coherent state and follows its oscillation faithfully. The center of the wave packet moves between -2 and 2 with a period 2π. We started from a uniform distribution of initial conditions and chose $a = 0.5$, $\alpha = 1$, and $x_0 = 2$. The sampling time step is $\Delta t = 0.01$, and the number of bins is $N_b = 50$, each with width $\Delta x = 0.0461$.

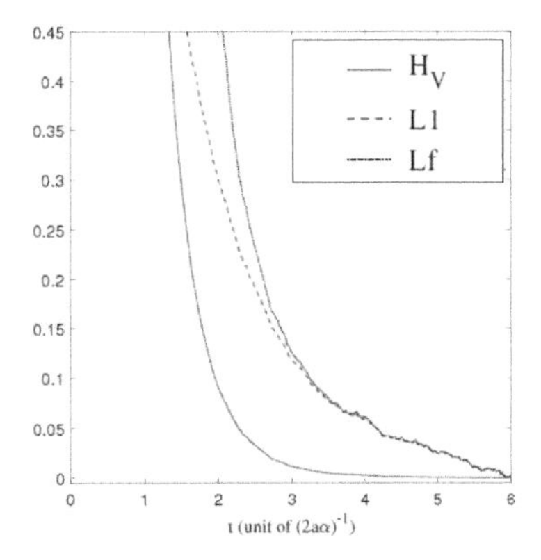

Figure 11. Time evolution of H_V (27), L_f (28), and L_1 (39), for a uniform initial probability distribution, showing the relaxation towards the distribution $|\Psi|^2$ of the coherent state of Equation (70). The simulation was performed for $\alpha = 1$, $a = 0.5$, and $\Delta t = 0.01$ and from 20,000 uniformly distributed initial conditions.

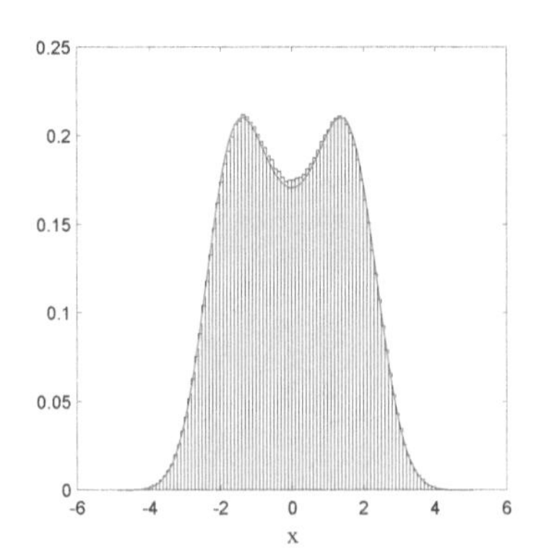

Figure 12. Histogram of the positions for a unique trajectory satisfying the Ito equation, Equation (16), for Equation (70). The full curve corresponds to the integration of $|\Psi|^2$ over one period. The center of the wave packet moves between -2 and 2 with a period 2π. Here, $a = 0.5$ and $\alpha = 1$. Total simulation time t is $t = 30,000$ and the sampling time step is $\Delta t = 0.01$. The initial position is $x_i = 1$, and the number of bins $N_b = 100$, each with width $\Delta x = 0.1$.

7.2. Onset of Equilibrium with a Dynamical Attractor in dBB Dynamics and Nelson Dynamics

If one wants to investigate the onset of equilibrium in dBB dynamics, one obviously has to consider non-static asymptotic distributions since in static cases the dBB dynamics freezes the trajectories (as the phase of the wave function is then position-independent). Even in the case of a coherent state (see Section 7.1), the distribution of dBB positions would never reach equilibrium because it moves as a whole (as the shape of a coherent state remains the same throughout time). In a sense, coherent states behave as solitary waves. Moreover, the absence of zeros in the wave function might explain why mixing does not occur. In Figure 13 we show the result of simulations of dBB trajectories in the case of a 2D harmonic oscillator for a quantum state consisting of a superposition of equally weighted products of states along X and Y, chosen among M energy (Fock) states ([49], Appendix 2A), with randomly chosen initial phases θ_{n_x,n_y}:

$$\Psi(x,y,t) = \frac{1}{\sqrt{M}} \sum_{n_x=0}^{\sqrt{M}-1} \sum_{n_y=0}^{\sqrt{M}-1} e^{i\,\theta_{n_x,n_y} - i\omega\left(n_x+n_y+1\right)t}\, \psi_{n_x}(x)\, \psi_{n_y}(y)\,. \tag{73}$$

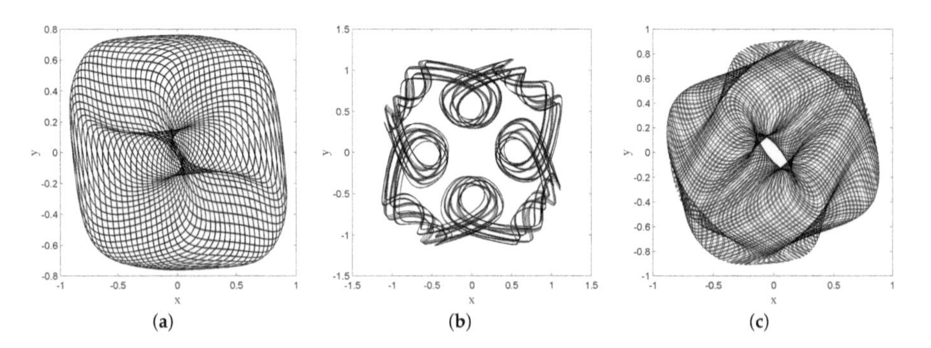

Figure 13. Plots showing three possible de Broglie–Bohm (dBB) trajectories for a single point particle in the case of (73) with $M = 2^2 = 4$. Each plot (**a**–**c**) is associated with different initial random phases (θ_{n_x,n_y} with n_x (n_y) taking the values 0, 1) and different initial positions.

We then compared the relaxation process for dBB with the quantum thermostat given by Nelson dynamics for $M = 4^2 = 16$ energy states. The results are shown in Figure 14 in which H_V (for the dBB and for the Nelson dynamics) and L_1 (for both the dBB and Nelson dynamics) are plotted at the (same) coarse-grained level. We started from a uniform distribution of positions; we took $\alpha = 0.1$. In both cases, the position distributions $\mathcal{P}$ and P converge to $|\Psi|^2$. Moreover, we recover an exponential decay for $\overline{H}_V$, as already observed in [26], even in the absence of stochastic (Brownian) noise *à la* Nelson. However, we observe that the convergence to equilibrium occurs faster in the presence of the quantum thermostat.

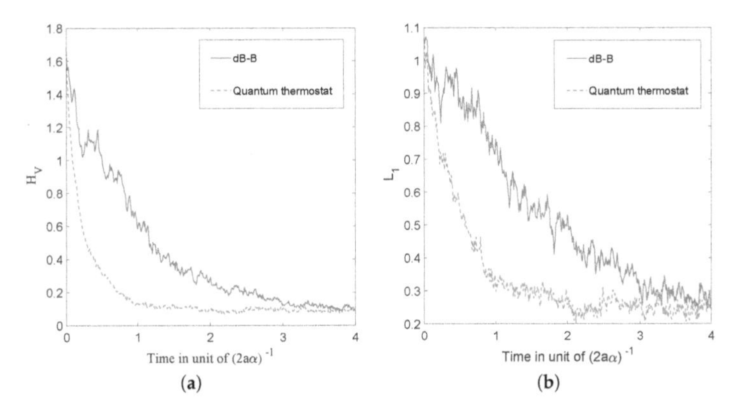

Figure 14. Plots of the evolution in time of the coarse-grained H-functions H_V (**a**) and L_1 (**b**) for the Nelson and dBB dynamics. The full line corresponds to the dBB dynamics and the dashed line corresponds to the quantum thermostat. We started from 10,000 initial positions uniformly distributed in a box of size 10×10; we chose $a = 0.5$, $\alpha = 0.1$, and $M = 4^2 = 16$ energy states.

8. Dynamical Model for Droplets and Double Quantization of the 2-D Harmonic Oscillator

In this section we shall focus on the description of droplets dynamics as described in [9,52], for a magnetized droplet moving in an isotropic 2-D harmonic potential. We shall show that dBB dynamics allows us to reproduce some of the main features of the experimental observations. In [9,52], it is reported that stable structures appear in the droplets dynamics whenever a double quantisation condition is satisfied. The Hamiltonian of the isotropic 2-D harmonic oscillator being invariant under rotations, we may indeed impose a double quantisation constraint, requiring that the energy states of the 2D quantum harmonic oscillator are also eigenstates of the angular momentum. In polar coordinates, these states (which are parameterized by two quantum numbers, the energy number n and the magnetic number m) are expressed as follows [58]:

$$\psi_{n,m}(r,\theta,t) = \sqrt{\frac{a}{\pi}\frac{k!}{(k+|m|)!}}\, e^{-\frac{ar^2}{2}} \left(\sqrt{a}\,r\right)^{|m|} \mathcal{L}_k^{|m|}\left[a\,r^2\right] e^{-i\omega(n+1)t+im\theta} \tag{74}$$

where $\mathcal{L}_k^{|m|}$ are the generalized Laguerre polynomials and $k = \frac{n-|m|}{2}$. Note that these solutions are linear combinations of the product of Fock states in x and y.

A first experimental result reported in [9] is the following: trajectories are chaotic and nearly unpredictable unless the spring constant of the harmonic potential takes quantized values that are strongly reminiscent of energy quantization (under the condition that, during the experiment, the size of the orbits is fixed once and for all). For quantized energies—in our case given by $E_n = (n+1)\hbar\omega$, for some "effective" value of $\hbar$ to be determined from actual experiments—stable orbits, to which one can attribute yet another quantum number, appear, this time for the angular momentum, which is strongly reminiscent of the magnetic number (the eigenvalue of the orbital momentum, perpendicular to the surface of the vessel, is given by the product of $\hbar$ and m). In [9] it is shown, for instance, that for the first excitation ($n = 1$, $m = \pm 1$) droplet orbits are circular or oval, turning clockwise or anti-clockwise depending on the sign of m. At the second energy level ($n = 2$, $m = -2, 0, +2$), ovals appear again for $m = \pm 2$ and lemniscates for an average value of the angular momentum $<m> = 0$. At the fourth energy level ($n = 4$, $m = -4, -2, 0, 2, 4$), trefoils appear (for $m = \pm 2$).

We simulated dBB trajectories, always considering a superposition of one of the aforementioned doubly quantized eigenstates $\psi_{n,m}$ with the ground state:

$$\Psi(r,\theta,t) = \xi_0 e^{-i\varphi_0} \psi_{0,0}(r,\theta,t) + \sum_{j=0}^{n} \xi_{j+1} e^{-i\varphi_{j+1}} \psi_{n,-n+2j}(r,\theta,t) \tag{75}$$

where φ_j and ξ_j are real numbers with $0 < \xi_0 \ll \xi_{j\neq 0}$. Computing the guidance relation of Equation (11) for a single eigenstate (74), one ends up with a value for ∇S for which the trajectories are circles of radius R around the origin, with tangential velocities proportional to m/R. In particular, the dynamics is frozen when $m = 0$.

Mixing the wave function with the ground state, however, generates a periodic (in time) component in the dBB velocity field, which turns circular orbits into ovals when ξ_0 is small enough, and eventually generates more complex structures, such as "rosaces" instead. We also tuned the energy difference between the ground state and the excited states such that two timescales characterize the dynamics. These are the "centrifugal" period, necessary for drawing a full circle around the origin, which varies as m/R^2, and the "Bohr" period which varies like $T/(n+1)$, where T is the classical period of the oscillator. Tuning these parameters, we were able to simulate dBB trajectories very similar to those reported in [9]. For instance, we found circles and ovals (see Figure 15a,b) for $(n,m) = (1,1)$ and $(n,m) = (2,2)$. Note that the lemniscate cannot be obtained with a superposition of the ground state and the $(n,m) = (2,0)$ state for which dBB velocities are necessarily purely radial, contrary to the suggestion made in [9]. Instead, it should be generated with a superposition of the ground state with $(n,m) = (2,+2), (2,-2)$, and $(2,0)$ in which the weights of the $m = +2$ and -2 components are slightly different (see Figure 15c). Figure 16 shows further detail of the evolution along this trajectory. Tuning the energy, we were also able to generate a trefoil and a "rosace" (see Figure 17).

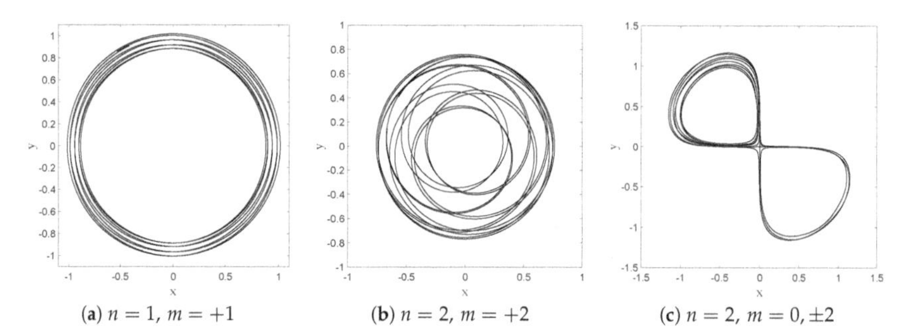

Figure 15. dBB trajectories obtained for a single point particle in a superposition of eigenstates (75). Each plot is associated with a different combination (n,m), as indicated. In the (**a**,**b**) graphs, we imposed $a = 1$ and, respectively, $\omega = 1, \frac{\xi_0}{\xi_2} = 0.05$ and $\omega = 0.5, \frac{\xi_0}{\xi_3} = 0.05$; for (**c**), we imposed $a = 3, \omega = 0.5, \frac{\xi_0}{\xi_3} = 0.0708, \frac{\xi_0}{\xi_2} = 0.0456$, and $\frac{\xi_0}{\xi_1} = 0.0773$.

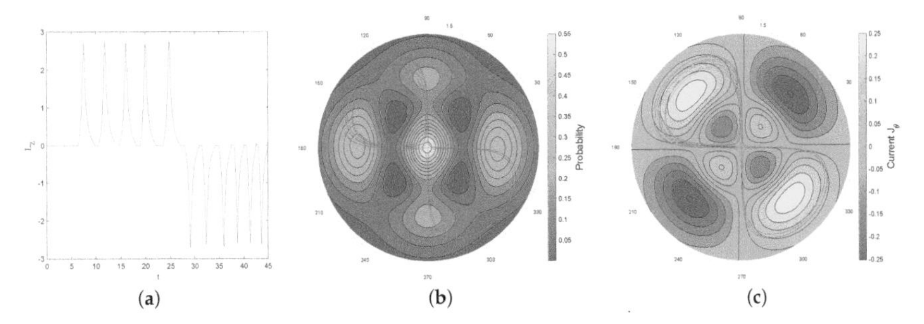

Figure 16. Plots of three quantities associated to the lemniscate in Figure 15c. (**a**) shows the L_z-component of the angular momentum, and the polar plots (**b**,**c**) show the probability density $|\psi|^2$ (**b**) and the θ-component of the probability current (5) along the trajectory (**c**).

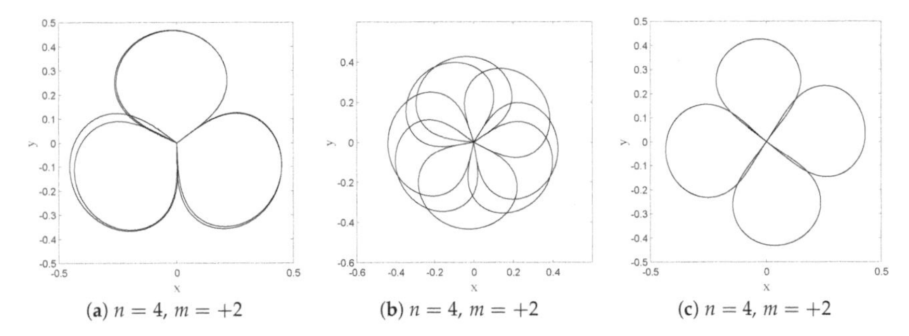

Figure 17. dBB trajectories obtained for a single point particle in a superposition of eigenstates (75). Plots (**a**,**b**) correspond to $\omega = 0,7$ and $\omega = 1$, respectively. Case (**c**) is obtained after multiplying the amplitude of the $(n, m) = (4, 2)$ state by a complex phase ($e^{(0.3i)}$). We took $a = 1$ in all cases.

It is worth noting, however, that chaos is omnipresent in the dBB dynamics for this system, in the sense that the trajectories exhibit an extreme sensitivity to the initial conditions, which explains why these dBB orbits mimicking stable droplets orbits are in general unstable. For instance, Figure 18 shows intermittent transitions between an oval trajectory and a lemniscate (as has also been reported in [9]), for a superposition of the ground state with the $(n, m) = (2, +2), (2, -2)$, and $(2, 0)$ states. Preliminary results furthermore show that the trajectories are also unstable under Nelson dynamics, i.e., in the presence of "noise," whenever this noise (parameterized by α in (16)) exceeds a critical value. Note that many experiments involving droplets are characterized by a lack of stability and predictability. For instance, the appearance of interferences similar to those obtained in a double slit experiment (see [49,59] for a description *à la* Nelson of the double slit experiment) has been attributed to "air currents" in [60]. Therefore, although our approach might not explain every detail of the double quantization reported in [9], it does reproduce many of its essential features, and we believe it would be very interesting to deepen this analogy. For instance, having access to the empirical values of the weights of the ground state, or of the effective values of $\hbar$ and of the mass in the case of droplets [56], would allow us to test our model in real detail.

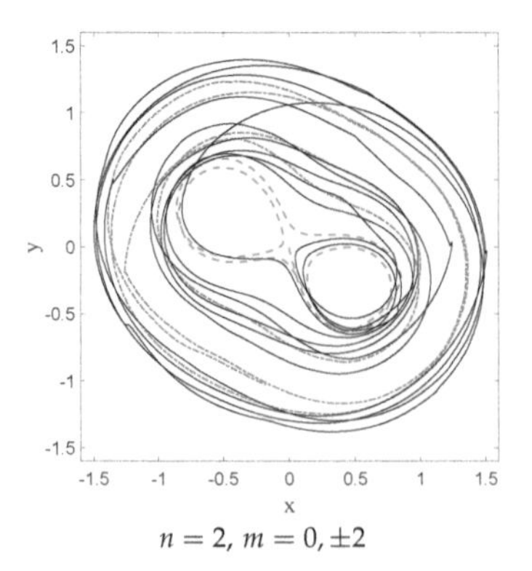

$n = 2$, $m = 0, \pm 2$

Figure 18. dBB trajectories obtained for a single point particle in a superposition of eigenstates (75) (with $m = -2, 0, 2$) showing intermittent transitions between two types of trajectories. The relevant parameter values are $\omega = 0.2$, $a = 1$, $\frac{\xi_0}{\xi_3} = 0.0342$, $\frac{\xi_0}{\xi_2} = 0.2547$, and $\frac{\xi_0}{\xi_1} = 0.0505$.

Another experiment, reported in [61], during which both the position of the droplet and the excitation of the bath are monitored, and where a superposition between two distinct modes of the bath is reported, could also provide more insight and might offer some means to test the validity of our model: using exactly the same observation device, but this time in the case where the droplet undergoes a 2-D isotropic potential, would allow one to check whether the modes of the bath are similar to the (n, m) quantum modes which we associate with the quantized droplets trajectories.

9. Conclusions and Open Questions

In this paper we studied stochastic, Nelson-like dynamics and dBB dynamics, with the aim of simulating the dynamics of droplets. The stochastic approach has the merit that it explicitly takes into account the influence of noise on the dynamics [59,62]. In contrast to experiments where noise is considered to be a parameter that should be minimized, here, noise is considered to be a relevant parameter for the dynamics (see also [53]). For instance, as we have shown, it plays an essential role in the relaxation towards equilibrium and in the ergodicity of the dynamics. In the dBB approach, on the other hand, the main ingredient is the chaotic nature of the dynamics [25]. Both models thus shed a different light on the dynamics and could possibly fit diverse sets of regimes in droplet dynamics. Note that in the limit where the amplitude of the Brownian motion in our Nelson dynamics tends to zero, the dynamics approaches dBB dynamics very closely. In sufficiently complex situations (e.g., when the mixing process due to the presence of zeros in the wave function becomes effective [26,40]), we expect the relaxation to equilibrium to be accompanied by chaotic rather than stochastic dynamics, as one has in Nelson dynamics (although Nelson dynamics with small but non-zero Brownian motion is hard to distinguish from dBB dynamics, it has the advantage that relaxation is guaranteed to occur, even in the absence of coarse graining and/or mixing).

Ultimately, experiments ought to indicate whether it is relevant, with respect to droplet phenomenology, to formalize the dynamical influence of noise à la Nelson (and/or dBB) as we did in the present paper. We have formulated several proposals in this sense in Sections 6.1 and 8. As emphasized throughout the paper, however, our models should be seen as a first step in the direction of a dynamical model, which remains to be formulated, combining Nelson's stochastic dynamics (and/or dBB dynamics) and memory effects. We think that the results of Section 8 show that this is a promising program for future research.

Finally, it is worth recalling some of the problems that arose when first de Broglie and then Bohm and Nelson developed their theories aimed at deriving quantum dynamics (statistics) as an emergent property, i.e., resulting from an underlying "hidden" dynamics.

The most severe problem is undoubtedly non-locality, which was recognized by Bohm [11,12] to be an irreducible feature of dBB dynamics (see also [19,63] for similar conclusions concerning Nelson-type dynamics). Today, under the influence of the work of John Bell [64] and his followers, it is widely recognized that quantum theory is irreducibly non-local, which makes it particularly difficult to mimic using classical models. Note that entanglement and non-locality (as well as decoherence, which is the corollary of entanglement [65]) only appear if we consider more than one particle at a time, which explains why we did not address these fundamental questions in the core of the paper, where a single droplet is described. It would be interesting to enlarge our model such that the presence of the environment can be taken into account. This would require incorporating the description of open quantum systems, for which a generalization of Bohmian dynamics has been developed in the past [66–69], but obviously this is beyond the scope of the present paper.

Another problem concerns the fact that the pilot wave is a complex function. This poses still unresolved problems in the case of Nelson dynamics because Nelson's diffusion process does not make it possible [70,71] to fix the phase of the wave function unequivocally (see [72] for an interesting proposal involving a multivalued wave function, also based on Zitterbewegung). In our approach, which is mainly of quantum inspiration, complex wave functions and imaginary phases appear spontaneously, but if we wish to scrutinize the link with the empirically observed modes at the surface of oil baths [9,52,56,61], it will be important to interpret the exact meaning of this complex phase. In the framework of his double solution program [73,74] de Broglie, and others, showed how to derive the Schrödinger equation from a Klein–Gordon equation in the non-relativistic limit. This is only possible provided the real wave bounces at an extremely high frequency (of the order of mc^2/h). A similar approach has been proposed in the context of droplets phenomenology in [75], where a complex Schrödinger equation is derived from the Klein–Gordon equation along these lines. Although such (interesting and promising) alternative studies of droplets solve the problem of the appearance of a complex phase in a classical context, it is worth noting that the phenomenological results outlined in Section 8, concerning the quantization of droplet orbits in the case of a harmonic potential [9,52], cannot be explained simply in terms of excited modes of the oil bath, because in these experiments only the droplet undergoes the harmonic potential, the oil bath being electromagnetically neutral. This difficulty actually concerns any classical model in which droplet dynamics is formulated in terms of classical modes of the bath only.

To conclude, in our view, the programs that aim at simulating droplet dynamics with quantum tools or at describing the emergence of quantum dynamics based on droplet dynamics, are still largely incomplete and raise challenging fundamental questions. This Pandora's box is now open and it will not be closed any time soon, but this is not something to be feared as it offers new and stimulating perspectives for future research in the field.

Author Contributions: Conceptualization, M.H., R.W., S.C. and T.D.; Formal Analysis, M.H., R.W. and T.D.; Funding Acquisition, R.W., S.C. and T.D.; Investigation, M.H. and R.W.; Methodology, M.H., R.W. and T.D.; Project Administration, R.W. and T.D.; Software, M.H. and S.C.; Supervision, T.D.; Validation, M.H. and R.W.; Visualization, M.H. and R.W.; Writing—Original Draft Preparation, M.H., R.W., S.C. and T.D.; Writing—Review & Editing, R.W. and T.D.

Funding: This research was funded by the John Templeton foundation (grant 60230).

Acknowledgments: The authors gratefully acknowledge funding and support from the John Templeton foundation (grant 60230, Non-Linearity and Quantum Mechanics: Limits of the No-Signaling Condition, 2016–2019) and an FQXi Physics of What Happens grant (Quantum Rogue Waves as Emergent Quantum Events, 2015–2017). RW would also like to thank FQXi for support through a mini-grant (Grant number FQXi-MGA-1819), which enabled him to elucidate crucial aspects of the relaxation to quantum equilibrium. SC also thanks the Foundational Questions Institute (fqxi.org) for its support though a mini-grant (Grant number FQXi-MGA-1705). T.D. also thanks Aurélien Drézet (CNRS-Grenoble) for drawing our attention to the paper of Kyprianidis and for useful comments on the draft of our paper during T.D.'s personal visit to Grenoble (3 May 2018). Sincere thanks to G. Grössing for his kind invitation to contribute to this special issue.

Conflicts of Interest: The authors declare no conflict of interest.

Appendix A. Numerical Simulations

Firstly, we discuss the case of the dBB dynamics. It is assumed that we have an analytical solution of the Schrödinger equation $\Psi(t, \mathbf{x})$. We want to compute the evolution of a given initial non-equilibrium density $\mathcal{P}(t_i, \mathbf{x})$ up to a final time t_f and for intermediate time events (we denote all these events by t_k, with $t_0 = t_i$ and $t_f = t_K$). In particular, we are interested in the coarse-grained non-equilibrium density

$$\overline{\mathcal{P}}(\mathbf{x}, t_k) = \frac{1}{\epsilon^3} \int_{\text{CG cell} \ni \mathbf{x}} d^3x \mathcal{P}(\mathbf{x}, t_k), \tag{A1}$$

which is defined in Equation (23).

Numerically, we replace that integral by a discrete sum over a finite set of points $\mathbf{x}^l$, which are uniformly distributed over the CG cells. In order to obtain the value of each $\mathcal{P}(\mathbf{x}^l, t_k)$ we use the Liouville relation

$$\frac{\mathcal{P}(\mathbf{x}^l, t_k)}{|\Psi(\mathbf{x}^l, t_k)|^2} = \frac{\mathcal{P}(\mathbf{x}_i^l, t_i)}{|\Psi(\mathbf{x}_i^l, t_i)|^2} \tag{A2}$$

where $\mathbf{x}_i^l$ is the position of the particle, which, when evolved according to Equation (11) from t_i up to t_k, gives $\mathbf{x}^l$.

In order to obtain $\mathbf{x}_i^l$ for each $\mathbf{x}^l$, we consider the time-reversed evolution with wave-function $\Psi^*(-t, \mathbf{x})$ and initial condition $\mathbf{x}^l$ at time $-t_k$. The position $\mathbf{x}^l$, if time evolved from $-t_k$ up to $-t_i$ according to Equation (11), will give the position $\mathbf{x}_i^l$. As there is usually no analytical solution of Equation (11), we use a Runge–Kutta (RK) algorithm [76] to obtain a numerical estimate of the position $\mathbf{x}_i^l$. In order to know if we can trust the result of the RK algorithm, we perform two realizations of the algorithm with different choices of a so-called tolerance parameter (the smaller the value of that tolerance parameter, the more precise the computation), say γ and $10^{-1}\gamma$. If the distance between the two positions is less than some chosen value δ, the result of the last iteration of the RK algorithm is trusted. Otherwise, we perform another iteration with $10^{-2}\gamma$ and we compare it to the previous realization of the RK algorithm. We repeat the procedure until the constraint on the distance between the two successive results of the RK algorithm is satisfied, or until we reach some minimal value of the tolerance parameter. In that case, the position $\mathbf{x}^l$ is considered as a bad position and it is discarded from the numerical integration of Equation (23). This method was used in [26].

That is one possible method but we could also adopt a more brute-force method: Randomly generate a set of N initial positions according to $\mathcal{P}(t_i, \mathbf{x})$ and let them evolve according to an Euler algorithm (that is, we divide the time-interval in small time-steps of length Δt and we increment the position by $\mathbf{v}(t)\Delta t$ at each time step). We record the positions of the N particles for each value of t_k, we count the number of particles in each CG cell for each time t_k (say n_{CG}), and we divide n_{CG} by N in order to define $\overline{\mathcal{P}}(\mathbf{x}, t_k)$. The first method turns out to be more efficient in the case of the dBB dynamics but it is not applicable in the presence of stochastic terms.

In the case of Nelson dynamics, we used the Euler–Maruyama method for stochastic processes to approximate the solution of the Ito equation, Equation (12). In the same way as Euler's method, the time T is divided into N small discrete time steps Δt. For each time t_i, we generated a random variable normally distributed $\Delta W_i = \sqrt{\Delta t}\mathcal{N}(0, 1)$. The integration scheme has the form

$$x_{i+1} = x_i + v(x_i, i\,\Delta t)\Delta t + \sqrt{\alpha}\,\Delta W_i. \tag{A3}$$

We invite the reader interested in the details to consult [77]. The remaining question is how to choose the time step Δt so that one can trust the result of the numerical simulations. One way to do this is the following. We know that the Born distribution remains invariant under Nelson's dynamics (equivariance). We therefore start with some value for Δt and decrease it until the result of the numerical simulation confirms this theoretical prediction. We then perform the numerical simulation for the non-equilibrium distribution with the value of Δt thus obtained.

References

1. Couder, Y.; Fort, E. Single-Particle Diffraction and Interference at a Macroscopic Scale. *Phys. Rev. Lett.* **2006**, *97*, 15410. [CrossRef] [PubMed]
2. Couder, Y.; Boudaoud, A.; Protière, S.; Fort, E. Walking droplets, a form of wave-particle duality at macroscopic scale? *Europhys. News* **2010**, *41*, 14–18. [CrossRef]

3. Couder, Y.; Protière, S.; Fort, E.; Boudaoud, A. Dynamical phenomena: Walking and orbiting droplets. *Nature* **2005**, *437*, 208. [CrossRef] [PubMed]
4. Harris, D.M.; Moukhtar, J.; Fort, E.; Couder, Y.; Bush, J.W. Wavelike statistics from pilot-wave dynamics in a circular corral. *Phys. Rev. E* **2013**, *88*, 011001. [CrossRef] [PubMed]
5. Bush, J.W.M. Pilot-wave hydrodynamics. *Annu. Rev. Fluid Mech.* **2015**, *47*, 269–292. [CrossRef]
6. Couder, Y.; Fort, E. Probabilities and trajectories in a classical wave-particle duality. *J. Phys. Conf. Ser.* **2012**, *361*, 012001. [CrossRef]
7. Bush, J.W.M. The new wave of pilot-wave theory. *Phys. Today* **2015**, *68*, 47–53. [CrossRef]
8. Eddi, A.; Sultan, E.; Moukhtar, J.; Fort, E.; Rossi, M.; Couder, Y. Information stored in Faraday waves: The origin of a path memory. *J. Fluid Mech.* **2011**, *674*, 433–463. [CrossRef]
9. Perrard, S.; Labousse, M.; Miskin, M.; Fort, E.; Couder, Y. Self-organization into quantized eigenstates of a classical wave-driven particle. *Nat. Commun.* **2014**, *5*, 3219. [CrossRef] [PubMed]
10. Durt, T. Do dice remember? *Int. J. Theor. Phys.* **1999**, *38*, 457–473. [CrossRef]
11. Bohm, D. A Suggested Interpretation of the Quantum Theory in Terms of "Hidden" Variables. I. *Phys. Rev.* **1952**, *85*, 166–179. [CrossRef]
12. Bohm, D. A Suggested Interpretation of the Quantum Theory in Terms of "Hidden" Variables. II. *Phys. Rev.* **1952**, *85*, 180–193. [CrossRef]
13. Nelson, E. *Dynamical Theories of Brownian Motion*; Princeton University Press: Princeton, NJ, USA, 1967; Volume 131, pp. 2381–2396.
14. Labousse, M. Etude D'une Dynamique à Mémoire de Chemin: une Expérimentation Théorique. Ph.D. Thesis, Université Pierre et Marie Curie UPMC Paris VI, Paris, France, 2014. (In French)
15. Fort, E.; Eddi, A.; Boudaoud, A.; Moukhtar, J.; Couder, Y. Path-memory induced quantization of classical orbits. *Proc. Natl. Acad. Sci. USA* **2010**, *107*, 17515–17520. [CrossRef]
16. Dubertrand, R.; Hubert, M.; Schlagheck, P.; Vandewalle, N.; Bastin, T.; Martin, J. Scattering theory of walking droplets in the presence of obstacles. *New J. Phys.* **2016**, *18*, 113037. [CrossRef]
17. Tadrist, L.; Shim, J.B.; Gilet, T.; Schlagheck, P. Faraday instability and subthreshold Faraday waves: surface waves emitted by walkers. *arXiv* **2017**, arXiv:1711.06791.
18. Bohm, D.; Vigier, J.P. Model of the causal interpretation of quantum theory in terms of a fluid with irregular fluctuations. *Phys. Rev.* **1954**, *96*, 208–216. [CrossRef]
19. Bohm, D.; Hiley, B. Non-locality and locality in the stochastic interpretation of quantum mechanics. *Phys. Rep.* **1989**, *172*, 93–122. [CrossRef]
20. Kyprianidis, P. The Principles of a Stochastic Formulation of Quantum Theory. *Found. Phys.* **1992**, *22*, 1449–1483. [CrossRef]
21. Valentini, A. Signal locality, uncertainty and the subquantum H-theorem. I. *Phys. Lett. A* **1991**, *156*, 5–11. [CrossRef]
22. Petroni, N.C. Asymptotic behaviour of densities for Nelson processes. In *Quantum Communications and Measurement*; Springer: New York, NY, USA, 1995; pp. 43–52.
23. Petroni, N.C.; Guerra, F. Quantum Mechanical States as Attractors for Nelson Processes. *Found. Phys.* **1995**, *25*, 297–315. [CrossRef]
24. Guerra, F. Introduction to Nelson Stochastic Mechanics as a Model for Quantum Mechanics. In *The Foundations of Quantum Mechanics—Historical Analysis and Open Questions*; Garola, C., Rossi, A., Eds.; Springer: Dordrecht, The Netherlands, 1995; pp. 339–355.
25. Efthymiopoulos, C.; Contopoulos, G.; Tzemos, A.C. Chaos in de Broglie–Bohm quantum mechanics and the dynamics of quantum relaxation. *Ann. Fond. Louis Broglie* **2017**, *42*, 133–160.
26. Valentini, A.; Westman, H. Dynamical origin of quantum probabilities. *Proc. R. Soc. A* **2005**, *461*, 253–272. [CrossRef]
27. Colin, S.; Struyve, W. Quantum non-equilibrium and relaxation to quantum equilibrium for a class of de Broglie–Bohm-type theories. *New J. Phys.* **2010**, *12*, 043008. [CrossRef]
28. Towler, M.; Russell, N.J.; Valentini, A. Time scales for dynamical relaxation to the Born rule. *Proc. R. Soc. A* **2011**, *468*, 990–1013. [CrossRef]
29. Colin, S. Relaxation to quantum equilibrium for Dirac fermions in the de Broglie–Bohm pilot-wave theory. *Proc. R. Soc. A* **2012**, *468*, 1116–1135. [CrossRef]

30. Contopoulos, G.; Delis, N.; Efthymiopoulos, C. Order in de Broglie–Bohm quantum mechanics. *J. Phys. A Math. Theor.* **2012**, *45*, 165301. [CrossRef]
31. Abraham, E.; Colin, S.; Valentini, A. Long-time relaxation in the pilot-wave theory. *J. Phys. A Math. Theor.* **2014**, *47*, 395306. [CrossRef]
32. De la Peña, L.; Cetto, A.M. *The Quantum Dice: An Introduction to Stochastic Electrodynamics*; Springer: Dordrecht, The Netherlands, 2013.
33. De la Peña, L.; Cetto, A.M.; Valdés Hernández, A. *The Emerging Quantum*; Springer: Basel, Switzerland, 2015.
34. Deotto, E.; Ghirardi, G.C. Bohmian mechanics revisited. *Found. Phys.* **1998**, *28*, 1–30. [CrossRef]
35. De Broglie, L. Interpretation of quantum mechanics by the double solution theory. *Ann. Fond. Louis Broglie* **1987**, *12*, 399–421.
36. Bacciagaluppi, G. Nelsonian mechanics revisited. *Found. Phys. Lett.* **1999**, *12*, 1–16. [CrossRef]
37. de Broglie, L. La mécanique ondulatoire et la structure atomique de la matière et du rayonnement. *J. Phys. Radium* **1927**, *8*, 225–241. (In French) [CrossRef]
38. Valentini, A. On the Pilot-Wave Theory of Classical, Quantum and Subquantum Physics. Ph.D Thesis, Scuola Internazionale Superiore di Studi Avanzati (SISSA), Trieste, Italy, 1992.
39. Valentini, A. Signal locality, uncertainty and the subquantum H-theorem. II. *Phys. Lett. A* **1991**, *158*, 1–8. [CrossRef]
40. Efthymiopoulos, C.; Kalapotharakos, C.; Contopoulos, G. Origin of chaos near critical points of quantum flow. *Phys. Rev. E* **2009**, *79*, 036203. [CrossRef] [PubMed]
41. Valentini, A. Inflationary cosmology as a probe of primordial quantum mechanics. *Phys. Rev. D* **2010**, *82*, 063513. [CrossRef]
42. Underwood, N.G.; Valentini, A. Quantum field theory of relic nonequilibrium systems. *Phys. Rev. D* **2015**, *92*, 063531. [CrossRef]
43. Jüngel, A. *Entropy Methods for Partial Differential Equations*; Springer: New York, NY, USA, 2016.
44. Bricmont, J. Bayes, Boltzmann and Bohm: Probabilities in Physics. In *Chances in Physics*; Bricmont, J., Ghirardi, G., Dürr, D., Petruccione, F., Galavotti, M.C., Zanghi, N., Eds.; Springer: Berlin/Heidelberg, Germany, 2001; pp. 3–21.
45. Gardiner, C.W. *Handbook of Stochastic Processes*; Springer: New York, NY, USA, 1985.
46. Risken, H. Fokker–Planck Equation. In *The Fokker–Planck Equation*; Risken, H., Frank, T., Eds.; Springer: Berlin/Heidelberg, Germany, 1996; pp. 63–95.
47. Petroni, N.C.; De Martino, S.; De Siena, S. Exact solutions of Fokker–Planck equations associated to quantum wave functions. *Phys. Lett. A* **1998**, *245*, 1–10. [CrossRef]
48. Brics, M.; Kaupuzs, J.; Mahnke, R. How to solve Fokker–Planck equation treating mixed eigenvalue spectrum? *Cond. Matter Phys.* **2013**, *16*, 13002. [CrossRef]
49. Hatifi, M.; Willox, R.; Colin, S.; Durt, T. Bouncing oil droplets, de Broglie's quantum thermostat and convergence to equilibrium. *arXiv* **2018**, arXiv:quant-ph/1807.00569v3.
50. Gray, R.M. *Probability, Random Processes, and Ergodic Properties*; Springer: New York, NY, USA, 2009.
51. Arnold, V.; Avez, A. *Problèmes Ergodiques de la Méchanique Classique*; Gauthier-Villars: Paris, France, 1967. (In French)
52. Labousse, M.; Oza, A.U.; Perrard, S.; Bush, J.W. Pilot-wave dynamics in a harmonic potential: Quantization and stability of circular orbits. *Phys. Rev. E* **2016**, *93*, 033122. [CrossRef] [PubMed]
53. Grössing, G. Sub-quantum thermodynamics as a basis of emergent quantum mechanics. *Entropy* **2010**, *12*, 1975–2044. [CrossRef]
54. Hestenes, D. The Zitterbewegung Interpretation of Quantum Mechanics. *Founds. Phys.* **1990**, *20*, 1213–1232. [CrossRef]
55. Colin, S.; Wiseman, H.M. The zig-zag road to reality. *J. Phys. A Math. Theor.* **2011**, *44*, 345304. [CrossRef]
56. Gilet, T. Quantumlike statistics of deterministic wave-particle interactions in a circular cavity. *Phys. Rev. E* **2016**, *93*, 042202. [CrossRef] [PubMed]
57. Bechhoefer, J.; Ego, V.; Manneville, S.; Johnson, B. An experimental study of the onset of parametrically pumped surface waves in viscous fluids. *J. Fluid Mech.* **1995**, *288*, 325–350. [CrossRef]
58. Cohen-Tannoudji, C.; Diu, B.; Laloe, F. *Quantum Mechanics*; Wiley: New York, NY, USA, 1977.
59. Hatifi, M.; Lopez-Fortin, C.; Durt, T. de Broglie's double solution: Limitations of the self-gravity approach. *Ann. Fond. Louis Broglie* **2018**, *43*, 63–90.

60. Pucci, G.; Harris, D.M.; Faria, L.M.; Bush, J.W.M. Walking droplets interacting with single and double slits. *J. Fluid Mech.* **2018**, *835*, 1136–1156. [CrossRef]
61. Sáenz, P.J.; Cristea-Platon, T.; Bush, J. Statistical projection effects in a hydrodynamic pilot-wave system. *Nature* **2017**, *14*, 3.
62. Durt, T. L. de Broglie's double solution and gravitation. *Ann. Fond. Louis Broglie* **2017**, *42*, 73–102.
63. Nelson, E. Review of stochastic mechanics. *J. Phys. Conf. Ser.* **2012**, *361*, 012011. [CrossRef]
64. Bell, J. On the Einstein–Podolsky–Rosen paradox. *Physics* **1964**, *1*, 195–200. [CrossRef]
65. Durt, T. Characterisation of an entanglement-free evolution. *arXiv* **2001**, arXiv:quant-ph/0109112.
66. Kostin, M. On the Schrödinger-Langevin Equation. *J. Chem. Phys.* **1972**, *57*, 3589–3591. [CrossRef]
67. Nassar, A. Fluid formulation of a generalised Schrodinger-Langevin equation. *J. Phys. A Math. General* **1985**, *18*, L509. [CrossRef]
68. Olavo, L.; Lapas, L.; Figueiredo, A. Foundations of quantum mechanics: The Langevin equations for QM. *Ann. Phys.* **2012**, *327*, 1391–1407. [CrossRef]
69. Nassar, A.B.; Miret-Artés, S. *Bohmian Mechanics, Open Quantum Systems and Continuous Measurements*; Springer: Basel, Switzerland, 2017.
70. Bacciagaluppi, G. A Conceptual Introduction to Nelson's Mechanics. In *Endophysics, Time, Quantum and the Subjective*; Buccheri, R., Saniga, M., Elitzur, A., Eds.; World Scientific: Singapore, 2005; pp. 367–388.
71. Wallstrom, T. Inequivalence between the Schrödinger equation and the Madelung hydrodynamic equations. *Phys. Rev. A* **1994**, *49*, 1613–1617. [CrossRef] [PubMed]
72. Derakhshani, M. A Suggested Answer to Wallstrom's Criticism (I): Zitterbewegung Stochastic Mechanics. *arXiv* **2015**, arXiv:quant-ph/1510.06391.
73. Bacciagaluppi, G.; Valentini, A. *Quantum Theory at the Crossroads: Reconsidering the 1927 Solvay Conference*; Cambridge University Press: Cambridge, UK, 2010.
74. Colin, S.; Durt, T.; Willox, R. de Broglie's double solution program: 90 years later. *Ann. Fond. Louis Broglie* **2017**, *42*, 19–71.
75. Borghesi, C. Equivalent quantum equations with effective gravity in a system inspired by bouncing droplets experiments. *arXiv* **2017**, arXiv:1706.05640.
76. Press, W.H.; Teukolsky, S.A.; Vetterling, W.T.; Flannery, B.P. *Numerical Recipes 3rd Edition: The Art of Scientific Computing*, 3rd ed.; Cambridge University Press: New York, NY, USA, 2007.
77. Higham, D.J. An algorithmic introduction to numerical simulation of stochastic differential equations. *SIAM Rev.* **2001**, *43*, 525–546. [CrossRef]

MDPI

Article

Nonlocality in Bell's Theorem, in Bohm's Theory, and in Many Interacting Worlds Theorising

Mojtaba Ghadimi, Michael J. W. Hall and Howard M. Wiseman *

Centre for Quantum Dynamics, Griffith University, Brisbane, Queensland 4111, Australia;
mghadimi@usc.edu.au (M.G.); michael.hall1@anu.edu.au (M.J.W.H.)
* Correspondence: h.wiseman@griffith.edu.au

Received: 2 July 2018; Accepted: 25 July 2018; Published: 30 July 2018

Abstract: "Locality" is a fraught word, even within the restricted context of Bell's theorem. As one of us has argued elsewhere, that is partly because Bell himself used the word with different meanings at different stages in his career. The original, weaker, meaning for locality was in his 1964 theorem: that the choice of setting by one party could never affect the outcome of a measurement performed by a distant second party. The epitome of a quantum theory violating this weak notion of locality (and hence exhibiting a strong form of nonlocality) is Bohmian mechanics. Recently, a new approach to quantum mechanics, inspired by Bohmian mechanics, has been proposed: Many Interacting Worlds. While it is conceptually clear how the interaction between worlds can enable this strong nonlocality, technical problems in the theory have thus far prevented a proof by simulation. Here we report significant progress in tackling one of the most basic difficulties that needs to be overcome: correctly modelling wavefunctions with nodes.

Keywords: Bell's theorem; Bohmian mechanics; nonlocality; many interacting worlds; wavefunction nodes

---

1. Introduction

This paper is based loosely on the talk given by its final author at the 2017 Symposium on Emergent Quantum Mechanics (EmQM17). That talk took as its principal inspiration the following questions (from the list provided by the organisers to give focus to the meeting):

- Is the universe local or nonlocal?
- What is the future of scientific explanation? Is scientific metaphysics, e.g., the notions of reality, causality, or physical influence, obsolete in mathematical accounts of the quantum world?

However, in view of the conference also being the David Bohm's Centennial Symposium, the speaker also briefly discussed some work relating to this question:

- What is David Bohm's legacy for the future of quantum physics?

This paper also concerns these two broad topics, but in the opposite ratio to that in the delivered talk. Section 2 of this paper addresses just the question "Is the universe local or nonlocal?" by examining the historical meaning of the term "local". (The second question listed above, which was addressed in the talk, and in References [1–3], is not addressed here). That section expresses the views of the invited contributor, not necessarily those of all the authors. Sections 3–6 concern Many Interacting Worlds (MIW), an approach to quantum mechanics inspired by Bohm's original hidden variables theory [4,5], and which thus we hope will be part of the answer to the last question above.

Having introduced the MIW approach in Section 3, we turn in Section 4 to how it can give rise to nonlocality; that is, how it can offer a positive answer to this (here abridged) focus question:

- Are nonlocal connections—e.g., "action-at-a-distance"—fundamental elements in a radically new conception of reality?

However, as Section 4 notes, there are still major difficulties faced in demonstrating these connections in MIW simulations. Section 5 introduces a new MIW technique—a higher-order interaction potential—to address one of these difficulties, namely, wavefunction nodes. Section 6 applies this idea to progress the MIW simulation of stable excited state nodes of a quantum particle in one dimension. An alternative method to address the node problem is presented and simulated in Appendix A. Section 7 concludes with a discussion of open challenges for MIW theorising.

2. Is the Universe Local or Nonlocal?

The answer to this EmQM17 focus question hinges, of course, on what one means by the term "local" (assuming that "nonlocal" is simply its complement). Perhaps surprisingly, it seems [2,6] that the word was not used in the context of interpreting EPR quantum correlations prior to the 1964 paper of Bell [7]. In that paper, Bell proved his 1964 Bell's theorem; to quote [7]:

> In a theory in which parameters are added to quantum mechanics to determine the results of individual measurements, without changing the statistical predictions, there must be a mechanism whereby the setting of one measuring device can influence the reading of another instrument, however remote.

In other words, some quantum phenomena are incompatible with the joint assumption of predetermination (or causality; Bell used both terms) and locality (or separability; Bell used both terms). In the above quote, it is the negation of locality that is characterised, in a way consistent with Bell's earlier definition of locality; to quote [7]:

> It is the requirement of locality, or more precisely that the result of a measurement on one system be unaffected by operations on a distant system with which it has interacted in the past, that creates the essential difficulty.

Thus, Bell intended to be (somewhat) precise about what he meant by locality. Unfortunately he did not give a general mathematical definition, nor did he define terms like "unaffected" or (in the first quote and elsewhere) "influence".

In the theorem he proves that the role "locality" plays is the following. The assumption of predetermination means that an arbitrarily long time before any measurements are performed, there existed in the world a collection of hidden variables, λ, that, together with the future measurement settings, determines the future outcomes. Adding the "vital assumption [of locality]" implies that the outcome A of one party (Alice, say) cannot depend on the settings b of a distant party (Bob, say), but only by her own setting a. That is, in a theory θ with local predetermination of outcomes, there exists a function, $A_\theta(a, \lambda)$ such that $A = A_\theta(a, \lambda)$, and likewise for Bob. From this, Bell was able to derive his famous theorem, that there exist sets of measurement on entangled quantum systems whose results cannot be explained by any such model. (Note that Bell made an implicit assumption in 1964, related to freedom of choice, that $P(\lambda|a, b) = P(\lambda)$. Here we follow Bell, as such an assumption is certainly necessary [8].)

In the above formulation, "locality" has a precise meaning only in the context that predetermination has already been assumed. This is fine for the purpose to which Bell puts it. If one wanted to broaden the definition so that it applied independently of the assumption of predetermination then, it has been argued [2,6,9,10], the natural reading of Bell's verbal definition would be as follows. An arbitrary theory contains initial variables λ, which may or may not be hidden, and which may or may not be sufficient to determine all outcomes, and is described (in the limited context we are considering) by theoretical probabilities $P_\theta(A, B|a, b, \lambda)$. The theory is local if there exists a function $P_\theta(A|a, \lambda)$ such that

$$\forall\, b,\ P_\theta(A|a, b, \lambda) = P_\theta(A|a, \lambda). \tag{1}$$

and likewise for Bob.

Others have argued [11–14], to the contrary, that this definition does not fit with the other use Bell makes of the concept of locality in his 1964 paper, which is the first paragraph of his Section II. There Bell implies that, according to the EPR argument [15], locality plus perfect correlations of outcomes implies predetermination. This shows, if nothing else, that Bell did think, in 1964, that locality was an assumption that had meaning prior to the assumption of predetermination. However, the obvious meaning, Equation (1), does not work in the EPR argument. Even one of Bell's most ardent admirers [13] was forced to admit this [14]:

> It is simply not clear how to translate Bell's words here (about locality) into a sharp mathematical statement in terms of which the EPR argument might be rigorously rehearsed. ...[I]t must be admitted that Bell's recapitulation of the EPR argument in this paragraph leaves something to be desired.

Regardless, is there really a problem here? Bell does not say that he believes that locality plus perfect correlations implies predetermination. He merely says, at the beginning of that paragraph, "...the EPR *argument* is as follows" (my emphasis). In the preceding (opening) paragraph he is even weaker: "The paradox of Einstein, Podolsky and Rosen was *advanced as an argument* ..." (my emphasis again). In a follow up paper in 1971 [16], Bell is weaker still. He gives three motivations for the assumption of predetermination (in a section entitled, unambiguously, "Motivations"). An EPR-style argument is the third motivation; he does not claim any logical deduction from locality but merely appeals to the intuition of the reader for the reasonableness of predetermination by hidden variables (see Reference [6] for details). The 1969 paper of CHSH [17] seems actively skeptical of the EPR argument, saying it "*led them to infer* that quantum mechanics is not a complete theory" (my emphasis). It is thus clear that, at this time, the EPR argument was far from being regarded as a rigorous proof for the necessity of hidden variables. (For a discussion of the extent to which the EPR argument is such a proof, see Reference [18].)

The situation with regard to the EPR argument changed rapidly, at least in Bell's mind, after he formulated, in 1976 [19], a concept that does allow one to infer predetermination from perfect correlations. This was the concept of "local causality", stated most succinctly in a later paper [20]

> A consequence ...of "local causality" [is] the outcomes [in the two labs] having no dependence on one another nor on the settings of the remote [measurement], but only on the local [measurement settings] and on the past causes.

In the situation considered above, a theory is "locally causal" only if there exists a function $P_\theta(A|a,\lambda)$ such that

$$\forall\, b, B,\ P_\theta(A|a,B,b,\lambda) = P_\theta(A|a,\lambda), \tag{2}$$

and likewise for Bob. Note the appearance of B as a conditional variable on the left hand side, which distinguishes this concept from "locality" in Equation (1). Moreover, this assumption obviates the need to consider determinism at all—it leads directly to the Bell inequalities that quantum mechanics violates. Thus, Bell gave what I have called [2] Bell's second Bell's theorem, in 1976 [19]:

> Quantum mechanics... gives certain correlations which ...cannot be [reproduced by] a locally causal theory.

Bell clearly (and, I think [2,3], rightly) thought local causality to be a more natural concept than locality as per Equation (1), as he never used the latter concept again. Regrettably, however, he did not abandon the word "locality". Rather, beginning even in 1976 [19], he sometimes used "local" as short-hand for "locally causal", and, a few years later, was apparently convinced that local causality was the concept that he (and EPR) had always used [21] (for details, see [2]). However, at least in his final word on the subject [20], Bell showed his preference unequivocally for the terminology "local causality" over "locality".

Thus we may return to the primary question posed above—*Is the universe local or nonlocal?* If by "locality" one means "locally causal", the concept Bell promoted for most of his career in quantum foundations [22], then the answer (barring more exotic possibilities such as "superdeterminism" [20], retrocausality [23], and the subjectivity of macroreality [24]) is that the universe is nonlocal; it violates local causality. To avoid confusion, we might agree to say that the universe is Bell-nonlocal [2]. If, on the other hand, one adopts the definition of "local" indicated by Bell's 1964 paper and commonly used in text books [25,26], then the answer is that we cannot say whether the universe is local or nonlocal. Operational quantum mechanics satisfies this weaker sense of locality, simply because it does not feature signalling faster than light, and denies the need for giving any account for quantum correlations beyond an operational one. We can only say that the universe is nonlocal, in this strict sense, if we make some other assumptions about its nature, such as determinism.

3. David Bohm's Legacy: Permission to Theorise Radically New Conceptions of Reality

The most famous quantum theory which does make the assumption of determinism is of course David Bohm's [4,5]. Indeed, Bohm's theory was an inspiration to Bell who summarised the result of his 1964 paper in the introduction-cum-abstract as

> [A] hidden variable interpretation of elementary quantum theory [4,5] has been explicitly constructed. That particular interpretation has ... a grossly nonlocal structure. This is characteristic, according to the result to be proved here, of any such theory which reproduces exactly the quantum mechanical predictions.

Why Bell considered Bohm's interpretation to be "grossly nonlocal", rather than nonlocal *simpliciter*, is unclear. Perhaps it was because the theory is nonlocal even in situations where there is an obvious local hidden variable theory, as in the EPR-correlations [15], or the EPR-Bohm correlations [27].

Unlike operational quantum mechanics, Bohm's theory is a precise and universal physical theory. Restricting to the case of interacting nonrelativistic scalar particles for simplicity of discussion, it takes the universe to be described by a universal wavefunction $\Psi(\mathbf{q})$, obeying Schrödinger's equation, where $\mathbf{q}$ is the vectorised list of the coordinates of all the particles. However, it also postulates a single point in configuration space, $\mathbf{x}$, which encodes the *real* positions of all these particles. This "marvellous point" [28] or "world-particle" [29] has a deterministic equation of motion $\dot{\mathbf{x}} = \mathbf{v}_\Psi(\mathbf{x})$ guided vicinally by $\Psi(\mathbf{q})$. (Note that "vicinal" is a synonym of "local" in the latter's quotidian sense, introduced here to avoid any possible confusion with "local" in the technical sense defined in Section 2.) Here, "vicinal guiding" means that the world-particle's velocity $\mathbf{v}_\Psi(\mathbf{x})$ depends on $\Psi(\mathbf{q})$ and finitely many derivatives, evaluated at $\mathbf{q} = \mathbf{x}$. However, vicinal in configuration space is not vicinal in 3D space—the positions of Bohmian particles in one region of 3D space can affect the motion of an arbitrarily distant particle if entanglement is present. Since it is the position of Bohmian particles that encodes what an experimenter decides to measure, this gives rise to the 'gross' nonlocality of Bohmian mechanics which Bell noted in 1964.

Bohm's original proposal [4,5] actually used a second-order dynamical equation $\ddot{\mathbf{x}} = \mathbf{a}_{\text{vicinal}}(\mathbf{x}) + \mathbf{a}_\Psi(\mathbf{x})$. The 3D-vicinal acceleration $\mathbf{a}_{\text{vicinal}}(\mathbf{x})$ is given by Newton's laws, involving inter-particle potentials which drop off with 3D-distance. The nonlocal effects in Bohm's theory arise from a separate, 3D-nonvicinal, quantum acceleration $\mathbf{a}_\Psi(\mathbf{x})$. Bohm's publication of an explicitly nonlocal theory seems to have made it acceptable for other physicists to publish realist approaches to quantum mechanics in direct opposition to the Copenhagen interpretation. This included: de Broglie in 1956 [30], reviving his unpublished idea from 30 years earlier which prefigured much of Bohm's work and used the first-order dynamics described earlier; and Everett in 1957 [31], introducing the relative state interpretation, more popularly known as the many worlds interpretation.

Taking inspiration from both Bohm and Everett, two of us plus Deckert introduced in 2014 [29] what we called the Many Interacting Worlds (MIW) approach to quantum mechanics. We suggested that it might be possible to reproduce quantum phenomena without a universal wavefunction $\Psi(\mathbf{q})$

(except to define initial conditions). In its place we postulated an enormous, but countable, ensemble $\mathbf{X} = \{\mathbf{x}_j : j\}$ of points $\mathbf{x}_j$ in configuration space (similar ideas have been proposed earlier by a number of authors [32–35], but they considered a continuum of worlds, which, in our view, leads to some of the same conceptual issues that Everett's interpretation faces—see also [36]). Each point is a world-particle, just as Bohmian mechanics postulates, and the dynamics is intended to reproduce a deterministic Bohmian trajectory for each world-particle. However, the trajectory of a world-particle is guided not by a wavefunction, but by the locations of nearby world-particles in configuration space: $\ddot{\mathbf{x}}_j = \mathbf{a}_{\text{vicinal}}(\mathbf{x}) + \mathbf{a}_{\text{MIW}}(\mathbf{N}_j)$, where $\mathbf{N}_j \subset \mathbf{X}$ is the set of world-particles in the vicinity (somehow defined) of world-particle $\mathbf{x}_j$, and $\mathbf{a}_{\text{MIW}}$ is some fixed (j-independent) function describing the interaction of many worlds.

In the MIW approach, probabilities arise only because observers are ignorant of which world $\mathbf{x}_j$ they actually occupy, and so assign an equal weighting to all worlds compatible with the macroscopic state of affairs they perceive, in accordance with Laplace's principle. In a typical experiment, where the outcome is indeterminate in operational quantum mechanics, the final configurations of the worlds in the MIW approach can be grouped into different subsets, $\mathbf{X}_o \subset \mathbf{X}$ of world-particles, still extremely large in number, based on shared macroscopic properties corresponding to the different possible outcomes o. In Everett's approaches, these groups correspond to branches of the universal wavefunction, and one has to argue that the square modulus of the coefficient of each branch somehow manifests as the correct probability for the experimenter. In the MIW approach, the operational quantum probabilities will be equal to the number of worlds in each group divided by the number of worlds at the start.

4. Nonlocality in the Many Interacting Worlds Approach

4.1. General Considerations

Just as Bohmian dynamics is vicinal in configuration space but gives rise to nonlocality in 3D space, so too is the type of MIW dynamics just described. Indeed, it must be nonlocal because, like Bohmian mechanics, it is deterministic. However, we can be more specific about how this nonlocality, in Bell's 1964 sense, arises in the MIW approach.

Consider a branch Ψ_E (in the Everettian sense) of the universal wavefunction Ψ. In the MIW approach, such a branch will correspond to a subset $\mathbf{X}_E \subset \mathbf{X}$ of world-particles, still extremely large in number, that are close together (on a macroscopic scale) in configuration space and mostly far (on a macroscopic scale) from the rest of $\mathbf{X}$. Thus, to a good approximation, this set of world-particles will evolve autonomously, on the time scale of interest.

Now consider a particular world-particle, $\mathbf{x}_j \in \mathbf{X}_E$, and a particular part of the vector $\mathbf{x}_j$, comprising a lower-dimensional vector $\mathbf{b}_j$, that contains the variables that encode the macroscopic fact of the decision by an experimenter, Bob, about what experiment to perform. That is, the decision is the same in every world under consideration here. However, there is no reason in the theory for $\mathbf{b}_j$ to be correlated with the other variables in $\mathbf{x}_j$, until the experiment is actually performed. Once it is performed, the value of $\mathbf{b}_j$ will have a direct (second-order in time) effect on the vicinal (in the 3D sense) variables in $\mathbf{x}_j$, say $\mathbf{B}_j$. Then, in the MIW approach, the change in $\mathbf{B}_j$ will cause a change (again second-order in time) in $\mathbf{x}_k \in \mathbf{N}_j$ (note that $\mathbf{N}_j$ is only a tiny subset of $\mathbf{X}_E$). Now say that in Ψ_E, there is entanglement between the system corresponding to $\mathbf{B}_j$, and a distant system corresponding to some other observables $\mathbf{A}_j$. Then the change in $\mathbf{x}_k$ will be a change not just in $\mathbf{B}_k$ but also in $\mathbf{A}_k$. This is because the MIW acceleration function $\mathbf{a}_{\text{MIW}}$ pays no heed to vicinality in the 3D sense—it cares only about the fact that the worlds $\mathbf{x}_k \in \mathbf{N}_j$ are in the vicinity of the world $\mathbf{x}_j$ in configuration space. Finally, the same interaction will cause a (second-order in time) change in $\mathbf{A}_j$, the distant observables in the original world j, since for at least some of the $\mathbf{x}_k \in \mathbf{N}_j$ we will have $\mathbf{x}_j \in \mathbf{N}_k$.

Thus it is clear in principle how the MIW approach can give rise to nonlocality in Bell's 1964 sense, with Bob's choice $\mathbf{b}_j$ causing a change in the variables $\mathbf{A}_j$, variables which correspond to the outcome observed by Alice in her distant, perhaps even space-like separated, lab, all in the same world

j. It might seem that Bob could use this to signal to Alice faster than light. However, that is not so, for essentially the same reason that signal-locality is respected in Bohmian mechanics. Alice and Bob are ignorant of which world j they inhabit. They know only the macroscopic configuration $\mathbf{X}_E$ in which their world must lie. A successful MIW theory would thus exhibit nonlocality at the hidden level of an individual world, but not at the observable level, averaging over all the worlds in $\mathbf{X}_E$.

4.2. Simulations

Can we explicitly simulate this in our MIW approach, showing, for example, the violation of a Bell inequality, or even just the EPR paradox, without signalling? Here we run up against the limitations in the development of our approach.

In our 2014 paper [29], we introduced a toy MIW theory to describe a very simple universe, comprising a single particle in 1D. We showed analytically that, in the limit of a large number of worlds, it gave the correct ground state distribution for a harmonic oscillator, and the correct description for the first and second moments of free particle evolution. We showed numerically that it could reproduce, at least qualitatively, the double-slit interference phenomenon. Furthermore, we argued that it can plausibly reproduce other generic quantum phenomena such as barrier tunnelling and reflection.

To demonstrate nonlocal correlations between measurements on distant quantum systems, we obviously require more than one particle. One might think one would require particles to model the measuring apparatuses as well as having at least two entangled particles. However, since the position of a particle is reified in the MIW approach, similarly to Bohm's theory [4,5], a particle can always represent its own measured position (this differs from its momentum, as any measurement of momentum will rely on the action of the quantum potential in Bohm's theory, or the interworld interaction potential in our approach). Moreover, if we consider a particle on a spring, with an externally controllable spring constant k, then either its initial position or its initial momentum can be encoded in position at a nominated time. This is because such evolution turns momentum information into position information at time $T/4 = (\pi/2)\sqrt{m/k}$, and back into position information at time $T/2$. (Alternatively, the same can be achieved even by free evolution with an externally controlled mass m, an assumption which does not violate any of the MIW framework. By reducing the mass, the time taken for the momentum to be encoded into the position can be made arbitrarily small; by increasing the mass, the position remains as it was initially to an arbitrarily good approximation. Thus, either position or momentum could be measured at a nominated time.) In this way, nonlocal correlations can be explored with a world of just two 1D particles. This scenario could be used not only to demonstrate the EPR paradox, but also Bell nonlocality, as proven by Bell himself in 1986 [37].

Recently, we have generalised MIW for a 2D particle, which is equivalent to two 1D particles, with some numerical success [38]. However, that study was restricted to finding ground states—stationary states with no nodes in the wavefunction. The same paper did consider finding the first excited states for a 1D particle, but only for symmetric potentials for which the node at $x = 0$ can be put in by hand. This restriction was necessary because, it is now clear, the toy MIW model of Reference [29] cannot reproduce stationary excited states at all. More generally, it cannot be expected to quantitatively reproduce dynamical quantum evolution in which nodes appear and disappear. In the proposal of Bell [37], to violate a Bell inequality by free evolution of two entangled particles, the required initial wavefunction has two nodes and can be expected to develop more during evolution. Thus, to study nonlocality in the MIW approach we certainly need to go beyond the toy model of Reference [29]. Progress in this direction is the topic of the next two sections.

5. MIW Beyond the Toy Model

The MIW toy model, for a nonrelativistic 1D quantum particle of mass m moving in a potential $V(q)$, has N worlds each containing a 1D particle. Denoting the positions and momenta of these world-particles by $X = (x_1, x_2, \ldots, x_N)$ and $P = (p_1, p_2, \ldots, p_N)$, their motion is described in the toy model by a Hamiltonian of the form [29]

$$H(X,P) = \sum_{n=1}^{N} \left(\frac{(p_n)^2}{2m} + V(x_n) \right) + U_{\text{toy}}(X). \tag{3}$$

The first term is just the sum of N classical Hamiltonians, one for each world. It is the second term, a potential energy responsible for interactions between the worlds, which accounts for quantum phenomena:

$$U_{\text{toy}}(X) = \frac{\hbar^2}{8m} \sum_{n=1}^{N} \left(\frac{1}{x_{n+1} - x_n} - \frac{1}{x_n - x_{n-1}} \right)^2. \tag{4}$$

For evaluation purposes we formally define $x_0 := -\infty$ and $x_{N+1} = \infty$, and the ordering $x_1 < x_2 < \cdots < x_N$ has been assumed (this ordering is preserved under time-evolution, due to the repulsive nature of the potential). The motion of each world-particle is determined by the usual Hamiltonian equations of motion, i.e.,

$$\dot{x}_n = \frac{\partial H}{\partial p_n} = m^{-1} p_n, \qquad \dot{p}_n = -\frac{\partial H}{\partial x_n} = -V'(x_n) - \frac{\partial U_{\text{toy}}}{\partial x_n}. \tag{5}$$

Note that the interworld potential $U_{\text{toy}}(X)$ vanishes in the classical limit $\hbar = 0$, and also for the case $N = 1$. In either of these cases, each world evolves independently, according to Newton's laws. More generally, however, for $\hbar \neq 0$ and $N > 1$ the interworld potential $U_{\text{toy}}(X)$ leads to forces on each world that act to reproduce quantum phenomena such as Ehrenfest's theorem, spreading of wave packets, tunneling through a barrier, and interference effects [29]. For the case of a 1D oscillator, $V(q) = \frac{1}{2} m\omega^2 q^2$, it has further been shown, in the limit $N \to \infty$, that the average energy per world of the MIW ground state converges to the quantum groundstate energy $\frac{1}{2}\hbar\omega$ [29], and that the corresponding stationary distribution of worlds samples the usual quantum Gaussian probability distribution [29,39].

The form of $U_{\text{toy}}(X)$ in Equation (4) above is a sum of three-body terms, leading to a force on the n-th world that depends on the positions of the two neighbouring worlds on either side. However, while this form is sufficient to reproduce the quantum phenomena noted above, we have found that it is too simple to model the behaviour of quantum wave functions with nodes. Hence, as noted in the previous section, we must turn to more complex forms of the interworld potential, involving interaction between greater numbers of neighbouring worlds.

Fortunately, there is a great deal of freedom in choosing this potential in the MIW approach. It is possible this freedom could be curtailed via suitable physically-motivated axioms (such as, for example, requiring Ehrenfest's theorem to hold). Here, however, we take a nonaxiomatic approach, to show how the interworld potential of the toy model can be straightforwardly generalised to allow direct interactions between an arbitrary number of worlds, in a manner corresponding to greater accuracy in approximation of the Bohmian acceleration. This greater accuracy supports a corresponding expectation of being able to successfully model wave function nodes.

In particular, it was shown in section II.D of Reference [29] that a suitable general form for the interworld potential in the 1D case is obtained by replacing $U_{\text{toy}}(X)$ in the Hamiltonian (3) by

$$U(X) = \frac{\hbar^2}{8m} \sum_{n=1}^{N} \left(\frac{P'_n}{P_n} \right)^2, \tag{6}$$

where P_n and P'_n are approximations to a probability density $P(q)$ at $q = x_n$, and where $P(q)$ corresponds to some smoothing of the empirical distribution of the world positions $\{x_1, x_2, \ldots, x_N\}$. Equation (6) may be regarded as an approximation of the quantum potential in Bohm's theory [4,5] (for which $P(q) = |\Psi(q)|^2$), and hence is expected to reproduce quantum evolution more and more closely as this approximation is improved. The toy model potential in Equation (4) is the simplest such approximation [29].

To go to higher order approximations, we tried two different methods. The first we call the rational smoothing method and the second the equivariance method. In rational smoothing we take a systematic approach to approximate $P(q)$ and its derivative by ratios of polynomials, where the order of these polynomials determines the number of neighbours that each world directly interacts with. This approach is developed in the next subsection, followed by an example. It will be applied to the description of nodes in Section 6.2. rational smoothing is less computational resource-intensive compared to the equivariance Method. The latter is discussed in Appendix A.

5.1. *Constructing Generalised Interworld Potentials*

If $P(q)$ is some smooth probability density as above, which approximates the distribution of world positions, $\{x_1, x_2, \dots, x_N\}$, then the cumulative distribution $C(x) := \int_{-\infty}^{x} dq\, P(q)$ must satisfy (at least approximately)

$$C(x_n) = \int_{-\infty}^{x_n} dq\, P(q) = u_n := \frac{n - \frac{1}{2}}{N} \tag{7}$$

Thus, the area under $P(q)$ is divided into N neighbourhoods, each containing one world and having area $1/N$. It will be assumed for simplicity that $P(q)$ is nonzero almost everywhere, implying that $C(x)$ is strictly monotonic and hence invertible.

To systematically approximate the quantum force at x_n to a given accuracy, we first define the inverse of the cumulative distribution by $y(u) := C^{-1}(u)$. It follows immediately from Equation (7) that

$$x_n = y(u_n). \tag{8}$$

Further, differentiating $u = C(y)$ with respect to y and using $d/dy \equiv (1/y')(d/du)$, gives $P(y) = 1/y'$, $P'(y) = -y''/(y')^3$, and hence that

$$\frac{P'(y)}{P(y)} = -\frac{y''}{(y')^2}. \tag{9}$$

Our aim is now to approximate this last expression by a function of the world positions, and then substitute this approximation into the right hand side of Equation (6) to obtain a corresponding form for the interworld potential $U(X)$.

In particular, expanding $y(u_{n+c})$ in a Taylor series about $y(u_n)$ gives the approximation

$$x_{n+c} - x_n = y(u_{n+c}) - y(u_n) \approx \sum_{l=1}^{L} \frac{1}{l!} \left(\frac{c}{N}\right)^l y_n^{(l)} \tag{10}$$

to accuracy $O(1/N^L)$, where $y_n^{(l)}$ denotes the lth derivative of $y(u)$ at $u = u_n$. The first L derivatives of y can therefore be approximated to this accuracy by choosing a set of coefficients $\{\alpha_{cl}\}$ such that

$$\sum_c \alpha_{cl}\, c^{l'} = l!\, \delta_{ll'}, \qquad l, l' = 1, 2, \dots, L. \tag{11}$$

It is shown how to construct suitable $\{\alpha_{cl}\}$ below. Equations (10) and (11) immediately yield the approximations

$$\frac{y_n^{(l)}}{N^l} \approx \sum_c \alpha_{cl}\, (x_{n+c} - x_n), \tag{12}$$

accurate to $O(1/N^L)$, for $l = 1, 2, \dots, L$. Finally, substitution into Equation (9) and hence into Equation (6) gives the corresponding interworld potential,

$$U_{\alpha,L}(X) = \sum_{n=1}^{N} U_{\alpha,L,n}(X) = \frac{\hbar^2}{8m} \sum_{n=1}^{N} \left\{ \frac{\sum_c \alpha_{c2}(x_{n+c} - x_n)}{[\sum_c \alpha_{c1}(x_{n+c} - x_n)]^2} \right\}^2 . \tag{13}$$

Note that this interworld potential is translation invariant, and scales as $1/\lambda^2$ under $x_n \to \lambda x_n$. It follows that analogues of Ehrenfest's theorem and wavepacket spreading hold for all such potentials [29].

An advantage of Hamiltonian based methods is that we can use a symplectic numerical integrator to make sure that the total energy is conserved [40]. The quantum force for the mth world-particle can be calculated from Equations (5) and (13) as:

$$f_m = -\frac{\partial U_{\alpha,L}(X)}{\partial x_m} = -\sum_{n=1}^{N} \frac{\partial U_{\alpha,L,n}(X)}{\partial x_m}. \tag{14}$$

5.2. Examples

To obtain an explicit example of the interworld potential in Equation (13), let M denote an $L \times C$ matrix with coefficients $M_{lc} = c^l$, and A denote the $C \times L$ matrix with coefficients α_{cl}, where c ranges over some set of C integers. Equation (11) can then be written in the matrix form

$$MA = \Delta := \text{diag}[1!, 2!, \ldots, L!]. \tag{15}$$

The existence of a solution requires $C \geq L$. Further, since the values of α_{c1} and α_{c2} are required in Equation (13), one must have $L \geq 2$. The corresponding solution is then $A = M^{-1}\Delta$, where M^{-1} denotes the inverse of M for $C = L$, and a pseudo-inverse of M for $C > L$.

It follows that the simplest interworld potential constructed in this way corresponds to $C = L = 2$. Labelling the two values of c by $c = \pm 1$ corresponds, via Equation (12), to approximating the derivatives of $y(x_n)$ via the values of the nearest neighbours $x_{n\pm 1}$. Ordering the values of c as $-1, 1$, the corresponding coefficients $\{\alpha_{cl}\}$ in Equation (13) are then given by

$$A = M^{-1}\Delta = \begin{pmatrix} -1 & 1 \\ 1 & 1 \end{pmatrix}^{-1} \begin{pmatrix} 1 & 0 \\ 0 & 2 \end{pmatrix} = \begin{pmatrix} -1/2 & 1 \\ 1/2 & 1 \end{pmatrix}. \tag{16}$$

Surprisingly, this is not actually equivalent to the toy model of Reference [29], even though it involves the same number of neighbours in the potential.

The simplest higher-order interworld potential corresponds to $C = L = 3$, but odd values of C necessarily introduce an unphysical left-right asymmetry, with each world being coupled to different numbers of neighbouring worlds on either side via $U_{\alpha,L,n}(X)$. To preserve symmetry we therefore next consider the case $C = L = 4$. Labelling the four values of c by $c = \pm 1, \pm 2$, the derivatives of $y(x_n)$ in Equation (12) can be calculated with values of the nearest and next nearest neighbours $x_{n\pm 1}$ and $x_{n\pm 2}$. If we order the values of c as $-2, -1, 1, 2$, the coefficients $\{\alpha_{cl}\}$ in Equation (13) are then given by

$$\begin{aligned} A = M^{-1}\Delta &= \begin{pmatrix} -2 & -1 & 1 & 2 \\ 4 & 1 & 1 & 4 \\ -8 & -1 & 1 & 8 \\ 16 & 1 & 1 & 16 \end{pmatrix}^{-1} \begin{pmatrix} 1 & 0 & 0 & 0 \\ 0 & 2 & 0 & 0 \\ 0 & 0 & 6 & 0 \\ 0 & 0 & 0 & 24 \end{pmatrix} \\ &= \begin{pmatrix} 1/12 & -1/12 & -1/2 & 1 \\ -2/3 & 4/3 & 1 & -4 \\ 2/3 & 4/3 & -1 & -4 \\ -1/12 & -1/12 & 1/2 & 1 \end{pmatrix}. \end{aligned} \tag{17}$$

6. Numerical Results

To test our higher-order methods against the original toy model, we apply them to the problem of finding the ground and the first excited state of a harmonic oscillator. It is the latter, containing a node, where the toy model fails and it is necessary to use a higher-order interworld potential. In this section we use dimensionless configuration coordinates

$$X_n := \sqrt{2m\omega/\hbar}\; x_n \tag{18}$$

and dimensionless times

$$T := \omega t/2\pi, \tag{19}$$

where ω is the harmonic oscillator frequency.

6.1. Toy Model

Figure 1b shows the result of applying the toy model in Equations (3) and (4) to the ground state of a harmonic oscillator, corresponding to $V(q) = \frac{1}{2}m\omega^2 x^2$. In this test we distributed 50 world-particles with x_n determined by inverting Equation (7) for the groundstate probability density $P_t^{(0)}(q)$ (Figure 1a). We then evolved these under the single-particle harmonic oscillator potential $\frac{1}{2}m\omega^2 x_n^2$ and the quantum interworld potential. We see that as expected for a stationary quantum potential, the classical and quantum forces cancel each other and the world-particles stay stationary. The slight oscillations for the world-particles near the boundary is due to differences between the toy-model and Bohmian potentials in areas with high curvature and low sampling. These differences imply an exact stationary state for the MIW potential that has slightly different values for the x_n for any finite value of N [29].

However, if we distribute world-particles based on the probability density of the first excited state of a harmonic oscillator (Figure 2a) and apply the same model, the world-particles will not stay stationary as expected (Figure 2b). We also implemented the nearest-neighbour interworld potential defined by Equation (17), and found very similar behaviour to the original toy model. That is, it fails to support stationary configurations corresponding to excited energy eigenstates.

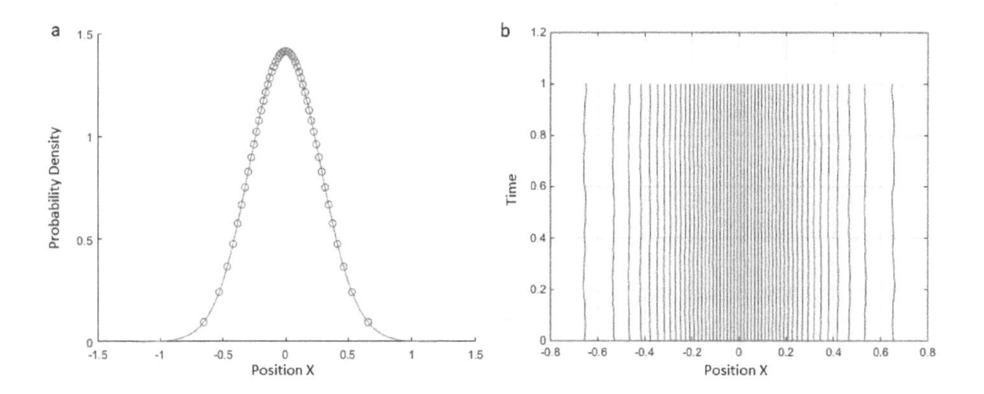

Figure 1. (**a**) 50 world-particles (that is, 50 different worlds describing a single particle) are distributed based on the probability density of the ground state of a harmonic oscillator. (**b**) Trajectories of the 50 world-particles. As expected, the classical and quantum forces approximately cancel each other and the world-particles stay approximately stationary. The slight oscillations near the boundary are due to approximation of the Bohmian potential by the toy-model potential.

The unsuccessful simulation results in Figure 3 show that the problem of the nodes cannot be fixed by increasing the sampling. In the case of a node, because the probability density is zero, there is always a second order curve between the two world-particles adjacent to the node that can never be correctly estimated by toy model approximation. This leads to a poor approximation of the Bohmian force for the world-particles near the node and instead of staying stationary the world-particles move towards the gap and fill the gap.

The general problem with areas of low probability density, and in particular in the region of the node of the quantum state at $x = 0$ in Figure 2a, is that sampling is low and nearest-neighbour approximations are not valid for calculating the quantum potential. If the probability is low but non-zero, theoretically, we can increase the number of worlds until we reach a good sampling in those areas. To test the simulation with higher sampling, we tried 5000 worlds. To reduce the simulation time, we only focused on the 20 world-particles around the node (Figure 3a). To apply the correct boundary condition for truncated area, we kept five world-particles near each boundary artificially fixed.

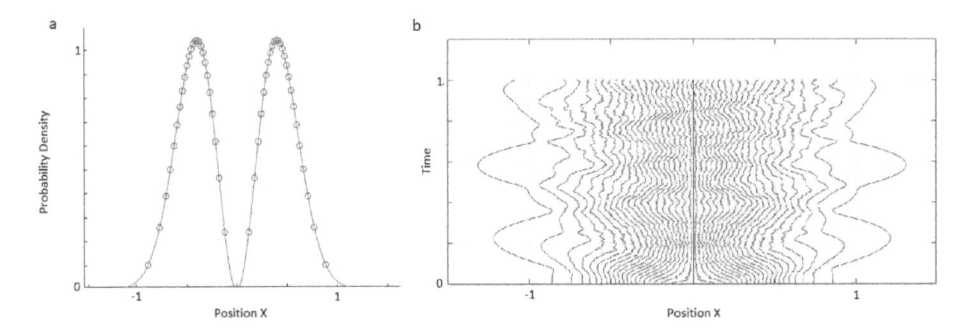

Figure 2. (**a**) 40 world-particles are distributed based on the probability density of the first excited state of a harmonic oscillator. (**b**) Trajectories for the first excited state of a harmonic oscillator using the potential in Equation (4). The world-particles do not stay stationary. Particularly those near the node move towards the middle and fill the gap.

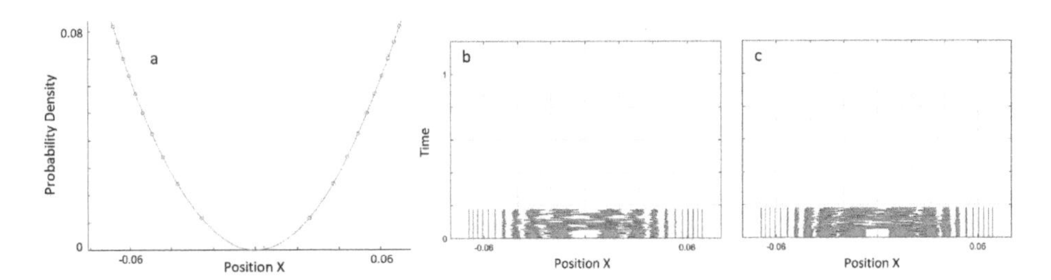

Figure 3. (**a**) 5000 world-particles are distributed based on the probability density of the first excited state of a harmonic oscillator. Only 20 world-particles around the node in the middle are shown. (**b**) Trajectories of the world-particles for the initial distribution in (**a**). To apply the correct boundary condition, we kept five world-particles, near each boundary, fixed and simulated the dynamics of the remaining 10 world-particles in the middle. time-steps are 10^{-8}. Since the nearest neighbours of the node do not stay near the starting point and move to the middle of the node after approximately 0.1 of a period, we did not continue the simulation for a full period. (**c**) The same test as (**b**) with time-steps of 10^{-9}. Thus, the failure of the simulation is not an artefact of large time-steps.

6.2. Higher Order Potential

To test a higher order approximation, we applied the rational smoothing model in Equation (13) for the case $L = 4$ (equivalent to a 5-world interaction), to the first excited state of a harmonic oscillator with 5000 worlds. This corresponds to the example in Equation (17).

In the first test we kept all the worlds artificially stationary except for the two worlds around the node. Figure 4 shows that these two middle worlds stay stationary which means that the quantum potential for those is well approximated.

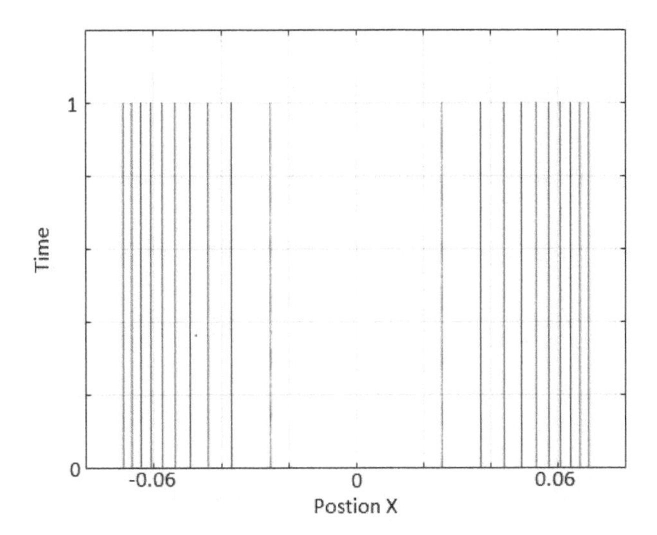

Figure 4. Simulation of only two worlds adjacent to the node for the first excited state of a harmonic oscillator. The initial positions were set by considering 5000 worlds in total to describe the excited state. To apply the boundary condition, the rest of the world-particles were kept fixed. For the evolution, rational smoothing with $L = 4$ (equivalent to 5-world approximation) is used. Time step is 10^{-9}. The two world-particles stay stationary as expected.

In the second test we tried the same scenario but this time simulated 10 worlds around the node and kept the rest stationary. Figure 5 shows that the simulation was successful in the sense that world-particles stayed close to the starting position and did not move towards the gap i.e., the low density region, corresponding to a node of the associated excited state, is preserved by the interaction, and the evolution is approximately stationary. The obvious difference compared to Figure 4 is the oscillations around the starting point. The oscillations appear chaotic, as might be expected for a system of nonlinearly coupled harmonic oscillators. These oscillations might also be due to the time step being too long. We were not able to test this conjecture because the test in Figure 5 took a few days on our desktop computer and, due to time constraints, we could not run it with smaller time steps.

We repeated the same test with 7-worlds approximation ($L = 6$). Figure 6 shows that this higher order approximation decreases oscillation compared to the results in Figure 5. Hence, the simulation of a wave function with nodes, which failed for the toy model (Figures 2 and 3), is seen to become more and more accurately modelled in the MIW approach as the number of directly interacting worlds is increased (Figures 5 and 6).

Similar convergence might also be possible with the equivariance Method presented in Appendix A. However, the results of the simplest test, with a potential involving five interacting worlds, were not as positive as those of the rational smoothing explored in this section. They are also more numerically intensive, so we did not pursue it further.

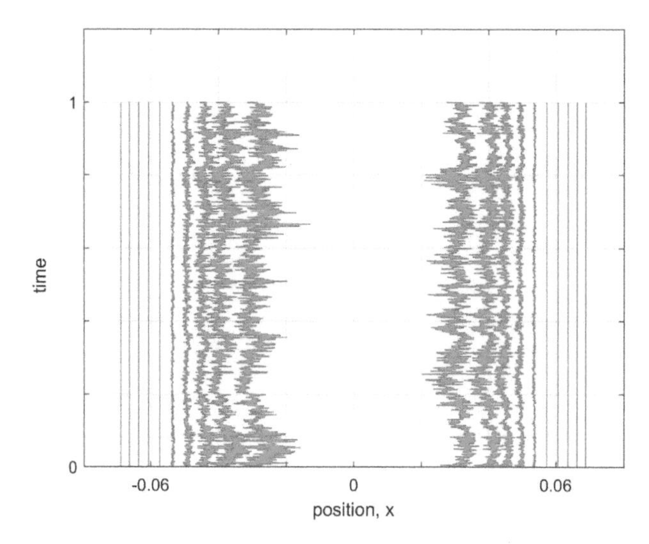

Figure 5. Simulation of 10 worlds (rather than two, as in Figure 4), five on either side of the node for the first excited state of a harmonic oscillator. Other details are as in Figure 4.

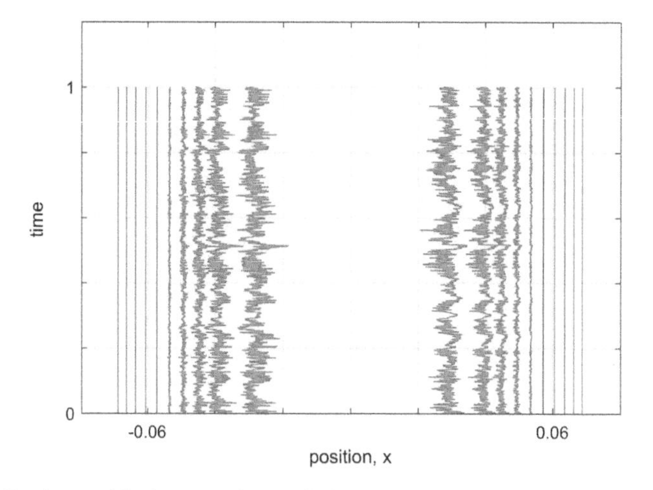

Figure 6. Simulation of the first excited state of a harmonic oscillator with rational smoothing, $L = 6$, (7-world approximation). Other details are as in Figure 5. The oscillations are much smaller than those in the $L = 4$ approximation in Figure 5.

7. Conclusions

Bell's theorem of 1964 showed that any deterministic interpretation of quantum mechanics must be nonlocal (Section 2). Bohm's theory of 1952 is the example *par excellence* of a nonlocal, deterministic theory (Section 3). The theoretical approach we introduced in 2014—Many Interacting Worlds—is also deterministic and, if it is to succeed in replicating Bohmian mechanics (and thus all quantum phenomena), must be nonlocal (Section 4). It is conceptually clear how an interworld potential can lead to nonlocality in Bell's 1964 sense, and, perhaps surprisingly, this could in principle be simulated with a universe of only two particles, in 1D (one spatial dimension), if an externally controlled spatially localized potential is allowed, as discussed in Section 4.

Unfortunately, this proof-of-principle simulation of Bell-nonlocality cannot be done using the toy model for an interworld potential introduced in Reference [29]. The reason is that it cannot deal with wavefunctions having nodes, and nodes are certainly necessary for modelling the entangled states and measurements necessary for violating a Bell inequality [37]. When one prepares a distribution of worlds, in one dimension, corresponding to a stable excited quantum state, the dynamics of the toy model causes the gap between worlds where the node should collapse, and no stationary configuration is reached. Dealing with nodes is a problem in many quantum simulation methods based on Bohmian mechanics [41]. Nodes should not be a problem for interpretations involving a continuum of worlds [32–35], as they are formulated to be exactly equivalent to quantum mechanics. However, as remarked in Section 3, our view is that these interpretations do not solve the conceptual problems of the Everettian many-worlds interpretation.

Here we showed that this problem of nodes in our discrete MIW approach may not be fundamental, but rather may be a result of using a too simple form for the inter-world interaction potential. By using a higher-order approximation to define our interworld potential from Bohm's quantum potential, we were able to show that a gap in the configuration of worlds, corresponding to the node of the first excited harmonic oscillator energy eigenstate, can remain open for at least a full harmonic oscillator period, and perhaps indefinitely. The world configurations were not exactly stationary, but rather had high frequency irregular oscillations. However, by increasing the order of the approximation, the size of the oscillations could be reduced.

Our simulations considered only the dynamics near the node, with more distant worlds artificially held fixed. Whether the node would remain stable if all worlds were allowed to evolve according to the MIW dynamics is an open question. In addition, our simulations were restricted to one particle in 1D. The MIW approach has been successfully used to simulate one particle in 2D (or, equivalently, two particles in 1D) [38,42] and 3D [42], but only to find the ground state configuration. Combining these research directions to be able to simulate stable excited states for two particles in 1D, and beyond, is a challenge for future work. Finally, to realise a simulation of the Bell experiment described in Section 4 would require simulating not just stationary nodes but also *dynamical* nodes, that may appear and then disappear in an instant, which are notoriously difficult to deal with in Bohmian-inspired numerical approaches [41]. Thus, much work remains to be done, but the positive results reported here are encouraging.

Author Contributions: H.M.W. and M.J.W.H. conceived the project to demonstrate Bell-nonlocality in the MIW approach. M.J.W.H. developed the rational smoothing method. M.G. developed the equivariance method and performed the simulations, with discussions involving all authors. H.M.W. drafted Sections 1–4 and 7, M.J.W.H. drafted Section 5, M.G. drafted Section 6 and the Appendix. All authors contributed to the final form of the paper.

Funding: This research was funded by the Foundational Questions Institute grant number FQXi-RFP-1519.

Acknowledgments: We thank Dirk-André Deckert for his formative and continuing role in MIW theorising with us, and the hospitality of Ludwig Maximilian University in facilitating the collaboration. H.M.W. and M.J.W.H. acknowledge Evan Gamble for valuable discussions on simulating excited states in the MIW approach, some years ago now. H.M.W. also thanks the Fetzer Franklin fund for support to attend the David Bohm Centennial Symposium at which some of the ideas in Section 2 solidified in dialogue with other invitees.

Conflicts of Interest: The authors declare no conflict of interest. The funding sponsors had no role in the design of the study; in the collection, analyses, or interpretation of data; in the writing of the manuscript, and in the decision to publish the results.

Abbreviations

The following abbreviations are used in this manuscript:

MIW	Many Interacting Worlds
EPR	Einstein, Podolsky, and Rosen
1D	One (spatial) dimension
2D	Two (spatial) dimensions
3D	Three (spatial) dimensions

Appendix A. A Equivariance Method

Here we introduce an alternative higher order method to rational smoothing, motivated by the equivariance property of Bohmian mechanics [11]. In particular, for a given configuration of worlds, $\{x_1, x_2, \ldots, x_n\}$, we construct a smooth polynomial probability density, $P_n(x)$, in the region of each world x_n, and use this to approximate the Bohmian potential. The coefficients of the polynomial are determined by requiring that area under $P_n(x)$ between x_n and x_{n+1} is equal to the constant value $1/N$ for each world, analogously to Equation (7). The accuracy of this equal-probability or "equivariance" method will increase with the degree of the polynomial.

Here we will illustrate the equivariance method for the case where $P_n(x)$ is third-order, corresponding to direct interactions between sets of five adjacent worlds. Thus,

$$P_n(x) = a_n + b_n x + c_n x^2 + d_n x^3. \tag{A1}$$

The equivariance method then requires that

$$\int_{x_n}^{x_{n+1}} P_n(x)dx = \text{ const.} = N^{-1}. \tag{A2}$$

To determine the coefficients of $P_n(x)$ in Equation (A1), we use four equations below based on the positions of the five worlds, $x_{n-2}, x_{n-1}, x_n, x_{n+1}$ and x_{n+2}:

$$\int_{x_{n-2}}^{x_{n-1}} P_n(x)dx = N^{-1} \tag{A3}$$

$$\int_{x_{n-1}}^{x_n} P_n(x)dx = N^{-1} \tag{A4}$$

$$\int_{x_n}^{x_{n+1}} P_n(x)dx = N^{-1} \tag{A5}$$

$$\int_{x_{n+1}}^{x_{n+2}} P_n(x)dx = N^{-1} \tag{A6}$$

Substituting $P_n(x)$ from Equation (A1) and evaluating the integrals we get

$$\begin{aligned}
a_n\,(x_{n-1} - x_{n-2}) + b_n\,\tfrac{1}{2}(x_{n-1}^2 - x_{n-2}^2) + c_n\,\tfrac{1}{3}(x_{n-1}^3 - x_{n-2}^3) + d_n\,\tfrac{1}{4}(x_{n-1}^4 - x_{n-2}^4) &= N^{-1} \\
a_n\,(x_n - x_{n-1}) + b_n\,\tfrac{1}{2}(x_n^2 - x_{n-1}^2) + c_n\,\tfrac{1}{3}(x_n^3 - x_{n-1}^3) + d_n\,\tfrac{1}{4}(x_n^4 - x_{n-1}^4) &= N^{-1} \\
a_n\,(x_{n+1} - x_n) + b_n\,\tfrac{1}{2}(x_{n+1}^2 - x_n^2) + c_n\,\tfrac{1}{3}(x_{n+1}^3 - x_n^3) + d_n\,\tfrac{1}{4}(x_{n+1}^4 - x_n^4) &= N^{-1} \\
a_n\,(x_{n+2} - x_{n+1}) + b_n\,\tfrac{1}{2}(x_{n+2}^2 - x_{n+1}^2) + c_n\,\tfrac{1}{3}(x_{n+2}^3 - x_{n+1}^3) + d_n\,\tfrac{1}{4}(x_{n+2}^4 - x_{n+1}^4) &= N^{-1}.
\end{aligned} \tag{A7}$$

If we define matrix K_n as

$$K_n = \begin{pmatrix} (x_{n-1} - x_{n-2}) & \tfrac{1}{2}(x_{n-1}^2 - x_{n-2}^2) & \tfrac{1}{3}(x_{n-1}^3 - x_{n-2}^3) & \tfrac{1}{4}(x_{n-1}^4 - x_{n-2}^4) \\ (x_n - x_{n-1}) & \tfrac{1}{2}(x_n^2 - x_{n-1}^2) & \tfrac{1}{3}(x_n^3 - x_{n-1}^3) & \tfrac{1}{4}(x_n^4 - x_{n-1}^4) \\ (x_{n+1} - x_n) & \tfrac{1}{2}(x_{n+1}^2 - x_n^2) & \tfrac{1}{3}(x_{n+1}^3 - x_n^3) & \tfrac{1}{4}(x_{n+1}^4 - x_n^4) \\ (x_{n+2} - x_{n+1}) & \tfrac{1}{2}(x_{n+2}^2 - x_{n+1}^2) & \tfrac{1}{3}(x_{n+2}^3 - x_{n+1}^3) & \tfrac{1}{4}(x_{n+2}^4 - x_{n+1}^4) \end{pmatrix}, \tag{A8}$$

we can rewrite Equation (A7) as

$$K_n \begin{pmatrix} a_n \\ b_n \\ c_n \\ d_n \end{pmatrix} = \begin{pmatrix} N^{-1} \\ N^{-1} \\ N^{-1} \\ N^{-1} \end{pmatrix}. \tag{A9}$$

Therefore:

$$\begin{pmatrix} a_n \\ b_n \\ c_n \\ d_n \end{pmatrix} = K_n^{-1} \begin{pmatrix} N^{-1} \\ N^{-1} \\ N^{-1} \\ N^{-1} \end{pmatrix}. \tag{A10}$$

We used Matlab to invert K_n and find coefficients a_n, b_n, c_n, and d_n. The resulting equations were too lengthy to include here. Substituting these coefficients into Equation (A1), we can find the probability.

To evaluate the quantum force, we use an interworld interaction potential of the form of Equation (6) and P_n as in Equation (A1) (with a_n, b_n, c_n and d_n taken from (A10)) to give

$$U = \frac{\hbar^2}{8m} \sum_{n=1}^{N} \left(\frac{b_n + 2c_n x_n + 3d_n x_n^2}{a_n + b_n x_n + c_n x_n^2 + d_n x_n^3} \right)^2. \tag{A11}$$

This may be compared with the corresponding interaction potential $U_{\alpha,L}$ in Equation (13) obtained via rational smoothing. The quantum force on each particle is evaluated similarly to Equation (14). We used Matlab to evaluate the analytical derivatives of these terms for the simulation. The resulting equation is too lengthy to include here (*c.* 360,000 characters for $N = 5$).

Figure A1 shows the result of applying the equivariance method to the first neighbours of the node in the first excited state of a harmonic oscillator. It shows that these world-particles do not move into the gap from their original position, but rather undergo some oscillatory motion. This is poorer behaviour than the corresponding rational smoothing simulation shown in Figure 4. The simulations were also slower because of the complexity of the analytical form of the force law, mentioned above. For these reasons we have not pursued this method further.

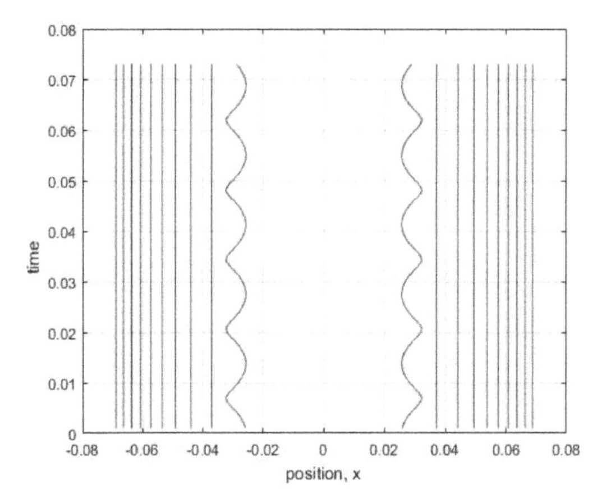

Figure A1. The first excited state of a harmonic oscillator is simulated using 5-world approximation in equivariance Mmethod. Five-thousand worlds are used and only the two world-particles next to the node are simulated. The rest are kept stationary, similar to Figure 4 for the rational smoothing case.

References

1. Wiseman, H.M. Bell's theorem still reverberates. *Nature* **2014**, *510*, 467–469. [CrossRef] [PubMed]
2. Wiseman, H.M. The two Bell's theorems of John Bell. *J. Phys. A* **2014**, *47*, 424001. [CrossRef]

3. Wiseman, H.M.; Cavalcanti, E.G. *Causarum Investigatio* and the Two Bell's Theorems of John Bell. In *Quantum [Un]speakables II: Half a Century of Bell's Theorem*; Bertlmann, R., Zeilinger, A., Eds.; Springer: Cham, Switzerland, 2017; pp. 119–142.
4. Bohm, D. A Suggested Interpretation of the Quantum Theory in Terms of "Hidden" Variables. I. *Phys. Rev.* **1952**, *85*, 166. [CrossRef]
5. Bohm, D. A Suggested Interpretation of the Quantum Theory in Terms of "Hidden" Variables. II. *Phys. Rev.* **1952**, *85*, 180. [CrossRef]
6. Wiseman, H.M.; Rieffel, E.G. Reply to Norsen's paper "Are there really two different Bell's theorems?". *Int. J. Quantum Found.* **2015**, *1*, 85–99.
7. Bell, J.S. On the Einstein-Podolsy-Rosen paradox. *Phys. Phys. Fiz. 1* **1964**, *1*, 195–200. [CrossRef]
8. Hall, M.J.W. *At the Frontier of Spacetime*; Asselmeyer-Maluga, T., Ed.; Springer: Cham, Switzerland, 2016; Chapter 11, pp. 189–204.
9. Jarrett, J. On the physical significance of the locality conditions in the Bell argument. *Noûs* **1984**, *18*, 569–589. [CrossRef]
10. Howard, D. Einstein on Locality and Separability. *Stud. Hist. Philos. Sci.* **1985**, *16*, 171–201. [CrossRef]
11. Dürr, D.; Goldstein, S.; Zanghì, N. Quantum Equilibrium and the Origin of Absolute Uncertainty. *J. Stat. Phys.* **1992**, *67*, 843–907. [CrossRef]
12. Maudlin, T. *Quantum Non-Locality and Relativity*; Blackwell: Oxford, UK, 1994.
13. Norsen, T. EPR and Bell Locality. *AIP Conf. Proc.* **2006**, *844*, 281–293.
14. Norsen, T. Are there really two different Bell's theorems? *Int. J. Quantum Found.* **2015**, *1*, 65–84.
15. Einstein, A.; Podolsky, B.; Rosen, N. Can Quantum-Mechanical Description of Physical Reality Be Considered Complete? *Phys. Rev.* **1935**, *47*, 777–780. [CrossRef]
16. Bell, J.S. Introduction to the Hidden-Variable Question. In *Foundations of Quantum Mechanics*; Academic: New York, NY, USA, 1971; pp. 171–181.
17. Clauser, J.F.; Horne, M.A.; Shimony, A.; Holt, R.A. Proposed Experiment to Test Local Hidden-Variable Theories. *Phys. Rev. Lett.* **1969**, *23*, 880–884. [CrossRef]
18. Wiseman, H.M. Quantum discord is Bohr's notion of non-mechanical disturbance introduced to counter the Einstein–Podolsky–Rosen argument. *Ann. Phys.* **2013**, *338*, 361–374. [CrossRef]
19. Bell, J.S. The Theory of Local Beables. *Epistemol. Lett.* **1976**, *9*, 11–24.
20. Bell, J.S. La nouvelle cuisine. In *Between Science and Technology*; Sarmeljin, A., Kroes, P., Eds.; Elsevier/North-Holland: Amsterdam, the Netherlands, 1990; pp. 97–115.
21. Bell, J.S. Bertlmann's socks and the nature of reality. *J. Phys. Colloq.* **1981**, *42*, 41–62.:1981202. [CrossRef]
22. Bell, M.; Gottfried, K.; Veltman, M. (Eds.) *John S. Bell on the Foundations of Quantum Mechanics*; World Scientific: Singapore, 2001.
23. Price, H. Toy Models for Retrocausality. *Stud. Hist. Philos. Mod. Phys.* **2008**, *39*, 752–761. [CrossRef]
24. Deutsch, D.; Hayden, P. Information flow in entangled quantum systems. *Proc. R. Soc. Lond. A* **1999**, *456*, 1759–1774. [CrossRef]
25. Nielsen, M.A.; Chuang, I.L. *Quantum Computation and Quantum Information*; Cambridge University Press: Cambridge, UK, 2000.
26. Schumacher, B.; Westmoreland, M. *Quantum Processes, Systems, and Information*; Cambridge University Press: Cambridge, UK, 2010.
27. Bohm, D. *Quantum Theory*; Prentice-Hall: New York, NY, USA, 1951.
28. Albert, D.Z. *After Physics*; Harvard University Press: Cambridge, MA, USA, 2015.
29. Hall, M.J.W.; Deckert, D.A.; Wiseman, H.M. Quantum Phenomena Modeled by Interactions between Many Classical Worlds. *Phys. Rev. X* **2014**, *4*, 041013. [CrossRef]
30. De Broglie, L. *Une Tentative D'interprétation Causale et non Linéaire de la Mécanique Ondulatoire: La théorie de la Double Solution*; Gauthier-Villars: Paris, France, 1956. (In French)
31. Everett, H. "Relative State" Formulation of Quantum Mechanics. *Rev. Mod. Phys.* **1957**, *29*, 454–462. [CrossRef]
32. Holland, P. Computing the wavefunction from trajectories: particle and wave pictures in quantum mechanics and their relation. *Ann. Phys.* **2005**, *315*, 505–531. [CrossRef]
33. Poirier, B. Bohmian mechanics without pilot waves. *Chem. Phys.* **2010**, *370*, 4–14. [CrossRef]

34. Parlant, G.; Ou, Y.C.; Park, K.; Poirier, B. Classical-like trajectory simulations for accurate computation of quantum reactive scattering probabilities. *Comput. Theor. Chem.* **2012**, *990*, 3–17. [CrossRef]
35. Schiff, J.; Poirier, B. Communication: Quantum mechanics without wavefunctions. *J. Chem. Phys.* **2012**, *136*, 031102. [CrossRef] [PubMed]
36. Sebens, C.T. Quantum Mechanics as Classical Physics. *Philos. Sci.* **2015**, *82*, 266–291. [CrossRef]
37. Bell, J.S. EPR correlations and EPW distributions. In *New Techniques and Ideas in Quantum Measurement Theory*; Greenberger, D.M., Ed.; New York Academy of Sciences: New York, NY, USA, 1986; Volume 480, p. 263.
38. Herrmann, H.; Hall, M.J.W.; Wiseman, H.M.; Deckert, D.A. Ground states in the Many Interacting Worlds approach. *arXiv* **2017**, arXiv:1712.01918v1.
39. McKeague, I.W.; Levin, B. Convergence of empirical distributions in an interpretation of quantum mechanics. *Ann. Appl. Probab.* **2016**, *2540*, 2555. [CrossRef] [PubMed]
40. Ruth, R. A Canonical Integration Technique. *IEEE Trans. Nucl. Sci.* **1983**, *30*, 92669. [CrossRef]
41. Babyuk, B.; Wyatt, R.E. Coping with the node problem in quantum hydrodynamics: The covering function method. *J. Chem. Phys.* **2004**, *121*, 9230. [CrossRef] [PubMed]
42. Sturniolo, S. Computational applications of the many-interacting-worlds interpretation of quantum mechanics. *Phys. Rev. E* **2018**, *97*, 053311. [CrossRef] [PubMed]

MDPI

Article

Spooky Action at a Temporal Distance

Emily Adlam

Basic Research Community for Physics, Leipzig 04315, Germany; eadlam90@gmail.com

Received: 25 November 2017; Accepted: 9 January 2018; Published: 10 January 2018

Abstract: Since the discovery of Bell's theorem, the physics community has come to take seriously the possibility that the universe might contain physical processes which are spatially nonlocal, but there has been no such revolution with regard to the possibility of *temporally* nonlocal processes. In this article, we argue that the assumption of temporal locality is actively limiting progress in the field of quantum foundations. We investigate the origins of the assumption, arguing that it has arisen for historical and pragmatic reasons rather than good scientific ones, then explain why temporal locality is in tension with relativity and review some recent results which cast doubt on its validity.

Keywords: quantum foundations; nonlocality; retrocausality; Bell's theorem

1. Introduction

Since the discovery of Bell's theorem [1], the physics community has broadly come to take seriously the possibility that the universe might contain physical processes which are spatially nonlocal. However, there has been no such revolution with regard to "temporal locality", i.e., the assumption that the probabilities attached to the outcomes of a measurement performed at a given time depend only on the state of the world at that time. Indeed, temporal locality remains almost ubiquitous in the way that scientists think about science and about what constitutes a reasonable scientific hypothesis.

An assumption so widespread and yet so infrequently justified is in serious danger of becoming a dogma. While it is true that temporal locality has previously been recognised as problematic by parts of the physics community, we argue that this recognition is not sufficiently widespread and that the assumption is actively limiting progress in the field of quantum foundations. In this article, we investigate the origins of this way of thinking about physics, arguing that it has become dominant for historical and pragmatic reasons rather than good scientific ones. We then explain why temporal locality is in tension with relativity, and review some recent results which cast doubt on the status of temporal locality in modern physics.

2. Temporal Locality

2.1. Definition

In seeking to set out a definition of temporal locality, a natural starting point is the standard mathematical definition of spatial locality [2,3]:

Definition 1. ***Spatial Locality:*** *Suppose that two observers, Alice and Bob, perform measurements on a shared physical system: Alice performs a measurement with setting a and obtains a measurement outcome A, while Bob performs a measurement with measurement setting b and obtains a measurement outcome B. Let λ be the joint state of the shared system prior to the two measurements. Then:*

$$p(A, B|a, b, \lambda) = p(A|a, \lambda)p(A|b, \lambda)$$

We can straightforwardly apply this language to the temporal case:

Definition 2. ***Temporal Locality:*** *Suppose that two observers, Alice and Bob, perform measurements on a shared physical system. At some time t_a, Alice performs a measurement with measurement setting a and at some time $t_a + \delta$ she obtains a measurement outcome A; likewise, at some time t_b, Bob performs a measurement with measurement setting b and at some time $t_b + \delta$ he obtains a measurement outcome B. Let $\lambda(t_a)$ be the state of the world at time t_a and let $\lambda(t_b)$ be the state of the world at time t_b. Then:*

$$p(A, B|a, b, \lambda(t_a), \lambda(t_b)) = p(A|a, \lambda(t_a))p(A|b, \lambda(t_b))$$

The central idea of this definition is that in a temporally local world there would be "no action at a temporal distance", i.e., all influences on a measurement outcome would be mediated by the state of the world immediately prior to the measurement. Of course, the definition does not lead to any specific theoretical constraints without some specification of what is included in "the state of the world at time t", but in this article we will not single out any unique way of characterising this state: instead, we will set out a range of options, acknowledging that there are a number of related concepts floating around in modern physics which might reasonably be subsumed under the heading of temporal locality.

It is helpful to approach this range of possibilities by describing some different ways in which physics might *fail* to be temporally nonlocal. First, a theory might fail to be temporally local by postulating non-Markovian laws, meaning that the results of a measurement at a given time can depend on facts about earlier times even if there is no record of those facts in the state of the world immediately prior to the measurement. Note that this is possible only within a theory in which the state of the world at time t, if such a thing exists, does not always contain complete information about everything that has happened before t. Alternatively, a theory might fail to be temporally local by being retrocausal, meaning that the results of measurements at a given time may depend in part on information about the future. We reinforce that retrocausality does not immediately imply temporal nonlocality: a retrocausal theory is temporally nonlocal only if it tells us that the result of a measurement can depend on facts about the future even if there is no record of those facts in the state of the world immediately prior to the measurement. Therefore this type of temporal nonlocality is possible only within a theory in which the state of the world at time t, if such a thing exists, does not always contain complete information about everything that happens after t—in particular, it must not be the case that the state of the world immediately prior to the measurement already contains a record of the future outcome of the measurement, as for example in theories which are deterministic in the traditional sense, meaning that the state of the world at a given time determines everything that happens at later times. Finally, a theory might fail to be temporally local by being atemporal, meaning that the course of history is determined "all at once" by external, global laws of nature, in much the same way as the rules of the game of sudoku apply to the whole grid at once rather than dictating the entries column by column from left to right. In such a theory, the result of a measurement at a given time may depend on global facts even if there is no record of those facts in the state of the world immediately prior to the measurement, and thus an atemporal theory will usually be temporally nonlocal, unless of course the theory tells us that the state of the world at time t always contains complete information about the history of the entire universe. Each of these alternatives singles out a different sense of temporal (non-)locality, and all three raise interesting possibilities for new ways of thinking about physics.

2.2. Motivation

Although physicists are certainly aware that the assumption of temporal locality is problematic, as a methodological principle it remains very widespread in the field. Although, presumably, some physicists would fight to the death for temporal locality, it seems likely that many others retain it simply because they regard it as a harmless simplification. However, we argue that the assumption is by no means harmless: temporal locality is deeply woven into many of the key results on which our present understanding of the interpretation of quantum theory is founded, and unpicking it would require a radical reinterpretation of the significance of those results.

In particular, much recent work in quantum foundations has been based within the "ontological models" framework introduced by Spekkens in [4], where it is supposed that every system has a single real "ontic state", which determines the probabilities for the outcomes of any measurement on that system. An ontological model thus consists of a space Λ of ontic states λ, a set of probability distributions $\mu^P(\lambda)$ giving the probability that the system ends up in the state λ when we perform the preparation procedure P, a set of response functions $\vec{\xi}_{M,O}(\lambda)$ giving the probability that we obtain outcome O when we perform measurement M on a system whose ontic state is λ, and a set of column-stochastic matrices T^X representing the way in which the ontic state is transformed when some operation X is applied to the system. Note that talk of "ontic states" does not imply that we are postulating the existence of hidden variables, because the "ontic state" could simply be the quantum state [5]. It should also be reinforced that one can make use of the formalism of ontological models without necessarily interpreting it as an attempt at a faithful representation of reality—Spekkens himself prefers to regard it as a classification schema which enables us to give precise mathematical definitions for concepts like contextuality [6]. Nonetheless, it seems to be the case that this formalism, or something close to it, *is* often regarded as a description of reality, and indeed as the only possible way of describing reality—for example, in [7], it is claimed that any model in which correlations are not explained by appeal to ontic states should not really be regarded as a realist model at all.

The ubiquity of this method of analysis matters, because the ontological models framework is explicitly temporally local. Not only that, temporal locality is the founding principle of the approach: the entire project of constructing an ontological model is premised on the assumption that measurement results must depend only on the information available in the ontic state at the time of the measurement. Consequently, temporal locality is the keystone of a number of influential results parsed in the language of ontological models, such as Spekkens' generalized proofs of contextuality [4], the Colbeck–Renner theorem [8], Hardy's theorem [9], and the Pusey–Barrett–Rudolph (PBR) theorem [10].

As a case study, let us consider the PBR theorem, which states that no model in which the quantum state is not an "element of reality" can reproduce all the quantitative predictions of quantum mechanics. Now, the term "element of reality" is a reference to a definition set out by Harrigan and Spekkens [11], but although this definition refers only to instantaneous facts, the proof of the PBR theorem depends implicitly on assumptions not only about states at a given time, but about the persistence of those states over time: PBR write that if there exists a set of four preparation procedures which all have some probability of preparing the same ontic state, then when this state is prepared, "*the measuring device is uncertain which of the four possible preparation methods was used, and on these occasions it runs the risk of giving an outcome that quantum theory predicts should occur with probability 0*" [10]. This makes it clear that the argument also requires the assumption that the outcome of the measurement can depend on facts about the system's preparation only via the mediation of an intervening state, so the PBR theorem should really be glossed as follows: *either* the quantum state is ontological, *or* some quantum measurement results must depend in a temporally nonlocal way on events at other times. In this context, then, the assumption of temporal locality is decidedly non-trivial—for example, anyone who wishes to push back against the ontological picture of quantum states should certainly be raising questions about this assumption.

Moreoever, most mainstream interpretations of quantum mechanics, including the Everett interpretation, spontaneous collapse models and the de Broglie Bohm approach, are prima facie temporally local. (We do not mean to suggest that these models could not be phrased in a temporally nonlocal way, nor even to assert that this has not already been done somewhere in the literature, but it does seem to be the case that temporal nonlocality is not a central feature of any of these interpretations). This suggests that fully embracing temporal nonlocality might open up untapped possibilities for the interpretation of quantum theory, and hence the whole landscape of quantum foundations becomes markedly different when temporal nonlocality is taken seriously.

3. Origins

Given that temporal locality plays such a key role in our modern understanding of quantum theory, it is important to understand the intellectual history of this idea. In this section, we argue that a number of historical and psychological factors are likely to have contributed to its prominence; indeed, temporal locality is, in a sense, built into the very structure of physics. Consider the long tradition of presenting theories in terms of their kinematics (the space of physical states postulated by the theory) and their dynamics (the set of laws by which these states evolve, according to the theory). This distinction can be traced back at least to Newton, who may have been the first to make a clear distinction between laws and initial conditions [12], and since Newton's time the formulation has become widespread: it is now almost mandatory to introduce a new physical theory by setting out a space of physical states and a set of evolutionary laws [4]. However, a physical state is, almost by definition, that which carries information from one time to another by means of its dynamical evolution, and thus by employing this mode of presentation we are already very close to assuming that information about one time can influence the results of measurements at other times only via a mediating physical state, thus ruling out temporal nonlocality almost by fiat. Temporal locality is thus very deeply ingrained in the way physicists are taught to think about physics.

There are also straightforward pragmatic reasons why temporal locality should have gained such ascendancy in science. After all, we ourselves are local agents—if we wish to influence events at a spatial or temporal distance we must do so via some spatiotemporally continuous process of mediation—and the fact that these constraints are, for us, so immediate and insurmountable naturally leads us to imagine that the laws of nature must be subject to similar constraints. The empirical results of quantum mechanics, such as the violation of Bell's inequality, have give us convincing reasons to question the resulting attachment to *spatial* locality, but *temporal* nonlocality has not thus far been subject to the same level of analysis and hence lives on in the ways we think and talk about quantum mechanics. Furthermore, as scientists, our primary practical interest is in formulating laws which enable us to predict the future given our knowledge of the present state of the world, and it is easy to move from the fact that most of the laws proposed by physicists have this form to the conclusion that the true underlying laws of nature must take the same form. However, it would be naive to suppose that the true laws of nature look exactly like the type of laws that human agents are most interested in formulating: as Wharton puts it: "There's one last anthropocentric attitude that needs to go, the idea that the computations we perform are the same computations performed by the universe". Assuming that our point of view is not central to the universe, it would be highly suspicious if the laws of nature were to be arranged so conveniently for us.

It also seems likely that certain elements of temporal locality have their origin in the viewpoint known in academic philosophy as "presentism", which holds that the only things which are real are the things which exist now [13–15]. A realist about science will clearly want to insist that measurement results can depend only on things that are real, and hence a realist who subscribes to presentism is compelled to believe that measurement results can depend only on facts about the world immediately prior to the measurement. Presentism is a very old philosophical idea, appearing in the writings of Aristotle and St. Augustine, and playing an important role in Buddhist philosphy, although with the advent of special relativity it has gone somewhat out of vogue as an explicit philosophical thesis: much has been written on the question of whether or not relativity makes presentism untenable [16–19], but whether or not the two can be formally reconciled, they are certainly in tension with one another. Nonetheless, although there are few modern physicists who would self-describe as presentists, the intuitive picture of the present as somehow specially privileged remains hard to shake, and it is likely that some element of this way of thinking contributes to the general conviction that scientific theories should respect temporal locality.

We reinforce that although these historical and psychological observations go some way towards explaining why our scientific theories tend to be temporally local, they do not offer any epistemic justification for thinking that the world actually *is* temporally local. Of course, it may be the case that some epistemic justification can be provided, but if such a justification exists it is certainly not commonly known and hence cannot be regarded as the main reason why our theories exhibit this feature. This indicates that the prominence of temporal locality in our standard approaches to physics may not be entirely rational and perhaps deserves greater scrutiny than it has thus far received.

The Pragmatic Argument

At this juncture, a defender of temporal locality might wish to suggest a different type of justification, using pragmatic rather than epistemic arguments. In particular, one might worry that if we accept that events at this moment may depend on events at any point in the past or future, it will become very difficult to track all the variables which might be relevant to the outcome of an experiment, and the whole scientific enterprise will be under threat. Indeed, similar objections were raised by Einstein concerning *spatial* nonlocality [20]:

> *An essential aspect of this arrangement of things in physics is that they lay claim, at a certain time, to an existence independent of one another, provided these objects "are situated in different parts of space". Unless one makes this kind of assumption about the independence of the existence (the "being-thus") of objects which are far apart from one another in space ... physical thinking in the familiar sense would not be possible. It is also hard to see any way of formulating and testing the laws of physics unless one makes a clear distinction of this kind.*

However, despite Einstein's concerns, it has not proven to be impossible to formulate a theory which allows for spatial nonlocality, because the nonlocal relations between events are governed by laws which enable us to identify regularities in patterns of dependence even between spatially separated events. Likewise, in principle it would not be impossible to move forward with a theory which allows for temporal nonlocality, provided that events at a time depend on events at other times in some regular, formalisable way—indeed, we already have a way of tracking patterns of dependence both temporally and spatially, since the quantum state gives a concise summary of all the information about the history of a system which we know to be relevant to the results of future measurements performed on that system. Therefore the assumption of temporal locality is not forced upon us by practical considerations, and it behoves us to consider the possibility that an explicitly temporally nonlocal theory might enable us to identify and track further regularities.

4. Relativity

In addition to these general concerns about epistemic rationality, there are also more specific technical reasons to be sceptical about temporal locality. In particular, as we describe in this section, temporal locality is in tension with both special and general relativity.

4.1. Special Relativity

The astute reader will already have noticed a problem with our definition of temporal locality: special relativity tells us there is no unique, observer-independent fact about what constitutes the state of the world at a given time [21], and hence the category "the state of the world at time t" is not even well-defined. It is possible to dodge this problem if we are working with a theory which is also *spatially* local, since the probabilities for the outcome at time $t + \delta t$ of a measurement which begins at a time t will then depend only on the state of the world at a fixed spacetime point, i.e., the spatiotemporal location at which the measurement begins, which is well-defined even in a relativistic context. However, the combination of spatial nonlocality, temporal locality and special relativity is straightforwardly inconsistent, since an instance of spatial nonlocality becomes an instance of temporal nonlocality under a change of reference frame.

This fact has consequences for many approaches to the interpretation of quantum mechanics. It is the main stumbling block for the de Broglie–Bohm pilot wave interpretation of quantum mechanics, which combines spatial nonlocality with temporal locality and consequently fails to be relativistically covariant in its standard form [22]. Similarly, Tumulka recently put forward what was intended to be a relativistically covariant version of the Ghirardi–Rimini–Weber (GRW) spontaneous collapse model [23], based on Bell's GRW flash ontology in which the point-like collapse events rather than the quantum states are regarded as fundamental [24], but it was subsequently pointed out by Gisin and Esfield that it is not possible to give a Lorentz invariant account of the temporal development of the flashes, so this model is relativistically invariant only if "one limits oneself to considering possible entire distributions of flashes, renouncing an account of the coming into being of the actual distribution of the flashes" [25]. A theory with laws governing entire distributions of flashes, rather than the temporal coming-into-being of the flashes, would be temporally nonlocal in the atemporal sense, and hence it seems that a temporally nonlocal approach is more or less mandatory if one wishes to achieve a Lorentz invariant version of the GRW flash ontology. A similar dilemma arises in the context of causal set theory, which we will examine in detail in Section 6.2. We would conjecture that this point is true more generally: to achieve relativistic covariance in an interpretation of quantum mechanics, it will usually be the case that one has to abandon the notion of temporally local "coming-into-being." (The Everett interpretation might be raised as a counterexample, but since the evolution postulated by the Everettian view takes place on configuration space rather than spacetime, it remains unclear what one should say about temporal locality and "coming-into-being" in that theory).

4.2. General Relativity

This point is even clearer in General Relativity (GR), where a solution to the Einstein field equations is not a state at a given time but rather an entire spacetime, a full history of a universe. It is tempting in view of this fact to argue that general relativity forces us to take an atemporal, temporally nonlocal viewpoint, but such a conclusion is not inevitable, because it has been shown that Einstein's equations are *compatible* with a well-posed initial value problem. We can split the Einstein equations into a set of constraint equations (the equations for which both indices are spatial) and a set of evolution equations (the equations for which one index is temporal); then, given a smooth three-manifold Σ and a set of initial conditions on that manifold which satisfy the constraint equations, there exists a unique globally hyperbolic solution to Einstein's equations—obtained by evolving the conditions on Σ forwards and/or backwards using the evolution equations—for which Σ is a Cauchy surface, meaning that the conditions on this surface determine the future and past uniquely [26–28]. This makes it possible to interpret general relativity as a temporally local theory with a kinematical state space restricted to the set of states which satisfy the constraint equations and a dynamics given by the evolution equations.

Of course, this reformulation will work in our actual universe only if the initial state of the actual universe does indeed satisfy the constraint equations. Do we know that this is the case? Arguably, yes—we know that on any hypersurface embedded in a spacetime which satisfies the Einstein equations, the conditions on the hypersurface must satisfy the constraint equations [26–28], so if we take it for granted that the universe as a whole satisfies the Einstein equations, then we can conclude that the initial state of the universe must satisfy the constraint equations. Nonetheless, something may be learned from considering the possibility of universes where the initial conditions do not satisfy the constraint equations. A key feature of the initial value formulation is that the constraint equations must be preserved under the evolution equations: if we vary the constraint equations while keeping the evolution equations the same, then in general we will find that initial states belonging the the kinematical state space will be taken to states outside the kinematical state space by dynamical evolution. This means that dynamics and kinematics are not fully independent in the initial value formulation of general relativity.

As Wharton points out [29], the traditional view of temporally local time evolution would have it that the laws of nature really do work like an initial value problem: the universe is presented with some initial state and must evolve it forward to produce a final state, just like a computer presented with an initial value and programmed to predict some value at some later time. The computer is not allowed to refuse the given value on the grounds that this is not the type of problem it prefers to deal with, and likewise, the dynamical laws of nature are not allowed to pick and choose the initial state on which they operate. Thus, even though general relativity can be given an initial value formulation, it is not at all clear that the traditional picture of temporally local time evolution can be maintained within this formulation, and thus general relativity may fit more naturally within a temporally nonlocal picture. (We pause here to note that general relativity is not the only theory in which a difficulty of this kind arises; the set of allowed kinematical states in Maxwell's electrodynamics is also subject to constraint equations, about which, one presumes, similar arguments could be made. We do not regard this as a weakness of our argument; indeed, it may be regarded as a further piece of evidence in favour of the view that the laws of physics are in fact temporally nonlocal).

4.2.1. Objection: The Independence of Dynamics and Kinematics

One might object to the argument given in Section 4.2 on the grounds that the independence of kinematics and dynamics on which the computational picture of the universe depends was never really realised even in pre-relativistic physics—after all, in any reasonable theory, the set of kinematical states must take a mathematical form such that the action of the dynamical laws is well-defined on every state in the set, so kinematics and dynamics can never be wholly independent. Moreover, the kinematics of a theory often makes ineliminable reference to dynamical quantities—witness the appearance of velocities in the characterisation of Newtonian phase space [30,31]. To which we say, first, so much the worse for the traditional view of time evolution! We will return to this theme in Section 6.1; however, we will also note here that the dependence in general relativity is of a more problematic kind. In Newtonian dynamics, a state belongs to the allowed kinematical set if and only if the action of the dynamical laws is well-defined on that state. This is a simple mathematical feature which can straightforwardly be regarded as a property of an individual state: it is, at least prima facie, a *temporally local* property. On the other hand, in general relativity a state belongs to the allowed kinematical set provided that it can be taken only into other members of the allowed kinematical set under allowed dynamical evolutions. How is this set defined? Can we simply say that the actual initial state of the world is chosen arbitrarily and the allowed kinematical set is then simply equal to the maximal set of states into which this state can be taken by allowed dynamical evolutions? This would restore the original picture in which the universe is presented with an initial state that it is not allowed to refuse. However, such a move would be a reasonable only if it is the case that a generic initial state will in this way give rise to an allowed kinematical set governed by a set of constraint equations which are not only comparably simple (by some appropriate measure of simplicity) to the actual constraint equations but which also can be unified with the actual dynamical equations in such a way as to produce a GR-like theory which is comparably simple (by some appropriate measure of simplicity) to the actual theory of GR; otherwise it would seem an implausible coincidence that we ended up with a universe governed by the simple, elegant laws of general relativity from an arbitrarily selected initial state. The argument thus hinges on a technical question whose answer is not presently known so for now we will content ourselves with noting that the equations of general relativity were derived in large part by appeal to the criterion of simplicity [32,33], and so it would seem quite surprising if there were a multitude of equally simple theories which would split into two parts to give the same dynamical equations but different constraint equations. If this move cannot be made, it seems as though the "initial" state must have been singled out on the basis that it would give rise to a particularly simple allowed kinematical set—which means that the choice of initial state actually depends on the state of the universe later in its evolution, so temporal nonlocality seems to be sneaking in through the back door.

4.2.2. Objection: Modality

One might also object to this argument of Section 4.2 on the grounds that there is something fishy about the modal step. Indeed, the argument is superficially similar to a well-known argument due to

Gödel, in which he argued that time cannot be absolute in general relativity because "the compatibility with the laws of nature of worlds in which there is no distinguished absolute time ... throws some light on the meaning of time also in the worlds in which an absolute time can be defined" [34]. This argument is regarded as problematic: in particular, the modal step has been challenged by Earman, who pointed out that it is not clear that "absoluteness", must be an *essential* property of time, and therefore perhaps we should simply say that the status of time varies along with the distribution of matter, so time is absolute in worlds where there is a distinguished absolute time and not absolute in worlds where there can be no distinguished absolute time [35,36]. In the same way, one might object to our appealing to worlds with different constraint equations but the same dynamical equations on the grounds that perhaps the dynamical equation should be varied along with the allowed kinematical set in such a way to ensure a suitably simple GR-like theory. However, the difficulty with the modal step of Gödel's argument stems precisely from the fact that in general relativity, spacetime and matter are not independent and therefore it is not reasonable to expect that questions about the nature of spacetime can be entirely divorced from facts about the constitution of matter in a particular universe; by contrast, on the traditional conception of time evolution, kinematics and dynamics are supposed to be independent, and therefore if this picture is correct it should be permissible to draw conclusions on the basis of holding the dynamics constant and varying kinematics, as we have done here.

4.3. Objection: Spacetimes That Are Not Globally Hyperbolic

Finally, one might worry that since we have only discussed the Cauchy problem in globally hyperbolic spacetimes, our argument might fail to go through if one allows the possibility of spacetimes that are not globally hyperbolic. We will not consider this case in detail here, but it seems likely that allowing spacetimes which are not globally hyperbolic would actually strengthen our argument. It is known that some but not all spacetimes with closed timelike curves admit a well-posed initial value problem, [37] yet a number of physicists have had the intuition that the laws of nature should not allow the existence of closed timelike curves, and to achieve this within the initial value formulation whilst not ruling out spacetimes which are not globally hyperbolic, it is necessary to place further constraints on the initial conditions to ensure that no closed timelike curves can be produced under the dynamical evolution [38]. Alternatively, one might want to allow closed timelike curves under the stipulation that they must be self-consistent, meaning that they do not produce "grandfather paradoxes" or comparable physical contradictions; and, again, this requires us to place constraints on the initial conditions to ensure that the local initial state can be extended to be part of a global solution which is well-defined throughout the non-singular regions of spacetime [38]. Either way, the specification of the allowed initial conditions once again makes reference to what are most naturally construed as global properties of an entire solution rather than temporally local properties of the initial state, which seems to support the temporally nonlocal viewpoint.

5. Three Options for Temporal Nonlocality

In this section we return to the three types of temporal nonlocality that we identified in Section 2.1, and review some relevant recent results.

5.1. Non-Markovian Laws

If the laws of nature do indeed prescribe a temporally local time evolution for the universe, this evolution must have the Markov property—that is, it must be possible to determine the probabilities for future evolution solely from the present state, without needing to know anything about the history [39]. However, we have good reason to be cautious about the Markov property in the context of quantum theories, because Montina has shown that any ontological Markovian theory of quantum mechanics requires a number of variables which grows exponentially with the physical size [40]. Montina concludes: "In order to avoid the exponential growth of the number of ontic variables, we have one possibility, to discard some hypotheses of the theorem. In our opinion, the Markovian property is the only one sacrificable".

Indeed, on reflection, it is clear that there is something rather strange about regarding quantum mechanics as a Markov process. As a general rule, Markov processes lose information over time, because details of the system's history which fail to be recorded are subsequently no longer accessible. However, the dynamics of unitary quantum mechanics is reversible, so if we take the quantum formalism literally, we conclude that in the absence of measurement, information about the past is never strictly lost—it just gets more and more spread out due to decoherence. Moreover much of this information will end up being stored in global properties of highly entangled systems which cannot be reduced to collections of properties of individual systems [41–43], so under the Markovian assumption we are forced to say that the information ends up stored in a "state" which is nonetheless not the state of any specific thing. However, if the formalism tells us that no information about the past is ever lost (except possibly in a measurement process) and also that most of this information usually cannot be attached to any single system or any particular physical location, then are we really saying anything particularly meaningful when we assert that the information is nonetheless all stored in the present state of the world? Under these circumstances, it is certainly more ontologically economically and arguably also more natural to say simply that measurement results at the present time depend *directly* on the history of the system, without any need for mediation via a nebulous state-like entity.

5.2. Retrocausality

Recently there has been renewed interest in "retrocausal" approaches to quantum theory [44–48], including a striking result due to Leifer and Pusey [7], expanding on an argument by Price [44], which demonstrates that if we insist on a certain kind of time-symmetry, quantum mechanics must allow for retrocausality, i.e., we must say that an experimenter's decision to choose a certain measurement setting can influence the properties of particles in the past.

This increase in support for retrocausality is exciting in that it represents an attempt to move away from standard paradigms for the laws of nature. However, the invocation of retrocausality may also be a retrograde step, if the notion is employed as a way of salvaging temporal locality even in the face of increasing evidence against it. To see this, we must disambiguate several different ways of thinking about retrocausality. One important distinction is introduced in ref [47], where the author distinguishes between theories which are retrocausal only in the sense of invoking "reverse causality", i.e., a simple global reversal of the direction of time, and retrocausal theories which allow causal influences from both the past and the future. But for our purposes, it is important to make a second distinction within this latter category, distinguishing between retrocausal theories which incorporate both backwards and forwards causal mediation, and retrocausal theories in which the universe is solved "all at once" without causal mediation in either direction. The first type of theory is perhaps best exemplified by the two-state vector interpretation of quantum mechanics [49], where measurement results at a given time still depend only on the state of the world at the time of the measurement, but this state now includes a "forward-evolving" state carrying information into the future from the past, and a "backward-evolving" state carrying information into the past from the future. This type of retrocausality, as in ref [45], still depends crucially on mediating states which carry information through time and thus such retrocausal theories look a lot like temporally local theories. However, retrocausal theories of the "all at once" type are naturally interpreted as temporally nonlocal, since although there is certainly a sense in which events in the future will have an influence on events in the past within such models, this influence need not be mediated by a record of those future events in the state of the world at the time of the measurement.

The distinction between these different ways of thinking about retrocausality is seldom made explicit in the literature on the subject, and thus it is not always straightforward to deduce into which camp various types of models are intended to fall. On the one hand, a number of recent models work by imposing global constraints and solving a constraint satisfaction problem across time [46,47], which tends to suggest the atemporal picture. On the other hand, it is common to motivate these models by arguing that retrocausality offers a means of salvaging *spatial* locality—the apparent nonlocality of the Bell experiments is to be explained by invoking a spatially local causal influence mediated via the *future* [44]—and this argument seems more at home within a picture in which influences from both the past and the future are mediated via a temporally and spatially local state. Similarly, the arguments of [7] are based on an assumption that Leifer and Pusey refer to as "λ-mediation", which asserts that any correlations between a preparation and a measurement made on a system should be mediated by the ontic state of the system—by which, presumably, they mean the ontic state immediately prior to the measurement. The term "ontic state" is deliberately used in a general way here so as to make the result applicable to a wide range of theories, in much the same way as we refrained from specifying in detail what constitutes "the state of the world at time t" in our definition of temporal locality, but it seems clear that any reasonable precisification of these two notions would imply that the state of the world at time t includes, at the least, all the ontic states of all systems which exist at time t (if and when such ontic states exist), and therefore "λ-mediation" is essentially identical to the assumption of temporal locality. Leifer and Pusey acknowledge that their mathematical formulation of λ-mediation cannot be precisely correct for a theory which includes retrocausality, but they hope to replace it by something salvaging the notion that measurement results depend only on the present ontic state, which they regard as "a core feature of a realist theory". Thus even within the retrocausality community the intuitive picture of mediation via a temporally local ontic state appears to persist.

However, this type of temporally local retrocausality requires a very finely tuned balancing act, because the backwards-evolving state must contain instructions which are compatible with the instructions from the forwards-evolving state—for example, the backwards-evolving state cannot enforce that a given particle must have some particular property if the forwards-evolving state enforces that it must have some mutually exclusive property. Formally, this balance is maintained because the future events which determine the backwards-evolving state are themselves determined by the forwards-evolving state at that time, but once we acknowledge this we are implicitly moving away from the picture of states evolving in fixed temporal directions and towards a kind of global coordination across time. In this picture, the assumption of temporal locality begins to seem highly artificial, and talk of a "backwards-evolving state" [49] or "influences that travel back in time" [7] look like rhetorical devices designed to preserve the appearance of temporal locality in a theory whose underlying structure is really temporally nonlocal.

We therefore suggest that the arguments of [7], along with other arguments that have been put forward in favour of retrocausality in quantum mechanics [44,46], are best interepreted as pointing us towards the atemporal type of temporal locality. Of course, this may be what has been intended by the proponents of retrocausality all along; if so, it would likely work in their favour to make this position clear. Indeed, it seems plausible that one major contribution to the reluctance of the wider physics community to take retrocausal theories seriously results from an implicit awareness of the tension that arises from attempting to balance information contained in forward and backward evolving states, and thus a significant conceptual barrier would be removed by moving explicitly to temporally nonlocal retrocausality.

5.3. Atemporal Laws

5.3.1. The Lagrangian Schema

The paradigm of temporal locality is closely linked to what Smolin has dubbed the "Newtonian schema" [50], which is the assumption that "the universe is a computational mechanism that takes some initial state as an input and generates future states as an output" [29]. Wharton points out that even within classical physics, an alternative approach was available to us in the form of the "Lagrangian schema", in which an experimental situation is described by a *Lagrangian*, a scalar function of various local parameters and their local derivative; the value of the Lagrangian for a given history is referred to as its "action", and the classical action principle (an example of a "variational principle") requires us to obtain predictions by choosing a set of boundary conditions and then extremizing the action [51]. For example, when the experimental situation under consideration is a ray of light travelling some unknown path, the Lagrangian is equal to the time taken to travel a given path, the boundary conditions are the initial and final positions of the ray of light, and thus we obtain Fermat's principle, which states that light will always take the path which minimizes the total time of travel [51]. The two pictures can be related to one another via the Euler–Lagrange equations, which allow us to obtain a set of dynamical equations of the Newtonian type from any Lagrangian.

Furthermore, there exists a Lagrangian-schema formulation of quantum physics—namely, the path integral approach to quantum field theory, which generalizes the classical action principle by requiring us to calculate the probability of an event as a sum over contributions from all possible histories including that event; the contribution of a history is proportional to $e^{iS/\hbar}$, where S is the action for the history, equal to the time integral of the Lagrangian along the history. Like classical Lagrangian methods, the path integral formalism is a powerful calculational tool, and indeed, certain interpretative approaches advocate treating path integrals as the fundamental object of the theory [52,53].

Since the action is a property of an entire history rather than a feature of moment-by-moment temporal evolution, a naive interpretation of these Lagrangian-schema versions of our physical theories leads naturally to a temporally nonlocal view, and in particular, to the atemporal variant of temporal nonlocality. Of course, we cannot argue that the mere existence of the Lagrangian schema forces us to adopt this atemporal view, as for simple mechanical systems, the extremization of the action is both necessary and sufficient for the satisfaction of the Euler–Lagrange equations, and hence for such systems the two pictures are exactly equivalent. Even in more complex systems, the extremization is still always *sufficient* for the satisfaction of the Euler–Lagrange equations [54,55], so we can always pass from the Lagrangian schema to the Newtonian schema. However, let us observe that it is by no means inevitable that world should have been governed by laws of nature that admit these two formulations: a set of dynamical equations can be derived as the Euler–Lagrange equations of a variational principle if and only if, after transformation into a certain canonical form, the right-hand sides of all the equations are derivable by differentiation from a single function H [54,56,57]. If nature is really best described by something like the Newtonian schema, the fact that the actual laws of nature are indeed so derivable is simply an inexplicable coincidence, whereas this property can easily be explained if we postulate that the Lagrangian picture is in fact closer to reality and therefore all dynamical laws are necessarily the Euler–Lagrange equations of some variational principle.

A more formal version of this argument must wait upon an answer to the complementary question about the necessary and sufficient conditions under which "atemporal" laws of nature admit a formulation in terms of temporally directed dynamical laws; we hope to address this technical question in future work, but at present we must acknowledge the possibility that the answer will make the Lagrangian schema seem equally in need of explanation, in which case the comparison would yield no clear argument in favour of either approach. Nonetheless, it is sufficient for our purposes here to note that there exists a well-developed atemporal alternative to the Newtonian schema. Historically, the Lagrangian schema has been regarded as a mere mathematical tool, whereas the Newtonian picture of states evolving forward in time is treated as an approximate description of reality; however, there does not seem to be an obvious justification for this preference, other than a preexisting prejudice in favour of temporal locality. If we look past this prejudice, there seem to be good reasons to consider taking the Lagrangian schema seriously as a possible description of reality, and if we do so, we must necessarily take atemporal variants of temporal nonlocality seriously as well.

5.3.2. New Models

The existence of the Lagrangian-schema formulations of both classical physics and quantum field theory makes it reasonable to argue that our best physical theories, as they currently stand, might be interpreted in a temporally nonlocal way. However, we can go further. Historically, the development of our physical theories has been constrained by the fact that research was by and large conducted within the Newtonian schema, and there has always been an expectation that new theories can be parsed in this framework—witness the great importance that was placed in the early history of General Relativity on showing that it could be given an initial-value formulation [27]. Thus we may well find that new avenues open up if we move to working directly on atemporal theories and indeed allow ourselves to consider theories which may ultimately turn out not to be susceptible to a Newtonian-schema formulation at all. Indeed, as noted in Section 5.2, a number of such models are already under development [46–48]. These toy models demonstrate that atemporal models are capable of reproducing in a natural way a number of prima facie puzzling features of quantum theory, such as the close resemblance between quantum wavefunction collapse and Bayesian updating [48], and thus provide motivation for further research into models of this kind.

6. Dynamics and Kinematics

We earlier identified the distinction between dynamics and kinematics as an important contribution to the status of temporal locality in physics. In this section we review recent work on this subject and discuss some resulting insights for the status of temporal locality in modern physics.

6.1. Spekkens on Dynamics vs. Kinematics

Spekkens has singled out the distinction between kinematics and dynamics as a potentially problematic feature of our standard physical paradigms: in [58], he argues that when new experimental data appears to falsify our existing theory, we can always choose freely whether to respond by altering the kinematics or the dynamics, and he gives a number of illustrative examples. He thus concludes that the distinction between kinematics and dynamics is doing no explanatory work in our theories, and appeals to ontological parsimony to motivate his call for physicists to move past this particular paradigm.

While we concur that the kinematics/dynamics split is problematic, we note that care must be taken with this line of argument to avoid slipping into conventionalism about the whole of science. Spekkens asserts that his approach "does not force us to operationalism", because he advocates only the rejection of distinctions which we can freely transform away without changing empirical predictions, and which are therefore not doing explanatory work. However, as Quine has shown, it can reasonably be argued that physical theories have empirical consequences only taken as a whole, and that consequently we always have freedom to choose which element of a theory to change in response to new empirical evidence, which would suggest that by Spekkens' criterion no distinction in any scientific theory is doing explanatory work [59,60]. Thus Spekkens' line of argument would seem to lead to the conclusion that we should simply give up on trying to formulate theories whose ontologies are endowed with nontrivial structure, a conclusion which realists about science will surely wish to avoid.

To do so, we must understand why Spekkens' argument has particular relevance in the context of the kinematical-dynamical distinction. In particular, let us reinforce that the distinction between kinematics and dynamics is not simply an individual element of some specific theory; the fact that it has become de rigueur to present new theories in this framework has made the kinematical-dynamical split into a meta-principle which physicists educated in this tradition may well regard as a defining feature of any meaningful scientific theory. By pointing out that the kinematical-dynamical distinction is not forced on us by any empirical evidence, Spekkens demotes it from a meta-principle back to a specific ontological hypothesis which should be subject to the same scrutiny and criticism as any other ontological hypothesis. The argument can then be understood as follows: as realists we choose to attach credence to certain ontologies, despite underdetermination by the empirical evidence, on the grounds of theoretical virtues like simplicity and explanatory power, and the same sorts of assessments should be applied to the distinction between kinematics and dynamics. Since theoretical distinctions in general do not have empirical content in and of themselves, it is no good insisting on a distinction between kinematics and dynamics in advance of specifying a particular ontology: we must evaluate the theoretical virtues of complete ontologies, some of which may incorporate such a split, others of which may not.

For clarity, at this point, we must mention a different way of thinking about the kinematics/dynamics distinction that has arisen through recent work on the philosophy of special relativity. In this tradition, "what it means for a phenomenon to be kinematical ... is that it is nothing but a specific instance of some generic feature of the world ... (and that) there is nothing more to learn from that particular phenomenon, neither about the specific system in which it occurs nor about the generic feature it instantiates". In other words, the dynamics/kinematics distinction is regarded as a stipulation about which things need to be explained and which things can be taken for granted. This is clearly quite a different concept from the notion of kinematics and dynamics that we have thus far referred to in this article, and to which Spekkens' argument pertains. Certainly it presupposes much less—in particular, such a distinction would still be perfectly meaningful within a theory which does not postulate a space of states and a set of evolutionary laws, whereas the more traditional way of distinguishing between kinematics and dynamics would be inapplicable in such a case. Nonetheless, we conjecture that similar arguments can be made about this more general distinction. Brown points out that "the distinction between kinematics and dynamics is not fundamental" and cites Pauli as making the same point in 1921, and, again, once this point is accepted it seems unreasonable to demand that all new theories must be presented in the framework of kinematics and dynamics: we may well find it heuristically useful to employ such a distinction in any particular case, but the judgement of its usefulness must be made in context, not in advance of the specification of a theory. We leave a more detailed development of this line of argument to future work.

6.2. Example: Causal Set Theory

As an example of a theory in which the kinematics/dynamics distinction may be less useful, consider the case of causal set theory, an approach to quantum gravity which holds that spacetime is fundamentally discrete. The "state space" of this theory is the space of causal sets—that is, sets of spacetime events with a partial order event defined over them. A causal set is, essentially, an entire history of a universe, with time and space being emergent from the partial order between pointlike events. However, it is not sufficient for the theory to simply specify this kinematical state space, because without further restrictions we will find that the majority of causal sets do not give rise to any low-dimensional emergent spacetime (a spacetime is said to emerge from a causal set iff it *faithfully approximates* the causal set—that is, we can embed the causal set into the spacetime in such a way that the causal relations are preserved (x lies before y in the partial order iff the embedding of x is in the past lightcone of the embedding of y), and on average one element of the causal set is mapped onto each Planck-sized volume of the spacetime, and the spacetime does not have structure at scales below the mean spacing of the events) so we must add in some way of singling out the permissible causal sets.

The standard way of doing this is to impose "dynamics", such as the classical sequential growth model in which elements are probabilistically added to the set one by one [61]. Proponents of causal set theory like to advertise it as an advantage of their approach that this dynamics provides us with a relativistically covariant notion of "becoming", allowing us to rescue the notion of temporal becoming and hence salvage our intuitive notion of time [62]. However, this claim cannot quite be taken at face value, because we encounter a difficulty akin to Smart's objection to the A-theory of time. Smart famously pointed out that if time really passes, we ought to be able to specify the rate at which it passes, which would require a second time-dimension with respect to which the passage of ordinary time can be measured [63]; and likewise in causal set theory, talking about the growth of the causal set seems to presuppose an external time dimension in which this growth can take place. The difficulty is all the more pertinent since the founding principle of the theory is that spacetime supervenes on the causal set [64] and thus proposing a dimension of time external to the causal set would seem to undermine the whole project. Rideout and Sorkin attempt to get around this by arguing that the birthing of events should be regarded as "constituting" time rather than occurring in time [61], but this seems like overkill: since spacetime supervenes on the causal set, a complete causal set already "constitutes" time, without any need to add in a process of growth. Furthermore, to ensure that the growth process satisfies general covariance, it is necessary to impose the requirement of discrete general covariance on the dyamics, meaning that the probability of reaching a particular final causet is independent of the path taken to reach that final causet—i.e., the probability does not depend on the order in which the elemets of the causet were "birthed". It is standard to interpret this by saying that there is no fact of the matter about which path was taken—the choice of path is pure gauge [61]—but this makes it implausible to regard the growth of the causal set as a real physical process, since probabilities are ultimately attached to the causal sets themselves rather than to the transitions that occur during the supposed growth [65,66]. Wütrick and Callender argue that these considerations simply show that modern physics requires us to adopt a 'novel and exotic' notion of becoming in which we are generally prohibited from saying which elements of the causal set exist at any stage of its growth [66]. However, this novel notion of "becoming" is so far removed from our intuitive notion of becoming that it is doubtful whether it can really be said to salvage our intuitive notion of time; moreoever, given that the dynamics cannot be taken literally, there is prima facie no way in which the growth of the causal set could even serve to explain why we have the subjective experience of becoming. The growth model, in fact, does not seem to add anything to the theory in terms of explanatory power: insofar as the dyamics succeeds in explaining why certain causal sets are permissible while others are not, and/or why the world is constituted by one causal set rather than another, the real explanatory work is done by the final probability distribution over causal sets rather than by the process of growth.

Thus it seems that all the growth picture is really doing is making causal set theory subjectively more palatable by soothing our uneasiness about attaching probabilities to entire courses of history, and, of course, allowing the causal set theorists to express their theory in the traditional framework of 'kinematics vs. dynamics". Thus, although it is *possible* to make a distinction between kinematics and dynamics within causal set theory, in this particular case the distinction seems not to be very useful and may in fact be holding us back from understanding the theory properly: perhaps the causal set theorists would do better to embrace the global nature of their theory and explicitly attach probabilities to entire causal sets, retaining the "growth" dynamics only as a calculational tool or perhaps even getting rid of it entirely in favour of a different way of calculating the relevant probabilties.

We have singled out the causal set approach here because the awkwardness of attempting to distinguish between kinematics and dynamics is particularly clear in this context, but we would contend that similar points apply more generally. Theories should not be forced into the kinematics-dynamics framework if they are not a natural fit for that framework: this practice imposes an artifical form of temporal locality on theories which are not inherently temporally local in their mathematical structure, which is likely to impede both understanding and also further theoretical progress.

7. Temporal Bell Inequalities and Entanglement in Time

The stark differences between contemporary attitudes to spatial and temporal locality can largely be traced back to the existence of Bell's inequalities and the fact that quantum mechanics is known to violate them [67], an experimentally verified fact which has led the physics community to at take seriously the possibility that spatially nonlocal processes may exist. Of course, the implication is not undisputed; although a number of experimental loopholes in Bell's theorem have been closed in recent years [68–70], there remain untested assumptions, such as the possibility that the choices of measurement on the two sides of the apparatus are not truly independent [71]. Furthermore, proponents of the Everett interpretation claim their approach can account for the Bell statistics in a spatially local way, and antirealists can avoid spatial nonlocality simply by denying that there exists any underlying process, local or otherwise, which accounts for the measurement statistics.

However, each of these ways around the conclusion of the theorem requires us to accept a fairly extreme proposition of one type or another, so it is fair to say that, conditional on a set of assumptions which seem very plausible to many people, the violation of Bell's inequality does indeed imply the existence of spatial nonlocality.

Thus it is very natural to consider whether some analogous set of equations are violated in the temporal case. The first point to be made is that the derivation of Bell's theorem assumes both spatial and temporal locality. If we relax the assumption of temporal locality, then we could say, for example, that the result of the measurement may depend directly on the state of the system being measured at times other than the time of measurement, including *future* times: as we noted above, the proponents of retrocausality have used this possibility to explain the violation of the Clauser–Horne–Shimony–Holt (CHSH) inequality via a local interaction which is mediated via the future [44]. Thus, it is not really fair to say that we have better evidence for spatial nonlocality than temporal nonlocality: we have exactly the same evidence for both. However, physicists have largely chosen to respond to this evidence by discarding spatial locality and retaining temporal locality (or indeed by arguing that we can salvage both), and therefore to have convincing evidence that points specifically to temporal locality, we would need not just a temporal analogue of Bell's inequalities, but a stronger result which shows that spatial nonlocality is not enough to explain the empirical results of quantum mechanics: the quantum world must be temporally nonlocal as well.

There exist several inequalities—mostly governing sequences of measurements performed on a single quantum system—which have been referred to as "temporal Bell's inequalities", and we will now consider whether any of them might be capable of providing the right sort of evidence. First, it should be clear that in the context of repeated measurements on a single quantum system, the assumption of temporal locality alone will not allow us to derive anything, since we can always choose to retain temporal locality by assuming that the entire history of a system is recorded in its present state. Thus, to obtain meaningful results, some further assumption p must be made, so we are never going to obtain a result stronger than "if this inequality is violated, then either $\neg p$, or quantum mechanics is not temporally local". For this to provide a convincing argument in favour of temporal locality, p would need to be an assumption so plausible that many people would be willing to abandon temporal locality before abandoning p.

In the case of the Leggett–Garg inequalites, the additional assumption is "macrorealism", which is the claim that a macroscopic object is at any given time in a definite ontic state and it is possible to determine which state it is in without changing the state or the subsequent system dynamics. Macrorealism is a strong assumption—too strong, in fact, for our purposes, because the only measurements referenced in the Leggett–Garg inequalities are measurements which reveal the definite ontic state of the system at the time of the measurement. As noted in Section 5.2, it is reasonable to assume that the state of the world at time t includes all the definite ontic states of all systems which exist at time t, and thus by definition the measurements referenced in the Leggett–Garg inequalities are only allowed to depend on the state of the world at the time of the measurement, which makes temporal locality irrelevant: whether or not the world is temporally nonlocal in general, for this specific type of measurement there is no freedom for the measurement result to depend on anything other than the present state of the world. There exist later reformulations of the Leggett–Garg inequalities which replace macrorealism with a weaker assumption, but most of these reformulations retain the assumption of "operational eigenstate realism", that is, the assumption that quantum systems are necessarily in states in which the quantity being measured has a definite value which is revealed deterministically by the measurement, and again this makes the assumption of temporal nonlocality irrelevant [72]. A similar issue arises for the set of temporal Bell inequalities derived in [73], and used to demonstrate the phenomenon of "entanglement in time". Here the derivation depends on temporal locality and also "realism", defined as the assumption that measurement results are determined by hidden properties that the particles carry prior to and independently of observation; one assumes that the state of the world at the time of the measurement would be expected to includ these "hidden properties" and, thus under this assumption measurement results can depend only on the state of the world at the time of the measurement, so the auxiliary assumption already implies temporal locality, in the sense in which we have used the term.

Thus, if we are to derive a temporal Bell's inequality has something to say about temporal locality in particular, we should look for an assumption p which does not itself imply temporal locality. A possible candidate is put forward in [74]: here, the derivation of the inequalities is based on the assumption that the results of measurements on a system of dimension d should be simulable by an ordered set of classical systems with no more than $\log_2(d)$ bits of communication between any consecutive pair of systems. It is helpful to split this assumption into two parts: first, the total amount of information about its history that can be carried forward in time by a quantum state of fixed dimension is upper bounded by $\log_2(d)$ bits; and second, the result of a measurement on the system is statistically independent of all information about its history which is not stored in its present state, i.e., the measurements in question are temporally local. Ref. [74] use this assumption to derive a bound on the minimum dimension of a system which can solve a certain sort of sequential problem, and then show that the problem can be solved by quantum systems of dimension smaller than this bound, indicating that quantum mechanics does not satisfy their assumption. This is exactly the kind of result needed to provide an argument for temporal locality: if we find it sufficiently unpalatable to postulate that quantum states may carry information greater than $\log_2(d)$ forward in time, we will have to conjecture instead that the later measurement results depend directly on earlier measurement settings and outcomes without being mediated via information carried forward in the state, and thus we may regard the violation of this inequality as a direct demonstration of temporal nonlocality at work in quantum mechanics, in the same way that a Bell experiment is a direct demonstration of spatial nonlocality at work in quantum mechanics. Admittedly, it may not be the case that there are many people who find the bound $\log_2(d)$ more intuitively plausible than temporal locality, but at least the result seems to be of the right form.

The Problem of Records

It may seem that the existence of records of past measurement results must always stymie any attempt to use the violation of some inequality to prove that the world must be both spatially and temporally nonlocal, since even if we do make an assumption like that of [74] to the effect that a given system can only carry a bounded amount of information forward in time, a proponent of temporal locality could always claim that a given result depends on the record of a given measurement result stored *elsewhere* in the present state of the world, rather than directly on the past events constituting the measurement. After all, in practice such records are very difficult (perhaps impossible!) to erase, and in any case, if a past measurement result could be permanently erased so that no record of it existed in the state of the world at the time of the next measurement, then we would never be able to observe the violation of the relevant inequality, since we could never have all the necessary results available to be compared at the same time.

We suggest the best way of resolving this difficulty is to adopt a halfway position inspired by our discussion in subsection "The Pragmatic Argument". The problem that we are facing can be understood as a particular instance of the general problem identified by Einstein: if spatial locality is simply abandoned wholesale, it becomes impossible to identify and control all the factors which might possibly influence the results of an experiment, and thus we lose the ability to draw meaningful conclusions from experimental results. Therefore, as noted in subsection "The Pragmatic Argument", to make progress we must assume that there are limits on spatial nonlocality. The most straightforward approach is to assume that the world is only as spatially nonlocal as quantum mechanics says it is, because then, provided the system being measured is in a sufficiently pure state, we can justify the assumption that the result of the measurement is independent both of records stored elsewhere in the world and of the state of the observer's brain. The resulting inequality will still be theory dependent, but at the very least the violation of such an inequality, assuming we are not willing to abandon the assumption that quantum states of dimension d may only carry $\log_2(d)$ bits of information forward in time, would force us to say either that the world must be temporally nonlocal or it must be more spatially nonlocal than quantum mechanics currently suggests.

An alternative would be to assume that we need only worry about spatial and/or temporal entanglement when the systems concerned can be connected via some reasonably simple spatiotemporal path, as for example in the case of two entangled particles which have interacted locally at some point in their shared past. This would have the advantage of removing any dependence on the present formalism of quantum mechanics, which may be desirable given that we do not know how much of that formalism would survive the move to a temporally nonlocal context, but on the other hand to make the criterion precise we would likely need a reasonably concrete proposal for an alternative theory, and at present only toy models are available to us for this purpose.

8. Conclusions

There already exists a small body of interesting work examining the possibility of what might be interpreted as temporally nonlocal approaches to quantum theory, although most of it has not yet reached the mainstream. Wharton, advocating the view that "the universe (runs) not as a computer, but as a global four-dimensional problem that (is) solved all at once" [29], has made progress with retrocausal models [75–77]; the consistent histories approach offers an approach to formulating laws of nature which constrain entire histories rather than moment-by-moment evolution [78,79], although there are a number of significant conceptual difficulties to be resolved, not least the question of what the probabilities prescribed by the theory are probabilities for [80]; and Ref. [81] puts forward a theoretical model, in which *"one particle at N times is ... equivalent to N (entangled) particles at one time"*, which, by emphasizing the parallel between spatial nonlocality and time-evolution, seems to lead naturally to a temporally nonlocal view. Similarly, there exist interpretations of quantum mechanics whose ontology consists entirely of pointlike events, such as the GRW flash ontology [23,24] or Kent's solution to the Lorentzian quantum reality problem [82], and one possible interpretation of these approaches would be to say that they have done away with the need for an ontic state as the carrier of information from the past to the future and hence should be regarded as temporally nonlocal.

This existing work is very promising, but we would argue that it does not go far enough. These approaches have been postulated as part of the project of interpreting the existing framework of quantum mechanics (and/or quantum field theory), and yet, once we accept that the universe may be generically nonlocal across both time and space, it becomes at least plausible that quantum theory as we know it is simply the local limit of a global theory which applies constraints across the whole of space and time. This means there is scope to be more ambitious: temporal nonlocality may ultimately point us not just to a new interpretation of quantum mechanics but to a new theory altogether.

Acknowledgments: Thanks to Jeremy Butterfield for helpful suggestions on an early draft of this article, and to anonymous reviewers for detailed and insightful comments. Thanks also to the Franklin Fetzer fund and the organizers of the EMQM17 workshop.

Conflicts of Interest: The authors declare no conflict of interest.

References

1. Bell, J.S. Free variables and local causality. In *Speakable and Unspeakable in Quantum Mechanics;* Cambridge University Press: Cambridge, UK, 1987.
2. Bell, J.S. La Nouvelle Cuisine. In *Speakable and Unspeakable in Quantum Mechanics;* Cambridge University Press: Cambridge, UK, 2004; pp. 232–248.
3. Jarrett, J.P. On the Physical Significance of the Locality Conditions in the Bell Arguments. *Noûs* **1984**, *18*, 569–589.
4. Spekkens, R.W. Contextuality for preparations, transformations, and unsharp measurements. *Phys. Rev. A* **2005**, *71*, 052108.
5. Leifer, M.S. Is the quantum state real? An extended review of ψ-ontology theorems. *Quanta* **2014**, *3*, 67–155.
6. Spekkens, R.W. (Perimeter Institute, Waterloo, Canada) Private communication, 2013.

7. Leifer, M.S.; Pusey, M.F. Is a time symmetric interpretation of quantum theory possible without retrocausality? *Proc. R. Soc. Lond. A Math. Phys. Eng. Sci.* **2017**, *473*, 20160607.
8. Colbeck, R.; Renner, R. The Completeness of Quantum Theory for Predicting Measurement Outcomes. In *Quantum Theory: Informational Foundations and Foils*; Springer: Berlin/Heidelberg, Germany, 2016; pp. 497–528.
9. Hardy, L. Are quantum states real? *Int. J. Mod. Phys. B* **2013**, *27*, 1345012.
10. Pusey, M.F.; Barrett, J.; Rudolph, T. On the reality of the quantum state. *Nat. Phys.* **2012**, *8*, 475–478.
11. Harrigan, N.; Spekkens, R.W. Einstein, incompleteness, and the epistemic view of quantum states. *Found. Phys.* **2010**, *40*, 125–157.
12. Wigner, E.P. Events, Laws of Nature, and Invariance Principles. *Science* **1964**, *145*, 995–999.
13. Bigelow, J. Presentism and properties. *Philos. Perspect.* **1996**, *10*, 35–52.
14. Tallant, J.; Ingram, D. Nefarious presentism. *Philos. Q.* **2015**, *65*, 355–371.
15. Deasy, D. What is presentism? *Noûs* **2017**, *51*, 378–397.
16. Saunders, S. How Relativity Contradicts Presentism. *R. Inst. Philos. Suppl.* **2002**, *50*, 277–292.
17. Putnam, H. Time and Physical Geometry. *J. Philos.* **1967**, *64*, 240–247.
18. Savitt, S.F. There's No Time Like the Present (in Minkowski Spacetime). *Philos. Sci.* **2000**, *67*, 574.
19. Hinchliff, M. The Puzzle of Change. *Philos. Perspect.* **1996**, *10*, 119–136.
20. Einstein, A. Quantum Mechanics and Reality. *Dialectica* **1948**, *2*, 320–324.
21. Einstein, A. *Relativity: The Special and General Theory*; Henry Holt: New York, NK, USA, 1920.
22. Goldstein, S. Bohmian Mechanics. In *The Stanford Encyclopedia of Philosophy*; Zalta, E.N., Ed.; Metaphysics Research Lab, Stanford University: Stanford, CA, USA, 2016.
23. Tumulka, R. A Relativistic Version of the Ghirardi Rimini Weber Model. *J. Stat. Phys.* **2006**, *125*, 821–840.
24. Bell, J.S.; Aspect, A. Are there quantum jumps? In *Speakable and Unspeakable in Quantum Mechanics*, 2nd ed.; Cambridge Books Online; Cambridge University Press: Cambridge, UK, 2004; pp. 201–212.
25. Esfeld, M.; Gisin, N. The GRW flash theory: A relativistic quantum ontology of matter in space-time? *Philos. Sci.* **2014**, *81*, 248–264.
26. Wald, R. *General Relativity*; University of Chicago Press: Chicago, IL, USA, 2010.
27. Ringström, H. *The Cauchy Problem in General Relativity*; European Mathematical Society: Zurich, Switzerland, 2009.
28. Foures-Bruhat, Y. Théorème d'existence pour certains systèmes d'équations aux dérivées partielles non linéaires. *Acta Math.* **1952**, *88*, 141–225. (In French)
29. Wharton, K. The Universe is not a Computer. In *Questioning the Foundations of Physics*; Aguirre, A.F.B., Merali, G., Eds.; Springer: Berlin/Heidelberg, Germany, 2015; pp. 177–190.
30. Bingham, G.P. A Note on Dynamics and Kinematics. *Haskins Lab. Stat. Rep. Speech Res.* **1988**, *93*, 247.
31. McCauley, J. *Classical Mechanics: Transformations, Flows, Integrable and Chaotic Dynamics*; Cambridge University Press: Cambridge, UK, 1997.
32. Norton, J.D. 'Nature is the Realisation of the Simplest Conceivable Mathematical Ideas': Einstein and the Canon of Mathematical Simplicity. *Stud. Hist. Philos. Sci. Part B Stud. Hist. Philos. Mod. Phys.* **2000**, *31*, 135–170.
33. Norton, J.D. Eliminative Induction as a Method of Discovery: Einstein's Discovery of General Relativity. In *The Creation of Ideas in Physics: Studies for a Methodology of Theory Construction*; Leplin, J., Ed.; Kluwer Academic Publishers: Dordrecht, The Netherlands, 1995; pp. 29–69.
34. Gödel, K. An Example of a New Type of Cosmological Solutions of Einstein's Field Equations of Gravitation. *Rev. Mod. Phys.* **1949**, *21*, 447–450.
35. Earman, J. *Bangs, Crunches, Whimpers, and Shrieks: Singularities and Acausalities in Relativistic Spacetimes*; Oxford University Press: Oxford, UK, 1995.
36. Smeenk, C.; Wuthrich, C. Time Travel and Time Machines. In *The Oxford Handbook of Time*; Callender, C., Ed.; Oxford University Press: Oxford, UK, 2009.
37. Friedman, J.L. The Cauchy Problem on Spacetimes That Are Not Globally Hyperbolic. In *The Einstein Equations and the Large Scale Behavior of Gravitational Fields*; Chruściel, P.T., Friedrich, H., Eds.; Birkhäuser Verlag: Basel, Switzerland, 2004; p. 331.
38. Friedman, J.; Morris, M.S.; Novikov, I.D.; Echeverria, F.; Klinkhammer, G.; Thorne, K.S.; Yurtsever, U. Cauchy problem in spacetimes with closed timelike curves. *Phys. Rev. D* **1990**, *42*, 1915–1930.
39. Gillespie, D. *Markov Processes: An Introduction for Physical Scientists*; Elsevier: Amsterdam, The Netherlands, 1991.

40. Montina, A. Exponential complexity and ontological theories of quantum mechanics. *Phys. Rev. A* **2008**, *77*, 022104.
41. Timpson, C. *Quantum Information Theory and the Foundations of Quantum Mechanics*; Oxford University Press: Oxford, UK, 2013.
42. Healey, R. *The Philosophy of Quantum Mechanics: An Interactive Interpretation*; Cambridge University Press: Cambridge, UK, 1991.
43. Morganti, M. A New Look at Relational Holism in Quantum Mechanics. *Philos. Sci.* **2009**, *76*, 1027–1038.
44. Price, H. Does time-symmetry imply retrocausality? How the quantum world says "Maybe"? *Stud. Hist. Philos. Sci. Part B Stud. Hist. Philos. Mod. Phys.* **2012**, *43*, 75–83.
45. Sutherland, R.I. Causally symmetric Bohm model. *Stud. Hist. Philos. Sci. Part B Stud. Hist. Philos. Mod. Phys.* **2008**, *39*, 782–805.
46. Wharton, K.B. A Novel Interpretation of the Klein-Gordon Equation. In *Quantum Theory: Reconsideration of Foundations-4*; Adenier, G., Khrennikov, A.Y., Lahti, P., Man'ko, V.I., Eds.; American Institute of Physics: College Park, MD, USA, 2007; pp. 339–343.
47. Price, H. Toy models for retrocausality. *Stud. Hist. Philos. Mod. Phys.* **2008**, *39*, 752–761.
48. Wharton, K. Quantum states as ordinary information. *Information* **2014**, *5*, 190–208.
49. Aharonov, Y.; Vaidman, L. The Two-State Vector Formalism of Quantum Mechanics. In *Time in Quantum Mechanics*; Muga, J.G., Mayato, R.S., Egusquiza, I.L., Eds.; Springer: Berlin/Heidelberg, Germany, 2002; pp. 369–412.
50. Smolin, L. The unique universe. *Phys. World* **2009**, *22*, 21.
51. Brizard, A. *An Introduction to Lagrangian Mechanics*; World Scientific: Singapore, 2008.
52. Hartle, J.B. The spacetime approach to quantum mechanics. *Vistas Astron.* **1993**, *37*, 569–583.
53. Sorkin, R.D. Quantum dynamics without the wavefunction. *J. Phys. A Math. Gen.* **2007**, *40*, 3207–3221.
54. Butterfield, J. Some Aspects of Modality in Analytical Mechanics. In *Formal Teleology and Causality*; Stöltzner, M., Weingartner, P., Eds.; Mentis: Paderborn, Germany, 2004.
55. Papastavridis, J. *Analytical Mechanics: A Comprehensive Treatise on the Dynamics of Constrained Systems: For Engineers, Physicists, and Mathematicians*; Oxford University Press: Oxford, UK, 2002.
56. Lanczos, C. *The Variational Principles of Mechanics*; Mathematical Expositions, University of Toronto Press: Toronto, ON, Canada, 1949.
57. Ostrogradsky, M. Mémoire sur les equations différentielles relative au problème des Isopérimètres. *Mem. Acad. St. Petersbourg* **1850**, *6*, 385–571. (In French)
58. Spekkens, R.W. The paradigm of kinematics and dynamics must yield to causal structure. In *Questioning the Foundations of Physics*; Springer: Berlin/Heidelberg, Germany, 2015; pp. 5–16.
59. Quine, W.V. Main Trends in Recent Philosophy: Two Dogmas of Empiricism. *Philos. Rev.* **1951**, *60*, 20–43.
60. Schnädelbach, H. Two Dogmas of Empiricism. Fifty Years After. *Grazer Philos. Stud.* **2003**, *66*, 7–12.
61. Rideout, D.P.; Sorkin, R.D. Classical sequential growth dynamics for causal sets. *Phys. Rev. D* **2000**, *61*, 024002.
62. Sorkin, R.D. Relativity Theory Does Not Imply that the Future Already Exists: A Counterexample. In *Relativity and the Dimensionality of the World*; Petkov, V., Ed.; Springer: Berlin/Heidelberg, Germany, 2007; p. 153.
63. Smart, J.J.C. The River of Time. *Mind* **1949**, *58*, 483–494.
64. Sorkin, R. Geometry from Order: Causal Sets. Available online: http://www.einstein-online.info/en/spotlights/causal_sets/ (accessed on 9 January 2018).
65. Arageorgis, A. Spacetime as a Causal Set: Universe as a Growing Block? *Belgrad. Philos. Annu.* **2016**, *29*, 33–55.
66. Wüthrich, C.; Callender, C. What Becomes of a Causal Set? *Br. J. Philos. Sci.* **2017**, *68*, 907–925.
67. Bell, J.S. On the problem of hidden variables in quantum mechanics. *Rev. Mod. Phys.* **1966**, *38*, 447.
68. Aspect, A.; Grangier, P.; Roger, G. Experimental Tests of Realistic Local Theories via Bell's Theorem. *Phys. Rev. Lett.* **1981**, *47*, 460–463.
69. Kielpinski, D.; Meyer, V.; Sackett, C.A.; Itano, W.M.; Monroe, C.; Wineland, D.J. Experimental violation of a Bell's inequality with efficient detection. *Nature* **2001**, *409*, 791–794.
70. Hensen, B.; Bernien, H.; Dréau, A.E.; Reiserer, A.; Kalb, N.; Blok, M.S.; Ruitenberg, J.; Vermeulen, R.F.L.; Schouten, R.N.; Abellán, C.; et al. Experimental loophole-free violation of a Bell inequality using entangled electron spins separated by 1.3 km. *Nature* **2015**, *526*, 682–686.

71. Hossenfelder, S. Testing superdeterministic conspiracy. *J. Phys. Conf. Ser.* **2014**, *504*, 012018.
72. Timpson, C.; Maroney, O. Quantum-vs. macro-realism: What does the Leggett-Garg inequality actually test? *The British Journal for the Philosophy of Science*. 2013. Available online: https://ora.ox.ac.uk/objects/uuid:c2c31bfa-f9d3-4bc6-9853-79fcd79917f7/datastreams/ATTACHMENT1 (accessed on 9 January 2018)
73. Brukner, C.; Taylor, S.; Cheung, S.; Vedral, V. Quantum Entanglement in Time. *arXiv* **2004**, arXiv:quant-ph/0402127.
74. Brierley, S.; Kosowski, A.; Markiewicz, M.; Paterek, T.; Przysiężna, A. Nonclassicality of temporal correlations. *Phys. Rev. Lett.* **2015**, *115*, 120404.
75. Price, H.; Wharton, K. A Live Alternative to Quantum Spooks. *arXiv* **2015**, arXiv:1510.06712.
76. Almada, D.; Ch'ng, K.; Kintner, S.; Morrison, B.; Wharton, K.B. Are Retrocausal Accounts of Entanglement Unnaturally Fine-Tuned? *Int. J. Quantum Found.* **2015**, *2*, 1–16.
77. Price, H.; Wharton, K. Disentangling the Quantum World. *Entropy* **2015**, *17*, 7752–7767.
78. Dowker, F.; Kent, A. On the consistent histories approach to quantum mechanics. *J. Stat. Phys.* **1996**, *82*, 1575–1646.
79. Halliwell, J.J. A Review of the Decoherent Histories Approach to Quantum Mechanics. In *Fundamental Problems in Quantum Theory*; Greenberger, D.M., Zelinger, A., Eds.; New York Academy of Sciences: New York, NY, USA, 1995; p. 726.
80. Okon, E.; Sudarsky, D. On the Consistency of the Consistent Histories Approach to Quantum Mechanics. *Found. Phys.* **2014**, *44*, 19–33.
81. Aharonov, Y.; Popescu, S.; Tollaksen, J. Each Instant of Time a New Universe. In *Quantum Theory: A Two-Time Success Story*; Struppa, D.C., Tollaksen, J.M., Eds.; Springer: Berlin/Heidelberg, Germany, 2014; p. 21, ISBN 978-88-470-5216-1.
82. Kent, A. Solution to the Lorentzian quantum reality problem. *Phys. Rev. A* **2014**, *90*, 012107.

Article

A New Class of Retrocausal Models

Ken Wharton

Department of Physics and Astronomy, San José State University, San José, CA 95192-0106, USA; kenneth.wharton@sjsu.edu

Received: 27 April 2018; Accepted: 24 May 2018; Published: 26 May 2018

Abstract: Globally-constrained classical fields provide a unexplored framework for modeling quantum phenomena, including apparent particle-like behavior. By allowing controllable constraints on unknown past fields, these models are retrocausal but not retro-signaling, respecting the conventional block universe viewpoint of classical spacetime. Several example models are developed that resolve the most essential problems with using classical electromagnetic fields to explain single-photon phenomena. These models share some similarities with Stochastic Electrodynamics, but without the infinite background energy problem, and with a clear path to explaining entanglement phenomena. Intriguingly, the average intermediate field intensities share a surprising connection with quantum "weak values", even in the single-photon limit. This new class of models is hoped to guide further research into spacetime-based accounts of weak values, entanglement, and other quantum phenomena.

Keywords: Retrocausation; weak values; Stochastic Electrodynamics

1. Introduction

In principle, retrocausal models of quantum phenomena offer the enticing possibility of replacing the high-dimensional configuration space of quantum mechanics with ordinary spacetime, without breaking Lorentz covariance or utilizing action-at-a-distance [1–6]. Any quantum model based entirely on spacetime-localized parameters would obviously be much easier to reconcile with general relativity, not to mention macroscopic classical observations. (In general, block-universe retrocausal models can violate Bell-type inequalities because they contain hidden variables λ that are constrained by the future measurement settings (a, b). These constraints can be mediated via continuous influence on the particle worldlines, explicitly violating the independence assumption $P(\lambda|a,b) = P(\lambda)$ utilized in Bell-type no-go theorems.)

In practice, however, the most sophisticated spacetime-based retrocausal models to date only apply to a pair of maximally entangled particles [3,7–9]. A recent retrocausal proposal from Sen [10] is more likely to extend to more of quantum theory, but without a retrocausal mechanism it would have to use calculations in configuration space, preparing whatever initial distribution is needed to match the expected final measurement. Sutherland's retrocausal Bohmian model [11] also uses some calculations in configuration space. Given the difficulties in extending known retrocausal models to more sophisticated situations, further development may require entirely new approaches.

One obvious way to change the character of existing retrocausal models is to replace the usual particle ontology with a framework built upon spacetime-based fields. Every quantum "particle", after all, is thought to actually be an excitation of a quantum field, and every quantum field has a corresponding classical field that could exist in ordinary spacetime. The classical Dirac field, for example, is a Dirac-spinor-valued function of ordinary spacetime, and is arguably a far closer analog to the electrons of quantum theory than a classical charged particle. This point is even more obvious when it comes to photons, which have no classical particle analog at all, but of course have a classical analog in the ordinary electromagnetic field.

This paper will outline a new class of field-based retrocausal models. Field-based accounts of particle phenomena are rare but not unprecedented, one example being the Bohmian account of photons [12,13], using fields in configuration space. One disadvantage to field-based models is that they are more complicated than particle models. However, if the reason that particle-based models cannot be extended to more realistic situations is that particles are too simple, then moving to the closer analog of classical fields might arguably be beneficial. Indeed, many quantum phenomena (superposition, interference, importance of relative phases, etc.) have excellent analogs in classical field behavior. In contrast, particles have essentially only one phenomenological advantage over fields: localized position measurements. The class of models proposed here may contain a solution to this problem, but the primary goal will be to set up a framework in which more detailed models can be developed (and to show that this framework is consistent with some known experimental results).

Apart from being an inherently closer analog to standard quantum theory, retrocausal field models have a few other interesting advantages to their particle counterparts. One intriguing development, outlined in detail below, is an account of the average "weak values" [14,15] measured in actual experiments, naturally emerging from the analysis of the intermediate field values. Another point of interest is that the framework here bears similarities to Stochastic Electrodynamics (SED), but without some of the conceptual difficulties encountered by that program (i.e., infinite background energy, and a lack of a response to Bell's theorem) [16,17]. Therefore, it seems hopeful that many of the successes of SED might be applied to a further development of this framework.

The plan of this paper is to start with a conceptual framework, motivating and explaining the general approach that will be utilized by the specific models. Section 3 then explores a simple example model that illustrates the general approach, as well as demonstrating how discrete outcomes can still be consistent with a field-based model. Section 4 then steps back to examine a large class of models, calculating the many-run average predictions given a minimal set of assumptions. These averages are then shown to essentially match the weak-value measurements. The results are then used to motivate an improved model, as discussed in Section 5, followed by preliminary conclusions and future research directions.

2. Conceptual Framework

Classical fields generally have Cauchy data on every spacelike hypersurface. Specifically, for second order field equations, knowledge of the field and its time derivative everywhere at one time is sufficient to calculate the field at all times. However, the uncertainty principle, applied in a field framework, implies that knowledge of this Cauchy data can never be obtained: No matter how precise a measurement, some components of the field can always elude detection. Therefore, it is impossible to assert that either the preparation or the measurement of a field represents the precise field configuration at that time. This point sheds serious doubt on the way that preparations are normally treated as exact initial boundary conditions (and, in most retrocausal models, the way that measurements are treated as exact final boundary conditions).

In accordance with this uncertainty, the field of Stochastic Electrodynamics (SED) explores the possibility that in addition to measured electromagnetic (EM) field values, there exists an unknown and unmeasured "classical zero-point" EM field that interacts with charges in the usual manner [16,17]. Starting from the assumption of relativistic covariance, a natural gaussian noise spectrum is derived, fixing one free parameter to match the effective quantum zero-point spectrum of a half-photon per EM field mode. Using classical physics, a remarkable range of quantum phenomena can be recovered from this assumption. However, these SED successes come with two enormous problems. First, the background spectrum diverges, implying an infinite stress energy tensor at every point in spacetime. Such a field would clearly be in conflict with our best understanding of general relativity, even with some additional ultraviolet cutoff. Second, there is no path to recovering all quantum phenomena via locally interacting fields, because of Bell-inequality violations in entanglement experiments.

Both of these problems have a potential resolution when using the Lagrangian Schema [3] familiar from least-action principles in classical physics. Instead of treating a spacetime system as a computer program that takes the past as an input and generates the future as an output, the Lagrangian Schema utilizes both past and future constraints, solving for entire spacetime structures "all at once". Unknown past parameters (say, the initial angle of a ray of light constrained by Fermat's principle of least time) are the *outputs* of such a calculation, not inputs. Crucially, the action S that is utilized by these calculations is a covariant scalar, and therefore provides path to a Lorentz covariant calculation of unknown field parameters, different from the divergent spectrum considered by SED. The key idea is to keep the action extremized as usual ($\delta S = 0$), while also imposing some additional constraint on the total action of the system. One intriguing option is to quantize the action ($S = nh$), a successful strategy from the "old" quantum theory that has not been pursued in a field context, and would motivate $\delta S = 0$ in the first place. (Here, the action S is the usual functional of the fields throughout any given spacetime subsystem, calculated by integrating the classical Lagrangian density over spacetime.)

Constraining the action does not merely ensure relativistic covariance. When complex macroscopic systems are included in the spacetime subsystem (i.e., preparation and measurement devices), they will obviously dominate the action, acting as enormous constraints on the microscopic fields, just as a thermal reservoir acts as a constraint on a single atom. The behavior of microscopic fields would therefore depend on what experimental apparatus is considered. Crucially, the action is an integral over spacetime systems, not merely spatial systems. Therefore, the future settings and orientations of measurement devices strongly influence the total action, and unknown microscopic fields at earlier times will be effectively constrained by those future devices. Again, those earlier field values are literally "outputs" of the full calculation, while the measurement settings are inputs.

Such models are correctly termed "retrocausal". Given the usual block universe framework from classical field theory and the interventionist definition of causation [18–21], any devices with free external settings are "causes", and any constrained parameters are "effects" (including field values at spacetime locations before the settings are chosen). Such models are retrocausal but not retro-signaling, because the future settings constrain unknown past field parameters, hidden by the uncertainty principle. (These models are also forward-causal, because the preparation is another intervention.) It is important not to view causation as a process—certainly not one "flowing" back-and-forth through time—as this would violate the block universe perspective. Instead, such systems are consistently solved "all-at-once", as in action principles. Additional discussion of this topic can be found in [2,4,22].

The retrocausal character of these models immediately provides a potential resolution to both of the problems with SED. Concerning the infinite-density zero point spectrum, SED assumes that all possible field modes are required because one never knows which ones will be relevant in the future. However, a retrocausal model is not "in the dark" about the future, because (in this case) the action is an integral that *includes* the future. The total action might very well only be highly sensitive to a bare few field modes. (Indeed, this is usually the case; consider an excited atom, waiting for a zero-point field to trigger "spontaneous" emission. Here, only one particular EM mode is required to explain the eventual emission of a photon, with the rest of the zero point field modes being irrelevant to a future photon detector.) As is shown below, it is not difficult to envision action constraints where typically only a few field modes need to be populated in the first place, resolving the problem of infinities encountered by SED. Furthermore, it is well-known that retrocausal models can naturally resolve Bell-inequality violations without action-at-a-distance, because the past hidden variables are naturally correlated with the future measurement settings [4,23]. (Numerous proof-of-principle retrocausal models of entanglement phenomena have been developed over the past decade [3,7–10].)

Unfortunately, solving for the exact action of even the simplest experiments is very hard. The macroscopic nature of preparation and measurement that makes them so potent as boundary constraints also makes them notoriously difficult to calculate exactly—especially when the relevant changes in the action are on the order of Planck's constant. Therefore, to initially consider such models, this paper will assume that any constraint on the total action manifests itself as certain rules constraining how microscopic fields are allowed to interact with the macroscopic devices. (Presumably, such rules would include quantization conditions, for example only allowing absorption of EM waves in packets of energy $\hbar\omega$.) This assumption will allow us to focus on what is happening between devices rather than in the devices themselves, setting aside those difficulties as a topic for future research.

This paper will proceed by simply exploring some possible higher-level interaction constraints (guided by other general principles such as time-symmetry), and determining whether they might plausibly lead to an accurate explanation of observed phenomena. At this level, the relativistic covariance will not be obvious; after all, when considering intermediate EM fields in a laboratory experiment, a special reference frame is determined by the macroscopic devices which constrain those fields. However, it seems plausible that if some higher-level model matches known experiments then a lower-level covariant account would eventually be acheivable, given that known experiments respect relativistic covariance.

The following examples will be focused on simple problems, with much attention given to the case where a single photon passes through a beamsplitter and is then measured on one path or the other. This is precisely the case where field approaches are thought to fail entirely, and therefore the most in need of careful analysis. In addition, bear in mind that these are representative examples of an entire class of models, not one particular model. It is hoped that, by laying out this new class of retrocausal models, one particular model will eventually emerge as a possible basis for a future reformulation of quantum theory.

3. Constrained Classical Fields

3.1. Classical Photons

Ordinary electromagnetism provides a natural analog to a single photon: a finite-duration electromagnetic wave with total energy $\hbar\omega$. Even in classical physics, all of the usual uncertainty relations exist between the wave's duration and its frequency ω; in the analysis below, we assume long-duration EM waves that have a reasonably well-defined frequency, in some well-defined beam such as the TEM_{00} gaussian mode of a narrow bandwidth laser. By normalizing the peak intensity I of this wave so that a total energy of $\hbar\omega$ corresponds to $I=1$, one can define a "Classical Photon Analog" (CPA).

Such CPAs are rarely considered, for the simple reason that they seem incompatible with the simple experiment shown in Figure 1a. If such a CPA were incident upon a beamsplitter, some fraction T of the energy would be transmitted and the remaining fraction $R = 1 - T$ would be reflected. This means that detectors A and B on these two paths would never see what actually happens, which is a full $\hbar\omega$ amount of energy on either A or B, with probabilities T and R, respectively. Indeed, this very experiment is usually viewed as proof that classical EM is incorrect.

Notice that the analysis in the previous paragraph assumed that the initial conditions were exactly known, which would violate the uncertainty principle. If unknown fields existed on top of the original CPA, boosting its total energy to something larger than $\hbar\omega$, it would change the analysis. For example, if the CPA resulted from a typical laser, the ultimate source of the photon could be traced back to a spontaneous emission event, and (in SED-style theories) such "spontaneous" emission is actually *stimulated* emission, due to unknown incident zero-point radiation. This unknown background would then still be present, boosting the intensity of the CPA such that $I>1$. Furthermore, every beamsplitter has a "dark" input port, from which any input radiation would also end up on the same two detectors, A and B. In quantum electrodynamics, it is essential that one remember to put an input vacuum state on such dark ports; the classical analog of this well-known procedure is to allow for possible unknown EM wave inputs from this direction.

The uncertain field strengths apply to the outputs as well as the inputs, from both time-symmetry and the uncertainty principle. Just because a CPA is measured on some detector A, it does not follow that there is no additional EM wave energy that goes unmeasured. Just because nothing is measured on detector B does not mean that there is no EM wave energy there at all. If one were to insist on a perfectly energy-free detector, one would violate the uncertainty principle.

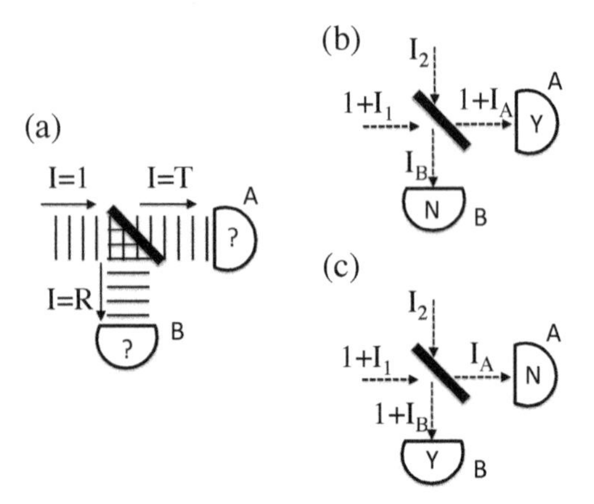

Figure 1. (**a**) A classical photon analog encounters a beamsplitter, and is divided among two detectors, in contradiction with observation. (**b**) A classical photon analog, boosted by some unknown peak intensity I_1, encounters the same beamsplitter. Another beam with unknown peak intensity I_2 enters the dark port. This is potentially consistent with a classical photon detection in only detector A ("Y" for yes, "N" for no), so long as the output intensities I_A and I_B remain unobserved. (The wavefronts have been replaced by dashed lines for clarity.) (**c**) The same inputs as in (**b**), but with outputs consistent with classical photon detection in only detector B, where the output intensities I_A and I_B again remain unobserved.

By adding these unknown input and output fields, Figure 1b demonstrates a classical beamsplitter scenario that is consistent with an observation of one CPA on detector A. In this case, two incoming beams, with peak intensities $1 + I_1$ and I_2, interfere to produce two outgoing beams with peak intensities $1 + I_A$ and I_B. The four unknown intensities are related by energy conservation, $I_1 + I_2 = I_A + I_B$, where the exact relationship between these four parameters is determined by the unknown phase difference between the incoming beams. Different intensities and phases could also result in the detection of exactly one CPA on detector B, as shown in Figure 1c. These scenarios are allowed by classical EM and consistent with observation, subject to known uncertainties in measuring field values, pointing the way towards a classical account of "single-photon" experiments. This is also distinct from prior field-based accounts of beamsplitter experiments [13]; here there is no need to non-locally transfer field energy from one path to another.

Some potential objections should be addressed. One might claim that quantum theory does allow certainty in the total energy of a photon, at the expense of timing and phase information. However, in quantum field theory, one can only arrive at this conclusion after one has renormalized the zero-point values of the electromagnetic field—the very motivation for I_1 and I_2 in the first place. (Furthermore, when hunting for some more-classical formulation of quantum theory, one should not assume that the original formulation is correct in every single detail.)

Another objection would be to point out the sheer implausibility of any appropriate beam I_2. Indeed, to interfere with the original CPA, it would have to come in with just the right frequency, spatial mode, pulse shape, and polarization. However, this concern makes the error of thinking of all past parameters as logical inputs. In the Lagrangian Schema, the logical inputs are the known constraints at the beginning and end of the relevant system. The unknown parameters are logical *outputs* of this Schema, just as the initial angle of the light ray in Fermat's principle. The models below aim to generate the parameters of the incoming beam in I_2, as constrained by the entire experiment. In action principles, just because a parameter is coming into the system at the temporal beginning does not mean that it is a logical input. In retrocausal models, these are the parameters that are the effects of the constraints, not causes in their own right. (Such unknown background fields do not have external settings by which they can be independently controlled, even in principle, and therefore they are not causal interventions.)

Even if the classical field configurations depicted in Figure 1 are possible, it remains to explain why the observed transmission shown in Figure 1b occurs with a probability T, while the observed reflection shown in Figure 1c occurs with a probability R. To extract probabilities from such a formulation, one obviously needs to assign probabilities to the unknown parameters, $P(I_1)$, $P(I_2)$, etc. However, use of the Lagrangian Schema requires an important distinction, in that the probabilities an agent would assign to the unknown fields would depend on that agent's information about the experimental geometry. In the absence of any information whatsoever, one would start with a "a priori probability distribution" $P_0(I_2)$—effectively a Bayesian prior that would be (Bayesian) updated upon learning about any experimental constraints. Any complete model would require both a probability distribution P_0 as well as rules for how the experimental geometry might further constrain the allowed field values.

Before giving an example model, one further problem should be noted. Even if one were successful in postulating some prior distribution $P_0(I_1)$ and $P_0(I_2)$ that eventually recovered the correct probabilities, this might very well break an important time symmetry. Specifically, the time-reverse of this situation would instead depend on $P_0(I_A)$ and $P_0(I_B)$. For that matter, if both outgoing ports have a wave with a peak intensity of at least $I = 1$, then the only parameters sensitive to which detector fires are the unobserved intensities I_A and I_B. Both arguments encourage us to include a consideration of the unknown outgoing intensities I_A and I_B in any model, not merely the unknown incoming fields.

3.2. Simple Model Example

The model considered in this section is meant to be an illustrative example of the class of retrocausal models described above, illustrating that it is possible to get particle-like phenomena from a field-based ontology, and also indicating a connection to some of the existing retrocausal accounts of entanglement.

One way to resolve the time-symmetry issues noted above is to impose a model constraint whereby the two unobserved incoming intensities I_1 and I_2 are always exactly equal to the unobserved outgoing intensities I_A and I_B (either $I_1 = I_A$ or $I_1 = I_B$). If this constraint is enforced, then assigning a probability of $P_0(I_1)P_0(I_2)$ to each diagram does not break any time symmetry, as this quantity will always be equal to $P_0(I_A)P_0(I_B)$. One simple rule that seems to work well in this case is the a priori distribution

$$P_0(I_Z) = Q\frac{1}{\sqrt{I_Z}} \quad (\textit{where } I_Z > \epsilon). \tag{1}$$

Here, I_Z is any of the allowed unobserved background intensities, Q is a normalization constant, and ϵ is some vanishingly small minimum intensity to avoid the pole at $I_Z = 0$. (While there may be a formal need to normalize this expression, there is never a practical need; these prior probabilities will be restricted by the experimental constraints before being utilized, and will have to be normalized again.) The only additional rule to recover the appropriate probabilities is that $I_1 \gg \epsilon$. (This might be motivated by the above analysis that laser photons would have to be triggered by background fields, so the known incoming CPA would have to be accompanied by a non-vanishing unobserved field.)

To see how these model assumptions lead to the appropriate probabilities, first consider that it is overwhelmingly probable that $I_2 \approx \epsilon$. Thus, in this case, we can ignore the input on the dark port of the beamsplitter. However, with only one non-vanishing input, there can be no interference, and both outputs must have non-vanishing intensities. The only way it is possible for detector A to fire, given the above constraints, is if $I_1 = I_B = R/T$ in Figure 1b (such that $I_2 = I_A = 0$). The only way it is possible for detector B to fire, in Figure 1c, is if $I_1 = I_A = T/R$.

With this added information from the experimental geometry, one would update the prior distribution $P_0(I_1)$ by constraining the only allowed values of I_1 to be R/T or T/R (and then normalizing). The relative probabilities of these two cases is therefore

$$\frac{P(A)}{P(B)} = \frac{\frac{1}{\sqrt{R/T}} P_0(I_2)}{\frac{1}{\sqrt{T/R}} P_0(I_2)} = \frac{T}{R}, \tag{2}$$

yielding the appropriate ratio of possible outcomes.

Taking stock of this result, here are the assumptions of this example model:

- The a priori probability distribution on each unknown field intensity is given by Equation (1)—to be updated for any given experiment.
- The unknown field values are further constrained to be equal as pairs, $\{I_1, I_2\} = \{I_A, I_B\}$.
- I_1 is non-negligible because it accompanies a known "photon".
- The probability of each diagram is given by $P_0(I_1)P_0(I_2)$, or equivalently, $P_0(I_A)P_0(I_B)$.

Note that it does not seem reasonable to assign the prior probability to the total incoming field $(1 + I_1)$, because Equation (1) should refer to the probability given no further information, not even the knowledge that there is an incoming photon's worth of energy on that channel. (The known incoming photon that defines this experiment is an addition to the a priori intensity, not a part of it.) Given these assumptions, one finds the appropriate probabilities for a detected transmission as compared to a detected reflection.

There are several other features of this example model. Given Equation (1), it should be obvious that the total energy in most zero-point fields should be effectively zero, resolving the standard SED problem of infinite zero-point energy. In addition, this model would work for any device that splits a photon into two paths (such as a polarizing cube), because the only relevant parameters are the classical transmission and reflection, T and R.

More importantly, this model allows one to recover the correct measurement probabilities for two maximally entangled photons in essentially the same way as several existing retrocausal models in the literature [3,7,8]. Consider two CPAs produced by parametric down-conversion in a nonlinear crystal, with identical but unknown polarizations (a standard technique for generating entangled photons). The three-wave mixing that classically describes the down-conversion process can be strongly driven by the presence of background fields matching one of the two output modes, M1, even if there is no background field on the other output mode, M2. (Given Equation (1), having essentially no background field on one of these modes is overwhelmingly probable.) Thus, in this case, the polarization of M2 necessarily matches the polarization of the unknown background field on M1 (the field that strongly drives the down-conversion process).

Now, assume both output photons are measured by polarizing cubes set at arbitrary polarization angles, followed by detectors. With no extra background field on M2, the only way that M2 could satisfy the above constraints at measurement would be if its polarization was *already* exactly aligned (modulo $\pi/2$) with the angle of the future polarizing cube. (In that case, no background field would be needed on that path; the bare CPA would fully arrive at one detector or the other.) However, we have established that the polarization of M2 was selected by the background field on M1, so the background field on M1 is also forced to align with the measurement angle on M2 (modulo $\pi/2$). In other words, solving the whole experiment "all at once", the polarization of both photons is effectively constrained to match one of the two future measurement angles.

This is essentially what happens in several previously-published retrocausal models of maximally entangled particles [3,7,8]. In these models, the properties of both particles (spin or polarization, depending on the context) are constrained to be aligned with one of the two future settings. The resulting probabilities are then entirely determined by the mis-matched particle, the one doesn't match the future settings. However, this is just a single-particle problem, and in this case the corresponding classical probabilties (R and T, given by Malus's Law at the final polarizer) are enforced by the above rules, matching experimental results for maximally entangled particles. The whole picture almost looks as if the measurement on one photon has collapsed the other photon into that same polarization, but in these models it was clear that the CPAs had the correct polarization all along, due to future constraints on the appropriate hidden fields.

3.3. Discussion

The above model was presented as an illustrating example, demonstrating one way to resolve the most obvious problems with classical photon analogs and SED-style approaches. Unfortunately, it does not seem to extend to more complicated situations. For example, if one additional beamsplitter is added, as in Figure 2, no obvious time-symmetric extension of the assumptions in the previous section lead to the correct results. In this case, one of the two dark ports would have to have non-negligible input fields. Performing this analysis, it is very difficult to invent any analogous rules that lead to the correct distribution of probabilities on the three output detectors.

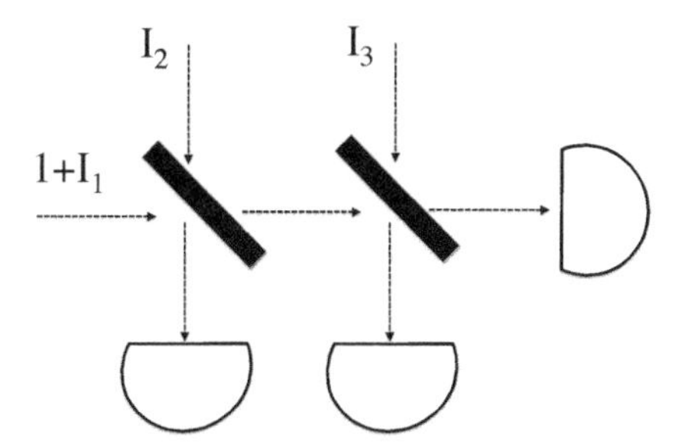

Figure 2. A classical photon analog encounters two beamsplitters, and is divided among three detectors. The CPA is boosted by some unknown peak intensity I_1, and each beamsplitter's dark port has an additional incident field with unknown intensity.

In Section 5, we show that it is possible to resolve this problem, using different assumptions to arrive at another model which works fine for multiple beamsplitters. However, before proceeding, it is worth reviewing the most important accomplishment so far. We have shown that it is possible to give a classical field account of an apparent single photon passing through a beamsplitter, matching known observations. Such models are generally thought to be impossible (setting aside nonlocal options [13]). Given that they *are* possible—if using the Lagrangian Schema—the next-level concern could be that such models are simply implausible. For phenomena that look so much like particle behavior, such classical-field-based models might seem to be essentially unmotivated.

The next section addresses this concern in two different ways. First, the experiments considered in Section 4 are expanded to include clear wave-like behavior, by combining two beamsplitters into an interferometer. Again, the input and output look like single particles, but now some essential wave interference is clearly occurring in the middle. Second, the averaged and post-selected results of these models can be compared with "weak values" that can be measured in actual experiments [14,15]. Notably, the results demonstrate a new connection between the average intermediate classical fields and experimental weak values. This correspondence is known in the high-field case [24–28], but here they are shown to apply even in the single-photon regime. Such a result will boost the general plausibility of this classical-field-based approach, and will also motivate an improved model for Section 5.

4. Averaged Fields and Weak Values

Even without a particular retrocausal model, it is still possible to draw conclusions as to the long-term averages predicted over many runs of the same experiment. The only assumption made here will be that every relevant unknown field component for a given experiment (both inputs and outputs) is treated the same as every other. In Figure 1, this would imply an equality between the averaged values $<I_1>=<I_2>=<I_A>=<I_B>$, each defined to be the quantity I_Z.

Not every model will lead to this assumption; indeed, the example model above does not, because the the CPA-accompanying field I_1 was treated differently from the dark port field I_2. However, for models which do not treat these fields differently, the averages converge onto parameters that can actually be measured in the laboratory: weak values [14,15]. This intriguing correspondence is arguably an independent motivation to pursue this style of retrocausal models.

4.1. Beamsplitter Analysis

Applying this average condition on the simple beamsplitter example of Figure 1b,c yields a phase relationship between the incoming beams, in order to retain the proper outputs. If θ is the phase difference between I_1 and I_2 before the beamsplitter, then taking into account the relative $\pi/2$ phase shift caused by the beamsplitter itself, a simple calculation for Figure 1b reveals that

$$\langle 1+I_A\rangle = I_Z + T - \left\langle 2\sqrt{RT(1+I_1)(I_2)}\sin\theta\right\rangle \tag{3}$$

$$\langle I_B\rangle = I_Z + R + \left\langle 2\sqrt{RT(1+I_1)(I_2)}\sin\theta\right\rangle . \tag{4}$$

Given the above restrictions on the average values, this is only possible if there exists a non-zero average correlation

$$C \equiv \left\langle \sqrt{(1+I_1)(I_2)}\sin\theta\right\rangle \tag{5}$$

between the inputs, such that $C = -\sqrt{R/4T}$. The same analysis applied to Figure 1c reveals that in this case $C = \sqrt{T/4R}$. (This implies some inherent probability distribution $P(I_1, I_2, \theta) \propto 1/|C|$ to yield the correct distribution of outcomes, which will inform some of the model-building in the next section.) In this case, there are no intermediate fields to analyze, as every mode is either an input or an output. To discuss intermediate fields, we must go to a more complicated scenario.

4.2. Interferometer Analysis

Consider the simple interferometer shown in Figure 3. For these purposes, we assume it is aligned such that the path length on the two arms is exactly equal. For further simplicity, the final beamsplitter is assumed to be 50/50. Again, the global constraints imply that either Figure 3a or Figure 3b actually happens. A calculation of the average intermediate value of I_x yields the same result as Equation (3), while the average value of I_y is the same as Equation (4). For Figure 3a, further interference at the final beamsplitter then yields, after some simplifying algebra,

$$\langle 1+I_A\rangle = (0.5+\sqrt{RT}) + I_Z + (T-R)\left\langle \sqrt{(1+I_1)(I_2)}\sin\theta\right\rangle \tag{6}$$

$$\langle I_B\rangle = (0.5-\sqrt{RT}) + I_Z - (T-R)\left\langle \sqrt{(1+I_1)(I_2)}\sin\theta\right\rangle . \tag{7}$$

The first term on the right of these expressions is the outgoing classical field intensity one would expect for a single CPA input, with no unknown fields. Because of our normalization, it is also the expected probability of a single-photon detection on that arm. The second term is just the average unknown field I_Z, and the final term is a correction to this average that is non-zero if the incoming unknown fields are correlated. Note that the quantity C defined in Equation (5) again appears in this final term.

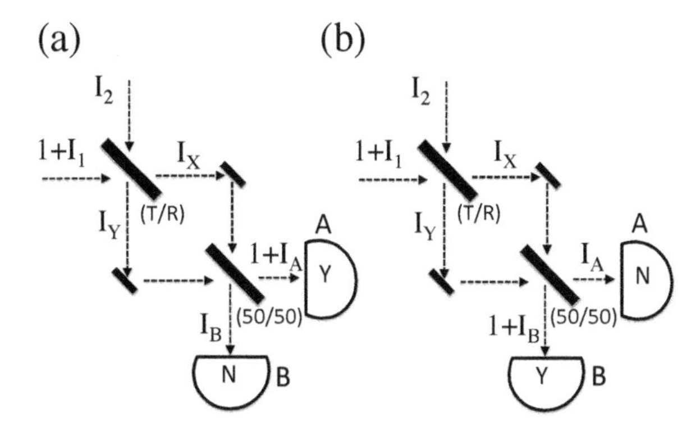

Figure 3. (**a**) A classical photon analog, boosted by some unknown peak intensity I_1, enters an interferometer through a beamsplitter with transmission fraction T. An unknown field also enters from the dark port. Both paths to the final 50/50 beamsplitter are the same length; the intermediate field intensities on these paths are I_X and I_Y. Here, detector A fires, leaving unmeasured output fields I_A and I_B. (**b**) The same situation as (**a**), except here detector B fires.

To make this end result compatible with the condition that $\langle 1 + I_A \rangle = 1 + I_Z$, the correlation term C must be constrained to be $C = (0.5 - \sqrt{RT})/(T - R)$. For Figure 3b, with detector B firing, this term must be $C = -(0.5 + \sqrt{RT})/(T - R)$. (As in the beamsplitter case, the quantity $1/|C|$ happens to be proportional to the probability of the corresponding outcome, for allowed values of C.) Notice that as the original beamsplitter approaches 50/50, the required value of C diverges for Figure 3b, but not for Figure 3a. That is because this case corresponds to a perfectly tuned interferometer, where detector A is certain to fire, but never B. (This analysis also goes through for an interferometer with an arbitrary phase shift, and arbitrary final beamsplitter ratio; these results will be detailed in a future publication.)

In this interferometer, once the outcome is known, it is possible to use C to calculate the average intensities $<I_X>$ and $<I_Y>$ on the intermediate paths. For Figure 3a, some algebra yields:

$$\langle I_X \rangle = I_Z + \frac{\sqrt{T}}{\sqrt{T}+\sqrt{R}} \tag{8}$$

$$\langle I_Y \rangle = I_Z + \frac{\sqrt{R}}{\sqrt{T}+\sqrt{R}}. \tag{9}$$

For Figure 3b, the corresponding average intermediate intensities are

$$\langle I_X \rangle = I_Z + \frac{\sqrt{T}}{\sqrt{T}-\sqrt{R}} \tag{10}$$

$$\langle I_Y \rangle = I_Z - \frac{\sqrt{R}}{\sqrt{T}-\sqrt{R}}. \tag{11}$$

Remarkably, as we are about to see, the non-I_Z portion of these calculated average intensities can actually be measured in the laboratory.

4.3. Weak Values

When the final outcome of a quantum experiment is known, it is possible to elegantly calculate the (averaged) result of a weak intermediate measurement via the real part of the "Weak Value" equation [14]:

$$\langle \mathbf{Q} \rangle_{weak} = Re\left(\frac{<\Phi|\mathbf{Q}|\Psi>}{<\Phi|\Psi>} \right). \tag{12}$$

Here, $|\Psi>$ is the initial wavefunction evolved forward to the intermediate time of interest, $|\Phi>$ is the final (measured) wavefunction evolved backward to the same time, and $\mathbf{Q}$ is the operator for which one would like to calculate the expected weak value. (Note that weak values by themselves are not retrocausal; post-selecting an outcome is not a causal intervention. However, if one takes the backward-evolved wavefunction $|\Phi>$ to be an element of reality, as done by one of the authors here [29], then one does have a retrocausal model—albeit in configuration space rather than spacetime.) Equation (12) yields the correct answer in the limit that the measurement $\mathbf{Q}$ is sufficiently weak, so that it does not appreciably affect the intermediate dynamics. The success of this equation has been verified in the laboratory [26], but is subject to a variety of interpretations. For example, $\langle \mathbf{Q} \rangle_{weak}$ can be negative, seemingly making a classical interpretation impossible.

In the case of the interferometer, the intermediate weak values can be calculated by recalling that it is the square root of the normalized intensity that maps to the wavefunction. (Of course, the standard wavefunction knows nothing about I_Z; only the prepared and detected photon are relevant in a quantum context.) Taking into account the phase shift due to a reflection, the wavefunction between the two beamsplitters is simply $|\Psi> = \sqrt{T}|X> + i\sqrt{R}|Y>$, where $|X>(|Y>)$ is the state of the photon on the upper (lower) arm of the interferometer.

The intermediate value of $|\Phi>$ depends on whether the photon is measured by detector A or B. The two possibilities are:

$$|\Phi_A> \quad = \quad \frac{1}{\sqrt{2}}\left(-i|X> + |Y>\right), \tag{13}$$

$$|\Phi_B> \quad = \quad \frac{1}{\sqrt{2}}\left(|X> - i|Y>\right). \tag{14}$$

Notice that, in this case, the reflection off the beamsplitter is associated with a negative $\pi/2$ phase shift, because we are evolving the final state in the opposite time direction.

These are easily inserted into Equation (12), where $\mathbf{Q} = |X><X|$ for a weak measurement of I_X, and $\mathbf{Q} = |Y><Y|$ for a weak measurement of I_Y. (Given our normalization, probability maps to peak intensity.) If the outcome is a detection on A, this yields

$$\langle I_X \rangle_{weak} = \frac{\sqrt{T}}{\sqrt{T}+\sqrt{R}}, \tag{15}$$

$$\langle I_Y \rangle_{weak} = \frac{\sqrt{R}}{\sqrt{T}+\sqrt{R}}. \tag{16}$$

If instead the outcome is a detection on B, one finds

$$\langle I_X \rangle_{weak} = \frac{\sqrt{T}}{\sqrt{T}-\sqrt{R}}, \tag{17}$$

$$\langle I_Y \rangle_{weak} = \frac{-\sqrt{R}}{\sqrt{T}-\sqrt{R}}. \tag{18}$$

Except for the background average intensity I_Z, these quantum weak values are precisely the same intermediate intensities computed in the previous section.

The earlier results were framed in an essentially classical context, but these weak values come from an inherently quantum calculation, with no clear interpretation. Some of the strangest features of weak values are when one gets a negative probability/intensity, which seem to have no classical analog whatsoever. For example, whenever detector B fires, either Equation (17) or Equation (18) will be negative. (Recall that if $T = R$, then B never fires.) Nevertheless, a classical interpretation of this negative weak value is still consistent with the earlier results of Equations (10) and (11), because those cases also include an additional unknown intensity I_Z. It is perfectly reasonable to have classical destructive interference that would decrease the average value of I_Y to below that of I_Z; after all, the latter is just an unknown classical field.

One objection here might be that for values of $T \approx R$, the weak values of Equations (17) and (18) could get arbitrarily large, such that I_Z would have to be very large as well to maintain a positive intensity for both Equations (10) and (11). However, consider that if I_Z were *not* large enough, then there would be no classical solution at all, in contradiction to the Lagrangian Schema assumptions considered above (requiring a global solution to the entire problem). Furthermore, if the weak values get very large, that is only because the outcome at B becomes very improbable, meaning that I_Z would rarely have to take a large value. As we show in the next section, there are reasonable a priori distributions of I_Z which would be consistent with this occasional restriction.

Such connections between uncertain classical fields and quantum weak values are certainly intriguing, and also under current investigation by at least one other group [30]. However, while it may be that the unknown-classical-field framework might help make some conceptual sense of quantum weak values, the main point here is simply that these two perspectives are mutually consistent. Specifically, the known experimental success of weak value predictions seems to equally support the unknown-field formalism presented above. It remains to be seen whether (and why) these two formalisms always seem to give compatible answers in every case, but this paper will set that question aside for future research.

For the purposes of this introductory paper, the final task will be to consider whether the above results indicate a more promising model of these experiments.

5. An Improved Model

Given the intriguing connection to weak values demonstrated in the previous section, it seems worth trying to revise the example model from Section 3. In Section 4, the new assumption which led to the successful result was that every unknown field component (I_1, I_2, I_A, I_B), should be treated on an equal footing, not singling out I_1 for accompanying a known photon. (Recall the average value of each of these was assumed to be some identical parameter I_Z.) Meanwhile, the central idea of the model in Section 3 is that time-symmetry could be enforced by demanding an exact equivalence between the two input fields (I_1, I_2) and the two output fields (I_A, I_B).

One obvious way to combine all these ideas is to instead demand an equivalence between all four of these intensities—not on average, but on every run of the experiment. This might seem to be in conflict with the weak value measurements, which are not the same on every run, but only converge to the weak values after an experimental averaging. However, these measurements are necessarily weak/noisy, so these results are inconclusive as to whether the underlying signal is constant or varying. (Alternatively, one could consider a class of models that on average converge to the below model, but this option will also be set aside for the purposes of this paper.)

With the very strict constraint that each of (I_1, I_2, I_A, I_B) are always equal to the same intensity I_Z, the only two free parameters are I_Z and the relative initial phase θ (between the two incoming modes $1 + I_1$ and I_2). In addition, θ and I_Z must be correlated, depending on the experimental parameters, in order to fulfill these constraints. For the case of the beamsplitter (Figure 1b,c), this amounts to removing all the time-averages from the analysis of Section 4.1. This leads to the conditions

$$\frac{1}{\sqrt{I_{ZA}^2 + I_{ZA}}} = -\sin\theta\sqrt{\frac{4T}{R}}, \tag{19}$$

$$\frac{1}{\sqrt{I_{ZB}^2 + I_{ZB}}} = \sin\theta\sqrt{\frac{4R}{T}}. \tag{20}$$

Here, I_{ZA} is the value of I_Z needed for an outcome on detector A (as in Figure 1b), and I_{ZB} is the value of I_Z needed for an outcome on detector B (as in Figure 1c). Both are functions of θ.

This model requires a priori probability distributions $P_0(I_Z)$ and $P_0'(\theta)$ (the prime is to distinguish these two functions). The hope is that these distributions can then be restricted by the global constraints such that the correct outcome probabilities are recovered. To implement the above constraints, instead of integrating over the two-dimensional space $[I_Z, \theta]$, the correlations between I_Z and θ essentially make this a one-dimensional space, which can be calculated with a delta function:

$$\frac{\int P_0(I_Z)P_0'(\theta)\delta(I_Z - I_{ZA})dI_Z d\theta}{\int P_0(I_Z)P_0'(\theta)\delta(I_Z - I_{ZB})dI_Z d\theta} = \frac{P(outcome\, A)}{P(outcome\, B)}. \tag{21}$$

It is very hard to imagine any rule whereby $P_0'(\theta)$ would not start out as a flat distribution—all relative phases should be equally a priori likely. The earlier observation that the appropriate probability was always proportional to $1/|C|$ (in both the beamsplitter and the interferometer geometries) motivates the following guess for an a priori probability distribution for background fields:

$$P_0(I_Z) \propto \frac{1}{\sqrt{I_Z^2 + I_Z}}, \tag{22}$$

assuming the normalization where $I = 1$ corresponds to a single classical photon. This expression diverges as $I_Z \to 0$, which is appropriate for avoiding the infinities of SED, although some cutoff would be required to form a normalized distribution. (Again, it is unclear whether an a priori assessment of relative likelihood would actually have to be normalized, given that in any experimental instance there would only be some values of I_Z which were possible, and only these probabilities would have to be normalized.)

Inserting Equation (22) into Equation (21), along with a flat distribution for $P_0'(\theta)$, the beamsplitter conditions from Equations (19) and (20) yield

$$\frac{\int_{\pi}^{2\pi} -\sin\theta\sqrt{4T/R}d\theta}{\int_0^{\pi} \sin\theta\sqrt{4R/T}d\theta} = \frac{T}{R}, \tag{23}$$

as desired. Here, the limits on θ come from the range of possible solutions to Equations (19) and (20). A similar successful result is found in the above case of the interferometer, because $1/|C|$ is again proportional to the outcome probability. This model also works well for the previously-problematic case of multiple beamsplitters shown in Figure 2. Now, because the incoming fields (I_1, I_2, I_3) are all equal, this essentially splits into two consecutive beamsplitter problems, and the probabilities of these two beamsplitters combine in an ordinary manner.

Summarizing the assumptions behind this improved model:

- The unknown field values are constrained to all be equal: $I_1 = I_2 = I_A = I_B$.
- The *apriori* probability distribution on each unknown field intensity is given by Equation (22)—but must be updated for any given experiment.
- The relative phase between the incoming fields is a priori completely unknown—but must be updated for any given experiment.

However, there is still a conceptual difficulty in this new model, in that all considered incoming field modes are constrained to be equal intensities, but we have left the *unconsidered* modes equal to zero. (Meaning, the modes with the wrong frequencies, or coming in the wrong direction, etc.). If literally all zero-point modes were non-zero, it would not only change the above calculations, but it would run directly into the usual infinities of SED. Thus, if this improved model were to be further developed, there would have to be some way to determine certain groups of background modes that were linked together through the model assumptions, while other background modes could be neglected.

This point is also essential if such a revised model is to apply to entangled particles. For two down-converted photons with identical polarizations, each measured by a separate beamsplitter, there are actually four relevant incoming field modes: the unknown intensity accompanying each photon, as well as the unknown intensity incident upon the dark port of each beamsplitter. If one sets all four of these peak intensities to the same I_Z, one does not recover the correct joint probabilities of the two measurements. However, if two of these fields are (nearly) zero, as described in Section 3.2, then the correct probabilities are recovered in the usual retrocausal manner (see Section 3.2 or [3,7,8]). Again, it seems that there must be some way to parse the background modes into special groups.

The model in this section is meant to be an example starting point, not some final product. Additional features and ideas that might prove useful for future model development will now be addressed in the final section.

6. Summary and Future Directions

Retrocausal accounts of quantum phenomena have come a long way since the initial proposal by Costa de Beauregard [31]. Notably, the number of retrocausal models in the literature has expanded significantly in the past decade alone [3,7–11,22,32–40], but more ideas are clearly needed. The central novelties in the class of models discussed here are: (1) using fields (exclusively) rather than particles; and (2) introducing uncertainty to even the initial and final boundary constraints. Any retrocausal model must have hidden variables (or else there is nothing for the future measurement choices to constrain), but it has always proved convenient to segregate the known parameters from the unknown parameters in a clear manner. Nature, however, may not respect such a convenience. In the case of realistic measurements on fields, there is every reason to think that our best knowledge of the field strength may not correspond to the actual value.

Although the models considered here obey classical field equations (in this case, classical electromagnetism), they only make sense in terms of the Lagrangian Schema, where the entire experiment is solved "all-at-once". Only then does it make sense to consider incoming dark-port fields (such as I_2), because the global solution may require these incoming modes in order have a solution. However, despite the presence of such fields at the beginning of the experiment (and, presumably, before it even begins), they are not "inputs" in the conventional sense; they are literally outputs of the retrocausal model.

The above models have demonstrated a number of features and consequences, most notably:

- Distributed classical fields can be consistent with particle-like detection events.
- There exist simple constraints and a priori field intensity distributions that yield the correct probabilities for basic experimental geometries.
- Most unobserved field modes are expected to have zero intensity (unlike in SED).
- The usual retrocausal account for maximally entangled photons still seems to be available.
- The average intermediate field values, minus the unobserved background, is precisely equal to the "weak value" predicted by quantum theory (in the cases considered so far).
- Negative weak values can have a classical interpretation, provided the unobserved background is sufficiently large.

This seems to be a promising start, but there are many other research directions that might be inspired by these models. For example, consider the motivation of action constraints, raised in Section 2. If the total action is ultimately important, then any constraint or probability rule would have to consider the contribution to the action of the microscopic intermediate fields. Even the simple case of a CPA passing through a finite-thickness beamsplitter has a non-trivial action. (A single free-field EM wave has a vanishing Lagrangian density at every point, but two crossing or interfering waves generally do not). It certainly seems worth developing models that constrain not only the inputs and outputs, but also these intermediate quantities (which would have the effect of further constraining the inputs and outputs).

Another possibility is to make the incoming beams more realistic, introducing spatially-varying noise, not just a single unknown parameter per beam. It is well-known that such spatial noise introduces bright speckles into laser profiles, and in some ways these speckles are analogous to detected photons—in terms of both probability distributions as well as their small spatial extent (compared to the full laser profile). A related point would be to introduce unknown *matter* fields, say some zero-point equivalent of the classical Dirac field, which would introduce further uncertainty and effective noise sources into the electromagnetic field. These research ideas, and other related approaches, are wide open for exploration.

Certainly, there are also conceptual and technical problems that need to be addressed, if such models are to be further developed. The largest unaddressed issue is how a global action constraint applied to macroscopic measurement devices might lead to specific rules that constrain the microscopic fields in a manner consistent with observation. (In general, two-time boundary constraints can be shown to lead to intermediate particle-like behavior [41], but different global rules will lead to different intermediate consequences.) The tension between a covariant action and the special frame of the measurement devices also needs to be treated consistently. Another topic that is in particular need of progress is an extension of retrocausal entanglement models to handle partially-entangled states, and not merely the maximally-entangled Bell states.

Although the challenges remain significant, the above list of accomplishments arising from this new class of models should give some hope that further accomplishments are possible. By branching out from particle-based models to field-based models, novel research directions are clearly motivated. The promise of such research, if successful, would be to supply a nearly-classical explanation for all quantum phenomena: realistic fields as the solution to a global constraint problem in spacetime.

Acknowledgments: The author would like to thank Justin Dressel for very helpful advice, Aephraim Steinberg for unintentional inspiration, Jan Walleczek for crucial support and encouragement, Ramen Bahuguna for insights concerning laser speckles, and Matt Leifer for hosting a productive visit to Chapman University. This work is supported in part by the Fetzer Franklin Fund of the John E. Fetzer Memorial Trust.

Conflicts of Interest: The author declares no conflict of interest.

References

1. Sutherland, R.I. Bell's theorem and backwards-in-time causality. *Int. J. Theor. Phys.* **1983**, *22*, 377–384. [CrossRef]
2. Price, H. *Time's Arrow & Archimedes' Point: New Directions for the Physics of Time*; Oxford University Press: Oxford, UK, 1997.
3. Wharton, K. Quantum states as ordinary information. *Information* **2014**, *5*, 190–208. [CrossRef]
4. Price, H.; Wharton, K. Disentangling the quantum world. *Entropy* **2015**, *17*, 7752–7767. [CrossRef]
5. Leifer, M.S.; Pusey, M.F. Is a time symmetric interpretation of quantum theory possible without retrocausality? *Proc. R. Soc. A* **2017**, *473*, 20160607. [CrossRef] [PubMed]
6. Adlam, E. Spooky Action at a Temporal Distance. *Entropy* **2018**, *20*, 41. [CrossRef]
7. Argaman, N. Bell's theorem and the causal arrow of time. *Am. J. Phys.* **2010**, *78*, 1007–1013. [CrossRef]
8. Almada, D.; Ch'ng, K.; Kintner, S.; Morrison, B.; Wharton, K. Are Retrocausal Accounts of Entanglement Unnaturally Fine-Tuned? *Int. J. Quantum Found.* **2016**, *2*, 1–14.
9. Weinstein, S. Learning the Einstein-Podolsky-Rosen correlations on a Restricted Boltzmann Machine. *arXiv* **2017**, arXiv:1707.03114. [CrossRef]
10. Sen, I. A local ψ-epistemic retrocasual hidden-variable model of Bell correlations with wavefunctions in physical space. *arXiv* **2018**, arXiv:1803.06458. [CrossRef]
11. Sutherland, R.I. Lagrangian Description for Particle Interpretations of Quantum Mechanics: Entangled Many-Particle Case. *Found. Phys.* **2017**, *47*, 174–207. [CrossRef]
12. Bohm, D.; Hiley, B.J.; Kaloyerou, P.N. An ontological basis for the quantum theory. *Phys. Rep.* **1987**, *144*, 321–375. [CrossRef]
13. Kaloyerou, P. The GRA beam-splitter experiments and particle-wave duality of light. *J. Phys. A* **2006**, *39*, 11541. [CrossRef]
14. Aharonov, Y.; Albert, D.Z.; Vaidman, L. How the result of a measurement of a component of the spin of a spin-1/2 particle can turn out to be 100. *Phys. Rev. Lett.* **1988**, *60*, 1351. [CrossRef] [PubMed]
15. Dressel, J.; Malik, M.; Miatto, F.M.; Jordan, A.N.; Boyd, R.W. Colloquium: Understanding quantum weak values: Basics and applications. *Rev. Mod. Phys.* **2014**, *86*, 307. [CrossRef]

16. Boyer, T.H. A brief survey of stochastic electrodynamics. In *Foundations of Radiation Theory and Quantum Electrodynamics*; Springer: Berlin/Heidelberg, Germany, 1980; pp. 49–63.
17. de La Pena, L.; Cetto, A.M. *The qUantum Dice: An Introduction to Stochastic Electrodynamics*; Springer Science & Business Media: Berlin, Germany, 2013.
18. Woodward, J. *Making Things Happen: A Theory of Causal Explanation*; Oxford University Press: Oxford, UK, 2005.
19. Price, H. Agency and probabilistic causality. *Br. J. Philos. Sci.* **1991**, *42*, 157–176. [CrossRef]
20. Pearl, J. *Causality*; Cambridge University Press: Cambridge, UK, 2009.
21. Menzies, P.; Price, H. Causation as a secondary quality. *Br. J. Philos. Sci.* **1993**, *44*, 187–203. [CrossRef]
22. Price, H. Toy models for retrocausality. *Stud. Hist. Philos. Sci. Part B* **2008**, *39*, 752–761. [CrossRef]
23. Leifer, M.S. Is the Quantum State Real? An Extended Review of ψ-ontology Theorems. *Quanta* **2014**, *3*, 67–155. [CrossRef]
24. Dressel, J.; Bliokh, K.Y.; Nori, F. Classical field approach to quantum weak measurements. *Phys. Rev. Lett.* **2014**, *112*, 110407. [CrossRef] [PubMed]
25. Dressel, J. Weak values as interference phenomena. *Phys. Rev. A* **2015**, *91*, 032116. [CrossRef]
26. Ritchie, N.; Story, J.G.; Hulet, R.G. Realization of a measurement of a "weak value". *Phys. Rev. Lett.* **1991**, *66*, 1107. [CrossRef] [PubMed]
27. Bliokh, K.Y.; Bekshaev, A.Y.; Kofman, A.G.; Nori, F. Photon trajectories, anomalous velocities and weak measurements: A classical interpretation. *New J. Phys.* **2013**, *15*, 073022. [CrossRef]
28. Howell, J.C.; Starling, D.J.; Dixon, P.B.; Vudyasetu, P.K.; Jordan, A.N. Interferometric weak value deflections: Quantum and classical treatments. *Phys. Rev. A* **2010**, *81*, 033813. [CrossRef]
29. Aharonov, Y.; Vaidman, L. The two-state vector formalism: An updated review. In *Time in Quantum Mechanics*; Springer: Berlin/Heidelberg, Germany, 2008; pp. 399–447.
30. Sinclair, J.; Spierings, D.; Brodutch, A.; Steinberg, A. Weak values and neoclassical realism. **2018**, in press.
31. De Beauregard, O.C. Une réponse à l'argument dirigé par Einstein, Podolsky et Rosen contre l'interprétation bohrienne des phénomènes quantiques. *C. R. Acad. Sci.* **1953**, *236*, 1632–1634. (In French)
32. Wharton, K. A novel interpretation of the Klein-Gordon equation. *Found. Phys.* **2010**, *40*, 313–332. [CrossRef]
33. Wharton, K.B.; Miller, D.J.; Price, H. Action duality: A constructive principle for quantum foundations. *Symmetry* **2011**, *3*, 524–540. [CrossRef]
34. Evans, P.W.; Price, H.; Wharton, K.B. New slant on the EPR-Bell experiment. *Br. J. Philos. Sci.* **2012**, *64*, 297–324. [CrossRef]
35. Harrison, A.K. Wavefunction collapse via a nonlocal relativistic variational principle. *arXiv* **2012**, arXiv:1204.3969. [CrossRef]
36. Schulman, L.S. Experimental test of the "Special State" theory of quantum measurement. *Entropy* **2012**, *14*, 665–686. [CrossRef]
37. Heaney, M.B. A symmetrical interpretation of the Klein-Gordon equation. *Found. Phys.* **2013**, *43*, 733–746. [CrossRef]
38. Corry, R. Retrocausal models for EPR. *Stud. Hist. Philos. Sci. Part B* **2015**, *49*, 1–9. [CrossRef]
39. Lazarovici, D. A relativistic retrocausal model violating Bell's inequality. *Proc. R. Soc. A* **2015**, *471*, 20140454. [CrossRef]
40. Silberstein, M.; Stuckey, W.M.; McDevitt, T. *Beyond the Dynamical Universe: Unifying Block Universe Physics and Time as Experienced*; Oxford University Press: Oxford, UK, 2018.
41. Wharton, K. Time-symmetric boundary conditions and quantum foundations. *Symmetry* **2010**, *2*, 272–283. [CrossRef]

Article

A Lenient Causal Arrow of Time?

Nathan Argaman

Department of Physics, Nuclear Research Center-Negev, P.O. Box 9001, Be'er Sheva 84190, Israel; argaman@mailaps.org

Received: 29 March 2018; Accepted: 15 April 2018; Published: 18 April 2018

Abstract: One of the basic assumptions underlying Bell's theorem is the causal arrow of time, having to do with temporal order rather than spatial separation. Nonetheless, the physical assumptions regarding causality are seldom studied in this context, and often even go unmentioned, in stark contrast with the many different possible locality conditions which have been studied and elaborated upon. In the present work, some retrocausal toy-models which reproduce the predictions of quantum mechanics for Bell-type correlations are reviewed. It is pointed out that a certain toy-model which is ostensibly superdeterministic—based on denying the free-variable status of some of quantum mechanics' input parameters—actually contains within it a complete retrocausal toy-model. Occam's razor thus indicates that the superdeterministic point of view is superfluous. A challenge is to generalize the retrocausal toy-models to a full theory—a reformulation of quantum mechanics—in which the standard causal arrow of time would be replaced by a more lenient one: an arrow of time applicable only to macroscopically-available information. In discussing such a reformulation, one finds that many of the perplexing features of quantum mechanics could arise naturally, especially in the context of stochastic theories.

Keywords: Bell's theorem; the causal arrow of time; retrocausality; superdeterminism; toy-models

1. Introduction

Bell's theorem is one of the most profound revelations of modern physics. In the Einstein–Podolsky–Rosen article [1], and in Bell's original proof [2], the discussion is based on notions of *locality*, but in a later review [3] Bell clarified that the relevant requirement of locality, often called Bell-locality, follows from the assumption of relativistic *causality*. The original "locality," Bell stated, is in fact simply an abbreviation of "local causality." It is perhaps natural that the original papers, more than the later reviews [3–5], were analyzed meticulously. Moreover, in all of Bell's writings [6,7] and in the overwhelming majority of the accompanying literature, the causal arrow of time is taken for granted, rather than identified as a physical assumption.

What does this assumption mean? Many physicists accept Hume's approach, which defines the concepts of "cause" and "effect" so that a cause always precedes its effect. However, for mathematical models of natural phenomena, it is also natural to take the inputs of a mathematical model as "causes" and the outputs as "effects," which corresponds to regarding the things we can control as causes (see, e.g., [8]). In this context, assuming the causal arrow of time means simply that the model accepts inputs from the past, such as initial conditions, and generates outputs that correspond to later times. Similarly, in such models an external field at a time t affects only the values of variables pertaining to later times. Many of the mathematical models used in physics are precisely of this type, e.g., the standard application of Newton's equations, with initial conditions taken as inputs, or the description of quantum wavefunctions evolving according to Schroedinger's equation between measurements, with Born's rule and the collapse postulate applied at the times of measurements. However, some established mathematical models, such as the stationary-action principle of classical mechanics, do not conform to this rule: at the mathematical level, they are in violation of the causal arrow [9] (to avoid confusion, we will use the terms "inputs" and outputs" rather than "causes" and "effects," whether or not a model obeys the causal arrow of time). A reasonable definition of the causal arrow of time for a physical theory is that it is possible to formulate it in terms of a mathematical model with inputs and outputs which conform to the causal arrow stated above.

For deterministic theories, modeled with differential equations, one can specify a concrete solution by choosing as inputs either initial or final conditions (or a combination of both). One can then reserve the identification of causes with inputs (and effects with outputs) only for models which conform to the arrow of time, obtaining consistency with Hume's approach. However, the situation is less clear for stochastic theories, such as quantum mechanics (QM). In general, it is not guaranteed that a mathematical model which does not conform to the causal arrow of time will have a reformulation which does.

Bell's writings indicate that, while he was interested in stochastic theories, he consistently accepted Hume's approach. For example, in [3], Bell contrasted local causality, which allows for stochastic probabilities, with local determinism, taking for granted that at the mathematical level the past affects the future rather than vice versa. Bell did often mention light cones, but did not pause to explain why the past light cone of an event, rather than its future light cone, is where one may find the inputs affecting it. When considering the possibility that relativistic causality could fail, he discussed a preferred frame of reference in which the causal arrow would still hold [10]. Bell applied relativistic causality to two separated particles which had originated together—the setup of the Einstein–Podolsky–Rosen article—and derived the condition of Bell-locality. Subsequently, many different notions of locality were identified, e.g., "parameter independence" vs. "outcome independence." In fact, a recent article lists no less than eight different statements/definitions of locality [11] (see also [12]). Nonetheless, the logical argumentation requires that we should pay at least as much attention to different definitions of causality as we pay to different notions of locality.

That the causal arrow of time is an essential assumption of Bell's no-go theorem, and hence should be called into question, is a point which was raised repeatedly in the literature [13–16]. When considering theories or models which violate the standard arrow of time, the time-reversal symmetry of microscopic physical theories is often used to argue that introducing any time-asymmetry should be avoided altogether. In the present work, the possibility of introducing such an asymmetry into the theory will be considered, with the aim of reproducing macroscopic phenomena. Specifically, an asymmetric, low-entropy-in-the-past condition will be applied. A directionality of time is expected to result from such asymmetry, but it need not be as strict as the standard causal arrow of time.

For this to work reasonably, one needs to distinguish between microscopic and macroscopic degrees of freedom, and to have the information carried by the macroscopic degrees of freedom constrained by an arrow of time. The microscopic degrees of freedom exhibit fluctuations which are affected by inputs from the future, but any attempt at amplifying these fluctuations and bringing them up to the macroscopic level must fail to produce any macroscopic information regarding the future inputs. This corresponds well to known facts concerning quantum fluctuations, the impossibility of using Bell-correlations for signaling, and the disturbance of a quantum system by measurement.

It thus appears that a most promising direction is to pursue retrocausal reformulations of quantum mechanics (QM). Reformulations are powerful tools in advancing theoretical physics, e.g., the Lagrangean and Hamiltonian reformulations of Newton's equations played essential roles in the development of QM. In fact, QM had two equivalent formulations to begin with—wave mechanics and matrix mechanics—and some of the most important subsequent advances were based on reformulations, as exemplified by path integrals. Additional examples include Bohmian mechanics [17,18], which motivated Bell in his original research on no-go theorems [19], and its stochastic version—diffusing particles guided by the quantum wavefunction, which was suggested by Bohm himself [20] and further developed by others [21]. Improving our understanding of QM appears to require such a radical retrocausal reformulation of the fundamental theory. This could have ramifications in contexts such as quantum computation and/or quantum gravity.

Notice that this approach is complementary to but quite distinct from the experimental approach, which is by and large the main activity following from Bell's theorem, e.g., [22]. The empirical adequacy of QM has by now been abundantly supported but can never reach the certitude of a mathematical theorem. The proven theorem states that the predictions of QM (i.e., the probabilities for the output parameters, given the input parameters) cannot be reproduced by a model or theory conforming to the condition of Bell-locality. Thus, our discussion will relate to models or theories which violate this mathematical condition by allowing retrocausality. Questions regarding the real state of the system, such as those which concerned Einstein [1], and which arise naturally when the theory is compared with experimental procedures, do not arise in this context of model-construction. In particular, the issue of counterfactual definiteness [23] is not relevant, because the discussion does not refer to the question whether or not the real system has a definite property, but instead to the much simpler question of whether or not the model or theory has a prediction for that property. The relationship between reality and the parameters and predictions of the theory is adequately handled by standard QM (and its interpretations), and need not be addressed when considering reformulations.

As a first step in this direction, it is appropriate to discuss retrocausal toy-models, i.e., mathematical models which reproduce the predictions of QM for the specific case considered in Bell's theorem. Two of the toy-models available will be reviewed. The first [16] was presented as a mathematical formulation of the retrocausal ideas expressed, e.g., by Cramer [14]. The second [24] was originally presented in a somewhat different context, associated with superdeterminism, i.e., the denial of the free-variable status of the inputs of QM, which is generally associated with the free will of the experimenters. It will be argued that while the latter model has a distinct technical advantage, the former presentation is more relevant as a basis for a discussion of future scientific theories.

The two retrocausal toy-models discussed share the undesirable feature of QM known as the measurement problem. On the other hand, they differ in that the "dynamics" in one is stochastic, while it is deterministic in the other. Can they be generalized to encompass all quantum phenomena, in a way which would allow an understanding of both nonlocality and quantum measurements? This possibility will be qualitatively discussed, assuming the stochastic option for the dynamics. In this context, it seems that many of the other mysterious aspects of QM, such as the exponentially large size of the Hilbert space required to describe n particles, and the dynamics involving unitary evolution punctuated by collapse, follow naturally. This discussion could be compared and contrasted with the questions regarding the relative sizes of ontic and epistemic spaces in the context of the ontological models framework, which is based on assuming strict causality. The difficulties that that framework faces [25] serve as further motivation for considering the retrocausal alternative.

The discussion of toy-models is the subject of Section 2, the pathway towards a general retrocausal reformulation of QM is discussed in Section 3, and conclusions are provided in Section 4.

2. Retrocausal Toy-Models

Bell's theorem concerns entangled particles, e.g., pairs of distant photons with polarizations entangled in a singlet state, as in some of the most remarkable early experiments [26] (originally, pairs of spin-half particles were considered). The predictions of QM for polarization measurements on the constituent photons of such a pair are given by the probabilities

$$p_{a,b}(A,B) = \frac{1}{4}[1 + A\,B\cos(2a - 2b)], \quad (1)$$

where a and b are angles, defined modulo π, specifying the orientations of the beam-splitting polarizers involved in the measurements, and $A, B = \pm 1$ represent the results of the measurements. $A = 1$ represents the first photon having a polarization along a, $A = -1$ represents a perpendicular polarization, and the polarization of the second photon with respect to the orientation b is similarly represented by B. In principle, even when one restricts QM to a description of the polarizations of a pair of photons, one has an additional input variable c specifying how the photons are prepared, and thus determining the initial wavefunction. However, in the present work, we will only refer to the singlet state, and thus c is a constant and can be dropped. The probabilities (1) are "local" in the no-signaling sense, i.e., the marginal probabilities are independent of the remote inputs: $p_{a,b}(A) \equiv \sum_B p_{a,b}(A,B) = \frac{1}{2}$ is independent of b, and similarly $p_{a,b}(B)$ is independent of a. However, they are "nonlocal" in the sense of Bell, i.e., one cannot write $p_{a,b}(A,B)$ as a product of two separate factors, $p_{a,\lambda}(A)$ and $p_{b,\lambda}(B)$, where λ is an additional parameter (or set of variables) describing the "state" the particles had in the past, which is taken to be independent of the inputs a and b.

2.1. A Simplistic Toy-Model

The idea of retrocausality is to recognize this last restriction as a physical assumption—the causal arrow of time—which may be inappropriate for a microscopic theory. If one allows λ to depend on a and b, the difficulty is resolved, as demonstrated by the following simplistic model [16]:

$$p_{a,b}(\lambda) = \frac{1}{4}\left[\delta(\lambda - a) + \delta\left(\lambda - a - \frac{\pi}{2}\right) + \delta(\lambda - b) + \delta\left(\lambda - b - \frac{\pi}{2}\right)\right], \tag{2}$$

where λ is an additional angle, defined modulo π, which represents the initial polarization of the photons. In this model, the predictions for the polarization measurements follow from the standard Malus' law,

$$p_{a,\lambda}(A) = \begin{cases} \cos^2(\lambda - a) & A = 1 \\ \sin^2(\lambda - a) & A = -1 \end{cases}, \tag{3}$$

and similarly for $p_{b,\lambda}(B)$. Combining these using

$$p_{a,b}(A,B) = \int d\lambda \; p_{a,b}(\lambda) p_{a,\lambda}(A) p_{b,\lambda}(B) \tag{4}$$

reproduces the predictions of QM, Equation (1). A toy-model which approaches this simplistic one in the appropriate limit ($\gamma \to 0$) is discussed in [27].

An attractive feature of this model is that one can consider what would happen if the value of λ were measured at the source. In the experiments [26], the singlet pair of photons was emitted by an atomic $(J=0) \to (J=1) \to (J=0)$ cascade. One can envision measuring the orientation of the angular momentum of the atom during the brief time it is in the intermediate state of the cascade, thereby inferring the initial polarization of the photons. In order to perform such a measurement, one must specify the direction in which the angular momentum is measured, e.g., whether it is a measurement of $\hat{J}_x$ or $\hat{J}_y$ (the results of this measurement and those of each one of the later photon-polarization measurements would be correlated per the predictions of QM). Clearly, this would constitute a "which path" measurement [28], and would disturb the system in a manner which ruins the entanglement between the two photons.

2.2. Hall's Toy-Model

According to (2), the variable λ carries much information regarding the values of a and b, or at least one of them. It is of interest to note that this "flow of information from the future to the past" is much more limited in other toy-models. A very efficient model, in this respect, was given in [24]. The variant of it pertaining to photons, rather than spin-half particles, is:

$$p_{a,b}(\lambda) = \frac{1}{\pi}\,\frac{1 + \acute{A}\acute{B}\cos(2a - 2b)}{1 + \acute{A}\acute{B}(1 - z)}, \tag{5}$$

where $\acute{A} = \text{sign}[\cos(2a - 2\lambda)]$, $\acute{B} = \text{sign}[\cos(2b - 2\lambda)]$ and $z = \frac{2}{\pi}|2a - 2b|$ are abbreviations (the denominator never vanishes, because $\acute{A}\acute{B} = 1$ when $z = 0$ and $\acute{A}\acute{B} = -1$ when $z = 2$). This model is deterministic in the sense that a and λ determine one possible value for A,

$$p_{a,\lambda}(A) = \delta_{A,\acute{A}}, \tag{6}$$

and similarly for B and $p_{b,\lambda}(B)$. Again, using Equation (4) to combine Equations (5) and (6) trivially reproduces the predictions of QM, Equation (1).

In the case of Equation (5), the "past" variable λ carries very little information on a and b, less than 0.07 Bits [29]. This can justifiably be seen as a definite advantage over the simplistic model of Equation (2). On the other hand, Refs. [24,29] can be criticized for not discussing retrocausality explicitly (although its relevance was briefly acknowledged by the author in [30]). In fact, the possibility that the variable λ depends on the inputs a and b, embodied in (5), is presented in [24] in a manner that does not imply a violation of the causal arrow of time.

2.3. Criticism of the Superdeterministic Approach

How could this come about? As described in the introduction, Bell emphasized the locality assumption, $p_{a,b,\lambda}(A, B) = p_{a,\lambda}(A)p_{b,\lambda}(B)$. In 1976 he was criticized by Shimony et al. [31] for not emphasizing the measurement-independence assumption

$$p_{a,b}(\lambda) = p(\lambda), \tag{7}$$

as well. Bell replied that he *had* (belatedly) made explicit the assumption that a and b were free variables, and that "this means that the values of such variables have implications only in their future light cones" [32]. He also emphasized that his work should not be understood as a philosophical discussion concerning the real world but as an analysis of the kinds of mathematical models or theories which may be applicable. Obviously, the mathematical notion of free variables, which pertains to such theories, is not specific to situations in which the past and/or the future are relevant. Thus, the measurement-independence assumption relies on two separate assumptions:

(i) a and b are free variables;

(iI) The causal arrow of time, with λ associated with a time earlier than that of a and b.

The first implies that a and b are independent of λ, and the second that λ is independent of a and b, giving a full justification of (7). However, saying that they are independent of each other might be misleading, because the word "independent" is used here with different mathematical meanings. In particular, mutual statistical independence is applicable only if the free-variable status of a and b is revoked, and they are replaced by random variables.

If indeed a and b are treated as random variables, like λ, then the probability distribution $p_{a,b}(\lambda)$ is replaced by a conditional probability, $p(\lambda|a, b)$ (Bell's notation, $\{\lambda|a, b\}$, could refer to either of these). The measurement independence condition (7) then reads $p(\lambda|a, b) = p(\lambda)$, and it no longer follows from the arrow of time alone—a conditional dependence of λ on a and b could also arise from a forwards-in-time dependence of a and/or b on λ, or from a common cause in the past. The latter is indeed the possibility considered in [31], which describes a conspiracy involving a person who has "concocted" a list of correlation data, an apparatus manufacturer, and the secretaries of two physicists who are to perform the experiments. When the j-th measurement is to be performed, each of the secretaries whispers the pre-listed setting to the corresponding physicist, who sets his apparatus accordingly, and the result registered by each apparatus is pre-programmed to correspond to the concocted list, rather than to an actual measurement. In this manner, any correlations can result, including those of (1), with no violations of local causality, but this is achieved conspiratorially.

The agreement of empirical observations with QM, whether a and b are selected at random or are determined by any other arrangement (say, a double-blind experimental procedure), provides strong evidence that it is appropriate to treat the orientations as free variables, and to avoid serious discussion of the possibility that they could be predetermined. Indeed, it was argued from the outset (in the concluding paragraphs of [31]) that the "enterprise of discovering the laws of nature" by "scientific experimentation" necessarily involves the assumption that "hidden conspiracies of this sort do not occur." Accordingly, the purpose of the discussion was not to cast doubt on the free-variable status of a and b, but merely to point out that a complete statement of Bell's theorem must include a reference to this status, assumption (i) above. Indeed, from that point on, Bell emphasized that the proof of his theorem assumes this, and engaged in brief discussions of the possibility of "superdeterministic" theories, in which there are no such free variables [4,5,32], concluding that even in such cases pseudorandom variables which are "sufficiently free for the purpose at hand" should be available. Subsequently, violations of measurement independence, Equation (7), were often perceived as equivalent to violations of assumption (i), involving free variables or "free will," and the similarly critical assumption (ii), involving the arrow of time, very often remained unmentioned.

As described above, the model of Equations (5) and (6) is presented in [24] in this manner—the violation of measurement independence is emphasized, the "free will" issue is discussed, and the role of the causal arrow of time is not; neither are the unscientific/conspiratorial concern and the notion of pseudorandom variables. In fact, in a more complete presentation of the model, the author finds it necessary to add a variable μ associated with the overlap of the backwards lightcones of a and b, and determining them (Section 5.1 of [29]). For an explicit version of the toy-model, it is suggested that μ simply consists of the values of a and b, determining them in a decidedly artificial manner. In addition, it was found necessary to invoke a correlation-does-not-imply- causation argument to explain how the ostensibly retrocausal dependence in Equation (5) is consistent with causality. When considering Equations (5) and (6) as a retrocausal rather than a free-variable-status-denying model, these additional steps are superfluous. Thus, one may add Occam's razor as a further argument to prefer the former over the latter, i.e., to interpret violations of measurement independence as violations of (ii) rather than of (i). The details given above are in the context of Refs. [24,29], but on the basis of the arguments given, the Occam's razor argument is expected to hold quite generally for superdeterministic models. Nevertheless, such models are often regarded as the leading alternative to standard QM, by both experimentalists (e.g., [33]) and leading theorists (e.g., [34]).

3. Toward a General Retrocausal Theory

Can the retrocausal toy-models discussed above be generalized to a full retrocausal theory of all quantum phenomena, including a resolution of the measurement problem? The development of such a theory would be revolutionary, as it would go beyond all previous developments in the foundations of QM: path integrals [35], Bohmian mechanics [17,18], histories approaches (e.g., [36]), stochastic mechanics [21], stochastic quantization [37], etc. In contemplating such a development, promoters of retrocausation often advocate a fully time-reversal-symmetric approach [15]. However, to bring microscopic theories in line with macroscopic phenomena, it is common in other contexts to break this symmetry by invoking a low-entropy condition in the remote past. It is therefore of interest to consider how a "fixed past" initial condition would affect different types of theories, as sketched in Figure 1 (see also [38]).

Consider a theoretical description of some degrees of freedom in spacetime, such as the modes of an electromagnetic field, and consider an external perturbation (possibly an oscillating dipole) which affects the dynamics of these degrees of freedom for a limited time, beginning at a time t_1. Furthermore, consider the possibility that the values of all the degrees of freedom described by the theory are given at a time t_0 in the past (if the dynamics is described by a second-order differential equation in time, the time derivatives of these values are also considered here as given). For a theory with deterministic dynamics, full specification of the field configuration at time t_0 would determine the field configuration for all times up to t_1, at which the external perturbation is applied. However, if the theory describing the dynamics of the fields is stochastic, this is no longer necessarily true and the probabilities for the fields at times before t_1 may be retrocausally affected by the external perturbation. In the case in which the dynamics allows for only small fluctuations around the classical behavior, such as the quantum vacuum fluctuations around a zero-field solution, one may expect a weaker form of causality to still hold.

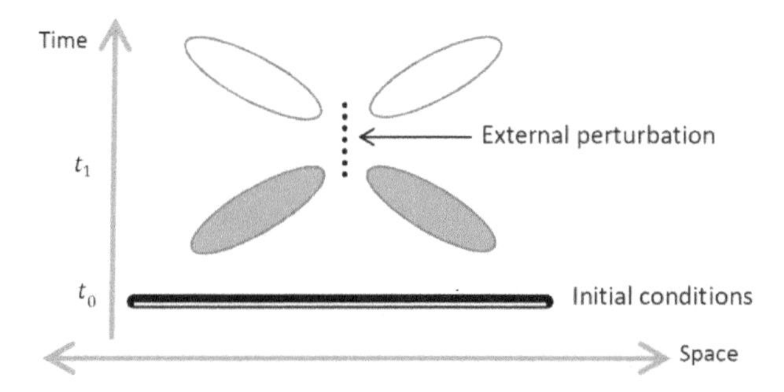

Figure 1. Sketch of a spacetime region permeated by fluctuating fields, with an external perturbation applied at times near t_1, and with the field configuration at an early time, t_0, fixed as initial conditions. If the fields are described by a deterministic theory, the field configurations before t_1 are unaffected by the external perturbation; in contrast, for stochastic theories the probability distribution of the fields at times between t_0 and t_1, indicated by the filled ovals, may depend on the perturbation. Nevertheless, the firing rate of a "detector" placed at or near these ovals must not depend on the perturbation (or else the no-signaling condition would be violated).

If the theory in question can also model the process of measurement, it must be capable of describing the inner workings of a detector, which could be placed within the regions marked by the open or the filled ovals in the sketch (times later than or earlier than t_1, respectively). The detector itself would consist of additional stochastic degrees of freedom and would presumably have an initial condition corresponding to a metastable state. Fluctuating out of the metastable state would correspond to a "click" in the detector, and the probability of such a fluctuation would depend on the fluctuations on the field to which the detector is coupled. Dissipative aspects of the detector could be modeled by coupling it to a "bath" of many "thermalized" degrees of freedom, subject to appropriate initial conditions of their own, in analogy to the Caldeira–Leggett description of quantum dissipation [39]. An "ideal" detector would correspond to the case for which the "click" is sufficiently dissipative to be "irreversible," so that coupling of further degrees of freedom to the detector would allow for copying or "cloning" of the information regarding whether or not a "click" has occurred. This would stand in contrast to the no-cloning condition which is expected to hold in general for stochastic theories (see, e.g., [40]).

If the detector operates at times after t_1 (the empty ovals), it is natural for the probability of detection to depend on the details of the external perturbation. However, if the detector operates at times earlier than t_1, it is necessary to require that the corresponding probability is independent of the external perturbation in order to avoid the possibility of signaling to the past. In other words, although the fluctuating fields are subject to rules which allow for retrocausality, the external perturbation and the detection events must be related in a manner which is subject to the causal arrow of time. As this causal arrow applies only to the "macroscopically available" or "clone-able" information, and does not affect the microscopic "hidden" degrees of freedom, it is perhaps appropriate to characterize it as "lenient." The fixed initial conditions are associated with low entropy, and therefore may be expected to break the symmetry of the theory in just the manner required to meet such a lenient causality requirement.

Note that we are here considering the possibility that an "agent" which is external to the theory may control the inputs and may use the information provided by the measured outputs, and that such an agent is subject to the arrow of time as well. For example, the agent may decide whether or not to apply the external perturbation according to whether or not a detection event has occurred at an earlier time. Signaling into the past must thus indeed be strictly impossible, as it would allow construction of causal loops—the well-known inconsistency arguments of the grandfather paradox (a.k.a. the bilking argument). However, these arguments involve only the information accessible at the macroscopic level, so they precisely allow the type of lenient causality described here, in which the microscopic degrees of freedom are affected retrocausally [15].

A theory of this type, if developed, would provide a natural explanation for many of the perplexing features of QM. The description above bears similarities to the analysis of quantum measurement by environment-induced superselection, or einselection [41]. It would be natural to supplement it by an alternative description which would represent only the macroscopically-available information regarding a certain subset of the degrees of freedom up to a time t. As the external perturbations applied at later times to these degrees of freedom are to be treated as unknown, this mathematical description would have to represent a large number of possibilities, exponentially large in the number of degrees of freedom involved. Furthermore, by definition it would have to evolve in an information-preserving manner as t is changed, except for the moments at which there is a change in the macroscopically-available information. This corresponds precisely to the evolution of quantum wavefunctions, which form exponentially large Hilbert spaces, and exhibit unitary evolution punctuated by "collapse" events (see also [42]).

For retrocuasal theories of the type considered here, it is expected that a direct generalization of the deduction of the lenient causality condition would lead to the slightly stronger condition of "information causality" [43]. This latter condition was put forward as a physical principle within an axiomatic approach, i.e., with the hope that all of the features of QM could be deduced from it. This was partly successful, as it was demonstrated that information causality implies Tsirelson's bound [44], a generalization of a Bell inequality which holds in the quantum realm. In this sense, it is expected that time-asymmetric retrocausal theories of the type considered here would, through the mathematical arguments of [43], provide an explanation for the fact that all quantum phenomena obey Tsirelson's inequality. Note that information causality is here to be deduced rather than assumed, and thus the fact that not all aspects of QM can be generated by it does not lead to any objections in the present context.

4. Summary and Discussion

The present work, like several other presentations at the EmQM17 David Bohm centennial conference [45–49], advocates the relaxation of the arrow-of-time assumption of Bell's theorem. In the first part, the role this assumption plays in the proof of the theorem was considered, and contrasted with the free-variable assumption, which is associated with the free will of the experimenters. Concrete toy-models which violate these assumptions were discussed. In the second part, an admittedly speculative discussion of the possibility of developing a retrocausal reformulation of QM which would describe all quantum phenomena in spacetime (rather than a Hilbert space), and would be free of the measurement problem, was given.

In the proof of Bell's theorem, the arrow-of-time assumption enters together with the free-variable status of the measurement settings, leading to the mathematical condition of measurement independence, Equation (7) above. Unfortunately, the arrow of time is usually taken for granted, rather than identified as a physical assumption, and relaxation of the measurement-independence condition is then associated with superdeterminism, i.e., denial of the free-variable status of the settings, rather than with retrocausality.

It was pointed out from the outset that for superdeterminism to provide a resolution of the difficulty exposed by Bell's theorem, one must assume that the measurement settings are produced in a conspiratorial manner, one which would undermine the scientific method [31]. In contrast, accepting violations of the mathematical causal-arrow-of-time condition was necessary already in the context of the stationary-action principle of classical mechanics. Thus, while both face the difficulty of overcoming the prejudices we have developed based on our experiences in the macroscopic world, there is a clear preference for the latter over the former. Nevertheless, the superdeterministic approach has received much attention recently, and several concrete examples of superdeterministic toy-models have been put forward. This may be due to the argument concerning the grandfather paradox, but such paradoxes are easily avoided if one takes retrocausality to affect only hidden variables [15], such as the which-path variables of standard QM [28] (those whose measurement precludes the observation of interference effects).

As a further argument against superdeterminism, one of the available toy-models [24] has been considered, and it was shown that it consists of a complete retrocausal toy-model, to which additional steps of argumentation have been added in order to transform it into a superdeterministic toy-model. Occam's razor thus rules against the superdeterministic approach and for the retrocausal interpretation of such toy-models. When interpreted in this manner, this particular toy-model of Ref. [24] has a distinct advantage over an earlier retrocausal toy-model [16], in that the microscopic degrees of freedom carry a very limited amount of information into the past.

Developing a general retrocausal theory of quantum phenomena, one that would be free of the measurement problem and not limited to the scope of a toy-model, is a grand challenge of quantum foundations. Whereas it is appropriate to discuss fully time-symmetric theories in this context [15], the possibility of breaking the symmetry by imposing fixed boundary conditions in the past was considered above. A theory with fully fixed initial conditions and deterministic dynamical rules cannot exhibit retrocausality of the type required by Bell's theorem. In contrast, a theory with time-symmetric stochastic dynamical rules would have its symmetry broken by the imposition of initial conditions, in a manner which may lead to a "lenient" arrow of time of the type observed macroscopically—an arrow of time applicable only to macroscopically available information.

Such a stochastic time-asymmetric approach is expected to enjoy two further advantages: (a) Constructing the corresponding "epistemic" state of knowledge, i.e., a mathematical representation for the macroscopically-available information up to a time t, would necessarily result in states which are exponentially complex for many degrees of freedom. These states would "evolve" with t in an information-preserving manner, except at the times of measurements, at which additional information becomes available. The correspondence to the complexity of quantum wavefunctions and their unitary/collapse evolution is clear. (b) The spatiotemporal "flow" of entropy/information in such theories is expected to lead to the information causality condition, and thus to Tsirelson's bound.

In closing, it is appropriate to quote from the concluding paragraph of Bell's last review of his theorem [5]:

> The unlikelihood of finding a sharp answer to this question [the measurement problem] reminds me of the relation of thermodynamics to fundamental theory. The more closely one looks at the fundamental laws of physics the less one sees of the laws of thermodynamics. The increase of entropy emerges only for large complicated systems, in an approximation depending on "largeness" and "complexity." Could it be that causal structure emerges only in something like a "thermodynamic" approximation, where the notions "measurement" and "external field" become legitimate approximations? Maybe that is part of the story, but I do not think it can be all. Local commutativity does not for me have a thermodynamic air about it. It is a challenge now to couple it with sharp internal concepts, rather than vague external ones.

Developing a fundamental retrocausal stochastic theory may resolve the issue, as the condition of local commutativity need only apply to the corresponding epistemic states, representing the macroscopically-available information. It is expected that the process of mathematically constructing such epistemic states would provide the "sharp internal concepts" required by Bell to meet this challenge.

Acknowledgments: This work was presented at the EmQM17 David Bohm centennial conference, and its publication was supported by the Fetzer Franklin Fund of the John E. Fetzer Memorial Trust.

Conflicts of Interest: The author declares no conflict of interest.

References

1. Einstein, A.; Podolsky, B.; Rosen, N. Can quantum-mechanical description of physical reality be considered complete? *Phys. Rev.* **1935**, *47*, 777–780. [CrossRef]
2. Bell, J.S. On the Einstein-Podolsky-Rosen paradox. *Physics* **1964**, *1*, 195–200. [CrossRef]
3. Bell, J.S. The theory of local beables. *Epistemol. Lett.* **1976**, *9*, 11–24, Reproduced in *Dialetica* **1985**, *39*, 86–96. [CrossRef]
4. Bell, J.S. Bertlmann's socks and the nature of reality. *J. Phys.* **1981**, *42*, C2-41–C2-62. [CrossRef]
5. Bell, J.S. La nouvelle cuisine. In *Between Science and Technology*; Sarlemijn, A., Kroes, P., Eds.; Elsevier/North-Holland: New York, NY, USA; Amsterdam, The Netherlands, 1990; pp. 97–115.
6. Bell, J.S. *Speakable and Unspeakable in Quantum Mechanics: Collected Papers on Quantum Philosophy*, 2nd ed.; Cambridge University Press: Cambridge, UK, 2004.
7. Bell, J.S.; Bell, M.; Gottfried, K.; Veltman, M. *John S. Bell on the Foundations of Quantum Mechanics*; World Scientific: Singapore, 2001.
8. Price, H. Agency and causal asymmetry. *Mind* **1992**, *101*, 501–520. [CrossRef]
9. Wharton, K. Time-symmetric boundary conditions and quantum foundations. *Symmetry* **2010**, *2*, 272–283. [CrossRef]
10. Bell, J.S. How to teach special relativity. *Prog. Sci. Cult.* **1976**, *1*, 1–13.
11. Wiseman, H.M. The two Bell's theorems of John Bell. *J. Phys. A Math. Theor.* **2014**, *47*, 424001. [CrossRef]
12. Norsen, T. Are there really two different Bell's theorems? *Int. J. Quantum Found.* **2015**, *1*, 65–84.
13. Costa de Beauregard, O. Une réponse à l'argument dirigé par Einstein, Podolsky et Rosen contre l'interprétation bohrienne des phénomènes quantiques. *C. R. Acad. Sci.* **1953**, *236*, 1632–1634. (In French)
14. Cramer, J.G. Generalized absorber theory and the Einstein-Podolsky-Rosen paradox. *Phys. Rev. D* **1980**, *22*, 362–376. [CrossRef]
15. Price, H. *Time's Arrow and Archimedes' Point: New Directions for the Physics of Time*; Oxford University Press: Oxford, UK, 1996.
16. Argaman, N. Bell's theorem and the causal arrow of time. *Am. J. Phys.* **2010**, *78*, 1007–1013. [CrossRef]
17. Bohm, D. A suggested interpretation of the quantum theory in terms of "hidden" variables. I. *Phys. Rev.* **1952**, *85*, 166–179. [CrossRef]
18. Bohm, D. A suggested interpretation of the quantum theory in terms of "hidden" variables. II. *Phys. Rev.* **1952**, *85*, 180–193. [CrossRef]
19. Bell, J.S. On the impossible pilot wave. *Found. Phys.* **1982**, *12*, 989–999. [CrossRef]
20. Bohm, D. *Causality and Chance in Modern Physics*; D. van Nostrand Co., Inc.: Princeton, NJ, USA, 1957.
21. Nelson, E. Review of stochastic mechanics. *J. Phys. Conf. Ser.* **2011**, *361*, 012011. [CrossRef]
22. Rosenfeld, W.; Burchardt, D.; Garthoff, R.; Redeker, K.; Ortegel, N.; Rau, M.; Weinfurter, H. Event-ready Bell test using entangled atoms simultaneously closing detection and locality loopholes. *Phys. Rev. Lett.* **2017**, *119*, 010402. [CrossRef] [PubMed]
23. Stapp, H.P. Nonlocal character of quantum theory. *Am. J. Phys.* **1997**, *65*, 300–303. [CrossRef]
24. Hall, M.J.W. Local deterministic model of singlet state correlations based on relaxing measurement independence. *Phys. Rev. Lett.* **2010**, *105*, 250404, Erratum in **2016**, *116*, 219902, doi:10.1103/PhysRevLett.116.219902. [CrossRef] [PubMed]
25. Leifer, M.S. Is the quantum state real? An extended review of ψ-ontology theorems. *Quanta* **2014**, *3*, 67–155. [CrossRef]
26. Aspect, A.; Dalibard, J.; Roger, J. Experimental test of Bell's inequalities using time-varying analyzers. *Phys. Rev. Lett.* **1982**, *49*, 1804–1807. [CrossRef]
27. Almada, D.; Ch'ng, K.; Kintner, S.; Morrison, B.; Wharton, K.B. Are retrocausal accounts of entanglement unnaturally fine-tuned? *Int. J. Quantum Found.* **2016**, *2*, 1–16.
28. Buks, E.; Schuster, R.; Heiblum, M.; Mahalu, D.; Umansky, V. Dephasing due to which path detector. *Phys. B Condens. Matter* **1998**, *249–251*, 295–301. [CrossRef]

29. Hall, M.J.W. The significance of measurement independence for Bell inequalities and locality. In *At the Frontier of Spacetime*; Asselmeyer-Maluga, T., Ed.; Springer: New York, NY, USA, 2016.
30. Hall, M.J.W. Relaxed Bell inequalities and Kochen-Specker theorems. *Phys. Rev. A* **2011**, *84*, 022102. [CrossRef]
31. Shimony, A.; Horne, M.A.; Clauser, J.F. Comment on "The theory of local beables". *Epistemol. Lett.* **1976**, *13*, 1–8, Reproduced in *Dialectica* **1985**, *39*, 97–102. [CrossRef]
32. Bell, J.S. Free variables and local causality. *Epistemol. Lett.* **1977**, *15*, 79–84, Reproduced in *Dialectica* **1985**, *39*, 103–106. [CrossRef]
33. Handsteiner, J.; Friedman, A.S.; Rauch, D.; Gallicchio, J.; Liu, B.; Hosp, H.; Kofler, J.; Bricher, D.; Fink, M.; Leung, C.; et al. Cosmic Bell test: Measurement settings from milky way stars. *Phys. Rev. Lett.* **2017**, *118*, 060401. [CrossRef] [PubMed]
34. 't Hooft, G. *The Cellular Automaton Interpretation of Quantum Mechanics*; Springer: Cham, Switzerland, 2016.
35. Feynman, R.P.; Hibbs, A.R.; Styer, D.F. *Quantum Mechanics and Path Integrals*, Emended ed.; Dover: Mineola, NY, USA, 2010.
36. Griffiths, R.B. *Consistent Quantum Theory*; Cambridge University Press: Cambridge, UK, 2002.
37. Damgaard, P.; Huffel, H. Stochastic quantization. *Phys. Rep.* **1987**, *152*, 227–398. [CrossRef]
38. Argaman, N. On Bell's theorem and causality. *arXiv* **2008**.
39. Caldeira, A.O.; Leggett, A.J. Quantum tunnelling in a dissipative system. *Ann. Phys.* **1983**, *149*, 374–456. [CrossRef]
40. Daffertshofer, A.; Plastino, A.R.; Plastino, A. Classical no-cloning theorem. *Phys. Rev. Lett.* **2002**, *88*, 210601. [CrossRef] [PubMed]
41. Zurek, W.H. Decoherence, einselection, and the quantum origins of the classical. *Rev. Mod. Phys.* **2003**, *75*, 715–775. [CrossRef]
42. Wharton, K. Quantum states as ordinary information. *Information* **2014**, *5*, 190–208. [CrossRef]
43. Pawlowski, M.; Paterek, T.; Kaszlikowski, D.; Scarani, V.; Winter, A.; Zukowski, M. Information causality as a physical principle. *Nature* **2009**, *461*, 1101–1104. [CrossRef] [PubMed]
44. Cirel'son, B.S. Quantum generalizations of Bell's inequality. *Lett. Math. Phys.* **1980**, *4*, 93–100. [CrossRef]
45. Price, H. Two Paths to the Parisian Zigzag. Available online: http://emqm17.org/presentations/Huw-Price/ (accessed on 30 March 2018).
46. Wharton, K. Live Options for Spacetime-Based Physics. Available online: http://emqm17.org/presentations/Kenneth-Wharton/ (accessed on 30 March 2018).
47. Adlam, E. A Tale of Two Anachronisms. Available online: http://emqm17.org/presentations/emily-adlam/ (accessed on 30 March 2018).
48. Walleczek, J. Nonlocality or Local Retrocausality?—The Non-Signalling Theorem in Ontological Quantum Mechanics. Available online: http://emqm17.org/presentations/Jan-Walleczek-2/ (accessed on 30 March 2018).
49. Adlam, E. Spooky action at a temporal distance. *Entropy* **2018**, *20*, 41. [CrossRef]

Article

Fundamental Irreversibility: Planckian or Schrödinger–Newton?

Lajos Diósi

Wigner Research Centre for Physics, H-1525 Budapest 114, P.O. Box 49, Hungary; diosi.lajos@wigner.mta.hu

Received: 18 May 2018; Accepted: 22 June 2018; Published: 27 June 2018

Abstract: The concept of universal gravity-related irreversibility began in quantum cosmology. The ultimate reason for universal irreversibility is thought to come from black holes close to the Planck scale. Quantum state reductions, unrelated to gravity or relativity but related to measurement devices, are completely different instances of irreversibilities. However, an intricate relationship between Newton gravity and quantized matter might result in fundamental and spontaneous quantum state reduction—in the non-relativistic Schrödinger–Newton context. The above two concepts of fundamental irreversibility emerged and evolved with few or even no interactions. The purpose here is to draw a parallel between the two approaches first, and to ask rather than answer the question: can both the Planckian and the Schrödinger–Newton indeterminacies/irreversibilities be two faces of the same universe. A related personal note of the author's 1986 meeting with Aharonov and Bohm is appended.

Keywords: fundamental irreversibility; space-time fluctuations; spontaneous state reduction

---

1. Introduction

Standard micro-dynamical equations, whether classical or quantum, are deterministic and reversible. They can, nonetheless, encode various options of irreversibility even at the fundamental level. Here, I am going to discuss two separate concepts of fundamental irreversibility, which are quite certain to overlap in the long run. The first option concerns space-time (gravity); it is relativistic, hallmarked by mainstream cosmologists and field theorists (including immortal ones). The second option is rooted in the explicit irreversibility of von Neumann measurement in non-relativistic quantum mechanics; its story is perhaps more diffusive than that of the first. The standard and linear story of Planck scale irreversibility is recapitulated in Section 2. I choose a personal account for the parallel story of the conjectured Newton-gravity-related non-relativistic irreversibility of macroscopic quantum mechanics in Section 3. I stop both stories in the 1980s when the same structure of heuristic master equations was proposed for the two options of fundamental irreversible dynamics—with different interpretations and regimes of significance, of course. Towards their reconciliation, Section 4 offers some thoughts and concludes in an open-ended fashion.

2. Irreversibility at Planck Scale

At the dawn of quantum-gravity research, Bronstein [1–3] discovered by heuristic calculations that the precise structure of space-time, contrary to the precise structure of electromagnetism, is unattainable if we rely on the quantized motion of test bodies. Subsequent decades raised stronger and famous arguments concerning space-time blurriness, unpredictability, its role in universal loss of information, of quantum coherence, and of microscopic reversibility in general. Wheeler [4] found that smooth space-time changes into a foamy structure of topological fluctuations at the Planck scale. Bekenstein [5] gave the first exact quantitative proposal toward fundamental irreversibility, claiming that black holes have entropy:

$$S = \frac{k_B}{4\ell_{Pl}^2} \times (\text{black hole surface area}), \tag{1}$$

where k_B is Boltzmann's constant; ℓ_{Pl} is the Planck length. This was confirmed by Hawking [6] who showed that black holes do, indeed, emit the corresponding thermal radiation. A short time later, he summarized the situation by stating the unpredictability of quantum-gravity at the Planck scale, leading him to propose that quantum field theory is fundamentally irreversible [7]. Accordingly, the unitary scattering operator $\hat{S}$ should be replaced by the more general superscattering operator $\$$ acting on the initial density operator $\hat{\rho}_{in}$ instead of the initial state vector:

$$\hat{\rho}_{out} = \$\hat{\rho}_{in} \neq \hat{S}\hat{\rho}_{in}\hat{S}^{\dagger}. \tag{2}$$

To resolve the detailed irreversible (non-unitary) dynamics beyond Hawking's superscattering, Ellis et al. [8] proposed a simple quantum-kinetic (master) equation, which Banks, Susskind and Peskin [9] generalized as follows:

$$\frac{d\hat{\rho}}{dt} = -\frac{i}{\hbar}[\hat{H},\hat{\rho}] - \frac{1}{2\hbar^2}\int\int [\hat{Q}(x),[\hat{Q}(y),\hat{\rho}\,]]h(x-y)d^3x d^3y, \tag{3}$$

where $\hat{H}$ is the Hamiltonian, $\hat{Q}(x)$ is a certain quantum field, and $h(x-y)$ is a positive symmetric kernel. The transparent structure allowed the authors to point out a substantial difficulty: non-conservation of energy-momentum.

3. Irreversibility in the Schrödinger–Newton Context

In the early 1970s, being a student fascinated already by quantum theory, I missed a dynamical formalism of the state vector collapse from it. If I were a student and aware of the related literature, I would have read the phenomenological model by Bohm and Bub [10]. However, I was not aware of it, and started to think on my own. If you open a textbook, you will read about the expansion of the time-dependent state vector $|t\rangle$ in terms of the energy eigenstates $|n\rangle$ of eigenvalues E_n, respectively. However, I wrote it with a little modification:

$$|t\rangle = \sum_n c_n \exp\left(-\frac{i}{\hbar}E_n(1+\delta)t\right)|n\rangle, \tag{4}$$

because I observed that by allowing a small randomness δ of the time flow, the average density matrix becomes gradually diagonal in the energy basis:

$$\overline{|t\rangle\langle t|} \longrightarrow \sum_n |c_n|^2 |n\rangle\langle n|, \tag{5}$$

exactly as if someone measured the energy. I made a prototype dynamical model of non-selective von Neumann measurements. A question remained unanswered: where does randomness of time come from? The hint should have come from the sadly forgotten Bronstein [1–3], but it came from Károlyházy after he gave department seminars in 1973 on his earlier work [11] where he used a Planck scale uncertainty of classical space-time and a very vague model of massive body's state vector collapse based upon it. Unfortunately, I had to do experimental particle physics for a decade.

Returning to theory, I showed [12] that the Newtonian limit of standard reversible semi-classical gravity, the so-called Schrödinger–Newton equation [13], obtains sensible solitonic wave functions for the massive (e.g., nano-) objects' center-of-mass. This determined my approach, i.e., to put *non-relativistic* flesh on the toy dynamics (4) and (5) of state vector reduction. The uncertainty δ of time flow should come from the Newtonian limit of the metric tensor element g_{00}, which is, in fact, the Newton potential ϕ. The unpredictability $\delta\phi$ of the Newton potential should depend on G and $\hbar$, but not on c. The choice was the following spatially correlated white-noise:

$$\overline{\delta\phi(x,t)\delta\phi(y,s)} = \frac{\hbar G}{|x-y|}\delta(t-s). \tag{6}$$

The random part of the Newton potential couples to the mass density operator $\hat{f}(x)$ via the interaction $\int \phi(x,t)\hat{f}(x)d^3x$, yielding the following master equation for the density operator:

$$\frac{d\hat{\rho}}{dt} = -\frac{i}{\hbar}[\hat{H},\hat{\rho}] - \frac{G}{2\hbar}\int\int [\hat{f}(x),[\hat{f}(y),\hat{\rho}\,]]\frac{1}{|x-y|}d^3xd^3y. \tag{7}$$

This dynamic is mimicking the (non-selective) von Neumann measurement of massive object's positions. It predicts the *spontaneous reduction* (decay) of Schrödinger cat states (see the same result in [13] by Penrose).

Before journal publication [14], I showed this result to Yakir Aharonov (read Appendix A). He warned me about the energy-momentum non-conservation. This came as a surprise to me as I had not read [9].

4. Planck Scale or Schrödinger–Newton Context?

Irreversibility at the Planck scale seems plausible within standard physics because of evaporating black holes (Section 2). Non-relativistic Schrödinger–Newton irreversibility (Section 3) is a conjecture, although its derivation is not much more heuristic than that of Planckian's. For both options, the same structure of master equations was proposed to encode the irreversible dynamics of the density operator. Planck scale irreversibilities from Equation (3) become significant for certain fundamental elementary particles. Contrary to that, Equation (7) predicts irreversibility for massive non-relativistic objects in the Schrödinger–Newton context. Whether the two underlying concepts are compatible at all is unknown. Whether the Newtonian unpredictabilities/fluctuations are the non-relativistic limit of the Planckian's? That is difficult to answer.

Let me mention, nonetheless, two examples where relativistic phenomenologies, different from the line of Section 2, turned out to reduce to the Schrödinger–Newton uncertainty (6) non-relativistically. Unruh [15] proposed a possible uncertainty relation between the metric and Einstein tensors, respectively. In the Newtonian limit, speed of light c cancels out and we are left with just the white-noise uncertainties (6), as pointed out in [14]. Penrose discussed the fundamental conflict between general relativity and quantization. To resolve it, heuristically at least, he also found the necessity of space-time's fundamental blurriness, guessed it non-relativistically and determined its equivalent with expression (6) up to a factor of 2 (a discrepancy which has recently been resolved by [16]).

Against questioning a possible transmutation of Planck scale uncertainties into the non-relativistic Schrödinger–Newton regime, I have an elementary argument. Consider the Schrödinger-equation for the center-of-mass of a big body such as $M = 1$ kg, with velocity 1 km/s which is fairly non-relativistic. Calculate the de Broglie wave length: $\lambda = (2\pi\hbar/mv) = 4.16 \times 10^{-36}$ m. This is smaller than the Planck length $\ell_{Pl} = 1.62 \times 10^{-35}$ m by about one order of magnitude. Since standard physics breaks down anyway at the Planck scale, we can no longer trust in the Schrödinger equation for the motion of our massive non-relativistic body. Planck scale space-time uncertainties have thus developed uncertainties in the Schrödinger dynamics of non-relativistic massive bodies. So far so good. However, will c be cancelled out so that we obtain the effective Schrödinger–Newton uncertainty (6) and (7) and the corresponding spontaneous reduction for massive objects [13,14]?

5. Concluding Remarks

Two independent theories of relativistic and non-relativistic fundamental irreversibility, both related to the conflict between gravity and quantization, are in the scope of this work. One was conceived and would be relevant in cosmology. The other one was born from the quantum measurement problem and would modify the quantum mechanics of massive bodies even in the lab. Their conceptions have been outlined in Sections 2 and 3, respectively, including their basics without details and later developments. Such a restricted presentation sufficed to expose the issue at the center of this work in Section 4: what is the relationship between the Planckian and the Schrödinger–Newton unpredictability of our space-time? The question remains unanswered, but our purpose has been to highlight it. In particular, we pointed out that Planckian unpredictability survives non-relativistically—for massive macroscopic quantized degrees of freedom.

Funding: This research was funded by the Hungarian Scientific Research Fund grant number 124351 and the EU COST Action grant number CA15220. **Acknowledgments:** The Fetzer Franklin Fund is acknowledged for generously covering my costs to attend the Emergent Quantum Mechanics 2017 conference (26–28 October, University of London) and to publish this work in open access.

Conflicts of Interest: The author declares no conflict of interests.

Appendix A

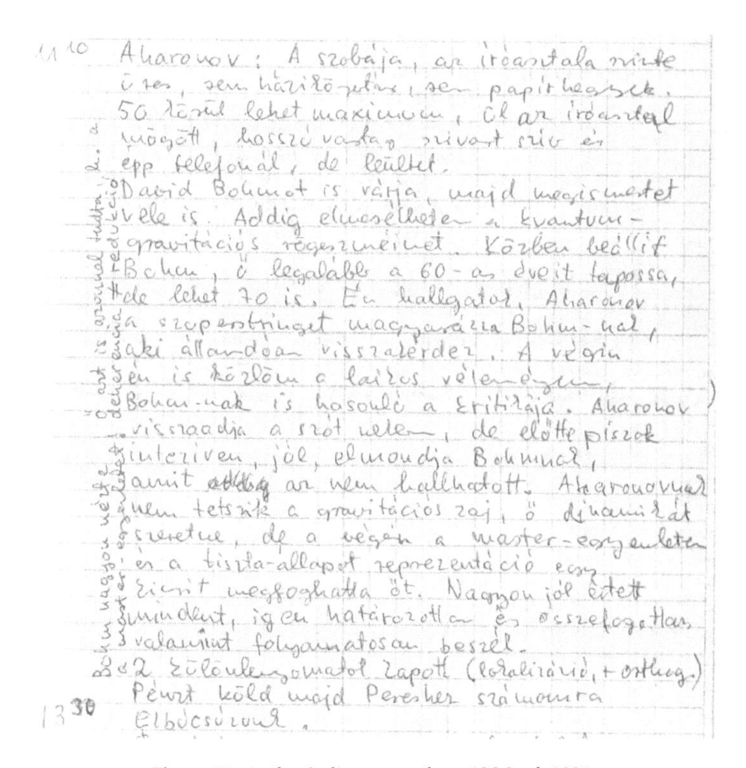

11^10 Aharonov: A szobája, az íróasztala szinte
üres, sem házikönyvtár, sem papírhegyek.
50 körül lehet maximum, ül az íróasztal
mögött, hosszú vastag szivart szív és
épp telefonál, de leültet.
David Bohmot is várja, majd megismertet
vele is. Addig elmesélhetem a kvantum-
gravitációs rögeszméimet. Közben beállít
Bohm, ő legalább a 60-as éveit tapossa,
de lehet 70 is. Én hallgatok, Aharonov
a szuperstringet magyarázza Bohm-nak,
aki állandóan visszakérdez. A végén
én is közlöm a laikus véleményem,
Bohm-nak is hasonló a kritikája. Aharonov
visszaadja a szót nekem, de előtte piszok
intenzíven, jól, elmondja Bohmnak,
amit ~~addig~~ az nem hallhatott. Aharonovnak
nem tetszik a gravitációs zaj, ő dinamikát
szeretne, de a végén a master-egyenleten
és a tiszta-állapot reprezentáció egy
kicsit megfoghatta őt. Nagyon jól értett
mindent, igen határozottan és összefogottan
valamint folyamatosan beszél.
2 különlenyomatot kapott (lokalizáció, + orthog.)
Pénzt küld majd Pereshez számomra
13^30 Elbúcsúzunk.

Figure A1. Author's diary page, from 18 March 1986.

It was Asher Peres who asked Yakir to receive the unknown theorist from Hungary. Below is the translation of my notes (Figure A1).

> 11[10] Aharonov: His office and desk are almost empty, no personal library, no paper piles. He is at most 50 or so. He sits behind the desk, smokes a long fat cigar, makes a phone call, and asks that I take a seat.

We await David Bohm, who I will also be introduced to. Until then, I can unfold my quantum-gravity idée fix. David Bohm arrives. He is at least in his 60s, but could be 70. I am listening as Aharonov explains the superstring to Bohm who is repeatedly asking questions. Finally, I also communicate my layman's views; Bohm's criticism is also akin. Aharonov allows me to speak, but first tells Bohm with hellish intensively what he could not have heard. Aharonov dislikes gravitational noise; he prefers dynamics. However, at the end, my master equation and the pure state representation may have caught him a bit. He understood everything very well, he spoke steadily, with real firmness and organization.

He got two offprints (localization + orthog.)

Peres will send money for me.

13^{30} We say goodbye.

Left margin: Bohm looked at the master equation intently! Immediately, he also knew that decoherence $\neq$ reduction.

References

1. Bronstein, M. Quantentheorie schwacher Gravitationsfelder. *Phys. Z. Sowjetunion* **1936**, *9*, 140–157. (In German)
2. Bronstein, M.P. Kvantovanie gravitatsionnykh voln. *Zh. Eksp. Theor. Fiz.* **1936**, *6*, 195–236. (In Russian)
3. Gorelik, G.M. Matvei Bronstein and quantum gravity: 70th anniversary of the unsolved problem. *Usp. Fiz. Nauk* **2005**, *48*, 1039–1053. [CrossRef]
4. Wheeler, J.A. *Geometrodynamics*; Academic Press: New York, NY, USA, 1962.
5. Bekenstein, J.D. Black holes and entropy. *Phys. Rev. D* **1973**, *7*, 2333. [CrossRef]
6. Hawking, S.W. Particle creation by black holes. *Commun. Math. Phys.* **1975**, *43*, 199–220. [CrossRef]
7. Hawking, S.W. The unpredictability of quantum gravity. *Commun. Math. Phys.* **1982**, *87*, 395–415. [CrossRef]
8. Ellis, J.; Hagelin, S.; Nanopoulos, D.V.; Srednicki, M. Search for violations of quantum mechanics. *Nucl. Phys. B* **1984**, *241*, 381–405. [CrossRef]
9. Banks, T.; Susskind, L.; Peskin, M.E. Difficulties for the evolution of pure states into mixed states. *Nucl. Phys. B* **1984**, *244*, 125–134. [CrossRef]
10. Bohm, D.; Bub, J. A proposed solution of the measurement problem in quantum mechanics by a hidden variable theory. *Rev. Mod. Phys.* **1966**, *38*, 453–469. [CrossRef]
11. Karolyhazy, F. Gravitation and quantum mechanics of macroscopic objects. *Nuovo Cim.* **1966**, *42*, 390–402. [CrossRef]
12. Diósi, L. Gravitation and quantum-mechanical localization of macro-objects. *Phys. Lett. A* **1984**, *105*, 199–202. [CrossRef]
13. Penrose, R. On gravity's role in quantum state reduction. *Gen. Relativ. Gravit.* **1996**, *28*, 581–600. [CrossRef]
14. Diósi, L. A universal master equation for the gravitational violation of quantum mechanics. *Phys. Lett. A* **1987**, *120*, 377–381. [CrossRef]
15. Unruh, W.G. Steps towards a quantum theory of gravity. In *Quantum Theory of Gravity*; Christensen, S.M., Ed.; Adam Hilger Ltd.: Bristol, UK, 1984; pp. 234–242.
16. Tilloy, A.; Diósi, L. Principle of least decoherence for Newtonian semi-classical gravity. *Phys. Rev. D* **2017**, *96*, 104045. [CrossRef]

MDPI

Article

Quantum Thermodynamics at Strong Coupling: Operator Thermodynamic Functions and Relations

Jen-Tsung Hsiang [1,*,†] and Bei-Lok Hu [2,†]

[1] Center for Field Theory and Particle Physics, Department of Physics, Fudan University, Shanghai 200433, China

[2] Maryland Center for Fundamental Physics and Joint Quantum Institute, University of Maryland, College Park, MD 20742-4111, USA; blhu@umd.edu

* Correspondence: cosmology@gmail.com; Tel.: +86-21-3124-3754

† These authors contributed equally to this work.

Received: 26 April 2018; Accepted: 30 May 2018; Published: 31 May 2018

Abstract: Identifying or constructing a fine-grained microscopic theory that will emerge under specific conditions to a known macroscopic theory is always a formidable challenge. Thermodynamics is perhaps one of the most powerful theories and best understood examples of emergence in physical sciences, which can be used for understanding the characteristics and mechanisms of emergent processes, both in terms of emergent structures and the emergent laws governing the effective or collective variables. Viewing quantum mechanics as an emergent theory requires a better understanding of all this. In this work we aim at a very modest goal, not quantum mechanics as thermodynamics, not yet, but the thermodynamics of quantum systems, or quantum thermodynamics. We will show why even with this minimal demand, there are many new issues which need be addressed and new rules formulated. The thermodynamics of small quantum many-body systems strongly coupled to a heat bath at low temperatures with non-Markovian behavior contains elements, such as quantum coherence, correlations, entanglement and fluctuations, that are not well recognized in traditional thermodynamics, built on large systems vanishingly weakly coupled to a non-dynamical reservoir. For quantum thermodynamics at strong coupling, one needs to reexamine the meaning of the thermodynamic functions, the viability of the thermodynamic relations and the validity of the thermodynamic laws anew. After a brief motivation, this paper starts with a short overview of the quantum formulation based on Gelin & Thoss and Seifert. We then provide a quantum formulation of Jarzynski's two representations. We show how to construct the operator thermodynamic potentials, the expectation values of which provide the familiar thermodynamic variables. Constructing the operator thermodynamic functions and verifying or modifying their relations is a necessary first step in the establishment of a viable thermodynamics theory for quantum systems. We mention noteworthy subtleties for quantum thermodynamics at strong coupling, such as in issues related to energy and entropy, and possible ambiguities of their operator forms. We end by indicating some fruitful pathways for further developments.

Keywords: quantum thermodynamics; strong coupling; operator thermodynamic functions

---

1. Quantum and Thermodynamics—Why?

Why is a paper on this subject matter appearing in this special issue of Entropy? The short answer is the same reason why EmQM appears in this journal Entropy, which is generally considered as treating topics in statistical mechanics: emergence. The long answer, serving as a justification for our dwelling on quantum thermodynamics in the realm of emergent quantum mechanics, is as follows. One way to see quantum mechanics as emergent is by analogy with hydrodynamics and thermodynamics, probably the two best known emergent theories because we know exactly what the collective variables are (thermodynamic functions and relations), the laws they obey (the four laws), how they are related to the basic constituents (molecular dynamics) and the mediating theory from which we can derive both the fundamental and the emergent theories (kinetic theory).

In analogy to emergent gravity [1–5], one of the present authors has championed the thesis of "general relativity as geometro-hydrodynamics" [6,7]. Since Verlinde's "gravity is entropic force" popularization of Jacobson's "Einstein equation of state" thesis [8–10], gravity as thermodynamics has caught a wider attention in the quantum gravity community. However, a new challenge arises. The question one of us posed for enthusiasts of this theme is the following: Since both gravity and thermodynamics are old subjects established centuries before the advent of quantum mechanics, and both can make sense and stand alone at the classical level without quantum theory, well, what exactly is quantum doing here?—What is the role of quantum in emergent gravity? Do we really need quantum if we view "gravity as thermodynamics"?

1.1. Quantum in "Gravity as Thermodynamics"

This question, "Wither the Quantum?", as Hu calls it, puts the spotlight on quantum, in how it contributes to the emergent phenomena which gives us both gravity and thermodynamics. One answer to this is to also consider quantum mechanics as emergent. For example, in the emergent theories of Adler and 't Hooft [11,12] probability theory, stochastic and statistical mechanics as a slate play a pivotal role, just as they do for thermodynamics and hydrodynamics arising from molecular dynamics. The same applies to emergent gravity: classical gravity captured by general relativity is an effective theory emergent from some fundamental theories of the basic constituents of spacetime functioning at the sub-Planckian scale. How these basic constituents interact, how their interaction strength varies with energy, how at some specific scale(s) some set(s) of collective variables and the law(s) governing them emerge, and in succession, leading to the physics at the lowest energy as we know it in today's universe is perhaps just as interesting as the manifestation of the relevant physics at the different scales familiar to us—from molecules to atoms to nucleons to quarks and below. Putting aside gravity for now in this investigation, we wish to see a deeper connection between micro and macro, quantum and thermo.

1.2. David Bohm: Quantum in Classical Terms

Here, Bohm's philosophical influence is evident. His pilot wave theory may not offer a better description or explanation of quantum phenomena, but the view that quantum mechanics is not a fundamental theory any more than a classical wave theory is, provides an inspiration for asking a deeper layer of questions: If we view quantum theory functioning in the capacity as thermodynamics, we should ask: What are the fundamental constituents, the laws governing them, and how quantum mechanics emerges from the sub-structures and theories depicting them (Long before we get to this point, many readers may have raised this objection: This is obviously nonsense: The second law of thermodynamics ostensibly shows the effects of an arrow of time, while quantum mechanics is time-reversal invariant. Well, if the mechanical processes which we can observe are in the underdamped regime where the dissipative effects are not strong enough, they would appear to obey time-reversal symmetry. This is not an outlandish explanation: For most physical systems, in the open system perspective, quantum phenomena in the system appears within the decoherence time which is many orders of magnitude shorter than the relaxation time, as is the case in many well-controlled environments (e.g., cavity QED). Or, if the system is near a nonequilibrium critical point. On this issue, cosmology, despite its seemingly remote bearing, may actually enter in a basic way, in terms of the origin of the arrow of time, and the mere fact that nonequilibrium conditions prevail in an expanding universe.).

1.3. Quantum as Thermodynamics?

In quantum thermodynamics, we may not see much in terms of what fundamental theory quantum mechanics emerges from (Adler and 't Hooft may have their answers: trace dynamics and cellular automaton, for instance, respectively), but even the juxtaposition or crossing of what is traditionally considered as governing the two opposite ends in the macro/micro and classical/quantum spectrum may reveal some deeper meaning in both. Macroscopic quantum phenomena is another such arena. For example, in small quantum systems, at low temperatures, or when the system is strongly coupled to its environment, is there a lower limit to the validity of the laws of thermodynamics, which play such an important role in our understanding of the macro world? Under what conditions will macroscopic entities show quantum phenomena? Is there an upper limit to quantum mechanics governing the meso domain? Is there a limit to quantum commanding the macro world? The above explains the philosophical issues which motivated us to take up a study of quantum thermodynamics. We are also of the opinion that useful philosophical discourses of any subject matter should be based on the hard-core scientific knowledge of that subject, down to all the nitty-gritty details of each important topic that makes up that body of knowledge. Thus we start with the basic demands in the formulation of quantum thermodynamics and try to meet them in a rigorous, no-nonsense way. The specific goal of this paper is to define the operator thermodynamic quantities and spell out their relations for quantum many-body systems in thermal equilibrium.

Quantum Thermodynamics

Quantum thermodynamics is a fast developing field, emergent from quantum many body physics and nonequilibrium statistical mechanics. Simply, it is the study of the thermodynamic properties of quantum many-body systems. Quantum now refers not just to the particle spin-statistics (boson vs. fermion) aspects in traditional quantum statistical mechanics, but also includes in the present era the quantum phase aspects, such as quantum coherence, quantum correlations, and quantum entanglement, where quantum information enters. The new challenges arise from several directions not falling under the assumptions of traditional classical thermodynamics: finding the quantum properties of small systems, at zero or very low temperature, strongly coupled to an environment, which could have non-ohmic spectral densities and colored noise, while the system evolves following a non-Markovian dynamics (with memory).

1.4. This Work

In this paper we discuss the issues and the technical challenges encountered in the first stage in the construction of a viable theory of quantum thermodynamics, where the system is strongly coupled with a heat bath. We wish to present in a systematic way how to introduce the operator thermodynamic functions and construct their relations. Here, we treat this problem in an equilibrium setting. There are other ways to construct such a theory, such as pursued in the so-called "eigenvalue thermalization" program [13–22], which treats the system and the environment as a closed system, or along the lines expounded in the fluctuation theorems [23–25] where the system under an external drive is allowed to evolve in a nonequilibrium albeit controlled manner [26,27], or in a fully nonequilibrium open quantum system dynamics program (see, e.g., [28,29] and references therein).

2. Quantum Thermodynamics at Strong Coupling: Background

2.1. New Challenges in Quantum Thermodynamics

Small quantum many-body systems strongly coupled to a heat reservoir at low temperatures are the new focuses of interest for quantum thermodynamics [30–32]. Under these hitherto lesser explored conditions, one needs to re-examine the meaning of the thermodynamic functions, the viability of the thermodynamic relations and the validity of the thermodynamic laws anew. Traditional thermodynamics is built on large systems weakly coupled to a reservoir [33], and for quantum systems, only the spin-statistics aspect is studied, as in quantum statistical mechanics, leaving the important factors of quantum coherence, correlations, entanglement and fluctuations as new challenges for quantum thermodynamics. To see the difference strong coupling makes, we take as an example the definition of heat, the energy transferred between the system and the reservoir, for systems strongly coupled to a bath. Esposito et al. [34], for example, show that any heat definition expressed as an energy change in the reservoir energy plus any fraction of the system-reservoir interaction is not an exact differential when evaluated along reversible isothermal transformations, except when that fraction is zero. Even in that latter case the reversible heat divided by temperature, namely entropy, does not satisfy the third law of thermodynamics and diverges in the low temperature limit. For quantum systems, as pointed out by Ankerhold and Pekola [35], in actual measurements, especially for solid state structures, quantum correlations between system and reservoir may be of relevance not only far from but also close to and in thermal equilibrium. Even in the weak coupling regime, this heat flow is substantial at low temperatures and may become comparable to typical predictions for the work based on conventional weak coupling approaches. It further depends sensitively on the non-Markovian features of the reservoir. These observations exemplify the intricacies involved in defining heat for strongly-coupled systems and added complexities for quantum systems, especially at low temperatures.

This incertitude regarding heat translates to ambiguity in the definition of thermodynamic functions and the thermodynamic relations. For example, it was shown [36–41] that the expressions for the specific heat derived from the internal energies of a quantum-mechanical harmonic oscillator bilinearly coupled to a harmonic bath calculated by two different approaches can have dramatically different behavior in the low temperature regime. To illustrate this point, Gelin and Thoss [42] compared these two approaches of calculating the internal energy of the system, which give identical results if the system-bath coupling is negligible, but predict significantly differently for finite system-bath coupling. In the first approach, the mean energy of the system given by the expectation value of the system Hamiltonian is evaluated with respect to the total (system + bath) canonical equilibrium distribution. The second approach is based on the partition function of the system, Z_s, which is postulated to be given as the ratio of the total (system + bath) and the bath partition functions. Gelin and Thoss [42] introduce a bath-induced interaction operator $\hat{\Delta}_s$, which would account for the effects of finite system-bath coupling and analyze the two approaches for several different systems including several quantum and classical point particles and nonlinear system bath coupling. They found that Approach II leads to very different results from Approach I, their differences exist already within classical mechanics, provided the system-bath interaction is not bilinear and/or the system of interest consists of more than a single particle.

Similar ambiguity appears in the entropy of the quantum system in the same setup. In the first approach, the von Neumann entropy is chosen to be the entropy of the system, while in the second approach, the entropy is given by the derivative of the system's free energy with respect to the inverse temperature. Both definitions are equivalent in the limit of weak system-bath coupling. However, at finite coupling strength, it has been noticed that [36,37,43,44] the von Neumann entropy may not vanish when the bath temperature is close to zero, even for a simple quantum system that consists of harmonic oscillators. This nonvanishing behavior of the von Neumann entropy is related to the quantum entanglement between the system and the bath [45–48]. On the other hand, the entropy defined in the second approach of the same system gives an expected vanishing result, consistent with the physical picture described by [49]. These disparities between different choices of the same thermodynamic function of the system can be traced to the degree in which the system-bath interaction enters into or counted as the system.

In a small quantum system, quantum fluctuations of a physical observable of the system can grow to the same order of magnitude as its average value. Thus a description of the thermodynamics of such a system based on the mean value is insufficient. To include the effects of quantum fluctuations of the thermodynamic variables, we need their corresponding quantum operators, from which we can calculate its higher moments. Likewise for quantities related to the quantum correlations and quantum coherence of such a system. Constructing the operator form of a thermodynamic variable is nontrivial and ambiguous at times. Take as an example the internal energy of the system. As is mentioned earlier, it can be given either as the expectation value of the system's Hamiltonian operator, or as the partial derivative of the (effective) partition function of the system. In the former case, the operator of the internal energy is already chosen but with arbitrariness, while in the latter case, as will be discussed in Section 7.3, our capabilities of identifying the corresponding operator is limited by our knowledge of the thermodynamic variables based on either its expectation or its classical counterpart. Either way there exists non-uniqueness in the introduction of the thermodynamic operator form of the system variables.

Even with a successful extraction and identification of the operator form from the expectation value or the partial derivative of the system's partition, such a thermodynamic operator for the reduced system in general acts on the Hilbert space of the full composite (system + environment). This is not always a desirable feature because most of the time we are interested in what happens to the system under the influence of the environment, whose details are of no particular concern. A description of the full composite does not easily translate to useful information about the system. Thus it is more preferable to derive operators of the system which solely act on the Hilbert space of the system, without reference to the environment. This is what we try to accomplish here; it is different from what has been reported in the literature.

2.2. Goal and Findings of Present Work

In traditional weak coupling thermodynamics, knowing the thermodynamic potentials and their relations enables one to construct a theory of thermodynamics. We believe they are also necessary for the establishment of a theory of thermodynamics for quantum systems. The new challenge is twofold: constructing operator thermodynamic potentials and treating quantum systems strong coupled to their baths. This is our goal in this paper. We shall provide a quantum formulation of Jarzynski's [27] classical thermodynamics at strong coupling. In contrast to the study of the nonequilibrium dynamics of an open quantum system (called ONEq in [29,50,51]), the combined system + environment studied here is a closed system, called the composite, which is assumed to stay in a global thermal state. In such a configuration, even though the interaction between the system and the bath is non-negligible, the partition function of the composite can be defined. This facilitates the introduction of thermodynamic potentials in a way similar to the traditional vanishing-coupling thermodynamics. We will focus on thermodynamic functions such as internal energy, enthalpy, entropy, and free energies, but exclude the consideration of heat and work in this paper, as quantum work is not well understood and requires a separate treatment.

This paper is organized as follows: In Section 3, we give a quick summary of the familiar thermodynamic relations, which we call traditional or weak-coupling thermodynamics, if only to establish notations. We then consider interacting quantum systems with the help of the Hamiltonian of mean force [52–58] and discuss some conceptual and technical difficulties we may face when the coupling strength becomes strong. In Sections 4–6, we overview the equilibrium quantum formulations of thermodynamics at strong coupling, based on Gelin & Thoss' and Seifert's work. In Section 7, we present a equilibrium quantum formulation of Jarzynski's thermodynamics in the "bare" and "partial molar" representations, respectively in the same setting. We conclude in Section 8 with a short summary, a brief discussion of two issues involved and a suggestion for further developments.

3. Thermodynamic Functions, Hamiltonian of Mean Force

We first summarize the familiar traditional thermodynamic relations, if only to establish notations. We then consider interacting quantum systems with the help of the Hamiltonian of mean force (Hamiltonian of mean force is a useful yet not indispensable concept for this purpose. It is useful because in the same representation, the formal expressions associated with it resemble the counterparts in the traditional weak coupling thermodynamics). We outline two quantum formulations of thermodynamic functions and relations; one based on Gelin and Thoss [42] and the other on Seifert [26]. With the abundance of thermodynamic quantities, a word about notations is helpful: quantum expectation values or classical ensemble averages are denoted by math calligraphic, quantum operators associated with the variable O will carry an overhat $\hat{O}$.

3.1. Traditional (Weak-Coupling) Equilibrium Thermodynamic Relations

The pre-conditions for the traditional weak-coupling thermodynamic theory to be well-defined and operative for a classical or quantum system are very specific despite their wide ranging applicability: (a) A system **S** of relatively few degrees of freedom is in contact with a thermal bath of a large number or infinite degrees of freedom (We shall consider only heat but no particle transfer here and thus the thermodynamics refers only to canonical, not grand canonical ensembles); (b) the coupling between the system and the bath is vanishingly small; and (c) the system is eternally in a thermal equilibrium state by proxy with the bath which is impervious to any change in the system. In weak-coupling thermodynamics, the bath variables are not dynamical variables (Dynamical variables are those which are determined consistently by the interplay between the system and the bath through their coupled equations of motion); they only provide weak-coupling thermodynamic parameters such as a temperature in canonical ensemble, or, in addition, a chemical potential, in grand canonical ensemble.

The classical thermodynamic relations among the internal energy $\mathcal{U}$, enthalpy $\mathcal{H}$, Helmholtz free energy $\mathcal{F}$ and Gibbs free energy $\mathcal{G}$ in conjunction with the temperature T, entropy $\mathcal{S}$, pressure P and volume $\mathcal{V}$ are well-known. From the first law, $d\mathcal{U} = T\,d\mathcal{S} - P\,d\mathcal{V}$. With $\mathcal{U} = \mathcal{U}(\mathcal{S},\mathcal{V})$, we have

$$T = \left(\frac{\partial \mathcal{U}}{\partial \mathcal{S}}\right)_{\mathcal{V}}, \qquad P = -\left(\frac{\partial \mathcal{U}}{\partial \mathcal{V}}\right)_{\mathcal{S}}. \tag{1}$$

By virtue of the differential form of the internal energy, the enthalpy $\mathcal{H} = \mathcal{U} + P\mathcal{V}$ obeys $d\mathcal{H} = T\,d\mathcal{S} + \mathcal{V}\,dP$. Since it is a function of the entropy and pressure, we can identify

$$T = \left(\frac{\partial \mathcal{H}}{\partial \mathcal{S}}\right)_{P}, \qquad \mathcal{V} = \left(\frac{\partial \mathcal{H}}{\partial P}\right)_{\mathcal{S}}. \tag{2}$$

Likewise, for the Helmholtz free energy $\mathcal{F} = \mathcal{U} - T\mathcal{S}$, we have

$$d\mathcal{F} = -\mathcal{S}\,dT - P\,d\mathcal{V}, \qquad \text{whence}, \qquad \mathcal{S} = -\left(\frac{\partial \mathcal{F}}{\partial T}\right)_{\mathcal{V}}, \qquad P = -\left(\frac{\partial \mathcal{F}}{\partial \mathcal{V}}\right)_{T}, \tag{3}$$

so the Helmholtz free energy is a function of the temperature and the volume, $\mathcal{F} = \mathcal{F}(T,\mathcal{V})$. Finally, the Gibbs free energy, defined by $\mathcal{G} = \mathcal{H} - T\mathcal{S}$, obeys

$$d\mathcal{G} = -\mathcal{S}\,dT + \mathcal{V}\,dP, \qquad \text{whence}, \qquad \mathcal{S} = -\left(\frac{\partial \mathcal{G}}{\partial T}\right)_{P}, \qquad V = \left(\frac{\partial \mathcal{G}}{\partial P}\right)_{T}. \tag{4}$$

Thus $\mathcal{G} = \mathcal{G}(T,P)$. Many more relations can be derived from these three basic relations. These relations are mutually compatible based on differential calculus.

Next we turn to the weak-coupling thermodynamics of quantum systems (Hereafter, we will choose the units such that the Boltzmann constant $k_B = 1$ and the reduced Planck constant $\hbar = 1$. In addition, to distinguish them from strong-coupling thermodynamics, all quantities defined in the context of traditional (weak-coupling) thermodynamics are identified with a subscript Θ). The state of a quantum system in contact with a heat bath at temperature $T = \beta^{-1}$ with vanishing coupling is described by the density matrix $\hat{\rho}_s = e^{-\beta \hat{H}_s}/\mathcal{Z}_\Theta$ where $\mathcal{Z}_\Theta = \mathrm{Tr}_s\, e^{-\beta \hat{H}_s}$ is the canonical partition function. Here $\hat{H}_s$ is the Hamiltonian of the system and is assumed to be independent of the inverse temperature $\beta = T^{-1}$. The notation Tr with a subscript s or b represents the sum over the states of the system or the bath respectively. The density matrix $\hat{\rho}_s$ is a time-independent Hermitian operator and is normalized to unity, i.e., $\mathrm{Tr}_s\,\hat{\rho}_s = 1$ to ensure unitarity.

The free energy $\mathcal{F}_\Theta$ of a quantum system in a canonical distribution is $\mathcal{F}_\Theta = -\beta^{-1}\ln\mathcal{Z}_\Theta$. The quantum expectation value $\langle \hat{H}_s\rangle$ is identified with the internal energy $\mathcal{U}_\Theta$ of the quantum system, and can be found by

$$\mathcal{U}_\Theta = \langle \hat{H}_s\rangle = \frac{1}{\mathcal{Z}_\Theta}\,\mathrm{Tr}_s\big\{\hat{H}_s\,e^{-\beta\hat{H}_s}\big\} = -\frac{\partial}{\partial\beta}\ln\mathcal{Z}_\Theta = \mathcal{F}_\Theta + \partial_\beta\mathcal{F}_\Theta\,. \tag{5}$$

If we define the entropy $\mathcal{S}_\Theta$ of the system by

$$\mathcal{S}_\Theta = \beta^2\partial_\beta\mathcal{F}_\Theta\,, \tag{6}$$

then it will be connected with the internal energy by the relation $\mathcal{F}_\Theta = \mathcal{U}_\Theta - T\,\mathcal{S}_\Theta$. These two expressions imply that the entropy of the quantum system can be expressed in terms of the density matrix $\mathcal{S}_\Theta = -\,\mathrm{Tr}_s\{\hat{\rho}_s\ln\hat{\rho}_s\}$, which is the von Neumann entropy. The von Neumann entropy plays an important role in quantum information as a measure of quantum entanglement, and can be used to measure the non-classical correlation in a pure-state system. (Beware of issues at zero temperature as discussed in Section 6.) Here we note that both internal energy and the entropy of the system can be equivalently defined in terms of the expectation values of the quantum operators, or as the derivative of the free energy.

The heat capacity $\mathcal{C}_\Theta = \partial\mathcal{U}_\Theta/\partial T = -\beta^2\partial_\beta\mathcal{U}_\Theta$ is given by

$$\mathcal{C}_\Theta = -2\beta^2\partial_\beta\mathcal{F}_\Theta - \beta^3\partial_\beta\mathcal{F}_\Theta = -\beta\,\partial_\beta\mathcal{S}_\Theta = \beta^2\Big[\langle\hat{H}_s^2\rangle - \langle\hat{H}_s\rangle^2\Big] \geq 0\,. \tag{7}$$

It is always semi-positive. Up to this point, under the vanishing system-bath coupling assumption, all the quantum thermodynamic potentials and relations still resemble their classical counterparts.

3.2. Quantum System in a Heat Bath with Nonvanishing Coupling

In formulating the quantum thermodynamics at strong coupling, we immediately face some conceptual and technical issues. At strong coupling, the interaction energy between the system and the bath is not negligible, so the total energy cannot be simply divided as the sum of the energies of the system and the bath. This introduces an ambiguity in the definition of, for example, internal energy. We may have more than one way to distribute the interaction energy between the system and the bath. The same ambiguity also arises in the other thermodynamic functions such as enthalpy and entropy, thus affecting the relations among the thermodynamic functions. On the technical side, in the course of formulating quantum thermodynamics, the non-commutative natures of the quantum operators make formidable what used to be straightforward algebraic manipulations in the classical thermodynamics.

Let us illustrate the previous points by an example. Consider in general an interacting quantum system **C** whose evolution is described by the Hamiltonian $\hat{H}_c = \hat{H}_s + \hat{H}_i + \hat{H}_b$, where $\hat{H}_s$, $\hat{H}_b$ are the Hamiltonians of the system **S** and the bath **B**, respectively and $\hat{H}_i$ accounts for the interaction between them. Suppose initially the composite **C** = **S** + **B** is in a global thermal equilibrium state which is stationary, and thus has reversible dynamics, described by the density matrix $\hat{\rho}_c = e^{-\beta\hat{H}_c}/\mathcal{Z}_c$ at the inverse temperature β^{-1}. The quantity $\mathcal{Z}_c = \mathrm{Tr}_{sb}\,e^{-\beta\hat{H}_c}$ is the partition function of the composite for the global thermal state.

In the case of vanishing coupling between the system and the bath, we may approximate the total Hamiltonian $\hat{H}_c$ to the leading order by $\hat{H}_c \simeq \hat{H}_s + \hat{H}_b$. Since $[\hat{H}_s, \hat{H}_b] = 0$, we notice that

$$\frac{1}{\mathcal{Z}_b}\,\mathrm{Tr}_b\,e^{-\beta\hat{H}_c} \simeq \frac{1}{\mathcal{Z}_b}\,\mathrm{Tr}_b\{e^{-\beta\hat{H}_s}e^{-\beta\hat{H}_b}\} = e^{-\beta\hat{H}_s}\,, \tag{8}$$

with the partition function of the free bath being given by $\mathcal{Z}_b = \mathrm{Tr}_b\, e^{-\beta\hat{H}_b}$. Equation (8) implies that the reduced state $\hat{\rho}_r = \mathrm{Tr}_b\, \hat{\rho}_c$, which is also stationary, will assume a canonical form

$$\rho_r = \frac{e^{-\beta\hat{H}_s}}{\mathcal{Z}_s} = \frac{1}{\mathcal{Z}_c}\,\mathrm{Tr}_b\, e^{-\beta\hat{H}_c}\,, \qquad \text{with} \qquad \mathcal{Z}_s = \mathrm{Tr}_s\, e^{-\beta\hat{H}_s}\,, \tag{9}$$

that is, $\mathcal{Z}_c \simeq \mathcal{Z}_s\mathcal{Z}_b$ in the limit of vanishing system-bath coupling. In addition, (9) ensures the proper normalization condition $\mathrm{Tr}_s\, \rho_r = 1$. Thus in the weak limit of the system-bath interaction, the reduced density matrix of the interacting composite system in the global thermal state will take the canonical form, hence it to some degree justifies the choice of the system state in the context of quantum thermodynamics in the textbooks [59,60]. Hereafter we will denote the reduced density matrix of the system by $\hat{\rho}_s$.

When the interaction between the system and the bath cannot be neglected, the righthand side of (8) no longer holds. In addition, non-commutating nature among the operators $\hat{H}_s$, $\hat{H}_i$ and $\hat{H}_b$ prevents us from writing $e^{-\beta(\hat{H}_s+\hat{H}_i+\hat{H}_b)} \neq e^{-\beta(\hat{H}_s+\hat{H}_b)}e^{-\beta\hat{H}_i}$, due to $[\hat{H}_s, \hat{H}_i] \neq 0$ and $[\hat{H}_b, \hat{H}_i] \neq 0$ in general. In fact, according to the Baker-Campbell-Haussdorff (BCH) formula, the previous decomposition will have the form

$$\begin{aligned} e^{-\beta(\hat{H}_s+\hat{H}_b)}e^{-\beta\hat{H}_i} = \exp\Big\{ & -\beta(\hat{H}_s + \hat{H}_i + \hat{H}_b) + \frac{\beta^2}{2!}\,[\hat{H}_s + \hat{H}_b, \hat{H}_i] \\ & - \frac{\beta^3}{3!}\Big(\frac{1}{2}\,[[\hat{H}_s + \hat{H}_b, \hat{H}_i], \hat{H}_i] + \frac{1}{2}\,[\hat{H}_s + \hat{H}_b, [\hat{H}_s + \hat{H}_b, \hat{H}_i]]\Big) + \cdots \Big\}. \end{aligned} \tag{10}$$

The exponent on the righthand side typically contains an infinite number of terms. This makes algebraic manipulation of the strongly interacting system rather complicated, in contrast to its classical or quantum weak-coupling counterpart.

3.3. Hamiltonian of Mean Force

To account for non-vanishing interactions (in this paper, we apply the Hamiltonian of mean force to a quantum system in the global thermal state setup without any time-dependent driving force. See [53–55] for its use in nonequilibrium systems at strong coupling), one can introduce the Hamiltonian of mean force H_s^* for the system defined by [56–58]

$$e^{-\beta\hat{H}_s^*} \equiv \frac{1}{\mathcal{Z}_b}\,\mathrm{Tr}_b\, e^{-\beta\hat{H}_c}\,. \tag{11}$$

It depends only on the system operator but has included all the influences from the bath. In the limit $\hat{H}_i$ is negligible $\hat{H}_s^* \simeq \hat{H}_s$; otherwise, in general $\hat{H}_s^* \neq \hat{H}_s$. The corresponding partition function $\mathcal{Z}^*$ is then given by $\mathcal{Z}^* = \mathrm{Tr}_s\, e^{-\beta\hat{H}_s^*} = \mathcal{Z}_b^{-1}\,\mathrm{Tr}_{sb}\, e^{-\beta\hat{H}_c} = \mathcal{Z}_c/\mathcal{Z}_b$.

If one followed the procedure of traditional weak-coupling thermodynamics to define the free energy as $\mathcal{F} = -\beta^{-1}\ln\mathcal{Z}$, then the total free energy $\mathcal{F}_c$ of the composite system can be given by a simple additive expression $\mathcal{F}_c = \mathcal{F}^* + \mathcal{F}_b$, with $\mathcal{F}^* = -\beta^{-1}\ln\mathcal{Z}^*$ and $\mathcal{F}_b = -\beta^{-1}\ln\mathcal{Z}_b$. In addition, one can write the reduced density matrix $\hat{\rho}_s$ in a form similar to (9), with the replacement of $\hat{H}_s$ by $\hat{H}_s^*$,

$$\hat{\rho}_s = \frac{1}{\mathcal{Z}_c}\,\mathrm{Tr}_b\, e^{-\beta\hat{H}_c} = \frac{e^{-\beta\hat{H}_s^*}}{\mathcal{Z}^*}\,, \tag{12}$$

in the hope that the conventional procedures of weak-coupling thermodynamics will follow. However, in Section 7 or in [42], we see that at strong coupling even though we already have the (reduced) density matrix and the free energy of the system, we can introduce two different sets of thermodynamic potentials for the system. The thermodynamic potentials in each set are mathematically self-consistent, but they are not compatible with their counterparts in the other set, in contrast to the weak coupling limit, where both definitions are equivalent.

Two earlier approaches to introduce the thermodynamic potentials in a strongly interacting system had been proposed by Gelin and Thoss [42], and by Seifert [26]. We shall summarize the approach I in Gelin and Thoss' work below and present a more detailed quantum formulation of Seifert's approach following, both for a configuration that the composite is initially in a global thermal state without any external force. A recent proposal by Jarzynski [27] for classical systems will be formulated quantum-mechanically in the same setting in Section 7.

4. Quantum Formulation of Gelin and Thoss' Thermodynamics at Strong Coupling

The first approach, based on Approach I of Gelin & Thoss [42], is rather intuitive, because their definitions of the internal energy and the entropy are the familiar ones in traditional thermodynamics. They define the internal energy $\mathcal{U}_s$ of the (reduced) system by the quantum expectation value of the system Hamiltonian operator alone, $\mathcal{U}_s = \operatorname{Tr}_s\{\hat{\rho}_s\,\hat{H}_s\}$, and choose the entropy to be the von Neumann (vN) entropy $\mathcal{S}_s = \mathcal{S}_{vN} = -\operatorname{Tr}_s\{\hat{\rho}_s \ln \hat{\rho}_s\}$. These are borrowed from the corresponding definitions in weak-coupling thermodynamics.

They write the same reduced density matrix (9) in a slightly different representation to highlight the difference from the weak-coupling thermodynamics case,

$$\hat{\rho}_s = \frac{1}{\mathcal{Z}_c}\,e^{-\beta(\hat{H}_s+\hat{\Delta}_s)} = e^{-\beta(\hat{H}_s+\hat{\Delta}_s-\mathcal{F}_c)}\,, \qquad \text{with} \qquad \mathcal{Z}_c = \operatorname{Tr}_s e^{-\beta(\hat{H}_s+\hat{\Delta}_s)} = e^{-\beta\mathcal{F}_c} \tag{13}$$

where $\hat{\Delta}_s$ depends only on the system variables but includes all of the influence from the bath from their interaction. Comparing this with (11), we note that $\hat{\Delta}_s$ is formally related to the Hamiltonian of mean force by $\hat{\Delta}_s = \hat{H}_s^* - \hat{H}_s + \mathcal{F}_b$. Finally, they let the partition function of the system take on the value $\mathcal{Z}_c$, which is distinct from $\mathcal{Z}^*$. Thus the corresponding free energy will be given by $\mathcal{F}_c$ which contains all the contributions from the composite **C**.

Although in this approach the definitions of internal energy and entropy of the system are quite intuitive, these two thermodynamic quantities do not enjoy simple relations with the partition function $\mathcal{Z}_c$, as in (5). From (13), we can show (Here some discretion is advised in taking the derivative with respect to β because in general an operator will not commute with its own derivative. See details in Appendix A)

$$-\frac{\partial}{\partial\beta}\ln\mathcal{Z}_c = -\frac{1}{\mathcal{Z}_c}\frac{\partial}{\partial\beta}\operatorname{Tr}_s e^{-\beta(\hat{H}_s+\hat{\Delta}_s)} = \langle\hat{H}_s\rangle + \langle\hat{\Delta}_s\rangle + \beta\,\langle\partial_\beta\hat{\Delta}_s\rangle \neq \mathcal{U}_s\,, \tag{14}$$

with the corresponding free energy $\mathcal{F}_c = \mathcal{U}_s + \langle\hat{\Delta}_s\rangle + \beta\,\langle\partial_\beta\hat{\Delta}_s\rangle - \beta\,\partial_\beta\mathcal{F}_c$. Here $\langle\cdots\rangle$ represents the expectation value taken with respect to the density matrix $\hat{\rho}_c$ of the composite. For a system operator $\hat{O}_s$, this definition yields an expectation value equal to that with respect to the reduced density matrix $\hat{\rho}_s$, namely,

$$\langle\hat{O}_s\rangle_s = \operatorname{Tr}_s\{\hat{\rho}_s\,\hat{O}_s\} = \operatorname{Tr}_{sb}\{\hat{\rho}_c\,\hat{O}_s\} = \langle\hat{O}_s\rangle\,. \tag{15}$$

Likewise, the von Neumann entropy $\mathcal{S}_{vN}$ can be expressed in terms of the free energy $\mathcal{F}_c$ by

$$\mathcal{S}_{vN} = \beta^2\partial_\beta\mathcal{F}_c - \beta^2\langle\partial_\beta\hat{\Delta}_s\rangle\,, \tag{16}$$

which does not resemble the traditional form in (3). Additionally, we observe the entropy so defined is not additive, that is,

$$\mathcal{S}_{vN} + \mathcal{S}_b = -\operatorname{Tr}_s\{\hat{\rho}_s\ln\hat{\rho}_s\} - \operatorname{Tr}_b\{\hat{\rho}_s\ln\hat{\rho}_b\} \neq -\operatorname{Tr}_{sb}\{\hat{\rho}_c\ln\hat{\rho}_c\} = \mathcal{S}_c\,. \tag{17}$$

Here $\mathcal{S}_c$ and $\mathcal{S}_b$ are the von Neumann entropies of the composite and the free bath, respectively. Note that the $\hat{\rho}_b$ in this formulation is the density matrix of the free bath, not the reduced density matrix of the bath, namely, $\hat{\rho}_b \neq \mathrm{Tr}_s\,\hat{\rho}_c$. The reduced density matrix of the bath will contain an additional overlap with the system owing to their coupling.

When the internal energy of the system given by the expectation value of the system Hamiltonian operator $\hat{H}_s$, the specific heat $\mathcal{C}_s$ will take the form,

$$\mathcal{C}_s = -\beta^2\partial_\beta\langle\hat{H}_s\rangle = -\beta\,\partial_\beta\mathcal{S}_{vN} - \beta^2\Big[\langle\partial_\beta\hat{\Delta}_s\rangle - \partial_\beta\langle\hat{\Delta}_s\rangle\Big]\,. \tag{18}$$

In general $\langle\partial_\beta\hat{\Delta}_s\rangle \neq \partial_\beta\langle\hat{\Delta}_s\rangle$ since the reduced density matrix $\hat{\rho}_s$ also has a temperature dependence. We thus see in this case the heat capacity cannot be directly given as the derivative of the (von Neumann) entropy with respect to β, as in (7).

In short, in this formulation, the thermodynamic potentials of the system are defined in a direct and intuitive way. They introduce an operator $\hat{\Delta}_s$ to highlight the foreseen ambiguity when the system is strongly coupled with the bath. Formally we see that $\hat{\Delta}_s = -\beta^{-1}\ln\mathrm{Tr}_b\,e^{-\beta(\hat{H}_s+\hat{H}_i+\hat{H}_b)} - \hat{H}_s$. In the limit of weak coupling, $\hat{H}_i \approx 0$, the operator $\hat{\Delta}_s$ reduces to

$$\hat{\Delta}_s \approx -\beta^{-1}\ln\mathrm{Tr}_b\,e^{-\beta(\hat{H}_s+\hat{H}_b)} - \hat{H}_s = -\beta^{-1}\ln\mathcal{Z}_b\,. \tag{19}$$

Hence in this limit, $\hat{\Delta}_s$ reduces to a c-number and plays the role of the free energy $\mathcal{F}_b$ of the free bath. Observe $\hat{\Delta}_s \approx \mathcal{F}_b$ in the weak coupling limit annuls the expression in the square brackets in (18) and restores the traditional relation (7) between the heat capacity and the entropy. However, even in the weak coupling limit, the internal energy still cannot be given by (5). The disparity lies in the identification of $\mathcal{Z}_c$ as the partition function of the system. As is clearly seen from (14), in the weak coupling limit, we have

$$-\frac{\partial}{\partial\beta}\ln\mathcal{Z}_c \approx -\frac{1}{\mathcal{Z}_c}\frac{\partial}{\partial\beta}\mathrm{Tr}_s\Big\{e^{-\beta\hat{H}_s}\mathcal{Z}_b\Big\} = \langle\hat{H}_s\rangle_s + \langle\hat{H}_b\rangle_b\,. \tag{20}$$

This implies that $\mathcal{Z}_c$ is not a good candidate for the partition function of the system. A more suitable option would be $\mathcal{Z}_c/\mathcal{Z}_b$.

5. Quantum Formulation of Seifert's Thermodynamics at Strong Coupling

If we literally follow (11) and identify $\hat{H}_s^*$ as the effective Hamiltonian operator of the (reduced) system, we will nominally interpret that the reduced system assumes a canonical distribution. Thus it is natural to identify $\mathcal{Z}^*$ as the partition function associated with the reduced state of the system.

Suppose we maintain the thermodynamic relations regardless of the coupling strength between the system and the bath. From (5) to (6), we will arrive at expressions of the internal energy and entropy of the system. This is essentially Seifert's approach [26] to the thermodynamics at strong coupling. Here we will present the quantum-mechanical version of it for a equilibrium configuration without the external drive, that is, $\lambda = 0$ in Seifert's notion.

First, from (11), we have the explicit form of the Hamiltonian of mean force $\hat{H}_s^*$

$$\hat{H}_s^* = -\beta^{-1}\ln\mathrm{Tr}_b\Big\{\exp\Big[-\beta\hat{H}_s - \beta\hat{H}_i - \beta\big(\hat{H}_b - \mathcal{F}_b\big)\Big]\Big\}\,. \tag{21}$$

This is the operator form of $\mathcal{H}(\xi_s,\lambda)$ in Equation (5) of [26]. Noting the non-commutative characters of the operators. Since $[\hat{H}_s,\hat{H}_i] \neq 0$,

$$e^{-\beta(\hat{H}_s+\hat{H}_i+\hat{H}_b)} \not\to e^{-\beta\hat{H}_s}e^{-\beta(\hat{H}_i+\hat{H}_b)}\,. \tag{22}$$

If one prefers factoring out $e^{-\beta\hat{H}_s}$ from $e^{-\beta(\hat{H}_s+\hat{H}_i+\hat{H}_b)}$, one can use the Baker-Campbell-Haussdorff formula, outlined in Appendix A, to expand out the operator products to a certain order commensurate with a specified degree of accuracy.

Second, it is readily seen that $p^{\rm eq}(\xi_s|\lambda=0)$ in Equation (4) of [26] is the reduced density matrix $\hat{\rho}_s$ of the system (12). The (Helmholtz) free energy $\mathcal{F}$ in Seifert's Equation (7) is exactly the free energy of the reduced system $\mathcal{F}^*$ in Section 3.3.

With these identifications, it is easier to find the rest of the physical quantities in Seifert's strong coupling thermodynamics for the equilibrium configuration. We now proceed to derive the entropy and the internal energy, i.e., Equations (8) and (9) of [26], for quantum systems in his framework in the equilibrium setting.

In general, an operator does not commute with its derivative, so taking the derivative of an operator-valued function or performing integration by parts on an operator-valued function can be nontrivial. Their subtleties are discussed in Appendix A, where we show that the derivative of an operator function is in general realized by its Taylor's expansion in a symmetrized form (A5). However, when such a form appears in the trace, the cyclic property of the trace allow us to manipulate the derivative of a operator-valued function as that of a c-number function thus sidestepping the symmetrized ordering challenge. Hence from the thermodynamic relation (6), we have

$$\mathcal{S}_s = -\beta\mathcal{F}^* + \beta\,\mathrm{Tr}_s\{\hat{\rho}_s(\hat{H}_s^* + \beta\,\partial_\beta\hat{H}_s^*)\}\,. \tag{23}$$

Here we recall that even though the operators $\hat{H}_s^*$ and $\partial_\beta\hat{H}_s^*$ in general do not commute, the trace operation allowing for cyclic permutations of the operator products eases the difficulties in their manipulation. Since (12) implies the operator identity $\beta\hat{H}_s^* = \beta\mathcal{F}^* - \ln\hat{\rho}_s$, we can recast (23) to

$$\mathcal{S}_s = \mathrm{Tr}_s\{\hat{\rho}_s(-\ln\rho_s + \beta^2\partial_\beta\hat{H}_s^*)\} = \mathcal{S}_{vN} + \beta^2\langle\partial_\beta H_s^*\rangle \neq -\mathrm{Tr}_s\{\hat{\rho}_s\ln\hat{\rho}_s\} = \mathcal{S}_{vN}\,. \tag{24}$$

This is the quantum counterpart of Seifert's entropy, Equation (8) of [26]. This entropy is often called the "thermodynamic" entropy in the literature. Note that it is not equal to the von Neumann ("statistical") entropy $\mathcal{S}_{vN}$ of the system.

The internal energy can be given by the thermodynamic relation

$$\mathcal{U}_s = \mathcal{F}^* + \beta^{-1}\mathcal{S}_s\,. \tag{25}$$

Thus from (23), we obtain,

$$\mathcal{U}_s = \mathrm{Tr}_s\{\hat{\rho}_s(\hat{H}_s^* + \beta\,\partial_\beta\hat{H}_s^*)\} = \langle\hat{H}_s^*\rangle + \beta\,\langle\partial_\beta\hat{H}_s^*\rangle \neq \langle\hat{H}_s^*\rangle\,. \tag{26}$$

This deviation results from the fact that $\hat{H}_s^*$, introduced in (11) may depend on β. When we take this into consideration, we can verify that the internal energy can also be consistently given by Equation (5)

$$-\frac{\partial}{\partial\beta}\ln\mathcal{Z}^* = \frac{1}{\mathcal{Z}^*}\,\mathrm{Tr}_s\Big\{\Big(\hat{H}_s^* + \beta\,\partial_\beta\hat{H}_s^*\Big)\,e^{-\beta\hat{H}_s^*}\Big\} = \mathcal{U}_s\,. \tag{27}$$

In fact, we can show, by recognizing $\mathcal{Z}^* = \mathcal{Z}_c/\mathcal{Z}_b$, that

$$\mathcal{U}_s = \langle\hat{H}_s\rangle + \big[\langle\hat{H}_i\rangle + \langle\hat{H}_b\rangle - \langle\hat{H}_b\rangle_b\big] \neq \langle\hat{H}_s\rangle\,, \tag{28}$$

with $\langle\hat{H}_b\rangle_b \equiv \mathrm{Tr}_b\{\hat{\rho}_b\,\hat{H}_b\}$, $\langle\hat{H}_b\rangle \equiv \mathrm{Tr}_{sb}\{\hat{\rho}_c\,\hat{H}_b\}$, and $\langle\hat{H}_s\rangle \equiv \mathrm{Tr}_{sb}\{\hat{\rho}_c\,\hat{H}_s\} = \langle\hat{H}_s\rangle_s$. Equation (28) implies that the internal energy, defined by (27), accommodates more than mere $\langle\hat{H}_s\rangle_s$. The additional pieces contain contributions from the bath and the interaction. In particular, when the coupling between the system and the bath is not negligible, we have $\langle\hat{H}_b\rangle \neq \langle\hat{H}_b\rangle_b$ in general. As a matter of fact, even the internal energy defined by the expectation value of the system Hamiltonian operator in approach I of Gelin & Thoss' work also encompasses influence from the bath because the reduced density matrix $\hat{\rho}_s$ includes all the effects of the bath on the system.

Hitherto, we have encountered three possible definitions of internal energies, namely, $\langle \hat{H}_s \rangle$, $\langle \hat{H}_s^* \rangle$, and $\mathcal{U}_s$. As can be clearly seen from (26) and (28), they essentially differ by the amount of the bath and the interaction energy which are counted toward the system energy. This ambiguity arises from strong coupling between the system and the bath. When the system-bath interaction is negligibly small, we have $\langle \hat{H}_i \rangle \approx 0$, and since in this limit, the full density matrix of the composite is approximately given by the product of that of the system and the bath, we arrive at $\langle \hat{H}_b \rangle \approx \langle \hat{H}_b \rangle_b$, and these three energies become equivalent.

To explicate the physical content of $\langle \hat{H}_s^* \rangle$, from (12) we can write $\langle \hat{H}_s^* \rangle$ as

$$\langle \hat{H}_s^* \rangle = -\frac{1}{\beta} \operatorname{Tr}_s\{\hat{\rho}_s \ln \hat{\rho}_s\} + \mathcal{F}^* = \beta^{-1} \mathcal{S}_{vN} + \mathcal{F}^* , \qquad \text{or} \qquad \mathcal{F}^* = \langle \hat{H}_s^* \rangle - \beta^{-1} \mathcal{S}_{vN} . \tag{29}$$

This offers an interesting comparison with (25), where $\mathcal{F}^* = \mathcal{U}_s - \beta^{-1} \mathcal{S}_s$. It may appear that we can replace the pair $(\mathcal{U}_s, \mathcal{S}_s)$ by another pair $(\langle \hat{H}_s^* \rangle, \mathcal{S}_{vN})$, leaving $\mathcal{F}^*$ unchanged, thus suggesting an alternative definition of internal energy by $\langle \hat{H}_s^* \rangle$ and that of entropy by $\mathcal{S}_{vN}$. However, in so doing, the new energy and entropy will not satisfy a simple thermodynamic relation like (5) and (6). This is a good sign, as it is an indication that certain internal consistency exists in the choice of the thermodynamic variables.

We now investigate the differences between the two definitions of entropy. From (24), we obtain

$$T(\mathcal{S}_s - \mathcal{S}_{vN}) = \operatorname{Tr}_s\{\hat{\rho}_s(\beta\, \partial_\beta \hat{H}_s^*)\} = \beta\, \partial_\beta \operatorname{Tr}_s\{\hat{\rho}_s\, \hat{H}_s^*\} - \beta \operatorname{Tr}_s\{(\partial_\beta \hat{\rho}_s) \hat{H}_s^*\} . \tag{30}$$

The factor $\partial_\beta \hat{\rho}_s$ can be written as $\partial_\beta \hat{\rho}_s = \langle \hat{H}_c \rangle \hat{\rho}_s - \operatorname{Tr}_b\{\hat{\rho}_c \hat{H}_c\}$ with $\partial_\beta \mathcal{Z}_c = -\langle \hat{H}_c \rangle \mathcal{Z}_c$. We then obtain

$$T(\mathcal{S}_s - \mathcal{S}_{vN}) = \beta\, \partial_\beta \langle \hat{H}_s^* \rangle + \beta\big[\langle \hat{H}_c\, \hat{H}_s^* \rangle - \langle \hat{H}_c \rangle \langle \hat{H}_s^* \rangle\big] . \tag{31}$$

Thus, part of the difference between the two entropies result from the correlation between the full Hamiltonian $\hat{H}_c$ and the Hamiltonian of mean force $\hat{H}_s^*$. This correlation will disappear in the vanishing coupling limit because there is no interaction to bridge the system and the bath. We also note that in the same limit, $\langle \hat{H}_s^* \rangle \approx \langle \hat{H}_s \rangle$ becomes temperature-independent, and both definitions of the entropy turn synonymous.

Since the von Neumann entropy $\mathcal{S}_{vN}$ can be used as a measure of entanglement between the system and the bath at zero temperature, we often introduce the quantum mutual information I_{sb} to quantify how they are correlated,

$$I_{sb} = \mathcal{S}_{vN} + \mathcal{S}_b' - \mathcal{S}_c \geq 0 , \tag{32}$$

where $\mathcal{S}_b'$ is the von Neumann entropy associated with the reduced density matrix $\hat{\varrho}_b$ of the bath, in contrast to $\mathcal{S}_b$ we have met earlier. This mutual information can be related to the quantum relative entropy $S(\hat{\rho}_c \| \hat{\rho}_s \otimes \hat{\varrho}_b)$ by

$$S(\hat{\rho}_c \| \hat{\rho}_s \otimes \hat{\varrho}_b) = \operatorname{Tr}_{sb}\Big\{\hat{\rho}_c \ln \hat{\rho}_c - \hat{\rho}_c \ln \hat{\rho}_s \otimes \hat{\varrho}_b\Big\} = I_{sb} , \tag{33}$$

because $\hat{\varrho}_b = \operatorname{Tr}_s \hat{\rho}_c$. On the other hand, Equation (23) imply that the thermodynamic entropy $\mathcal{S}_s$ is additive $\mathcal{S}_s + \mathcal{S}_b = \mathcal{S}_c$, from which we find

$$I_{sb} = (\mathcal{S}_b' - \mathcal{S}_b) + (\mathcal{S}_{vN} - \mathcal{S}_s) . \tag{34}$$

This and (31) provide different perspectives on how the difference between the two system entropies is related to the system-bath entanglement, and how the system-bath coupling has a role in establishing such correlations.

Following the definitions of the internal energy (27) and the entropy (23), the heat capacity of the system still satisfies a familiar relation $\mathcal{C}_s = -\beta^2\,\partial_\beta \mathcal{U}_s = \beta^2\,\partial^2_\beta \ln \mathcal{Z}^* = -\beta\,\partial_\beta \mathcal{S}_s$. Compared with (18), with the help of (28), we clearly see their difference, caused by different definitions of internal energy, is given by

$$-\beta^2\,\partial_\beta\big(\mathcal{U}_s - \langle \hat{H}_s\rangle\big) = -\beta^2\,\partial_\beta\big[\langle \hat{H}_i\rangle + \langle \hat{H}_b\rangle - \langle \hat{H}_b\rangle_b\big]. \tag{35}$$

6. Issues of These Two Approaches: Entropy and Internal Energy

Both equilibrium quantum formulations for thermodynamics at strong coupling are based on plausible assumptions and are mathematically sound. In Approach I outlined in Section 4, one starts with intuitive definitions of the thermodynamics quantities, inspired by traditional thermodynamics for classical systems premised on vanishingly weak coupling between the system and the bath. This leads to modifications in the thermodynamic relations of the relevant thermodynamics quantities. In Approach II delineated in Section 5, one opts to maintain the familiar thermodynamic relations but is compelled to deal with a rather obscure interpretation of the thermodynamic potentials. Although both approaches in the vanishing system-bath coupling limit are compatible, as shown in Section 3.1, they in general entail distinct definitions of the thermodynamic functions. This disparity is amplified with strong system-bath coupling in the deep quantum regime, where quantum coherence plays an increasingly significant role. Thus, even though both approaches possess the same correct classical thermodynamic limit, they are not guaranteed to give unique physical results in the deep quantum regime, even for simple quantum systems, which are areas for interesting further investigations.

To highlight the issues more explicitly, we can apply these two methods to a simple and completely solvable model, namely, a Brownian oscillator linearly but strongly coupled with a large (or infinitely large, as modeled by a scalar field) bath. We will see both approaches at some point, or others that produce ambiguous or paradoxical results. We make a few observations in the following section.

6.1. Entropy

(1) It has been discussed in [36,43,44] that the von Neumann entropy $\mathcal{S}_{vN}$ will not approach to zero for the finite system-bath coupling in the limit of zero temperature, but the thermodynamic entropy $\mathcal{S}_s$, defined in Approach II, behaves nicely in the same limit.

(2) It has been shown [49] that if the composite is in a global thermal state, the discrete energy spectrum of the undamped oscillator will become a continuous one with a unique ground level. This supports physics described by the thermodynamic entropy $\mathcal{S}_s$.

(3) It has been argued [44–46] that the entanglement between the system and the bath prevents the von Neumann entropy from approaching zero at zero temperature. Without quantum entanglement between the system and the bath, the lowest energy level of the composite system will be given by the tensor product of the ground state of the unperturbed system and bath, that is, a pure state. In this case, the von Neumann entropy will go to zero as expected, and this is the scenario that occurred in traditional quantum/classical thermodynamics in the vanishing system-bath coupling limit.

6.2. Internal Energy

- It has been discussed [37–40] that the internal energy defined in Approach II can lead to anomalous behavior of the heat capacity in the low temperature limit. When the system, consisting of a quantum oscillator [40] or a free particle [37–39] is coupled to a heat bath modeled by a large number of quantum harmonic oscillators, the heat capacity of the system can become negative if the temperature of the bath is sufficiently low. If the internal energy defined in Approach I is used to compute the heat capacity, then it has been shown that the heat capacity remains positive for all nonzero temperatures but vanishes in the zero bath temperature limit, for a system with one harmonic oscillator [37], or a finite number of coupled harmonic oscillators [29]. This discrepancy may result from the fact that the internal energy defined in Approach II contains contributions from the interaction and the bath Hamiltonian.

It seems to imply that in the low-temperature, strong coupling regime, it remains an open question how to properly define the thermodynamic functions; being able to show the well-known behaviors in the classical thermodynamic limit is a necessary but not sufficient condition.

7. Quantum Formulation of Jarzynski's Strong Coupling Thermodynamics

We now provide a quantum formulation of Jarzynski's classical results [27] but for composite system **C** kept under thermal equilibrium. The Hamiltonian operator of the composite $\mathbf{C} = \mathbf{S} + \mathbf{B}$ is assumed to take the form

$$\hat{H}_c = \hat{H}_s + \hat{H}_i + \hat{H}_b + J \cdot \hat{A}_b \,, \tag{36}$$

Here in this paper, J will be some external, but constant c-number drive acting on the bath via a bath operator $\hat{A}_b$. It can be a constant pressure, as in Jarzynski's classical formulation, and then $\hat{A}_b$ will be an operator corresponding to V_b, conjugated to P. However, in general, $\hat{A}_b$ will be the operator of the quantity conjugated to J. This analogy, though formal, provides an alternative route to introduce the operator conjugated to J.

If the composite system is in thermal equilibrium at the temperature β^{-1}, its state is described by the density matrix operator $\hat{\rho}_c = e^{-\beta\hat{H}_c} / \mathcal{Z}_c$, where $\mathcal{Z}_c = \mathrm{Tr}_{sb}\{e^{-\beta\hat{H}_c}\}$, a c-number, is the partition function of the composite. For later convenience, we also define the corresponding quantities for the bath **B** when it is coupled to the system **S**, $\hat{\rho}_b = e^{-\beta(\hat{H}_b + J\cdot\hat{A}_b)} / \mathcal{Z}_b$ with the bath partition function $\mathcal{Z}_b = \mathrm{Tr}_b\{e^{-\beta(\hat{H}_b+J\cdot\hat{A}_b)}\}$. We introduce the Hamiltonian operator of mean force $\hat{H}_s^*$ by

$$e^{-\beta\hat{H}_s^*} \equiv \frac{1}{\mathcal{Z}_b}\,\mathrm{Tr}_b\Big\{e^{-\beta\hat{H}_c}\Big\}\,, \tag{37}$$

such that the reduced density matrix of the system **S** takes the form

$$\hat{\rho}_s \equiv \mathrm{Tr}_b\,\hat{\rho}_c = \frac{1}{\mathcal{Z}_s}\,e^{-\beta\hat{H}_s^*}\,, \qquad \text{with} \qquad \mathcal{Z}_s = \frac{\mathcal{Z}_c}{\mathcal{Z}_b} = \mathrm{Tr}_s\Big\{e^{-\beta\hat{H}_s^*}\Big\}\,. \tag{38}$$

The quantity $\mathcal{Z}_s$ can be viewed as an effective partition function of the system **S**. This is motivated by the observation that, in the absence of coupling between **S** and **B**, or in the weak coupling limit, the composite is additive so its partition function is the product of those of the subsystems, i.e., $\mathcal{Z}_c = \mathcal{Z}_s\mathcal{Z}_b$. The difference $\hat{H}_s^* - \hat{H}_s$ modifies the dynamics of the system **S** due to its interaction with the bath **B**.

In fact, by the construction, $e^{-\beta H_s^*}$, once sandwiched by the appropriate states of the system **S** and expressed in the imaginary-time path integral formalism (for further details regarding the connection with the influence action, please refer to [28,61,62]), is formally $e^{-S_{CG}}$, where S_{CG} is the coarse-grained effective action of the system **S**, wick-rotated to the imaginary time. Thus formally $\beta(\hat{H}_s^* - \hat{H}_s)$ is equivalent to the influence action in the imaginary time formalism.

Similar to the classical formulations, we may have two different representations of the operator $\hat{A}_s$ of the system.

7.1. "Bare" Representation

In the bare representation, we may define $\hat{A}_s = (\hat{H}_s^* - \hat{H}_s)/J$, and the internal energy operator $\hat{U}_s$ and the enthalpy operator $\hat{\mathfrak{H}}_s$, respectively, by $\hat{U}_s = \hat{H}_s$ and $\hat{\mathfrak{H}}_s = \hat{H}_s^*$, with expectation values given by $\mathcal{U}_s = \mathrm{Tr}_s\{\hat{\rho}_s\,\hat{U}_s\}$ and $\mathcal{H}_s = \mathrm{Tr}_s\{\hat{\rho}_s\,\hat{\mathfrak{H}}_s\} = \mathcal{U}_s + J\cdot\mathcal{A}_s$, corresponding to the internal energy and the enthalpy we are familiar with, respectively. Here $\mathcal{A}_s = \mathrm{Tr}_s\{\hat{\rho}_s\,\hat{A}_s\}$. The entropy is chosen to be the von Neumann entropy of the system

$$\mathcal{S}_s = \mathrm{Tr}_s\{\hat{\rho}_s\,\ln\hat{\rho}_s\} = \beta(\mathcal{H}_s - \mathcal{G}_s)\,. \tag{39}$$

The Gibbs free energy $\mathcal{G}_s$ is defined as $\mathcal{G}_s = -\beta^{-1} \ln \mathcal{Z}_s$. These definitions are in exact parallel to those in the classical formulation [27].

7.2. "Partial Molar" Representation

In contrast to the bare representation, we can alternatively define the operator $\hat{A}_s$ of the system **S** that corresponds to $\hat{A}_b$ of the bath **B** by $\hat{A}_s = \partial(\hat{H}_s^* - \hat{H}_s)/\partial J = \partial\hat{H}_s^*/\partial J$. The last equality results from the fact that $\hat{H}_s$ has no dependence on the external parameter J. Owing to the non-commutativity of operators, the micro-physics interpretation of the operator $\hat{A}_s$ is not so transparent. We first focus on its quantum expectation value $\mathcal{A}_s$

$$\mathcal{A}_s = \mathrm{Tr}_s\Big\{\hat{\rho}_s\,\hat{A}_s\Big\} = \frac{1}{\mathcal{Z}_s}\,\mathrm{Tr}_s\Big\{e^{-\beta\hat{H}_s^*}\,\frac{\partial\hat{H}_s^*}{\partial J}\Big\}. \tag{40}$$

As stressed earlier, since the operator does not commute with its derivative, care must be taken when we move the derivative around in an operator expression. However, from (A7), we learn that the righthand side of (40) can be identified as

$$\mathrm{Tr}_s\Big\{e^{-\beta\hat{H}_s^*}\,\frac{\partial\hat{H}_s^*}{\partial J}\Big\} = -\beta^{-1}\,\frac{\partial}{\partial J}\,\mathrm{Tr}_s\Big\{e^{-\beta\hat{H}_s^*}\Big\}, \tag{41}$$

and thus we have $\mathcal{A}_s = -\beta^{-1}\,\partial\ln\mathcal{Z}_s/\partial J$. The advantage of this expression is that the observation of $\mathcal{Z}_s = \mathcal{Z}_c/\mathcal{Z}_b$ enables us to write $\mathcal{A}_s$ as $\mathcal{A}_s = \mathcal{A}_c - \mathcal{A}_b$, if we have defined the corresponding expectation values for the composite **C** and the bath **B** by $\mathcal{A}_c = -\beta^{-1}\,\partial\ln\mathcal{Z}_c/\partial J$ and $\mathcal{A}_b = -\beta^{-1}\,\partial\ln\mathcal{Z}_b/\partial J$. In particular we can check that $\mathcal{A}_b$ indeed is the expectation value of the operator $\hat{A}_b$, that is, $\mathcal{A}_b = \mathrm{Tr}_b\{\hat{\rho}_b\,\hat{A}_b\} = \mathrm{Tr}_{sb}\{\hat{\rho}_b\,\hat{A}_b\}$. The latter expression can nicely bridge with $\mathcal{A}_c$ for the composite, $\mathcal{A}_c = \mathrm{Tr}_{sb}\{\hat{\rho}\,\hat{A}_b\}$. Thus the expectation value $\mathcal{A}$ is additive. Its value for the combined systems is equal to the sum of those of the subsystems, $\mathcal{A}_c = \mathcal{A}_s + \mathcal{A}_b$. In fact, this additive property holds for all the thermodynamics potentials introduced afterwards. This is a nice feature in Jarzynski's partial molar representation or in Seifert's approach.

From this aspect, we can interpret $\mathcal{A}_s$ as the change of $\mathcal{A}_b$ due to the intervention of the system **S**. For example, consider a photon gas inside a cavity box, one side of which is a movable classical mirror and is exerted by a constant pressure. Assume originally the photon gas and the mirror are in thermal equilibrium. In this cavity we now place a Brownian charged oscillator and maintain the new composite system in thermal equilibrium at the same temperature and the same pressure (The equilibration process in this example can be awfully complicated if we mind the subtleties regarding whether the photon gas can ever reach thermal equilibrium in a cavity whose walls are perfectly reflective and so on. For the present argument, we assume equilibration is possible and there is no leakage of the photons). Then we should note that there is a minute change in the mean position of the mirror before and after the Brownian charged oscillator is placed into the cavity. This change can also be translated to an effective or dynamical size of the charged oscillator due to its interaction with the photon gas, and thus is accounted for in $\mathcal{A}_s$ when J is identified as the constant pressure applied to the wall.

From this example, it is tempting to identify $J\cdot\hat{A}_s$ as some quantum work operator (its value depends on the interaction between the system and the bath and when this interaction is switched on. It is thus path-dependent in the parameter space of the coupling constant). Alternatively we may view it or its expectation as some additional "energy content" of the system **S** due to its interaction with the bath when the composite is acted upon by an external agent J, since $\hat{A}_s$ is related to $\hat{H}_s^* - \hat{H}_s$ [63,64]. Inspired by this observation and taking the hint from the definition of $\mathcal{A}_s$, we introduce the enthalpy of the system **S** by

$$\mathcal{H}_s \equiv -\frac{\partial}{\partial\beta}\ln\mathcal{Z}_s = \mathcal{H}_c - \mathcal{H}_b\,, \tag{42}$$

where we have identified the enthalpies of the composite $\mathbf{C} = \mathbf{S} + \mathbf{B}$ and the bath $\mathbf{B}$ as $\mathcal{H}_c = -\partial \ln \mathcal{Z}_c / \partial \beta$ and $\mathcal{H}_b = -\partial \ln \mathcal{Z}_b / \partial \beta$. We may rewrite them as $\mathcal{H}_c = \langle \hat{H}_s \rangle + \langle \hat{H}_i \rangle + \langle \hat{H}_b \rangle + J \cdot \mathcal{A}_c$ and $\mathcal{H}_b = \langle \hat{H}_b \rangle_b + J \cdot \mathcal{A}_b$. It implies that (1) the system enthalpy can be decomposed as

$$\mathcal{H}_s = \mathcal{H}_c - \mathcal{H}_b = \left[\langle \hat{H}_s \rangle + \langle \hat{H}_i \rangle + \langle \hat{H}_b \rangle - \langle \hat{H}_b \rangle_b \right] + J \cdot \mathcal{A}_s \,, \tag{43}$$

and (2) the internal energy $\mathcal{U}_s$ of the system $\mathbf{S}$ can be consistently inferred as

$$\mathcal{U}_s = \langle \hat{H}_s \rangle + \langle \hat{H}_i \rangle + \left[\langle \hat{H}_b \rangle - \langle \hat{H}_b \rangle_b \right] . \tag{44}$$

This is exactly the same internal energy obtained in Seifert's approach in the equilibrium setting. We can define the internal energy of the composite system and of the bath by $\mathcal{U}_c = \langle \hat{H}_s \rangle + \langle \hat{H}_i \rangle + \langle \hat{H}_b \rangle$ and $\mathcal{U}_b = \langle \hat{H}_b \rangle_b$, and then we may conclude $\mathcal{U}_s = \mathcal{U}_c - \mathcal{U}_b$. Thus the internal energy $\mathcal{U}_s$ also includes contributions that naïvely we will not ordinarily attribute to the system, such as $\langle \hat{H}_b \rangle - \langle \hat{H}_b \rangle_b$. Doing so will complicate the physical connotation of the internal energy of the system.

Up to this moment, we essentially write the thermodynamic quantities by the quantum expectation value and in terms of the partition functions. Thus it is appropriate to introduce the Gibbs free energies of the composite $\mathbf{C}$, the system $\mathbf{S}$, and the bath $\mathbf{B}$ by $\mathcal{G}_a = -\beta^{-1} \ln \mathcal{Z}_a$, where $a = c$, s and b, and they obey the additive property of the Gibbs energy, $\mathcal{G}_c = \mathcal{G}_s + \mathcal{G}_b$. Furthermore, in the composite, we note that

$$\beta(\mathcal{H}_c - \mathcal{G}_c) = \beta^2 \frac{\partial \mathcal{G}_c}{\partial \beta} = -\operatorname{Tr}_{sb}\{\hat{\rho}_c \ln \hat{\rho}_c\} . \tag{45}$$

From (45), we can consistently define the entropy $\mathcal{S}$ of the composite by $\mathcal{S}_c = \beta(\mathcal{H}_c - \mathcal{G}_c)$ and, similarly, the entropy $\mathcal{S}_b$ of the bath:

$$\mathcal{S}_b = \beta(\mathcal{H}_b - \mathcal{G}_b) = \beta^2 \frac{\partial \mathcal{G}_b}{\partial \beta} = -\operatorname{Tr}_b\{\hat{\rho}_b \ln \hat{\rho}_b\} . \tag{46}$$

The additive property of the free energy and the enthalpy implies that the entropy $\mathcal{S}_s$ of the system in this representation is also additive, $\mathcal{S}_s = \mathcal{S}_c - \mathcal{S}_b$, and is given by

$$\mathcal{S}_s = \beta(\mathcal{H}_s - \mathcal{G}_s) = \beta^2 \frac{\partial \mathcal{G}_s}{\partial \beta} = -\operatorname{Tr}_{sb}\{\hat{\rho}_c \ln \hat{\rho}_c\} + \operatorname{Tr}_b\{\hat{\rho}_b \ln \hat{\rho}_b\} \neq -\operatorname{Tr}_s\{\hat{\rho}_s \ln \hat{\rho}_s\} , \tag{47}$$

Note it is not equal to the von Neumann entropy, which is defined as the entropy of the system in the "bare" representation.

7.3. Operator Forms of the Thermodynamic Functions

In trying to formulate a set of laws to describe the thermodynamics of a quantum system (even the existence of such a theory, under certain appropriate conditions, is not a matter of presumption or prescription, but by demonstration and proof) it would be most convenient if we could define operators of the thermodynamic functions in such a way that the quantum expectation values of those operators give the familiar expressions for the thermodynamic functions. As we see it, this is the paramount challenge in the formal establishment of quantum thermodynamics as a viable theory. The laws of thermodynamics have been understood in terms of the mean values of the relevant operator quantities. For a system where the fluctuations of the thermodynamic functions become comparable to the corresponding mean values, the thermodynamic laws governing the mean values need be supplanted by laws governing their quantum fluctuations or higher order quantum correlations. A case in point for classical systems where fluctuations are as important as the mean values is near the critical point of the system. The truly quantum properties would impact on the quantum thermodynamics for small quantum systems in the regimes of strong couplings to its environment, and at low temperatures, where quantum coherence effects take center stage. Having the operator forms of these thermodynamic potentials allows us to calculate the higher-order quantum correlations of those quantities existent in larger fluctuations.

In the following sections, we will attempt to identify the operator form of the thermodynamic function for the reduced system.

7.3.1. Enthalpy and Energy Operators: Caution

In fact, we may deduce the operator form of the quantities introduced earlier. For example, we may intuitively define the enthalpy operator $\hat{\mathfrak{H}}$ of the composite by $\hat{\mathfrak{H}}_c = \hat{H}_s + \hat{H}_i + \hat{H}_b + J \cdot \hat{A}_b$, and then it is clear to see that the expectation value $\mathcal{H}_c$ is related to this operator by $\mathcal{H}_c = \mathrm{Tr}_{sb}\{\hat{\rho}_c\,\hat{\mathfrak{H}}_c\} = \langle\hat{\mathfrak{H}}_c\rangle$. Likewise, the enthalpy operator $\hat{\mathfrak{H}}_b$ of the bath **B** can be defined by $\hat{\mathfrak{H}}_b = \hat{H}_b + J \cdot \hat{A}_b$, and its expectation value gives $\mathcal{H}_b = \mathrm{Tr}_b\{\hat{\rho}_b\,\hat{\mathfrak{H}}_b\} = \langle\hat{\mathfrak{H}}_b\rangle_b$. Moreover, the internal energy operator $\hat{\mathcal{U}}_c$ of the composite system and the expectation value can be chosen such that $\hat{\mathcal{U}}_c = \hat{H}_s + \hat{H}_i + \hat{H}_b$ such that $\mathcal{U}_c = \mathrm{Tr}_{sb}\{\hat{\rho}_c\,\hat{\mathcal{U}}_c\} = \langle\hat{\mathcal{U}}_c\rangle$. For the bath, the internal energy operator $\hat{\mathcal{U}}_b$ is, intuitively, $\hat{\mathcal{U}}_b = \hat{H}_b$ with expectation values $\mathcal{U}_b = \mathrm{Tr}_b\{\hat{\rho}_b\,\hat{\mathcal{U}}_b\} = \langle\hat{\mathcal{U}}_b\rangle_b$ that is consistent with the expressions of the internal energy discussed earlier.

Despite their intuitively appealing appearances, these operator forms of the enthalpies and internal energies are not very useful. Inadvertent use of them may result in errors. For example, we cannot define the enthalpy operator of system **S** simply by the difference of $\hat{\mathfrak{H}}_c$ and $\hat{\mathfrak{H}}_b$, since

$$\hat{\mathfrak{H}}_s \stackrel{?}{=} \hat{\mathfrak{H}}_c - \hat{\mathfrak{H}}_b = \hat{H}_s + \hat{H}_i\,. \tag{48}$$

This result in (48) is nonsensical because (1) the righthand side still explicitly depends on the bath degree of freedom; (2) we cannot take its trace with respect to the state of the system, $\hat{\rho}_s$; and thus (3) the expectation value will not be $\mathcal{H}_s$. This is because the operators defined this way act on Hilbert spaces different from that of $\hat{\rho}_s$; $\hat{\mathfrak{H}}_c$ is an operator in the Hilbert space of the composite while $\hat{\mathfrak{H}}_b$ is an operator in the Hilbert space of the bath. Neither operator acts exclusively in the Hilbert space of the system. Thus, extreme care is needed when manipulating the operator forms of the thermodynamical potentials. What one needs to do is to seek the local forms of these operators, i.e., operators which act only on the Hilbert space of the system. This can be done in parallel to Jarzynski's classical formulation.

7.3.2. System Enthalpy Operator: Approved

We first inspect the internal energy operator. Since the averaged internal energy of the composite system is given by $\mathcal{U}_c = \mathrm{Tr}_{sb}\{e^{-\beta\hat{H}_c}\,(\hat{H}_s + \hat{H}_i + \hat{H}_b)\}/\mathcal{Z}_c$, we can rewrite the expressions inside the trace into

$$\mathcal{U}_c = \mathrm{Tr}_s\left\{\hat{\rho}_s\,\hat{\mathcal{Z}}_i^{-1}\,\mathrm{Tr}_b\left[e^{+\beta\hat{H}_s}e^{-\beta\hat{H}_c}\,(\hat{H}_s + \hat{H}_i + \hat{H}_b)\right]\right\}, \tag{49}$$

in a way analogous to Jarzynksi's classical formulation. Here we have used the fact that $\mathcal{Z}_c = \mathcal{Z}_s\mathcal{Z}_b$ and the identity for the operator $\hat{\mathcal{Z}}_i$

$$\hat{\mathcal{Z}}_i \equiv e^{+\beta\hat{H}_s}\,\mathrm{Tr}_b\left\{e^{-\beta\left(\hat{H}_s+\hat{H}_i+\hat{H}_b+J\cdot\hat{A}_b\right)}\right\} = \mathcal{Z}_b\,e^{+\beta\hat{H}_s}e^{-\beta\hat{H}_s^*}\,, \qquad \Leftrightarrow \qquad \frac{e^{+\beta\hat{H}_s^*}}{\mathcal{Z}_b} = \hat{\mathcal{Z}}_i^{-1}\,e^{+\beta\hat{H}_s}\,.$$

If we define an internal energy operator $\hat{\mathcal{U}}_i$ by

$$\hat{\mathcal{U}}_i = \hat{\mathcal{Z}}_i^{-1}\,\mathrm{Tr}_b\left[e^{+\beta\hat{H}_s}e^{-\beta\hat{H}_c}\,(\hat{H}_s + \hat{H}_i + \hat{H}_b)\right] = \mathcal{Z}_b^{-1}e^{+\beta\hat{H}_s^*}\,\mathrm{Tr}_b\left[e^{-\beta\hat{H}_c}\,(\hat{H}_s + \hat{H}_i + \hat{H}_b)\right], \tag{50}$$

then we obtain a new representation of $\mathcal{U}_c$

$$\mathcal{U}_c = \mathrm{Tr}_s\{\hat{\rho}_s\,\hat{U}_i\}\,. \tag{51}$$

Equation (50) is an operator expression of the internal energy of the composite system, on account of the non-commutativity of the operators, but its expectation value is taken with respect to the system's density matrix $\hat{\rho}_s$. With the help of (A9), this is equivalent to Equation (28) of [26] in the $J = 0$ case. In addition, we note that $\hat{Z}_i$ is an operator, not a c number. Since $\mathrm{Tr}_s\,\hat{\rho}_s = 1$, we may define the operator $\hat{U}_s$ by

$$\hat{U}_s = \hat{U}_i - \mathcal{U}_b\,, \tag{52}$$

such that $\mathrm{Tr}_s\{\hat{\rho}_s\,\hat{U}_s\} = \mathrm{Tr}_s\{\hat{\rho}_s\,\hat{U}_i\} - \mathrm{Tr}_s\{\hat{\rho}_s\,\mathcal{U}_b\} = \mathcal{U}_c - \mathcal{U}_b = \mathcal{U}_s$. The advantage of (50), (52) is that, unlike those introduced in the previous subsection, they are all operators in the Hilbert space of the system **S**. Indeed, using the identity operator $\hat{I}_s$ in the Hilbert space of the system **S** we can also define $\hat{U}_b$ as $\hat{U}_b = \mathcal{U}_b\,\hat{I}_s$.

In the same fashion, we may rewrite $\mathcal{A}_c$ by

$$\mathcal{A}_c = \mathrm{Tr}_s\left\{\frac{1}{Z}\,\mathrm{Tr}_s\left[e^{-\beta\hat{H}}\,\hat{A}_b\right]\right\} = \mathrm{Tr}_s\left\{\hat{\rho}_s\,\hat{Z}_i^{-1}\,\mathrm{Tr}_b\left[e^{+\beta\hat{H}_s}e^{-\beta\hat{H}}\,\hat{A}_b\right]\right\}. \tag{53}$$

Thus we can define

$$\hat{A}_i = \hat{Z}_i^{-1}\,\mathrm{Tr}_b\left[e^{+\beta\hat{H}_s}e^{-\beta\hat{H}}\,\hat{A}_b\right], \tag{54}$$

so that $\mathcal{A}_c = \mathrm{Tr}_s\{\hat{\rho}_s\,\hat{A}_i\}$. We then can have a local form for the $\hat{A}_s$ given by $\hat{A}_s = \hat{A}_i - \hat{A}_b$ in close resemblance to its classical expression in [27], if we re-define $\hat{A}_b$ as $\hat{A}_b = \mathcal{A}_b\,\hat{I}_s$. The expectation value of $\hat{A}_s$ is then given by $\mathrm{Tr}_s\{\hat{\rho}_s\,\hat{A}_s\} = \mathrm{Tr}_s\{\hat{\rho}_s\,\hat{A}_i\} - \mathcal{A}_b = \mathcal{A}_c - \mathcal{A}_b = \mathcal{A}_s$.

Now we proceed with constructing a local form of the enthalpy operator of the system. From (52) and the definition of the operator $\hat{A}_s$, we claim that the local form $\hat{\mathfrak{H}}_s$ is

$$\hat{\mathfrak{H}}_s = \hat{U}_s + J\cdot\hat{A}_s\,. \tag{55}$$

We can straightforwardly show that $\mathrm{Tr}_s\{\hat{\rho}_s\,\hat{\mathfrak{H}}_s\} = \mathrm{Tr}_s\{\hat{\rho}_s\,\hat{U}_s\} + J\cdot\mathrm{Tr}_s\{\hat{\rho}_s\,\hat{A}_s\} = \mathcal{U}_s + J\cdot\mathcal{A}_s$. Thus we have succeeded in constructing the operators that correspond to $\mathcal{A}_s, \mathcal{U}_s, \mathcal{H}_s$ in forms local in the Hilbert space of the system **S**. However, as can be seen from their expressions, their meanings are not transparent a priori. They are determined a posteriori because we would like their expectation values to take certain forms. This can pose a question about the uniqueness of these operators (A similar issue is also raised in [58] for the classical formulation. However, in this context it is not clear whether this ambiguity can be fixed by calculating the cumulants associated with these operators. If there exist physical, measurable observables that correspond to the expectation values of the moments of these operators, then one may entertain the possibility of using them to uniquely determine these operators.). At least for a given reduced density matrix $\hat{\rho}_s$ of the system, we can always attach a system operator $\hat{\Lambda}_s$ that satisfies $\mathrm{Tr}_s\,\hat{\rho}_s\hat{\Lambda}_s = 0$ to the definitions of those local operators, that is, any system operator that has a zero mean. The choice of $\hat{\Lambda}_s$ may not be unique in the sense that in the basis $\{|n\rangle\}$ that diagonalizes $\hat{\rho}_S$, we can write $\mathrm{Tr}_S\{\hat{\rho}_S\hat{\Lambda}_S\} = 0$ as

$$\mathrm{Tr}_S\{\hat{\rho}_S\hat{\Lambda}_S\} = \sum_{m,n}\langle n|\hat{\rho}_S|m\rangle\langle m|\hat{\Lambda}_S|n\rangle = \sum_n\left(\hat{\rho}_S\right)_{nn}\left(\hat{\Lambda}_S\right)_{nn} = 0\,. \tag{56}$$

It says that the vectors that are respectively composed of the diagonal elements of $\hat{\rho}_S$ and $\hat{\Lambda}_S$ are orthogonal, but it does not place any restriction on the off-diagonal elements of $\hat{\Lambda}_S$ on this basis.

8. Conclusions

8.1. Summary

In this paper we provide quantum formulations of equilibrium thermodynamic functions and their relations for Jarzynski's classical thermodynamics at strong coupling [27] without the consideration of heat and work. The combined system + environment, called the composite, is assumed initially to be in a global thermal state. In such a configuration, even though the interaction between the system and the bath is non-negligible, the partition function of the composite is well defined. This facilitates the introduction of thermodynamic potentials in a way similar to the traditional vanishing-coupling thermodynamics. Such a configuration allows for the introduction of enthalpy by introducing an external agent (keeping at constant pressure in Jarzynski's case) acting on the conjugate bath operator. The effect can be represented by an equivalent effect on the system, which then appears in the expression of the enthalpy of the system. There are two ways to capture this effect, called the "bare" and "partial molar" representations by Jarzynski. We have worked out a quantum formulation for each of these two representations of Jarzynski's classical thermodynamics. In addition, we attempt to identify the operator forms of the thermodynamic functions, which can be potentially useful in directly addressing the effects of quantum fluctuations on thermodynamics at strong coupling, where the fluctuations can reach the same order of magnitude as the corresponding mean values.

8.2. Issues

We mention two outstanding issues of the two representations or approaches, namely, internal energy and entropy of the system at strong coupling in the global thermal state. When the quantum versions of these two approaches are applied to a small quantum system that strongly couples with a low-temperature bath, some nonintuitive results have been reported in the literature [29,36–40,43,44]. If the von Neumann entropy is adopted as the system entropy, then when the system and the bath are entangled, this entropy will not approach zero for a simple system such as a harmonic oscillator in the zero temperature bath, contradicting the result in [49], where it has been shown that the ground state of such a composite system is non-degenerate in general, thus implying vanishing entropy at zero temperature. On the other hand, if the system's internal energy is defined as the derivative of the partition function of the system, the heat capacity derived therefrom can take on negative values in the low temperature regime when the system consists of free particles or coupled harmonic oscillators. This anomalous behavior, not seen when the internal energy defined as the expectation value of the system Hamiltonian, may be traced to an excessive inclusion of the interaction and the bath contributions, as shown in (44), in the definition of the internal energy of the system.

8.3. Further Developments

To complete a theory for equilibrium quantum thermodynamics, we need to include the considerations of heat and work. For understanding quantum work, the physical meaning of the operator $\hat{A}_s$ conjugate to J is a key issue. At the next level of investigation, reaching out to nonequilibrium conditions, also at strong coupling, we posit that the operator $\hat{\Delta}_s$, introduced by Gelin and Thoss, and the Hamiltonian operator of mean force can be related to the influence action or the coarse-grained effective action [28,61,62] of the system when they are sandwiched by the states of the system and formulated in the finite temperature imaginary-time path integral method. This and an earlier observation we made for the partition function point out a way to extend the present equilibrium quantum thermodynamics at strong coupling for a closed system in a global thermal state to a nonequilibrium setting, by employing the real-time closed-time-path formalism used in [28,29,61,62]. This is the goal of our next paper in this series [65].

Author Contributions: J.-T.H. and B.-L.H. contributed equally to the conceptualization, methodology, investigation and writing of this paper.

Funding: This research received no external funding.

Acknowledgments: B.-L.H. wishes to thank the organizers of the EmQM-Bohm2017 conference for their invitation and hospitality. We thank Chris Jarzynski for explaining the key points in his work, and Chung-Hsien Chou and Yigit Subaşı for helpful discussions. B.-L.H. visited the Fudan University Physics Theory Center when this work commenced, while J.-T.H. visited the University of Maryland Physics Theory Center when this work was concluded. The support of visits from these centers are gratefully acknowledged.

Conflicts of Interest: The authors declare no conflict of interest.

Appendix A. Handling Operator Products in Quantum Thermodynamics

In deriving various thermodynamic relations in the context of quantum thermodynamics, we often come up with expressions involving exponential of the sum of two operators, say $\hat{\lambda}$ and $\hat{\mu}$. Unlike its *c*-number counterpart, in general such an exponential cannot be written as a product of exponential of the respective operators, that is,

$$e^{\hat{\lambda}+\hat{\mu}} \stackrel{?}{=} e^{\hat{\lambda}}\, e^{\hat{\mu}}\,, \tag{A1}$$

because these two operators $\hat{\lambda}$, $\hat{\mu}$ may not commute. From Baker-Campbell-Haussdorff formulas, the righthand side of (A1) in fact is given by

$$e^{\hat{\lambda}}e^{\hat{\mu}} = \exp\left(\hat{\lambda}+\hat{\mu}+\frac{1}{2!}\left[\hat{\lambda},\hat{\mu}\right]+\frac{1}{3!}\left\{\frac{1}{2}\left[\left[\hat{\lambda},\hat{\mu}\right],\hat{\mu}\right]+\frac{1}{2}\left[\hat{\lambda},\left[\hat{\lambda},\hat{\mu}\right]\right]\right\}+\cdots\right), \tag{A2}$$

for any two operators $\hat{\lambda}$, $\hat{\mu}$. However, in the special case that $[\hat{\lambda},\hat{\mu}]=0$, the equality in (A1) indeed is valid. The other useful expression in the Baker-Campbell-Hausdorff formulas is

$$e^{\hat{\lambda}}\,\hat{\mu}\,e^{-\hat{\lambda}} = \hat{\mu}+\left[\hat{\lambda},\hat{\mu}\right]+\frac{1}{2}\left[\hat{\lambda},\left[\hat{\lambda},\hat{\mu}\right]\right]+\cdots. \tag{A3}$$

This is particularly useful in deriving the unitary transformation of $\hat{\mu}$ by the operator $e^{\hat{\lambda}}$.

We also often come to a situation that we need to take a derivative of an exponential of the operator. This is less straightforward than is expected due to the fact that the operator in the exponent may not commute with its own derivative. For example, consider an operator $\hat{O}(\chi)$ of the form $\hat{O}(\chi)=\alpha(\chi)\,\hat{X}+\beta(\chi)\,\hat{P}$, with α, β being functions of χ, but the operators $\hat{X}$, $\hat{P}$ of the canonical variables have no explicit χ dependence. We immediately see the trivial result $[\hat{O}(\chi),\,\hat{O}(\chi)]=0$, but $[\hat{O}(\chi),\partial_\chi\hat{O}(\chi)]=(\alpha\dot{\beta}-\dot{\alpha}\beta)\,[\hat{X},\hat{P}]\neq 0$, where the overhead dot represents the derivative with respect to χ. This introduces complications in taking the derivative of, say, $e^{-\hat{O}(\chi)}$ with respect to χ. If we realize an operator function in terms of its Taylor's expansion, then

$$e^{-\hat{O}(\chi)} = \sum_{k=0}^{\infty}\frac{(-1)^k}{k!}\,\hat{O}^k(\chi)\,. \tag{A4}$$

Taking the derivative of (A4) with respective to χ yields

$$\begin{aligned}
&\partial_\chi e^{-\hat{O}}\\
&=-\sum_{k=1}^{\infty}\frac{(-1)^{k-1}}{(k-1)!}\left\{\frac{1}{k}\left[(\partial_\chi\hat{O})\underbrace{\hat{O}\cdots\cdots\hat{O}}_{(k-1)\text{ terms}}+\hat{O}(\partial_\chi\hat{O})\underbrace{\hat{O}\cdots\cdots\hat{O}}_{(k-2)\text{ terms}}+\cdots\cdots+\underbrace{\hat{O}\cdots\cdots\hat{O}}_{(k-1)\text{ terms}}(\partial_\chi\hat{O})\right]\right\}\\
&=-\left[(\partial_\chi\hat{O})\,e^{-\hat{O}}\right]_{\text{sym}},
\end{aligned} \tag{A5}$$

where we define the symmetrized product $(\hat{O}_1\hat{O}_2\cdots\hat{O}_k)_{\text{sym}}$ as a generalization of the anti-commutator by

$$(\hat{O}_1\hat{O}_2\cdots\hat{O}_k)_{\text{sym}} = \frac{1}{\#\text{ of perm.}}\sum_{\#\text{ of perm.}}\hat{O}_{\sigma_1}\hat{O}_{\sigma_2}\cdots\hat{O}_{\sigma_k}\,, \tag{A6}$$

with σ being the permutations of 1, 2, $\cdots$, k. Thus the expression, say, $(\partial_\beta\hat{H}_s^*)\,e^{-\beta\hat{H}_s^*}$, and similar derivative expressions will be understood in this manner as a symmetrized product of $\partial_\beta\hat{H}_s^*$ and the Taylor-expanded $e^{-\beta\hat{H}_s^*}$, shown in (A5).

However, if the derivative like (A5) is taken within a trace, then the complicated expression (A5) will reduce to a simple form

$$\mathrm{Tr}\{\partial_\chi e^{-\hat{O}}\} = -\sum_{k=1}^{\infty} \frac{(-1)^{k-1}}{(k-1)!}\,\mathrm{Tr}\Big\{(\partial_\chi \hat{O})\,\hat{O}^{k-1}\Big\} = -\,\mathrm{Tr}\Big\{(\partial_\chi \hat{O})\,e^{-\hat{O}}\Big\}, \tag{A7}$$

due to the cyclic property of the trace formula. Hence in general we have $\partial_\chi e^{-\hat{O}} = -\big[(\partial_\lambda \hat{O})\,e^{-\hat{O}}\big]_{\mathrm{sym}}$, but its trace gives

$$\mathrm{Tr}\{\partial_\chi e^{-\hat{O}}\} = -\,\mathrm{Tr}\Big\{\Big[(\partial_\lambda \hat{O})\,e^{-\hat{O}}\Big]_{\mathrm{sym}}\Big\} = -\,\mathrm{Tr}\{(\partial_\chi \hat{O})\,e^{-\hat{O}}\}, \tag{A8}$$

as if the operator $\hat{O}$ is a *c*-number. Note here we have assumed the traces applied in (A7) and (A8) are not a partial trace; otherwise the same symmetrization procedure is still needed.

A special case of (A7) is $\partial_\chi e^{-\chi\hat{O}}$, in which $\hat{O}$ has no explicit dependence on χ. Then it is straightforward to perform the differentiation, and since $[\hat{O}, e^{-\chi\hat{O}}] = 0$, we obtain

$$\partial_\chi e^{-\chi\hat{O}} = -\hat{O}\,e^{-\chi\hat{O}}. \tag{A9}$$

Next we give an explicit application of (A7). In particular, we focus on the expression

$$-\frac{1}{\mathcal{Z}_c}\frac{\partial}{\partial\beta}\,\mathrm{Tr}_s\, e^{-\beta(\hat{H}_s+\hat{\Delta}_s)}, \qquad \text{with} \qquad \mathcal{Z}_c = \mathrm{Tr}_s\, e^{-\beta(\hat{H}_s+\hat{\Delta}_s)}. \tag{A10}$$

Carrying out the differentiation of (A10) gives

$$= \frac{1}{\mathcal{Z}_c}\,\mathrm{Tr}_s\Big\{\big(\hat{H}_s + \hat{\Delta}_s + \beta\,\partial_\beta\hat{\Delta}_s\big)e^{-\beta(\hat{H}_s+\hat{\Delta}_s)}\Big\} = \langle\hat{H}_s\rangle + \langle\hat{\Delta}_s\rangle + \beta\,\langle\partial_\beta\hat{\Delta}_s\rangle. \tag{A11}$$

The first two terms after the first equal sign in (A11) is the consequence of (A9), while the third term results from (A8) due to the trace.

References

1. Sakharov, A.D. Vacuum quantum fluctuations in curved space and the theory of gravitation. *Dokl. Akad. Nauk SSSR* **1967**, *177*, 70.
2. Volovik, G.E. *The Universe in a Hhelium Droplet*; Clarendon Press: Wotton-under-Edge, UK, 2003.
3. Volovik, G.E. Fermi-point scenario for emergent gravity. In Proceedings of the Conference from Quantum to Emergent Gravity: Theory and Phenomenology (PoS(QG-Ph)), Trieste, Italy, 11–15 June 2007.
4. Wen, X.-G. *Quantum Field Theory of Many-Body Systems*; Oxford University Press: Oxford, UK, 2004.
5. Levin, M.; Wen, X.-G. Fermions, strings, and gauge fields in lattice spin models. *Phys. Rev. B* **2003**, *67*, 245316.
6. Hu, B.L. General relativity as geometro-hydrodynamics. In Proceedings of the Second Sakharov International Conference, Lebedev Physical Institute, Moscow, Russia, 20–24 May 1996; arXiv:gr-qc/9607070.
7. Hu, B.L. Emergent/quantum gravity: Macro/micro structures of spacetime. *J. Phys. Conf. Ser.* **2009**, *174*, 012015, doi:10.1088/1742-6596/174/1/012015.
8. Jacobson, T. The Einstein equation of state. *Phys. Rev. Lett.* **1995**, *75*, 1260–1263.
9. Padmanabhan, T. Thermodynamical aspects of gravity: New insights. *Rep. Prog. Phys.* **2010**, *73*, 046901.
10. Verlinde, E.P. On the origin of gravity and the laws of Newton. *arXiv* **2010**, arXiv:1001.0785.
11. Adler, S.L. *Quantum Theory As an Emergent Phenomenon*; Cambridge University Press: Cambridge, UK, 2004.
12. 't Hooft, G. The cellular automaton interpretation of quantum mechanics. *AIP Conf. Proc.* **2007**, *957*, 154–163.
13. Deutsch, J.M. Quantum statistical mechanics in a closed system. *Phys. Rev. A* **1991**, *43*, 2046.
14. Srednicki, M. Chaos and quantum thermalization. *Phys. Rev. E* **1994**, *50*, 888.
15. Goldstein, S.; Lebowitz, J.L.; Tumulka, R.; Zanghì, N. Canonical typicality. *Phys. Rev. Lett.* **2006**, *96*, 050403.
16. Popescu, S.; Short, A.J.; Winter, A. Entanglement and the foundations of statistical mechanics. *Nat. Phys.* **2006**, *2*, 754.
17. Linden, N.; Popescu, S.; Short, A.J.; Winter, A. Quantum mechanical evolution towards thermal equilibrium. *Phys. Rev. E* **2009**, *79*, 061103.
18. Short, A.J.; Farrelly, T.C. Quantum equilibration in finite time. *New J. Phys.* **2012**, *14*, 013063.

19. Reimann, P. Foundation of statistical mechanics under experimentally realistic conditions. *Phys. Rev. Lett.* **2008**, *101*, 190403.
20. Polkovnikov, A.; Sengupta, K.; Silva, A.; Vengalattore, M. Nonequilibrium dynamics of closed interacting quantum systems. *Rev. Mod. Phys.* **2011**, *83*, 863.
21. Cazalilla, M.A.; Rigol, M. Focus on dynamics and thermalization in isolated quantum many-body systems. *New J. Phys.* **2010**, *12*, 55006.
22. Gogolin, C.; Eisert, J. Equilibration, thermalization, and the emergence of statistical mechanics in closed quantum systems. *Rep. Prog. Phys.* **2016**, *79*, 056001.
23. Jarzynski, C. Nonequilibrium equality for free energy differences. *Phys. Rev. Lett.* **1997**, *78*, 2690.
24. Crooks, G.E. Entropy production fluctuation theorem and the nonequilibrium work relation for free energy differences. *Phys. Rev. E* **1999**, *60*, 2721.
25. Campisi, M.; Hänggi, P.; Talkner, P. Quantum fluctuation relations: Foundations and applications. *Rev. Mod. Phys.* **2011**, *83*, 771.
26. Seifert, U. First and second law of thermodynamics at strong coupling. *Phys. Rev. Lett.* **2016**, *116*, 020601.
27. Jarzynski, C. Stochastic and macroscopic thermodynamics of strongly coupled systems. *Phys. Rev. X* **2017**, *7*, 011008.
28. Weiss, U. *Quantum Dissipative Systems*, 4th ed.; World Scientific: Singapore, 2012.
29. Hsiang, J.-T.; Chou, C.H.; Subaşı, Y.; Hu, B.L. Quantum thermodynamics from the nonequilibrium dynamics of open systems: Energy, heat capacity and the third law. *Phys. Rev. E* **2018**, *97*, 012135.
30. Gemmer, J.; Michel, M.; Mahler, G. *Quantum Thermodynamics—Emergence of Thermodynamic Behavior within Composite Quantum Systems*, 2nd ed.; Springer: Berlin, Germany, 2004.
31. Kosloff, R. Quantum thermodynamics: A dynamical viewpoint. *Entropy* **2013**, *15*, 2100–2128.
32. Esposito, M.; Ochoa, M.A.; Galperin, M. Quantum thermodynamics: A nonequilibrium Green's function approach. *Phys. Rev. Lett.* **2015**, *114*, 080602.
33. Spohn, H.; Lebowitz, J.L. Irreversible thermodynamics for quantum systems weakly coupled to thermal reservoirs. *Adv. Chem. Phys.* **1978**, *38*, 109–142.
34. Esposito, M.; Ochoa, M.A.; Galperin, M. Nature of heat in strongly coupled open quantum systems. *Phys. Rev. B* **2015**, *92*, 235440.
35. Ankerhold, J.; Pekola, J.P. Heat due to system-reservoir correlations in thermal equilibrium. *Phys. Rev. B* **2014**, *90*, 075421.
36. Nieuwenhuizen, T.M.; Allahverdyan, A.E. Statistical thermodynamics of quantum Brownian motion: Construction of perpetuum mobile of the second kind. *Phys. Rev. E* **2002**, *66*, 036102.
37. Hänggi, P.; Ingold, G.-L. Quantum Brownian motion and the third law of thermodynamics. *Acta Phys. Pol. B* **2006**, **37**, 1537.
38. Hänggi, P.; Ingold, G.-L.; Talkner, P. Finite quantum dissipation: The challenge of obtaining specific heat. *New J. Phys.* **2008**, *10*, 115008.
39. Ingold, G.-L.; Hänggi, P.; Talkner, P. Specific heat anomalies of open quantum systems. *Phys. Rev. E* **2009**, *79*, 061105.
40. Hasegawa, H. Specific heat anomalies of small quantum systems subjected to finite baths. *J. Math. Phys.* **2011**, *52*, 123301.
41. Pucci, L.; Esposito, M.; Peliti, L. Entropy production in quantum Brownian motion. *J. Stat. Mech.* **2013**, *2013*, P04005.
42. Gelin, M.F.; Thoss, M. Thermodynamics of a subensemble of a canonical ensemble. *Phys. Rev. E* **2009**, *79*, 051121.
43. O'Connell, R.F. Does the third law of thermodynamics hold in the quantum regime? *J. Stat. Phys.* **2006**, *124*, 15–23.
44. Hörhammer, C.; Büttner, H. Information and entropy in quantum Brownian motion–Thermodynamic entropy versus von Neumann entropy. *J. Stat. Phys.* **2008**, *133*, 1161–1174.
45. Jordan, A.N.; Büttiker, M. Entanglement energetics at zero temperature. *Phys. Rev. Lett.* **2004**, *92*, 247901.
46. Hilt, S.; Lutz, E. System-bath entanglement in quantum thermodynamics. *Phys. Rev. A* **2009**, *79*, 010101.
47. Esposito, M.; Lindenberg, K.; van den Broeck, C. Entropy production as correlation between system and reservoir. *New J. Phys.* **2010**, *12*, 013013.

48. Deffner, S.; Lutz, E. Nonequilibrium entropy production for open quantum systems. *Phys. Rev. Lett.* **2011**, *107*, 140404.
49. Hanke, A.; Zwerger, W. Density of states of a damped quantum harmonic oscillator. *Phys. Rev. E* **1995**, *52*, 6875.
50. Hsiang, J.-T.; Hu, B.L. Nonequilibrium steady state in open quantum systems: Influence action, stochastic equation and power balance. *Ann. Phys.* **2015**, *362*, 139–169.
51. Hsiang, J.-T.; Hu, B.L. 'Hot entanglement'?: A nonequilibrium quantum field theory scrutiny. *Phy. Lett. B* **2015**, *750*, 396–400.
52. Kirkwood, J.G. Statistical mechanics of fluid mixtures. *J. Chem. Phys.* **1935**, *3*, 300–313.
53. Jarzynski, C. Nonequilibrium work theorem for a system strongly coupled to a thermal environment. *J. Stat. Mech.* **2004**, *2004*, P09005.
54. Strasberg, P.; Schaller, G.; Lambert, N.; Brandes, T. Nonequilibrium thermodynamics in the strong coupling and non-Markovian regime based on a reaction coordinate mapping. *New J. Phys.* **2016**, *18*, 073007.
55. Strasberg, P.; Esposito, M. Stochastic thermodynamics in the strong coupling regime: An unambiguous approach based on coarse graining. *Phys. Rev. E* **2017**, *95*, 062101.
56. Campisi, M.; Talkner, P.; Hänggi, P. Fluctuation theorem for arbitrary open quantum systems. *Phys. Rev. Lett.* **2009**, *102*, 210401.
57. Hilt, S.; Thomas, B.; Lutz, E. Hamiltonian of mean force for damped quantum systems. *Phys. Rev. E* **2011**, *84*, 031110.
58. Talkner, P.; Hänggi, P. Open system trajectories specify fluctuating work but not heat. *Phys. Rev. E* **2016**, *94*, 022143.
59. Reif, F. *Fundamentals of Statistical and Thermal Physics*; McGraw-Hill: New York, NY, USA, 1965.
60. Huang, K. *Statistical Mechanics*, 2nd ed.; Wiley: New York, NY, USA, 1987.
61. Grabert, H.; Schramm, P.; Ingold, G.L. Quantum Brownian motion: The functional integral approach. *Phys. Rep.* **1988**, *168*, 115–207.
62. Calzetta, E.; Hu, B.L. *Nonequilibrium Quantum Field Theory*; Cambridge University Press: Cambridge, UK, 2008.
63. Philbin, T.G.; Anders, J. Thermal energies of classical and quantum damped oscillators coupled to reservoirs. *J. Phys. A* **2016**, *49*, 215303.
64. Hörhammer, C.; Büttner, H. Thermodynamics of quantum Brownian motion with internal degrees of freedom: The role of entanglement in the strong-coupling quantum regime. *J. Phys. A* **2005**, *38*, 7325.
65. Hsiang, J.-T.; Hu, B.L. Thermodynamic functions and relations for nonequilibrium quantum systems. Unpublished work, 2018.

MDPI

Article

The Definition of Entropy for Quantum Unstable Systems: A View-Point Based on the Properties of Gamow States

Osvaldo Civitarese [1,*] and Manuel Gadella [2]

1 Department of Physics, University of La Plata, and IFLP-CONICET, 49 y 115. c.c.67, La Plata 1900, Argentina
2 Departamento de Física Teórica, Atómica y Óptica and IMUVA, Universidad de Valladolid, Paseo Belén 7, 47011 Valladolid, Spain; manuelgadella1@gmail.com
* Correspondence: osvaldo.civitarese@fisica.unlp.edu.ar

Received: 23 November 2017; Accepted: 1 March 2018; Published: 28 March 2018

Abstract: In this paper, we review the concept of entropy in connection with the description of quantum unstable systems. We revise the conventional definition of entropy due to Boltzmann and extend it so as to include the presence of complex-energy states. After introducing a generalized basis of states which includes resonances, and working with amplitudes instead of probabilities, we found an expression for the entropy which exhibits real and imaginary components. We discuss the meaning of the imaginary part of the entropy on the basis of the similarities existing between thermal and time evolutions.

Keywords: entropy and time evolution; resonances in quantum systems; the Friedrichs model; complex entropy

1. Introduction

The definition of entropy and its interpretation in terms of the evolution to equilibrium of isolated systems was a crucial step in understanding the link between mechanical and thermal features in classical mechanics [1]. The notion of probability applies, both in classical phase-space as well as in quantum mechanics, and from this the connection between entropy and the number of degrees of freedom of a system has been established [2]. The main difference between classical and quantum mechanical counting of states is, of course, the existence of the exclusion principle (for fermions) and other symmetry restrictions (both for fermions and bosons) imposed to quantum states. In both cases, fermions and bosons, the definition of the probability assigned to a state remains valid. This is not the case for states with complex energies, where the time evolution is non-oscillatory. States with complex energy, such as the Gamow states [3], are well described in the theory of scattering [4] and found as solutions of the analytical continuation of quantum relativistic and non-relativistic equations [5]. Several problems arise in dealing with these states, particularly their non-normalizability [6,7]. Most of these difficulties are removed with the use of amplitudes, which are the solutions of the equations and/or with the corresponding propagators, instead of working with their modulus. A suitable tool to work with Gamow states, in order to extract their thermodynamical information, is the path integration. In performing the path-integration we shall be dealing with amplitudes instead of probabilities, a concept which cannot be applied to states with complex energy.

In the present article, we are going to show that a comprehensive scheme leading to the definition of entropy for resonances can be rigorously designed by adopting path integration techniques. We shall discuss this method as well as its application to a model for resonances which is analytically solvable.

The paper is organized as follows. In Section 2 we revisit the conventional definition of entropy and relate it to time dependent operations, such as time inversion and time displacement.

Section 3 deals with the identification of resonances in quantum physical systems and illustrate their time dependence. These properties are then shown to be found over solid mathematical basis; e.g., we construct the decaying states in the framework of rigged Hilbert spaces [8]. Section 4 is devoted to the notion of complex entropy and in Section 5 we investigate the possible connection between our definition of complex entropy and the class of time operators [9]. Our final remarks and conclusions are presented in Section 6.

2. Entropy and time Evolution

In the context of quantum mechanics in the Heisenberg picture, the time evolution of a system is governed by its Hamiltonian. Each operator obeys the following equation of motion:

$$[H, \mathcal{O}] = -i\hbar\dot{\mathcal{O}}\,. \tag{1}$$

In classical mechanics the commutator (1) is replaced by a Poisson bracket and the corresponding time evolution is determined by the classical Hamilton–Jacobi equations. From a perspective other than thermodynamics, the evolution of a system is determined by the extreme of its free energy. For the moment we shall assume that the number of particles of the system is constant, this is why it makes sense to refer to the Helmholtz free energy $F = E - TS$. In such a circumstance, the change of the entropy S with respect to the energy, at constant temperature, is given by the equation

$$\frac{1}{T} = \frac{\partial S}{\partial \overline{E}}\,, \tag{2}$$

where $\overline{E}$ is the mean value of the energy and T is the absolute temperature at which the extreme of the free energy is reached (e.g., at the equilibrium). Though (2) seems to belong to a class of equations of motion different from (1), the difference is only apparent, since both equations fix physical values at equilibrium. Then, we may establish a correspondence between a class of operators and the entropy, as the associated observable. We shall return to this point later.

It was Boltzmann who realized that the number of degrees of freedom of a classical system is proportional to the logarithm of the number Ω of micro-states of the system, from where one derives the relation between the number of degrees of freedom and the entropy, i.e., $S = -k\log\Omega$. The way, in which the entropy evolves as a physical system approaches to the equilibrium, is given by a celebrated theorem due to Boltzmann, the $\mathbb{H}$-theorem [2]. The $\mathbb{H}$-theorem states that if $P_r(t)$ is the probability that a system is in the state r at time t and if we define $\mathbb{H} := \sum_r P_r(t)\log P_r(t)$, where the sum extends to all possible states of the system; then, $d\mathbb{H}/dt \leq 0$. The consequence is clear, since the entropy is given by $S = -k\mathbb{H}$, so that $dS/dt \geq 0$. The entropy monotonically increases with time until the system reaches the thermodynamic equilibrium.

The same time evolution is expressed by means of the quantum evolution operator e^{-itH}, so that if O is the operator representing a given quantum observable at time $t = 0$, the operator at time t is given by $O(t) = e^{itH}Oe^{-itH}$ (Unless otherwise stated, we take $\hbar = 1$ everywhere in the text.). Therefore, it should exist a direct connection between both descriptions of the evolution to equilibrium. However, from the time dependence of the observables of a system, one cannot always extract the direction of the evolution. The time reversal operation inverses the sense of time, so that it performs the operation $t \mapsto -t$. In classical mechanics, this means that the time reversal operation reverses momenta, velocities, etc, so that it reverses the velocities of the charges. This produces a change of the sign in the magnetic field, while leaves invariant the electric field.

In quantum mechanics, the time reversal operation is represented by the action of an operator, $\mathcal{T}$, on the space of wave functions. According to Wigner [10], time reversal is an operation such that the following operations performed sequentially give the identity:

$$\text{time displacement by } t \times \text{ time reversal } \times \text{ time displacement by } t \times \text{ time reversal}\,.$$

The above operations result on the identity if

time displacement by $t \times$ time reversal $=$ time reversal $\times$ time displacement by $-t$.

This point of view implies that the time reversal operator $\mathcal{T}$ has to be anti-linear in the sense that for any linear combination of states ψ_i and two complex numbers λ_i, $i = 1, 2$, $\mathcal{T}(\lambda_1\psi_1 + \lambda_2\psi_2) = \lambda_1^* \mathcal{T}(\psi_1) + \lambda_2^* \mathcal{T}(\psi_2)$, where the star denotes complex conjugation. In fact, if $\psi(x,t)$ is the wave function for some quantum pure state at time t, we have that $\mathcal{T}\psi(x,t) = \psi^*(x,-t)$ [11]. In addition, Wigner showed that in the construction of projective representations of the Poincaré group, extended with time inversion and parity, four independent choices exist for the time reversal operator. One is the just mentioned operator $\mathcal{T}$ and the other three require a doubling of the representation space [10,12]. From a conceptual point of view, we are faced to a difficult question, namely: If equilibrium appears in a particular instant of the time evolution of a system and is governed by a Hamiltonian, which is the operator that obeys Equation (1) and has the entropy given by the associated observable so that Equation (2) is fulfilled? One may also think that equilibrium is just a manifestation of the violation of the time-reversal symmetry, as shown by the behaviour of the entropy as a time dependent observable, as follows from the $\mathbb{H}$-theorem.

3. Resonances in Quantum Systems

As is well known, unstable quantum states are very frequent in Nature. They are characterized by two parameters: E_R and Γ, which are the real and imaginary parts of the energy, respectively. The quantity Γ is the inverse of the state half-life. Usually, one may consider that unstable quantum states are produced by the capture of a particle by a center of forces and its subsequent decay, a situation which is conveniently described by quantum scattering. The process of capture is often ignored as one is mainly concerned with the process of decay [5,8]. They are detected experimentally by the presence of some scattering features, such as a sharp bump in the cross section or a sudden change in the value of phase shifts. Due to this fact, unstable quantum states are usually called resonances. We shall use this denomination hereafter.

After this characterization of resonances in the context of scattering theory, they can be identified with poles in the analytic continuation of the S-matrix, provided that some smooth conditions be satisfied [13]. If this analytic continuation is performed in the energy representation the S matrix becomes a function of a complex energy defined on a two sheeted Riemann surface [13]. Resonances appear as pairs of complex conjugate poles located on the second Riemann sheet at the points $z_R = E_R \pm i\Gamma/2$, where $E_R > 0$ is the resonance energy and $\Gamma > 0$ the inverse of the half life, as said before.

The description of a quantum scattering process requires of two Hamiltonians. One is the free Hamiltonian H_0 that gives the free evolution of states. The other is a total Hamiltonian $H = H_0 + V$, where V is the potential which produces the scattering. In the case of having resonances due to scattering, the potential V determines the forces that produce the capture and the later decay of the resonant particle.

A particularly interesting model for quantum resonances is the Friedrichs model [14–16]. In the simplest formulation of the Friedrichs model a bound state interacts with an external field. As the result of this interaction, the bound state becomes unstable and, therefore, it is interpreted as a resonance. In the language of the Hamiltonian pair $\{H_0, H\}$, we have that

$$H_0 = \omega_0\, a^\dagger\, a + \int_0^\infty \omega\, b_\omega^\dagger\, b_\omega\, d\omega\,, \quad V = \lambda \int_0^\infty f(\omega)[a^\dagger\, b_\omega + a\, b_\omega^\dagger]\, d\omega\,. \tag{3}$$

We see that H_0 is the sum of two terms. In the former, $a^\dagger$ and a are, respectively, the creation and annihilation of a bound state of energy $\omega_0 > 0$. The integral term in H_0 is the simplest representation of a field in the energy representation, where $b_\omega^\dagger$ and b_ω are, respectively, the creation and annihilation operators of a state in the continuum with energy $\omega > 0$. Thus, H_0 has a non-degenerated continuous spectrum, $[0, \infty)$, plus a discrete eigenvalue $\omega_0 > 0$ imbedded in the continuum. The potential V intertwines discrete and continuous spectrum, where $f(\omega)$ is a regular function called the form factor and λ a real coupling constant. When the interaction V is switched on, the bound state becomes a resonance with complex energy given by

$$z_R = E_R - i\Gamma/2\,. \tag{4}$$

Observations on the resonance decay show that the decay rate is approximately given by $e^{-t\Gamma/2}$. Now, if any bound state is represented by a square integrable wave function, is this the same for unstable quantum states (resonances)? Let us assume that a resonance state is represented by a vector state ψ. The survival amplitude is defined as $\langle\psi|e^{-itH}|\psi\rangle$ and the survival probability, $P(t)$, as the modulus square of the survival amplitude, i.e., $P(t) = |\langle\psi|e^{-itH}|\psi\rangle|^2$. If ψ is to represent a resonance state, we expect that $P(t) \approx e^{-t\Gamma}$ for all values of t.

However, this is not the case. In general, one may prove that there exist states ψ for which $P(t)$ is approximately $e^{-t\Gamma}$ for most of observational values of time. These states may serve as resonance states. However, simple theoretical considerations show that deviations from the exponential decay law must exist for small and large values of time. These deviations are essential, i.e., they are a consequence of quantum theory, in particular of the semi-boundedness of the Hamiltonian, and not the product of noise or other interactions [17]. There exists some experimental evidence on the existence of such deviations [18,19].

As a matter of fact, $P(t) = |\langle\psi|e^{-itH}|\psi\rangle|^2$ shows a similar behaviour at very small times for all scattering states ψ. A simple calculation shows that $P'(0) = 0$, where $P'(t)$ is the derivative of $P(t)$ with respect to t. This has a subtle consequence known as the Zeno effect: sequential (repeated) measurements of the decay probability at very short intervals of time may prevent a decaying system to decay [20]. (This is the origin of the deviations of the purely exponential decay law for very short times, since if $P(t) = e^{-t\Gamma}$, then $P'(0) = -\Gamma \neq 0$.)

Nevertheless, a wide range of experiments on decaying systems show that the exponential decay is a good approximation for most purposes. Then, the consideration of states that have a purely exponential decay should be in order. These states can be rigorously constructed as eigenvectors, ψ^D, of H with eigenvalue z_R as in (4), $H\psi^D = z_R\,\psi^D$. Each of the eigenvectors ψ^D is called a decaying Gamow vector and has the property that $e^{-itH}\psi^D = e^{-iE_Rt}\,e^{-\Gamma t/2}\psi^D$, i.e., it decays exponentially as $t \longmapsto \infty$. Since ψ^D is an eigenvector of a self adjoint Hamiltonian with complex eigenvalue, then ψ^D cannot belong to the Hilbert space where H is densely defined and self-adjoint. Instead, it belongs to the dual $\Phi^\times$ of a rigged Hilbert space $\Phi \subset \mathcal{H} \subset \Phi^\times$.

If a normalizable vector ψ is taken to represent a resonance state, one may write $\psi = \psi^D + \psi^B$, where ψ^B accounts for the deviations from exponential law for very short and very large times [5,17]. Except for these two regimes of time, ψ^D is a good approximation for ψ. However, as ψ^D is not normalizable in the usual sense, one finds methodological difficulties to define mean values of observables on ψ^D [6,7]. These difficulties will re-emerge as one attempts to assign a value to the entropy for quantum decaying systems.

4. Complex Entropy

It should be clearly stated that any quantum unstable system should obey the laws of thermodynamics. The point is that a precise formulation of these laws for quantum decaying states has not been formulated yet, up to our knowledge. Also, quantum statistical mechanics should extend its scope so as to embrace these kind of systems.

Based on this idea, one may ask for a suitable definition of the entropy for quantum unstable systems. At least three approaches have been proposed. A first approach was proposed by the Brussels group, it relies on the construction of an entropy operator, defined as a monotonic function of the time operator [9,21], which can be rigorously defined from a mathematical point of view under reasonable physical properties, see [22]. This way has not been fully explored yet.

A second approach was suggested by Kobayashi and Shimbori [23,24]. There the entropy for a quantum unstable state, described by a pole of the form $z_R = E_R - i\Gamma/2$, is a sum of a contribution of the entropy of the real part, E_R, and a contribution from the imaginary part Γ, so that $S = S(E_R) + S(\Gamma)$ [24]. This keeps the entropy as a real function of the resonance pole. In fact, real and imaginary parts of the complex resonance energy z_R are treated as if they were two independent systems. In this picture, decaying processes transfer entropy from $S(\Gamma)$ to $S = S(E_R)$ and the rate of this transference depends on time. Each part has its own temperature, which suggests a notion of complex temperature.

We advocate a third approach, which does not make use of the entropy operator and avoids any possible reference to complex temperatures. Following a suggestion in [23], we assume that quantum decaying states are in thermodynamic equilibrium, provided that the half life be sufficiently large, or equivalently, that the imaginary part of its energy, Γ, be sufficiently small.

Then, in order to give a definition of the entropy for quantum unstable states, we need a universal model of resonances for which mathematical operations could be performed as much as possible. This is given by the Friedrichs model described in the previous section. In the Friedrichs model, resonances are produced after the interaction of a discrete bound state with a continuum of states with a much larger degeneracy, so that it may be taken as a good example of a situation amenable to a statistical description based on the canonical ensemble representation, where the continuum is playing the role of the environment interacting with the isolated discrete state. For simplicity, we may consider a Friedrichs model with one resonance only, although more complicated models could be used for the same purpose [16]. As is well known, the canonical entropy is given by the formula:

$$S = k\left(1 - \beta\frac{\partial}{\partial\beta}\right)\log Z\,, \tag{5}$$

where $Z = \mathrm{Tr}\, e^{-\beta H}$ is the partition function corresponding to the total hamiltonian H and $\beta = 1/(kT)$, where T is the absolute temperature and k the Boltzmann constant.

In order to evaluate $Z = \mathrm{Tr}\, e^{-\beta H}$, it seems reasonable to use a generalized basis of vectors which includes the Gamow state ψ^D. This is given by ψ^D and the so called generalized outgoing eigenvectors of the total Hamiltonian $\{|\omega^+\rangle\}$, with $H|\omega^+\rangle = \omega\,|\omega^+\rangle$, for all $\omega \geq 0$. Then, the partition function would have taken the following form:

$$Z = \mathrm{Tr}\, e^{-\beta H} = \langle\psi^D|e^{-\beta H}|\psi^D\rangle + \int_0^\infty \langle\omega^+|e^{-\beta H}|\omega^+\rangle\, d\omega\,. \tag{6}$$

However, this formula is not computable, as brackets of the form $\langle\psi^D|\psi^D\rangle$ or $\langle\omega^+|\omega^+\rangle$ are not well defined [7,16,25].

Then, we have to circumvent this problem by using a different technique based on the use of path integrals to calculate partition functions as introduced by Feynman and Hibbs [26]. In our approach, we have adopted path integration in order to write the partition function using a basis of coherent states. Thus, we construct coherent states in the following form: creation, $A^\dagger_{\mathrm{IN}}$, and annihilation, A_{OUT}, operators for the Gamow state ψ^D may be constructed for the second quantized Friedrichs model as described in [16]. Then, for any complex number α, we define the coherent state $|\alpha\rangle$ as:

$$|\alpha\rangle = \exp[\alpha\, A^\dagger_{\mathrm{IN}} - \alpha^*\, A_{\mathrm{OUT}}]\,|0\rangle\,, \tag{7}$$

where $|0\rangle$ is the vacuum state and the star denotes complex conjugation. Then, an evaluation of the partition function, although somehow cumbersome, can be done. Details are given in [27–29]. We arrive to the following result for the entropy of a Gamow state with complex energy $z_R = E_R - i\Gamma/2$:

$$S = k \left[1 - \ln \left(\beta \sqrt{E_R^2 + \frac{\Gamma^2}{4}} \right) - i \arctan \left(\frac{\Gamma}{2E_R} \right) \right] . \tag{8}$$

The result for S is complex and this fact requires of some comments. Firstly, the method used to obtain the above formula is a straightforward generalization of a similar method, which uses path-integration and coherent states, developed to obtain an approximation to the entropy of the harmonic oscillator [29] avoiding the use of probabilities. For the case of the harmonic oscillator it yields $S \approx k(1 - \log(\beta\hbar\omega))$. Note that this is exactly the result that we obtain in the limit $\Gamma \mapsto 0$ and $E_R = \hbar\omega$. Secondly, since quantum resonances have complex energies with a different physical interpretation of real and imaginary parts, it is not a surprise that the same situation arises for the entropy. The resonance in the Friedrichs model is produced by the interaction of the bound state with the external field that plays the role of an external bath [30]. With this idea in mind, one may interpret the real part of (8) as the system entropy and its imaginary part as the entropy transferred from the resonance to the external bath.

There is another approach, described in [29], which leads to a complex entropy for an unstable quantum state. It is based on the fact that the total Hamiltonian has the form $H = z_R\, A^\dagger_{\rm IN}\, A_{\rm OUT}$, plus a much smaller background term which is neglected. Then, by using the property that the trace is invariant under cyclic permutations and formulas like

$$[H, A^\dagger_{\rm IN}(\tau)] = \frac{\partial}{\partial\tau} A^\dagger_{\rm IN}(\tau) = z_R\, A^\dagger_{\rm IN}\,;\; [H, A_{\rm OUT}(\tau)] = \frac{\partial}{\partial\tau} A_{\rm OUT} = -z_R\, A_{\rm OUT}\,, \tag{9}$$

and for operators of the form $O(\tau) = e^{\tau H}\, O\, e^{-\tau H}$ with $\tau = \beta$, we obtain the desired result, of which (8) could be considered as a reasonable approximation. Note that the definition for $O(\tau)$ has a great similarity with the definition for the time evolution of an operator, as suggested before (see Section 2).

5. Time-Temperature Plane

Let us consider a quantum observable O and define $O(\tau)$ as in the previous section, right after formula (9), where $\tau = \beta = 1/(kT)$, being T the absolute temperature. $O(\tau)$ denotes the thermal evolution of the observable. On the other hand, if we consider the time evolution of the observable O under a Hamiltonian H, we have that $\tau = -it$. The transformation from the first to the second is sometimes called the Wick rotation [1].

This suggests the possibility of a description of time evolution for non-equilibrium systems using the dependence on both variables time and inverse temperature. The picture would be a complex plane in which the real part is given by the inverse of the temperature and the imaginary part by time. Similar notions have been applied to introduce dual-thermal degrees of freedom and close-path integrals [31].

Time operators have been defined for different purposes and different contexts [32,33]. For instance, assume that H is a densely defined Hamiltonian on a given Hilbert space $\mathcal{H}$. This system has an internal time operator $\mathbb{T}$ if for any density operator ρ in a domain dense in the space of Hilbert-Schmidt operators on $\mathcal{H}$, we have that $e^{-itH}\, \mathbb{T}\, e^{itH}\, \rho = (\mathbb{T} + tI)\rho$, for any t real, where I is the identity. Not any Hamiltonian system may have an internal time operator [34].

One of the interesting aspects of time operators is the possibility of constructing Liapunov quantum variables, i.e., variables monotonic on time. One may understand the role of such variables as indicators of the approach to equilibrium for complex systems, particularly for a definition of an entropy operator. Here, we refer to a construction proposed in [9] and valid for a large class of situations. A necessary condition for the use of the procedure outlined in [9] is that the Hamiltonian be unbounded and have an absolutely continuous spectrum, usually the half line $[0, \infty)$, if this condition is fulfilled H is said to be semi-bounded.

The idea of a time operator emerges from the comparison between the position-momentum and energy-time uncertainty relations. However, and due to the semi-boundedness of the Hamiltonian in non-relativistic quantum mechanics, a commutation relation of the type $[H, \mathbb{T}] = iI$ cannot hold. In any case, if the Hamiltonian is semi-bounded with absolutely continuous spectrum, its corresponding Liouvillian $L = H \otimes I - I \otimes H$ has a continuous spectrum that covers the whole real axis. In this case, it may be possible to define a time operator $\mathbb{T}$ as the conjugate of the Liouvillian operator, $[L, \mathbb{T}] = -iI$. Note that these operators have to be defined on $\mathcal{H} \otimes \mathcal{H}$, i.e., the space of Hilbert-Schmidt operators on $\mathcal{H}$.

If this were the case, one may define the entropy as some monotonic function of the time operator, as done in [9], i.e., $S = f(\mathbb{T})$. Attempts to define a time operator, and hence an entropy operator for unstable decaying systems are on the course.

6. Final Remarks

In classical mechanics, the approach to equilibrium is a manifestation of the time reversal symmetry breaking. This is formulated via the Boltzmann $\mathcal{H}$ theorem according to which the entropy monotonically increases up to a critical point, usually a maximum, at equilibrium. In classical electrodynamics, the retarded solutions of the Maxwell equations are privileged over the advanced solutions, thus showing a time asymmetry.

In quantum mechanics, the decay of unstable systems such as quantum resonances gives a sense of time reversal symmetry breaking. One finds a need for a proper formulation describing this situation in a similar context as in classical mechanics, whenever possible. Then, it seems necessary to define the notion of entropy for quantum decaying systems.

We have introduced an idea toward a proper definition of this entropy based on the use of Gamow states as state vectors for resonances. However, a naive presentation using standard tools of quantum mechanics yields to inconsistencies due to the ill definition of some formal averages. We have shown that the use of path integration over coherent states, which have been constructed with the help of creation and annihilation operators of Gamow vectors, gives a reasonable outcome. The resulting entropy is complex, with an imaginary part which gives an account for the interactions of decaying states with their surroundings.

We have discussed the formal similarities between thermal and time evolution of states. Concerning quantum decaying states, we have introduced a representation, (9), that gives the thermal evolution of the creation and annihilation operators for the Gamow states. These are differential equations that admit the following solutions:

$$A_{\rm IN}^{\dagger}(\tau) = e^{\tau z_R}\, A_{\rm IN}^{\dagger}\,, \quad \text{and} \quad A_{\rm OUT}(\tau) = e^{-\tau z_R}\, A_{\rm OUT}\,. \tag{10}$$

Identities (10) are useful in order to obtain an expression for the complex entropy valid for the quantum unstable state created by $A_{\rm IN}^{\dagger}$ and annihilated by $A_{\rm OUT}$. Here, we choose $\tau = \beta = 1/(kT)$, as given in [29]. In this case, and since the resonance energy E_R is taken to be positive, the highest the temperature, the smaller $A_{\rm IN}^{\dagger}(\tau)$ and the larger $A_{\rm OUT}(\tau)$.

A completely different interpretation comes when $\tau = -it$, i.e., when we consider the time evolution of Gamow states. Now, $A_{\rm IN}^{\dagger}(t) = e^{-itE_R}\, e^{-\Gamma t/2}\, A_{\rm IN}^{\dagger}$ and $A_{\rm OUT}(t) = e^{itE_R}\, e^{\Gamma t/2}\, A_{\rm OUT}$, so that the creation operator for a Gamow state decays with time while the annihilation operator grows with time. This is called the Wick rotation.

A future perspective could be a definition of the entropy operator as a function of the time operator. This is defined on an algebras of observables where Gamow states play a role as functionals over this algebra. A prototype of this algebraic model has been constructed [35] and the investigation is on the course.

Acknowledgments: Partial financial support in acknowledged to the National Research Council of Argentina (CONICET) PIP 616 , to the Agencia Nacional de Promocion Cientifica y Tecnica de Argentina (ANPCYT), the Project MTM2014-57129-C2-1-P of the Spanish Ministerio de Economía y Competitividad and the Project VA057U16, awarded by the Junta de Castilla y León (Spain).

Author Contributions: This paper is a fully collaborative work and both authors have equally contributed to all technical details as well as to the final presentation of the manuscript.

Conflicts of Interest: The authors declare no conflict of interest.

References

1. Reichl, L. *A Modern Course in Statistical Physics*, 4th ed.; Wiley-VCH: Weinheim, Germany, 2016.
2. Huang, K. *Statistical Mechanics*, 2nd ed.; Wiley: New York, NY, USA, 1987.
3. Gamow, G. On the Quantum theory of atomic nucleus. *Z. Phys.* **1928**, *51*, 204–212.
4. Newton, R.G. *Scattering Theory of Waves and Particles*, 2nd ed.; Springer: New York, NY, USA, 1982.
5. Civitarese, O.; Gadella, M. Physical and mathematical aspects of Gamow states. *Phys. Rep.* **2004**, *396*, 41–113.
6. Berggren, T. Expectation value of an operator in a resonant state. *Phys. Lett. B* **1996**, *373*, 1–4.
7. Civitarese, O.; Gadella, M.; Id Betán, R. On the mean value of the energy for resonance state. *Nucl. Phys. A* **1999**, *660*, 255–266.
8. Bohm, A.; Gadella, M. *Dirac Kets, Gamow Vectors and Gelfand Triplets*; Springer Lecture Notes in Physics; Springer: Berlin, Germany, 1989; Volume 348.
9. Misra, B.; Prigogine, I.; Courbage, M. Lyapunov variable: Entropy and measurement in quantum mechanics. *Proc. Natl. Acad. Sci. USA* **1979**, *76*, 4768–4772.
10. Wigner, E.P. *Group Theoretical Concepts and Methods in Elementary Particle Physics*; Gordon and Breach: New York, NY, USA, 1994.
11. Gadella, M.; de la Madrid, R. Resonances and time reversal operator in rigged Hilbert spaces. *Int. J. Theor. Phys.* **1999**, *38*, 93–113.
12. Wigner, E.P. *Symmetries and Reflections*; Indiana University Press: Bloomington, Indiana, 1967.
13. Bohm, A. *Quantum Mechanics: Foundations and Applications*; Springer: New York, NY, USA, 1986.
14. Friedrichs, K.O. On the perturbation of continuous spectra, Commun. *Pure Appl. Math.* **1948**, *1*, 361–406.
15. Horwitz, L.P.; Marchand, J.P. The decay scattering system. *Rocky Mtn. J. Math.* **1971**, *1*, 225–252.
16. Gadella, M.; Pronko, G.P. The Friedrichs model and its use in resonance phenomena. *Fortschr. Phys.* **2011**, *59*, 795–859.
17. Fonda, L.; Ghirardi, G.C.; Rimini, A. Decay theory of unstable quantum systems. *Rep. Prog. Phys.* **1978**, *41*, 587–631.
18. Fischer, M.C.; Gutierrez-Medina, B.; Raizen, M.G. Observation of the quantum Zeno and anti-Zeno effects in an unstable system. *Phys. Rev. Lett.* **2001**, *87*, 40402.
19. Rothe, C.; Hintschich, S.L.; Monkman, A.P. Violation of the exponential-decay law at long times. *Phys. Rev. Lett.* **2006**, *96*, 163601.
20. Misra, B.; Sudarshan, E.C.G. Zeno's paradox in quantum theory. *J. Math. Phys.* **1977**, *18*, 756–763.
21. Karpov, E.; Ordóñez, G.; Petrosky, T.; Prigogine, I. Microscopic entropy. *Int. J. Quantum Chem.* **2004**, *98*, 69–77.
22. Sorger, U.; Suchanecki, Z. Nonlocalization Properties of Time Operator Transformations. *Int. J. Theor. Phys.* **2015**, *54*, 787–800.
23. Kobayhashi, T.; Shimbori, T. Statistical mechanics for unstable states in Gelfand triplets and investigations of parabolic potential barriers. *Phys. Rev. E* **2001**, *63*, 56101.
24. Kobayhashi, T. New aspects in physics of Gelfand triplets. *Int. J. Theor. Phys.* **2003**, *42*, 2265–2283.
25. Antoniou, I.; Prigogine, I. Intrinsic irreversibility and integrability of dynamics. *Phys. A Stat. Mech. Appl.* **1993**, *192*, 443–464.
26. Feynman, R.P.; Hibbs, A.R. *Quantum Mechanics and Path Integrals*; Mc-Graw Hill: New York, NY, USA, 1965.
27. Civitarese, O.; Gadella, M. An approximation to the entropy for quantum decaying states. *Int. J. Geom. Methods Mod. Phys.* **2013**, *10*, 1360009.
28. Civitarese, O.; Gadella, M. On the concept of entropy for quantum decaying systems. *Found. Phys.* **2013**, *43*, 1275–1294.

29. Civitarese, O.; Gadella, M. On the entropy for unstable fermionic and bosonic states. *Phys. A Stat. Mech. Its Appl.* **2014**, *404*, 302–314.
30. Antoniou, I.; Gadella, M.; Karpov, E.; Prigogine, I.; Pronko, G. Gamow algebras. *Chaos Solitons Fractals* **2001**, *12*, 2757–2775.
31. Kamenev, A. *Field Theory of Non-Equilibrium Systems*; Cambridge University Press: Cambridge, UK, 2011.
32. Antoniou, I.; Suchanecki, Z. Time operators associated to dilations of Markov processes. In *Evolution Equations: Applications to Physics, Industry, Life Sciences and Economics (Levico Terme, 2000)*; Progress in Nonlinear Differential Equations and Their Applications; Birkhäuser: Basel, Switzerland, 2003; Volume 55, pp. 13–23.
33. Suchanecki, Z.; Antoniou, I. Time operators, innovations and approximations. *Chaos Solitons Fractals* **2003**, *17*, 337–342.
34. Gómez Cubillo, F.; Suchanecki, Z.; Villullas, S. On lambda and time operators: The inverse intertwining problem revisited. *Int. J. Theor. Phys.* **2011**, *50*, 2074–2083.
35. Castagnino, M.; Gadella, M.; Betan, R.; Laura, R. Gamow functionals on operator algebras. *J. Phys. A Math. Gen.* **2001**, *34*, 10067–10083.

Article

What Constitutes Emergent Quantum Reality? A Complex System Exploration from Entropic Gravity and the Universal Constants

Arno Keppens

Space Pole, Circular Avenue, 1180 Brussels, Belgium; arnok@oma.be; Tel.: +32-2373-0412

Received: 29 March 2018; Accepted: 30 April 2018; Published: 2 May 2018

Abstract: In this work, it is acknowledged that important attempts to devise an emergent quantum (gravity) theory require space-time to be discretized at the Planck scale. It is therefore conjectured that reality is identical to a sub-quantum dynamics of ontological micro-constituents that are connected by a single interaction law. To arrive at a complex system-based toy-model identification of these micro-constituents, two strategies are combined. First, by seeing gravity as an entropic phenomenon and generalizing the dimensional reduction of the associated holographic principle, the universal constants of free space are related to assumed attributes of the micro-constituents. Second, as the effective field dynamics of the micro-constituents must eventually obey Einstein's field equations, a sub-quantum interaction law is derived from a solution of these equations. A Planck-scale origin for thermodynamic black hole characteristics and novel views on entropic gravity theory result from this approach, which eventually provides a different view on quantum gravity and its unification with the fundamental forces.

Keywords: quantum ontology; sub-quantum dynamics; micro-constituents; emergent space-time; emergent quantum gravity; entropic gravity; black hole thermodynamics

---

1. Introduction

Important attempts to devise an emergent quantum (gravity) theory require space-time to be discretized at the Planck scale [1]. The identification of the discrete micro-constituents of space-time is therefore one of the biggest research questions in present-day physics. Yet, if space-time is indeed an effective field, emerging from the interaction of its micro-constituents only, then quantizing some aspect of general relativity will not help us identify its fundamental degrees of freedom—by analogy, we would arrive at a theory of phonons rather than a description of the underlying atoms of the condensate [2–4]. For that reason, in correspondence with Oriti [5], in this work "we consider the emergence of continuum space and time from the collective behavior of discrete, pre-geometric atoms of quantum space, and (analogously consider) space-time as a kind of condensate".

Yet, by viewing the conjectured pre-geometric atoms of quantum space as the ontological micro-constituents of our emergent reality, its effective macro-dynamics, including space and time, is expected to benefit from a complex (nonlinear) sub-quantum dynamical systems approach for its appropriate understanding in terms of the fundamental degrees of freedom. According to Ladyman et al. [6] "a complex system is an ensemble of many elements which are interacting in a disordered way, resulting in robust organization and memory". The necessary qualitative conditions, although being not necessarily jointly sufficient, for the emergence of a complex dynamic that shows spontaneous yet persistent ordering can be correspondingly defined as "numerosity" (an ensemble of many fungible elements) and "interaction" (through direct nonlinear causality) [6].

This work hence attempts to provide a parsimonious complex systems approach, as a kind of toy model, for identifying space-time's ontological micro-constituents and their interaction, i.e., their sub-quantum dynamics. Motivated by Occam's razor, it is here assumed that only one type of such micro-constituents exists, and that a single interaction law connects them relationally [7]. This assumption entails that effective space-time, matter, gravity, and the other fundamental forces should emerge from the interaction, through their fundamental degrees of freedom as dynamical attributes, of the single-type micro-constituents. A number of analogue gravity models or condensed matter approaches to quantum gravity already adopt this strategy, but typically lack background-independence in their interactions [4,8]. This background-independence however is required for interactions that induce (and thus precede) the emergence of any space-time that could serve as a reference metric.

In order to arrive at a background-independent micro-constituent interaction (law) that reproduces general relativity's dynamical space-time (including gravity) in its effective field behavior, we adopt and combine two strategies. First, motivated by the works of Jacobson [2], Padmanabhan [9], and Verlinde [10] (or see Padmanabhan [11] for more recent progress), we will conceive of gravity as a thermodynamic phenomenon or an emergent entropic force. These authors have demonstrated how Einstein's field equations can be considered to originate from space-time's thermodynamic degrees of freedom at a causal (black hole or holographic) horizon. In this work however, in order to identify the micro-constituents of space-time and their relation with common physical quantities, the dimensional reduction of the holographic principle as presented by 't Hooft [12] is generalized to non-holographic reference surfaces. It is shown that the universal constants of free space can then be related to attributes of the atoms of quantum space.

Second, a reverse-engineering argument, somewhat characteristic for complex dynamical systems approaches and encouraged by Hu [13] for emergent quantum gravity research, is used to put forward an approximation of the background-independent interaction law that connects the conjectured single-type micro-constituents of space-time: as the emergent effective field dynamics of the micro-constituents must eventually obey Einstein's relativistic field equations [14], a micro-constituent interaction law that yields the required diffeomorphism invariant field behavior can be obtained from a solution of these equations. The resulting interaction law is however formulated within the emergent relativistic space-time framework itself, and not in a fundamental pre-space-time framework. The latter option is very much complicated by the involvement of some sort of "external time" that is tied to the pre-space-time dynamics of the micro-theory [4]. This flaw seems familiar—and acceptable—when looking at the analogous issue in perturbative string theory, see for instance Huggett and Vistarini [15].

Together, these two strategies thus allow identifying—in a first rudimentary way—the micro-constituents of space-time and their basic interaction. The explicit constituent-based complex systems approach presented in this work additionally allows deriving black hole thermodynamics in a way that is believed to be more direct and intuitive than previous accounts [16–18] and related aspects of entropic gravity, the latter even for non-holographic reference surfaces. Both phenomena are reproduced in terms of space-time's micro-constituents and the number of fundamental (thermodynamic) degrees of freedom at their availability on the surface of reference. This complex toy model of quantum reality is therefore anticipated to point the way towards a more mature emergent theory of quantum gravity, while a generalization of the constituent-based origin of the gravitational field finally hints at a unification of the fundamental forces.

2. Constituent Identification

We initiate our complex systems-based toy model of emergent reality with a rudimentary attempt to identify space-time's ontological micro-constituents. It is thereby assumed that only one type of such micro-constituents exists, which entails that effective space-time, matter, gravity, and the other fundamental forces should emerge from the interaction, through their attributes, of these single-type micro-constituents only. This also entails that the universal constants of free space, like the speed of light in vacuum c, the gravitational constant G, the (reduced) Planck constant $\hbar$, and the Boltzmann constant k_B, are expected to be in some way all related to the attributes of the micro-constituents.

A direct connection between the universal constants of free space and associated space-time constituent properties is therefore derived in the following.

As space-time (curvature) and gravitational effects are unified by Einstein's relativistic field equations, it seems evident to first establish a relationship between the mass m or energy E enclosed within a certain space-time volume V on the one hand and an invariable property (say G_0) of each of the n_V individual space-time constituents within that volume on the other hand:

$$m \propto n_V G_0. \tag{1}$$

Let us denote this mass and energy defining attribute, G_0, which should obviously be related to the gravitational constant, as a micro-constituent's "gravitational presence" (this choice is elucidated later on). Yet, masses also experience their mutual full extent from a distance, i.e., without shared knowledge of their respective n_V. We must, therefore, relate the "information" about the amount of micro-constituents within the volume V to some "information" on its surface $A = \partial V$, which is the kind of dimensional reduction that was proposed by 't Hooft [12] in his holographic principle. This principle is generalized to non-holographic surfaces here with the following premise: the amount of micro-constituents n_V contained within an enclosed space-time volume V is proportional to the amount of micro-constituents n_A that overlaps with the surface $A = \partial V$ of that volume: $n_V \propto n_A$. As a result, one can rewrite Equation (1) as:

$$m \propto n_A G_0. \tag{2}$$

Relating the above to common physical quantities can be achieved by use of straightforward dimensional analysis. By simply rearranging the unit dimensions of G one has:

$$m \propto \frac{c^3}{G} \Delta t. \tag{3}$$

By combination of Equations (2) and (3), and thereby taking $\Delta t = t_P$ to explicitly relate the constituents to the Planck scale (and unit system), one can identify each mass as follows:

$$m \equiv n_A G_0 t_P \tag{4}$$

with $G_0 \propto c^3/G$ from Equation (3). Equation (4) implies:

$$\begin{array}{rcl} m_0 & = & G_0 l_P / c \\ E_0 & = & G_0 l_P c \end{array} \tag{5}$$

so that we can write $m = n_A m_0$ and $E = n_A E_0$ with m_0 and E_0 the rather abstract unit mass and unit energy that are associated with the exchange of a single space-time micro-constituent through the surface A, respectively. In the following, n_A is replaced by n, as always the micro-constituents on the reference surface are intended.

Up to this point, our analysis has been limited to linear relationships in terms of the numbers of micro-constituents. This changes when considering temperature T and entropy S that both depend on a system's thermodynamic degrees of freedom. Motivated by the entropic gravity argumentation from Padmanabhan [9] and Verlinde [10] for holographic surfaces, yet keeping our non-holographic premise and Equation (2) in mind, we here apply the equipartition theorem to the generalized reference surface A (assuming that it also holds approximately for non-trivial energy distributions in quantum systems). The equipartition theorem then states that the energy nE_0 of V, because of its representation by the n micro-constituents at the surface A of V, is equally distributed over all degrees of freedom N on A, or $E = nE_0 = Nk_B T/2$, which immediately results in:

$$T = \frac{2nE_0}{Nk_B}. \tag{6}$$

The connection between temperature and entropy as conjugate thermodynamic variables through $T = \Delta E/\Delta S$, which is discretized because of the finite-sized micro-constituents, moreover yields:

$$\Delta S = \frac{k_B N}{2}\frac{\Delta n}{n}. \tag{7}$$

By direct integration for constant N, i.e., over the reference surface A, Equation (7) becomes:

$$S = \frac{k_B N}{2}\ln(n) \tag{8}$$

so that, on the Planck unit scale, $S_P = k_B \ln(2)$ bit or $S_P = k_B$ nit (as required by definition) only when $n = N = 2$. This entails that a surface enclosing a single Planck mass exchanges two space-time micro-constituents with the outer environment during a single Planck time interval or $\sim 10^{43}$ constituents over a second. The entropy associated with a single constituent occupying one fundamental degree of freedom $S\ (n = 1, N = 1)$ obviously equals zero, yet one can define $S_0 = S\ (n = 2, N = 1) = k_B/2$ nit as a unit simplification, wherefrom, upon insertion into Equations (6) and (8) respectively:

$$T = \frac{n}{N}\frac{E_0}{S_0} = \frac{n}{N}T_0 \tag{9}$$

and

$$S = S_0 N \ln(n). \tag{10}$$

Comparison with the Boltzmann formula $S = k_B \ln(\Omega)$ shows that the number of microstates Ω that corresponds with a given macrostate encompassing N surface degrees of freedom for n micro-constituents is given by $\Omega = n^N$ as one would expect.

By combining $m_P = 2G_0 l_P/c$ with the Planck definitions of mass $m_P = \sqrt{\hbar c/G}$ and length $l_P = \sqrt{\hbar G/c^3}$ [19], one obtains:

$$\begin{array}{rcl} G & = & c^3/2G_0 \\ \hbar & = & 2G_0 l_P^2 \end{array}. \tag{11}$$

As summarized in Table 1, the above allows translating the universal constants of free space into four attributes of space-time's micro-constituents and corresponding constituent units. Note that products of constituent units of complementary variables, like time and energy or position and momentum, immediately yield $G_0 l_P^2 = \hbar/2$. This result suggests a direct connection between the discreteness of the micro-constituents, forcing measurement outcomes to refer to an integer amount of constituents, and the Heisenberg uncertainty relations [20].

Table 1. Translation (first column) of universal constants of free space into space-time constituent attributes (second column) and its effect on the definition of basic units (third column).

Constants Translation	Constituent Attributes	Constituent Units
$\hbar = 2G_0 l_P^2 \to l_P$	Size	$l_0 = l_P$
$c \to c$	Velocity	$t_0 = t_P = l_P/c$
$G = c^3/2G_0 \to G_0$	Gravitational presence	$m_0 = G_0 l_P/c = m_P/2$
$k_B = 2S_0 \to S_0$	Unit entropy	$S_0 = S_P/2\ (T_0 = T_P)$

3. Constituent Interaction

Inventing a valid constant translation and unit redefinition can be done in numerous ways and is therefore not highly remarkable. The translation developed above however aims at getting as close as possible to the very nature of reality by considering the attributes that are allocated to individual micro-constituents of space-time as its basis. The next step in our search for a complex theory of quantum gravity would then be to connect the constituent properties defined in Table 1 by an interaction law that yields an effective dynamics in agreement with present-day physics theories. From a gravitational perspective, the emergent effective field dynamics must obey Einstein's field equations of general relativity [14]. Motivated by Hu [13], a relational micro-constituent interaction law that yields diffeomorphism invariant fielding behavior, yet formulated within the emergent relativistic space-time framework, can therefore be derived from a solution of these equations.

In the weak field approximation (neglecting the exact Schwarzschild solution to simplify the discussion), where the metric tensor is defined as a small perturbation ($\ll 1$) on the Minkowski metric due to a mass M, the line element ds at a distance R from M is given by [14]:

$$ds^2 \approx \left(1 - \frac{2GM}{c^2R}\right)c^2dt^2 - \left(1 + \frac{2GM}{c^2R}\right)dl^2 \tag{12}$$

with $dl^2 = dx^2 + dy^2 + dz^2$. As the micro-constituents move at the speed of light (see Table 1), the effective space-time constituent speed, denoted as c', is then given by $ds = 0$ or:

$$c' \equiv \frac{dl}{dt} \approx c\left(1 - \frac{2GM}{c^2R}\right). \tag{13}$$

In constituent units, this becomes:

$$c' \approx c\left(1 - \frac{l_P}{R}n_M\right) \equiv c(1 - \rho_r) \tag{14}$$

whereby $\rho_r \equiv n_M l_P/R = n_M/R_P$ is defined as the "radial constituent density" i.e., the amount of micro-constituents exchanged by M through the surface $4\pi R^2$ relative to the distance R from M in units l_P, which reflects gravity's spherical isotropy.

Equation (14) shows that the constituent speed as measured in a non-inertial coordinate system at distance R from M indeed decreases with declining R [21,22]. Stated differently, there exists an effective index of refraction $\eta \approx (1 - \rho_r)^{-1}$ with ρ_r representing an effective local constituent density (field). According to the same non-inertial coordinate system, the space-time constituents must therefore undergo an acceleration a_0 given by $dc'/dt \approx 2GM/R^2$ or $dc'/dt \approx c^2 l_P n_M/R^2$ in constituent units, wherefrom:

$$a_0 \approx \frac{4\pi c^2}{l_P}\frac{n_M}{N} \tag{15}$$

provided that $N = A/l_P^2 = 4\pi R^2/l_P^2 = 4\pi R_P^2$ here. This identity however has been derived by Padmanabhan for any diffeomorphism invariant theory [23,24]. By the very conception of mass in Equation (4), n_M refers to the number of space-time constituents intersecting a spherical surface with radius R, entailing that N must indeed equal the number of fundamental degrees of freedom on this same surface in constituent units. Most importantly, Equation (15) translates the presence of a remote massive object M into a local experience (and interaction) of gravitational presences at distance R from M, i.e., into a function of the amount of micro-constituents n_M relative to the number of degrees of freedom N at their availability (also see next section). There is no reference to any prior geometry, or in other words Equation (15) is a background-independent constituent interaction law.

Black hole thermodynamics follows straightforwardly [25]: A spherical surface with radius R_S enclosing a compound massive object M will have $c' \to 0$ when its radial constituent density $\rho_r = n_M l_P/R_S \to 1$ according to Equation (14). This means that the escape velocity from M equals c at $R_S = n_M l_P$, which exactly matches the Schwarzschild radius $R_S = 2GM/c^2$ in constituent units. The corresponding number of degrees of freedom of the spherical reference surface at R_S is hence given by $N_S = 4\pi R_S^2/l_P^2 = 4\pi n_M^2$, entailing that $\Delta S_{BH} = 2\pi k_B n_M \Delta n_M$ from Equation (7). Integration yields

$$S_{BH} = \pi k_B n_M^2 = \frac{k_B N_S}{4} \tag{16}$$

in agreement with Hawking's black hole entropy expression [26]. The Bekenstein–Hawking black hole radiation temperature T_{BH} can be determined most easily from Equation (9):

$$T_{BH} = \frac{n_M}{N_S} T_0 = \frac{T_0}{4\pi n_M} \tag{17}$$

which is identical to the result obtained by inserting the constant translations proposed in the previous section into the regular Bekenstein–Hawking expression [27,28]. This constituent-based origin for thermodynamic black hole characteristics is however considered to be more direct and intuitive than earlier accounts [16–18].

4. Entropic Gravity

Based predominantly on the works by Padmanabhan [9] and Verlinde [10], we attempt to relate the previous outcomes back to the interpretation of gravity as an entropic force, yet generalized to non-holographic reference surfaces. Adopting Verlinde's classical approach first, consider the force F induced by a mass $M = n_M m_0$ onto a mass $m = n_m m_0$ (and vice-versa) at distance R, which is according to Newton's law and in constituent units given by:

$$F = \frac{G_0 l_P^2 c}{2R^2} n_m n_M. \tag{18}$$

This force induces an acceleration a_m on m of the size F/m or:

$$a_m = \frac{2\pi c^2}{l_P} \frac{n_M}{N} \tag{19}$$

which differs from Equation (15) only by a factor of two, as one would expect for a calculation that omits relativity's temporal perturbation of the space-time metric [22]. Equation (19), however, immediately reproduces the Unruh temperature expression upon insertion of Equation (6) [29]. This straightforward connection in constituent units again supports the idea to regard gravity as a thermodynamic phenomenon or an emergent entropic force, as suggested before.

According to Verlinde, one can write the gravitational pull induced by M on m also as [10]:

$$F = \left(\frac{\Delta E}{\Delta R}\right)_m = \left(\frac{\Delta E}{\Delta S}\right)_m \left(\frac{\Delta S}{\Delta R}\right)_m \tag{20}$$

with immediately from Equation (6) for the reference surface temperature induced by m:

$$\left(\frac{\Delta E}{\Delta S}\right)_m = \frac{2G_0 l_P c}{k_B} \frac{n_m}{N}. \tag{21}$$

Also according to Verlinde, the last factor in Equation (20), being the entropy variation ΔS at the location of m that corresponds to a variation in the distance ΔR between the two masses, can be considered from the Bekenstein conjecture [27]: The effective distance shift that is needed to add one unit of entropy $\Delta S = k_B$ to the holographic reference surface at m equals the Compton wavelength $\hbar/mc = 2l_P/n_m$ wherefrom (with subscript B to denote the Bekenstein-based approach):

$$\left(\frac{\Delta S}{\Delta R}\right)_B = \frac{k_B n_m}{2l_P}. \tag{22}$$

However, inserting Equations (21) and (22) into Equation (20) only yields Equations (18) and (19) apart from an unexplained factor $2\pi n_M/n_m$ or $4\pi n_M/n_m$ with respect to the general relativistic Equation (15). Such dissimilarity, which must be due to the Bekenstein conjecture, has also been observed by Verlinde in regular units [10]. Verlinde nevertheless uses his version of Equation (22) to relate the classical gravitational acceleration with a mass-induced entropy gradient. The same result (still by a factor $2\pi n_M/n_m$) is immediately obtained here by inserting the latter identity into Equation (19):

$$a_{m,B} = \frac{4\pi c^2}{k_B N}\frac{n_M}{n_m}\left(\frac{\Delta S}{\Delta R}\right)_B. \tag{23}$$

For a general description that is not bound to a holographic scenario, Equation (8) instead of the Bekenstein conjecture should be used as a starting point for determining the distance-dependent entropy gradient that is induced by the mass M. In that case, with n_M being independent of R:

$$\left(\frac{\Delta S}{\Delta R}\right)_C = \frac{k_B}{2}\frac{8\pi R}{l_P^2}\ln(n_M) = \frac{2S}{R} \tag{24}$$

whereby the subscript C stresses the constituent-based approach, so that:

$$a_{m,C} = \pi c^2 \frac{n_M}{N}\frac{R_P}{S}\left(\frac{\Delta S}{\Delta R}\right)_C \tag{25}$$

One can immediately reproduce the results by Padmanabhan [9] and Verlinde [10] by insertion of the Schwarzschild solutions $R_S = n_M l_P$ and $S_{BH} = \pi k_B n_M^2$ into Equations (24) and (25) respectively, yielding (with subscript S for Schwarzschild):

$$\left(\frac{\Delta S}{\Delta R}\right)_S = \frac{2S_{BH}}{R_S} = \frac{2\pi k_B n_M}{l_P} \tag{26}$$

which indeed differs from Equation (22) by a factor $4\pi n_M/n_m$ as anticipated, and consequently for the entropy-induced acceleration:

$$a_{m,S} = \frac{c^2}{k_B N}\left(\frac{\Delta S}{\Delta R}\right)_S. \tag{27}$$

The entropic interpretation of gravitational pull can however be simplified by definition of an "informational constituent density" $\rho_i = n_M/N$, which is like a temperature according to Equation (9), as the amount of micro-constituents n_M that is exchanged by M relative to the number of degrees of freedom N at their availability on a spherical reference surface at distance R. Taking into account again that $N = 4\pi R_P^2$, the gradient of ρ_i as experienced by m is given by:

$$\frac{\Delta\rho_i}{\Delta R} = \frac{\Delta}{\Delta R}\left(\frac{n_M l_P^2}{4\pi R^2}\right) = -\frac{2\rho_i}{R}. \tag{28}$$

Note the similarity with the entropic gradient in Equation (24). As a result, the gravitational acceleration is very straightforwardly considered as being induced by an informational constituent density gradient also in Equation (19):

$$a_m = -\pi c^2 R_P \frac{\Delta\rho_i}{\Delta R}. \tag{26}$$

For the relativistic space-time constituents interacting through Equation (15), this means that:

$$a_0 \approx -2\pi c^2 R_P \frac{\Delta\rho_i}{\Delta R} = -c^2\frac{\Delta\rho_r}{\Delta R}. \tag{30}$$

corresponding elegantly with a gravitational potential $\varphi = c^2 n_M/R_P$.

The interpretation of entropic gravity by Padmanabhan [9] and Verlinde [10] in terms of a temperature-induced entropy change on a holographic screen due to a mass m (the Bekenstein conjecture), which causes an entropy gradient, which causes acceleration, is thus replaced here by an interpretation of gravitational pull in terms of micro-constituent density gradients: Each mass can be experienced by a remote mass, due to the experience of an effective (informational) constituent density gradient, which can be expressed as a temperature or entropy gradient, and which causes an acceleration. Although technical differences are small, the latter interpretation is believed to provide an improved conceptual understanding of emergent quantum gravity in terms of space-time's micro-constituents and the fundamental degrees of freedom at their availability. Further entropic gravity generalizations by Padmanabhan [9] and Verlinde [10] still hold true, while a covariant Lagrangian version has been provided by Hossenfelder [30]. Relating the micro-constituent-based interpretation of entropic gravity as presented here to promising studies in entropic cosmology [31–35] is subject of ongoing research.

5. Discussion

From the necessary conditions for the emergence of a complex dynamical system, it has been conjectured that reality is identical to a sub-quantum dynamics of indistinguishable yet ontological micro-constituents that are connected by a single interaction law. In order to arrive at a first toy-model identification of these micro-constituents, two strategies have been combined. First, it is obvious that masses, which can only consist of constituent collections, require a means to fully experience each other from a distance, i.e., some kind of information about the presence and extent of each mass must be remotely available. This kind of dimensional reduction of information has been achieved from a micro-constituent-based generalization of the holographic principle within a thermodynamic interpretation of gravity. The generalization allowed identifying Planck-scale constituent attributes from the universal constants of free space, like G and $\hbar$, that can be seen as unit conversion constants as a result. Second, as the effective field dynamics of the constituents must eventually obey Einstein's field equations, a sub-quantum interaction law, although formulated within the emergent relativistic space-time framework, has been derived from an approximate solution of these equations.

Generalizing the workings of the holographic principle to all reference surfaces, however, also called for a corresponding generalization of the Bekenstein conjecture, which assesses the entropy change at a black hole's surface upon mass aggregation. This conjecture has been used to connect the gravitational acceleration near a holographic surface to an entropy gradient by Padmanabhan [9] and Verlinde [10]. In this work, however, relating the experience of a distant mass to the entropy (gradient) has been achieved for non-holographic surfaces from the number of micro-constituents that are distributed over the surfaces' fundamental degrees of freedom. Taking a Schwarzschild surface as reference immediately reproduced the holographic entropic gravity results and provided a constituent-based origin for thermodynamic black hole characteristics. The interpretation of gravity in terms of an effective constituent density gradient is believed to provide a more straightforward understanding towards an emergent quantum gravity theory.

The general conclusion "that acceleration is related to an entropy gradient" [9] or a constituent density gradient also calls for a more general interpretation of the fundamental forces. If reality is indeed identical to a single type of space-time micro-constituents interacting through the proposed law (or similar), than this assumption entails that not only effective space-time and gravity, but also the other fundamental forces should emerge from the interaction of the micro-constituents. Unruh's argument that every acceleration induces a temperature was inverted by Padmanabhan [9] and Verlinde [10] to state that gravitational acceleration or inertia is induced by a temperature-induced entropy gradient, but can hence also be understood to be generally reversible, indicating that every fundamental acceleration (or force) is induced by an effective constituent density gradient.

In line with the common interpretation of Einstein's field equations, one could indeed imagine that a composite body (i.e., a space-time constituent collection) experiencing no net force whatsoever must be located within an isotropic space-time constituent density distribution, while every "force" that disturbs the isotropy, as a "space-time curvature" effect on the surrounding micro-constituent density distribution, is compensated for by a macroscopic acceleration, as effectively induced by a sub-quantum micro-constituent dynamics according to Equation (30), to a geodesic trajectory. This view corresponds with the idea that according to general relativity gravity is not a force in the classical sense as objects do not couple to the gravitational field; objects just exist and, if not differently constrained, follow geodesic trajectories [36].

Differences between the Standard Model matter and force particles must in this view emerge from different types of 'clustering' of the space-time micro-constituents, while no specific clustering configuration seems to be required for the emergence of space-time and gravity. Note that correspondingly every part of the universe can be attributed mass and energy, but not any other Standard Model attribute that requires a specific constituent configuration. The strength gap between the gravitational pull and the other fundamental forces that involve clustered space-time anisotropies is therefore anticipated. In agreement with experiment, this gap however should narrow when the number of background constituents increases up to a high-energy level where the constituent density discrepancy becomes vague or disappears.

The biggest open question towards unification of the fundamental forces within this line of research is then whether the interaction according to the law proposed in Equation (30) also allows for different types of micro-constituent clustering behavior that yield Standard Model physics, or whether other constituent attributes and interaction laws are required. Yet, for the accustomed probability wave dynamics within quantum mechanics, one could expect that each constituent cluster shows an internal micro-constituent dynamics that can be assessed by the use of wave characteristics, which are merely descriptive choices in function of an observer's Eigen-time. These descriptive choices could be quantized in terms of a wavelike Gibbs ensemble probability density function for the cluster's micro-constituents. Thereby taking into account the finite extent l_P of the constituents, one arrives at a canonical quantization that relates to quantum mechanics' probability density function. This function is denoted "densité de présence" in French, wherefrom the (gravitational) "presence" attribute specification in this work.

Funding: The Fetzer Franklin Fund is acknowledged for awarding a travel grant to present this work at the Emergent Quantum Mechanics 2017 conference (26–28 October, University of London) and for covering the costs to publish this work in open access. The John Templeton Foundation is acknowledged for awarding a visiting scholarship to the Space and Time after Quantum Gravity research group at the University of Geneva, allowing for extensive discussion of this work with Niels Linnemann, Karen Crowther, Baptiste Le Bihan, and Christian Wüthrich. Thanks for helpful thoughts also go to Antoine Acke and Erik Verlinde.

Conflicts of Interest: The author declares no conflict of interest. The founding sponsors had no role in the design of the study; in the collection, analyses, or interpretation of data; in the writing of the manuscript, and in the decision to publish the results.

References

1. Oriti, D. *Approaches to Quantum Gravity: Toward a New Understanding of Space, Time and Matter*, 1st ed.; Cambridge University Press: Cambridge, UK, 2009.
2. Jacobson, T.A. Thermodynamics of space-time: The Einstein equation of state. *Phys. Rev. Lett.* **1995**, *75*, 1260–1263. [CrossRef] [PubMed]
3. Barceló, C.; Visser, M.; Liberati, S. Einstein gravity as an emergent phenomenon? *Int. J. Mod. Phys. D* **2001**, *10*, 799–806. [CrossRef]
4. Crowther, K. *Effective Spacetime: Understanding Emergence in Effective Field Theory and Quantum Gravity*, 1st ed.; Springer International Publishing: Basel, Switzerland, 2016.
5. Oriti, D. Disappearance and emergence of space and time in quantum gravity. *Stud. Hist. Philos. Mod. Phys.* **2014**, *46*, 186–199. [CrossRef]
6. Ladyman, J.; Lambert, J.; Wiesner, K. What is a complex system? *Eur. J. Philos. Sci.* **2012**, *3*, 33–67. [CrossRef]
7. Berghofer, P. Ontic structural realism and quantum field theory: Are there intrinsic properties at the most fundamental level of reality? *Stud. Hist. Philos. Mod. Phys.* **2017**, *9*, 1–13. [CrossRef]
8. Barceló, C.; Liberati, S.; Visser, M. Analogue gravity. *Living Rev. Relativ.* **2011**, *14*, 1–159. [CrossRef] [PubMed]

9. Padmanabhan, T. A new perspective on gravity and dynamics of space-time. *Int. J. Mod. Phys. D* **2005**, *14*, 2263–2269. [CrossRef]
10. Verlinde, E. On the origin of gravity and the laws of Newton. *J. High Energy Phys.* **2010**, *29*, 1–27. [CrossRef]
11. Padmanabhan, T. Emergent gravity paradigm: Recent progress. *Mod. Phys. Lett. A* **2015**, *30*, 1–21. [CrossRef]
12. Hooft, G.T. Dimensional reduction in quantum gravity. *arXiv*, 1993.
13. Hu, B.L. Emergent/quantum gravity: Macro/micro structures of spacetime. *J. Phys. Conf. Ser.* **2009**, *174*, 1–16. [CrossRef]
14. Einstein, A. Die Feldgleichungen der Gravitation. *Sitzungsber. Königl. Preuss. Akad. Wiss. Berlin* **1915**, *48*, 844–847. (In German)
15. Huggett, N.; Vistarini, T. Deriving general relativity from string theory. *Proc. Philos. Sci. Assoc.* **2014**, *14*, 1–12. [CrossRef]
16. Strominger, A.; Vafa, C. Microscopic origin of the Bekenstein–Hawking entropy. *Phys. Lett. B* **1996**, *379*, 99–104. [CrossRef]
17. Rovelli, C. Black hole entropy from loop quantum gravity. *Phys. Rev. Lett.* **1996**, *77*, 3288–3291. [CrossRef] [PubMed]
18. Ashtekar, A.; Baez, J.; Corichi, A.; Krasnov, K. Quantum geometry and black hole entropy. *Phys. Rev. Lett.* **1998**, *80*, 904. [CrossRef]
19. Planck, M. Über irreversible Strahlungsvorgänge. *Sitzungsber. Königl. Preuss. Akad. Wiss. Berlin* **1900**, *306*, 69–122. (In German) [CrossRef]
20. Heisenberg, W. Über den anschaulichen Inhalt der quantentheoretischen Kinematik und Mechanik. *Z. Phys.* **1927**, *43*, 172–198. [CrossRef]
21. Einstein, A. Einfluss der Schwerkraft auf die Ausbreitung des Lichtes. *Annal. Phys.* **1911**, *35*, 898–908. [CrossRef]
22. Einstein, A. *Relativity: The Special and General Theory*, 1st ed.; Henry Holt & Co.: New York, NY, USA, 1920.
23. Padmanabhan, T. Equipartition of energy in the horizon degrees of freedom and the emergence of gravity. *Mod. Phys. Lett. A* **2010**, *25*, 1129–1136. [CrossRef]
24. Padmanabhan, T. Surface density of spacetime degrees of freedom from equipartition law in theories of gravity. *Phys. Rev. D* **2010**, *81*, 1–12. [CrossRef]
25. Bardeen, J.M.; Carter, B.; Hawking, S.W. The four laws of black hole mechanics. *Commun. Math. Phys.* **1973**, *31*, 161–170. [CrossRef]
26. Hawking, S.W. Particle creation by black holes. *Commun. Math. Phys.* **1975**, *43*, 199–220. [CrossRef]
27. Bekenstein, J.D. Black holes and entropy. *Phys. Rev. D* **1973**, *7*, 2333–2346. [CrossRef]
28. Hawking, S.W. Black hole explosions? *Nature* **1974**, *248*, 30–31. [CrossRef]
29. Unruh, W.G. Notes on black-hole evaporation. *Phys. Rev. D* **1976**, *14*, 870–892. [CrossRef]
30. Hossenfelder, S. A Covariant version of Verlinde's emergent gravity. *Phys. Rev. D* **2017**, *95*, 1–8. [CrossRef]
31. Cai, Y.F.; Liu, J.; Li, H. Entropic cosmology: A unified model of inflation and late-time acceleration. *Phys. Lett. B* **2010**, *690*, 213–219. [CrossRef]
32. Easson, D.A.; Frampton, P.H.; Smoot, G.F. Entropic accelerating universe. *Phys. Lett. B* **2011**, *696*, 273–277. [CrossRef]
33. Koivisto, T.S.; Mota, D.F.; Zumalacárregui, M. Constraining entropic cosmology. *JCAP* **2011**, *2*, 1–27. [CrossRef]
34. Basilakos, S.; Solà, J. Entropic-force dark energy reconsidered. *Phys. Rev. D* **2014**, *90*, 1–11. [CrossRef]
35. Verlinde, E. Emergent gravity and the dark universe. *SciPost Phys.* **2017**, *2*, 1–41. [CrossRef]
36. Maudlin, T. On the unification of physics. *J. Philos.* **1996**, *93*, 129–144. [CrossRef]

Article

Generalized Lagrangian Path Approach to Manifestly-Covariant Quantum Gravity Theory

Massimo Tessarotto [1,2] and Claudio Cremaschini [3,*]

1 Department of Mathematics and Geosciences, University of Trieste, Via Valerio 12, 34127 Trieste, Italy; maxtextss@gmail.com
2 Institute of Physics, Faculty of Philosophy and Science, Silesian University in Opava, Bezručovo nám.13, CZ-74601 Opava, Czech Republic
3 Institute of Physics and Research Center for Theoretical Physics and Astrophysics, Faculty of Philosophy and Science, Silesian University in Opava, Bezručovo nám.13, CZ-74601 Opava, Czech Republic
* Correspondence: claudiocremaschini@gmail.com

Received: 10 January 2018; Accepted: 8 March 2018; Published: 19 March 2018

Abstract: A trajectory-based representation for the quantum theory of the gravitational field is formulated. This is achieved in terms of a covariant Generalized Lagrangian-Path (GLP) approach which relies on a suitable statistical representation of Bohmian Lagrangian trajectories, referred to here as *GLP-representation*. The result is established in the framework of the manifestly-covariant quantum gravity theory (CQG-theory) proposed recently and the related CQG-wave equation advancing in proper-time the quantum state associated with massive gravitons. Generally non-stationary analytical solutions for the CQG-wave equation with non-vanishing cosmological constant are determined in such a framework, which exhibit Gaussian-like probability densities that are non-dispersive in proper-time. As a remarkable outcome of the theory achieved by implementing these analytical solutions, the existence of an emergent gravity phenomenon is proven to hold. Accordingly, it is shown that a mean-field background space-time metric tensor can be expressed in terms of a suitable statistical average of stochastic fluctuations of the quantum gravitational field whose quantum-wave dynamics is described by GLP trajectories.

Keywords: quantum mechanics; generalized Lagrangian paths; covariant quantum gravity; emergent space-time; Gaussian-like solutions

PACS: 03.65.Ca; 03.65.Ta

---

1. Introduction

The search for a theory of quantum gravity that is consistent both with the principles of quantum mechanics [1] as well as with the postulates of the classical Einstein theory of General Relativity (GR) [2–4] has represented so far one of the most challenging and hard-to-solve conceptual problems of mathematical and theoretical physics alike. The crucial issue is about the possibility of achieving in the context of either classical or quantum relativistic theories, and in particular for a quantum theory of gravity, a truly coordinate- (i.e., frame-) independent representation, namely which satisfies, besides the general covariance principle, also the so-called principle of manifest covariance. In fact, although the choice of special coordinate systems is always legitimate for all physical systems either discrete or continuous, including in particular classical and quantum gravity, the intrinsic objective nature of physical laws makes them frame-independent.

However, for these principles to actually apply, a background space-time picture must hold. This means, more precisely, that a suitable classical curved space-time $\{\mathbf{Q}^4, \widehat{g}\}$ must exist with respect to which both general covariance principle and principle of manifest covariance can be prescribed. As a consequence, when parameterized with respect to a coordinate system $r \equiv \{r^\mu\}$ the same space-time must be endowed with a well-defined (i.e., uniquely prescribed and hence deterministic) symmetric metric tensor $\widehat{g}$, represented equivalently in terms of its covariant $\widehat{g} \equiv \{\widehat{g}_{\mu\nu}\}$ and countervariant $\widehat{g} \equiv \{\widehat{g}^{\mu\nu}\}$ forms , which is referred to in the following as the "*background*" field tensor. In particular, $\mathbf{Q}^4$ can be identified with a time-oriented four dimensional Riemann space-time. Thus, although the precise choice of the same background space-time itself remains in principle arbitrary, as a consequence of the principle of manifest covariance it should always be possible to represent all quantum observables (of the theory), including the corresponding quantum Hamiltonian operator and quantum canonical variables/operators (see below), in 4-tensor form. This requires to cast them exclusively as 4-tensor fields with respect to the group of local point transformations (LPT group)

$$r \equiv \{r^\mu\} \to r' \equiv \{r'^\mu\} = r'(r) \tag{1}$$

mapping $\{\mathbf{Q}^4, \widehat{g}\}$ in itself [5].

In such a framework $\widehat{g}$ is considered as a classical (i.e., deterministic) tensor field, to be identified as the metric tensor field of $\{\mathbf{Q}^4, \widehat{g}\}$ which—as such—determines the geometric properties of the same space-time. This means more precisely that:

Prescription a: Its covariant and countervariant components, i.e., respectively, $\widehat{g}_{\mu\nu}$ and $\widehat{g}^{\mu\nu}$, must lower and raise tensor indices of arbitrary tensor fields and also prescribe the standard connections (Christoffel symbols) appearing in the covariant derivatives.

Prescription b: It determines the Ricci tensor, the Ricci 4-scalar and the coupling contained in the stress–energy tensor due to external sources, in the sequel, respectively, identified with the symbols $\widehat{R}_{\mu\nu} \equiv R_{\mu\nu}(\widehat{g})$, $\widehat{R} \equiv R(\widehat{g}) \equiv \widehat{g}^{\alpha\beta}\widehat{R}_{\alpha\beta}$ and $\widehat{T}_{\mu\nu} = T_{\mu\nu}(\widehat{g})$.

Prescription c: Consequently, $\widehat{g}$ can be identified with a particular solution of the Einstein field equations

$$\widehat{R}_{\mu\nu} - \frac{1}{2}\left[\widehat{R} - 2\Lambda\right]\widehat{g}_{\mu\nu} = \frac{8\pi G}{c^4}\widehat{T}_{\mu\nu}, \tag{2}$$

where as usual Λ denotes the cosmological constant.

Prescription d: $\widehat{g}$ determines uniquely *the Riemann distance s*, or *proper-time*, on the space-time $\{\mathbf{Q}^4, \widehat{g}\}$ by means of the 4-scalar equation

$$ds^2 = \widehat{g}_{\mu\nu}(r,s)dr^\mu dr^\nu. \tag{3}$$

One notices that, in accordance with [5], here $dr^\mu \equiv dr^\mu(s)$ and ds identify, respectively, the 4-tensor displacement and its corresponding 4-scalar line-element (arc length), both evaluated along a suitable worldline. For this purpose, the latter is identified with an arbitrary geodetics $r(s) \equiv \{r^\mu(s)\}$ belonging to $\{\mathbf{Q}^4, \widehat{g}\}$ that crosses an arbitrary 4-position $r^\mu \equiv r_o^\mu$, and hence fulfills the initial condition $r^\mu(s_o) = r_o^\mu$, at some proper time s_o (which for definiteness can always be set $s_o = 0$). Consequently, the definition of proper time remains unambiguous and unique also for arbitrary finite values of $s \in I$ (with $I \equiv \mathbb{R}$ the real axis), being identified with the arc length along the (unique) geodetics $r(s) \equiv \{r^\mu(s)\}$ joining $r^\mu(s_o) = r_o^\mu$ with an arbitrary 4-position r_1^μ, i.e., such that $r^\mu(s_1) = r_1^\mu$ for a given s_1 is assumed to exist. For example, the proper time can always be defined along an appropriate observer geodetics.

Prescription e: One notices that in principle the background metric tensor might be taken of the form $\widehat{g}(r,s) \equiv \{\widehat{g}_{\mu\nu}(r,s)\}$, i.e., allowed to depend explicitly also on the proper time s. In the following, however, we shall restrict the treatment to the customary case in which the background metric tensor solution of the Einstein field equations is purely dependent only on the 4-position r^μ, namely is of the form

$$\widehat{g} = \widehat{g}(r), \tag{4}$$

which identifies a *stationary metric tensor*.

Next, let us consider the prescription holding for the Lagrangian continuum coordinates $g \equiv \{g^{\mu\nu}\}$ and the conjugate momentum operator $\pi \equiv \{\pi^{\mu\nu}\}$, again both to be considered as 4-tensor fields with respect to the group of local point transformations in Equation (1):

Prescription f: As a consequence of the stationarity assumption in Equation (4), for all sets $(r,s) \in \{\mathbf{Q}^4, \widehat{g}\} \times I$ tensor decompositions of the form

$$\begin{cases} g(r,s) = \widehat{g}(r) + \delta g(r,s), \\ \pi(r,s) = \delta\pi(r,s), \end{cases} \tag{5}$$

will be assumed to hold for the quantum gravity theory, with $\delta g(r,s) \equiv \{\delta g_{\mu\nu}(r,s)\}$ and $\delta\pi(r,s) \equiv \{\delta\pi_{\mu\nu}(r,s)\}$ denoting the corresponding *quantum fluctuations*, represented by a coordinate displacement field and momentum operator which by assumption may depend explicitly on the variables (r,s).

A promising new scenario for quantum gravity fulfilling these requirements has recently been established in [6–11]. This is realized by the theory of manifestly-covariant quantum gravity, denoted as CQG-theory, which is based on the manifestly-covariant canonical quantization (*g-quantization*) of the classical Hamiltonian state $\{g(r,s), \pi(r,s)\}$. It must be clarified that in the present treatment the concept of manifest covariance means that CQG-theory is realized by a formulation in which all classical and quantum Hamiltonian field variables or operators, including continuum coordinates, conjugate momenta and Hamiltonian densities transform as 4-tensors, i.e., fulfill covariance tensor transformation laws with respect to the group of local point transformations in Equation (1). Although a manifestly-covariant theory of this type need not necessarily be unique, the involved notion of manifest covariance given here is certainly unambiguously determined when the background space-time $\{\mathbf{Q}^4, \widehat{g}\}$ is prescribed. On the other hand, an alternative route is also available. This is based on the preliminary introduction of a non-canonical mapping in which the classical (and hence also the quantum) Hamiltonian state $\{g(r,s), \pi(r,s)\}$ is mapped by means of a diffeomorphism onto a suitable set of non-canonical variables

$$\{g(r,s), \pi(r,s)\} \Leftrightarrow \{\eta(r,s), \chi(r,s)\}, \tag{6}$$

in which, however, $\eta(r,s) \equiv \{\eta_{\alpha\beta}(r,s)\}$ and $\chi(r,s) \equiv \{\chi_{\alpha\beta}(r,s)\}$ are not represented by 4-tensor variables. When expressed in terms of the transformed variables $\{\eta(r,s), \chi(r,s)\}$ CQG-theory does not lose obviously the property of covariance (its equations remain covariant with respect to the LPT-group) although its variables (i.e., $\{\eta(r,s), \chi(r,s)\}$) are not represented by 4-tensors. Such a notion will be referred to as property of plain covariance of the theory. The distinction between the two notions of covariance (manifest or plain) is, however, important. In fact manifest covariance represents a stronger condition for the realization of a quantum theory of gravitational field with respect to literature approaches which, instead, may or may not rely on weaker notions of covariance such as that of plain covariance (see also subsequent discussion in Section 2).

As such, CQG-theory is endowed with a number of key features, since: A) it preserves the background metric tensor $\widehat{g}(r)$ which is identified with a classical field tensor; B) it satisfies the *quantum unitarity principle,* i.e., the quantum probability is conserved; C) it is *constraint-free*, in the sense that the quantum Lagrangian variables $g \equiv g(r,s)$ are identified with independent tensor fields; D) it is *non-perturbative* so that the quantum fluctuations $\delta g(r,s)$ and $\delta\pi(r,s)$ need not be regarded as asymptotically "small" in some appropriate sense with respect to the background metric tensor $\widehat{g}(r)$. Its foundations (for a detailed discussion see [9]) lie on the preliminary establishment of a variational formulation of GR achieved in the context of a covariant DeDonder–Weyl-type approach to continuum field-Hamiltonian dynamics [12–19] in which the background space-time $\{\mathbf{Q}^4, \widehat{g}\}$ is considered prescribed [7,8].

In the following, we intend to shed further light on key aspects of the CQG-theory which are intimately related with its consistent realization. These include in particular two crucial "*tests of consistency*" for CQG-theory which should actually be regarded as mandatory physical prerequisites for any quantum theory of gravity fulfilling both the principles of general and manifest covariance.

The first one is that, although quantum corrections may in principle occur [7,8], it must be possible to preserve the functional form of the Einstein field equations consistent with the so-called *emergent gravity picture*. More precisely, the latter equations should follow uniquely from quantum theory itself *without performing the semiclassical continuum limit* (namely obtained letting in particular $\hbar \rightarrow 0$; see for example [20] where the derivation of the Einstein field equation was discussed in the context of loop quantum gravity). This property will be referred to here as "*first-type emergent-gravity paradigm*".

The second test of consistency, to be investigated here, refers instead to the validity of an emergent-gravity picture also for the deterministic background metric tensor $\widehat{g}(r)$, in the sense that the same $\widehat{g}(r)$ should be prescribed by means of a suitably-defined quantum/stochastic expectation value of the quantum state. This property will be denoted here as "*second-type emergent-gravity paradigm*". A basic requirement needed for its verification is manifestly the determination of a suitable class of particular solutions of the quantum wave-function, i.e., the CQG-wave equation for the quantum state $\psi(g,r,s)$ earlier pointed out in [10].

With this hindsight in mind, in the following Eulerian and Lagrangian representations are preliminarily distinguished for the CQG-wave equation and its corresponding set of quantum hydrodynamic equations (QHE). The latter are implied by the Madelung representation [21] of the quantum wave function written in Eulerian form $\psi \equiv \psi(g,r,s)$, namely distinguishing the dependences in terms of the Lagrangian coordinates $g \equiv \{g_{\mu\nu}\}$ and the parameters (r,s) as

$$\psi(g,r,s) = \sqrt{\rho(g,r,s)} \exp\left\{\frac{i}{\hbar} S^{(q)}(g,r,s)\right\}. \tag{7}$$

Here, the real fields $\left\{\rho, S^{(q)}\right\} \equiv \left\{\rho(g,r,s) = |\psi(g,r,s)|^2, S^{(q)}(g,r,s)\right\}$ identify the quantum fluid 4-scalar fields written in Eulerian form, namely the quantum probability density function (PDF) and the quantum phase-function. In particular, the intent of the investigation concerns the introduction of a trajectory-based or Lagrangian representation of CQG-theory (see subsequent Sections 4 and 5), to be distinguished from the Eulerian one (see Section 3) and referred to here as *Generalized Lagrangian-path approach* to CQG-theory. This goal is obtained by means of an appropriate parameterization of the corresponding set of quantum hydrodynamic equations, following in turn from the CQG wave-equation and based on the Madelung representation recalled above (see Equation (7)). More precisely, this concerns the investigation of:

- *Goal #1*: Explicit solutions of the CQG-quantum hydrodynamic equations satisfying suitable physical requirements.
- *Goal #2*: The "emergent" character of the classical background space-time metric tensor $\widehat{g}(r)$, to be determined in terms of quantum theory. Accordingly, the background metric tensor $\widehat{g}(r)$ should be identified with a suitably-defined quantum expectation value of the quantum state, i.e., weighted in terms of the corresponding quantum probability density (PDF).
- *Goal #3*: The existence of either stationary or, more generally, non-stationary solutions with respect to the proper-time s, i.e., explicitly dependent on s, for the quantum state ψ expressed via the Madelung representation (see Equation (7)).
- *Goal #4*: The search of Gaussian-like or Gaussian realizations for the quantum PDF ρ.
- *Goal #5*: The search of separable solutions of the quantum Hamilton-Jacobi (H-J) equation in terms of the quantum phase-function $S^{(q)}$ and the investigation of their qualitative properties and in particular their asymptotic behavior for $s \rightarrow +\infty$.

For the tasks indicated above, in close similarity with non-relativistic quantum mechanics (see [22,23]), two choices are in principle available. The first one is based on the introduction of deterministic Lagrangian trajectories $\{g(s), s \in I\}$, or Lagrangian-Paths (LP), analogous to those adopted in the context of the Bohmian representation of non-relativistic quantum mechanics [24–31].

This provides a Bohmian interpretation (of CQG-theory) which is ontologically equivalent to CQG-theory itself [32].

Hence, the tensor field $\delta g(s) \equiv \delta g_L(s)$ is uniquely determined by means of a map of the type

$$s \rightarrow \delta g_L(s) \equiv \delta g_L(r(s), s), \tag{8}$$

with $r = r(s)$ denoting the parameterization in terms of geodetic curves associated with the classical background field tensor $\widehat{g}(r) \equiv \widehat{g}(r(s))$ (see *Prescription d* above, [9] and related discussion in Section 4), so that, in terms of $g_L(s) \equiv g(s)$, it follows that $\{g(s), s \in I\} \equiv \{g_L(s) = \widehat{g}(r(s)) + \delta g_L(s), s \in I\}$. The second choice, instead, and the one which is at the basis of the GLP trajectory-based approach (or *GLP-representation*) adopted here, is achieved in terms of suitable stochastic, i.e., intrinsically non-unique, Lagrangian trajectories which are referred to here as *Generalized Lagrangian Paths* (GLP). Such a notion, which is inspired and extends to CQG-theory the analogous approach earlier developed for non-relativistic quantum mechanics [22], is based on a suitable generalization of the concept of LP (see Section 5 below). In such a context, each deterministic LP $\{g(s), s \in I\}$ is replaced with a continuous statistical ensemble of *stochastic GLP trajectories* $\{G(s), s \in I\}$. More precisely, introducing in analogy with Equation (5) the tensor decomposition

$$G(s) = \widehat{g}(r(s)) + \delta G(s), \tag{9}$$

with $\delta G(s) \equiv \{\delta G_{\mu\nu}(r(s), s)\}$ being a suitable tensor field denoted as *GLP-displacement* to be later defined, each GLP trajectory

$$\{G(s), s \in I\} \equiv \{\widehat{g}(r(s)) + \delta G(s), s \in I\} \tag{10}$$

is parameterized in terms of the displacement field, to be considered as a stochastic field tensor,

$$\Delta g = \delta g(s) - \delta G(s), \tag{11}$$

with $\Delta g \equiv \{\Delta g_{\mu\nu}\}$ denoting a suitable constant second-order tensor field referred to here as *stochastic displacement field tensor*. For definiteness, it is required that its covariant components at proper-times s and s_o, $\Delta g_{\mu\nu}(s) = \delta g_{\mu\nu}(s) - \delta G_{\mu\nu}(s)$ and $\Delta g_{\mu\nu}(s_o) = \delta g_{\mu\nu}(s_o) - \delta G_{\mu\nu}(s_o)$, are prescribed so that for all $s, s_o \in I$

$$\Delta g_{\mu\nu}(s) = \Delta g_{\mu\nu}(s_o). \tag{12}$$

Then, this implies that its counter-variant components $\Delta g^{\mu\nu}(s)$ and $\Delta g^{\mu\nu}(s_o)$ can be equivalently determined in terms of the prescribed field tensors $\widehat{g}_{\mu\nu}(r) \equiv \widehat{g}(r(s))$ or $\widehat{g}_{\mu\nu}(r_o) \equiv \widehat{g}(r(s_o))$, so that one also has necessarily for all $s, s_o \in I$ that:

$$\Delta g^{\mu\nu}(s) = \Delta g^{\mu\nu}(s_o). \tag{13}$$

Consequently, each GLP trajectory is actually represented by a configuration-space curve of the type $\{G(s), s \in I\} \equiv \{\widehat{g}(r(s)) + \delta g(s) - \Delta g, s \in I\}$, so that upon varying the stochastic displacement field tensor Δg it actually defines a statistical ensemble of trajectories. In terms of them, i.e., by parameterizing the CQG wave-function $\psi(g, r, s)$ (or equivalently the corresponding quantum fluid fields) in terms of the GLP-displacement $\delta G(s) = \delta g(s) - \Delta g$, the *GLP-representation* of CQG-theory is then achieved. This amounts to introduce the composed mapping $\psi(g, r, s) \rightarrow \psi(G(s), \Delta g, \widehat{g}, r, s)$, where $\psi(G(s), \Delta g, \widehat{g}, r, s)$ denotes the *GLP-parameterized quantum wave-function* in which the dependence in terms of the displacement tensor field Δg is explicitly allowed.

As shown in Section 5, the adoption of the GLP parameterization for CQG-theory actually leaves unchanged the underlying axioms established in [10], thus providing a Lagrangian representation of CQG-theory which is ontologically equivalent to CQG-theory itself. The remarkable new aspects of the GLP formalism, however, are that it will be shown: First, to determine a solution method for the CGQ-wave equation, to be referred to here as *GLP-approach*, permitting the explicit construction of physically-relevant particular realizations of the CGQ-quantum state $\psi(s)$. Second, to realize quantum solutions which are consistent with the emergent-gravity picture. In particular, for this purpose, the background field tensor will be shown to be determined equivalently either in terms of quantum expectation values or via a suitably-prescribed stochastic average of the quantum field tensor $g_{\mu\nu}$. This includes the determination of particular solutions of the CQG-wave equation which, consistent with Goal #1, satisfy the following physical requirements:

- *Requirement #1:* the quantum wave-function $\psi(s)$ is *dynamically consistent*, namely for which the PDF $\rho(g,r,s) \equiv |\psi(g,r,s)|^2$ associated with the quantum wave-function $\psi(g,r,s)$ is globally prescribed and summable in the quantum configuration space U_g in such a way that the corresponding probability $|\psi|^2 d(g)$ is similarly globally conserved for arbitrary subsets of the quantum configuration space U_g. As discussed below a prerequisite for meeting such a requirement is the validity of suitable Heisenberg inequalities earlier determined in [11].
- *Requirement #2:* $\psi(s)$ exhibits the *explicit dependence in terms of a stochastic observable,* so to yield a so-called Stochastic-Variable Approach to quantum theory [22,33–35]. In the context of CQG-theory this should be generally identified with a 4-tensor field depending on the physical quantum observable $g_{\mu\nu}(r,s)$ and realizing a stochastic variable endowed with a stochastic probability density, i.e., dependent on a suitable stochastic field. Such a stochastic field will be identified in the following with the second-order real and observable stochastic displacement field tensor $\Delta g = \{\Delta g_{\mu\nu}\}$ defined by Equation (11) which by assumption depends functionally on $g_{\mu\nu}$ (and hence $\delta g_{\mu\nu}(r,s)$ too).
- *Requirement #3:* the PDF ρ is endowed with a *Gaussian-like* behavior and is *non-dispersive* in character, namely in the sense of assuming that in the subset of the proper-time axis I in which ψ is defined, its probability density $|\psi|^2$ can be identified for all $s \in I \equiv \mathbb{R}$ with a Gaussian-like PDF depending on Δg and $\widehat{g}$, and thus by itself realizes a stochastic function. These particular solutions of the CQG-wave equation are generally non-stationary and are required to preserve their Gaussian-like character, and therefore to be non-dispersive, i.e., free of any spreading behavior during the proper-time quantum dynamical evolution.
- *Requirement #4:* the quantum wave function holds for arbitrary realizations of the deterministic background metric tensor $\widehat{g}(r)$ and in particular in the case of vacuum solutions of the Einstein field equations.

Requirements #1–#4 are physically motivated. More precisely, the first one is needed to warrant the validity of the quantum unitarity principle, i.e., the conservation of quantum probability. The second requirement, instead, is instrumental for the present theory. In fact, as clarified below, the existence of the stochastic tensor observable $\Delta g(g)$ is mandatory for the development of a GLP-approach in the context of CQG-theory. The third requirement is related to the issue about the physical origin of the cosmological constant [36]. The existence of Gaussian-like solutions for the quantum PDF $\rho(s)$ is mandatory in order to establish the connection between the CQG-theory and the Einstein field equations and to identify its precise quantum origin in terms of the Bohm vacuum interaction [32,37,38]. Finally, Requirement #4 is intimately related to the principle of manifest covariance and the deterministic character of the background metric tensor $\widehat{g}(r)$.

As a further remark, one notices that Requirements #2 and #3 are qualitatively similar to those set at the basis of the GLP-approach developed for non-relativistic quantum mechanics. These led to the identification and proof of existence of non-dispersive Gaussian-like, or even properly Gaussian, particular solutions of the Schroedinger equation originally conjectured by Schrödinger himself in 1926 [39]. It is therefore natural to conjecture that analogous properties should hold in the context of the CQG-theory. As a remarkable conceptual outcome of the GLP theory, it is then shown that the discovery of analytical solutions satisfying physical Requirements #1–#4 allows for the investigation of theoretical aspects of the quantization of the gravitational field which go beyond the framework of so-called first-quantization, toward inclusion of second-quantization effects. This refers to quantum interactions of the gravitational field with itself which are intrinsically proper-time dependent contributions generated by the quantum wave dynamics retained in the solution of the same background metric tensor. In particular, in this work the existence of an emergent gravity phenomenon is displayed, which establishes a precise relationship between the background metric tensor $\widehat{g}_{\mu\nu}$ and the quantum field $g_{\mu\nu}$. In detail, it is shown that $\widehat{g}_{\mu\nu}$ can be represented as a mean-field background space-time metric tensor provided by a statistical moment of the Gaussian (or more generally Gaussian-like) PDF ρ. Hence, from the physical point of view $\widehat{g}_{\mu\nu}$ can be effectively interpreted as arising from a statistical average of stochastic fluctuations of the quantum gravitational field $g_{\mu\nu}$ whose quantum-wave dynamics is described by GLP trajectories.

In detail, the structure of the paper is as follows. First, a qualitative comparison between CQG-theory and literature approaches to quantum gravity and its Bohmian formulation is proposed in Section 2. The Eulerian representation of CQG-theory is then presented in Section 3. Subsequently, the Lagrangian-path and Generalized Lagrangian-path representations are pointed out in Sections 4 and 5, together with their Bohmian, i.e., deterministic, and correspondingly stochastic interpretations. Next, consistent with the axioms of CQG-theory, in Section 6, the establishment of the stochastic probability density attached with the stochastic displacement field tensor Δg is achieved. This is shown to be necessarily identified with the initial quantum PDF. In connection with such a prescription, in the same section the problem is posed of the construction of generalized Gaussian particular solutions for the quantum PDF $\rho(g,r,s)$. Subsequently, in Section 7 the search of separable solutions of the corresponding quantum H-J equation is investigated. As a result, asymptotic conditions are investigated warranting the quantum phase function to be expressed in terms of polynomials of Δg. Finally, in Section 8, the main conclusions of the paper are drawn, while Appendices A and B contain mathematical details of the calculations.

2. Quantum Gravity Theories and Bohmian Formulation in Literature

This section is intended to provide a summary of the relevant conceptual features of CQG theory, together with an exhaustive discussion of literature works dealing with quantum gravity theories and corresponding Bohmian formulations. The aim of such a comparison with previous literature is twofold. From one side, we intend pointing out the main differences and significant progresses of CQG-theory from alternative approaches to quantum gravity. From the other side, we are interested in stating which are the common aspects of the present approach with other quantum theories of the gravitational field, and in which sense CQG-theory and the literature formulations discussed here can be reconciled or regarded as complementary. A review of the mathematical foundations of CQG theory and its Hamiltonian structure is treated separately in Section 3.

We start by noting that, according to [40] quantization methods, in both quantum mechanics and quantum gravity, can be classified in two classes, denoted, respectively, as the canonical and the covariant approaches. These differ in the way in which both the quantum state and the space-time are treated. In fact, the canonical quantization approach is based, first on the preliminary introduction of (3 + 1) or (2 + 2)-decompositions (or foliations [41–43]) for the representation of the space-time and, second, on the adoption of a quantum state represented in terms of non-4-tensor continuum fields. As such, by construction these theories are not covariant with respect to the LPT-group (1). Nevertheless, they still may retain well-definite covariance properties with respect to appropriate subgroups of local point transformations. For example, in the case of the (3 + 1)-decomposition covariance is warranted with respect to arbitrary point transformations which preserve the same foliation. In the covariant approaches, instead, typically all physical quantities including the quantum state are represented exclusively by means of 4-tensor fields. so that the property of manifest covariance remains fulfilled. Consequently, for these approaches, covariant quantization involves the assumption of some sort of classical background space-time structure on which a quantum gravity theory is constructed, for example identified with the flat Minkowski space-time. To realize such a strategy, however, it turns out that the quantum state is typically represented in terms of superabundant variables. Thus, in such cases, covariant quantization may also require the treatment of suitable constraint conditions.

Let us briefly analyze both approaches in more detail, considering first the canonical approach. A choice of this type is exemplified by the one adopted by Dirac and based on the Dirac constrained dynamics [44–48]. Dirac Hamiltonian approach to quantum gravity is not manifestly covariant, in reference both to transformation properties with respect to local as well as non-local point transformations (see discussion in [5]). In this picture in fact the field variable is identified with the metric tensor $g_{\mu\nu}$, but the corresponding "generalized velocity" is defined as $g_{\mu\nu,0}$, namely with respect to the "time" component of the 4-position. This choice necessarily violates the principle of manifest covariance [7,8]. Consequently, in Dirac's canonical theory, the canonical momentum remains identified with the manifestly non-tensorial quantity $\pi^{\mu\nu}_{Dirac} = \frac{\partial L_{EH}}{\partial g_{\mu\nu,0}}$, where L_{EH} is the Einstein-Hilbert variational Lagrangian density.

The same kind of ingredients is at the basis of the approach developed by Arnowitt, Deser and Misner (ADM theory, 1959–1962 [49]). In addition, in the ADM case, manifest covariance is lost because of the adoption of Lagrangian and Hamiltonian variables which are not 4-tensors. In fact, ADM theory is based on the introduction of a (3 + 1)-decomposition of space-time which, by construction, is foliation dependent, in the sense that it relies on a peculiar choice of a family of GR frames for which "time" and "space" transform separately, so that space-time is effectively split into the direct product of a one-dimensional time and a three-dimensional space subsets, respectively [50]. A quantum gravity theory constructed upon the ADM Hamiltonian formulation of gravitational field leads to postulating a quantum wave equation of Wheeler-DeWitt type [51]. The latter one is expressed as an evolution Schrödinger-like equation advancing the dynamics of the wave function with respect to the coordinate-time t of the ADM foliation, which is not an invariant parameter. In addition, in the absence of background space-time, the same equation carries a conceptual problem related in principle to the definition of the same coordinate time, which is simultaneously the dynamical parameter and a component of space-time which must be quantized by solving the wave equation. This marks a point of difference with respect to CQG theory and CQG-wave equation (see Equation (16) below), which represents a dynamical evolution equation with respect to an invariant (i.e., 4-scalar) proper-time s defined on the prescribed background space-time, without introduction of any kind of space-time foliation.

Another important approach is the one exemplified by the choice of so-called Ashtekar variables, originally identified respectively with a suitable self-dual spinorial connection (the generalized coordinates) and their conjugate momenta (see [52,53]). Ashtekar variables provide an alternative canonical representation of General Relativity, and this choice is at the basis of the so-called "loop representation of quantum general relativity" [54] usually referred to as "loop quantum gravity" (LQG) and first introduced by Rovelli and Smolin during 1988–1990 [55,56] (see also [57]). Nevertheless, the Ashtekar variables can also be shown to be by construction intrinsically manifestly non-tensorial in character. The basic consequence is that also the canonical representation of Einstein field equations based on these variables, as well as ultimately also LQG itself, violates the principle of manifest covariance. In contrast, in the framework of CQG-theory the choice of Hamiltonian state and quantum variables satisfies manifest covariance, whereby the dynamical variables are expressed by means of 4-tensor quantities.

However, despite these considerations, it must be stressed that both the canonical approach and CQG-theory can be regarded also complementary from a certain point of view, this because they exhibit distinctive physical properties associated with two canonical Hamiltonian structures underlying General Relativity itself. The corresponding Hamiltonian flows, however, are different, being referred to an appropriate coordinate-time of space-time foliation in the canonical approach, and to a suitable invariant proper-time in the present theory. Consequently, the physical interpretation of quantum theories of General Relativity build upon these Hamiltonian structures remain distinctive.

The CQG-theory in fact reveals the possible existence of a discrete spectrum of metric tensors having non-vanishing momenta at quantum level, while canonical approaches deal with the quantum discretization of single space-time hypersurfaces implied by space-time foliation.

Let us now consider the covariant approaches to quantum gravity [58–60]. In this case, the usual strategy is to split the space-time metric tensor $g_{\mu\nu}$ in two parts according to the decomposition of the type $g_{\mu\nu} = \eta_{\mu\nu} + h_{\mu\nu}$, where $\eta_{\mu\nu}$ is the background metric tensor defining the space-time geometry (usually identified with the flat background), and $h_{\mu\nu}$ is the dynamical field (deviation field) for which quantization applies. From the conceptual point of view there are some similarities between the literature covariant approaches and the manifestly-covariant quantum gravity theory adopted here. The main points of contact are: (1) the adoption of 4-tensor variables, without invoking any space-time foliation; (2) the implementation of a first-quantization approach, in the sense that there exists by assumption a continuum classical background space-time with a geometric connotation, over which the relevant quantum fields are dynamically evolving; and (3) the adoption of superabundant variables, which in the two approaches are identified with the sets $(\eta_{\mu\nu}, h_{\mu\nu})$ and $(\widehat{g}_{\mu\nu}, g_{\mu\nu})$ respectively.

It is important nevertheless to emphasize the relevant differences existing with respect to literature covariant approaches. First, CQG-theory is intrinsically non-perturbative in character, so that the background metric tensor can be identified with an arbitrary continuum solution of the Einstein equations (not necessarily the flat space-time), while a priori the canonical variable $g_{\mu\nu}$ is not required to be necessarily a perturbation field. On the other hand, a decomposition of the type in Equation (5) resembling the one invoked in covariant literature approaches can always be introduced a posteriori for the implementation of appropriate analytical solution methods, like GLP theory proposed here or the analytical evaluation of discrete-spectrum quantum solutions discussed in [10]. Second, the present theory is constructed starting from the DeDonder–Weyl manifestly-covariant approach [12,13]. Consequently, CQG-theory is based on a variational formulation which relies on the introduction of a synchronous variational principle for the Einstein equations first reported in [7]. This represents a unique feature of manifestly-covariant quantum gravity theory, since previous literature is actually based on the adoption of asynchronous variational principles, i.e., in which the invariant volume element is considered variational rather than prescribed. As shown in [7], it is precisely the synchronous principle which allows the distinction between variational and extremal (or prescribed) metric tensors, and the consequent introduction of non-vanishing canonical momenta. The same feature has also made possible the formulation of manifestly-covariant classical Lagrangian, Hamiltonian and Hamilton–Jacobi theories of General Relativity and the corresponding subsequent manifestly-covariant quantum theory. Third, in CQG-theory superabundant unconstrained variables are implemented, while the same covariant quantization holds with respect to a four dimensional space-time, with no extra-dimensions being required for its prescription.

Finally, regarding covariant quantization, a further interesting comparison concerns the Batalin-Vilkovisky formalism originally developed in [61–64]. This method is usually implemented for the quantization of gauge field theories and topological field theories in Lagrangian formulation [65–67], while the corresponding Hamiltonian formulation can be found in [61]. Further critical aspects of the Batalin-Vilkovisky formalism can be found for example in [68]. In the case of the gravitational field it has been formerly applied in the context of perturbative quantum gravity to treat constraints arising from initial metric decomposition (i.e., in reference with the so-called gauge-fixing and ghost terms). Its basic features are the adoption of an asynchronous Lagrangian variational principle of General Relativity [7], the use of superabundant canonical variables and the consequent introduction of constraints. These features mark the main differences with CQG-theory, which is non-perturbative, constraint-free and follows from the synchronous Lagrangian variational principle defined in [7].

In view of these considerations, CQG-theory can be said to realize at the same time both a canonical and a manifestly-covariant quantization method, in this way establishing a connection with former canonical and covariant approaches. Nevertheless, a number of conceptual new features of the present theory depart in several ways from previous literature. This conclusion is supported by the analytical results already established by CQG-theory and presented in [10,11], which concern the existence of invariant discrete-energy spectrum for the quantum gravitational field, the graviton mass estimate associated with a non-vanishing cosmological constant and the validity of Heisenberg inequalities.

Extending further these results, in the following a trajectory-based representation of CQG-theory is developed, which permits the analytical construction of generally non-stationary solutions of the CQG-wave equation. Previous efforts to construct Bohmian representations of canonical quantum gravity and applications to cosmology have been pursued in the past literature. These are typically based on the Wheeler–DeWitt quantum equation. Relevant progresses in this directions can be found for example in [69–72], where conceptual features/differences characterizing the Bohmian approach to quantum gravity (in terms of trajectories) with respect to previous customary approaches were clearly stated. The GLP representation of CQG-theory proposed in the present paper shares the conceptual advantages of adopting a Bohmian approach to quantum physics. On the other hand, it differs from the mentioned literature in that it is built upon CQG-theory, which is manifestly covariant contrary to the Wheeler–DeWitt equation, and more important because it has a stochastic character, namely in the sense that single Bohmian trajectories are replaced by ensembles of stochastic trajectories with prescribed probability density.

3. Eulerian Representation

In this section, the basic formalism of CQG-theory formulated in [9,10] is recalled. Starting point of CQG-theory is the realization of the quantum-wave function which, for an arbitrary prescribed background space-time $(\mathbf{Q}^4, \widehat{g})$, determines the CQG-quantum state. In analogy with non-relativistic quantum mechanics, this can be first prescribed in the so-called Eulerian form. In this picture, the state is assumed to to depend on two sets of independent variables, respectively, represented by suitable configuration-space Lagrangian variables and, second, by the space-time coordinates and time. In the present case these are identified with the continuum field variables (Lagrangian coordinates) $g \equiv \{g_{\mu\nu}\}$ and, respectively, by the 4-position $r \equiv \{r^\mu\}$ and the background space-time proper-time s, so that the wave function takes generally the form $\psi \equiv \psi(g, r, s)$, where a possible explicit dependence in terms of the background metric tensor $\widehat{g}$ is understood. Regarding the notations, first $g = \{g_{\mu\nu}\}$ spans the quantum configuration space U_g of the same wave-function, i.e., the set on which the associated quantum PDF $\rho(g, r, s) = |\psi(g, r, s)|^2$ is prescribed. Second, $g = \{g_{\mu\nu}\}$ is realized by means of real symmetric tensors, so that U_g is a $10-$dimensional real vector space, namely $U_g \subseteq \mathbb{R}^{10}$. Third, in the whole time-axis $I \equiv \mathbb{R}$, $r \equiv \{r^\mu\}$ denotes the instantaneous 4-position of suitably-prescribed space-time trajectories $r = r(s)$, while the explicit $s-$dependence includes also the possible dependence (of ψ) in terms of the corresponding tangent 4-vector, i.e., $t(s) \equiv \{t^\mu(s)\} \equiv \frac{dr^\mu(s)}{ds}$. Here, $\frac{d}{ds}$ identifies the total covariant s-derivative operator

$$\frac{d}{ds} \equiv \left.\frac{d}{ds}\right|_r + \left.\frac{d}{ds}\right|_s, \tag{14}$$

with $\left.\frac{d}{ds}\right|_r \equiv \left.\frac{\partial}{\partial s}\right|_r$ and $\left.\frac{d}{ds}\right|_s \equiv t^\alpha \nabla_\alpha$ being the covariant s-derivatives performed at constant $r \equiv \{r^\mu\}$ and constant s respectively.

A realization of the parameterization $\psi \equiv \psi(g, r, s)$ is provided by the geodetics of the metric field tensor $\widehat{g} \equiv \widehat{g}(r)$, namely the integral curves of the initial-value problem [9,10]

$$\begin{cases} \frac{dr^\mu(s)}{ds} = t^\mu(s), \\ \frac{Dt^\mu(s)}{Ds} = 0, \\ r^\mu(s_o) = r^\mu_o, \\ t^\mu(s_o) = t^\mu_o, \end{cases} \tag{15}$$

with $\left(r_o^\mu, t_o^\mu\right)$ denoting, respectively, arbitrary initial 4-position of $\{\mathbf{Q}^4, \widehat{g}\}$ and a corresponding (arbitrary) tangent 4-vector, while the standard connections in the covariant derivative $\frac{D}{Ds}$ are prescribed again in terms the background metric tensor $\widehat{g}(r)$. Since each point $r^\mu \equiv r^\mu(s)$ can be crossed by infinite arbitrary geodetics having different tangent 4-vectors $t^\mu \equiv t^\mu(s)$ it follows that the wave-function parameterization $\psi \equiv \psi(g,r,s)$ may generally depend explicitly on the choice of the geodetics, i.e., on t^μ too. In particular, $\psi(g,r,s)$ will be assumed to contain the following smooth dependences:

(1) *Explicit g-dependence:* $\psi(g,r,s)$ is assumed to be a $C^{(2)}$, smoothly differentiable complex function of the continuum Lagrangian variables $g = \{g_{\mu\nu}\}$.

(2) *Explicit and implicit s-dependences:* $\psi(g,r,s)$ may depend both explicitly and implicitly on s. The implicit dependence occurs via $r(s)$ and $t(s)$ and therefore also in terms of the prescribed metric tensor through its explicit spatial dependence $\widehat{g}(r)$. These s-dependences will all be assumed to realize in terms of $\psi(g,r,s)$ a $C^{(1)}$, smoothly differentiable function of s.

The next step is the identification of the quantum-wave equation which determines the CQG state $\psi(g,r,s)$. This task is achieved by means of the *CQG-wave equation* [10]. Written again in the Eulerian form, the latter is realized by the initial-value problem

$$\left\{ \begin{array}{c} i\hbar\frac{\partial}{\partial s}\psi(g,r,s) = [H_R, \psi(g,r,s)] \equiv H_R\psi(g,r,s), \\ \psi(g,r(s_o) = r_o, s_o) = \psi_o(g,r_o), \end{array} \right. \tag{16}$$

where in the first equation the squared-brackets denote the quantum commutator in standard notation, while the operator $\frac{\partial}{\partial s}$ appearing in the scalar Equation (16) in the Eulerian representation coincides with $\frac{\partial}{\partial s} \equiv \frac{d}{ds}$, being $\frac{d}{ds}$ the total covariant s-derivative in Equation (14) again prescribed in terms of the background metric tensor $\widehat{g}(r)$. Notice that the initial-value problem in Equation (16) can be represented equivalently in terms of the initial quantum fluid fields

$$\left\{ \begin{array}{c} \rho(g,r(s_o) = r_o, s_o) = \rho_o(g,r_o), \\ S^{(q)}(g,r(s_o) = r_o, s_o) = S_o^{(q)}(g,r_o). \end{array} \right. \tag{17}$$

As such, provided $\psi(g,r,s)$ is suitably smooth the solution of Equation (16) is unique. Thus, Equation (16) realizes a hyperbolic evolution equation, i.e., a first-order PDE with respect to the proper time s. In the same equation H_R denotes the quantum Hamiltonian operator characteristic of CQG-theory, to be expressed in terms of the relevant quantum momentum operator, namely $\pi_{\mu\nu}^{(q)} = -\frac{i\hbar}{\alpha L}\frac{\partial}{\partial g^{\mu\nu}}$. Here, the partial derivative is performed keeping constant all remaining variables appearing in $\psi(g,r,s)$, while L and α are, respectively, a suitably-defined 4-scalar scale-length and a dimensional 4-scalar parameter related to the universal constant $\kappa = \frac{c^3}{16\pi G}$ (see again [10]). Then, the quantum Hamiltonian operator H_R takes the form

$$H_R \equiv T_R^{(q)} + V(g,r,s), \tag{18}$$

with $T_R^{(q)}(x,\widehat{g}) \equiv \frac{1}{2\alpha L}\pi_{\mu\nu}^{(q)}\pi^{(q)\mu\nu}$ and $V(g,r,s)$ being, respectively, the effective kinetic energy operator and the effective potential energy

$$\left\{ \begin{array}{c} V(g,r,s) \equiv \sigma V_o(g) + \sigma V_F(g,r,s), \\ V_o(g) \equiv \alpha L h \left[g^{\mu\nu}\widehat{R}_{\mu\nu} - 2\Lambda\right], \\ V_F \equiv \frac{\alpha L}{k} h L_F(g,r,s), \end{array} \right. \tag{19}$$

with $V_o(g)$ and $V_F(g,r,s)$ identifying the vacuum and external effective contributions to the effective potential $V(g,r,s)$. Here, the notation is given according to [10]. Thus, all hatted quantities are evaluated with respect to the background metric tensor $\widehat{g}$ only while the multiplicative 4-scalar gauge function σ is taken to be $\sigma = -1$. In addition, $V \equiv V(g,r,s)$ itself is determined up to an arbitrary additive gauge transformation of the form $V \rightarrow V' = V - \frac{d}{ds}F(g,r(s)s)\Big|_g$, being $F(g,r(s)s)$ a 4-scalar function of the form

$$F(g,r(s)s) = g_{\mu\nu}G^{\mu\nu}(\widehat{g},r(s),s) + F_1(\widehat{g},r(s),s), \tag{20}$$

with $G^{\mu\nu}(\widehat{g},r(s),s)$ and $F_1(\widehat{g},r(s),s)$ denoting, respectively, a 4-tensor and a 4-scalar smoothly differential real gauge, i.e., arbitrary, functions. A characteristic element of CQG-theory is the quantity $h \equiv h(g)$ first introduced in [7]. The prescription of $h(g)$ is obtained in terms of a polynomial function of $g \equiv \widehat{g} + \delta g$, with δg being an in principle arbitrary variational displacement so that according to the same reference (see also [7]):

$$h(g) = 2 - \frac{1}{4}\left(\widehat{g}^{\alpha\beta} + \delta g^{\alpha\beta}\right)\left(\widehat{g}^{\mu\nu} + \delta g^{\mu\nu}\right)\widehat{g}_{\alpha\mu}\widehat{g}_{\beta\nu}. \tag{21}$$

As a final remark, we notice that the Eulerian CQG-state defined by the complex function $\psi(g,r,s)$ can always be cast in the form of an exponential representation via the Madelung representation recalled above. Elementary algebra [10,11] then shows that, based on the quantum-wave Equation (16), the same quantum fluid fields necessarily fulfill the corresponding set of Eulerian CQG-quantum hydrodynamic equations. In the Eulerian representation, upon identifying again $\frac{d}{ds}$ with the total covariant $s-$derivative operator in Equation (14), these are realized respectively by the continuity and quantum Hamilton-Jacobi equations:

$$\begin{cases} \frac{d\rho(g,r,s)}{ds} + \frac{\partial}{\partial g_{\mu\nu}}\left(\rho(g,r,s)V_{\mu\nu}(g,r,s)\right) = 0, \\ \frac{dS^{(q)}(g,r,s)}{ds} + H_c(g,r,s) = 0, \end{cases} \tag{22}$$

which represent a set of evolution PDEs for the quantum fluid fields $\rho(g,r,s)$ and $S^{(q)}(g,r,s)$. Notice that, in the previous equations, $V_{\mu\nu}(q,s)$ and $H_c(g,r,s)$ denote, respectively, the tensor "velocity" field $V_{\mu\nu}(g,r,s) = \frac{1}{\alpha L}\frac{\partial S^{(q)}(g,r,s)}{\partial g^{\mu\nu}}$ and the effective quantum Hamiltonian density

$$H_c(g,r,s) = \frac{1}{2\alpha L}\frac{\partial S^{(q)}(g,r,s)}{\partial g^{\mu\nu}}\frac{\partial S^{(q)}(g,r,s)}{\partial g_{\mu\nu}} + V_{QM}(g,r,s) + V(g,r,s), \tag{23}$$

with

$$T = \frac{1}{2\alpha L}\frac{\partial S^{(q)}(g,r,s)}{\partial g^{\mu\nu}}\frac{\partial S^{(q)}(g,r,s)}{\partial g_{\mu\nu}} \tag{24}$$

being the effective kinetic energy. In addition, $V(g,r,s)$ and $V_{QM}(g,r,s)$ identify, respectively, the effective potential density (19) and the Bohm effective quantum potential

$$V_{QM}(g,r,s) \equiv -\frac{\hbar^2}{8\alpha L}\frac{\partial \ln\rho(g,r,s)}{\partial g^{\mu\nu}}\frac{\partial \ln\rho(g,r,s)}{\partial g_{\mu\nu}} - \frac{\hbar^2}{4\alpha L}\frac{\partial^2 \ln\rho(g,r,s)}{\partial g_{\mu\nu}\partial g^{\mu\nu}}, \tag{25}$$

or equivalently $V_{QM}(g,r,s) \equiv \frac{\hbar^2}{8\alpha L}\frac{\partial \ln\rho(g,r,s)}{\partial g^{\mu\nu}}\frac{\partial \ln\rho(g,r,s)}{\partial g_{\mu\nu}} - \frac{\hbar^2}{4\alpha L}\frac{\partial^2\rho(g,r,s)}{\rho\partial g_{\mu\nu}\partial g^{\mu\nu}}$.

4. Lagrangian Path (Bohmian) Representation

It is well known that in the non-relativistic framework the Bohmian interpretation of quantum mechanics provides the corresponding trajectory-based Lagrangian Path representation (*LP-representation*) of the Schroedinger quantum-wave equation (see [73,74] for a review of the topic). The intrinsic similarity with the CQG-wave equation suggests that an analogous Lagrangian representation is possible also for the same equation, so that as a consequence, a "Bohmian" trajectory-based interpretation can be achieved in the context of CQG-theory too. In both cases, in fact, the Lagrangian representation is based on the introduction of a suitable family of configuration-space trajectories, or Lagrangian Paths (LP), which for each "point" of the appropriate quantum configuration space are unique. In the context of CQG-theory, the LP-representation involves the introduction for all $s \in I$ of the correspondence (8), with $\delta g_{\mu\nu} \equiv \delta g_{L\mu\nu}(s) \in U_g$ belonging to a suitable curve $\{g_L(s), \forall s \in I\}$ of the configuration space U_g denoted as *Lagrangian path*. Consequently, each LP is identified with a well-defined characteristics associated with the tensor velocity field $V_{\mu\nu}(g, r, s)$. For definiteness, based on the tensor decomposition (5), the LP-representation involves parameterizing all quantum fields, and in particular the quantum state, in terms of $g_{L\mu\nu}(s)$ thus letting $\psi \equiv \psi(g_{L\mu\nu}(s), r(s), s)$. As such, $\delta g_{L\mu\nu}(s)$ is constructed in such a way that its "tangent" coincides with the local value of the tensor velocity field $V_{\mu\nu}$, namely so that they fulfill the initial-value problem

$$\left\{ \begin{array}{c} \frac{D}{Ds} g_{L\mu\nu}(s) = V_{\mu\nu}(g_L(s), s), \\ g_{\mu\nu}(s_o) = g_{\mu\nu}^{(o)}. \end{array} \right. \tag{26}$$

Here, $\frac{D}{Ds}$ identifies the LP-derivative (or covariant s-derivative) realized by the operator

$$\frac{D}{Ds} \equiv \left.\frac{d}{ds}\right|_{\delta g_{L\mu\nu}(s)} + V_{\mu\nu}(g_L(s), s) \frac{\partial}{\partial \delta g_{L\mu\nu}}, \tag{27}$$

where the two terms on the R.H.S. of Equation (27) identify respectively the covariant s-derivative performed at constant $\delta g_{L\mu\nu} \equiv \delta g_{L\mu\nu}(s)$, namely

$$\left.\frac{d}{ds}\right|_{\delta g_{L\mu\nu}(s)} \equiv \left.\frac{D}{Ds}\right|_{\delta g_{\mu\nu}} = \left[\left.\frac{\partial}{\partial s}\right|_r + t^\alpha \nabla_\alpha \right]_{\delta g_{L\mu\nu}}, \tag{28}$$

and the convective derivative performed with respect to the Lagrangian coordinates $\delta g_{L\mu\nu}(s)$ while keeping constant $\widehat{g}(r) \equiv \{\widehat{g}_{\mu\nu}(r(s))\}$. In view of Equation (5), Equation (26) can be written as

$$\left\{ \begin{array}{c} \frac{D}{Ds} \delta g_{L\mu\nu}(s) = V_{\mu\nu}(\widehat{g}(r) + \delta g_L(s), s), \\ \delta g_{L\mu\nu}(s_o) = \delta g_{\mu\nu}^{(o)}. \end{array} \right. \tag{29}$$

Consequently, Equation (29) can be integrated to give

$$\delta g_{L\mu\nu}(s) = \delta g_{\mu\nu}^{(o)} + \int_{s_o}^{s} ds' V_{\mu\nu}(\widehat{g}(r) + \delta g_L(s'), s'), \tag{30}$$

which determines the LP itself, namely the trajectory $\{g_L(s), \forall s \in I\} \equiv \{g_L(s) \equiv \widehat{g}(r) + \delta g_L(s), \forall s \in I\}$. However, if $H_{\mu\nu} \equiv H_{\mu\nu}(\widehat{g}(r))$ denotes an arbitrary smoothly-differentiable tensor function of $\widehat{g}(r)$, it is obvious that also the arbitrary additive tensor quantity of the form $\frac{D}{Ds}\left[\delta g_{L\mu\nu}(s) + H_{\mu\nu}(\widehat{g}(r))\right]$ satisfies identically Equation (29). Since uniqueness of the solution $\delta g_{L\mu\nu}(s)$ given by Equation (30) is warranted by prescribing $\delta g_{\mu\nu}^{(o)}$, the mapping

$$g_{L\mu\nu}(s_o) = g_{\mu\nu}^{(o)} \Leftrightarrow g_{L\mu\nu}(s) \tag{31}$$

identifies a classical dynamical system (CDS), i.e., a diffeomeorphism mutually mapping in each other two arbitrary points $g_{L\mu\nu}(s_o)$ and $g_{L\mu\nu}(s)$ which belong to the same LP. Consequently, the Liouville theorem warrants that the Jacobian determinant of the transformation (31) is

$$\left| \frac{\partial \delta g_L(s)}{\partial \delta g_L(s_o)} \right| = \exp \left\{ \int_{s_o}^{s} ds' \frac{\partial V_{\mu\nu}(g_L(s'), s')}{\partial g_{L\mu\nu}(s')} \right\}. \tag{32}$$

The Lagrangian representation of CQG-theory is then achieved by means of the formal replacement $g \rightarrow g_L(s)$ to be made in the quantum wave-function, i.e., introducing in the CQG-wave Equation (16) the LP-parameterization $\psi = \psi(g_L(s), s)$ and similarly for the quantum fluid fields, namely

$$\left\{\rho, S^{(q)}\right\} \equiv \left\{\rho(g_L(s), s), S^{(q)}(g_L(s), s)\right\}. \tag{33}$$

As a result, in terms of the tensor velocity field in the LP-representation, namely $V_{\mu\nu}(g_L(s), s) \equiv \frac{1}{\alpha L}\frac{\partial S^{(q)}(g_L(s), r(s), s)}{\partial \delta g_L^{\mu\nu}(s)}$, the quantum hydrodynamic Equation (22) can be set at once in the corresponding Lagrangian form. To obtain them one notices preliminarily that

$$\frac{D}{Ds}S^{(q)}(g_L(s), s) \equiv \frac{d}{ds}S^{(q)}(g_L(s), s) + V_{\mu\nu}(g_L(s), s)\frac{\partial S^{(q)}(g_L(s), s)}{\partial g_{L\mu\nu}(s)}, \tag{34}$$

with $\frac{D}{Ds}$ and $\frac{d}{ds}$ identifying respectively the LP-derivative in Equation (27) and the total covariant s-derivative operator in Equation (14). Consequently, the LP-representation of the quantum fluid Equation (22) is given by the PDEs

$$\begin{cases} \frac{D}{Ds}\rho(g_L(s), s) = -\rho(g_L(s), s)\frac{\partial V_{\mu\nu}(g_L(s), s)}{\partial g_{L\mu\nu}(s)}, \\ \frac{D}{Ds}S^{(q)}(g_L(s), s) = V_{\mu\nu}(g_L(s), s)\frac{\partial S^{(q)}(g_L(s), s)}{\partial g_{L\mu\nu}(s)} - H_c(g_L(s), s), \end{cases} \tag{35}$$

where $H_c(g_L(s), s)$ identifies the effective quantum Hamiltonian density in Equation (23) parameterized in terms of $g_L(s)$. Thus, in particular, the continuity equation (first part of Equation (35)) can be formally integrated to give the LP-parameterized integral continuity equation

$$\rho(g_L(s), s) = \rho(g_L(s_o), s_o)\exp\left\{-\int_{s_o}^{s} ds'\frac{\partial V_{\mu\nu}(g_L(s'), s')}{\partial g_{L\mu\nu}(s')}\right\}, \tag{36}$$

with $\rho(g_L(s_o), s_o) \equiv \rho(g_L(s_o), r(s_o) = r_o, s_o)$ denoting the initial quantum PDF, namely

$$\rho(g_L(s_o), r(s_o) = r_o, s_o) = \rho_o(g_L(s_o), r_o). \tag{37}$$

Together with Liouville theorem in Equation (32) this implies therefore the conservation laws

$$d(g_L(s))\rho(g_L(s), s) = d(g_L(s_o))\rho(g_L(s_o), s_o), \tag{38}$$

$$\int_{U_g} d(g_L(s))\rho(g_L(s), s) = \int_{U_g} d(g_L(s_o))\rho(g_L(s_o), s_o) = 1, \tag{39}$$

which warrant, consistent with the quantum unitarity principle, the conservation of the quantum probability in U_g.

We conclude this section noting that from a mathematical viewpoint the Lagrangian formulation of CQG-theory is actually realized solely by the LP-parameterized quantum hydrodynamic Equations (35). Therefore, the Lagrangian and Eulerian quantum hydrodynamic equations are manifestly equivalent. This suggests that a Bohmian interpretation of the Lagrangian-path representation of the CQG-theory is in principle possible. However, just as in the case of the Schroedinger equation (see related discussion in [22]), a basic difficulty of such an interpretations lies in the uniqueness feature, and consequently the intrinsic deterministic character, of each LP. Such a property, in fact, appears potentially in contradiction with the notion of quantum measurement holding in the context of CQG-theory and the validity of Heisenberg inequalities [11].

5. Generalized Lagrangian Path Representation

The considerations indicated above lead us to introduce the notion of Generalized Lagrangian Path (GLP) and of the corresponding GLP-representation obtained in this way for the quantum wave-function and quantum fluids fields. As anticipated above (see Introduction) this is achieved by means of the introduction of a suitable set of intrinsically non-unique and stochastic trajectories, to be referred to as generalized Lagrangian paths (GLPs), in terms of which the quantum wave-equation, as well as the corresponding set of quantum fluid fields and quantum hydrodynamic equations, can be parameterized. In the context of CQG-theory the mathematical problem of formulating its GLP-representation involves the introduction for all $s \in I$ of a suitable correspondence of the type

$$s \to \delta G_L(s), \tag{40}$$

referred to as *GLP-map*. Then, upon invoking the tensor decomposition (Equation (9)), a GLP is the curve $\{G_L(s), \forall s \in I\}$ of the quantum configuration space U_g which is defined by Equation (10) and is realized by the ensemble of "points" of U_g spanned by the tensor field $G_L(s) \equiv G(s)$ and obtained varying $s \in I$. The underlying basic idea is therefore to replace a single LP, prescribed in terms of a solution of the initial-value problem in Equation (26), with an infinite set of stochastic trajectories, each one identified with a single GLP and characterized by a unique choice of a suitable stochastic tensor $\Delta g = \{\Delta g_{\mu\nu}\}$. This effectively involves introducing a parameter-dependent mapping of the type

$$\{g_L(s), \forall s \in I\} \to \{G_L(s), \forall s \in I\}, \tag{41}$$

whose realization depends on the prescription of $\Delta g = \{\Delta g_{\mu\nu}\}$. Then, the GLP-map in Equation (41) is realized by means of the following two requirements.

- *GLP Requirement #1* - The first one is realized by prescribing $\delta G_{L\mu\nu}(s)$ in terms of the displacement tensor $\delta g_{L\mu\nu}(s)$ which is determined according to Equation (5). This yields therefore the identity

$$G_{L\mu\nu}(s) = \widehat{g}_{\mu\nu}(r) + \delta g_{L\mu\nu}(s) - \Delta g_{\mu\nu}, \tag{42}$$

with Δg denoting the *stochastic displacement 4-tensor*

$$\Delta g = g - G_L(s) \equiv \delta g - \delta G_L(s). \tag{43}$$

Notice that, here, $g_{\mu\nu} = g_{L\mu\nu}(s)$, and hence $\delta g_{\mu\nu} \equiv \delta g_{L\mu\nu}(s)$. Consequently, it is understood that Δg must be endowed with a suitable stochastic PDF to be suitably prescribed. In this regards, taking Δg as an independent stochastic variable, it is natural to assume that the same PDF should be a stationary and spatially uniform probability distribution, i.e., a function independent of r, s as well as $\delta g_L(s)$, but still allowed to depend in principle on the prescribed metric tensor $\widehat{g}_{\mu\nu}(r)$. More precisely, this means assuming the same PDF to be realized in terms of a smoothly differentiable and strictly positive function of the form

$$f = f(\Delta g, \widehat{g}). \tag{44}$$

Hence, the corresponding notion of stochastic average for an arbitrary smooth function $X(\Delta g, r, s)$ is prescribed in terms of the weighted integral

$$\langle X(\Delta g, r, s)\rangle_{stoch} \equiv \int_{U_g} d(\Delta g) X(\Delta g, r, s) f(\Delta g, \widehat{g}), \tag{45}$$

to be performed on the configuration space U_g. In particular, besides the prescription (44), $f(\Delta g, \widehat{g})$ should be prescribed so that the following stochastic averages are also fulfilled:

$$\begin{cases} \langle 1 \rangle_{stoch} \equiv \int\limits_{U_g} d(\Delta g) f(\Delta g, \widehat{g}) = 1, \\ \langle \Delta g_{\mu\nu} \rangle_{stoch} \equiv \int\limits_{U_g} d(\Delta g) \Delta g_{\mu\nu} f(\Delta g, \widehat{g})) = \pm \widehat{g}_{\mu\nu}(r), \\ \sigma_{\Delta g}^2 \equiv \left\langle (\Delta g - \langle \Delta g \rangle_{stoch})^2 \right\rangle_{stoch} \equiv \\ \int\limits_{U_g} d(\Delta g) \left(\Delta g - \langle \Delta g \rangle_{stoch}\right)^2 f(\Delta g, \widehat{g}) = r_{th}^2, \end{cases} \tag{46}$$

with $(\Delta g - \langle \Delta g \rangle_{stoch})^2 \equiv [\Delta g_{\mu\nu} - \langle \Delta g_{\mu\nu} \rangle_{stoch}] [\Delta g^{\mu\nu} - \langle \Delta g^{\mu\nu} \rangle_{stoch}]$ and $\sigma_{\Delta g}$ denoting the standard deviation of Δg to be identified with the dimensionless 4-scalar parameter $r_{th}^2 > 0$. Notice in addition that, here, for consistency with the same assumption in Equation (44), r_{th}^2 must be assumed to be a non-vanishing constant, i.e., independent of both (r, s).

- *GLP Requirement #2*- The second one is obtained requiring that $\Delta g = \{\Delta g_{\mu\nu}\}$ is constant for all $s \in I$ and for an arbitrary Lagrangian Path, i.e., it is prescribed so that identically for all $s, s_o \in I$ it occurs that

$$\Delta g_{\mu\nu}(s) = \Delta g_{\mu\nu}(s_o). \tag{47}$$

Notice that here $\frac{D}{Ds}\delta g_{L\mu\nu}(s) = \frac{D}{Ds}\delta G_{L\mu\nu}(s) \equiv V_{\mu\nu}(G_L(s), \Delta g, s))$, with $V_{\mu\nu}(G_L(s), \Delta g, s)$ being the tensor velocity field in the GLP-representation, namely

$$V_{\mu\nu}(G_L(s), \Delta g, s) = \frac{1}{\alpha L} \frac{\partial S^{(q)}(G_L(s), \Delta g, s)}{\partial \delta g_L^{\mu\nu}(s)}, \tag{48}$$

while $\frac{D}{Ds}$ is the Lagrangian derivative defined above (see Equation (27)). As a result, the constraint condition (47) necessarily implies also that

$$\frac{D}{Ds}\Delta g_{\mu\nu} \equiv \frac{D}{Ds}\delta g_{L\mu\nu}(s) - \frac{D}{Ds}\delta G_{L\mu\nu}(s) \equiv 0. \tag{49}$$

As a consequence of Requirements #1 and #2, for all $s \in I$, the correspondence in Equation (40) is uniquely established, in the sense that, for each determination of the stochastic displacement Δg, $G_L(s, \Delta g) \equiv G_L(s)$ belongs to a uniquely-prescribed curve $\{G_L(s), \forall s \in I\}$, identifying a GLP which spans the quantum configuration space U_g. More precisely, a generic GLP $\{G_L(s), \forall s \in I\}$ is identified with the integral curve determined by the GLP-initial-value problem

$$\begin{cases} \frac{D}{Ds}\delta G_{L\mu\nu}(s) = V_{\mu\nu}(G_L(s), \Delta g, s), \\ \delta G_{L\mu\nu}(s_o) = \delta g_{\mu\nu}^{(o)} - \Delta g_{\mu\nu}. \end{cases} \tag{50}$$

In addition, here, the map $G_L(s_o) \Leftrightarrow G_L(s)$ defines again a classical dynamical system with Jacobian determinant

$$\left| \frac{\partial G_L(s)}{\partial G_L(s_o)} \right| = \exp\left\{ \int\limits_{s_o}^{s} ds' \frac{\partial V_{\mu\nu}(G_L(s') + \Delta g, s')}{\partial g_{L\mu\nu}(s')} \right\}. \tag{51}$$

The ensemble of integral curves $\{G_L(s), \forall s \in I\}$ obtained by varying Δg in U_g identifies therefore an infinite set of GLP which are associated with the tensor velocity field $V_{\mu\nu}(G_L(s) + \Delta g, s)$. One notices, however, that by construction,

$$V_{\mu\nu}(G_L(s) + \Delta g, s) = V_{\mu\nu}(g_L(s), s). \tag{52}$$

Thus, the same infinite set of GLP is actually associated with the same local value of the tensor velocity field $V_{\mu\nu}(g_L(s), s)$. Thus, in contrast with the LP defined above (in terms of Equation (26)), this means that the GLP which are associated with the local tensor velocity field $V_{\mu\nu}(g_L(s),s)$ are non-unique (and actually infinite), each one being determined by Δg. Precisely because the same trajectories are stochastic and hence non-unique, such a feature is in principle compatible with the possible interpretation of the GLPs as *physical quantum trajectories in the configurations space* U_g. Nevertheless, the prerequisite for making actually possible such an interpretation is, ultimately, the ontological equivalence of the GLP-parameterization for the quantum state ψ with the "standard" Eulerian representation of the same quantum wave-function. In other words, the adoption of the GLP and in particular the prescription of the stochastic PDF $f(\Delta g)$ associated with the same constant stochastic displacement tensor Δg (see Equation (44)), should be possible leaving unchanged the axioms of CQG-theory.

For definiteness, let us now pose the problem of introducing explicitly the parameterization of the quantum fluid fields and the related GLP-representation of the QHE. In principle, this can simply be obtained from the corresponding LP-parameterization indicated above noting that $\delta g(s) = \Delta g + \delta G(s)$. However, in formal analogy with the GLP-approach to non-relativistic quantum mechanics earlier indicated, a more general parameterization in terms of the stochastic displacement tensor field Δg, to be referred in the sequel as GLP-parameterization, is possible. This involves assuming that the CQG-wave function may be of the type

$$\psi = \psi(G_L(s), \Delta g, s), \tag{53}$$

i.e., to include also an explicit dependence in terms of $\Delta g \equiv \{\Delta g_{\mu\nu}\}$. Therefore, the corresponding GLP-parameterization of the quantum fluid fields is taken of the form

$$\left\{\rho, S^{(q)}\right\}_{(s)} \equiv \left\{\rho(G_L(s), \Delta g, s), S^{(q)}(G_L(s), \Delta g, s)\right\}. \tag{54}$$

Nevertheless, the quantum hydrodynamic Equation (22), when expressed in the GLP-parameterization, remain formally analogous to those obtained in the LP-parameterization (see Equation (35)), so that the same equations must determine the map

$$\left\{\rho, S^{(q)}\right\}_{(s_o)} \equiv \left\{\rho_o, S_o^{(q)}\right\} \to \left\{\rho, S^{(q)}\right\}_{(s)}, \tag{55}$$

with $\left\{\rho_o, S_o^{(q)}\right\}$ being suitable initial quantum fluid fields. Hence, for consistency, these should be again assumed of the form

$$\left\{\rho_o, S_o^{(q)}\right\} \equiv \left\{\rho_o(G_L(s_o), \Delta g), S_o^{(q)}(G_L(s_o), \Delta g)\right\}. \tag{56}$$

In detail, in the GLP-representation the quantum hydrodynamic Equation (35) are now realized by the PDEs

$$\begin{cases} \frac{D}{Ds}\rho(G_L(s), \Delta g, s) = -\rho(G_L(s), \Delta g, s)\frac{\partial V_{\mu\nu}(G_L(s), \Delta g, s)}{\partial g_{L\mu\nu}(s)}, \\ \frac{D}{Ds}S^{(q)}(G_L(s), \Delta g, s) = K_c(G_L(s), \Delta g, s), \end{cases} \tag{57}$$

representing, respectively, the GLP-parameterized quantum continuity and H-J equations, where

$$K_c(G_L(s), \Delta g, s) = V_{\mu\nu}(G_L(s), \Delta g, s)\frac{\partial S^{(q)}(G_L(s, \Delta g), \Delta g, s)}{\partial g_{L\mu\nu}(s)} - H_c(G_L(s), \Delta g, s) \tag{58}$$

and $H_c(G_L(s), \Delta g, s)$ identifies now the effective quantum Hamiltonian density in Equation (23) expressed in terms of the GLP-parameterization. Thus, from Equation (23), it follows that

$$H_c(G_L(s), \Delta g, s) = T(G_L(s), \Delta g, s) - V(G_L(s), \Delta g, s) - V_{QM}(G_L(s), \Delta g, s), \tag{59}$$

with $T \equiv T(G_L(s), \Delta g, s)$, $V \equiv V(G_L(s), \Delta g, s)$ and $V_{QM} \equiv V_{QM}(G_L(s), \Delta g, s)$ denoting now in terms of the GLP-parameterization respectively the effective kinetic energy and classical potential density given by Equations (24), (19) and the Bohm effective quantum potential in Equation (25). Thus, regarding the representation of the effective potential energy V, and in particular its vacuum contribution $V_o \equiv V_o(G_L(s), \Delta g, s)$ (see Equation (19)), to be used in the context of the GLP-approach, one notices that the displacement 4-tensor δg entering the expression of the variational parameter in Equation (21) remains non-unique. One notices that, due to its arbitrariness, the displacement 4-tensor can always be identified with $\delta g \equiv \Delta g$, being Δg the stochastic constant displacement field tensor introduced above (see Equation (11)), so that actually $h(g)$ can be conveniently represented as

$$h(\widehat{g} + \Delta g) = 2 - \frac{1}{4}\left(\widehat{g}^{\alpha\beta} + \Delta g^{\alpha\beta}\right)\left(\widehat{g}^{\mu\nu} + \Delta g^{\mu\nu}\right)\widehat{g}_{\alpha\mu}\widehat{g}_{\beta\nu}, \tag{60}$$

while the vacuum effective potential becomes:

$$V_o(G_L(s), \Delta g, s) \equiv \sigma\alpha L h(\widehat{g} + \Delta g)\left[\left(\widehat{g}_{pq}(s) + \Delta g_{pq}\right)\widehat{g}^{pq}(r) - 2\right]\Lambda. \tag{61}$$

Useful implications of the GLP-representation in Equations (53)–(54) follow by inspection of the GLP-quantum continuity equation (see first equation in Equation (57)) obtained above. The first one follows by noting that the same equation implies also

$$\frac{D}{Ds}\ln\rho(G_L(s), \Delta g, s) = -\frac{\partial V_{\mu\nu}(G_L(s), \Delta g, s)}{\partial g_{L\mu\nu}(s)}, \tag{62}$$

so that its formal integration generates the map $\rho(G_L(s_o), \Delta g, s_o) \rightarrow \rho(G_L(s), \Delta g, s)$, with $\rho(G_L(s), \Delta g, s)$ denoting the proper-time evolved quantum PDF, namely

$$\rho(G_L(s), \Delta g, s) = \rho(G_L(s_o), \Delta g, s_o)\exp\left\{-\int_{s_o}^{s} ds' \frac{\partial V_{\mu\nu}(G_L(s'), \Delta g, s')}{\partial g_{L\mu\nu}(s')}\right\}. \tag{63}$$

Notice that the integration on the R.H.S is performed along the GLP-trajectory $\{G_L(s, \Delta g), \forall s \in I\}$, i.e., for a prescribed constant stochastic displacement 4-tensor Δg, while $\rho(G_L(s_o), \Delta g, s_o)$ identifies the initial, and in principle still arbitrary, PDF. The second implication concerns the quantum H-J equation itself. In fact, the formal solution in Equation (63) permits to cast it in terms of an (implicit) equation for the GLP-parameterized quantum phase-function $S^{(q)}(G_L(s), \Delta g, s)$ only. Consequently, provided an explicit realization is reached for the GLP-trajectory $\{G_L(s), \forall s \in I\}$, by solving the initial-value problem (50), the same H-J equation should uniquely determine the corresponding solution $S^{(q)}(G_L(s), \Delta g, s)$ as a real function of Δg and s only. A notable feature worth to be stressed here is about the prescription of the same initial PDF $\rho(G_L(s_o), \Delta g, s_o)$. This manifestly generally differs from the one considered above in the case of the LP-parameterization (see Equation (37)), where no explicit $\Delta g-$dependences was assumed. In fact, consistent with the GLP-parameterization introduced above (see Equation (54)), this is now taken of the form (56). This means that it may include in particular an admissible choice for the initial PDF provided by a probability density of the form

$$\rho(G_L(s_o), \Delta g, s_o) = \rho_o(\Delta g + \widehat{g}(r_o)), \tag{64}$$

with $\rho(\Delta g + \widehat{g}(r_o))$ to be determined as indicated below.

6. GLP Approach: Determination of the Stochastic PDF for Δg and of the Quantum PDF

The problem addressed in this section is twofold. First, it concerns the identification of the stochastic probability density $f(\Delta g, \widehat{g}_{\mu\nu})$ which is associated with the stochastic displacement tensor field $\Delta g \equiv \{\Delta g_{\mu\nu}\}$ and is consistent with the requirements indicated above, i.e., Equation (44), together with the aforementioned constraint conditions in Equation (46). Second, it deals with the prescription of the CQG-probability density, in particular the initial one ρ_o, to be adopted in the GLP-parameterization, see Equation (17) as well as Equations (54) and (64) above. In fact, both prescriptions should be actually regarded as mandatory prerequisites for the consistency of the GLP-representation and its ontological equivalence with the corresponding Eulerian representation of CQG-theory. In this section, we intend to show that the two issues are actually intrinsically related.

In particular we aim to prove that the initial quantum PDF can be prescribed in such a way that it coincides with a shifted Gaussian PDF, such a choice being consistent with the principle of entropy maximization (PEM), i.e., determined so to maximize the initial Boltzmann–Shannon entropy associated with the initial PDF. Consequently, the same initial PDF is shown to satisfy suitable symmetry properties (see Proposition #1). Furthermore the problem is posed of the determination of the quantum expectation values evaluated with respect to the GLP-parameterized quantum PDF. As a result, for arbitrary observables which are identified with ordinary tensor functions, equivalent representations of the GLP-quantum expectation values are pointed out (Proposition #2). A notable related implication refers to the physical interpretation of CQG-theory arising in such a context which is analogous to the so-called emergent gravity picture of quantum gravity. This follows by noting that, by suitable prescription of the initial quantum PDF, the background metric tensor $\widehat{g}(r(s))$ is uniquely determined, at any arbitrary proper-time s, in terms of an appropriate expectation value of the quantum PDF (see Proposition #3).

6.1. Prescription of the Stochastic PDF

The two topics indicated above actually have a unique solution. This follows at once provided the axiomatic setting of CQG-theory is invoked. Let us consider, in fact, the problem of the determination of $f(\Delta g, \widehat{g})$. In the context of CQG-theory, as in the case of Quantum Mechanics (see related discussion in [22]), the independent prescription of $f(\Delta g, \widehat{g})$ potentially may amount to the introduction of an additional axiom, thus possibly giving rise to additional conceptual difficulties related to the notions of quantum measurement and quantum expectation values. To overcome this issue, while leaving unaffected the axioms of CQG-theory earlier introduced in [10] and, at the same time, warranting the ontological equivalence indicated above, the only possible choice for $f(\Delta g, \widehat{g})$ is that it coincides with the initial quantum PDF ρ_o. This means also, of course, that ρ_o must be necessarily of the type (64), namely such that

$$f(\Delta g, \widehat{g}) \equiv \rho_o(\Delta g \pm \widehat{g}(r_o)), \tag{65}$$

and therefore fulfilling also the constraint conditions indicated above (see Equation (46)). Incidentally, as explained below, from the conceptual viewpoint, this choice exhibits remarkable features.

6.2. The Initial Quantum PDF ρ_o and Its Invariance Property

The first one, as a specific application of the GLP formalism, concerns the prescription itself of the initial quantum PDF. In fact, in validity of the identification in Equation (65), the constraints in Equation (46) included in *Requirement #1* indicated above actually uniquely prescribe the form of the initial PDF $\rho_o(\Delta g \pm \widehat{g}(r_o))$. In fact, let us introduce for definiteness the Boltzmann-Shannon entropy associated with the same PDF, which is provided by the functional

$$S(\rho_o(\Delta g + \widehat{g}(r_o))) = -\int_{U_g} d(\Delta g)\rho_o(\Delta g + \widehat{g}(r_o)) \ln \rho_o(\Delta g + \widehat{g}(r_o)), \tag{66}$$

with $\rho_o(\Delta g, \widehat{g}_{\mu\nu}(r)) \equiv f(\Delta g + \widehat{g}_{\mu\nu}(r))$ being assumed to satisfy the same constraint equations indicated above (i.e., Equation (46). Then, one can show that the PDF $\rho_o(\Delta g + \widehat{g}(r_o))$ which fulfills the so-called Principle of Entropy Maximization (PEM, Jaynes 1957), namely maximizes $S(\rho_o(\Delta g + \widehat{g}(r_o)))$

when subject to the same constraints, is unique. Straightforward algebra shows that in the whole configuration domain U_g it coincides with the PDF

$$\rho_o(\Delta g \pm \widehat{g}(r_o)) = \frac{1}{\pi^5 r_{th}^{10}} \exp\left\{ -\frac{(\Delta g \pm \widehat{g}(r_o))^2}{r_{th}^2} \right\} \equiv \rho_G(\Delta g \pm \widehat{g}(r_o)), \tag{67}$$

with $\rho_G(\Delta g \pm \widehat{g}(r_o))$ denoting a shifted Gaussian PDF in which both r_{th}^2 and $(\Delta g \pm \widehat{g}(r_o))^2$ are 4-scalars, and in particular r_{th}^2 is a constant independent of (r,s), while

$$(\Delta g \pm \widehat{g}(r_o))^2 \equiv (\Delta g \pm \widehat{g}(r_o))_{\mu\nu} (\Delta g \pm \widehat{g}(r_o))^{\mu\nu}. \tag{68}$$

Therefore, we conclude that the Gaussian PDF in Equation (67) realizes *the most likely PDF*, i.e., the one which, when subject to the constraints (46), maximizes the Boltzmann-Shannon entropy $S(\rho_o(\Delta g + \widehat{g}(r_o)))$ in Equation (66).

Let us now denote with

$$\rho_G(\Delta g \pm \widehat{g}(r)) = \frac{1}{\pi^5 r_{th}^{10}} \exp\left\{ -\frac{(\Delta g \pm \widehat{g}(r))^2}{r_{th}^2} \right\} \tag{69}$$

the Gaussian PDF in Equation (67) evaluated for a generic 4-position $r(s)$ generally different from the initial one $r_o \equiv r(s_o)$. Then, it is possible to show that a formal solution $\rho(G_L(s), \Delta g, s)$ of the quantum continuity equation can more generally be taken of the form

$$\rho(G_L(s), \Delta g, s) = \rho_G(\Delta g \pm \widehat{g}(r)) \exp\left\{ -\int_{s_o}^{s} ds' \frac{\partial V_{\mu\nu}(G_L(s'), \Delta g, s')}{\partial g_{L\mu\nu}(s')} \right\}. \tag{70}$$

Let us display for this purpose an invariance property of the initial PDF. The following proposition is proven to hold.

Proposition 1. Invariance of the Gaussian PDF $\rho_G(\Delta g \pm \widehat{g}(r))$

The following two propositions hold:

P1$_1$) The Gaussian PDF $\rho_G(\Delta g \pm \widehat{g}(r))$ prescribed by Equation (69) satisfies the invariance condition

$$\frac{D}{Ds} \ln \rho_G(\Delta g \pm \widehat{g}(r)) = 0. \tag{71}$$

P1$_2$) Equation (70) realizes a particular solution of the quantum continuity equation in Equation (57).

Proof. To prove the invariance property in Equation (71) in proposition P1$_1$, one first notices that $(\Delta g \pm \widehat{g}(r))^2 \equiv (\Delta g)^2 \pm 2\Delta g_{\mu\nu} \widehat{g}^{\mu\nu}(r) + 4$, where

$$\begin{cases} (\Delta g(s))^2 = (\Delta g(s_o))^2, \\ \Delta g_{\mu\nu}(s) \widehat{g}^{\mu\nu}(r) = \Delta g_{\mu\nu}(s_o) \widehat{g}^{\mu\nu}(r). \end{cases} \tag{72}$$

Consequently, it follows that identically $\frac{D}{Ds}(\Delta g(s))^2 \equiv 0$, while due to the second equation in (72)

$$\frac{D}{Ds} \Delta g_{\mu\nu}(s) \widehat{g}^{\mu\nu}(r) = \Delta g_{\mu\nu}(s_o) \frac{D}{Ds} \widehat{g}^{\mu\nu}(r), \tag{73}$$

where one has that identically $\frac{D}{Ds} \widehat{g}^{\mu\nu}(r) \equiv 0$. Hence, Equation (71) necessarily holds. This implies in turn that Equation (70) is indeed a particular solution of the quantum continuity equation, as can be easily verified by algebraic calculation after substitution in the same equation. This proves Proposition P1$_2$. □

6.3. GLP-Quantum and Stochastic Expectation Values

The second implication of Equation (65) concerns the prescription of the quantum and stochastic expectation values of arbitrary observables which are identified with ordinary tensor functions.

Indeed, first, since $\Delta g \equiv \{\Delta g_{\mu\nu}\}$ is an observable, $\rho_o(\Delta g)$ remains in turn an observable too. Second, the *quantum expectation values* of quantum observables can be determined explicitly, *without performing a separate stochastic average.* In fact, let us consider for definiteness a generic observable which is represented by an ordinary s-dependent real function $X(s) \equiv X(G_L(s), \Delta g, s)$. According to the GLP-representation its quantum expectation value is given by the configuration-space weighted integral (hereon referred to as *GLP-quantum expectation value*):

$$\langle X(s)\rangle = \int_{U_g} d(\delta G_L)\rho(G_L, \Delta g, s)X(G_L, \Delta g, s), \tag{74}$$

where the integration is performed with respect to $\delta G_L \equiv \delta G_L(s)$, keeping constant both $\delta g_{L\mu\nu}(s)$ and the background metric tensor $\widehat{g}(r) \equiv \widehat{g}(r(s))$ in terms of $\rho(G_L, \Delta g, s) \equiv \rho(G_L(s), \Delta g, s)$, the latter being prescribed according to Equation (70). One can show that the following equivalent representations of $\langle X(s)\rangle$ hold.

Proposition 2. Equivalent representations of the GLP-quantum expectation value $\langle X(s)\rangle$

In validity of Proposition 1 and Equation (74), the following equivalent representations of the GLP-quantum expectation value $\langle X(s)\rangle$ hold:

(1) First, $\langle X(s)\rangle$ can be expressed by means of the expectation value in terms of the initial quantum PDF. This yields

$$\langle X(s)\rangle = \int_{U_g} d(\delta G_L(s_o))\rho_G(\Delta g \pm \widehat{g}(r))X(G_L(s), \Delta g, s), \tag{75}$$

where the integration is performed on the initial values of the tensor field $\delta G_L(s_o)$ instead of $\delta G_L(s)$. In the same integral both $\delta g_{L\mu\nu}(s)$ and $\widehat{g}(r(s))$ are again kept constant.

(2) Second, the same integral can also be equivalently performed in terms of the integration variable $\Delta g \equiv \{\Delta g_{\mu\nu}\}$ instead of the initial fields $\delta G_L(s_o)$, thus yielding

$$\langle X(s)\rangle = \int_{U_g} d(\Delta g)\rho_o(\Delta g \pm \widehat{g}(r_o))X(G_L(s), \Delta g, s) \equiv \langle X(s), \widehat{g}(r_o)\rangle_{\Delta g}, \tag{76}$$

where $\langle X(s), \widehat{g}(r_o)\rangle_{\Delta g}$ identifies the stochastic average of $X(G_L(s), \Delta g, s)$, performed in terms of the stochastic PDF $\rho_o(\Delta g \pm \widehat{g}(r_o))$ while again keeping constant $\delta g_{L\mu\nu}(s)$ and $\widehat{g}(r(s))$.

3) Finally, the integral in Equation (76) can also be equivalently performed in terms of the integral

$$\langle X(s)\rangle = \int_{U_g} d(\Delta g)\rho_o(\Delta g \pm \widehat{g}(r(s)))X(G_L(s), \Delta g, s) \equiv \langle X(s), \widehat{g}(r(s))\rangle_{\Delta g}, \tag{77}$$

where $\langle X(s), \widehat{g}(r(s))\rangle_{\Delta g}$ identifies the stochastic average of $X(G_L(s), \Delta g, s)$, performed in terms of the stochastic PDF $\rho_o(\Delta g \pm \widehat{g}(r))$ while keeping constant $\delta g_{L\mu\nu}(s)$ and $\widehat{g}(r)$.

Proof. Consider first Equation (75). Its proof follows by noting that the integral in Equation (74) can be equivalently represented in terms of the inverse mapping $\delta G_L(s) \rightarrow \delta G_L(s_o)$. This implies, in fact, the differential identity

$$d(\delta G_L(s)) = d(\delta G_L(s_o))\left|\frac{\partial \delta G_L(s)}{\partial \delta G_L(s_o)}\right|, \tag{78}$$

where, thanks to Liouville theorem the Jacobian determinant $\left|\frac{\partial \delta G_L(s)}{\partial \delta G_L(s_o)}\right|$ can be shown to be

$$\left|\frac{\partial \delta G_L(s)}{\partial \delta G_L(s_o)}\right| = \exp\left\{\int_{s_o}^{s} ds' \frac{\partial V_{\mu\nu}(G_L(s'),\Delta g, s')}{\partial g_{L\mu\nu}(s')}\right\}. \tag{79}$$

Next, by invoking the solution of the quantum continuity Equation (70), conservation of probability warrants that

$$d(\delta G_L)\rho(G_L(s),\Delta g, s) = d(\delta G_L(s_o))\rho_o(\Delta g \pm \widehat{g}(r(s))), \tag{80}$$

which in turn implies Equation (75). The proof of Equation (76) is obtained in a similar way by noting that (see Equation (43)) $\Delta g_{\mu\nu} = \delta g_{L\mu\nu}(s_o) - \delta G_{L\mu\nu}(s_o)$ so that the same integral (75) can also be equivalently performed in terms of the integration variable $\Delta g \equiv \{\Delta g_{\mu\nu}\}$ while keeping constant $\delta g_{L\mu\nu}(s_o)$ and $\widehat{g}(r)$. Hence, it follows that

$$d(\delta G_L(s_o)) = d(\Delta g)\left|\frac{\partial \delta G_L(s_o)}{\partial \Delta g}\right| = d(\Delta g), \tag{81}$$

since the Jacobian determinant $\left|\frac{\partial \delta G_L(s_o)}{\partial \Delta g}\right|$ is by construction identically equal to 1. Hence, the differential identity (80) necessarily holds, thus yielding also Equation (76). Finally, the proof of Equation (77) follows from Equation (76) being an immediate consequence of Proposition 1. □

6.4. *Generalized Gaussian PDF and Emergent Gravity Interpretation*

Let us examine the implications of the previous Propositions 1 and 2. The first one concerns the determination of the proper-time evolved quantum PDF $\rho(G_L(s),\Delta g, s)$, to be based on Proposition 1 (see the conservation Equation (71)) and Equation (63). This is given by Equation (70). Notice that, although $\rho_G(\Delta g \pm \widehat{g}(r))$ is a shifted Gaussian PDF, $\rho(G_L(s),\Delta g, s)$ is generally not so. Its precise realization depends in fact on the quantum phase-function $S^{(q)}(G_L(s),\Delta g, s)$, i.e., the corresponding solution of the quantum H-J equation (in Equation (57)). As a result, the tensor velocity field $V_{\mu\nu}(G_L(s),\Delta g, s)$ at this stage is still unknown, thus leaving still undetermined the precise functional form of $\rho(G_L(s),\Delta g, s)$, so that in general the proper-time evolved PDF $\rho(G_L(s),\Delta g, s)$, in contrast to the initial PDF, may be generally not Gaussian any more. For this reason, Equation (70) will be referred to in the following as *Generalized Gaussian PDF*.

The second implication, which is also relevant for the physical interpretation of the GLP-approach, concerns the following statement.

Proposition 3. Determination of $\widehat{g}(r)$ (Emergent gravity)
The generalized Gaussian PDF (70) for all $r \in \{\mathbf{Q}^4, \widehat{g}\}$ admits for the stochastic displacement 4-tensor $\Delta g_{\mu\nu}$ the following GLP-quantum/stochastic expectation value (in which both $\delta g_{L\mu\nu}(s)$ and $\widehat{g}(r(s))$ are again kept constant in the integration):

$$\langle \Delta g_{\mu\nu}\rangle \equiv \langle \Delta g_{\mu\nu}\rangle_{\Delta g} = \int_{U_g} d(\Delta g)\rho_G(\Delta g \pm \widehat{g}(r))\Delta g_{\mu\nu} = \mp \widehat{g}_{\mu\nu}(r). \tag{82}$$

Proof. The proof follows as an immediate consequence of Proposition 2 and in particular thanks to Equation (77). □

The consequence is that, in the whole space-time and for all proper-times s (i.e., for arbitrary $(r \equiv r(s), s)$), the local value of the background metric tensor $\widehat{g}(r)$ is prescribed by means of the GLP-quantum expectation value of the stochastic displacement 4-tensor $\Delta g_{\mu\nu}$, i.e., $\langle \Delta g_{\mu\nu}\rangle$, or equivalently by means of the corresponding stochastic average $\langle \Delta g_{\mu\nu}\rangle_{\Delta g}$ evaluated in terms of the stochastic PDF $\rho_G(\Delta g \pm \widehat{g}(r))$. In this regard one notices that for the validity of Proposition 3 the initial PDF must be identified with the stochastic PDF $f(\Delta g, \widehat{g})$, with the latter satisfying the constraint conditions (46). This implies the existence of an emergent gravity phenomenon, in the sense that the background metric tensor $\widehat{g}(r) \equiv \widehat{g}(r(s))$ "emerges" from the quantum gravitational field $g_{\mu\nu}$ as the quantum/stochastic expectation value of the stochastic quantum displacement tensor $\Delta g_{\mu\nu}$ which characterizes the covariant GLP theory.

The conclusion provides a physical interpretation of CQG-theory. Indeed, consistent with the second-type emergent-gravity paradigm referred to above (see Introduction), the background space-time appears through a mean-field gravitational tensor as the result of a suitable ensemble average of an underlying quantum/stochastic virtual space-time whose quantum-wave dynamics is described by GLP trajectories. A notable aspect of the conclusion is, however, that the representation of the proper-time evolved PDF provided by Equation (70) is of general character. In fact, Equation (82) holds independent also of the precise prescription of the classical/quantum effective potential in the quantum Hamiltonian operator. Therefore, *the emergent-gravity interpretation of $\widehat{g}(r)$ is an intrinsic characteristic feature of the GLP-representation* developed here for CQG-theory, whereby the background metric tensor $\widehat{g}(r)$ can be effectively interpreted as arising from the stochastic fluctuations of GLP trajectories having a suitable stochastic probability distribution identified with a Gaussian or more generally Gaussian-like PDF. It follows that $\widehat{g}(r)$ can be then obtained exactly as a statistical moment in terms of weighted integral over the stochastic tensor $\Delta g_{\mu\nu}$.

In this sense, the concept of emergent gravity proposed here has similarities with the analogous one to be found in the literature, namely the conjecture that the geometrical properties of space-time should reveal themselves as a mean field description of microscopic stochastic or quantum degrees of freedom underlying the classical solution [75,76]. However, the physical context proposed here differs from the customary one adopted in the literature, whereby according to the common emergent gravity paradigm the Einstein field equations of gravity should have an emergent character in that, in validity of suitable assumptions, they can be shown to arise from a thermodynamic approach to space-time [77,78]. Nevertheless, the explicit construction of particular solutions of the GLP-parameterized quantum continuity and H-J equations indicated above (see Equation (57)) remains necessary and requires the introduction of suitable representations both for the quantum phase-function $S^{(q)}(G_L(s), \Delta g, s)$ and the quantum effective potential $V(G_L(s), \Delta g, s)$ (see next Section).

7. GLP Approach: Polynomial Decomposition of the Quantum Phase Function

Based on these premises, we can now implement the GLP formalism and proceed constructing particular solutions of the quantum H-J equation (see second part of Equation (57)). More precisely, the goal here is to look for solutions of the quantum phase function expressed in the GLP-parameterization, i.e., $S^{(q)}(G_L(s), \Delta g, s)$, which are expressed by means of polynomial decompositions in terms of power series of the stochastic tensor Δg. For definiteness, in the sequel the case is considered in which the following pre-requisites apply:

(A) "Harmonic" polynomial decomposition of $S^{(q)}(G_L(s), \Delta g, s)$, i.e., the same quantum-phase function is expressed in terms of a second-degree polynomial of the form

$$S^{(q)}(G_L(s), \Delta g, s) = \frac{a_{pq}^{\alpha\beta}(s)}{2} \Delta g_{\alpha\beta} \Delta g^{pq} + b_{\alpha\beta}(s) \Delta g^{\alpha\beta} + c(s), \tag{83}$$

with $a_{\mu\nu}^{\alpha\beta}(s)$, $b_{\mu\nu}(s)$ and $c(s)$ denoting, respectively, suitable real 4-tensors and a 4-scalar functions of s to be determined in terms of the same H-J equation. As shown below, this implies that the effective kinetic energy $T(G_L(s), \Delta g, s)$ defined by Equation (24) and the Bohm effective quantum potential $V_{QM}(G_L(s), \Delta g, s)$ prescribed according to Equation (25) are both realized by means of polynomials of second degree in Δg.

(B) An analogous "Harmonic" polynomial decomposition holds for $V(G_L(s), \Delta g, s)$: namely, that a polynomial representation of analogous type should apply also for the total quantum effective potential density appearing in the quantum H-J equation (see Equation (19)). The latter, to be generally considered of the form $V(G_L(s), \Delta g, s)$, should therefore admit a polynomial representation of the type

$$V(G_L(s), \Delta g, s) = \frac{A_{pq}^{\alpha\beta}(s)}{2} \Delta g_{\alpha\beta} \Delta g^{pq} + B_{\alpha\beta}(s) \Delta g^{\alpha\beta} + C(s), \tag{84}$$

where the tensor coefficients $A_{\mu\nu}^{\alpha\beta}(s)$, $B_{\mu\nu}(s)$ and $C(s)$ are considered here functions of s alone to be suitably determined.

7.1. Implications of the Polynomial Decomposition for $S^{(q)}(G_L(s), \Delta g, s)$

Let us investigate in detail the consequences of the prescription (83) set on the quantum phase-function $S^{(q)}(G_L(s), \Delta g, s)$. One notices, first, that this property permits to identify uniquely the proper-time evolved quantum PDF in terms of a Gaussian PDF, which means that, apart for a proper-time dependent factor, in such a case the PDF $\rho(G_L(s), \Delta g, s)$ becomes intrinsically non-dispersive in character. In this regard the following statement holds.

Proposition 4. Determination of the Gaussian PDF $\rho(G_L(s), \Delta g, s)$
In validity of the harmonic polynomial decomposition in Equation (83), the generalized Gaussian PDF in Equation (70) takes the form of the Gaussian PDF

$$\rho(G_L(s), \Delta g, s) \equiv \rho_G(\Delta g + \widehat{g}(r)) \exp \left\{ -16 \int_{s_0}^{s} ds p^2(s') \frac{a(s')}{\alpha L} \right\}, \tag{85}$$

where $p(s')$ and $a(s')$ are the 4-scalar functions respectively prescribed by Equation (A19) and Equation (A18) in Appendix A.

Proof. The proof follows by noting that in this case the tensor velocity $V_{\mu\nu}(G_L(s), \Delta g, s)$ defined by Equation (48) becomes explicitly

$$V^{\mu\nu}(G_L(s), \Delta g, s) \equiv \frac{1}{\alpha L} \frac{\partial S^{(q)}(G_L(s), \Delta g, s)}{\partial g_{L\mu\nu}(s)} = \frac{a_{pq}^{\alpha\beta}(s)}{\alpha L} \frac{\partial \Delta g_{\alpha\beta}}{\partial g_{L\mu\nu}(s)} \Delta g^{pq} + \frac{1}{\alpha L} \frac{\partial \Delta g_{\alpha\beta}}{\partial g_{L\mu\nu}(s)} b^{\alpha\beta}(s). \tag{86}$$

Consequently, the divergence of the tensor velocity $\frac{\partial V^{\mu\nu}(\Delta g, s')}{\partial g_L^{\mu\nu}(s')}$, which enters the exponential occurring on the R.H.S. of Equation (63), delivers

$$\frac{\partial V^{\mu\nu}(\Delta g, s')}{\partial g_L^{\mu\nu}(s')} = \frac{1}{\alpha L} \frac{\partial^2 S^{(q)}(\Delta g, s')}{\partial g_L^{\mu\nu}(s') \partial g_{L\mu\nu}(s')} = \frac{a_{pq}^{\alpha\beta}(s)}{\alpha L} \frac{\partial \Delta g_{\alpha\beta}}{\partial g_{L\mu\nu}(s')} \frac{\partial \Delta g^{pq}}{\partial g_L^{\mu\nu}(s')}, \tag{87}$$

where the evaluation of the fourth order tensor $\frac{\partial \Delta g_{\alpha\beta}}{\partial g_{L\mu\nu}(s)}$ is reported in Appendix A (see, e.g., Equation (A2) together with Propositions (A1) and (A2)).

Hence, the previous equation implies in turn

$$\frac{\partial V^{\mu\nu}(\Delta g, s)}{\partial g_L^{\mu\nu}(s)} = p^2(s) \frac{a_{pq}^{\alpha\beta}(s)}{\alpha L} \delta_{\alpha\beta}^{\mu\nu} \delta_{\mu\nu}^{pq} \equiv 16 p^2(s) \frac{a(s)}{\alpha L}, \tag{88}$$

where the notation $\delta_{\alpha\beta}^{\mu\nu} \equiv \delta_\alpha^\mu \delta_\beta^\nu$ has been introduced. Consequently, the proper-time evolved quantum PDF in Equation (70) takes the form of Equation (85). □

Next, let us consider the evaluation of effective kinetic energy $T(G_L(s), \Delta g, s)$ defined by Equation (24) and of the Bohm potential given by Equation (25). Regarding $T(G_L(s), \Delta g, s)$, thanks again to Equation (83), direct evaluation delivers

$$T(G_L(s), \Delta g, s) = \frac{p^2(s)}{2\alpha L}\left[a_{\mu\nu}^{\alpha\beta}(s)a_{pq}^{\mu\nu}(s)\Delta g_{\alpha\beta}\Delta g^{pq} + b_{\mu\nu}(s)b^{\mu\nu}(s) + 2a_{\alpha\beta}^{\mu\nu}(s)b_{\mu\nu}(s)\Delta g^{\alpha\beta}\right]. \quad (89)$$

Concerning instead the Bohm potential, one notices that, by invoking Proposition 4 (i.e., Equation (85)), the two source terms on the R.H.S. of Equation (25) become, respectively,

$$\frac{\partial \ln \rho(G_L(s), \Delta g, s)}{\partial g_L^{\mu\nu}(s)} = -\frac{2}{r_{th}^2}p(s)\left(\Delta g_{\mu\nu} \pm \widehat{g}_{\mu\nu}(r)\right), \quad (90)$$

$$\frac{\partial^2 \ln \rho(G_L(s), \Delta g, s)}{\partial g_{L\mu\nu}(s)\partial g_L^{\mu\nu}(s)} = -\frac{8}{r_{th}^2}p^2(s). \quad (91)$$

Consequently, direct substitution in the same equation delivers for the Bohm potential the representation:

$$\begin{aligned} V_{QM}(G_L(s), \Delta g, s) \equiv & -\frac{\hbar^2}{8\alpha L}\left[\frac{2}{r_{th}^2}p(s)\left(\Delta g_{\mu\nu} \pm \widehat{g}_{\mu\nu}(r)\right)\right]\left[\frac{2}{r_{th}^2}p(s)\left(\Delta g^{\mu\nu} \pm \widehat{g}^{\mu\nu}(r)\right)\right] \\ & -\frac{\hbar^2}{4\alpha L}\left[-\frac{8}{r_{th}^2}p^2(s)\right], \end{aligned} \quad (92)$$

which can be equivalently written as

$$V_{QM}(G_L(s), \Delta g, s) \equiv -\frac{\hbar^2 p^2(s)}{2\alpha L r_{th}^4}\left(\Delta g_{\mu\nu}\Delta g^{\mu\nu} \pm 2\widehat{g}_{\mu\nu}(r)\Delta g^{\mu\nu} + 4\right) + \frac{2\hbar^2 p^2(s)}{\alpha L r_{th}^2}. \quad (93)$$

7.2. Implications of the Polynomial Decomposition for $V(G_L(s), \Delta g, s)$

That an explicit realization of the polynomial representation of the type in Equation (84) is actually possible for the effective classical potential density $V(G_L(s), \Delta g, s)$ given by Equation (19) follows by its definition. For definiteness, let us show how this task can be achieved for a specific realization, i.e., in the case of vacuum. The following proposition holds.

Proposition 5. Harmonic representation of the vacuum effective potential

The vacuum effective potential in Equation (19) in the harmonic polynomial representation in Equation (84) takes the form

$$V_o(g + \Delta g) = 2\sigma\alpha L\Lambda + \sigma\alpha L\Lambda\left[-\frac{1}{2}\Delta g_{\mu\nu}\Delta g^{\mu\nu} - \frac{1}{2}\Delta g^{\mu\nu}\widehat{g}_{\mu\nu}(r)\Delta g^{\alpha\beta}\widehat{g}_{\alpha\beta}(r)\right]. \quad (94)$$

Proof. In fact, from Equation (61), the vacuum effective potential $V_o\left(\widehat{g} + \Delta g\right)$ becomes

$$V_o(g + \Delta g) \equiv \sigma\alpha L\Lambda\left[2 - \frac{1}{4}\left(\widehat{g}_{\mu\nu}(r) + \Delta g_{\mu\nu}\right)\left(\widehat{g}^{\mu\nu}(r) + \Delta g^{\mu\nu}\right)\right]\left[\left(\widehat{g}_{pq}(r) + \Delta g_{pq}\right)\widehat{g}^{pq}(r) - 2\right]. \quad (95)$$

The harmonic representation is obtained dropping terms of order $(\Delta g)^3$ or higher. When this is done in the previous equation, Equation (94) is recovered at once. □

The form of the source term in Equation (94) suggests to seek for the tensor coefficient $a_{\mu\nu}^{\alpha\beta}(s)$ in Equation (83) a particular realization of the form

$$a_{pq}^{\alpha\beta}(s) = \frac{1}{2}\left[a_{(o)}(s)\delta_{pq}^{\alpha\beta} + a_{(1)}(s)\widehat{g}_{pq}(r)\widehat{g}^{\alpha\beta}(r)\right], \quad (96)$$

so that upon invoking Equation (A18), namely letting $a^{\alpha\beta}_{\mu\nu}(s)\delta^{\mu\nu}_{\alpha\beta} \equiv 4a(s)$, it follows $a(s) = \frac{1}{2}\left[a_{(o)} + a_{(1)}\right]$. Consequently, one finds that the tensor coefficients $a^{\alpha\beta}_{pq}(s)$ in Equation (83) can also be written as

$$a^{\alpha\beta}_{pq}(s) = \frac{1}{2}\left[2a(s)\delta^{\alpha\beta}_{pq} + a_{(1)}(s)\left(\widehat{g}_{pq}(r)\widehat{g}^{\alpha\beta}(r) - \delta^{\alpha\beta}_{pq}\right)\right]. \tag{97}$$

In addition, straightforward algebra yields the identities represented by Equations (A23)–(A29) which are reported in Appendix B.

7.3. Construction of the GLP-Equations

We now pose the problem of the construction of the set of ODEs, which, in validity of the Harmonic polynomial decompositions indicated above, determine a separable solution of the quantum H-J equation in Equation (57), and are thus equivalent to the same equation. In the case of the vacuum effective potential by equating all terms in the polynomial expansion, one obtains a set of ODEs for the 4-scalar coefficients $a_{(o)}(s)$, $a_{(1)}(s)$ and $c(s)$ and the 4-tensor $b_{\alpha\beta}(s)$, here referred to as *GLP-equations*. These are provided by the first-order ODEs:

$$\begin{cases} \frac{1}{4}\frac{d}{ds}a_{(o)}(s) = \frac{p^2(s)}{8\alpha L}a^2_{(o)}(s) - \frac{\hbar^2}{2\alpha L}\frac{1}{r^4_{th}}p^2(s) + \frac{1}{2}\sigma\alpha L\Lambda + G_{(o)}, \\ \frac{1}{4}\frac{d}{ds}a_{(1)}(s) = \frac{p^2(s)}{8\alpha L}\left(4a^2_{(1)}(s) + 2a_{(o)}(s)a_{(1)}(s)\right) + \frac{1}{2}\sigma\alpha L\Lambda + G_{(1)}, \\ \frac{d}{ds}b_{\alpha\beta}(s) = \frac{p^2(s)}{2\alpha L}\left[b_{\alpha\beta}a_{(o)}(s) + a_{(1)}(s)\widehat{g}_{\alpha\beta}(r)\widehat{g}^{\mu\nu}(r)b_{\mu\nu}(s)\right], \\ \frac{d}{ds}c(s) = \frac{p^2(s)}{2\alpha L}b_{\mu\nu}(s)b^{\mu\nu}(s) + \frac{2\hbar^2}{\alpha L}\frac{1}{r^2_{th}}p^2(s) + C_o(s), \end{cases} \tag{98}$$

where $G_{(o)}$, $G_{(1)}$ and $C_o(s)$ are in principle arbitrary 4-scalar gauge functions. These can be prescribed in such a way that there exists a stationary null solution for the 4-scalar coefficient $a(s) \equiv \widehat{a}(s)$, namely such that for all $s \in I$, $\widehat{a}(s) = 0$, and hence identically for all $s, s_o \in I$,

$$\begin{cases} \widehat{a}_{(o)}(s) \equiv \widehat{a}_{(o)}(s_o), \\ \widehat{a}_{(1)}(s) = \widehat{a}_{(1)}(s_o), \\ \widehat{a}_{(1)}(s_o) = -\widehat{a}_{(o)}(s_o), \end{cases} \tag{99}$$

which realizes a particular stationary solution of Equation (98). This requires suitably-identifying the gauge functions $G_{(o)}$ and $G_{(1)}$, which, for consistency with Equation (99), can always be prescribed in such a way that

$$\begin{cases} G_{(o)} = -\frac{1}{8\alpha L}\widehat{a}^2_{(o)}(s_o) + \frac{\hbar^2}{2\alpha L}\frac{1}{r^4_{th}} - \frac{1}{2}\sigma\alpha L\Lambda, \\ G_{(1)} = -\frac{1}{4\alpha L}\widehat{a}^2_{(o)}(s_o) - \frac{1}{2}\sigma\alpha L\Lambda \equiv 0, \end{cases} \tag{100}$$

so that the first two parts of Equation (98) can be written explicitly as

$$\begin{cases} \frac{1}{4}\frac{d}{ds}a_{(o)}(s) = \frac{1}{8\alpha L}\left[p^2(s)a^2_{(o)}(s) - 2\alpha^2L^2\Lambda\right] - \frac{\hbar^2}{2\alpha L}\frac{1}{r^4_{th}}\left[p^2(s) - 1\right], \\ \frac{1}{4}\frac{d}{ds}a_{(1)}(s) = \frac{p^2(s)}{8\alpha L}\left(4a^2_{(1)}(s) + 2a_{(o)}(s)a_{(1)}(s)\right) - \frac{1}{2}\alpha L\Lambda. \end{cases} \tag{101}$$

We finally notice that the previous equations can also be conveniently cast in dimensionless form. Noting that $[a] = \left[a_{(o)}\right] = \left[a_{(1)}\right] = [\hbar] = [\alpha]$, the dimensionless representation is obtained by means of the dimensionless variables

$$\begin{cases} \overline{a}_{(o)}(\theta) = \frac{a_{(o)}(\theta)}{\alpha}, \\ \overline{a}_{(1)}(\theta) = \frac{a_{(1)}}{\alpha}, \\ \overline{b}_{\alpha\beta}(\theta) = \frac{b_{\alpha\beta}(s)}{\alpha}, \\ \overline{c}(\theta) = \frac{c}{\alpha}, \\ \theta = \frac{2s}{L}, \\ \overline{\Lambda} = \Lambda L^2 \cong 9.408, \end{cases} \tag{102}$$

where $\overline{\Lambda}$ identifies in dimensionless units the experimental value of cosmological constant, here evaluated in terms of the Compton Length L which corresponds to the graviton-mass estimate given in [10]. Then, introducing the notations

$$\begin{cases} Y(\theta) \equiv \left(1 + \int\limits_{\theta_o}^{\theta} d\theta' \overline{a}(\theta')\right)^{1/2}, \\ Z(\theta) = \frac{\overline{a}_{(1)}(\theta)}{Y(\theta)^2}, \end{cases} \tag{103}$$

Equation (98) can be shown to be equivalent to the following set of ODEs for the coefficients $\overline{a}(\theta)$ and $\overline{a}_{(1)}(\theta)$:

$$\begin{cases} \frac{d^2}{d\theta^2} Y(\theta) = \frac{3}{16} \frac{Z^2(\theta)}{Y(\theta)} - \frac{3}{16} \frac{\overline{\Lambda}}{Y(\theta)} + \frac{\hbar^2}{4\alpha^2 r_{th}^4} \frac{Y(\theta)^2 - 1}{Y(\theta)^3}, \\ \frac{d}{d\theta} Z(\theta) = \frac{Z^2(\theta)}{Y(\theta)} - \frac{1}{2} \frac{\overline{\Lambda}}{Y(\theta)}, \end{cases} \tag{104}$$

which admit the stationary solution

$$\begin{cases} \overline{a}(s) = 0, \\ \overline{a}_{(o)}^2(s) - \frac{1}{2}\overline{\Lambda} = 0. \end{cases} \tag{105}$$

7.4. Small-Amplitude Solutions: Conditions of Validity

Now, we look for small-amplitude solutions of Equation (104). For definiteness, let us introduce the representations

$$\begin{cases} Y(\theta) = Y(\theta_o) + \delta Y(\theta), \\ Z(\theta) = Z(\theta_o) + \delta Z(\theta), \end{cases} \tag{106}$$

with $Y(\theta_o) = 1$, $Z(\theta_o) = \overline{a}_{(1)}(\theta_o) = \pm\sqrt{\frac{1}{2}\overline{\Lambda}}$ and $\delta Y(\theta)$, $\delta Z(\theta)$ denoting displacements such that for all $\theta \in I_{s_{\theta o}}^{(+)} \equiv [\theta_o, +\infty]$

$$\begin{aligned} 0 &< \delta Y(\theta) \ll 1, \\ 0 &< \left| \delta Z(\theta) / \sqrt{\frac{1}{2}\overline{\Lambda}} \right| \ll 1. \end{aligned} \tag{107}$$

These will be denoted as small-amplitude solutions. In this regard, the following proposition holds.

Proposition 6. Small-amplitude solutions of Equation (104)
For all $s \in I_{s_o}^{(+)} \equiv [s_o, +\infty]$ Equation (104) admit small-amplitude solutions.

Proof. In fact, upon linearization, Equation (104) implies

$$\begin{cases} \frac{d^2}{d\theta^2} \delta Y(\theta) = \frac{3}{16} \left[\pm 2\sqrt{\frac{1}{2}\overline{\Lambda}} \delta Z - \frac{1}{2}\overline{\Lambda} \delta Y(\theta) \right] + \frac{3}{16}\overline{\Lambda} \delta Y(\theta) + \frac{\hbar^2}{4\alpha^2 r_{th}^4} 2\delta Y(\theta), \\ \frac{d}{d\theta} \delta Z(\theta) = \pm 2\sqrt{\frac{1}{2}\overline{\Lambda}} \delta Z. \end{cases} \tag{108}$$

The two equations deliver, respectively, the solutions

$$\left\{ \begin{array}{l} \delta Z(\theta) = \delta Z(\theta_o) \exp \left\{ \pm 2\sqrt{\frac{1}{2}\overline{\Lambda}}(\theta - \theta_o) \right\}, \\ \delta a(\theta) = A \delta Z(\theta_o) \exp \left\{ \pm 2\sqrt{\frac{1}{2}\overline{\Lambda}}(\theta - \theta_o) \right\}, \end{array} \right. \tag{109}$$

with A denoting the constant coefficient

$$A = \frac{\frac{3}{4}\overline{\Lambda}}{1 - \frac{9}{32}\overline{\Lambda} - \frac{\hbar^2}{2\alpha^2 r_{th}^4}}. \tag{110}$$

Consequently, Equation (109) implies also that

$$\delta a_{(1)}(\theta) = \delta Z(\theta) \left[1 + \frac{A}{2} \right]. \tag{111}$$

□

Thus, we conclude that small-amplitude solutions of the GLP-equations in Equation (104) indeed exist which depend exponentially on proper time, the exponential factor being of the form $\exp\left\{-2\sqrt{\frac{1}{2}\overline{\Lambda}}(\theta - \theta_o)\right\}$ or $\exp\left\{2\sqrt{\frac{1}{2}\overline{\Lambda}}(\theta - \theta_o)\right\}$, respectively. These are referred to, respectively, as *decay* and *blow-up* small-amplitude solutions. In the two cases for $\theta - \theta_o \to +\infty$, these either decay to the constant solution or diverge exponentially. Therefore, quantum stationary solutions can be identified with asymptotic ones, i.e., as final states of decaying quantum solutions. Blow-up solutions, however, for finite times $\theta - \theta_o > 0$ necessarily violate the ordering assumptions (107) and, as such, Equations (109) and (111) are no longer applicable in such a case.

The investigation of the blow-up solutions requires therefore the proper consideration of the set of GLP-equations in Equation (104). One can show, however, that, if the following asymptotic orderings apply,

$$Y(\theta) \gg 1, \tag{112}$$

$$\left| Z(\theta) / \sqrt{\frac{1}{2}\overline{\Lambda}} \right| \gg 1, \tag{113}$$

then, in such a case, the asymptotic limits must apply

$$\lim_{\theta - \theta_o \to +\infty} Y(\theta) = +\infty, \tag{114}$$

$$\lim_{\theta - \theta_o \to +\infty} \frac{d}{d\theta} Y(\theta) = 0. \tag{115}$$

These imply in turn also the vanishing of the 4-scalar coefficient $p(s)$ (see Appendix A) in the proper-time limit $s - s_o \to +\infty$, i.e.,

$$\lim_{s - s_o \to +\infty} p(s) = 0. \tag{116}$$

The implication of Equation (116) is however the violation in the same limit of the Heisenberg inequality

$$\left\langle \left(\Delta g_{(\mu)(\nu)} \right)^2 \right\rangle \left\langle (\Delta \pi_{\mu\nu})^2 \right\rangle_1 \geq \frac{\hbar^2}{4}, \tag{117}$$

pointed out in [11], with $\left\langle \left(\Delta g_{(\mu)(\nu)} \right)^2 \right\rangle$ and $\left\langle (\Delta \pi_{\mu\nu})^2 \right\rangle_1$ denoting respectively

$$\left\langle \left(\Delta\pi_{\mu\nu}\right)^2\right\rangle_1 = \frac{\hbar^2}{4}\int_{U_g} d(g)\rho\frac{\partial\ln\rho}{\partial g^{\mu\nu}}\frac{\partial\ln\rho}{\partial g^{(\mu)(\nu)}}, \tag{118}$$

$$\left\langle \left(\Delta g_{\mu\nu}\right)^2\right\rangle = \int_{U_g} d(g)\rho\left(g_{\mu\nu}-\widetilde{g}_{\mu\nu}\right)\left(g_{(\mu)(\nu)}-\widetilde{g}_{(\mu)(\nu)}\right) = \frac{1}{10}r_{th}^2. \tag{119}$$

In fact, due to Equation (116), it follows that

$$\lim_{s-s_o\to+\infty}\left\langle \left(\Delta g_{\mu\nu}\right)^2\right\rangle = 0. \tag{120}$$

Instead, one can show that constant or small-amplitude decaying solutions satisfy the Heisenberg inequality in Equation (117) and as such realize physically admissible quantum solutions. Such a conclusion, therefore, rules out blow-up solutions from the class of physically-admissible solutions in the same limit.

8. Conclusions

In this paper, the basic principles of a new trajectory-based approach to manifestly-covariant quantum gravity (CQG) theory have been laid down. This provide new physical insight into the nature and behavior of the manifestly-covariant quantum-wave equation and corresponding equivalent set of quantum hydrodynamic equations that are realized by means of CQG-theory. For its similarity with the analogous Generalized Lagrangian Path approach holding in non relativistic quantum mechanics [22], this is referred to here as Generalized Lagrangian Path (GLP) approach (or representation) of CQG-theory.

The GLP approach presented here has been shown to be ontologically equivalent to the "standard" formulation of CQG-theory based on the Eulerian CQG-wave equation. This occurs because, provided the stochastic PDF $f(\Delta g, \widehat{g})$ is identified with the Gaussian PDF $\rho_G(\Delta g \pm \widehat{g}(r_o))$ defined above (see Equation (69)), it does not require any kind of addition/modification of the related fundamental axioms established in [10]. This feature permits one to effectively reconcile the Eulerian and Lagrangian descriptions of covariant quantum gravity, which are achieved respectively in terms of the Eulerian and GLP representations of CQG-wave equation and of the quantum wave-function. Nevertheless, it also provides a statistical generalization of the Bohmian interpretation of quantum gravity based on the notion of unique, i.e., deterministic, configuration-space Lagrangian trajectories belonging to the configuration space U_g spanned by the symmetric tensor field $g \equiv \{g_{\mu\nu}\}$. In fact, in the framework of GLP-theory, each Bohmian trajectory is associated with an infinite ensemble of stochastic Lagrangian trajectories associated with the stochastic tensor variable $\Delta g_{\mu\nu}$. Thus, GLP trajectories replace the customary deterministic Lagrangian trajectories (LPs) adopted in the original Bohmian approach, from which they inherently differ for their stochastic character. Consequently, it is shown that it is possible to replace each LP with a corresponding continuum set of stochastic GLP.

A further notable aspect of the GLP approach is, however, that it realizes at the same time also a solution method for the CQG-wave equation and the corresponding equivalent quantum hydrodynamic equations. This is obtained by means of the explicit parameterization of the same equations (and of the quantum wave-function) in terms of the stochastic displacement tensor $\Delta g_{\mu\nu}$ introduced here (see Equation (11)). As an application of the theory developed in this paper, the problem of constructing Gaussian or Gaussian-like solutions of the CQG-wave equation has been addressed. For this purpose, the case of vacuum fields, i.e., obtained in the absence of external classical sources but with the inclusion of a non-vanishing cosmological constant, has been considered. In this connection the explicit construction of solutions of the CQG-quantum hydrodynamic equations has been carried out in which the GLP-parameterized quantum wave function $\psi(G_L(s), \Delta g, s)$ is characterized by a globally-defined Gaussian-like or Gaussian PDF which satisfies identically the corresponding quantum continuity equation. As a notable result, the validity of the emergent-gravity picture has been demonstrated, referred to here as "*second-type emergent-gravity paradigm*". Accordingly, the background space-time metric tensor $\widehat{g}_{\mu\nu}(r)$ of CQG-theory has been identified in terms of a suitable quantum/stochastic expectation value of the quantum state, i.e., weighted in terms of the corresponding quantum PDF.

In addition, the problem of the construction of separable solutions of the quantum Hamilton-Jacobi (H-J) equation has been posed which satisfy at the same time also the requirements that the quantum wave function $\psi(G_L(s), \Delta g, s)$ is dynamically consistent, in the sense that the corresponding (GLP-parameterized) quantum PDF $\rho(G_L(s), \Delta g, s)$ associated with the quantum wave-function is globally conserved. The solution of the H-J equation has been based on the polynomial representations of the quantum effective potential. In particular, separable solutions for the GLP-parameterized quantum phase function $S(G_L(s), \Delta g, s)$ have been determined based on a harmonic (i.e., second degree) polynomial expansion with respect to the stochastic displacement tensor $\Delta g_{\mu\nu}$. The coefficients of the same expansion have been shown to satisfy an equivalent set of first-order evolution ODEs, denoted as GLP-equations. The same coefficients admit both stationary and non-stationary solutions with respect to the dependence on the background proper-time s. Non-stationary solutions include, in particular, the case of small-amplitude solutions which remain globally (i.e., for all s greater than the initial proper-time s_o) suitably close to the stationary ones. These have been identified here with particular solutions exponentially decaying (to the constant ones).

These conclusions show that particular solutions of the CQG-quantum wave-equation exist which are characterized by Gaussian quantum PDF. Remarkably, the same solutions can be either stationary, i.e., characterized by quantum wave-functions of the type $\psi = \psi(G_L(s), \Delta g)$, or non-stationary ones $\psi(G_L(s), \Delta g, s)$, namely depending explicitly on the proper-time s. This scenario is promising for its possible implications suggesting that the investigation of non-stationary solutions of the quantum wave-function may be actually an important and challenging subject of future research in quantum gravity, quantum cosmology and CQG-theory.

Acknowledgments: Work developed within the research project of the Albert Einstein Center for Gravitation and Astrophysics, Czech Science Foundation No. 14-37086G (M.T.). The publication of this work was supported by the Fetzer Franklin Fund of the John E. Fetzer Memorial Trust.

Author Contributions: The paper is a part of a long-standing collaboration between the two authors which ranges from Mathematical Physics and Quantum Mechanics to Quantum Gravity and Cosmology. The ideas expressed here are fruit of lengthy-debated joint discussions between the authors. The initial motivation of the work is related to the extension of the GLP-approach originally developed for Non-Relativistic Quantum Mechanics (Massimo Tessarotto and Massimo Tessarotto, 2016) to the context of CQG-theory.

Conflicts of Interest: The authors declare no conflict of interest.

Appendix A. Evaluation of $p(s)$ and Differential Iden-Tities

In this appendix, the proof of Equation (85) in Proposition #4 and the determination of the 4-scalar factor $p(s)$ are explicitly pointed out in the following propositions.

Proposition A1. Determination of the tensor field $\frac{\partial \Delta g_{\alpha\beta}}{\partial g_{L\mu\nu}(s')}$

Given validity of the polynomial representation in Equation (83), the tensor field $\frac{\partial \Delta g_{\alpha\beta}}{\partial g_{L\mu\nu}(s')}$ takes the form

$$\frac{\partial \Delta g_{\mu'\nu'}}{\partial g_{L\mu\nu}(s')} = -\frac{\partial \Delta g_{\mu'\nu'}}{\partial G_{L\mu\nu}(s')}, \tag{A1}$$

with

$$\frac{\partial \Delta g_{\mu'\nu'}}{\partial g_{L\mu\nu}(s')} = \delta^{\mu}_{\mu'}\delta^{\nu}_{\nu'}p(s), \tag{A2}$$

and $p(s)$ is the 4-scalar function determined by the integral equation

$$p(s) = \frac{1}{1 + \int_{s_o}^{s} ds' \frac{1}{\alpha L} a(s') g(s')}. \tag{A3}$$

Here, $a(s) \equiv \frac{1}{16} a_{\alpha\beta}^{pq}(s) \delta_{pq}^{\alpha\beta}$ and $a_{\alpha\beta}^{pq}(s)$ is the tensor introduced in the polynomial decomposition of the phase function $S^{(q)}$ given by Equation (83).

Proof. One first notices that, provided the quantum phase function is of the form $S^{(q)} = S^{(q)}(\Delta g, s')$, and noting that $\delta g_{L\mu\nu}^{(o)} = \delta G_{L\mu\nu}^{(o)} + \Delta g_{\mu\nu}$, then the LP-initial-value problem in Equation (30) delivers

$$\delta g_{L\mu\nu}(s) = \delta g_{L\mu\nu}^{(o)} + \int_{s_o}^{s} ds' \frac{1}{\alpha L} \frac{\partial S^{(q)}(\Delta g, s')}{\partial g_L^{\mu\nu}(s')}, \tag{A4}$$

or equivalently

$$\delta g_{L\mu\nu}(s) = \delta G_{L\mu\nu}^{(o)} + \Delta g_{\mu\nu} + \int_{s_o}^{s} ds' \frac{1}{\alpha L} \frac{\partial S^{(q)}(\Delta g, s')}{\partial g_L^{\mu\nu}(s')}. \tag{A5}$$

The last equation therefore implies also that the solution to the GLP-initial-value problem in Equation (50) is similarly

$$\delta G_{L\mu\nu}(s) = \delta g_{L\mu\nu}(s_o) - \Delta g_{\mu\nu} + \int_{s_o}^{s} ds' \frac{1}{\alpha L} \frac{\partial S^{(q)}(\Delta g, s')}{\partial g_L^{\mu\nu}(s')}. \tag{A6}$$

Then, differentiating Equation (A5) with respect to $\delta g_{L\mu\nu}(s)$ while keeping $\delta G_{L\mu\nu}(s_o) \equiv \delta G_{L\mu\nu}^{(o)}$ constant, yields

$$\delta_{\mu'}^{\mu} \delta_{\nu'}^{\nu} \equiv \frac{\partial g_{L\mu'\nu'}(s)}{\partial g_{L\mu\nu}(s)} = \frac{\partial \Delta g_{\mu'\nu'}}{\partial g_{L\mu\nu}(s)} + \frac{\partial \Delta g_{\alpha\beta}}{\partial g_{L\mu\nu}(s)} \frac{\partial}{\partial \Delta g_{\alpha\beta}} \int_{s_o}^{s} ds' \frac{1}{\alpha L} \frac{\partial S^{(q)}(\Delta g, s')}{\partial g_L^{\mu'\nu'}(s')}, \tag{A7}$$

where in the following we shall adopt the short notation $\delta_{\mu'\nu'}^{\mu\nu} \equiv \delta_{\mu'}^{\mu} \delta_{\nu'}^{\nu}$ and by construction

$$\frac{\partial S^{(q)}(\Delta g, s')}{\partial G_L^{\mu'\nu'}(s')} = -\frac{\partial S^{(q)}(\Delta g, s')}{\partial g_L^{\mu'\nu'}(s')}, \tag{A8}$$

and hence

$$\frac{\partial \Delta g_{\mu'\nu'}}{\partial G_L^{\mu\nu}(s)} = -\frac{\partial \Delta g_{\mu'\nu'}}{\partial g_L^{\mu\nu}(s)}. \tag{A9}$$

Consequently, if one performs the differentiation of Equation (A6) with respect to $G_{L\mu\nu}(s)$ while keeping $\delta g_{L\mu\nu}(s_o) \equiv \delta g_{L\mu\nu}^{(o)}$ as constant, it follows equivalently that

$$\delta_{\mu'}^{\mu} \delta_{\nu'}^{\nu} \equiv \frac{\partial G_{L\mu'\nu'}(s)}{\partial G_{L\mu\nu}(s)} = -\frac{\partial \Delta g_{\mu'\nu'}}{\partial G_{L\mu\nu}(s)} - \frac{\partial \Delta g_{\alpha\beta}}{\partial G_{L\mu\nu}(s)} \frac{\partial}{\partial \Delta g_{\alpha\beta}} \int_{s_o}^{s} ds' \frac{1}{\alpha L} \frac{\partial S^{(q)}(\Delta g, s')}{\partial g_L^{\mu'\nu'}(s')}. \tag{A10}$$

Therefore, from Equation (A7), denoting $\delta_{\mu'\nu'}^{\mu\nu} \equiv \delta_{\mu'}^{\mu} \delta_{\nu'}^{\nu}$, it follows

$$\delta_{\mu\nu}^{\mu'\nu'} = \frac{\partial \Delta g^{\mu'\nu'}}{\partial g_L^{\mu\nu}(s)} + \frac{\partial \Delta g^{\alpha\beta}}{\partial g_L^{\mu\nu}(s)} \frac{\partial}{\partial \Delta g^{\alpha\beta}} \int_{s_o}^{s} ds' \frac{1}{\alpha L} \frac{\partial S^{(q)}(\Delta g, s')}{\partial g_{L\mu'\nu'}(s')}, \tag{A11}$$

where due to the polynomial representation in Equation (83)

$$\frac{\partial S^{(q)}(g_L(s'),\Delta g,s')}{\partial g_{L\mu'\nu'}(s')} = \frac{\partial \Delta g_{pq}}{\partial g_{L\mu'\nu'}(s')}\left[a^{pq}_{p'q'}(s')\Delta g^{p'q'} + b^{pq}(s)\right], \tag{A12}$$

$$\frac{\partial}{\partial \Delta g^{\alpha\beta}}\frac{\partial S^{(q)}(g_L(s'),\Delta g,s')}{\partial g_{L\mu'\nu'}(s')} = a^{\mu\nu}_{\alpha\beta}(s')\frac{\partial \Delta g_{\mu\nu}}{\partial g_{L\mu'\nu'}(s')}. \tag{A13}$$

As a result Equation (A11) delivers

$$\delta^{\mu'\nu'}_{\mu\nu} = \frac{\partial \Delta g^{\mu'\nu'}}{\partial g^{\mu\nu}_L(s)} + \frac{\partial \Delta g^{\alpha\beta}}{\partial g^{\mu\nu}_L(s)}\int_{s_o}^{s} ds' \frac{a^{pq}_{\alpha\beta}(s')}{\alpha L}\frac{\partial \Delta g_{pq}}{\partial g_{L\mu'\nu'}(s')}, \tag{A14}$$

thus implying validity of Equation (A2). In fact, thanks to Equation (A2), we can write the previous equation as

$$\delta^{\mu'\nu'}_{\mu\nu} = \delta^{\mu'\nu'}_{\mu\nu}\, g(s) + \delta^{\alpha\beta}_{\mu\nu} g(s)\int_{s_o}^{s} ds' \frac{a^{pq}_{\alpha\beta}(s')}{\alpha L}\delta^{\mu'\nu'}_{pq}\, g(s'). \tag{A15}$$

Then, defining

$$a(s')\,\delta^{\mu'\nu'}_{\mu\nu} \equiv \delta^{\alpha\beta}_{\mu\nu} a^{pq}_{\alpha\beta}(s')\delta^{\mu'\nu'}_{pq} \tag{A16}$$

and substituting, after simplification we get that

$$g(s)\left[1 + \int_{s_o}^{s} ds' \frac{1}{\alpha L} a(s')g(s')\right] = 1, \tag{A17}$$

while straightforward algebra yields

$$a(s') = \frac{1}{16}\delta^{\alpha\beta}_{\mu\nu} a^{pq}_{\alpha\beta}(s')\delta^{\mu'\nu'}_{pq}\,\delta^{\mu\nu}_{\mu'\nu'} \equiv \frac{1}{16} a^{pq}_{\alpha\beta}(s')\delta^{\alpha\beta}_{pq}. \tag{A18}$$

Thus, provided $1 + \int_{s_o}^{s} ds' \frac{1}{\alpha L} a(s')g(s') \neq 0$, Equation (A3) follows. □

Proposition A2. Determination of the 4-scalar function $p(s)$

In validity of Equation (A3), it follows that

$$|p(s)| = \frac{1}{\left(1 + \frac{2}{\alpha L}\int_{s_o}^{s} ds' a(s')\right)^{1/2}}. \tag{A19}$$

Proof. In fact, if $p(s) \neq 0$, Equation (A17) implies

$$1 + \int_{s_o}^{s} ds' \frac{1}{\alpha L} a(s')g(s') = \frac{1}{p(s)}. \tag{A20}$$

Differentiating the same equation term by term with respect to s yields the ODE

$$\frac{1}{\alpha L} a(s)p(s) = -\frac{p'(s)}{p^2(s)}. \tag{A21}$$

This can be solved noting that $p(s_o) = 1$. Thus, one finds

$$\frac{1}{2p(s)^2} - \frac{1}{2} = \frac{1}{\alpha L}\int_{s_o}^{s} ds' a(s'), \tag{A22}$$

whose solution is given by Equation (A19). □

Appendix B. Differential Identities for the Tensor Coefficients $a_{pq}^{\alpha\beta}(s)$

In this appendix, the explicit calculations are reported for several useful identities invoked in Section 7. First, one notices that, invoking Equation (96), it follows that

$$a_{\mu\nu}^{\alpha\beta}(s)a_{pq}^{\mu\nu}(s) = \frac{1}{4}\left[a_{(o)}^{2}(s)\delta_{pq}^{\alpha\beta} + \left(4a_{(1)}^{2}(s) + 2a_{(o)}(s)a_{(1)}(s)\right)\widehat{g}_{pq}(s)\widehat{g}^{\alpha\beta}(s)\right], \tag{A23}$$

and similarly

$$4a_{(1)}^{2}(s) + 2a_{(o)}(s)a_{(1)}(s) = 2a_{(1)}^{2}(s) + 4a(s)a_{(1)}(s), \tag{A24}$$

$$\begin{aligned} &a_{(o)}^{2}(s) + 4a_{(1)}^{2}(s) + 2a_{(o)}(s)a_{(1)}(s) = \\ &= \left[a_{(o)}(s) + a_{(1)}(s)\right]^{2} + 3a_{(1)}^{2}(s) = 4a^{2}(s) + 3a_{(1)}^{2}(s). \end{aligned} \tag{A25}$$

The prescription in Equation (96) implies therefore:
(A) from Equation (84):

$$\begin{aligned} \frac{D}{Ds}S^{(q)}(\Delta g, s) &= \frac{1}{4}\Delta g_{\alpha\beta}\Delta g^{\mu\nu}\left[\frac{d}{ds}a_{(o)}(s)\delta_{\mu\nu}^{\alpha\beta} + \widehat{g}_{\mu\nu}(r)\widehat{g}^{\alpha\beta}(s)\frac{d}{ds}a_{(1)}(s)\right] \\ &\quad + \Delta g^{\alpha\beta}\frac{d}{ds}b_{\alpha\beta}(s) + \frac{d}{ds}c(s); \end{aligned} \tag{A26}$$

(B) from Equation (89):

$$\frac{\partial S^{(q)}(G_L(s), \Delta g, s)}{\partial g_L^{\mu\nu}(s)} = \frac{1}{2}\left[a_{(o)}(s)\delta_{\mu\nu}^{\alpha\beta} + a_{(1)}(s)\widehat{g}_{\mu\nu}(r)\widehat{g}^{\alpha\beta}(r)\right]\Delta g_{\alpha\beta}p(s) + b_{\mu\nu}(s)p(s). \tag{A27}$$

Hence, the quantum 4-tensor fluid velocity field can be represented as

$$V_{\mu\nu} = \frac{1}{2\alpha L}\left[a_{(o)}(s)\delta_{\mu\nu}^{\alpha\beta} + a_{(1)}(s)\widehat{g}_{\mu\nu}(r)\widehat{g}^{\alpha\beta}(r)\right]\Delta g_{\alpha\beta}p(s) + \frac{1}{\alpha L}b_{\mu\nu}(s)p(s), \tag{A28}$$

with the first term on the R.H.S., linearly proportional to Δg, representing the stochastic part of the quantum fluid velocity. Similarly one obtains that, in Equation (89), the following identities hold:

$$\left\{\begin{array}{c} a_{\mu\nu}^{\alpha\beta}(s)a_{pq}^{\mu\nu}(s) = \frac{1}{4}\left[a_{(o)}^{2}(s)\delta_{pq}^{\alpha\beta} + \left(4a_{(1)}^{2}(s) + 2a_{(o)}(s)a_{(1)}(s)\right)\widehat{g}_{pq}(s)\widehat{g}^{\alpha\beta}(s)\right], \\ 2a_{\alpha\beta}^{\mu\nu}(s)b_{\mu\nu}(s) = \left[a_{(o)}(s)\delta_{\mu\nu}^{\alpha\beta} + a_{(1)}(s)\widehat{g}_{\mu\nu}(r)\widehat{g}^{\alpha\beta}(r)\right]b^{\mu\nu}(s). \end{array}\right. \tag{A29}$$

References

1. Messiah, A. *Quantum Mechanics*; Dover Pubs: New York, NY, USA, 1999.
2. Einstein, A. *The Meaning of Relativity*; Princeton University Press: Princeton, NJ, USA, 2004.
3. Landau, L.D.; Lifschitz, E.M. Field Theory. In *Theoretical Physics*; Addison-Wesley: New York, NY, USA, 1957.
4. Misner, C.W.; Thorne, K.S.; Wheeler, J.A. *Gravitation*; W.H. Freeman and Company: New York, NY, USA, 1973.
5. Tessarotto, M.; Cremaschini, C. Theory of Nonlocal Point Transformations in General Relativity. *Adv. Math. Phys.* **2016**, *2016*, 9619326.
6. Cremaschini, C.; Tessarotto, M. Quantum theory of extended particle dynamics in the presence of EM radiation-reaction. *Eur. Phys. J. Plus* **2015**, *130*, 166.
7. Cremaschini, C.; Tessarotto, M. Synchronous Lagrangian variational principles in general relativity. *Eur. Phys. J. Plus* **2015**, *130*, 123.
8. Cremaschini, C.; Tessarotto, M. Manifest covariant Hamiltonian theory of general relativity. *Appl. Phys. Res.* **2016**, *8*, 2.

9. Cremaschini, C.; Tessarotto, M. Hamiltonian approach to GR—Part 1: Covariant theory of classical gravity. *Eur. Phys. J. C* **2017**, *77*, 329.
10. Cremaschini, C.; Tessarotto, M. Hamiltonian approach to GR—Part 2: Covariant theory of quantum gravity. *Eur. Phys. J. C* **2017**, *77*, 330.
11. Cremaschini, C.; Tessarotto, M. Quantum-Wave Equation and Heisenberg inequalities of covariant quantum gravity. *Entropy* **2017**, *19*, 339.
12. De Donder, Th. *Théorie Invariantive Du Calcul des Variations*; Gaultier-Villars & Cia: Paris, France, 1930. (In French)
13. Weyl, H. Geodesic fields in the calculus of variation for multiple integrals. *Ann. Math.* **1935**, *36*, 607.
14. Saunders, D.J. *The Geometry of Jet Bundles*; Cambridge University Press: Cambridge, UK, 1989.
15. Sardanashvily, G. *Generalized Hamiltonian Formalism for Field Theory*; World Scientific Publishing: Singapore, 1995.
16. Echeverría-Enríquez, A.; Muñoz-Lecanda, M.C.; Román-Roy, N. Geometry of Lagrangian first-order classical field theories. *Fortschr. Phys.* **1996**, *44*, 235.
17. Forger, M.; Paufler, C.; Romer, H. The poisson bracket for Poisson forms in multisymplectic field theory. *Rev. Math. Phys.* **2003**, *15*, 705.
18. Kisil, V.V. *p*-Mechanics as a physical theory: An introduction. *J. Phys. A Math. Gen.* **2004**, *37*, 183.
19. Struckmeier, J.; Redelbach, A. Covariant Hamiltonian field theory. *Int. J. Mod. Phys. E* **2008**, *17*, 435.
20. Han, M. Einstein equation from covariant loop quantum gravity in semiclassical continuum limit. *Phys. Rev. D* **2017**, *96*, 024047.
21. Madelung, E. Quantentheorie in hydrodynamischer form. *Z. Phys.* **1927**, *40*, 322. (In German)
22. Tessarotto, M.; Cremaschini, C. Generalized Lagrangian-path representation of non-relativistic quantum mechanics. *Found. Phys.* **2016**, *46*, 1022.
23. Tessarotto, M.; Mond, M.; Batic, D. Hamiltonian structure of the Schrödinger classical dynamical system. *Found. Phys.* **2016**, *46*, 1127.
24. Nelson, E. Derivation of the Schrödinger equation from Newtonian mechanics. *Phys. Rev.* **1966**, *150*, 1079.
25. Bouda, A. From a mechanical Lagrangian to the Schrödinger equation: A modified version of the quantum Newton law. *Int. J. Mod. Phys. A* **2003**, *18*, 3347.
26. Holland, P. Computing the wavefunction from trajectories: Particle and wave pictures in quantum mechanics and their relation. *Ann. Phys.* **2005**, *315*, 505.
27. Poirier, B. Bohmian mechanics without pilot waves. *Chem. Phys.* **2010**, *370*, 4.
28. Holland, P. Foreword. In *Quantum Trajectories*; Chattaraj, P., Ed.; Taylor & Francis/CRC: Boca Raton, FL, USA, 2010.
29. Poirier, B. Trajectory-based derivation of classical and quantum mechanics. In *Quantum Trajectories*; Hughes, K.H., Parlant, G., Eds; Daresbury Laboratory: Daresbury, UK, 2011.
30. Schiff, J.; Poirier, B. Communication: Quantum mechanics without wavefunctions. *J. Chem. Phys.* **2012**, *136*, 031102.
31. Parlant, G.; Ou, Y.C.; Park, K.; Poirier, B. Classical-like trajectory simulations for accurate computation of quantum reactive scattering probabilities. *Comput. Theoret. Chem.* **2012**, *990*, 3.
32. Bohm, D.; Hiley, B.J.; Kaloyerou, P.N. An ontological basis for the quantum theory. *Phys. Rep.* **1987**, *144*, 321.
33. Bohm, D. A suggested interpretation of the quantum theory in terms of "hidden" variables. I. *Phys. Rev.* **1952**, *85*, 166.
34. Bohm, D. A suggested interpretation of the quantum theory in terms of "hidden" variables. II. *Phys. Rev.* **1952**, *85*, 180.
35. Bohm, D. Reply to a criticism of a causal re-interpretation of the quantum theory. *Phys. Rev.* **1952**, *87*, 389.
36. Weinberg, S. The cosmological constant problem. *Rev. Mod. Phys.* **1989**, *61*, 1.
37. Grössing, G. On the thermodynamic origin of the quantum potential. *Physica A Stat. Mech. Appl.* **2009**, *388*, 811.
38. Dennis, G.; de Gosson, M.A.; Hiley, B.J. Bohm's quantum potential as an internal energy. *Phys. Lett. A* **2015**, *379*, 1224.
39. Schrödinger, E. Der stetige Übergang von der Mikro- zur Makromechanik. *Die Naturwisseschaften* **1926**, *14*, 664. (In German)
40. Ashtekar, A. Gravity and the quantum. *New J. Phys.* **2005**, *7*, 198.

41. Etienne, Z.B.; Liu, Y.T.; Shapiro, S.L. Relativistic magnetohydrodynamics in dynamical spacetimes: A new adaptive mesh refinement implementation. *Phys. Rev. D* **2010**, *82*, 084031.
42. Gheorghiu, T.; Vacaru, O.; Vacaru, S. Off-diagonal deformations of Kerr black holes in Einstein and modified massive gravity and higher dimensions. *Eur. Phys. J. C* **2014**, *74*, 3152.
43. Ruchin, V.; Vacaru, O.; Vacaru, S. On relativistic generalization of Perelman's W-entropy and thermodynamic description of gravitational fields and cosmology. *Eur. Phys. J. C* **2017**, *77*, 184.
44. Dirac, P.A.M. Generalized Hamiltonian dynamics. *Can. J. Math.* **1950**, *2*, 129.
45. Sundermeyer, K. *Constrained Dynamics*; Lecture Notes in Physics Series 169; Springer-Verlag: Berlin, Germany, 1982.
46. Sudarshan, E.C.G.; Mukunda, N. *Classical Dynamics: A Modern Perspective*; Wiley-Interscience Publication: New York, NY, USA, 1964.
47. Mukunda, N. Generators of symmetry transformations for constrained Hamiltonian systems. *Phys. Scr.* **1980**, *21*, 783.
48. Castellani, L. Symmetries in constrained hamiltonian systems. *Ann. Phys.* **1982**, *143*, 357.
49. Arnowitt, R.; Deser, S.; Misner, C.W. *Gravitation: An Introduction to Current Research*; Witten, L., Ed.; Wiley: New York, NY, USA, 1962.
50. Alcubierre, M. *Introduction to 3+1 Numerical Relativity*; Oxford University Press: Oxford, UK, 2008.
51. DeWitt, B.S. Quantum theory of gravity. *Phys. Rev.* **1967**, *60*, 1113.
52. Ashtekar, A. New variables for classical and quantum gravity. *Phys. Rev. Lett.* **1986**, *57*, 2244.
53. Ashtekar, A. New Hamiltonian formulation of general relativity. *Phys. Rev. D* **1987**, *36*, 1587.
54. Jacobson, T.; Smolin, L. Nonperturbative quantum geometries. *Nucl. Phys. B* **1988**, *299*, 295.
55. Rovelli, C.; Smolin, L. Knot theory and quantum gravity. *Phys. Rev. Lett.* **1988**, *61*, 1155.
56. Rovelli, C.; Smolin, L. Loop space representation of quantum general relativity. *Nucl. Phys. B* **1990**, *331*, 80.
57. Rovelli, C. Ashtekar formulation of general relativity and loop space nonperturbative quantum gravity: A Report. *Class. Quantum Gravity* **1991**, *8*, 1613.
58. Ashtekar, A.; Geroch, R. Quantum theory of gravitation. *Rep. Prog. Phys.* **1974**, *37*, 1211.
59. Weinberg, S. *Gravitation and Cosmology*; John Wiley: New York, NY, USA, 1972.
60. DeWitt, B.S. Covariant quantum geometrodynamics. In *Magic without Magic*; Wheeler, J.A., Klauder, J.R., Eds; W. H. Freeman: San Francisco, CA, USA, 1972.
61. Batalin, I.A.; Vilkovisky, G.A. Relativistic S-matrix of dynamical systems with boson and fermion constraints. *Phys. Lett. B* **1977**, *69*, 309.
62. Batalin, I.A.; Vilkovisky, G.A. Gauge algebra and quantization. *Phys. Lett. B* **1981**, *102*, 27.
63. Batalin, I.A.; Vilkovisky, G.A. Feynman rules for reducible gauge theories. *Phys. Lett. B* **1983**, *120*, 166.
64. Batalin, I.A.; Vilkovisky, G.A. Quantization of gauge theories with linearly dependent generators. *Phys. Rev. D* **1983**, *28*, 2567.
65. Mandal, B.P.; Rai, S.K.; Upadhyay, S. Finite nilpotent symmetry in Batalin-Vilkovisky formalism. *Eur. Phys. Lett.* **2010**, *92*, 21001.
66. Upadhyay, S.; Mandal, B.P. BV formulation of higher form gauge theories in a superspace. *Eur. Phys. J. C* **2012**, *72*, 2059.
67. Upadhyay, S. Perturbative quantum gravity in Batalin-Vilkovisky formalism. *Phys. Lett. B* **2013**, *723*, 470.
68. Fredenhagen, K.; Rejzner, K. Batalin-Vilkovisky formalism in the functional approach to classical field theory. *Comm. Math. Phys.* **2012**, *314*, 93.
69. Pinto-Neto, N.; Santos, G.; Struyve, W. Quantum-to-classical transition of primordial cosmological perturbations in de Broglie-Bohm quantum theory. *Phys. Rev. D* **2012**, *85*, 083506.
70. Pinto-Neto, N.; Falciano, F.T.; Pereira, R.; Santini, E.S. Wheeler-DeWitt quantization can solve the singularity problem. *Phys. Rev. D* **2012**, *86*, 063504.
71. Pinto-Neto, N.; Fabris, J.C. Quantum cosmology from the de Broglie-Bohm perspective. *Class. Quantum Gravity* **2013**, *30*, 143001.
72. Falciano, F.T.; Pinto-Neto, N.; Struyve, W. Wheeler-DeWitt quantization and singularities. *Phys. Rev. D* **2015**, *91*, 043524.
73. Holland, P.R. *The Quantum Theory of Motion*; Cambridge University Press: Cambridge, UK, 1993.
74. Wyatt, R. *Quantum Dynamics with Trajectories*; Springer-Verlag: Berlin, Germany, 2005.
75. Bhattacharya, S.; Shankaranarayanan, S. How emergent is gravity? *Int. J. Mod. Phys. D* **2015**, *24*, 1544005.

76. Padmanabhan, T. Emergent gravity paradigm: Recent progress. *Mod. Phys. Lett. A* **2015**, *30*, 1540007.
77. Jacobson, T. Thermodynamics of spacetime: The Einstein equation of state. *Phys. Rev. Lett.* **1995**, *75*, 1260.
78. Faizal, M.; Ashour, A.; Alcheikh, M.; Alasfar, L.; Alsaleh, S.; Mahroussah, A. Quantum fluctuations from thermal fluctuations in Jacobson formalism. *Eur. Phys. J. C* **2017**, *77*, 608.

Article

Experimental Non-Violation of the Bell Inequality

Tim N. Palmer

Department of Physics, University of Oxford, Oxford OX1 3PU, UK; tim.palmer@physics.ox.ac.uk

Received: 7 April 2018; Accepted: 2 May 2018; Published: 10 May 2018

Abstract: A finite non-classical framework for qubit physics is described that challenges the conclusion that the Bell Inequality has been shown to have been violated experimentally, even approximately. This framework postulates the primacy of a fractal-like 'invariant set' geometry I_U in cosmological state space, on which the universe evolves deterministically and causally, and from which space-time and the laws of physics in space-time are emergent. Consistent with the assumed primacy of I_U, a non-Euclidean (and hence non-classical) metric g_p is defined in cosmological state space. Here, p is a large but finite integer (whose inverse may reflect the weakness of gravity). Points that do not lie on I_U are necessarily g_p-distant from points that do. g_p is related to the p-adic metric of number theory. Using number-theoretic properties of spherical triangles, the Clauser-Horne-Shimony-Holt (CHSH) inequality, whose violation would rule out local realism, is shown to be undefined in this framework. Moreover, the CHSH-like inequalities violated experimentally are shown to be g_p-distant from the CHSH inequality. This result fails in the singular limit $p = \infty$, at which g_p is Euclidean and the corresponding model classical. Although Invariant Set Theory is deterministic and locally causal, it is not conspiratorial and does not compromise experimenter free will. The relationship between Invariant Set Theory, Bohmian Theory, The Cellular Automaton Interpretation of Quantum Theory and p-adic Quantum Theory is discussed.

Keywords: Bell theorem; fractal geometry; p-adic metric; singular limit; gravity; conspiracy; free will; number theory; quantum potential

---

1. Introduction

Recent experiments (e.g., [1]) have seemingly put beyond doubt the conclusion that the CHSH version

$$|\mathrm{Corr}(0,0) + \mathrm{Corr}(1,0) + \mathrm{Corr}(0,1) - \mathrm{Corr}(1,1)| \leq 2 \tag{1}$$

of the Bell Inequality is violated robustly for a range of experimental protocols and measurement settings. As a result, it is very widely believed that physical theory cannot be based on Einsteinian notions of realism and local causality ('local realism'). Here, $\mathrm{Corr}(X, Y)$ denotes the correlation between spin measurements performed by Alice and Bob on entangled particle pairs produced in the singlet quantum state, where $X = 0,1$ and $Y = 0,1$ correspond to pairs of freely-chosen points on Alice and Bob's celestial spheres, respectively.

Of course, in the *precise* form as written, (1) has not been shown to have been violated experimentally. In practice, the four correlations on the left-hand side of (1) are each estimated from a separate sub-ensemble of particles with measurements performed at different times and/or spatial locations. Hence, for example, the measurement orientation corresponding to $Y = 0$ for the first sub-ensemble cannot correspond to *precisely* the same measurement orientation $Y = 0$ for the second sub-ensemble; as a matter of principle, Bob cannot shield his apparatus from the effects of ubiquitous gravitational waves associated for example with distant astrophysical events. Hence, as a matter of principle, what is actually violated experimentally is not (1) but

$$|\mathrm{Corr}(0,0) + \mathrm{Corr}(1,0') + \mathrm{Corr}(0',1) - \mathrm{Corr}(1',1')| \leq 2, \tag{2}$$

where, relative to the Euclidean metric of space-time, $0 \approx 0'$ and $1 \approx 1'$ for X and Y.

Could the difference between $0 \approx 0'$, $1 \approx 1'$ on the one hand, and $0 = 0'$, $1 = 1'$ on the other, actually matter? More specifically, is there a plausible framework for physical theory where (1) is the singular [2] rather than the smooth limit of (2) as $0' \to 0$, $1' \to 1$ and, therefore, where (2) is in some sense physically distinct from (1), no matter how accurate are our finite-precision experiments? Intuitively, it would seem not, as Bell [3] himself argued with a form of 'epsilonic' analysis. Indeed, common sense might suggest to even contemplate such a possibility would be to entertain a theory that was not only grotesquely fine-tuned, but one that was inconsistent with the fact that the experimental violation of (2) does not require any precision in setting the polariser orientations.

The purpose of this paper is to argue that, in this respect, we are being fooled by our intuition. It is worthwhile beginning the discussion with a close analogy: the Penrose Impossible Triangle (sometimes known as the tribar). The triangle seems impossible because we intuitively assume that any two sides of the triangle necessarily become *close* at a common vertex. Relaxing this metric assumption makes it possible to construct such Penrose Triangles in 3D physical space: it is the projection into 2D of such a 3D structure that provides the illusion (but not the reality) of inconsistency.

The relevance of this example is to draw attention to the notion of distance. There is no doubt that space-time has a locally Euclidean metric. However, should we assume such a metric for state space? In conventional quantum theory based on complex Hilbert Space, this assumption is forced on us. However, motivated by both nonlinear dynamical systems theory and p-adic number theory, we outline in Section 2 a plausible and robust locally causal framework where the metric on state space is explicitly not Euclidean. This framework arises from the 'invariant set' postulate [4–6] that a certain fractal-like subset I_U of cosmological state space is primal in the sense that the universe itself can be considered a deterministic dynamical system evolving on I_U, and moreover that space-time and the laws of physics in space-time are emergent from the geometry of I_U. Within 'Invariant Set Theory', complex Hilbert states have finite frequentist probabilistic interpretations as incompletely defined trajectory segments on I_U, requiring squared amplitudes and complex phases take rational values. By implication, complex Hilbert states with irrational squared amplitudes or irrational complex phases have no status as probabilistically defined trajectory segments on I_U and are therefore 'non-ontic'. A key number theorem is introduced that establishes an incompatibility between rational angles and rational cosines and which completely underpins the viability of the invariant set postulate as a realistic causal basis for quantum physics. In Section 2, a metric g_p (where p is a large integer) is introduced on state space, which respects the fundamental primacy of I_U and with respect to which ontic and non-ontic Hilbert states are necessarily distant from one another.

After a warm-up discussion in Section 3 where the number theorem above is used to account for the non-commutativity of spin observables in Invariant Set Theory, in Section 4 we discuss the Bell Theorem. It is shown that the violation of (2) is generically robust to g_p-small-amplitude perturbations. However, the set of all inequalities encompassed by such perturbations does not and cannot include the Bell inequality (1) itself, whose violation would be needed to rule out local realism [7]. As shown, (1) is necessarily constructed from Hilbert states with irrational descriptors, i.e., non-ontic states not lying on I_U and therefore g_p distant from the ontic states lying on I_U. In this sense, (1) is neither satisfied nor violated in Invariant Set Theory: it is simply undefined. This is not so much a loophole as a gaping chasm in the Bell Theorem, allowing a new type of a locally causal theory as a candidate descriptor of quantum physics (and hence, potentially, a novel approach to synthesise quantum and gravitational physics). Invariant Set Theory has the added bonus that it is essentially a finite theory, in contradistinction with quantum theory, where the role of the infinitesimal appears to be foundational [8]. As discussed in Section 5, although deterministic, Invariant Set Theory is not conspiratorial and respects experimenter free will. In Section 6, Invariant Set Theory is compared with Bohmian Theory, 'tHooft's Cellular Automaton Interpretation of Quantum Theory and p-Adic Quantum Theory. Further discussion and analysis of the issues of robustness and local causality are provided in Section 7.

2. Invariant Set Theory

Results below summarise more detailed analysis given in [6]. As mentioned above, we posit some primal compact fractal-like geometry I_U in cosmological state space, on which the universe U as a self-contained locally causal deterministic system evolves [4,5]. Figure 1 illustrates the local fractal structure of I_U. On the left is shown, at some $(j-1)$th fractal iterate of I_U, a single state-space trajectory segment ('history') in some three-dimensional subspace of state space. At the jth iterate, this trajectory segment comprises a helix of $N \gg 0$ fine-scale trajectories and an additional $N+1$th trajectory (not shown) at the centre of the helix. The winding frequency ω of a jth iterate helical segment is assumed proportional to the energy E associated with the subsystem described by this subspace. In this sense, the deBroglie relationship $E = \hbar\omega$ reflects a key element of the invariant set postulate: that the laws of physics in space-time are manifestations of the geometry of the more primal I_U. At the $(j+1)$th iterate (not shown), the helical trajectory segments are themselves found to be helical. In general, a cross section through a $(j-1)$th trajectory segment comprises a Cantor set $\mathcal{C}$ comprising $p = N+1$ iterated disks (Figure 2).

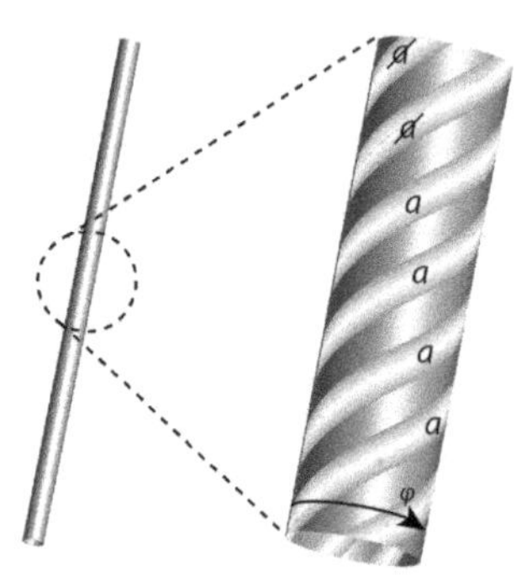

Figure 1. A state-space trajectory segment, which appears to be a simple line on some coarse scale, is in fact found to be, on magnification, a helix of trajectories. On further magnification, each of these helical trajectory segments is itself a helix of trajectories, and so on. A cross section through the original coarse-scale trajectory segment is a Cantor Set as illustrated below. At any particular level of magnification (i.e., fractal iterate), the trajectory segments can be labelled a or $\not{a}$ according to the regime to which they evolve.

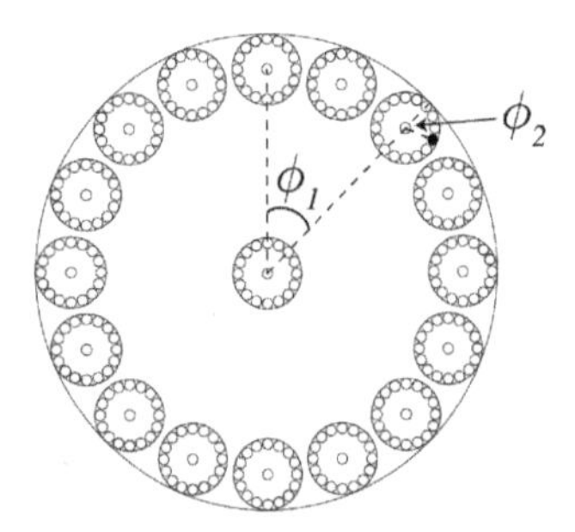

Figure 2. A Cantor Set $\mathcal{C}$, comprising $p = 17$ iterated disks: $N = 16$ iterated pieces around the edge of a disk and 1 at the centre of a disk. Here, a single disk at the $(j-1)$th fractal iteration comprises 17 jth-iterate disks, and each of these comprises 17 $(j+1)$th-iterate disks. An element of $\mathcal{C}$ can be represented by a sequence $\{\phi_1, \phi_2, \phi_3, \ldots\}$, where $\phi_i/2\pi = n/N \in \mathbb{Q}$.

We will now suppose that, at any fractal iterate, the N helical trajectory segments in Figure 1 can be labelled according to a process illustrated in Figure 3, associated with the divergence and nonlinear clustering of trajectories into two distinct state-space regimes or clusters labelled a and $\not{a}$ (the central $(N+1)$th trajectory is assumed to lie on the basin boundary and hence to evolve neither to a nor $\not{a}$). This divergence reflects the generic phenomenon of decoherence (essentially the butterfly effect) as the sub-system interacts with its environment. These clusters correspond to the measurement eigenstates of quantum theory.

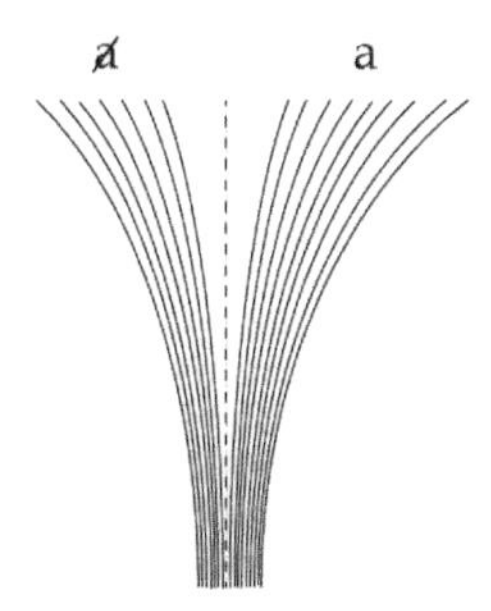

Figure 3. Here, $N = 16$ classical state-space trajectories diverge into two distinct regimes labelled a and $\not{a}$. In this example, seven of the 16 evolve to the $\not{a}$ regime and the other nine evolve to the a regime. In terms of the parameter θ described in the text, here $\cos\theta = 1/8 \in \mathbb{Q}$.

Within this geometric framework, complex Hilbert vectors can be used to provide some incomplete probabilistic description of reality. For example, consider the $(j+1)$th iterate disks inside a jth iterate disk and labelled ϕ_2 in Figure 2. Suppose that $N\cos^2(\theta/2)$ of these $(j+1)$th iterate disks are labelled a, and that *reality* corresponds to one of the N disks and is therefore either labelled a or $\not{a}$. Then, as discussed in [6] in more detail and with $\phi = \phi_2$, an incomplete representation of *reality* can be given probabilistically by the complex Hilbert vector

$$\cos\frac{\theta}{2}\,|a\rangle + e^{i\phi}\sin\frac{\theta}{2}\,|\not{a}\rangle. \tag{3}$$

In particular, it is necessary that $\phi/2\pi$ and $\cos^2(\theta/2)$ (and hence $\cos\theta$) are rational numbers. By contrast, a putative Hilbert vector where $\cos\theta \notin \mathbb{Q}$ or $\phi/2\pi \notin \mathbb{Q}$ cannot provide an incomplete representation of any trajectory segment on I_U and, therefore, in Invariant Set Theory, cannot correspond to an ontic state. More general tensor-product Hilbert states can also be used to provide incomplete representations of multi-variate properties of *reality*. Again, it is necessary that all squared amplitudes are rational, and all complex phase angles are rational multiples of 2π [6].

A crucial number theorem that completely underpins this framework is the following:

Theorem 1. *Let $\phi/\pi \in \mathbb{Q}$. Then, $\cos\phi \notin \mathbb{Q}$ except when $\cos\phi = 0, \pm\frac{1}{2}, \pm 1$.* [9,10]

Proof. Assume that $2\cos\phi = a/b$, where $a, b \in \mathbb{Z}$ have no common factors and $b \neq 0$. Since $2\cos 2\phi = (2\cos\phi)^2 - 2$, then $2\cos 2\phi = (a^2 - 2b^2)/b^2$. Now, $a^2 - 2b^2$ and b^2 have no common factors, since if p were a prime number dividing both, then $p|b^2 \implies p|b$ and $p|(a^2 - 2b^2) \implies p|a$, a contradiction. Hence, if $b \neq \pm 1$, then the denominators in $2\cos\phi, 2\cos 2\phi, 2\cos 4\phi, 2\cos 8\phi \ldots$ get bigger without limit. On the other hand, if $\phi/\pi = m/n$, where $m, n \in \mathbb{Z}$ have no common factors, then the sequence $(2\cos 2^k\phi)_{k\in\mathbb{N}}$ admits at most n values. Hence, we have a contradiction. Hence, $b = \pm 1$ and $\cos\phi = 0, \pm\frac{1}{2}, \pm 1$. □

We now define a metric g_p that respects the primacy of I_U where ontic states on I_U and non-ontic states off I_U are necessarily distant from one another (no matter how close they may appear from a Euclidean perspective). For all $x \in C$, $y \in C$ and $z \notin C$,

- $g_p(x,y)$ is Euclidean,
- $g_p(x,y) \leq 1$,
- $g_p(x,z) = g_p(y,z) = p$.

Hence, if $p \gg 1$, we can say that z is g_p-distant from both x and y. It is easily shown that g_p satisfies the axioms for a metric (e.g., the triangle inequality) on cosmological state space and that g_p is related to the p-adic metric of number theory [6].

3. The Sequential Stern-Gerlach Experiment

As a warm up to the Bell Theorem, we discuss in this Section one of the classic experiments designed to introduce students to non-commutativity of spin observables in quantum theory (e.g., [11]). Consider an ensemble of spin-1/2 particles prepared by the first of three Stern-Gerlach apparatuses (Figure 4a) with spins oriented in the direction $\hat{\mathbf{a}}$ in physical 3-space. The particles that are prepared spin-up by this first apparatus pass through a second Stern-Gerlach apparatus oriented in the direction $\hat{\mathbf{b}}$. The particles that are output along the spin-up channel of the second apparatus are then passed into a third Stern-Gerlach apparatus oriented in the direction $\hat{\mathbf{c}}$. The directions $\hat{\mathbf{a}}$, $\hat{\mathbf{b}}$ and $\hat{\mathbf{c}}$ correspond to points A, B and C on the celestial sphere $\mathbb{S}^2$ (Figure 4b). Typically, the directions $\hat{\mathbf{a}}$, $\hat{\mathbf{b}}$ and $\hat{\mathbf{c}}$ are designed to be coplanar, i.e. A, B and C lie on a great circle. However, this is impossible to achieve *precisely*: as a matter of principle, one cannot shield the experiment from the distorting effects of gravitational waves. Hence, as in Figure 4b, we assume that A, B and C are the vertices of some non-degenerate triangle $\triangle ABC$, where the angle γ is not equal to 180° precisely.

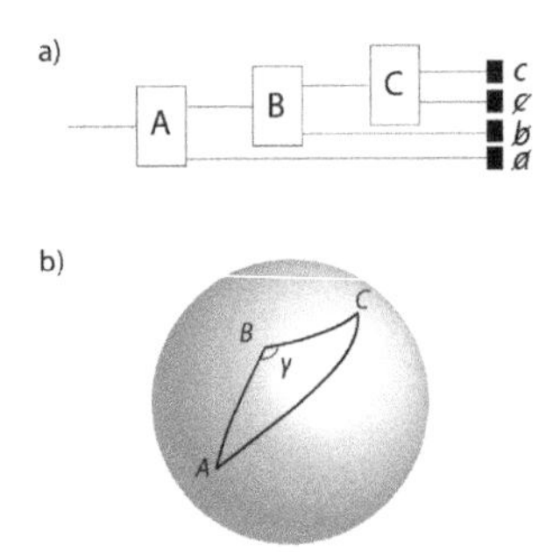

Figure 4. (**a**) a sequential Stern-Gerlach experiment where a particle is sent through three Stern-Gerlach devices, A, B and C; (**b**) A, B and C shown as directions on the celestial sphere. Although to experimental accuracy A, B and C may be coplanar, they are not coplanar precisely. In invariant set theory, we demonstrate the non-commutativity of spin observables by number theory.

We now show that if a particle was measured by the apparatus in the order A-B-C, then it could not have been measured in the order A-C-B. That is, the measurements in directions $\hat{\mathbf{b}}$ and $\hat{\mathbf{c}}$ cannot be performed simultaneously. In Invariant Set Theory, this result is derived by number theory. Consider a particle sent through the sequential Stern-Gerlach apparatus in the configuration A-B-C and where the detector corresponding to either c or $\not{c}$ was triggered. Then, in Invariant Set Theory, we require that all of $\cos\theta_{AB}$, $\cos\theta_{BC}$ and γ must be rational for the experiment to lie on I_U and hence correspond to *reality*. We now ask the question: what would the outcome have been for that particle had the configuration been A-C-B. For there to have been a definite outcome, we require, in addition, that $\cos\theta_{AC} \in \mathbb{Q}$. However, by the cosine rule for spherical triangles

$$\cos\theta_{AC} = \cos\theta_{AB}\cos\theta_{BC} + \sin\theta_{AB}\sin\theta_{BC}\cos\gamma. \tag{4}$$

The right-hand side is the sum of two terms. The first is rational since it is the product of two terms each of which, by construction, is rational. The second is the product of three terms the last of which, $\cos\gamma$, is irrational, except for the eight exceptions listed in the Theorem above. Since γ is only approximately equal to 180°, $\cos\gamma$ is irrational. Since θ_{AB}, θ_{BC} and γ are independent degrees of freedom defining the triangle $\triangle ABC$, there is no reason why $\sin\theta_{AB}$ and $\sin\theta_{BC}$ should conspire with $\cos\gamma$ to make the product $\sin\theta_{AB}\sin\theta_{BC}\cos\gamma$ rational. Hence, $\cos\theta_{AC}$ is the sum of a rational and an irrational and is therefore irrational. Hence, for the particle where A-B-C was measured, the counterfactual A-C-B is undefined and could not be an element of *reality*. Put another way, the state $U' \notin I_U$ of the universe associated with the configuration A-C-B is g_p distant from the state $U \in I_U$ associated with configuration A-B-C.

We can of course envisage performing two separate sequential Stern-Gerlach experiments (one on a Monday, the other on a Tuesday, say) where the order of the Stern-Gerlach apparatuses was A-B-C and A-C-B, respectively. For Monday's experiment, $\cos\theta_{AB}$ and $\cos\theta_{BC}$ are rational, and the angle subtended at B is a rational multiple of 2π. For Tuesday's experiment, $\cos\theta_{AC}$ and $\cos\theta_{BC}$ are rational, and the angle subtended at C is a rational multiple of 2π. As before, this would be impossible if the triangle $\triangle ABC$ was *precisely* the same on Monday and Tuesday. However, this will not be the case—background space-time ripples are necessarily different on Tuesday compared with Monday.

One potential objection should be answered before moving on to the Bell Theorem. In the discussion above, θ denoted a relative orientation in physical space, whereas in the discussion in Section 2, θ was merely a parameter whose cosine gave the probability of one measurement outcome rather than another. How is it that θ can now be interpreted as an orientation in physical space? The answer relates to the existence of spinorial structure on $\mathcal{C}$. To see this, the reader is directed to [6].

4. The Bell Inequality

Consider now the relationship between (1) and (2) from the perspective of Invariant Set Theory. As above, let $X = 0, 1$, $Y = 0, 1$ denote four random points on the sphere, three of which (relevant to the discussion below) are shown in Figure 5a. Let θ_{XY} denote the relative orientation between an X point and a Y point. Recall that complex Hilbert states can represent uncertain trajectory segments on I_U providing squared amplitudes are rational. Hence, $\mathrm{Corr}(X, Y) = -\cos\theta_{XY}$ requires $\cos\theta_{XY} \in \mathbb{Q}$.

Suppose Alice freely chooses $X = 0$ and Bob $Y = 0$ when measuring a particular entangled particle pair. Then, it must be the case that $\cos\theta_{00} \in \mathbb{Q}$. Could Alice and Bob have chosen $X = 1$ and $Y = 0$ when measuring this same particle pair, given that they actually chose $X = 0$, $Y = 0$? In other words, does the state U' of the universe in which this counterfactual experiment takes place also lie on I_U? To answer the question in the affirmative, we additionally require $\cos\theta_{10} \in \mathbb{Q}$. However, as with the Stern-Gerlach analysis, applying the cosine rule for spherical triangles, we have

$$\cos\theta_{10} = \cos\theta_{00}\cos\alpha_X + \sin\theta_{00}\sin\alpha_X\cos\gamma, \tag{5}$$

where α_X is the angular distance between $X = 0$ and $X = 1$. Now, it is always possible for Alice to send the particle which she has just measured in the $X = 0$ direction, back into her measuring apparatus to be again measured in the $X = 1$ direction. Hence, $\cos\alpha_X$ must be rational. Now, we also require the angle γ to be a rational multiples of 2π. This would be so if the three points $X = 0, 1$ and $Y = 0$ lay on a great circle *exactly*, so that $\gamma = 180^\circ$ *precisely*. However, as before, because of ubiquitous unshieldable gravitational waves, this cannot be the case. Hence, $\cos\theta_{01}$ is the sum of two terms, the first a rational and the second the product of three independent terms, the last of which is irrational. Being independent, these three terms cannot conspire to make their product rational. Hence, $\cos\theta_{01}$ is the sum of a rational and an irrational and must therefore be irrational. Hence, the state of the universe U' in which the counterfactual experiment takes place is not realistic and is g_p-distant from worlds on I_U. Hence, the counterfactual question cannot be answered in the affirmative: $\mathrm{Corr}(1, 0)$ is undefined. In general, it is never the case that all four correlations in (1) are definable on I_U—the Bell inequality is always undefined in Invariant Set Theory.

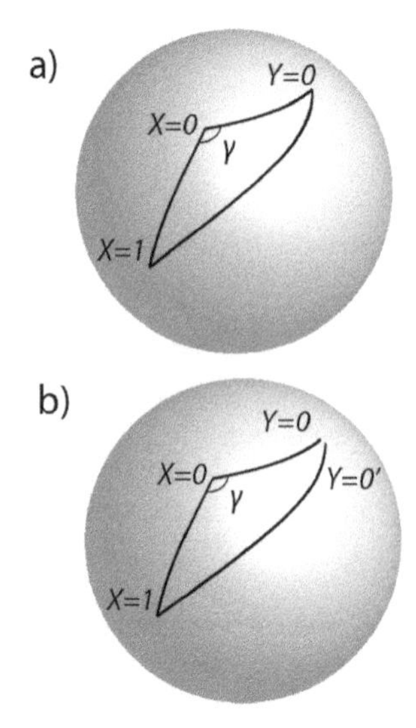

Figure 5. (**a**) in general, it is impossible for all the cosines of the angular lengths of all three sides of the spherical triangle to be rational, and the internal angles rational multiples of 2π. (**b**) what actually occurs when (2) is tested experimentally. Here, the cosines of the angular lengths of all sides are rational. In a precise sense, (b) is g_p distant from (a).

An experimenter might ask how one could set up an experiment with sufficient care to ensure that the corresponding Hilbert state descriptors were rational rather than irrational. The answer is that the experimenter need take no care: if an experiment is performable, i.e., corresponds to some $U \in I_U$, then by construction the descriptors must be rational. Physical perturbations (e.g., gravitational waves in space-time) only introduce uncertainty in the values of the rational descriptors and not in the fact that they are rational. Conversely, if the descriptor of a counterfactual state is irrational, then no amount of noise that respects the primacy of I_U can change it into an ontic state. This property provides an attractive finitist feature that is missing in conventional physical theories based on $\mathbb{R}$ or $\mathbb{C}$ and hence in theories that utilise the Euclidean state-space metric. Hence, in the real world of experiments, both $\cos\theta_{00}$ and $\cos\theta_{10'}$ in (2) are necessarily and robustly rational (Figure 5b), consistent with the fact that the individual sub-ensembles are measured at different times and/or locations, and that unshieldable gravitational waves ensure that orientations are not *precisely* the same when these different sub-ensembles are measured. Indeed, we can infer the existence of an effectively infinite family of orientations where all of $\cos\theta_{00}, \cos\theta_{10'}, \cos\theta_{0'1}, \cos\theta_{1'1'}$ in (2) are rational. However, by construction, none of the orientations so generated includes those associated with (1), which is therefore indeed the singular limit of and g_p-distant from (2). Just as the paradox of the Penrose 'impossible triangle' is resolved by realising that the sides of the triangle are not necessarily close near a vertex of the triangle, so too here. As discussed in [6], many of the familiar 'paradoxes' of quantum theory can be interpreted realistically and causally with g_p as the metric of state space.

5. Conspiracy and Free Will

As discussed, if the spins of an entangled particle pair are measured relative to $X = 0, Y = 0$, then by construction the spins of this particular particle pair could not have been measured relative to the directions $X = 1$ and $X = 0$, respectively. This is nothing to do with entanglement per se, but is rather a manifestation of quantum complementarity (associated with the non-commutativity of quantum observables). For example, as discussed in Section 3, if an experimenter performs a sequential Stern-Gerlach experiment in the order A-B-C, then, as a matter of principle, an experiment could not have been performed on the same particle in the order A-C-B. That is to say, if the state U of the universe associated with one experiment lies on I_U, then the corresponding state U' of the universe associated with the other experiment does not lie on I_U and is g_p distant from states on I_U.

This implies a violation of the principle of Measurement Independence (MI), usually framed in terms of some probability density $\rho(\lambda)$ of so-called 'hidden-variables' λ associated with the particles in question. In particular, MI is violated if

$$\rho(\lambda) \neq \rho(\lambda|m), \tag{6}$$

where m denotes some particular measurement, e.g., the A-B-C experiment or the $X = 0$, $Y = 0$ experiment. Another way of saying this is that the 'hidden variables' are contextual [12] and it is well known that contextual hidden variables could provide a route to negating the Bell Theorem. However, violation of MI is often seen as either implausibly conspiratorial [13] or inconsistent with experimenter free will [14]. As discussed below, Invariant Set Theory is neither implausibly conspiratorial nor inconsistent with experimenter free will.

It is important in the discussion below to recognise that, in Invariant Set Theory, the violation of MI is not imposed by fiat. Rather, it is a consequence of the postulate that I_U is a primal fractal-like geometry on which states of the universe evolve.

5.1. Nullifying the Notion of Conspiracy

Consider a specific and pertinent example. Suppose, in a Bell experiment, the measuring apparatuses are set according to the frequency ν of photons emitted by distant stars. If $\nu < \nu_0$, a reference frequency, suppose the $X = 0, Y = 0$ directions are chosen; if $\nu > \nu_0$, then the $X = 1, Y = 0$ directions are chosen. Let Λ_{00} denote some sample space of hidden variables associated with the choice $X = 0, Y = 0$ and so on. In Invariant Set Theory, if $\lambda \in \Lambda_{00}$, then the outcome of measurements in the $X = 0, Y = 0$ directions is well defined, but, by the discussion above, the outcome of measurements in the $X = 1, Y = 0$ directions is undefined. Conversely, if $\lambda \in \Lambda_{10}$, then the outcome of measurements in the $X = 1, Y = 0$ directions is well defined, but the outcome of measurements in the $X = 0, Y = 0$ directions is undefined.

The notion of conspiracy arises because we have two seemingly independent pieces of information that determine the type of measurement made: the frequency ν and the hidden variables λ. Since the photon is emitted long before the entangled particles, and in a different part of the universe, one could imagine that these quantities can be varied independently of one another. If this is so, then the possibility that $\nu < \nu_0$ but $\lambda \in \Lambda_{10}$ leads to inconsistency since the latter combination is associated with a state of the universe not lying on I_U. Hence, there must be some unpalatable 'conspiracy' between ν and λ, so the argument goes, to prevent such inconsistency.

However, this conclusion is incorrect. Firstly, a thousand years (say) before the experiment is performed, the photon does not and cannot 'know' that its frequency (and not, say, bits from a to-be-made sci-fi movie) will be the determinant of some future measurement settings. If the experimenters decide what whimsical process (photons or movie bits) they will use to set the measurement orientations at space-time event D, then the information needed to determine the nature of this decision will be completely delocalised on spacelike hypersurfaces in the causal past of D. Perturbing any one bit anywhere in the causal past of D could change the nature of this decision. This is the butterfly effect, generic to nonlinear systems evolving on fractal attractors, and the stuff of numerous sci-fi movies.

Similarly, it is incorrect to assume that the entangled particles 'know' whether their hidden variables belong to Λ_{00} or Λ_{10}. Consistent with both the quantum field-theoretic notion of a particle as a field excitation, and the invariant set premise that the laws of physics in space-time derive from a holistic geometry in state space, it is incorrect to think of λ as somehow internal and localised to the particles being measured. Consider the following analogy. Babies (almost always) either belong to the set Λ_{male} or Λ_{female} of babies who are male or female at birth. Information that determines which of the two sets a particular baby belongs is internal to the baby. This type of analogy describes classical hidden-variable theory (probe the particle to find its hidden variables and determine what spin it will have when measured in a particular direction), but does not describe the situation here. A more accurate analogy is this: at birth, all babies either belong to the set Λ_L or $\Lambda_{\not L}$ of humans that, as adults, either fall in love or don't. Information that determines to which of Λ_L or $\Lambda_{\not L}$ a particular baby belongs is clearly not internal to the baby. In particular, if L_B denotes the event where adult Bob falls in love, then the information λ that determines that baby Bob belongs to Λ_{L_B} is (as above) completely delocalised on spacelike hypersurfaces in the causal past of L_B and hence cannot, in principle, be known to baby Bob. Because λ is delocalised in this way, then when a counterfactual experiment is described as an alternative experiment on the same particle, i.e., on the same λ, effectively we are describing a hypothetical experiment where the measurement set up is changed, but the rest of the universe is held fixed.

In short, there is no implausible conspiracy between the so-called hidden variables and determinants of the processes which set the measurement observations. As discussed, information which determines these supposedly independent quantities are in fact highly intertwined on spacelike hypersurfaces in the causal past of the experiment.

5.2. *Free Will and Inaccessible Determinism*

The only way there can be a conflict between free will and determinism is if it was found to be possible to compute the future algorithmically, faster than the universe actually evolves. If a computer can predict my future actions reliably, then I am an automaton. Without this, determinism and free will are completely compatible with one another. Is it possible to compute the future with some faster-than-reality computational subset of the universe? No! If I_U was a strict fractal, then it would actually be non-computational [15,16]. However, if I_U is some finite fractal-like limit cycle, it will still have a property called computational irreducibility [17]: we cannot reliably predict which set of measurement settings will be chosen with a computationally simpler approximation of the full system. In particular, supressing just one bit of information on some initial spacelike hypersurface when integrating forward in time can lead to a completely different choice of measurement setting. This is again the butterfly effect and is generic for systems that evolve on fractal invariant sets in state space. This property of computational irreducibility can be considered as implying an 'inaccessible' form of determinism.

Sometimes the word 'pre-destination' is used as a synonym for determinism. For example, in a deterministic world, it was already pre-destined at the time of the dinosaurs that Alice would do this measurement and not that. For some, such pre-destination sounds implausible. What is sometimes forgotten with this example is that the information that determines that Alice would do this measurement and not that at the time of the dinosaurs is completely delocalised on the intersection of some spacelike hypersurface at the time of the dinosaurs with the causal past of the event, where Alice makes the measurement. That is to say, the information that determines Alice's measurement choice is completely inaccessible at the time of the dinosaurs: it is buried down at the Planck scale over regions of space spanning hundreds of millions of light years. Changing just one bit on a Planck-scale variable on this hypersurface could change what measurement Alice makes. For the third time, this is the butterfly effect and generic for systems that evolve on fractal invariant sets. Is that problematic for deterministic theories of physics?

Further discussion of the free-will issue is given in Section 7.

6. Relations to Other Approaches

In this section, we discuss the relationship between Invariant Set Theory and some other approaches to quantum physics.

6.1. Bohmian Theory

Since this paper was written as part of a celebration of the 100th birthday of David Bohm, it is worth commenting on possible links to the de-Broglie/Bohm interpretation of quantum theory [18]. Both Bohmian theory and Invariant Set theory are deterministic. However, Bohmian theory is necessarily non-local (i.e., not locally causal) whilst Invariant Set theory is not. The reason for this difference hinges around differences between the quantum potential, a differentiable potential function in configuration space, and I_U a fractal geometric object in state space. In particular, because there are no 'holes' in the quantum potential, Bohmian theory must be counterfactually complete and hence satisfy MI (and can therefore only violate the Bell Theorem by being nonlocal). By contrast, as discussed, the Invariant Set does generically have holes, is therefore not counterfactually complete, and hence does not satisfy MI and therefore is not required to satisfy the Bell inequalities despite being locally causal.

This raises the following tantalising possibility. Perhaps one should think of the Bohmian quantum potential as a 'coarse-grained' approximation to the fractal geometric structure of state space. As a simple illustration, it would be possible to mimic some aspects of the behaviour of the state vector of the Lorenz '63 system [19] by a stochastically forced motion in a double potential well. Of course, the fractal structure of the attractor would be 'smoothed out' in such a potential-well system. Thinking of the quantum potential as an approximation to some object in state space with rich geometric number-theoretic structure may provide a direction in which to move Bohmian theory forward.

6.2. The Cellular Automaton Interpretation of Quantum Mechanics

In both Invariant Set Theory and the Cellular Automaton Interpretation of Quantum Mechanics [20], it is concluded that quantum physics can be described by deterministic causal laws, and where the Bell Theorem is negated through a failure of counterfactual incompleteness [20]. However, the reasons for rejecting counterfactual completeness are different in these two approaches. The Cellular Automaton Interpretation rejects counterfactual definiteness by the assumption that the cosmological initial conditions are somehow special, in the sense that counterfactual perturbations to the initial conditions that would lead to the counterfactual measurements, are somehow excluded from the theory. In common with contemporary physics, 't Hooft separates the laws of physics from the initial conditions, i.e., treats them as separate. Treating the initial conditions as special in this way could be viewed as ad hoc, as well as being imprecise. What exactly is it about the initial conditions that prevents these quantum counterfactuals?

By contrast, in Invariant Set Theory, the initial conditions and the laws of physics are not independently specifiable items and that the Universe as a dynamical system evolves on the invariant set. Hence, by constructions, the cosmological initial conditions must lie on the invariant set. That is to say, there is no fundamental distinction between the laws of physics and the initial conditions in the sense that there is in standard theory. It is the structure of the invariant set that leads to counterfactual incompleteness i.e., where certain quantum counterfactual perturbations take a state lying on the invariant set and take it off the invariant set. Ultimately, the structure that leads to this property is number theoretic: that rational angles cannot have rational cosines, and vice versa.

6.3. p-Adic Quantum Theory

Over the years, there have been attempts to reformulate quantum theory, e.g., by replacing the complex coefficients of Hilbert states with p-adic numbers (there is a p-adic correspondence) to the complex numbers. A review of such approaches is given in [21]. One of the motivations for this was the idea that space-time may have some fractal structure on the Planck scale. Drawing on this work, Khrennikov [22] proposed a hidden-variable model where the probability distribution $\rho(\lambda)$ was defined on the p-adic numbers. Now, essentially because $-1 = 1 + 2 + 4 \ldots$ in 2-adic number theory, this allows for negative probabilities, which, although very esoteric, also arose for different reasons in Dirac's work on relativistic quantisation, and therefore have some basis in quantum physics. Khrennikov argued that the existence of such negative probabilities could negate the Bell Theorem.

It is important to distance the present work from such p-adic quantum theory. Here, we do not introduce p-adic numbers into Hilbert Space or (as a result) into probability distributions. Fundamentally, Invariant Set Theory is a deterministic theory and not a probabilistic theory. A key element of Invariant Set Theory is to reject the notion of the algebraically closed Hilbert Space as a state space for quantum physics, irrespective of whether the Hilbert Space is closed on the reals, the complexes or the p-adics. The reason for such rejection of this is ultimately so that physics can be described finitely [6]. In Invariant Set Theory, probabilities are defined by elementary frequentism and therefore are simply and intuitively rational numbers on the interval $[0,1]$. In fact, in the work presented in [6], maps on p-adic integers are defined to describe the deterministic laws that may underpin quantum physics. This is where algebraic closure can be reinstated—not on the Hilbert vectors or associated probabilities.

Indeed, the metric g_p that plays such a central role in this paper, whilst motivated by the p-adic metric and the fact that the set of p-adic integers is homeomorphic to Cantor sets, is not identical to it. g_p has the property that, if p is large, points which do not lie on I_U are necessarily distant from points that do lie on I_U. This reflects the fact that the p-adic distance between p-adic numbers that are not p-adic integers, and p-adic integers, is $\geq p$.

This comparison with p-adic quantum theory raises an important point. A key motivation for developing Invariant Set Theory (in addition to be able to describe physics finitely) was to make the Bell Theorem understandable. If a particular formalism leads one to trade one type of incomprehensible property (nonlocality) for another (negative probability), then, as far as the author is concerned, there is no compelling reason to adopt such a formalism. As discussed further in the next Section, Invariant Set Theory does not require any esoteric notions to explain the Bell Theorem (once one has embraced the key fact that space-time and state-space can have very different metrics).

7. Discussion

A theoretical framework has been outlined that asserts that no physical experiment has or will demonstrate that the Bell inequality (1) is violated—even approximately. In this framework, (1) is the singular limit of the experimentally tested (2), (1) is undefined and (2) is not approximately equivalent to (1). Key to this formulation, a non-Euclidean metric g_p is introduced on state space. g_p, related (but not entirely equivalent) to the p-adic metric of number theory, respects the primacy of an assumed fractal geometry I_U on which the universe is assumed to evolve and from which the laws of physics derive. Based on g_p, we can make an ontological distinction between Hilbert vectors with rational descriptors (rational squared amplitudes and rational complex phases) and irrational descriptors. Based on this framework, it is claimed that experiments do not rule out Einsteinian determinism and causality. Only in the singular classical limit at $p = \infty$ could experiments be used to rule out local causality.

Making an ontological distinction between Hilbert vectors with rational and irrational descriptors is likely to induce a sense of unease (indeed scepticism) amongst many readers; not least, the results above may appear to be inconsistent with the experimental fact that the violation of Bell-like inequalities is insensitive to the precise orientation of polarisers. To alleviate this sense of unease, consider the function $f(x)$ on $[0,1]$ such that $f(x) = x^2$ if x is rational, and $f(x) = 3$ (say), otherwise. This function is everywhere discontinuous and hence non-differentiable on the reals, and therefore $f(x)$ could hardly describe how experimental values vary smoothly with experimental parameter x. However, consider a physical theory T that demands that states of reality (that is to say, states of systems that can be probed by experiment or are otherwise amenable to observation) are only associated with rational values x and that irrational only arise in T when considering hypothetical counterfactual states that did not occur in reality. To have such a property, T would be a profoundly nonlinear theory. Based on this, T has the property that $f(x)$ is not only continuous but also (using the rational calculus) differentiable over the set of physically realistic values of x. Over a large number of parameter values x, an experimenter might return the values $f(x) = x^2 + \epsilon$, where ϵ denoted some random experimental error. The experiments would, by construction, never return the value $f(x) = 3$. As a result, a theoretician (unaware of T) might construct a linear theory T' on the reals where $f(x) = x^2$ for all $0 \leq x \leq 1$. Like T, T' would describe the results of experiments well. Being linear, T' would be analytically and computationally tractable, making it a convenient tool in practice. However, T' would incorrectly ascribe values to counterfactual states and this could lead to inconsistencies in the interpretation of T'. Some might argue that since, by construction, such inconsistencies have no implications for the real world of experimentation, one should 'shut up and calculate' with T' and not waste time searching for the deeper theory T. However, the failure of T' to describe the nonlinear structure in T may have implications elsewhere, e.g., when trying to extend T' to account for phenomena beyond the experiments that have so far been conducted (see below).

However, unlike T', T is an unrealistically fine-tuned theory [23,24], since rationals and irrationals lie arbitrarily close to each other on the real line with respect to the standard Euclidean metric. The notion that distances in physics should be necessarily described by the Euclidean metric is a deeply held intuition, since almost the first thing we learn as babies is a sense of spatial awareness (for the baby to get its hand close to a colourful toy, it has to learn to equate closeness with smallness of Euclidean distance). However, here we are considering distances in state space, not in space time, and where our intuitions may not apply. This raises the question about whether there is a state-space metric where realistic and counterfactual states are actually distant from one another. The toy model example here is too simple to allow such an alternate interpretation. However, using fractal geometry we have shown that there is a model where such states are indeed distant from one another, thus negating the fine-tuning argument.

As discussed in the Introduction, a rather beautiful example of how relying on intuition about distance can lead to inconsistency is provided by the Penrose Impossible Triangle. We claim that quantum theory (cf., T') is similarly inconsistent, even though it is wonderfully accurate and a convenient tool for analytic manipulation and computation. This inconsistency arises from the use of the Euclidean metric forced on us by the assumption that state space is the algebraically closed Hilbert Space. By weakening this assumption, allowing only Hilbert states with rational descriptors as elements of physical reality, the inconsistencies associated with the Bell and other no-go theorems [6], disappear. The key conclusion we can draw from this discussion is that, in Invariant Set Theory, there is no contradiction with the fact that the violations of (2) are insensitive to polariser orientation.

Let us now discuss a related issue. Let us fix the orientation of Bob's measuring device and ask whether, according to Invariant Set Theory, this in any way constrains Alice in orienting her measuring device. If Alice was somehow constrained, she would not only not be a free agent, she would somehow be remotely under the influence of Bob—clearly unacceptable in any theoretical framework purporting to reinstate the notion of Einsteinian local realism. The answer is that, for all practical purposes, Alice is under no such constraint. What 'for all practical purposes' means is that within *any* neighbourhood of Alice's celestial sphere, no matter how small, there exist orientations which Alice is free to choose from, providing p is sufficiently big. That is to say, the set of orientations from which Alice can choose is as dense as one likes, providing p is big enough. Colloquially, we can indeed say that Alice can set her measuring apparatus as she pleases—she is a free agent in any practical sense of the phrase. Nevertheless, whatever the size of p, Invariant Set Theory requires that the cosine of the relative orientation between Alice and Bob's measuring apparatuses must be rational. For large p, this is an utter irrelevance in the *design* of a Bell experiment. However, it is crucially important in the *interpretation* of a Bell experiment, since, as discussed, the counterfactual states needed to establish (1) will inevitably lie off I_U and are therefore g_p distant from the states measured by experiment. In conclusion, in making the statement that the orientation of Bob's polariser does not influence Alice's choice of polariser orientation, the contrast between the Euclidean metric of space-time and the metric g_p of state space becomes crucial.

The author believes that the ongoing failure to synthesise quantum and gravitational physics satisfactorily arises from the fact that quantum theory is inimical to the local realism of general relativity, and a synthesis between these two areas of physics will require a nonlinear theory of quantum physics, less like T' and more like T. As such, in the analysis above, it is plausible that p^{-1} defines the gravitational coupling constant (and so the largeness of p reflects the weakness of gravity). Indeed, the fact that gravitational waves provide an *in principle* unshieldable source of noise to ensure that rational angles can never have rational cosines (a central theorem to this paper) may be evidence of a deep link to the phenomenon of gravity, with experimental consequences for the dark universe and for quantum gravity. These more speculative notions have been developed elsewhere [25].

Acknowledgments: This research was supported by a Royal Society Research Professorship. The author thanks Andrei Khrennikov and Felix Tennie for helpful comments on an early draft of this paper and Eric Cavalcanti for helpful discussions on causal discovery and fine tuning.

Conflicts of Interest: The author declares no conflict of interest.

References

1. Shalm, A.E.A. Strong Loophole-Free Test of Local Realism. *Phys. Rev. Lett.* **2015**, *115*, 250402. [CrossRef] [PubMed]
2. Berry, M. Singular Limits. *Phys. Today* **2002**, *55*, 10–11. [CrossRef]
3. Bell, J. On the Einstein-Podolsky-Rosen Paradox. *Physics* **1964**, *1*, 195–200. [CrossRef]
4. Palmer, T. The invariant set postulate: A new geometric framework for the foundations of quantum theory and the role played by gravity. *Proc. R. Soc.* **2009**, *A465*, 3165–3185. [CrossRef]
5. Palmer, T. Lorenz, Gödel and Penrose: New perspectives on determinism and causality in fundamental physics. *Contemp. Phys.* **2014**, *55*, 157–178. [CrossRef]
6. Palmer, T. A Finite Theory of Qubit Physics. *arXiv* **2018**, arXiv:1804.01734.
7. Clauser, J.; Horne, M.; Shimony, A.; Holt, R. Proposed experiment to test local hidden variable theories. *Phys. Rev. Lett.* **1969**, *23*, 880–884. [CrossRef]
8. Hardy, L. Why is nature described by quantum theory? In *Science and Ultimate Reality: Quantum Theory, Cosmology and Complexity*; Barrow, J.D., Davies, P.C.W., Harper, C.L., Jr., Eds.; Cambridge University Press: Cambridge, UK, 2004.
9. Niven, I. *Irrational Numbers*; The Mathematical Association of America: Washington, DC, USA, 1956.
10. Jahnel, J. When does the (co)-sine of a rational angle give a rational number? *arXiv* **2010**, arXiv:1006.2938.
11. Sakurai, J. *Modern Quantum Mechanics*; Adison Wesley Longman: Boston, MA, USA, 1994.
12. Nieuwenhuizen, T.M. Is the contextuality loophole fatal for the derivation of Bell inequalities? *Found. Phys.* **2011**, *41*, 580–591. [CrossRef]
13. Bell, J. Free variables and local causality. *Dialectica* **1985**, *39*, 103. [CrossRef]
14. Wiseman, H.; Cavalcanti, E.G. Causarum Investigatio and the Two Bell's Theorems of John Bell. *arXiv* **2015**, arXiv:1503.06413.
15. Blum, L.; Cucker, F.; Shub, M.; Smale, S. *Complexity and Real Computation*; Springer: Berlin, Germany, 1997.
16. Dube, S. Undecidable problems in fractal geometry. *Complex Syst.* **1993**, *7*, 423–444.
17. Wolfram, S. *A New Kind of Science*; Wolfram Media: Oxfordshire, UK, 2002.
18. Bohm, D.; Hiley, B.J. *The Undivided Universe*; Routledge: Abington-on-Thames, UK, 2003.
19. Lorenz, E. Deterministic nonperiodic flow. *J. Atmos. Sci.* **1963**, *20*, 130–141. [CrossRef]
20. Hooft, G. *The Cellular Automaton Interpretation of Quantum Mechanics*; Springer: Berlin, Germany, 2016.
21. Dragovich, B.; Khrennikov, A.Y.; Kozyrev, S.V.; Volovich, I.V.; Zelenov, E.I. p-Adic Mathematical Physics: The First 30 Years. *p-Adic Numbers Ultrametr. Anal. Appl.* **2017**, *9*, 87–121. [CrossRef]
22. Khrennikov, A. The problem of hidden variables in quantum mechanics. *Phys. Lett. A* **1995**, *200*, 219–223. [CrossRef]
23. Wood, C.J.; Spekkens, R.W. The lesson of causal discovery algorithms for quantum correlations: Causal explanations of Bell-inequality violations require fine tuning. *New J. Phys.* **2015**, *17*, 033002. [CrossRef]

24. Cavalcanti, E. G. Classical causal models for Bell and Kochen-Specker Inequality Violations require fine tuning. *Phys. Rev. X* **2018**, *8*, 021018. [CrossRef]
25. Palmer, T. A Gravitational Theory of the Quantum. *arXiv* **2017**, arXiv:1709.00329.

Article

On Ontological Alternatives to Bohmian Mechanics

Thomas Filk [1,2]

[1] Institute for Physics, University of Freiburg, Hermann-Herder-Str. 3, D-79104 Freiburg, Germany; thomas.filk@physik.uni-freiburg.de
[2] Parmenides Foundation for the Study of Thinking, D-82049 Munich-Pullach, Germany

Received: 20 April 2018; Accepted: 13 June 2018; Published: 19 June 2018

Abstract: The article describes an interpretation of the mathematical formalism of standard quantum mechanics in terms of relations. In particular, the wave function $\psi(x)$ is interpreted as a complex-valued relation between an entity (often called "particle") and a second entity x (often called "spatial point"). Such complex-valued relations can also be formulated for classical physical systems. Entanglement is interpreted as a relation between two entities (particles or properties of particles). Such relations define the concept of "being next to each other", which implies that entangled entities are close to each other, even though they might appear to be far away with respect to a classical background space. However, when space is also considered to be a network of relations (of which the classical background space is a large-scale continuum limit), such nearest neighbor configurations are possible. The measurement problem is discussed from the perspective of this interpretation. It should be emphasized that this interpretation is not meant to be a serious attempt to describe *the* ontology of our world, but its purpose is to make it obvious that, besides Bohmian mechanics, presumably many other ontological interpretations of quantum theory exist.

Keywords: relational space; relational interpretation of quantum mechanics; measurement problem; non-locality

---

1. Introduction

Bohmian mechanics [1] became a refuge for scientists and philosophers of science in search of an interpretation of quantum theory, which offers a consistent ontology (for an introduction to Bohmian mechanics, see e.g., [2–4]). Its experimentally verifiable predictions agree with standard quantum mechanics almost by construction, but its ontology is based on degrees of freedom which, again by construction, are not directly accessible to experiments. There are other models that can offer an ontology, like the so-called collapse models of Ghirardi, Rimini and Weber [5,6] (for relativistic extensions see, e.g., [7]); however, these models predict a deviation from quantum mechanics for mesoscopic systems, for which the influence of collapse centers cannot be neglected. Similar models by Karolyhazy [8] and Penrose [9,10] attribute the physical collapse of the wave function to an influence of gravity, which effectively leads to similar deviations from quantum theory as the collapse models. Sooner or later, we should be able to decide by experiment whether or not these collapse models are correct. However, any experimental disagreement with Bohmian mechanics would also be a disagreement with standard quantum theory (at least for those observables for which measurable expectation values can be calculated in quantum theory).

David Bohm himself did not consider his model of quantum theory as *the* ontology of the world, but he emphasized on several occasions that this model is *about ontology* [11]. (Similar remarks can also be found in the first chapter of [12].) He proved that an ontological formulation of quantum theory is possible, despite so-called "no-go"-theorems by von Neumann [13] and others. The price which Bohmian mechanics has to pay is a non-local 'influenciability', not associated with energy or physical information (in the sense that this influenciabilty can be used to transmit signals), and the introduction of unobservable degrees of freedom. Whether or not non-local influenciability exists also for other interpretations of quantum theory is a matter of debate, however, as Bohmian mechanics defines an ontology, this influenciability has an ontological basis, in contrast, e.g., to interpretations of quantum theory, where quantum states represent our knowledge about the world (an example is QBism; for a review, see [14]; for criticism, see [15]).

Bohmian mechanics is based on an ontology that is close to classical Newtonian mechanics in the sense that there exist particles that propagate along well-defined trajectories, and, in addition, there exist fields (the so-called guidance fields or pilot waves), which can be derived from the solutions of Schrödinger's equation and which play a similar role as potentials in classical physics.

In this article, I will outline an interpretation of the quantum formalism which differs in essential ways from Bohmian mechanics: there are no particles that follow trajectories, but there are entities that have relations to spatial points, and a change of these relations leads to the impression of motion. Like Bohmian mechanics, this ontology is based on an unchanged quantum formalism, and, therefore, leads to the same predictions as quantum theory proper. However, I should emphasize that this difference in interpretations refers to Bohmian mechanics in its standard form (like, e.g., in [2,3]), not necessarily to the implicate order, which, according to the later ideas of Bohm, may underlie both quantum theory and relativity.

At first sight, some parts of this ontological model may look very artificial and seem far from being "natural". However, the message is not to sell this as the "right" ontology. The message is that, apart from Bohmian mechanics, other ontological interpretations of quantum theory are possible (I am convinced that there are many more models that "do the job") and that, therefore, none of these interpretations can claim to be *the* ontological interpretation of quantum theory.

Many details of this ontology have not been worked out completely, but it should be obvious that this is possible in principle (often in many different ways). Again, this is because my aim is not to propose a complete model but because I want to convince the reader that such models are possible. Some "mechanisms" have been copied from neural network theory, which is one of my research fields, but I am convinced that almost any field (engineering, biology, psychology, sociology, maybe even chemical networks, fluid dynamics, complex systems theory, electronics, etc.) can give rise to the mechanisms needed to fill the gaps between the general concepts and concrete realizations. For me, the lesson is twofold: (1) we shouldn't give up until we find a really "natural" and maybe even "beautiful" interpretation, (2) until then we can stick to the cooking recipe of quantum mechanics in the firm knowledge that ontological interpretations do exist.

The next section contains a very brief summary of the ingredients of Bohmian mechanics as compared to the relational model of this article. Most of the rest can be considered an elaboration on these ingredients. In Section 3, I will introduce the notion of a relational space and a relational location. In this section, I will also describe the necessary generalizations that are needed to apply this picture to quantum theory. Section 4 describes this relational framework for one-particle quantum mechanics, while, in Section 5, I will extend this formalism to many-particle quantum systems. Section 6 indicates what a generalization to quantum field theory may look like. A brief summary concludes this article.

Finally, I would like to mention that there exists a "relational interpretation of quantum theory", mainly due to Carlo Rovelli [16]. There may be parallels, but, according to my understanding of Rovelli's theory, at least the starting points are different. The notion of "relational" in Rovelli's interpretation refers more to what is observed *in relation to* an observer. In the interpretation given here, "relational" refers to objective, observer independent relations between certain entities. In order to distinguish my relational interpretation of quantum mechanics from Rovelli's interpretation, I will sometimes refer to my interpretation as the "micro-relational interpretation". Some of the ideas presented in this article have been published earlier (see [17–19]) but are presented from slightly different perspectives. In particular, this article includes more explicit examples from every-day life or classical physics (like neural networks) that exhibit similar relational structures. This includes examples for complex relations, entanglement and "collapse".

2. Bohmian Mechanics and the Microrelational Interpretation in a Nutshell

This section is a brief summary of the essential features of Bohmian mechanics on the one hand and the microrelational interpretation on the other. The aim is to emphasize the differences in both ontologies. Furthermore, it is minimalistic in the sense that I list the necessary features of both ontologies in order to agree with standard quantum theory.

In Bohmian mechanics, the wave function $\psi(x)$ *and* the particle haven an ontological character. The wave function satisfies Schrödinger's equation. The particle is guided by the field and its trajectory $x(t)$ is such that the probability density of finding the particle at a particular location x is proportional to $|\psi(x)|^2$. Bohmian mechanics specifies an equation of motion for this trajectory in terms of the wave function and it can be shown that, for these dynamics, the probability requirement holds (at least for quite general initial conditions). However, this particular equation of motion, which is deterministic and can be derived from a polar decomposition of the wave function, is not the only possibility to satisfy the probability requirement.

For multi-particle systems, the wave function is defined over configuration space. The trajectory of several particles becomes the trajectory of a single point in configuration space. In this way, entanglement is automatically built into the model.

In the microrelational interpretation, particles exist as entities that can have relations to other entities. In particular, they do not (for a given moment in time) have a fixed location in space, but their "location" is specified by a complex-valued relation, which is defined by the wave function $\psi(x)$. This wave function also satisfies Schrödinger's equation. Particles don't move, but their relations to spatial entities change. Probing the relation of a particle with a particular point x (or volume in space) induces an "all-or-nothing" change: with a probability (density) proportional to $|\psi(x)|^2$, the relation becomes 1 (the particle "is" at this particular point or in that volume); or it becomes zero for this point. This behavior may sound artificial, but, in Section 4.5, I discuss an every-day example that exhibits similar properties.

The microrelational interpretation doesn't need space (or space-time) to be relational, but defining also relations between spatial entities and letting the topological and metrical properties of space become large scale features of this relational space makes the whole picture more coherent.

Entanglement is interpreted as a relation between two particles such that the relations of one particle to spatial entities depend on the relations of the other particle to spatial entities. This leads to three different types of relations: relations among spatial entities (leading to, say, Euclidean space), relations between particles and spatial entities (they are defined by the wave function), and relations among particles (which lead to entanglement). Whether these three types of relations are fundamentally different (or just different manifestations of the same type of fundamental relation) is left open. The interpretation of entanglement as a 'nearest neighbor' relation makes it possible to keep locality (in a sense defined in Section 5.3), which in my opinion is a charming feature of a relational interpretation.

One tentative idea is that the relations underlying entanglement are the most fundamental ones (these relations may not be the same as the entanglement relations used in standard quantum mechanics, but relations that allow the two entities to share certain information such that in measurements quantum correlations are observed). The two other relations are then effective large scale limits of this fundamental relation.

The whole approach is not to be understood as a fully worked out theory or model but rather as a "program". The idea is not to fill out the details—I am convinced that this can be done in many different ways. The question is whether there are fundamental or logical limitations that would make such a program impossible. If not, many ontological interpretations of quantum theory may coexist and (unless we are able to probe space or space-time at the fundamental level) are indistinguishable with respect to their experimental predictions.

3. Relational Entities

Starting from a mathematical definition of "relation", I first introduce the notion of a relational space. This concept was favored by many philosophers of science, amongst others by Descartes [20] and Leibniz [21]. It is the antipode to the notion of an absolute space that is viewed like a "stage" for matter and which was favored, amongst others, by Newton. Finally, I discuss the notion of "being somewhere" with respect to a relational space, and I will extend this notion to complex-valued relations.

3.1. Mathematical Relations

Mathematically, a relation E on a set V is a subset of $V \times V$. We can represent V by a set of points and the relations E by (directed) lines. Any relation can be represented by a (directed) graph. If the relation E is symmetric (i.e., $(a,b) \in E \Rightarrow (b,a) \in E$ for all $a,b \in V$), it can be represented by an undirected graph (Figure 1). Any relation can be expressed by its adjacency matrix:

$$A_{xy} = \begin{cases} 1, & \text{if } (y,x) \in E, \\ 0, & \text{otherwise.} \end{cases} \tag{1}$$

In the following, I will exclude reflexive relations, i.e., the diagonal elements of the adjacency matrix are zero. This leaves us with $2^{V(V-1)/2}$ different undirected relational sets or $2^{V(V-1)}$ directed sets.

In Figure 1, all elements are uniquely specified by these relations. For example, such a specification could be:

- One node has four neighbors.
- One node has two neighbors.
- One node has only one neighbor, which has three neighbors.
- One node has only one neighbor, which has four neighbors.
- One node has three neighbors of which one has one neighbor.
- One node has three neighbors of which one has two neighbors.

Actually, this is the smallest non-trivial (having more than one node) connected (each node can be connected to each other node by a path along existing lines) undirected relational set for which this is possible. The requirement for this being possible is that the graph has no symmetry. This means that there exists no permutation of vertices that leaves the graph unchanged. In other words, there exists no permutation matrix acting on the elements of V that commutes with A. The probability for a random graph having a symmetry, i.e., that some of its elements are not uniquely identifiable by the relational properties of the graph, gets smaller with an increasing number of vertices.

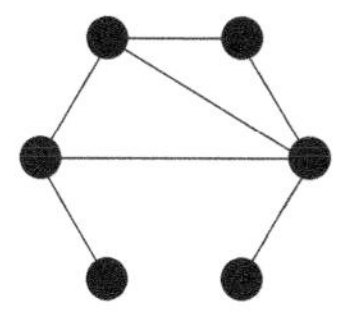

Figure 1. A set with six points and undirected relations. In this case, the relations allow a unique identification of the elements.

3.2. Relational Space

A relational space is defined as a set V of elements, which will be called spatial points, and an undirected relation E, which defines a "nearest neighbor" relation for spatial points. "Nearest neighbor" is not to be understood as "next to each other in an already existing space". Well-known examples of relational spaces are "co-authorship networks" in a scientific community (two scientists being "nearest neighbors" if they are co-authors of a scientific article), semantic networks (two words being related if they are listed as synonyms in a dictionary), protein networks in organisms (e.g., two proteins being related if a chemical reaction in this organism involves both proteins), etc. (for more examples and the statistical properties of such networks, see, e.g., [22]).

While in general such relational spaces are called networks, I will use the term "space" if this network is meant to be a model of the underlying structure of our three-dimensional space. In addition, even though I will refer to the elements of this space as "spatial points", the notion of a point should not be taken literally because, in a network, the shape of an object is defined by its relations. In addition, the representation of relational spaces by undirected graphs, for which the spatial points are depicted as nodes (often in a plane) with lines connecting these nodes, serves merely as an illustration of the relations. The location of these nodes as points in a plane has no intrinsic meaning whatsoever.

There are several ways to define a distance between the elements of a relational set. One possibility is the "length of the shortest connecting path", i.e., the minimal number of nearest neighbor steps which are needed to connect the two points. This is often called the mathematical distance. Another possibility is the "propagator distance" [23,24] that is motivated by physical arguments (the propagation of particles in a scaling limit) and involves statistical sums over all paths connecting two points. Once a distance has been defined for any two points, the dimension d of a point $a \in V$ is defined by the relation $\mathrm{Vol}(a) \propto r(a)^d$, where $\mathrm{Vol}(a)$ is the number of nodes within a distance of $r(a)$ from a point a. If this dimension is independent of the point a, we call it the dimension of the graph. The concept of a scalar curvature can be defined by deviations from this formula (in [25], geometrical concepts have been investigated in more detail for such relational spaces). Of course, these concepts are only meaningful for very large graphs (ideally for $V \to \infty$). We assume that such a relational space—almost flat and of dimension three—is given. The extension of this concept to relational space-time sets—so-called causal sets, for which the elements are events—will be discussed in Section 6.

3.3. "To Be" in a Relational Space

Having discussed the notion of a relational space, I now discuss the meaning of "Where *is* an object?" In a relational space, the location of a spatial point is given by the set of relations it has to all other spatial points. However, when we want to specify the location of a non-spatial object (e.g., an entity which we might associate with a charged particle, i.e., an entity which is different from a spatial point), we have more options. In particular, the type of relations that this entity has with spatial entities will, at least in general, be different from the type of relations between spatial entities.

In an absolute space, the location of an object is defined by $x(t)$, i.e., by specifying the spatial point x *at* which an object is at this particular moment. In a relational space, the location of an object is defined by a field $\chi(x,t)$ ($x \in V$), indicating the relations of this object to the spatial points (see Figure 2). For an undirected mathematical relation, this field only assumes two values, 0 and 1, depending on whether or not the relation exists, while, for a directed mathematical relation, this field can be considered as having two binary components specifying the "in"- and "out"-relations.

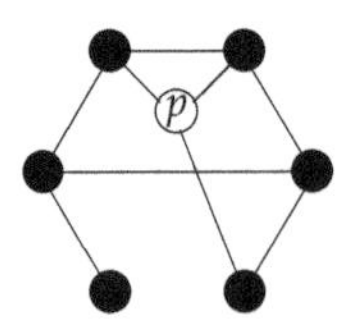

Figure 2. The location of an object p in a relational structure is defined by the spatial points to which it is related. Equivalently, one can specify this relation by the characteristic function of this set of spatial points. If the relations of the object to 'space' are directed, we can specify it by two characteristic functions.

In a relational space, it is also possible that an entity can be "in two different spatial regions simultaneously" (for an example, see Figure 3).

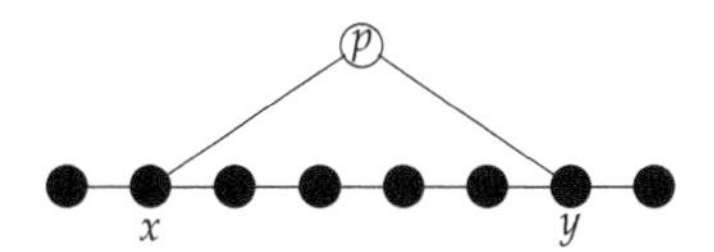

Figure 3. In a relational framework, a particle can be at two locations simultaneously. In the given example, the object p "is" at the points x and y simultaneously.

More general, if an object p has relations to many spatial points that are distributed over a region with large distances, we may say that this object is non-local.

3.4. Complex Valued Relations

Up to now, I have considered "yes-or-no"- relations, defined by binary functions $\chi : V \times V \to \{0, 1\}$. For the corresponding graphs, a line between two points is either present or not. However, one can generalize this concept by attributing weights or distances or other quantities to lines. In this way, one arrives at the notion of a network. (For me, a network is a graph with additional structures; however, not all authors make this distinction.)

In quantum mechanics, the state of a particle at a particular moment t can be characterized by its wave function $\psi(x, t)$, which is a complex-valued field over space. In the next section, I will distinguish between the relations among spatial points and the relations between some entity ("particle") and spatial points. For the latter, the relations are defined by the wave function. However, here I only want to emphasize that, in general, complex-valued relations are also possible.

The following examples of networks, in which the relations between nodes can be generalized from a binary value (being there or being absent) to complex values, just serves as an illustration that networks with complex-valued relations exist. In most of the examples, a link between two nodes is used for an exchange of information (in the broadest sense) or activity between the nodes. This information or activity can be coded as a complex number for at least two reasons: either because there is a flow in both directions, from node x to node y and vice versa, or because the activity has an amplitude and a phase. From classical wave models, it is well known that these two possibilities are not completely independent.

As a first example, consider a network of computer servers (this example will be extended in Section 4.5). In this case, two servers are said to be "related", if there exists a direct connection from one of the servers to the other. For a particular server p inside such a network, we can define a complex-valued, time-dependent relational structure by a function $\psi_p(q; t) : V \to \mathbb{C}$ (where V now denotes the set of computers), which characterizes the information exchanged between server p and another server q. As such, a function exists for all servers in the network, and we end up with $\psi(p, q; t) \equiv \psi_p(q; t) : V \times V \to \mathbb{C}$.

Another example of networks in which the relations between the constituents can be described by a complex-valued function are electric circuits with resistors, capacities and/or coils—the "relations" between nodes being the currents.

A different type of network, which I will sometimes use as an example, are neural networks (for an introduction to neural networks, see, e.g., [26]). One can define several types of complex-valued relations in such networks:

- In a neural network, the directed link between two nodes (which in this case are referred to as neurons) has a weight (the synaptic weight) which determines the transmission intensity of a signal. Negative weights indicate inhibitory influences. As the network is directed, the connection between two nodes is specified by two real-valued weights that can be combined into a complex number. These weights change over time as a result of learning.

- In so-called spiking neural networks, the signal consists of a firing rate (the number of spikes per unit of time) that is transmitted from one neuron to another. The time scales on which these firing rates change are much shorter than the time scales for changes in the synaptic weights, so that the synaptic weights can roughly be considered as constant. The connections (the synapses) between neurons are directed, but it often happens that connections exist in both directions. In addition, firing can occur in a synchronized way between clusters of neurons or asynchronous. Thus, the relative phases in spiking neurons can be important.
- On large scales (averaging over several hundreds of neurons), the activity in neural networks is sometimes described by a complex field (see, e.g., [27,28]). Together with David Bohm(!), the famous neuroscientist Karl Pribram developed a quantum field theoretic approach to consciousness [29], which was related to Bohm's ideas of an implicate and explicate order [30].

4. One-Particle Quantum Mechanics

We know that quantum mechanics, based on Schrödinger's equation, is only a non-relativistic limit of a theory that is considered to be more fundamental: quantum field theory. In Section 6, I will indicate how quantum field theory might be formulated in a relational setting. In this section, however, mainly for didactical purposes, I indicate how a relational re-interpretation of quantum mechanics might lead to an ontology of quantum theory.

4.1. *The Generalized Relational Structure of "Location"*

As we have seen in the previous section, in a relational framework, the position of an object is defined by the spatial points to which it is related. One of the consequences is that an object can be "at several spatial points simultaneously" (see Figure 3). Exactly this feature is one of the conundrums in the standard formulation of quantum mechanics. The wave function $\psi(x)$ of a particle does not mark a particular point of space as the position of that particle, but it defines a whole region of space in which the particle, if measured, can be found. According to the general interpretation of quantum mechanics (not Bohmian mechanics), this uncertainty in the position of a particle is not due to a lack of knowledge but intrinsic. An object like an electron "presents" itself as a particle, when a proper measurement is performed.

There are many ways to combine a discrete model of space(-time) with quantum mechanics (a by far not complete selection of approaches can be found in [31,32]). In the following, I will describe just one possible model (more details can be found in [17–19]). In this model, the connection between wave mechanics and a relational model is made by generalizing the concept of a relation.

In principle, the relational structure among spatial points can be anything which in a large-scale limit gives rise to the topological and geometrical properties of our three-dimensional space. However, for simplicity, I still assume the relational structure among spatial points as represented by an undirected graph, i.e., for two spatial points x and y, a connection is either present or absent. Even this simple structure can in principle yield the desired large-scale limit.

In the micro-relational interpretation, the relations between an object ("particle") and the set of spatial points will be generalized from a binary function to a complex-valued function $\psi : V \to \mathbb{C}$, and this complex function is the wave function of this object. Again, the micro-structure is not necessarily fixed: the requirement is that, in a large-scale limit, the relational structure between objects and spatial points yields the wave function; however, for simplicity, I assume that the micro-relations already have the complex values of the wave function. That such complex-valued relations can occur even in classical systems has been indicated in the last section.

I want to emphasize that, in the framework discussed here, the relational description of a single particle does not require a new mathematical formalism as compared to standard quantum theory (apart from the discretizations of space and time—and even these are not required). It is simply a different interpretation of the usual concept of a wave function (this will also hold for many-particle systems). In this re-interpretation, the absolute value of $\psi(x)$, i.e., $p(x) = |\psi(x)|^2$, should still give the probability(density) for finding a particle in a particular location when a measurement is performed. The changes with respect to the standard interpretation of quantum mechanics are minor: instead of speaking of a "probability amplitude" $\psi(x)$, I refer to this function as a complex-valued relation. When this relation is probed by a measurement, it changes according to a "winner takes it all" manner (the "collapse" of the wave function). I will come back to these points in Section 4.4.

4.2. The Dynamics of Relations

For a relational structure between an object (particle) p and spatial points, one can define the dynamics as follows. First of all, as we are considering a discretized space, we also discretize time and formulate the dynamics in terms of an iterative mapping which defines $\psi_p(x, t+1)$, the ψ_p-function at time-step $t+1$, as a linear function of $\psi_p(x, t)$. (I use the notation $\psi_p(x)$ to indicate that this is the wave function of an entity p.) A natural candidate for such a dynamics is the following equation:

$$\psi_p(x, t+1) = \psi_p(x, t) + \epsilon \left(\alpha \left(\sum_y A_{xy} \psi_p(y, t) \right) + \beta V(x) \psi_p(x, t) \right) . \tag{2}$$

The second term on the right-hand side (proportional to a constant ϵ) corresponds to the change of the generalized relation $\psi_p(x, t)$. There are two contributions: the first one describes the propagation of the relation from one spatial point to a neighbored point (expressed by the adjacency matrix A_{xy}), the second one describes an additional change of the relation due to a local potential. This second term may also depend on the valency or degree d_x of point x, i.e., the number of points it is related to. It is not hard to see that, under very general conditions, such an equation becomes a Schrödinger-type equation in a continuum limit. Notice in this context that, for undirected graphs, the matrix $D - A$ (where D is a diagonal matrix—the so-called degree matrix—with $D_{xx} = d_x$) is the graph Laplacian, i.e., the discretized analogue of the Laplace operator (see, e.g., [33]).

This specifies the dynamics of the relations of an object to the network of spatial points. However, how do relations change in general, e.g., how does the adjacency matrix A_{xy} of spatial points change? (Changes in the purely spatial relations imply changes in the geometrical properties of space, i.e., they may become relevant if we include gravity.) At this stage, I introduce a locality requirement: If x and y are not related at time t (i.e., $(x, y) \notin E$), they can only become related at time $t+1$ if there exists a point z such that $(x, z) \in E$ and $(y, z) \in E$ at time t (see Figure 4). This locality requirement is not mandatory, but it is quite satisfying from a philosophical point of view. As we will see later (Section 5.3), quantum correlations become local in this picture.

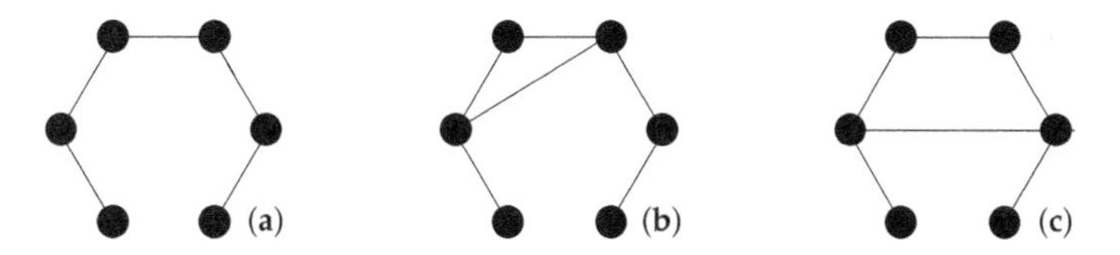

Figure 4. Propagation of relations: only via the intermediate step (**b**) can the additional relation in (**c**) be created from the relational space (**a**).

One may add further requirements, e.g., that an existing relation $(x, y) \in E$ at time t can only be removed at time $t+1$, if there exists a point z such that either (x, z) or (y, z) have changed their status as a relation from $t-1$ to t. This "deletion" of relations may become relevant in the context of the measurement problem (see Section 4.4).

4.3. The Double-Slit Experiment and "Sum over Histories"

For many scientists, the double-slit experiment is "the only mystery" of quantum mechanics (see, e.g., [34]). This may be arguable (in particular in view of the "mystic" effects related to entanglement), but the double-slit experiment has always been one of the paradigms of quantum theory. Therefore, it might be of interest to see how the double-slit phenomenon is explained in the relational setting (see Figure 5).

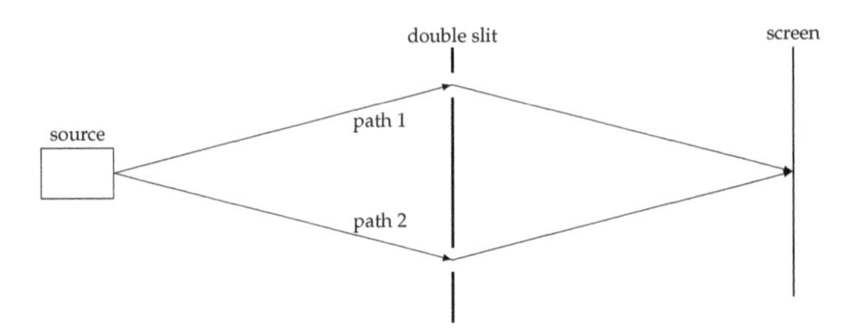

Figure 5. In the double-slit experiment, the total amplitude can be obtained by assuming that *aparticle* propagates along path 1 AND path 2. In the micro-relational interpretation, *the relations* of a particle propagate along path 1 and path 2.

The interference pattern is easily explained by assuming a wave $\psi(x)$ to propagate through the slits. The two parts of the wave behind the slits interfere and the intensity on the screen is obtained by the absolute square of the sum of these two parts. In a particle picture, Feynman's "summation over paths" can be interpreted as: a particle propagates along path 1 with an amplitude ψ_1 associated with this path, and it also propagates along path 2 with an amplitude ψ_2 associated with this process. The absolute square of the sum of these two amplitudes yields the probability of finding the particle at a particular spot on the screen.

In the micro-relational interpretation, we can re-interpret the "summation over paths" in the following way: the entity p (the particle) has relations that propagate along path 1 and it also has relations that propagate along path 2 (compare Figure 3). The absolute square of the sum of these (complex-valued) relations yields the probability of finding the particle at a particular spot on the screen.

More generally, we can re-interpret Feynman's "summation over paths"-representation for the propagator of a particle as "a relation propagates along path 1 AND a relation propagates along path 2 AND ...". This seems to be much less weird than "the particle propagates along path 1 AND it propagates along path 2 AND it propagates along path 3 ...".

4.4. Measurements

With respect to measurements, two fundamental concepts of quantum theory have to be explained: (1) probabilities are given by the absolute square of the scalar products of vectors that represent states, and (2) quantum states "collapse" into a new state (depending on the outcome of a measurement) as the result of a measurement. The first concept is known as Born's rule, the second is sometimes called the "collapse" postulate. With respect to Born's rule, we are mainly interested in probabilities for finding an entity at a spatial point (or in a spatial region), i.e., in probabilities proportional to $|\psi_p(x)|^2$.

Of course, we can simply postulate a mechanism which respects these two rules. Nobody will deny that it is easy to program a computer (essentially a classical system) to calculate probabilities from the absolute squares of complex functions and to update these complex functions according to the collapse postulate. However, I would like to include some examples of classical systems in which similar rules can be found.

In oscillating systems like the harmonic oscillator or simple waves, energies and intensities are given by the square of an amplitude. This also holds for alternating currents and voltages (the energy and electrical power being proportional to the product of these two). However, also in other processes, we encounter this relation: for instance, in diffusion processes or Brownian motion, the probability of finding a particle in a distance d from the origin of propagation is proportional to d^2. Thus, if a process is triggered by this particle (or by an intensity exceeding a given threshold) and if the 'relations' correspond to inverse distances, this process is triggered with a probability proportional to the square of these relations.

In addition, the collapse postulate is not completely unknown in classical physics: in some neural networks (e.g., in so-called Kohonen networks, see, e.g., [35]) the first neuron, which starts to fire as the result of an integrated input, sends inhibitory signals to all other neurons such that these will not fire. This mechanism—the first firing neuron inhibiting all other neurons—is sometimes called the "winner-takes-it-all" principle.

4.5. An "Every-Day" Example for Measurements and the Collapse

Instead of elaborating on possible realizations of the measurement process, I describe an every-day example which at first sight seems to have nothing to do with quantum theory. I hope that, in the end, the relationships will become obvious. This example relates to an example that has already been mentioned in Section 3.4.

You book a flight. What you get is an e-ticket. The essential information on that e-ticket is your name and the e-ticket number. Of course, it also tells you the flight number, the date and time of departure, the duration and additional information about your flight. What you need before you can enter the plane is a boarding-pass, which assigns to you your seat in the plane. The information that transforms your e-ticket into a boarding-pass is stored in some server at the airport or the airline.

Before things get too complicated, I consider a simplified system (which comes close to the situation a few years ago). There is a single server that has the information about your e-ticket. Distributed over the airport are several counters with printers (Figure 6). When you go to one of the counters and present your e-ticket number, you will get your boarding-pass.

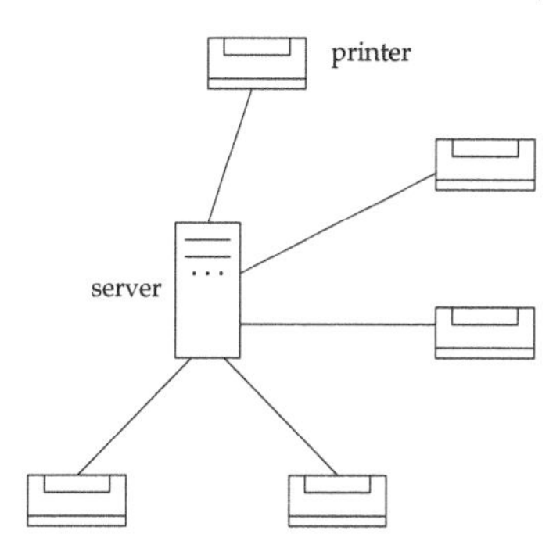

Figure 6. A server connected to a periphery of counters with printers is a model for a measurement in a relational system. A boarding pass exists only virtually as a program instruction in the server. Only when an e-ticket number is presented at a counter—this is the measurement—does the boarding pass become reality at the printer of this counter.

This boarding pass exists only once. You cannot go to a second counter and get a second boarding pass (you might get a second print-out, but it will be for the same seat number in the plane; in this sense, you can never get a second pass). In addition, when you go to a counter you never get "half a boarding-pass", or part of a boarding pass, and, for the rest, you have to go to a different counter. Its an all or nothing situation, and the "all" can only happen once.

In order to make the situation more similar to quantum theory, let us assume a fictitious world in which you will get your boarding-pass at a particular counter only with a certain probability. If at one counter you do not get it, you have to try a different counter, and again you will get it only with a certain probability. The server decides probabilistically at which counter you will get the boarding pass, but it will set this probability to zero for a counter where you already unsuccessfully tried to get your boarding-pass (and it will renormalize the probabilities for the other counters). Eventually, you will get your boarding-pass at one of the counters.

The similarity to quantum theory should now be obvious: the boarding-pass is an entity that exists only "virtually" before it becomes reality as the result of a measurement. The measurement consists of the presentation of your E-ticket at one of the many counters. Before you make this "measurement", the boarding-pass existed as a "potentia" (a virtual entity) at all counters simultaneously while, upon making the measurement, it becomes reality at only one of the counters. The counters represent certain locations where the boarding-pass can become reality. They correspond to the (discretized) spatial points.

Of course, you could come to the airport with your family and maybe many friends and present the e-ticket number at all possible counters simultaneously. Only at one of the counters will one of the members of your group get the boarding-pass.

5. Many-Particle Systems and Entanglement

5.1. General Remarks

Many-particle systems are a general problem for ontological theories. The many-particle wave function is defined in configuration space, i.e., in a $3N$-dimensional space. Thus, in contrast to electric and magnetic fields or the metric field of space-time, which also "guide" particles, this field does not have an ontology in ordinary space. This feature is often used as an argument against Bohmian mechanics. The counter-argument is that, for a system of N interacting particles, the potential in Newtonian mechanics is also defined in configuration space, and the corresponding force acting on one particle may depend on the positions of all other particles. Factorization only occurs for external potentials or forces acting on single particles and being independent of the positions of the other particles.

Before I elaborate further on this subject, I take up the example of the last Section 4.5. A server at an airport can not only handle a single boarding-pass but thousands of e-tickets and boarding-passes simultaneously. If you have made a booking for two persons, you can instruct the server to hand over both boarding-passes at the same counter. Such correlations among the probabilities, at which counters the boarding passes "come into reality", resemble entanglement correlations. As long as there is an information exchange within the server, entanglement correlations are no miracle. Even if there are several servers handling the boarding passes at an airport, there is no miracle if the information is shared by these servers, i.e., if there is an information exchange between them.

This example can be taken as a hint that entanglement correlations are the result of an immediate exchange between entities. These entities have to be directly related to each other. In other words, a relation between two entities can exhibit itself as an entanglement between these entities.

5.2. Relations for Two-Particle Systems

The simplest way to extend the one-particle picture to a two particle picture is to add two elements to the set of spatial points V, so that the set of elements now is $\{p_1, p_2\} \cup V$ (see Figure 7).

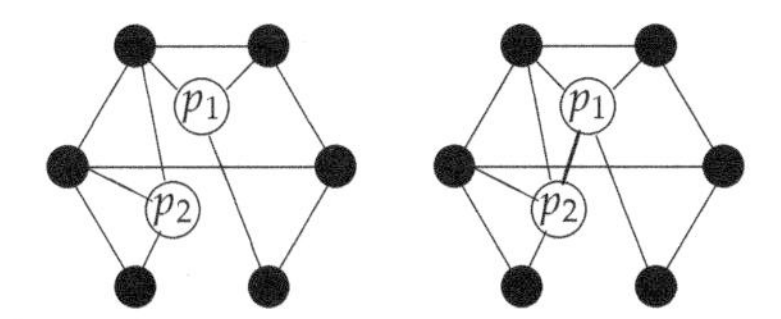

Figure 7. (**left**) two objects in a relational structure. Each object has its own set of relations to the spatial points. The relations factorize; (**right**) there can also be a direct relation between the two objects. This may lead to entanglement.

The generalization of this construction to several particles is straightforward. For n particles, the relational space consists of the elements $V_n = \{p_1, ..., p_n\} \cup V$ and a generalized relation is a subset of $V_n \times V_n$. In Figure 7, the relations are undirected, but, depending on the nature of these relations, some of them may also be directed.

We now have encountered three types of relations (see Figure 8):

1. Relations between spatial entities: these are considered to be non-directed and give rise, on a large scale, to the geometry of space.
2. Relations between "particles" and spatial entities: these relations maybe directed and give rise, on a large scale, to the wave function.
3. Relations between "particles": These relations are present if the particles are entangled. They allow for a direct information transfer between particles and characterize the form of entanglement.

When we compare this picture (e.g., Figure 7 or Figure 8) with our metaphor of servers and counters (or printers), the counters correspond to the spatial points *where*, in certain measurements, particles (boarding-passes) can be found. The algorithm that is stored in the server and which upon the presentation of the e-ticket sends the printing command to the periphery corresponds to the "virtual" entity before a measurement. The server (or the net of servers) just handles these virtual "many-particle algorithms". Relations between particles, i.e., entanglement, can be compared with certain constraints between the different algorithms, and relations between particles and spatial points can be compared with connections between servers and printers (allowing for a selective output of the boarding pass at exactly one of the printers). In our metaphor, we do not take into account direct relations between printers. Such relations would define a spatial "neighborhood" and eventually give rise to a topology and a geometry on the set of printers.

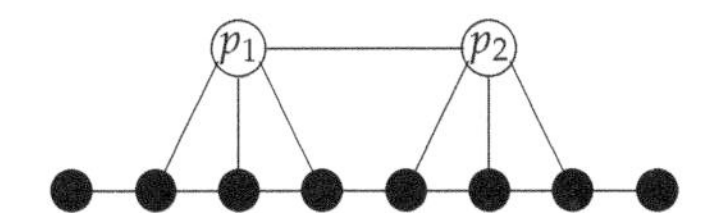

Figure 8. When there are several objects (in this case two), we have three different types of relations: (1) relations among spatial points, (2) relations between "particles" and spatial points, and (3) relations between the "particles".

There is an interesting point here: if entangled entities can exchange information, what mechanism restricts the degree of entanglement correlations to the Tsirelson bound [36] (see, e.g., [37–39])? There is no reason why correlations between systems that can exchange information are subject to a constraint that is much below the maximum possible correlation (Popescu–Rohrlich (PR) boxes [40] have maximum correlations), and it is easy to construct classical machines for which the correlations assume this PR-bound (see [41]). Of course, if the information exchanged is tailored according to the quantum formalism, this bound will be respected. However, it remains a general question why quantum correlations are subject to this bound.

5.3. *Local or Non-Local, That Is the Question*

One of the more speculative consequences of a relational space and, in particular, relational locations of objects in such a space, is the possibility of an ostensible superluminal propagation of influences (changes in relations) in the sense that actually this propagation of influences is subject to a locality principle (see Section 4.2), but, for an observer, it may look like an immediate, non-local influence or change of relations.

Of course, much depends on how we measure distances between the elements of relational sets. As already mentioned, apart from the simple mathematical concept of distance (number of links for the shortest path connecting two elements), one may use a propagator distance, which involves a summation over all paths connecting two elements. In [17,18], I have dealt with the consequences of such definitions.

Here, I would like to emphasize a slightly different point of view. Let us assume that distances in space are determined exclusively by the spatial relations and that these relations remain constant (e.g., consider a three-dimensional hypercubic lattice). Now, consider the situation of Figure 8: two entangled particles, each having relations to spatial points in regions that might be far away from each other. However, due to entanglement, these particles are directly related and can therefore "communicate" almost instantly. Two objects, which are entangled, are "nearest neighbors" and never far away from each other in the sense of relations. (There is a a similarity to the ideas behind the so-called ER = EPR conjecture of Maldacena and Susskind [42]: Two particles which are entangled (EPR, Einstein–Podolsky–Rosen entanglement, [43]) are connected by an Einstein–Rosen (ER) wormhole.)

There is a curious observation that supports this general idea: entanglement is always built-up locally; however, it can be destroyed non-locally. In order for two distant objects to be entangled, they either were directly (locally) involved in an interaction in the past (e.g., they were created in a decay process) or one of them interacted locally with a particle that already was entangled with the other (entanglement swapping). Both are local processes according to the definition given in Section 4.2. However, if two distant particles are entangled, this entanglement relation can be "broken" (they become separated) by a local interaction (e.g., a measurement) performed at only one of the particles. This asymmetry with respect to entanglement creation and entanglement destruction is nicely explained in the relational structure.

Violations of Bell's inequalities are sometimes taken as a proof that any ontological model of quantum theory has to be non-local. Only seldom is it explicitly stated that this conclusion is based on a classical (non-quantum) picture of space-time, e.g., a Minkowski space-time as a background. The ER = EPR conjecture as well as the micro-relational interpretation circumvent this assumption.

6. Relational Space-Time—Relational Events

The previous sections assumed a relational space and a relational notion of "location" for an object in such a space. In this section, I will briefly sketch a relational structure of space-time.

When dealing with space-time, the relevant "objects" (the elements of space-time) are events. If space-time is considered as "absolute" (e.g., Minkowski space-time), the events are located at particular space-time points. In a relational picture, the locations of events (space-time points) are defined by their relations to other events.

One starting point may be the model of causal sets (see, e.g., [44,45] and Figure 9, left). In this case, all relations are assumed to be time-like or light-like (depending on the details of the formalism). There are no space-like relations. The causal structure of space-time is built into the relational structure. I will assume such a relational structure for the space-time events that make up "empty space", i.e., which in a large-scale limit approaches a Minkowski space or any other vacuum solution of Einstein's equations.

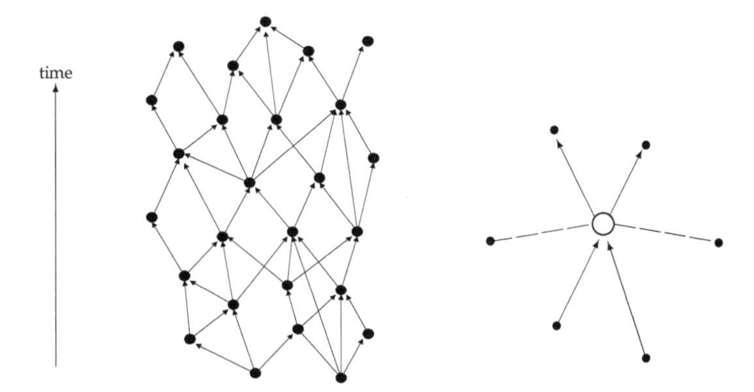

Figure 9. (**left**) the events making up the canvas of "space-time" are endowed with a causal structure; (**right**) a physical, object-related event can be related to the events of "space-time" in three different ways: It can be causally influenced by events in its past, it can influence events in its future and there may be "space-like" relations to events that are in the causal complement. The distinction between "space-like" events and time-like or light-like events depend on the real and imaginary parts of causal Green's functions.

With respect to the relations of an event like the emission of a photon by an electron, i.e., an event which involves entities like particles, I will choose a different structure. Having a quantum field theory in mind, I define space-like and time-like (including light-like) relations for object-related events (see Figure 9, right). The distinction between space-like and time-like relations will be that space-like relations are real-valued while time-like relations are complex-valued. The reason behind this definition is that, in quantum field theory, Green's functions have real and imaginary parts for time-like separated points but only real parts for space-like separated points; and the relations that I associate to an event are defined by the Green's functions.

Without going into details, I just consider the simple process of Coulomb scattering of two electrons in the lowest approximation (Figure 10). Two elementary events—the emission of a photon of one electron and the absorption of the photon of the other electron—constitute this process. Usually, the asymptotic states are characterized by their momenta, but, for simplicity, I consider the process as determined by four external events x_1, x_2, x_3, x_4 that correspond to two initial states of the electrons and two final states of the electrons, respectively. Suppressing all indices referring to the spin of the electrons and the polarization of the photons as well as factors of π and other normalization factors etc., the amplitude for this process can formally be expressed as

$$A(x_1, x_2, x_3, x_4) \propto \int \mathrm{d}y_1^4 \int \mathrm{d}y_2^4 \, S(x_1, y_1) S(x_2, y_2) G(y_1, y_2) S(y_1, x_3) S(y_2, x_4) \,. \tag{3}$$

Here, $S(x, y)$ denotes the electron propagator (from space-time point x to space-time point y) and $G(y_1, y_2)$ the propagator of the exchanged (virtual) photon. In general, the contributions from these propagators are complex functions. Each propagator defines a generalized relation between the event (say y_1) and other events (in this case y_2, x_1 and x_3). The fact that we have to integrate over the "location" y_1 of this event indicates that this event does not happen at a particular point, but, in principle, everywhere in space-time. At least, this is the usual interpretation of this integration: we have to sum over all histories, i.e., all positions for these events. In the mircro-relational picture, this integration is interpreted as a "sum" over all relations which one event, say "emission of a photon", has to all the other events of the space-time canvas. (Actually, as the exchange propagator for the photon between event y_1 and y_2 will not be on mass-shell, emission of a photon and absorption of a photon cannot be distinguished and should rather be interpreted as 'interaction with a virtual photon').

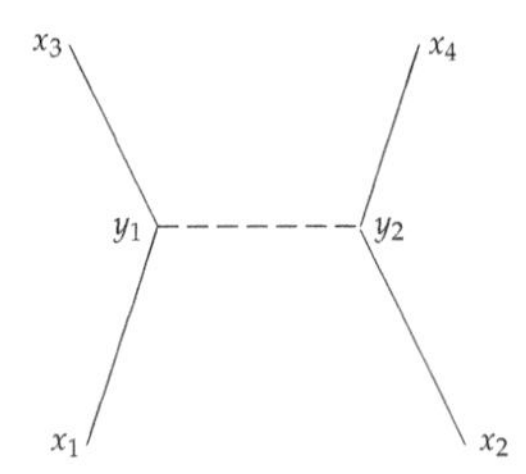

Figure 10. The lowest order approximation of a Coulomb scattering of two electrons by an exchange of a (virtual) photon. The points x_i are kept fixed while one has to integrate over all possible positions of the intermediate events at y_1 and y_2.

Thus, in the micro-relational interpretation, events do not have a particular location, but they have relations to all other events, space-time events and object-related events. The amplitude for a particular process in quantum field theory is just the remainder of the sum over all these relations. (For more details, see [18,19].)

7. Conclusions

I have argued that the concept of "locality" receives a completely different meaning when the positions or locations of entities (objects or events) are defined in a relational sense as compared to an absolute space or space-time. In particular, many counter-intuitive aspects of quantum theory appear less weird from this perspective. A relational space or space-time as well as a relational structure between particles might also be a way to circumvent the constraints given by Bell-type inequalities: the "elements of reality" and the requirement of locality are no-longer mutually exclusive.

I should add as a final remark that the ontological interpretation presented in this article is not necessarily opposed to Bohmian mechanics, at least not in the sense David Bohm interpreted his theory (see, e.g., [12]). The implicate order (or the structure underlying quantum theory and the theory of relativity) could be relational and the ideals outlined in this article may, in a large-scale continuum limit, lead to Bohmian mechanics.

Acknowledgments: I gratefully acknowledge stimulating discussions with Basil Hiley, Paavo Pylkkännen, Nikolaus von Stillfried, and Jan Walleczek during the London Conference EmQM17—David Bohm Centennial Symposium. I also acknowledge the generous hospitality of the Fetzer Franklin Fund during this conference.

Conflicts of Interest: The author declares no conflict of interest.

References

1. Bohm, D.J. A Suggested Interpretation of the Quantum Theory in Terms of 'Hidden' Variables I & II. *Phys. Rev.* **1952**, *85*, 166.
2. Cushing, J.T.; Fine, A.; Goldstein, S. *Bohmian Mechanics and Quantum Theory: An Appraisal;* Boston Studies in the Philosophy of Science 184; Springer: Dordrecht, The Netherlands, 1996.
3. Dürr, D.; Teufel, S. *Bohmian Mechanics: The Physics and Mathematics of Quantum Theory;* Springer: Berlin, Heidelberg, 2012.
4. Holland, P.R. *The Quantum Theory of Motion: An Account of the DeBroglie-Bohm Causal Interpretation of Quantum Mechanics;* Cambridge University Press: Cambridge, UK, 1993.
5. Ghirardi, G.; Rimini, A.; Weber, T. A Model for a Unified Quantum Description of Macroscopic and Microscopic Systems. In *Quantum Probability and Applications;* Accardi, L., von Waldenfels, W., Eds.; Springer: Berlin, Germany, 1985.
6. Ghirardi, G.; Rimini, A.; Weber, T. Unified Dynamics for microscopic and macroscopic systems. *Phys. Rev. D* **1985**, *34*, 470. [CrossRef]

7. Tumulka, R. A Relativistic Version of the Ghirardi-Rimini-Weber Model. *J. Stat. Phys.* **2006**, *125*, 821. [CrossRef]
8. Károlyházy, F. Gravitation and Quantum Mechanics of Macroscopic Objects. *Nuovo Cimento* **1966**, *42*, 1506. [CrossRef]
9. Penrose, R. *The Emperor's New Mind*; Oxford University Press: Oxford, UK, 1989.
10. Penrose, R. On the Gravitization of Quantum Mechanics 1: Quantum State Reduction. *Found. Phys.* **2014**, *44*, 557–575. [CrossRef]
11. Pylkkänen, P. (University of Helsinki, Helsinki, Finland). Private communication, 2018. Paavo Pylkkänen refers to a letter from D. Bohm to K. Popper, who had accused him of 'saving determinism'. Bohm wrote "This is not about determinism, this is about ontology".
12. Bohm, D.J.; Hiley, B. *The Undivided Universe: An Ontological Interpretation of Quantum Theory*; Routledge: London, UK, 1993.
13. Von Neumann, J. *Mathematical Foundations of Quantum Mechanics (orig. 1932)*, 1st ed.; Princeton University Press: Princeton, NJ, USA, 1955.
14. Fuchs, C.A. QBism, the Perimeter of Quantum Bayesianism. *arXiv* **2010**, arXiv:1003.5209.
15. Bacciagaluppi, G. A Critic Looks at QBism. In *New Directions in the Philosophy of Science*; Galavotti, M., Dieks, D., Gonzalez, W., Hartmann, S., Uebel, T., Weber, M., Eds.; The Philosophy of Science in a European Perspective; Springer: Cham, Switzerland, 2014; Volume 5.
16. Rovelli, C. Relational Quantum Mechanics. *Int. J. Theor. Phys.* **1998**, *35*, 1637–1678. [CrossRef]
17. Filk, T. The Problem of Locality and a Relational Interpretation of the Wave Function. In *Quantum Theory: Reconsideration of Foundations—3*; Adenier Khrennikov, G., Yu, A., Nieuwenhuizen, T.M., Eds.; American Institute of Physics: Melville, NY, USA, 2006; pp. 305–311.
18. Filk, T. Relational interpretation of the wave function and a possible way around Bell's theorem. *Int. J. Theor. Phys.* **2006**, 45, 1166–1180. [CrossRef]
19. Filk, T. Relational Events and the Conflict Between Relativity and the Collapse. In *Re-Thinking Time at the Interface of Physics and Philosophy—The Forgotten Present*; Parmenides Book Series *On Thinking*; von Müller, A., Filk, T., Eds.; Springer: Basel, Switzerland, 2015; pp. 67–91.
20. Descartes, R. *The Principles of Philosophy*; (orig. 1644); Kessinger Pub. Co.: Whitefish, MT, Canada, 2010.
21. Ariew, R. *G.W. Leibniz and Samuel Clarke: Correspondence*; Hackett Publishing Co. Inc.: Indianapolis, IN, USA; Cambridge, MA, USA, 2000.
22. Albert, R.; Barabasi, L. Statistical mechanics of complex networks. *Rev. Mod. Phys.* **2002**, *75*, 47–97. [CrossRef]
23. Billoire, A.; David, F. Scaling properties of randomly triangulated planar random surfaces: A numerical study. *Nucl. Phys. B* **1986**, *275*, 617. [CrossRef]
24. Filk, T. Equivalence of Massive Propagator Distance and Mathematical Distance on Graphs. *Mod. Phys. Lett. A* **1992**, *7*, 2637–2645. [CrossRef]
25. Filk, T. Random graph gauge theories as toy models for non-perturbative string theories. *Class. Quantum Grav.* **2000**, *17*, 4841–4854. [CrossRef]
26. Haykin, S. *Neural Networks and Learning Machines*, 3rd ed.; Prentice Hall International: Upper Saddle River, NJ, USA, 2008.
27. Griffith, J.S. A field theory of neural nets I: Derivation of field equations. *Bull. Math. Biophys.* **1963**, *25*, 111–120. [CrossRef] [PubMed]
28. Griffith, J.S. A field theory of neural nets II: Properties of field equations. *Bull. Math. Biophys.* **1965**, *27*, 187–195. [CrossRef] [PubMed]
29. Pribram, K. (Ed.) *Rethinking Neural Networks: Quantum Fields and Biological Data*; Lawrence Erlbaum Associates: Hillsdale, NJ, USA, 1993.
30. Bohm, D.J. *Wholeness and the Implicate Order*; Routledge: London, UK, 1980.
31. Clark, C. Quantum Theory in Discrete Spacetime. Preprint UCLA. 2010. Available online: http://dfcd.net/articles/discrete.pdf (accessed on 18 June 2018).
32. Šťovíček, P.; Tolar, J. Quantum Mechanics in Discrete Space-Time. *Rep. Math. Phys.* **1984**, *20*, 157. [CrossRef]
33. Chung, F.R.K. *Spectral Graph Theory*, 2nd ed.; American Mathematical Society: Providence, RI, USA, 1997.
34. Feynman, R.P.; Leighton, R.B.; Sands, M. *The Feynman Lectures on Physics*; Addison-Wesley Longman: Amsterdam, The Netherlands, 1965; Volume 3.

35. Kohonen, T. Self-Organized Formation of Topologically Correct Feature Maps. *Biol. Cybern.* **1982**, *43*, 59–69. [CrossRef]
36. Cirel'son, B.S. Quantum Generalizations of Bell's Inequality. *Lett. Math. Phys.* **1980**, *4*, 93. [CrossRef]
37. Grinbaum, A. Reconstruction of quantum theory. *Br. J. Philos. Sci.* **2007**, *58*, 387–408. [CrossRef]
38. Grinbaum, A. Quantum Correlations: Challenging the Tsirelson Bound. In *Quantum Interaction, Lecture Notes in Computer Science*; Atmanspacher, H., Filk, T., Pothos, E., Eds.; Springer: Basel, Switzerland, 2015; Volume 9535, pp. 3–11.
39. Popescu, S. Nonlocality beyond quantum mechanics. *Nat. Phys.* **2014**, *10*, 264–270. [CrossRef]
40. Popescu, S.; Rohrlich, D. Generic quantum nonlocality. *Phys. Lett.* **1992**, *166*, 293–297. [CrossRef]
41. Filk, T. A mechanical model of a PR-box. *arXiv* **2015**, arXiv:1507.06789.
42. Maldacena, J.; Susskind, L. Cool horizons for entangled black holes. *Fortsch. Phys.* **2013**, *61*, 781–811. arXiv:1306.0533.
43. Einstein, A.; Podolsky, B.; Rosen, N. Can quantum-mechanical description of physical reality be considered complete? *Phys. Rev.* **1935**, *47*, 777. [CrossRef]
44. Bombelli, L.; Lee, J.; Meyer, D.; Sorkin, R.D. Space-Time as a Causal Set. *Phys. Rev. Lett.* **1987**, *59*, 521. [CrossRef] [PubMed]
45. Sorkin, R.D. Light, Links and Causal Sets. *Proc. DICE2008 Meet. J. Phys. Conf. Ser.* **2009**, *174*, 012018. [CrossRef]

MDPI

Article

Analytical and Numerical Treatments of Conservative Diffusions and the Burgers Equation

Dimiter Prodanov †

Department of Environment, Health and Safety, Imec, 3001 Leuven, Belgium; dimiter.prodanov@imec.be
† Current address: Imec, Kapeldreef 75, 3001 Leuven, Belgium

Received: 15 April 2018; Accepted: 21 June 2018; Published: 25 June 2018

Abstract: The present work is concerned with the study of a generalized Langevin equation and its link to the physical theories of statistical mechanics and scale relativity. It is demonstrated that the form of the coefficients of the Langevin equation depends critically on the assumption of continuity of the reconstructed trajectory. This in turn demands for the fluctuations of the diffusion term to be discontinuous in time. This paper further investigates the connection between the scale-relativistic and stochastic mechanics approaches, respectively, with the study of the Burgers equation, which in this case appears as a stochastic geodesic equation for the drift. By further demanding time reversibility of the drift, the Langevin equation can also describe equivalent quantum-mechanical systems in a path-wise manner. The resulting statistical description obeys the Fokker–Planck equation of the probability density of the differential system, which can be readily estimated from simulations of the random paths. Based on the Fokker–Planck formalism, a new derivation of the transient probability densities is presented. Finally, stochastic simulations are compared to the theoretical results.

Keywords: stochastic differential equations; Monte Carlo simulations; Burgers equation; Langevin equation; fractional velocity

PACS: 02.30.Jr, 02.30.Uu, 02.50.Ga, 02.70.Uu

MSC: 60J65, 76R50, 65R20

---

1. Introduction

The Langevin equation was introduced in order to describe the motion of a test particle subjected to a fluctuating force and a viscous drag [1]. Its formulation was later generalized to encompass also other types of systems. The Langevin equation is also fundamental for the stochastic interpretation of Quantum Mechanics (QM) [2] and it also appears, in the form of a geodesic equation, in the scale relativity theory (SR) developed by Nottale [3]. The equation represents a substantial theoretical innovation because it was in fact the first stochastic differential equation. The formal theory of stochastic differential equations was developed much later by the works of Itô and Stratonovich (see, for example, [4] for introduction).

In contrast to the picture of diffusion as an uncorrelated random walk, the theory of dynamical systems makes it possible to treat diffusion as a deterministic dynamical process. There the Langevin dynamics can be also driven by chaotic but deterministic processes [5–7]. Emergence of diffusive behavior and Markovian evolution was also addressed by Gillespie [8]. The recent study of Tyran-Kaminska demonstrates that simple diffusion processes can emerge as weak limits of piecewise continuous processes constructed within a totally deterministic framework [7]. This is a finding which lends credence to the widely used techniques of Monte Carlo simulations using pseudo-random number generators.

A different way of looking at the Langevin equation is to specify a *fractal* driving process instead of the stochastic Wiener process. Examples can be given by the studies of deterministic diffusion, where generalized Takagi functions appear [9,10]. Using this approach, both fractal and linear behaviours of the diffusion coefficients can be demonstrated. Together, the studies mentioned so far demonstrate a fundamental interplay between emergent stochasticity, chaotic dynamics and fractality, which governs transport phenomena.

The term *generalized Langevin equation* is typically used in the physics literature to describe the system's memory effects conveyed by non-Markovian color noises [11]. The present paper will generalize the Langevin equation in a different way. The Markovian character of the driving signal will be preserved, but the signal will be assumed to have some properties, leading to fractal behaviour—notably a suitably dense set of points where its Hölder exponent is fractional. Furthermore, the linearity restriction of drift term will be relaxed and instead the drift will be assumed to be a smooth function of position and time.

Interpretations of quantum mechanics are drawing a reemerging attention in the light of the centennial anniversary of David Bohm. Part of the present work was presented as a poster at the Emergent Quantum Mechanics 2017 conference in London. Results of the present work have been derived using the machinery of stochastic mechanics. On the other hand, the paper does not make strong foundational claims; instead, it is concerned with some questions about the mathematical foundations of the scale relativity theory, its link to stochastic mechanics and the theory of the Burgers equation. To the author's knowledge, such a link to the Burgers equation was not recognized before.

The Burgers equation was initially formulated by Bateman while modeling the weakly viscous liquid motion [12]. It can be derived from the full Navier–Stokes equations under some simplifying assumptions. It was later studied extensively by Burgers as a cartoon model of turbulence [13]. Presently, the number of applications of the Burgers equation is very diverse. It has been used to model physical systems, such as surface perturbations, acoustic waves, electromagnetic waves, density waves, or traffic (see, for example, [14]). The stochastic representation of the Burgers equation can be traced back to the seminal works of Busnello et al. [15,16]. Later, Constantin and Iyer derived a probabilistic representation of the deterministic three-dimensional Navier–Stokes equations [17,18]. The result presented here complements the findings of these authors as incompressibility, and hence the harmonicity of the drift, in the Burgers equation is not required.

The paper starts by briefly presenting stochastic mechanics and scale relativity. Section 2 demonstrates a general result about stochastic representations of Hölder-continuous signals leading to the Langevin equation. Section 3 introduces Nelson's characterization of a stochastic process. Section 4 introduces the complex representation of the drift in stochastic mechanics and scale relativity. Section 5 establishes the connection with the Burgers equation. Based on the Fokker–Planck formalism, a new derivation of the transient probability densities is presented. Section 6 discusses the Burgers equation as a geodesic-type of equation. The Cole–Hopf transformations are discussed as solution techniques for the Burgers equation in Section 7. Moreover, it is demonstrated how complex Cole–Hopf transformations map the complex Burgers equation, derived in a variational setting, to the free Schroedinger equation. Finally, in Section 8, numerical simulations are compared with the theoretical results.

1.1. Stochastic Mechanics (SM)

In the 1930s, certain similarities between the equations of classical statistical mechanics and the Schrödinger equation were discovered. These findings led to the stochastic interpretation of quantum mechanics. In stochastic mechanics, quantum phenomena are described in terms of diffusions instead of wave functions. The main equation of motion is in fact the Langevin equation. The formal equations of stochastic mechanics were formulated at first by Fényes [19] and Weizel [20] and later taken up by Nelson [2]. Following this interpretation, the trajectories of the configuration, described by a Markov stochastic process, are regarded as physically real. Nelson's original formulation employed a stochastic version of the Newton's law and time reversibility of the process. Interestingly, the form of the stochastic acceleration had to be postulated.

A Lagrangian formulation of stochastic mechanics was achieved by Pavon in complex form [21]. However, the given presentation is far from intuitive. In his treatment, the stochastic Lagrangian is the classical Lagrangian evaluated on a complex-valued velocity field in place of the real-valued classical velocity, while the dynamics is given by a complex-valued stochastic differential equation, similar to the treatment of Nottale. The Lagrangian problem was formulated as a constrained optimization problem, where the dynamics acted as the constraint.

1.2. Scale Relativity Theory (SR)

The scale relativity theory extends the principle of relativity also to resolution scales [3,22,23]. The main tenet of the theory is that there is no preferred scale of description of the physical reality. Therefore, a physical phenomenon must be described simultaneously at all admissible scales. While this is consistent with calculus for differentiable signals, the situation changes if non-differentiable models, such as Brownian motion or Mandelbrot's multiplicative cascades [24], are addressed. For these cases, the scale of observation (or resolution) is present irreducibly in the local description of a phenomenon. This led Nottale to postulate the fractality of the underlying mathematical variety (i.e., a pseudo-manifold) describing the observables. It should be noted that, in Nottale's approach, only finite differences are admissible. The scale relativistic approach results in corrections of Hamiltonian mechanics that arise due to the non-differentiability of trajectories, which are treated as virtual paths. Nottale introduces a complex operator that he calls the scale derivative, which acts as a pseudo-derivative (see Section 4 for details).Using this tool, Nottale gives an informal derivation of the Schrödinger equation from the classical Newtonian equation of dynamics, via a quantization procedure that follows from an extension of Einstein's relativity principle called the scale relativity principle.

2. Stochastic Representation of Trajectories

If one considers the Brownian particle as a subsystem and the surrounding particles as an infinite dimensional thermal reservoir, the Langevin equation precisely models the situation where the subsystem suitably interacts with the thermal reservoir. The type of the effective random force can be identified with a Wiener process, which has continuous but non-differentiable paths almost everywhere. Mathematical descriptions of strongly nonlinear phenomena necessitate the relaxation of the global assumption of differentiability. In contrast, classical physics assumes global smoothness of the signals and continuity of their first two derivatives. Therefore, non-smooth phenomena, such as fractals slip through its conceptual net. This argument can be further elaborated as follows. Consider the measurement of a trajectory in time $x(t)$. Non-differentiability can occur in three scenarios:

1. divergence of the velocity, that is divergence of the difference quotient,
2. oscillatory singularity or
3. difference between forward and backward velocities.

While for scenarios (1) and (2) the velocities (i.e., derivatives) can not be defined mathematically, scenario (3) requires dropping only the assumption of continuity of the resulting velocity. That is, $x'_{+}(t) \neq x'_{-}(t)$ at the point of non-differentiability t. A simple example of such behavior is the signal $x(t) = |t|$ around the origin $t = 0$. While scenario (2) is excluded by the scale relativity theory, scenario (1) leads to scale dependence of the difference quotient. Examples of fractal functions, such as the mathematical Brownian motion paths, are typically of divergent length. This at best can be viewed as a mathematical idealization since in this case the work for moving a particle along its trajectory must be infinite. On the other hand, non-differentiability does not need to occur "everywhere" (i.e., with full Lebesgue measure) on a trajectory. In this case, the trajectory can be almost everywhere differentiable except on a certain dense set of points. Examples of these are the singular functions, such as the Salem-de Rham's functions [25] or the well known Cantor's function. Singular functions have finite lengths, therefore the exerted displacement work is also finite. This makes them promising candidates for conceptualization of non-smooth phenomena in physics.

The relationship between Nelson's and Nottale's approaches can be established in a formal way. For clarity of the argument, we focus on the one-dimensional case. First, let's establish the concept of stochastic embedding of a signal. In the following, we assume that the deterministic signal (i.e., trajectory) will be represented by an **equivalence class of stochastic paths** having the same expectation as the given deterministic signal. Mathematical notation and preliminaries for the subsequent treatment are presented in Appendix A. A possibly non-differentiable continuous trajectory is represented by a continuous Markov stochastic process evaluated in the virtual space of paths as follows:

Definition 1 (Markov Stochastic Embedding). *Consider a bounded deterministic signal $x(t)$ on the compact interval $T \subseteq \mathbb{R}$ representing time. Define the stochastic embedding $\mathcal{S}_\rho$ in the probability space $(T \otimes \Omega, \mathcal{F}, \rho)$, where ρ is the probability density, as the isomorphism*

$$\begin{aligned} \mathcal{S}_\rho : &\quad T \otimes \mathbb{R} \mapsto (T \otimes \Omega, \mathcal{F}, \rho), \\ \mathcal{S}_\rho : &\quad (t, x(t)) \mapsto X(t, \omega), \quad X \in T \otimes \Omega, \end{aligned}$$

under the constraint

$$\mathbb{E}^\omega X(t, \omega) = x(t),$$

where the random variables sampled at different times t are independent and identically distributed (i. i. d.) and $\mathcal{F}$ is a σ-algebra.

Note: the ω-index will be skipped from the notation wherever convenient for clarity. In addition, X_t and $X(t)$ will be used interchangeably. Deterministic signals are denoted by the lower case, while the stochastic by upper case letters.

The above definition implicitly assumes that $X_t \in \mathbb{L}^1(T \otimes \Omega, \mathcal{F}, \rho)$ and $\mathbb{E}^\omega X(t, \omega) < \infty$.

The name of the embedding is justified by the following Lemma:

Lemma 1. *The stochastic process under the above definition has the Markov property.*

Proof. By construction for fixed $t, \delta \in \mathcal{F}$

$$\mathbb{E}^\omega X_t = x(t), \quad \mathbb{E}^\omega X_{t+\delta} = x(t + \delta).$$

The conditional expectation is

$$\mathbb{E}^\omega (X_{t+\delta} | X_t) = \int_\Omega \xi \frac{\rho(\xi, X_t)}{\rho(X_t)} d\xi,$$

where $\xi \equiv X_{t+\delta}$ is used for notational convenience. However, by independence of the variables $\rho(X_{t+\delta}, X_t) = \rho(X_{t+\delta})\rho(X_t)$. Therefore, $\mathbb{E}^\omega X_{t+\delta} = \mathbb{E}^\omega (X_{t+\delta} | X_t)$. Since δ can be either positive or negative, the claim follows. □

Consider the nonlinear problem, where the phase-space trajectory of a system is represented by a Hölder function $x(t)$ (see Appendix A, Definition A2) and t is a real-valued parameter, for example time or curve length. Let us suppose that the continuous temporal evolution of a differential system can be represented by a generalization of the Langevin equation of the form

$$dx(t) = a(x, t)dt + B(x, t)dt^\beta, \ \beta < 1, \tag{1}$$

where $a(x,t)$ and B are bounded and measurable functions of the co-ordinates and furthermore $a(x,t)$ is continuous in both x and t. That is, for all ϵ, such that $0 \le \epsilon \le dt$

$$\Delta_{\epsilon}^{+}[x](t) = x(t+\epsilon) - x(t) = a(x,t)\epsilon + B(x,t)\epsilon^{\beta} + \mathcal{O}(\epsilon).$$

This can be recognized as the Hölder growth condition of order β, since $a(x,t)\epsilon$ is an $\mathcal{O}(\epsilon)$ term. The fractional exponent β is treated as a free parameter with value to be determined later.

The type of admissible functions coupled to the fractional exponent depends critically on the assumption of continuity of the reconstructed trajectory. This in turn demands for the fluctuations of the fractional term to be discontinuous. The proof technique is introduced in [26], while the argument is similar to the one presented by Gillespie [8].

Without loss of generality, set $a = 0$. Let $x_{t+\epsilon} = x_t + B(x_t,t) + \mathcal{O}(\epsilon^{\beta})$ and $|\Delta_{\epsilon}x| \le K\epsilon^{\beta}$. Fix the interval $[t, t+\epsilon]$ and choose a partition of points $\mathcal{P} = \{t_k = t + \epsilon k/N\}$

$$x_{t_k} = x_{t_{k-1}} + B(x_{t_{k-1}}, t_{k-1})\,(\epsilon/N)^{\beta} + \mathcal{O}\left((\epsilon/N)^{\beta}\right).$$

Therefore, by induction

$$\Delta_{\epsilon}x = x_{t+\epsilon} - x_t = \frac{1}{N^{\beta}} \sum_{k=0}^{N-1} B(x_{t_k}, t_k)\epsilon^{\beta} + \mathcal{O}\left(N^{1-\beta}\epsilon^{\beta}\right).$$

If we suppose that B is continuous in x, implying also continuity in t, after taking supremum limit on both sides

$$\limsup_{\epsilon\to 0} \frac{\Delta_{\epsilon}x}{\epsilon^{\beta}} = N^{1-\beta}B(x_t,t) = B(x_t,t).$$

Therefore, either $\beta = 1$ (which is forbidden by hypothesis) or else $B = 0$ so that $B(x,t)$ must oscillate from point to point if $\beta < 1$. Then, let's denote the set $\chi_{\beta} := \{B(x_t,t) \neq 0\}$.

The argument demonstrates that so-defined set is totally disconnected in the topology of the real line [26]. This allows for the choice of the algebra $\mathcal{F}$, since we can demand that $\Omega \subseteq \chi_{\beta}$ has for elements the semi-open intervals $[\tau_i, \tau_j),\quad \tau_{i,j} \in \chi_{\beta}$. Furthermore, the initial system in Equation (1) is equivalent to the finite existence of the fractional velocity $B(x,t) = v_{+}^{\beta}x(\tau_i) \neq 0$, since the differential system can be recognized as fractional Taylor series [26]. In other words, the events in the probability space are the observations of non-vanishing values of the fractional velocity of the signal.

From now on, let $\mathcal{P}_{\tau} \equiv \mathcal{P} \subseteq \mathcal{F}$. Without loss of generality, suppose that $\mathcal{O}\left(N^{1-\beta}\epsilon^{\beta}\right) \le 1$. The stochastic representation $x_t \mapsto (X_t(\omega), \rho)$ is such that

$$\mathbb{E}\,\frac{\Delta_{\epsilon}X}{\epsilon^{\beta}} - \mathcal{O}_{\epsilon} = \frac{1}{N^{\beta}} \sum_{k=0}^{N-1} \mathbb{E}\,B(X_{t_k}, t_k), \quad \forall N.$$

Therefore, we demand that $B(X_t,t)$ is $\mathcal{F}$-measurable and $\mathbb{L}^2(\Omega, T)$ as a technical condition.

By the Hölder condition, $|x_{t_k} - x_{t_{k-1}}| \le K_k\epsilon^{\beta}$ for some set of constants K_k. Then, by transfer,

$$|\mathbb{E}\,B(X_{t_k}, t_k) - \mathbb{E}\,B(X_{t_{k-1}}, t_{k-1})| \le K_k\epsilon^{\beta}.$$

Therefore, $\mathbb{E}\,B(X_{t_k}, t_k)$ exists and is bounded. By the same argument,

$$\mathbb{E}\,(\Delta_{\epsilon}X)^2 \le K_k^k\epsilon^{2\beta}.$$

Then, we proceed by induction. Let $K_s = \sup_i K_i^2$ from the above partition:

$$(\Delta_\epsilon x)^2 = \sum_{i,j=0}^{N-1} \Delta_i x \Delta_j x = \sum_{i=0}^{N-1} \Delta_i x^2 + 2 \sum_{i<j}^{N-1} \Delta_i x \Delta_j x \leq 3K_s\, \epsilon^{2\beta} N^{1-2\beta}.$$

Therefore, for the embedded variable,

$$\mathbb{E}\,(\Delta_\epsilon X)^2 = \mathbb{E} \sum_{i,j=0}^{N-1} \Delta_i X \Delta_j X = \mathbb{E} \sum_{i=0}^{N-1} \Delta_i X^2 + 2\mathbb{E} \sum_{i<j}^{N-1} \Delta_i X \Delta_j X \leq 3K_s\, \epsilon^{2\beta} N^{1-2\beta}.$$

Since $\Delta_i X \Delta_j X$ are independent by Lemma 1, $\mathbb{E}\Delta_i X \Delta_j X = \mathbb{E}\Delta_i X \mathbb{E}\Delta_j X \leq K_s$. Therefore, $Var[\Delta_\epsilon X] \leq 3K_s\, \epsilon^{2\beta} N^{1-2\beta}$.

The argument can be specialized to $\beta = 1/2$ where $Var[\Delta_\epsilon X] \leq K_s\, \epsilon^{2\beta}$. Therefore, the variance exists $\forall N$ and the Central Limit Theorem holds. Since by Lemma 1 the process is Markovian, it must follow that in limit $N \to \infty$ the random process is Wiener.

Now suppose that $a \neq 0$. Then, since $a(x)$ is continuous of bounded variation (BVC, see Appendix A), then a.e.,

$$\mathbb{E}a(X,t) = \mathbb{E}a(x+Z,t) = \mathbb{E}(a(x,t) + a'_x Z + o\,(Z)) = a(x,t), \quad Z = X_t - x_t$$

and

$$\mathbb{E}a(X,t)^2 = a(x,t)^2 + {a'_x}^2 \sigma^2, \quad \sigma^2 = \mathbb{E}Z^2,$$

with σ^2 existing by the previous argument. Therefore, $Var[\Delta_\epsilon X] \leq 3K_s\, \epsilon^{2\beta} N^{1-2\beta} - a(x,t)^2 \epsilon^2 \leq 3K_s\, \epsilon^{2\beta} N^{1-2\beta}$ by the same argument as in the previous case. Therefore, for $\beta = 1/2$, the limit of the random process is Wiener.

Let us denote the limit Wiener process by W_t. Using the stationarity and self-similarity of the increments $\Delta_\epsilon^+ W_t = \sqrt{\epsilon}\, N(0,1)$, where $N(0,1)$ is a standard Gaussian random variable. Therefore, for $\beta = 1/2$, the velocity can be regularized to a finite value if we take the expectation. That is,

$$v_+^\beta \mathbb{E} W_t = 0,$$

since $\Delta_\epsilon^+ \mathbb{E} W_t = 0$. However,

$$v_+^\beta\ \mathbb{E}|W_t| = \int_0^\infty \sqrt{\frac{2}{\pi}} e^{-z^2/2} dz = 1\,.$$

The estimate holds a.s. since $\mathbb{P}(W_t = 0) = 0$, where $\mathbb{P}$ denotes probability.

Finally, there is a function $b(X,t)$, such that $b(X,t)\xi = B(X,t)$, $\xi \sim N(0,1)$. This follows directly from the axiom of choice, since we can always choose $\xi = 1$. Therefore, the last equation can be treated as a definition of $b(X,t)$.

In summary, the following theorem can be formulated:

Theorem 1 (Gaussian stochastic embedding). *Suppose that $x(t)$ is β-differentiable of order $\beta = 1/2$ in the interval $T = [t, t+\epsilon]$ and*

$$dx(t) = a(x,t)dt + B(x,t)dt^\beta$$

for $0 < dt \leq \epsilon$, where $a(x,t)$ is continuous in both x and t and $B(x,t) \in \mathbb{L}^2(\Omega, T)$ is bounded but discontinuous. Furthermore, let χ_β be the set of change (Definition A6) of $f[T]$.

Then, $x(t)$ can be embedded in a probability space $(T \otimes \Omega, \mathcal{F}, \rho)$, such that

1. $\Omega \subseteq \chi_\beta$,
2. *X_t has i. i. d. Gaussian increments,*
3. *$\mathbb{E}X_t = x(t)$ and*

4. $v_+^{\beta}\mathbb{E}\left(|X_t||X_t = x\right) = v_+^{\beta}|x(t)| = |b(x,t)|$ *hold almost sure.*

Furthermore, the stochastic differential equation

$$dX_t = a(X_t, t)dt + b(X_t, t)dW_t$$

holds a.s. In the last equation, W_t is a standard Wiener process and

$$b(X,t)\xi := B(X,t), \quad \xi \sim N(0,1).$$

Such embedding can be also called a **consistent stochastic embedding**. This theorem allows for Nelson's characterization of the Langevin diffusion process.

3. Nelson's Characterization

The Langevin equation can describe equivalent quantum-mechanical systems in a path-wise manner. These are the so-called *conservative diffusions* of Carlen [27]. The existence of so-conceived QM particle paths was proven under certain reasonable conditions [27]. Starting from the generalized Langevin equation, the argument can be specialized to a Wiener driving process, which can be handled using the apparatus of Itô calculus. Consider the stochastic differential equation with continuous drift and diffusion coefficients

$$dX_t = a(X,t)dt + b(X,t)dW_t,$$

where $a(X,t)$ and $b(X,t)$ are smooth functions of the co-ordinates and dW_t are the increments of a Wiener process $dW_{dt} \sim N(0, dt)$ adapted to the past filtration $\mathbb{F}_{t>0}$ – i.e., starting from the initial state.

Let $\mathbb{E}X_t = x(t)$. Following Nelson [2], the forward and backward and *drift*, respectively *diffusion* coefficients, can be identified as the averaged velocities [28]:

$$a = \lim_{dt \to 0} \mathbb{E}\left(\frac{X_{t+dt} - X_t}{dt} \middle| X_t = x \right) = \frac{d}{dt}(x - b\sqrt{dt}), \tag{2}$$

$$|b| = \lim_{dt \to 0} \mathbb{E}\left(\frac{|X_{t+dt} - X_t|}{\sqrt{dt}} \middle| X_t = x \right) = v_+^{1/2}|x|. \tag{3}$$

The evolution of the density of the process can be computed from the forward Fokker–Planck equation

$$\frac{\partial}{\partial t}\rho + \frac{\partial}{\partial x}(a\rho) - \frac{1}{2}\frac{\partial^2}{\partial x^2}\left(b^2\rho\right) = 0, \tag{4}$$

which can be recognized as a conservation law for the probability current j:

$$\frac{\partial}{\partial t}\rho + \frac{\partial}{\partial x}j = 0, \quad j := a\rho - \frac{1}{2}\frac{\partial}{\partial x}b^2\rho.$$

Under the finite energy technical condition, there is a backwards process with the *same transition density*

$$dX_t = \hat{a}(X,t)dt + b(X,t)d\hat{W}_t,$$

which is adapted to the future filtration $\mathbb{F}_{t<T}$ – i.e., starting from the final state. This leads to the anticipative (i.e., anti-Itô or anticipative) stochastic integrals. This process has Fokker–Planck equation

$$\frac{\partial}{\partial t}\rho + \frac{\partial}{\partial x}(\hat{a}\rho) + \frac{1}{2}\frac{\partial^2}{\partial x^2}\left(b^2\rho\right) = 0. \tag{5}$$

Then, it follows that the Nelson's *osmotic velocity* can be defined from

$$a - \hat{a} = b^2 \frac{\partial}{\partial x} \log b^2 \rho + \phi(t),$$

where $\phi(t)$ is an arbitrary $\mathbb{C}^1$ function of time as $u := \frac{1}{2}(a - \hat{a})$ and the *current velocity* as

$$v := \frac{1}{2}(a + \hat{a})$$

so that a continuity equation holds for the density

$$\frac{\partial}{\partial t}\rho + \frac{\partial}{\partial x} v\rho = 0.$$

Furthermore, Pavon [21] has established that the entropy production over the whole space is

$$H'(t) := -\frac{d}{dt} \int_R \rho \log \rho \, dx^3 = -\frac{2}{b^2} \mathbb{E}\, uv.$$

Thus, for a Markov diffusion process,

$$\mathbb{E} \frac{1}{b^2} \int_s^r uv \, dt = \frac{1}{2}(H(s) - H(r))$$

for a constant b.

4. The Complex Velocity Operator in SR and SM Theories

Scale relativity treats velocity only as a difference quotient. This is a necessity due to the assumed non-differentiability of the trajectories. Non-differentiability leads to introduction of two velocity fields—forward and backward, depending on the direction of differentiation in time. These fields are assumed to be finite for small values of the time step dt but they diverge to infinity in the limit $dt \to 0$ in a standard analysis setting. Therefore, such velocity fields can be defined only up to a finite resolution underlying the physical phenomenon under study. The velocity fields are assumed to admit representation of the form of a sum of a "classical part" plus a correction of a resolution-dependent and diverging fractal part. The classical part corresponds to the **absolutely continuous** part of the trajectory, while the fractal part corresponds to the **singular** and possibly **oscillatory** parts. Since, at the level of physical description, there is no way to favor the forward rather than the backward velocity, the description should incorporate them on equal grounds, i.e., forming a bivariate vector field $\mathbb{R} \otimes \mathbb{R}$

$$v_+ := \frac{\Delta_{dt}^+ x}{dt} \quad \otimes \quad v_+ := \frac{\Delta_{dt}^- x}{dt}.$$

This bivariate vector field is represented by a complex-valued vector field [29] as $\mathbf{v} = V - iU \in \mathbb{R}^3$ with components given by $U := \frac{1}{2}(v_+ + v_-)$, $V := \frac{1}{2}(v_+ - v_-)$, where V is interpreted as the "classical" velocity and U is a new quasi-velocity quantity (i.e., the *osmotic* velocity in the terminology of Nelson). Under these assumptions, Nottale introduces a complexified material derivative, which is a pseudo-differential operator acting on scalar functions as

$$\mathcal{D}F = \partial_t F + (\mathbf{v} \cdot \nabla) F - i\sigma^2 \nabla^2 F,$$

where σ is a constant, quantifying the effect of changing the resolution scale.

Stochastic mechanics allows for a similar treatment of the complex of forward and backward diffusions. The drift, resp. diffusion coefficients can be further embedded in a complex space as proposed by Pavon [21]:

$$a \otimes \hat{a} \mapsto \mathcal{V} := v - iu,$$
$$X_{t+dt} \otimes X_{t-dt} \mapsto d\mathcal{X} = \frac{1}{2}\left(X_{t+dt} + X_{t-dt}\right) - i\frac{1}{2}\left(X_{t+dt} - X_{t-dt}\right),$$

so that the diffusion process becomes complex. It follows that

$$d\mathcal{X} = \mathcal{V}dt + \frac{1-i}{2} b\, dW_t + \frac{1+i}{2}\hat{b}\, d\hat{W}_t.$$

In the case when $b = \hat{b}$,

$$d\mathcal{X} = \mathcal{V}dt + \frac{1-i}{2} b\left(dW_t + id\hat{W}_t\right) = \mathcal{V}dt + \frac{e^{-\frac{i\pi}{4}}}{\sqrt{2}} b\left(dW_t + id\hat{W}_t\right).$$

Therefore, we can designate a new complex stochastic variable

$$Z_t := \frac{dW_t + id\hat{W}_t}{\sqrt{2}}.$$

Because of its double adaptation, Z_t retains its local martingale properties: that is, $\mathbb{E}Z_t = 0$. In this case, notably $Var\ Z_t = 0$, but $\mathbb{E}|Z_t|^2 = 1$, so that finally,

$$d\mathcal{X} = \mathcal{V}dt + \sqrt{-i} b\, dZ_t.$$

Therefore, a formal Itô differential can be introduced in exactly the same way

$$dF = \frac{\partial}{\partial t}F + d\mathcal{X}\frac{\partial}{\partial x}F + \frac{1}{2}\left[d\mathcal{X}^2\right]\frac{\partial^2}{\partial x^2}F \tag{6}$$

with quadratic variation $\left[d\mathcal{X}^2\right] = -ib^2 dt$. Therefore, in components,

$$dF = \left(\frac{\partial}{\partial t}F + \mathcal{V}\frac{\partial F}{\partial x} - \frac{ib^2}{2}\frac{\partial^2}{\partial x^2}F\right)dt + \sqrt{-i}b\frac{\partial F}{\partial x}\, dZ_t, \tag{7}$$

which generalize to

$$dF = \left(\partial_t F + (\mathcal{V}\cdot\nabla)F - \frac{ib^2}{2}\nabla^2 F\right)dt + \sqrt{-i}b\left(dZ_t \cdot \nabla\right)F$$

in three dimensions [28]. It is apparent that both theories share an identical algebraical structure, while SM can be considered as a stochastic representation of SR.

Remark 1. *Conceptually, the forward process can be interpreted as a prediction, while the backward process can be interpreted as a retrodiction.*

Note that, in the complex formulation of Pavon, the real part of the driving process Z_t corresponds to the forward (i.e., adapted to the past) process, while the imaginary part corresponds to the backward (i.e., adapted to the future) process. This is of course one of infinitely many choices, since the complex factor in the diffusion coefficient is a root of unity and hence represents a rotation in the complex plane.

The martingale property of the complex Wiener process conceptually means that the knowledge of the past and future of the process do not bias the outcome at the present time (i.e., at measurement). Note that the mapping is invertible since

$$X_{t+dt} = Re\,(d\mathcal{X}) + Im\,(d\mathcal{X})\,, \quad | \quad X_{t-dt} = Re\,(d\mathcal{X}) - Im\,(d\mathcal{X})\,.$$

From these formulas, it is apparent that the real part, or respectively the imaginary part of the resulting process do not have separate meanings, as they mix the predictive process with the retrodictive process. To illustrate the point, suppose that $F = F_r + iF_i$ and $a = a_r + ia_i$ and the original process $d\mathcal{X}$ is transformed as $F(\mathcal{X})$. Then, a straightforward calculation gives

$$Re(dF) = \left(\frac{\partial F_r}{\partial t} + a_r\frac{\partial}{\partial x}F_r - a_i\frac{\partial}{\partial x}F_i + \frac{b^2}{2}\frac{\partial^2}{\partial x^2}F_i\right)dt \\ + \frac{b}{\sqrt{2}}\left(dW_t\left(\frac{\partial}{\partial x}F_r + \frac{\partial}{\partial x}F_i\right) + d\hat{W}_t\left(\frac{\partial}{\partial x}Fr - \frac{\partial}{\partial x}F_i\right)\right), \quad (8)$$

$$Im(dF) = \left(\frac{\partial F_i}{\partial t} + a_r\frac{\partial}{\partial x}F_r + a_i\frac{\partial}{\partial x}F_i + \frac{b^2}{2}\frac{\partial^2}{\partial x^2}F_r\right)dt \\ - \frac{b}{\sqrt{2}}\left(dW_t\left(\frac{\partial}{\partial x}F_r - \frac{\partial}{\partial x}F_i\right) - d\hat{W}_t\left(\frac{\partial}{\partial x}Fr + \frac{\partial}{\partial x}F_i\right)\right). \quad (9)$$

5. The Real Stochastic Geodesic Equations

The appearance of the Wiener process entails the application of the fundamental Itô Lemma for the forward (i.e., adapted to the past, plus sign) or the backward processes (i.e., adapted to the future, minus sign), respectively. In differential notation, it reads

$$dF(X) = dX\frac{\partial}{\partial x}F \pm \frac{[dX^2]}{2}\frac{\partial^2}{\partial x^2}F, \quad (10)$$

where $[dX^2] = b^2 dt$ is the quadratic variation of the process. It can be seen that in this case the (forward) differential operator d acts as a material derivative.

The term *geodesic* will be interpreted as a solution of a variational problem [30,31]. A brief treatment is given in Appendix B. By application of Itô's Lemma, the forward geodesic equation can be obtained as:

$$\frac{\partial}{\partial t}a + a\frac{\partial}{\partial x}a + \frac{b^2}{2}\cdot\frac{\partial^2}{\partial x^2}a = 0. \quad (11)$$

This can be recognized as a Burgers equation with negative kinematic viscosity for the drift field [13].

The backward geodesic equation follows from the application of the Itô's lemma for the anticipative process

$$\frac{\partial}{\partial t}a' + a'\frac{\partial}{\partial x}a' - \frac{b^2}{2}\frac{\partial^2}{\partial x^2}a' = 0 \quad (12)$$

This can be recognized as a Burgers equation with positive kinematic viscosity for the drift field.

The solution of the Burgers equation is well known and can be given by the convolution integrals (Equation (44)) for the case of positive viscosity [13]. The case about the negative viscosity can not be easily solved using Fourier transform. Therefore, a different solution technique will be pursued. Time-varying solutions will be constructed from topological deformations of the stationary solutions.

In QM applications, $b = \hbar/2m$. Normalization $b = 1$ will be assumed further in most cases to simplify calculations.

5.1. Path-Wise Separable Solutions

In the first instance, one can solve the geodesic equation by supposing separability. By making the ansatz $a(x,t) = f(x)g(t)$, we arrive at the equation:

$$\frac{f''(x)}{2f(x)} + g(t)\,f'(x) + \frac{g'(t)}{g(t)} = 0.$$

This has the unique solution

$$a(x,t) = \frac{x + x_0}{t + T}. \tag{13}$$

The resulting Itô equation can be formulated as

$$dX = \frac{X + x_0}{t + T}dt + dW_t.$$

The stochastic differential equation for the drift is therefore

$$da = \frac{1}{t + T}dW_t,$$

which can be integrated exactly in Itô's sense as

$$a(t) = a_0 + \int_0^t \frac{dW_s}{s + T}, \quad a_0 = \frac{x_0}{T}. \tag{14}$$

Therefore,

$$X(t) = \frac{x_0}{T}(t + T) + (t + T)\int_0^t \frac{dW_s}{T + s}, \tag{15}$$

where T is the stopping time. Therefore, an exact numerical quadrature can be performed (Figure 1)

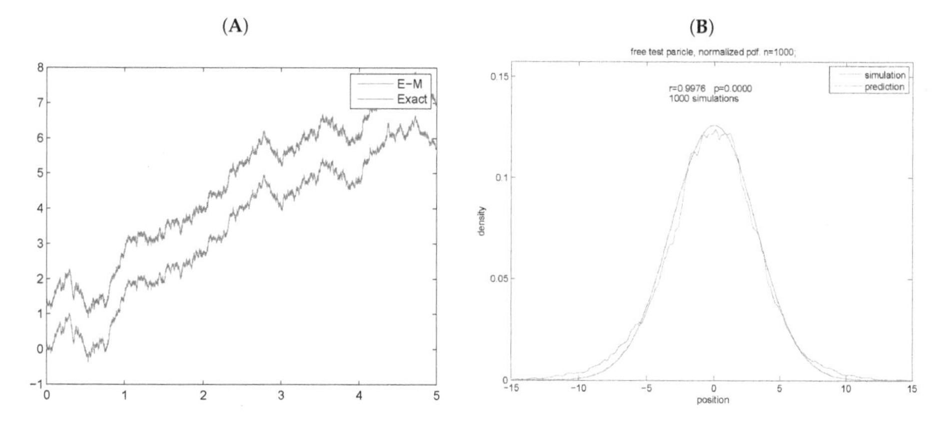

Figure 1. Virtual trajectories of the separable process. (**A**) virtual trajectories; (**B**) empirical vs. theoretical density. (**A**) exact simulation of separable process is compared with the Euler–Maruyama algorithm. E—Exact simulation, E–M—Euler–Maruyama simulation; An offset is added to the exact solution for appreciation. Time is given in arbitrary units; (**B**) the empirical transition density is estimated from $n = \log^2(N_s\,N)$ bins. Pearson's correlation is given as an inset—r = 0.9976.

The corresponding density can be obtained from the Fokker–Planck equation

$$\frac{\partial}{\partial t}\rho + \frac{\partial}{\partial x}\left(\frac{\rho\, x}{t+T}\right) - \frac{1}{2}\frac{\partial^2}{\partial x^2}\rho = 0$$

with solution

$$\rho(x,t) = \frac{1}{\sqrt{2\pi\,(T+t)}} \exp\left(\frac{x^2}{2(t+T)}\right). \tag{16}$$

It should be noted that, under time reversion, we arrive at the same solution, which however leads to a different Fokker–Planck equation

$$\frac{\partial}{\partial t}\rho + \frac{\partial}{\partial x}\left(\frac{\rho\, x}{t-T}\right) + \frac{1}{2}\frac{\partial^2}{\partial x^2}\rho = 0$$

with solution

$$\rho(x,t) = \frac{1}{\sqrt{2\pi\,(T-t)}} \exp\left(-\frac{x^2}{2(t-T)}\right),$$

which can be recognized as a Brownian bridge. The entropy of this density can be calculated as

$$H(t) = \frac{\log 2\pi\,(t-T) + 1}{2}.$$

5.2. Stationary Drift Fields

For time-homogeneous diffusion, the geodesic equation can be brought into the form

$$\frac{1}{2}\frac{\partial}{\partial x}\left(a^2 + \frac{\partial}{\partial x}a\right) = 0,$$

which can be integrated once to give

$$a^2 + \frac{\partial}{\partial x}a = -E.$$

The integration constant E can be identified with the energy. The resulting first order ordinary differential equation (ODE) can be solved as

$$a(x) = -\sqrt{E}\,\tan\left(\sqrt{E}\,x + c\right), \quad E > 0, \tag{17}$$

$$a(x) = \frac{1}{x+c}, \quad E = 0. \tag{18}$$

The solution for $E > 0$ was identified by Herman [32]. By translation, invariance of the coordinates, $c = 0$ is admissible. This observation will be used further for the transient solution. The link between the two solutions can be established as follows. Note that

$$a(x) = \sqrt{E}\,\cot\sqrt{E}\,x$$

is also a solution. Then,

$$\lim_{E\to 0}\sqrt{E}\,\cot\sqrt{E}\,x = \frac{1}{x},$$

which is the second solution.

The expectation of the trajectory can be obtained by solving the ODE

$$\frac{dx}{dt} = -\sqrt{E}\,\tan\sqrt{E}\,x$$

so that

$$\sqrt{E}\, x(t) = \arcsin e^{-Et+c}, \tag{19}$$

$$\sqrt{E}\, x(t) = \arccos e^{-Et+c}. \tag{20}$$

In accordance with so-developed theory for $c = 0$,

$$v_+^{1/2} x(t=0) = \pm \lim_{h \to 0+} E \frac{2\sqrt{h}\, e^{-Eh}}{\sqrt{1 - e^{-2Eh}}} = \pm \sqrt{2E}.$$

Furthermore, for $E = 0$,

$$\frac{dx}{dt} = \pm \frac{1}{x}$$

so that in the same way

$$x(t) = c \pm \sqrt{2t}.$$

The backward geodesic equation

$$\frac{1}{2} \frac{\partial}{\partial x} \left(a^2 - \frac{\partial}{\partial x} a \right) = 0$$

by the same method leads to

$$a(x) = -\sqrt{E} \tanh \left(\sqrt{E}\, x + c \right), \quad E > 0, \tag{21}$$

$$a(x) = -\frac{1}{x + c}, \quad E = 0. \tag{22}$$

5.3. Stationary Density Solutions

The stationary density $\rho(y)$ is a solution of the Fokker–Planck (i.e., forward Kolmogorov) equations parametrized by E:

$$\frac{1}{2} \frac{\partial^2}{\partial y^2} \rho \left(y^2 + 1 \right)^2 = 0, \quad E > 0, \tag{23}$$

$$\frac{1}{2} \frac{\partial^2}{\partial y^2} \rho\, y^4 = 0, \quad E = 0. \tag{24}$$

The case $E > 0$ leads to

$$\frac{\partial}{\partial x} \tan(x) \rho + \frac{1}{2} \frac{\partial^2}{\partial x^2} \rho = 0 \tag{25}$$

with stationary solution

$$\rho = \cos^2 \sqrt{E} x,$$

which can be valid on a bounded domain. The entropy of this solution in the domain $[-\pi/(2\sqrt{E}), \pi/(2\sqrt{E})]$ can be calculated as

$$H = \frac{\pi \log 4}{2} - \frac{\pi}{2}.$$

The case $E = 0$ leads to

$$\frac{\partial}{\partial x} \frac{\rho}{x} - \frac{1}{2} \frac{\partial^2}{\partial x^2} \rho = 0 \tag{26}$$

with a stationary solution

$$\rho = |x|,$$

which can be valid on a bounded domain.

5.4. Transient Drift Fields

The solution of the Burgers equation is well known and can be given by the convolution integrals (Equation (44)) for the case of positive viscosity [13]. The case of negative viscosity emerging here is more challenging and it will be solved by a deformation of the stationary solution, so that in limit the stationary solution is recovered:

$$\lim_{t\to\infty} a(t,x) = a(x), \quad E > 0.$$

The solution is sought in the form (neglecting scale factors)

$$a(t,x) = -\frac{\sin x}{\cos x + f(t),}$$

which results in a linear ODE for the unknown function $f(t)$:

$$\frac{2f'(t) + f(t)}{2\left(\cos x + f(t)\right)^2} \sin x = 0.$$

By variation of the parameters, the solution for $a(t,x)$ is given as

$$a(t,x) = -\sqrt{E}\frac{\sin\sqrt{E}x}{\cos\sqrt{E}x + ke^{-\frac{Et}{2}}}, \tag{27}$$

where the constant E represents an energy scale and k is an arbitrary constant. We can assume normalization, for example $k = \pm\sqrt{E}$, such that $a(t, \frac{\pi}{2E}) = \pm 1$. Plots are presented in Figure 2.

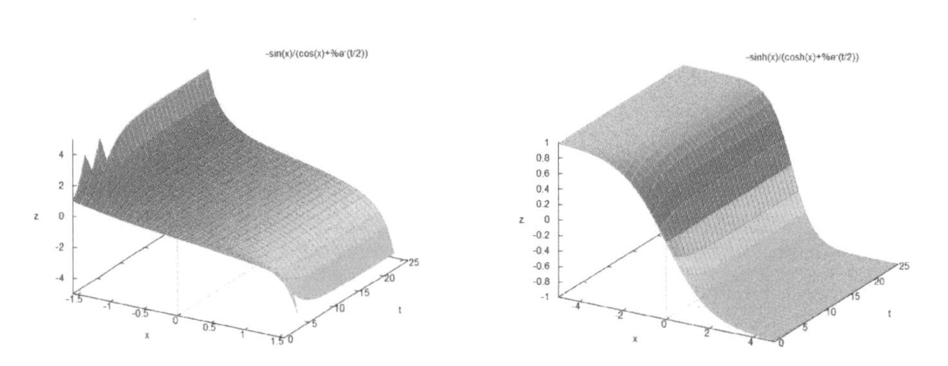

Figure 2. Time-varying drift fields for $E = 1, k = 1$. (**A**) forward drift; (**B**) backward drift.

The transformed Itô drift equation for $k = 1$ reads

$$da(t,x) = -E\frac{e^{-Et/2}\cos\sqrt{E}x + 1}{\left(e^{-Et/2} + \cos\sqrt{E}x\right)^2}dW_t = -\sqrt{E}\frac{e^{-Et/2}\cos\sqrt{E}x + 1}{\left(e^{-Et/2} + \cos\sqrt{E}x\right)^2}dW_{Et}.$$

It can be further noticed that rescaling in a pair of new variables $x' = \sqrt{E}x, \quad t' = Et$ leaves the ratio

$$z = \frac{x^2}{t} = \frac{x'^2}{t'}$$

invariant so that z becomes a similarity variable.

Furthermore, a formal forward Kolmogorov equation can be written in the $y = a(t,x)$ variable with $E = 1$ as

$$\frac{\partial}{\partial t}\rho - \frac{1}{2}\frac{\partial^2}{\partial y^2}\frac{\rho e^t \left(\cos(x) + e^{\frac{t}{2}}\right)^2}{\left(e^{\frac{t}{2}}\cos(x) + 1\right)^4} = 0, \quad x = \pm\arcsin\left(\frac{e^{-\frac{t}{2}}y\left(\sqrt{(e^t-1)\,y^2 + e^t} - 1\right)}{y^2+1}\right),$$

however its solution is challenging due to its mixed nonlinearity and will not be attempted here. Nevertheless, the analysis presented so far assures that asymptotically ρ can be obtained as a solution of the stationary equation.

The backward geodesic equation leads to the following solution :

$$a(t,x)' = -\sqrt{E}\frac{\sinh\sqrt{E}x}{\cosh\sqrt{E}x + ke^{-\frac{Et}{2}}}. \tag{28}$$

5.5. Asymptotic Density Solutions

The forward drift itself $a \equiv y$ (symbol changed) obeys the transformed stochastic differential equations

$$E > 0: \quad dy = \sqrt{E}\left(y^2+1\right)dW_t = \left(y^2+1\right)dW_{E\,t}, \tag{29}$$

$$E = 0: \quad dy = y^2\,dW_t. \tag{30}$$

The density ρ is a solution of the forward Kolmogorov equations parametrized by E:

$$\frac{\partial}{\partial E\,t}\rho = \frac{1}{2}\frac{\partial^2}{\partial y^2}\rho\left(y^2+1\right)^2, \quad E \neq 0, \tag{31}$$

$$\frac{\partial}{\partial t}\rho = \frac{1}{2}\frac{\partial^2}{\partial y^2}\rho\,y^4, \quad E = 0. \tag{32}$$

The solutions can be obtained using the Laplace transform $\mathcal{L}_s f(t) \mapsto \hat{f}(s)$. In this way, the partial differential equation can be transformed into an ODE for the Laplace variable:

$$-\frac{1}{2}\frac{\partial^2}{\partial y^2}\hat{\rho}y^4 + \hat{\rho}\,s = \rho(0,y), \tag{33}$$

$$-\frac{1}{2}\frac{\partial^2}{\partial y^2}\hat{\rho}\left(y^2+1\right)^2 + \hat{\rho}\,s = \rho(0,y). \tag{34}$$

To obtain the Green's function, we take homogeneous initial conditions a.e. The solutions in the time domain can be obtained by the inverse Laplace transformation:

$$\hat{\rho}(s,y) = \frac{Ae^{-\frac{\sqrt{2s}}{y}}}{\sqrt{s}\,y^3} \xrightarrow{\mathcal{L}_s^{-1}} \rho(t,y) = \frac{A}{\sqrt{t}\,y^3}e^{-\frac{1}{2ty^2}}, \tag{35}$$

$$\hat{\rho}(s,y) = \frac{e^{-\sqrt{2s/E-1}\,\arctan(y)}}{(y^2+1)^{\frac{3}{2}}} \xrightarrow{\mathcal{L}_s^{-1}} \rho(t,y) = \frac{e^{-\frac{\arctan^2 y-(Et)^2}{2Et}}}{\sqrt{\pi\,Et}\,\sqrt{(y^2+1)}^3}. \tag{36}$$

In the position space, the solution can be obtained using Grisanov's theorem [4]:

$$\rho(t,x) = \frac{|x|}{\sqrt{t}} \exp\left(-\frac{x^2}{2t}\right), \tag{37}$$

$$\rho(t,x) = \frac{|\cos(\sqrt{E}\,x)|}{\sqrt{\pi E\,t}} \exp\left(\frac{Et}{2} - \frac{x^2}{2t}\right). \tag{38}$$

The second equation is not acceptable from a physical point of view since $\lim_{t\to\infty} \rho(t,x)$ diverges. In the same way for the backward drift,

$$E > 0: \quad dy = \sqrt{E}\left(y^2 - 1\right) dW_t = \left(y^2 - 1\right) dW_{E\,t}, \tag{39}$$

$$E = 0: \quad dy = y^2\, dW_t, \tag{40}$$

$$\frac{\partial}{\partial E\,t}\rho = \frac{1}{2}\frac{\partial^2}{\partial y^2}\rho\left(y^2 - 1\right)^2, \quad E \neq 0, \tag{41}$$

with solutions

$$\rho(t,y) = \frac{1}{\sqrt{\pi\,Et}\,\sqrt{(y^2-1)}^3} \exp\left(-\frac{\operatorname{arctanh}^2 y + (Et)^2}{2E\,t}\right)$$

in the drift space and in position space

$$\rho(t,x) = \frac{\cosh(\sqrt{E}\,x)}{\sqrt{\pi E\,t}} \exp\left(-\frac{x^2}{2t} - \frac{Et}{2}\right),$$

respectively. This is acceptable from a physical point of view since $\lim_{t\to\infty} \rho(t,x) = 0$, which is a correct asymptotic behavior.

6. The Complex Stochastic Geodesic Equations

The complexification removes the restriction of positive definiteness of the E parameter so that the substitution $t \mapsto \pm Et$ becomes admissible by an appropriate cut along the complex plane.

In a similar way, for the complex case, we have

$$dX_t = -i\sqrt{E}\tanh\sqrt{E}X_t\,dt + \sqrt{-i}\,dZ_t,$$

which, under substitution, $y = \tanh x$ leads to

$$dy = \sqrt{-i}\sqrt{E}\left(y^2 - 1\right) dZ_t,$$

By the same methods as used above, the asymptotic density for the drift variable can be obtained as

$$\rho(t,y) = Re\frac{1}{\sqrt{\pi t}\,(y^2-1)^{\frac{3}{2}}} \exp\left(\frac{iEt}{2} - \frac{i\operatorname{arctanh}^2 y}{2Et}\right).$$

For the resulting density in the position space, it can be calculated that

$$\rho(t,x) = Re\frac{i\cosh(\sqrt{E}x)}{\sqrt{\pi Et}} \exp\left(\frac{iEt}{2} - \frac{ix^2}{2t}\right). \tag{42}$$

In a similar way, for the other solution,

$$dX = i\sqrt{E}\tan\sqrt{E}Xdt + \sqrt{-i}\,dZ_t,$$

which under substitution $y = \tan\sqrt{E}x$ leads to the drift equation

$$dy = \sqrt{-i}\sqrt{E}(1+y^2)\,dZ_t.$$

The drift density can be readily obtained as

$$\rho(t,y) = \frac{1}{\sqrt{\pi t}\,(y^2+1)^{\frac{3}{2}}}\exp\left(-\frac{iEt}{2} - \frac{i\arctan^2 y}{2Et}\right).$$

In the position space, the density is of the form

$$\rho(t,x) = Re\frac{|\cos(\sqrt{E}x)|}{\sqrt{\pi E t}}\exp\left(-\frac{ix^2}{2t} - \frac{iEt}{2}\right). \quad (43)$$

In either case, the densities asymptotically approach zero.

7. Real-Valued and Complex Cole–Hopf Transformations

The Burgers equation can be linearized by the Cole–Hopf transformation [33,34]. This mapping transforms the nonlinear Burgers equation into the linear heat conduction equation in the following way. Let

$$u = \frac{\partial}{\partial x}\log a.$$

Substitution into Equation (11) leads to

$$\frac{1}{2u^2}\left(u\frac{\partial^3 u}{\partial x^3} + 2u\frac{\partial^2 u}{\partial t\partial x} - \frac{\partial u}{\partial x}\frac{\partial^2 u}{\partial x^2} - 2\frac{\partial u}{\partial t}\frac{\partial u}{\partial x}\right) = 0.$$

This can be recognized as

$$\frac{\partial}{\partial x}\frac{1}{u}\left(\frac{\partial}{\partial t}u + \frac{1}{2}\frac{\partial^2}{\partial x^2}u\right) = 0,$$

which is equivalent to a solution of the equation

$$\frac{\partial}{\partial t}u + \frac{1}{2}\frac{\partial^2}{\partial x^2}u = 0.$$

It should be noted that if instead of the forward development (i.e., prediction) one takes the backward development (i.e., retrodiction), the usual form of the Burgers equation is recovered. This corresponds to the anticipative Wiener process, which is subject to the anticipative Itô calculus [17,35]:

$$\frac{\partial}{\partial t}\hat{a} + \hat{a}\frac{\partial}{\partial x}\hat{a} - \frac{1}{2}\frac{\partial^2}{\partial x^2}\hat{a} = 0.$$

In this case, the usual general solution can be revealed

$$\phi_0(x) = \exp\left(\frac{1}{2\nu}\int_0^x \hat{a}_0(u)du\right), \quad (44)$$

$$\hat{a}(x,t) = \frac{\partial}{\partial x}\log\frac{1}{2\sqrt{\pi\nu t}}\int_{-\infty}^{\infty}\phi_0(u)e^{-\frac{(x-u)^2}{4\nu t}}du = \frac{\int_{-\infty}^{\infty}\frac{x-u}{t}\phi_0(u)e^{-\frac{(x-u)^2}{4\nu t}}du}{\int_{-\infty}^{\infty}\phi_0(u)e^{-\frac{(x-u)^2}{4\nu t}}du}, \quad (45)$$

where $\nu = 1/2$ is the viscosity coefficient.

In the complex case, starting from the generalized Itô differential, the complex velocity field becomes

$$d\mathcal{V} = \left(\frac{\partial}{\partial t}\mathcal{V} + \mathcal{V}\frac{\partial}{\partial x}\mathcal{V} - \frac{ib^2}{2}\frac{\partial^2}{\partial x^2}\mathcal{V} \right) dt + \sqrt{-i}b\frac{\partial}{\partial x}\mathcal{V}\, dZ_t.$$

The geodesic equation reads

$$\mathbb{E}\, d\mathcal{V} = 0.$$

Therefore, by the martingale property, this is equivalent to

$$\frac{\partial}{\partial t}\mathcal{V} + \mathcal{V}\frac{\partial}{\partial x}\mathcal{V} - \frac{ib^2}{2}\frac{\partial^2}{\partial x^2}\mathcal{V} = 0,$$

which can be recognized as a generalized Burgers equation with imaginary kinematic viscosity coefficient. Applying the complex Cole–Hopf transformation as [36]

$$\mathcal{V} = -i\frac{\partial}{\partial x}\log \mathcal{U}, \quad -\pi < \arg \mathcal{U} < \pi$$

and specializing to $b = 1$ leads to

$$-\frac{\mathcal{U}\left(\frac{\partial^3}{\partial x^3}\mathcal{U}\right) - \left(\frac{\partial}{\partial x}\mathcal{U}\right)\left(\frac{\partial^2}{\partial x^2}\mathcal{U}\right) - 2i\left(\frac{\partial}{\partial t}\mathcal{U}\right)\left(\frac{\partial}{\partial x}\mathcal{U}\right) + 2i\mathcal{U}\left(\frac{\partial^2}{\partial t \partial x}\mathcal{U}\right)}{2\mathcal{U}^2} = 0,$$

which can be recognized as a gradient

$$-\frac{\partial}{\partial x}\frac{1}{\mathcal{U}}\left(i\frac{\partial}{\partial t}\mathcal{U} + \frac{1}{2}\frac{\partial^2}{\partial x^2}\mathcal{U} \right) = 0.$$

The last equation is equivalent to the solution of the free Schrödinger equation. On the other hand, the diffusion part is simply

$$-\sqrt{i}\left(\frac{\partial^2}{\partial x^2}\log \mathcal{U}\right) dZ_t = -\sqrt{i}\left(\frac{\partial}{\partial x}\frac{1}{\mathcal{U}}\frac{\partial}{\partial x}\mathcal{U}\right)$$

since $-i\sqrt{-i} = -\sqrt{i}$.

This corresponds with the arguments given in [37] that the coefficient of the stochastic noise should be purely imaginary. Calculations can be reproduced in the computer algebra system Maxima [38].

8. Numerical Results

The different types of solutions of the stochastic geodesic equation were simulated using the Euler–Maruyama algorithm. Simulations were performed in Matlab. An example of a simulation script is given in Appendix C.

8.1. Exact Simulations

The exact simulations of the separable process were compared to simulations computed by the Euler–Maruyama algorithm. Achieved correlation was 1.0 while the mean squared error was on the order of 1×10^{-8}.The comparison is presented in Figure 1. The empirical transition density was computed from $N_s = 1000$ simulations and correlated to the graph of Equation (16). The theoretical density was computed from Equation (16) for stopping time $T = 10$. Achieved correlation was 0.9976.

8.2. Free Diffusion

The normalized asymptotic transient density of the free particle distribution can be recognized as the Rayleigh's distribution (Figure 3)

$$R(x,t) = \frac{|x|}{2t} e^{-\frac{x^2}{2t}}.$$

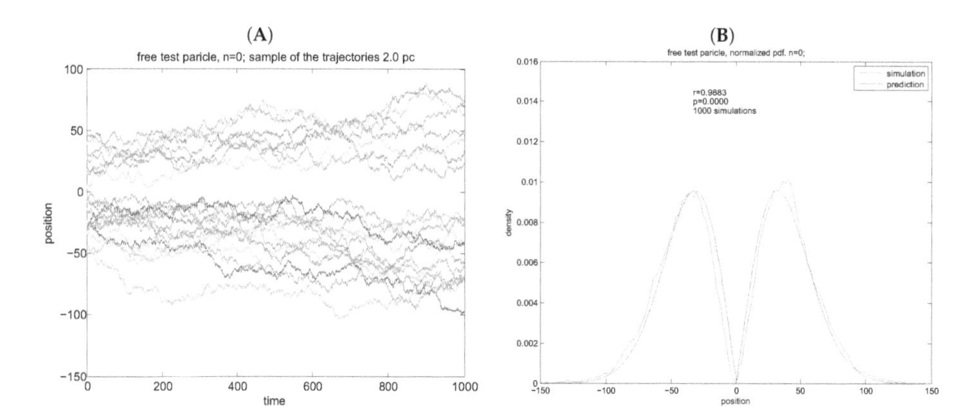

Figure 3. Simulations of free particles virtual trajectories. (**A**) virtual trajectories: free particles; (**B**) empirical vs. theoretical density. Simulations are based on $N = 10,000$ points in $N_s = 1000$ simulations. (**A**,**B**) width of potential well is $2L = 100$ units. The empirical *pdf* is estimated from $n = \log^2(N_s N)$ bins. Pearson's correlations are given as inset—r = 0.9883. Norming of the free particle transient results in Rayleigh density.

8.3. Particle in a Box

The third simulated case comprised a freely diffusing particle in a square potential well of size $2L$. The approach was based on Hermann [32]. Individual trajectories were simulated according to the fundamental equation using the scheme of Euler–Maruyama:

$$x_{n+1} = x_n - 2D\Delta t \frac{\pi n}{L} \tan\left(\frac{\pi n}{L} x_n - \pi \frac{n+1}{2}\right) + \sqrt{2D\Delta t}\, \Delta W_n,$$

where $\Delta W_n \sim N(0,1)$.

Restarting boundary conditions were used for the simulations to avoid distortions of the distribution. That is, if a simulated particle crossed the boundaries its position was reset to its original position.

The initial particle positions were sampled from a uniform distribution between $-L$ and L. The theoretical density for the particle in a box case is given by

$$\rho_s(x) = \frac{2}{L} \sin^2\left(n\pi\left(\frac{x}{L} + \frac{1}{2}\right)\right).$$

Results are based on $N = 10000$ points in $N_s = 1000$ simulations.

The empirical *pdf* is estimated from $n = \log(N_s N)^2$ bins. Pearson's correlations are given as insets: B – r = 0.9939, D – r = 0.9871. For both cases, the numerical precision correlates excellently with the analytical solutions (Figure 4).

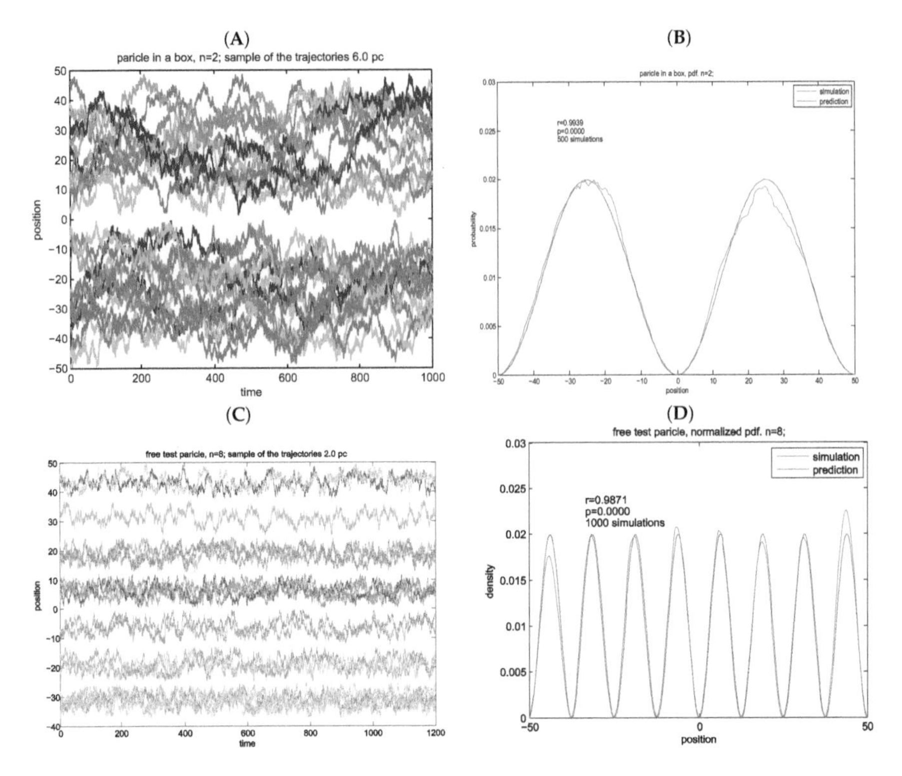

Figure 4. Simulations of particles in a box for two quantum numbers. (**A**) virtual trajectories: particle in a box, n = 2; (**B**) empirical vs. theoretical density; (**C**) test particles in a box, n = 8; (**D**) empirical vs. theoretical density. (**A**,**B**) width of potential well is $2L = 100$ units.

9. Discussion

This work was motivated in part by the premise that inherently nonlinear phenomena need development of novel mathematical tools for their description. The relaxation of the differentiability assumption opens new avenues in describing physical phenomena, as demonstrated by SM and SR, but also challenges existing mathematical methods, which are developed for smooth signals [2,3]. While this description can be achieved also by fractional differ–integrals, or by multi-scale approaches [39], the present work focused on a local description. The reason for this choice is that locality provides a direct way of physical interpretation of the obtained results. In this regard, Hölderian functions can be used as building blocks of such strongly nonlinear models, which give rise to singular [24,40] or non-differentiable models.

The second motivation of the present work was to investigate the potential of stochastic methods for simulations of quantum-mechanical and convection-diffusive systems. While the usual presentation of the stochastic mechanics typically used the Schrödinger equation as a solution device and paths were constructed from solutions of the Schrödinger equation, this is not necessary. McClendon and Rabitz simulated several quantum systems using the differential equations of Nelson's stochastic quantization as a starting point [41]. In the framework of scale relativity, Herman [32] and later Al Rashid et al. [42] simulated QM particle in a box using the Langevin equations. Later, Al-Rashid et al. [43] simulated the quantum harmonic oscillator extending Herman's approach. The approach presented here can be used as an alternative to numerical solutions of the Schrödinger equation. In this scenario, the density of the solution can be sampled from Monte Carlo simulations as demonstrated. Presented numerical approaches can be used, for example, for simulations of nanoparticles or quantum dots, which are mesoscopic objects and are expected to have properties intermediate between macroscopic and quantum systems [44]. This can be of interest, for example in sedimentation studies, where Langevin dynamics was proposed [45]. In principle, presented results can be extended towards asynchronous simulations using the Gillespie's algorithm [8]. This can be achieved using time steps distributed exponentially.

Obtained results can be also discussed in view of the fluctuation-dissipation relationships. The fluctuation-dissipation theorem relates the linear response relaxation of a system from a non-equilibrium state to the properties of fluctuations in equilibrium. This is an exact result in the case of the Ornstein–Uhlembeck process, where the drift term is linear. The geodesic treatment in the present work provides a different relationship between the drift and $a(x,t)$ and diffusivity $b(x,t)$. In the small perturbation regime around the equilibrium the geodesic process $x_{eq}(t)$ can be approximated by an Ornstein–Uhlembeck process for the fluctuation term ($\xi = \delta x(t)$), therefore an appropriate fluctuation-dissipation theorem can be formulated assuming that equipartition also holds.

A fact that is not fully addressed by both stochastic mechanics and scale relativity is why do the theories work only for (box) fractal dimension 2 of the paths. While Nottale gives an heuristic argument and claims that the prescription of a Wiener process may be generalized, he does not proceed to rigorously develop the argument. On the other hand, the stochastic mechanics fixes from the start the Wiener process as a driving noise. While this may look plausible in view of the traditions in the treatment of Brownian motion, it is a choice that should be justified as nowadays anomalous types of diffusion dynamics are also recognized and systematically investigated (overview in [46]). The answer to this question can be given more easily by an approach inspired by Nottale and is partially given by the argument given by Gillepsie [8]. The original argument in [8] contains an explicit assumption of existence of the second moment of the distribution, which amounts to assuming Hölder continuity of order $1/2$ as demonstrated here in Theorem 1. The theorem also corresponds to the result established for fractal interpolation computed via a chaos game where the limit random distribution has been identified with the Gaussian distribution [47,48].

While in SR particle 'trajectories' are considered to be only virtual, SM and the original formulation of Bohm's quantum mechanics treat them as physically real. It is noteworthy that recently Flack and Hiley [49] demonstrated that that a Bohm 'trajectory' is the average of an ensemble of actual individual stochastic Feynman paths. This is in line with the treatment of the problem by the stochastic mechanics and scale relativity and promotes the view that Bohm's quantum mechanics is a mean field theory of the stochastic mechanics.

Funding: The work has been supported in part by grants from Research Fund—Flanders (FWO), contract number VS.097.16N, Fetzer Franklin Memorial Trust and the NanoStreeM H2020 project, contract number 688194.

Acknowledgments: All calculations were assisted by the computer algebra system *Maxima*. The author would like to acknowledge Stephan LeBohec for critical reading and Laurent Nottale for inspirational discussions.

Conflicts of Interest: The funding sponsors had no role in the design of the study; in the collection, analyses, or interpretation of data; in the writing of the manuscript, and in the decision to publish the results.

Appendix A. Notations, General Definitions and Properties of Fractional Velocity

Definition A1 (Asymptotic $\mathcal{O}$ notation). *The notation $\mathcal{O}(x^\alpha)$ is interpreted as the convention that*

$$\lim_{x \to 0} \frac{\mathcal{O}(x^\alpha)}{x^\alpha} = 0$$

for $\alpha > 0$. The notation $\mathcal{O}_x$ will be interpreted to indicate a Cauchy-null sequence with no particular power dependence of x.

Definition A2. *We say that f is of (point-wise) Hölder class $\mathbb{H}^\beta$ if for a given x there exist two positive constants $C, \delta \in \mathbb{R}$ that for an arbitrary $y \in Dom[f]$ and given $|x-y| \le \delta$ fulfill the inequality $|f(x) - f(y)| \le C|x-y|^\beta$, where $|\cdot|$ denotes the norm of the argument.*

Definition A3. *Define the parametrized difference operators acting on a function $f(x)$ as*

$$\Delta_{\epsilon}^{\pm}[f](x) := sgn(\epsilon)\left(f(x+\epsilon) - f(x)\right).$$

The first one we refer to as forward difference operator, the second one we refer to as backward difference operator.

Definition A4. *Define Fractional Variation operators of order $0 \leq \beta \leq 1$ as*

$$v_{\epsilon\pm}^{\beta}[f](x) := \frac{\Delta_{\epsilon}^{\pm}[f](x)}{|\epsilon|^{\beta}}. \tag{A1}$$

This section follows the presentation given recently in [50].

Definition A5 (Fractional order velocity). *Define the* **fractional velocity** *of fractional order β as the limit*

$$v_{\pm}^{\beta}f(x) := \lim_{\epsilon \to 0} \frac{\Delta_{\epsilon}^{\pm}[f](x)}{|\epsilon|^{\beta}}, \tag{A2}$$

where $0 < \beta \leq 1$ are real parameters and $f(x)$ is real-valued function. A function for which at least one of $v_{\pm}^{\beta}f(x)$ exists finitely will be called β-differentiable at the point x.

In the above definition, we do not require upfront equality of left and right β-velocities. This amounts to not demanding continuity of the β-velocities in advance. Instead, continuity is a property, which is fulfilled under certain conditions.

Definition A6. *The set of points where the fractional velocity exists finitely and $v_{\pm}^{\beta}f(x) \neq 0$ will be denoted as the* **set of change** $\chi_{\pm}^{\beta}(f) := \left\{x : v_{\pm}^{\beta}f(x) \neq 0\right\}$.

Since the set of change $\chi_{+}^{\alpha}(f)$ is totally disconnected [26], some of the useful properties of ordinary derivatives, notably the continuity and the semi-group composition property, are lost.

Definition A7. *β-Regularized derivative of a function is defined as:*

$$\frac{d^{\beta\pm}}{dx}f(x) := \lim_{\epsilon \to 0} \frac{\Delta_{\epsilon}^{\pm}[f](x) - v_{+}^{\beta}f(x)\,\epsilon^{\beta}}{\epsilon}.$$

We will require as usual that the forward and backward regularized derivatives be equal for a uniformly continuous function.

In this section, we assume that the functions are BVC in the neighborhood of the point of interest. Under this assumption, we have

- Product rule

$$v_{+}^{\beta}[f\,g](x) = v_{+}^{\beta}f(x)\,g(x) + v_{+}^{\beta}g(x)\,f(x) + [f,g]_{\beta}^{+}(x),$$
$$v_{-}^{\beta}[f\,g](x) = v_{-}^{\beta}f(x)\,g(x) + v_{-}^{\beta}g(x)\,f(x) - [f,g]_{\beta}^{-}(x),$$

- Quotient rule

$$v_{+}^{\beta}[f/g](x) = \frac{v_{+}^{\beta}f(x)\,g(x) - v_{+}^{\beta}g(x)\,f(x) - [f,g]_{\beta}^{+}}{g^{2}(x)},$$
$$v_{-}^{\beta}[f/g](x) = \frac{v_{-}^{\beta}f(x)\,g(x) - v_{-}^{\beta}g(x)\,f(x) + [f,g]_{\beta}^{-}}{g^{2}(x)},$$

where

$$[f,g]_{\beta}^{\pm}(x) := \lim_{\epsilon \to 0} v_{\epsilon\pm}^{\gamma}[f](x)\; v_{\epsilon\pm}^{\beta-\gamma}[g](x),$$

wherever $[f,g]_{\beta}^{\pm}(x) \neq 0$.

For compositions of functions,

- $f \in \mathbb{H}^\beta$ and $g \in \mathbb{C}^1$

$$v_+^\beta f \circ g\,(x) = v_+^\beta f\,(g)\,(g'(x))^\beta,$$
$$v_-^\beta f \circ g\,(x) = v_-^\beta f\,(g)\,(g'(x))^\beta,$$

- $f \in \mathbb{C}^1$ and $g \in \mathbb{H}^\beta$

$$v_+^\beta f \circ g\,(x) = f'(g)\,v_+^\beta g\,(x),$$
$$v_-^\beta f \circ g\,(x) = f'(g)\,v_-^\beta g\,(x).$$

Basic evaluation formula [51]:

$$v_\pm^\beta f\,(x) = \frac{1}{\beta}\lim_{\epsilon\to 0}\epsilon^{1-\beta} f'(x \pm \epsilon).$$

Derivative regularization [52]:
Let $f(t,w) \in \mathbb{C}^2$ be composition with $w(x)$, a $1/q$-differentiable function at x, then

$$\frac{d^\pm}{dx} f(x,w) = \frac{\partial f}{\partial x} + \frac{d^\pm}{dx} w(x) \cdot \frac{\partial f}{\partial w} \pm \frac{1}{q!}[w^q]^\pm \cdot \frac{\partial^q f}{\partial w^q}, \tag{A3}$$

where

$$[w^q]^\pm = \left(v_\pm^{1/q} w\,(x)\right)^q$$

is the fractal q-adic (co-)variation.

Appendix B. The Stochastic Variation Problem

The study of stochastic Lagrangian variational principles has been motivated initially by quantum mechanics and optimal control problems. This section gives only a sketch for the treatment of the problem. The reader is directed to [21,30,31] for more details. In the simplest form, this is the minimization of the regularized functional assuming a constant diffusion coefficient b

$$S_\alpha(t_0,T) := \lim_{N\to\infty} \mathbb{E}\left((\mathcal{P}_N) \sum_{t=t_0}^{t=T} \frac{1}{2}\frac{(\Delta X_k)^2}{\Delta t_k} - \sigma\left(\alpha - \frac{1}{2}\right) b^2 \,\middle|\, X_k = x(\alpha t_k + (1-\alpha)t_{k+1}) \right)$$

for the partition $\mathcal{P}_N$ and $\sigma = \operatorname{sign}\left(\alpha - \frac{1}{2}\right)$.

Thus, suppose that $\alpha = 1$. Then, the increments can be interpreted as Itô integrals so that by the Itô isometry since finite summation and integration commute

$$\mathbb{E}\left(\frac{1}{2\Delta t_k}(\Delta X_k)^2 - \frac{1}{2}b^2 \,\middle|\, X_k = x(t_k)\right) =$$
$$\frac{1}{2\Delta t_k}\left(\int_{t_k}^{t_{k+1}} a ds\right)^2 + \frac{1}{\Delta t_k}\left(\int_{t_k}^{t_{k+1}} a ds\right)\mathbb{E}\left(\int_{t_k}^{t_{k+1}} b dw\right) + \frac{1}{2\Delta t_k}\mathbb{E}\left(\int_{t_k}^{t_{k+1}} b dw\right)^2 - \frac{1}{2}b^2 =$$
$$\frac{1}{2\Delta t_k}\left(\int_{t_k}^{t_{k+1}} a ds\right)^2 + \frac{1}{2\Delta t_k}\int_{t_k}^{t_{k+1}} b^2 ds - \frac{1}{2}b^2 = a\int_{t_k}^{t_{k+1}} a ds + \mathcal{O}(\Delta t_k).$$

Therefore, $S_\alpha(t_0,T)$ is minimal if the drift vanishes on $\mathcal{P}_N$. Suppose that X_t is varied by a small smooth function $\lambda\phi(t,x)$, where the smallness is controlled by λ, then the Itô lemma should be applied so that $\mathbb{E}(d\delta X_t|\mathcal{F}) = 0$ on the difference process $\delta X_t = \lambda\phi(t,x)dt + bdW_t$. Therefore,

$$\mathbb{E}\,(d\phi|\mathcal{F}) = \lambda dt\left(\frac{\partial}{\partial t}\phi + \phi\frac{\partial}{\partial x}\phi + \frac{b^2}{2}\frac{\partial^2}{\partial x^2}\phi\right) = 0 \tag{A4}$$

should hold. The same calculation can be performed for $\alpha = 0$ if the Itô integral is replaced by the anticipative Itô integral. In this case, $\sigma = -1$ and the integration is reversed

$$\mathbb{E}\left(\frac{1}{2\Delta t_k}(\Delta X_k)^2 + \frac{1}{2}b^2\middle| X_k = x(t_{k+1})\right) =$$

$$\frac{1}{2\Delta t_k}\left(\int_{t_{k+1}}^{t_k} ads\right)^2 + \frac{1}{\Delta t_k}\left(\int_{t_{k+1}}^{t_k} ads\right)\mathbb{E}\left(\int_{t_{k+1}}^{t_k} bdw\right) + \frac{1}{2\Delta t_k}\mathbb{E}\left(\int_{t_{k+1}}^{t_k} bdw\right)^2 + \frac{1}{2}b^2 =$$

$$\frac{1}{2\Delta t_k}\left(\int_{t_{k+1}}^{t_k} ads\right)^2 + \frac{1}{2\Delta t_k}\int_{t_{k+1}}^{t_k} b^2 ds + \frac{1}{2}b^2 = a\int_{t_k}^{t_{k+1}} ads + o\left(\Delta t_k\right).$$

In this case, the backward Itô formula also applies as

$$\mathbb{E}\left(d\phi|\mathcal{F}\right) = \lambda dt\left(\frac{\partial}{\partial t}\phi + \phi\frac{\partial}{\partial x}\phi - \frac{b^2}{2}\frac{\partial^2}{\partial x^2}\phi\right) = 0. \quad \text{(A5)}$$

Remark A1. *The treatment of Pavon [21] uses the symmetrized functional $S = S_0 + S_1$ together with a constraint on anti-symmetrized functional $S_0 - S_1$ in the present notation.*

Appendix C. Matlab Simulation Code

Listing 1: Exact simulation Matlab code.

```
clear all;
close all;

% Choose suitably small time step
dt = 1/2^10;
st=sqrt(dt);
sig=1;
% Stopping time
T=10;

t = 0:dt:T-dt;                    % time vector
% Set initial condition
rand('state',200);            % net random seed
nsim=1000;                                        % number of simulations
N =length(t);
% populate random vector
r = randn(N,nsim);

% seed initial conditions
x0 = T*(rand(nsim,1) -0.5)/2;

%% SDE
% Euler- Maruyama
%
y = zeros(N,nsim);
y(1,:) = x0';
for i = 1 :N-1
y(i+1,:) = y(i,:)+ dt*y(i,:)./(t(i)+T) + st*r(i,:);
end

%%%%%%%%%%
% Exact simluation
%
tau=t'+T;
w= zeros(N,nsim);
z= zeros(N,nsim);
for i=1:nsim
z= r(:,i)./tau;
w(:,i)= tau.*( cumsum(z)*st);
w(:,i)=[x0(i); w(1:end-1,i)+x0(i)*tau(1:end-1)/T];
```

and

References

1. Langevin, P. Sur la theórie du mouvement Brownien. *C. R. Acad. Sci.* **1908**, *146*, 530–533. (in French)
2. Nelson, E. Derivation of the Schrödinger equation from Newtonian mechanics. *Phys. Rev.* **1966**, *150*, 1079. [CrossRef]
3. Nottale, L. Fractals in the Quantum Theory of Spacetime. *Int. J. Mod. Phys. A* **1989**, *4*, 5047–5117. doi:10.1007/978-0-387-30440-3_228. [CrossRef]
4. Øksendal, B. *Stochastic Differential Equations*, 6th ed.; Springer: Berlin/Heidelberg, Germany, 2003. doi:10.1007/978-3-642-14394-6.
5. Beck, C. Dynamical systems of Langevin type. *Phys. A Stat. Mech. Appl.* **1996**, *233*, 419–440. doi:10.1016/s0378-4371(96)00254-3. [CrossRef]
6. Mackey, M.; Tyran-Kaminska, M. Deterministic Brownian motion: The effects of perturbing a dynamical system by a chaotic semi-dynamical system. *Phys. Rep.* **2006**, *422*, 167–222. doi:10.1016/j.physrep.2005.09.002. [CrossRef]
7. Tyran-Kamińska, M. Diffusion and Deterministic Systems. *Math. Model. Nat. Phenom.* **2014**, *9*, 139–150. doi:10.1051/mmnp/20149110. [CrossRef]
8. Gillespie, D.T. The mathematics of Brownian motion and Johnson noise. *Am. J. Phys.* **1996**, *64*, 225–240. [CrossRef]
9. Klages, R. *Deterministic Diffusion in One-Dimensional Chaotic Dynamical Systems*; Wissenschaft und Technik Verlag: Berlin, Germany, 1996.
10. Knight, G.; Klages, R. Linear and fractal diffusion coefficients in a family of one-dimensional chaotic maps. *Nonlinearity* **2010**, *24*, 227–241. doi:10.1088/0951-7715/24/1/011. [CrossRef]
11. Zwanzig, R. *Nonequilibrium Statistical Mechanics*; Oxford University Press: Oxford, UK, 2001.
12. Bateman, H. Some recent researches in the motion of fluids. *Mon. Weather Rev.* **1915**, *43*, 163–167. [CrossRef]
13. Burgers, J.M. *The Nonlinear Diffusion Equation*; Springer: Dordrecht, The Netherlands, 1974. doi:10.1007/978-94-010-1745-9.
14. Gurbatov, S.; Malakhov, A.; Saichev, A.; Saichev, A.I.; Malakhov, A.N.; Gurbatov, S.N. *Nonlinear Random Waves and Turbulence in Nondispersive Media: Waves, Rays, Particles (Nonlinear Science: Theory & Applications)*; John Wiley & Sons Ltd.: Hoboken, NJ, USA, 1992.
15. Busnello, B. A Probabilistic Approach to the Two-Dimensional Navier-Stokes Equations. *Ann. Probab.* **1999**, *27*, 1750–1780. doi:10.1214/aop/1022874814. [CrossRef]
16. Busnello, B.; Flandoli, F.; Romito, M. A probabilistic representation for the vorticity of a three-dimensional viscous fluid and for general systems of parabolic equations. *Proc. Edinb. Math. Soc.* **2005**, *48*, 295–336. doi:10.1017/s0013091503000506. [CrossRef]
17. Constantin, P.; Iyer, G. A stochastic Lagrangian representation of the three-dimensional incompressible Navier-Stokes equations. *Commun. Pure Appl. Math.* **2007**, *61*, 330–345. doi:10.1002/cpa.20192. [CrossRef]
18. Constantin, P.; Iyer, G. A stochastic-Lagrangian approach to the Navier-Stokes equations in domains with boundary. *Ann. Appl. Probab.* **2011**, *21*, 1466–1492. doi:10.1214/10-aap731. [CrossRef]
19. Fényes, I. Eine wahrscheinlichkeitstheoretische Begründung und Interpretation der Quantenmechanik. *Z. Phys.* **1952**, *132*, 81–106. doi:10.1007/bf01338578. (In German) [CrossRef]
20. Weizel, W. Ableitung der Quantentheorie aus einem klassischen, kausal determinierten Modell. *Z. Phys.* **1953**, *134*, 264–285. doi:10.1007/bf01330155. (in German) [CrossRef]
21. Pavon, M. Hamilton's principle in stochastic mechanics. *J. Math. Phys.* **1995**, *36*, 6774–6800. doi:10.1063/1.531187. [CrossRef]
22. Nottale, L. Scale Relativity and Schrödinger's equation. *Chaos Solitons Fractals* **1998**, *9*, 1051–1061. doi:10.1016/S0960-0779(97)00190-2. [CrossRef]
23. Nottale, L. Macroscopic Quantum-Type Potentials in Theoretical Systems Biology. *Cells* **2013**, *3*, 1–35. doi:10.3390/cells3010001. [CrossRef] [PubMed]
24. Mandelbrot, B.B. Intermittent turbulence in self-similar cascades: divergence of high moments and dimension of the carrier. In *Multifractals and 1/f Noise: Wild Self-Affinity in Physics (1963–1976)*; Springer: New York, NY, USA, 1999; pp. 317–357. doi:10.1007/978-1-4612-2150-0_15.

25. Salem, R. On some singular monotonic functions which are strictly increasing. *Trans. Am. Math. Soc.* **1943**, *53*, 427–439. [CrossRef]
26. Prodanov, D. Conditions for continuity of fractional velocity and existence of fractional Taylor expansions. *Chaos Solitons Fractals* **2017**, *102*, 236–244. doi:10.1016/j.chaos.2017.05.014. [CrossRef]
27. Carlen, E.A. Conservative diffusions. *Commun. Math. Phys.* **1984**, *94*, 293–315. doi:10.1007/bf01224827. [CrossRef]
28. Guerra, F. The Foundations of Quantum Mechanics—Historical Analysis and Open Questions. In *The Foundations of Quantum Mechanics—Historical Analysis and Open Questions: Lecce, 1993*; Garola, C., Rossi, A., Eds.; Springer: Dordrecht, The Netherlands, 1995; pp. 339–355. doi:10.1007/978-94-011-0029-8_28.
29. Nottale, L. *Fractal Space-Time And Microphysics: Towards A Theory Of Scale Relativity*; World Scientific: Singapore, 1993.
30. Zambrini, J.C. Variational processes and stochastic versions of mechanics. *J. Math. Phys.* **1986**, *27*, 2307–2330. doi:10.1063/1.527002. [CrossRef]
31. Yasue, K. Stochastic calculus of variations. *J. Funct. Anal.* **1981**, *41*, 327–340. doi:10.1016/0022-1236(81)90079-3. [CrossRef]
32. Hermann, R. Numerical simulation of a quantum particle in a box. *J. Phys. A Math. Gen.* **1997**, *30*, 3967. [CrossRef]
33. Hopf, E. The partial differential equation $u_t + uu_x = \mu u_{xx}$. *Commun. Pure Appl. Math.* **1950**, *3*, 201–230. doi:10.1002/cpa.3160030302. [CrossRef]
34. Cole, J.D. On a quasi-linear parabolic equation occurring in aerodynamics. *Quart. Appl. Math.* **1951**, *9*, 225–236. [CrossRef]
35. Dunkel, J. Relativistic Brownian Motion and Diffusion Processes. Ph.D. Thesis, Augsburg University, Augsburg, Germany, 2008.
36. Lage, J.L.; Kulish, V.V. On the Relationship between Fluid Velocity and de Broglie's Wave Function and the Implications to the Navier–Stokes Equation. *Int. J. Fluid Mech. Res.* **2002**, *29*, 13. doi:10.1615/interjfluidmechres.v29.i1.30. [CrossRef]
37. Tsekov, R. On the Stochastic Origin of Quantum Mechanics. *Rep. Adv. Phys. Sci.* **2017**, *1*, 1750008. doi:10.1142/s2424942417500086. [CrossRef]
38. *Cole Hopf Transformations in Maxima*; MDPI: Switzerland, 2018; doi:10.5281/zenodo.1285516. [CrossRef]
39. Cresson, J. Scale calculus and the Schrödinger equation. *J. Math. Phys.* **2003**, *44*, 4907–4938. doi:10.1063/1.1618923. [CrossRef]
40. Sornette, D. Brownian representation of fractal quantum paths. *Eur. J. Phys.* **1990**, *11*, 334. [CrossRef]
41. McClendon, M.; Rabitz, H. Numerical simulations in stochastic mechanics. *Phys. Rev. A* **1988**, *37*, 3479–3492. doi:10.1103/physreva.37.3479. [CrossRef]
42. Al-Rashid, S.N.T.; Habeeb, M.A.Z.; Ahmad, K.A. Application of Scale Relativity (ScR) Theory to the Problem of a Particle in a Finite One-Dimensional Square Well (FODSW) Potential. *J. Quantum Inf. Sci.* **2011**, *1*, 7–17. doi:10.4236/jqis.2011.11002. [CrossRef]
43. Al-Rashid, S.N.T.; Habeeb, M.A.Z.; Ahmed, K.A. Application of Scale Relativity to the Problem of a Particle in a Simple Harmonic Oscillator Potential. *J. Quantum Inf. Sci.* **2017**, *7*, 77–88. doi:10.4236/jqis.2017.73008. [CrossRef]
44. Botez, I.C.; Agop, M. Order-Disorder Transition in Nanostructures via Non-Differentiability. *J. Comp. Theor. Nanosci.* **2015**, *12*, 1746–1755. doi:10.1166/jctn.2015.3953. [CrossRef]
45. Ganguly, S.; Chakraborty, S. Sedimentation of nanoparticles in nanoscale colloidal suspensions. *Phys. Lett. A* **2011**, *375*, 2394–2399. doi:10.1016/j.physleta.2011.04.018. [CrossRef]
46. Metzler, R.; Klafter, J. The restaurant at the end of the random walk: recent developments in the description of anomalous transport by fractional dynamics. *J. Phys. A Math. Gen.* **2004**, *37*, R161. [CrossRef]
47. Puente, C.; López, M.; Pinzón, J.; Angulo, J. The Gaussian Distribution Revisited. *Adv. Appl. Probab.* **1996**, *28*, 500. doi:10.2307/1428069. [CrossRef]
48. Puente, C. The exquisite geometric structure of a Central Limit Theorem. *Fractals* **2003**, *11*, 39–52. doi:10.1142/s0218348x03001458. [CrossRef]
49. Flack, R.; Hiley, B. Feynman Paths and Weak Values. *Entropy* **2018**, *20*, 367. doi:10.3390/e20050367. [CrossRef]

50. Prodanov, D. Fractional Velocity as a Tool for the Study of Non-Linear Problems. *Fractal Fract.* **2018**, *2*, 4. doi:10.3390/fractalfract2010004. [CrossRef]
51. Prodanov, D. Characterization of strongly nonlinear and singular functions by scale space analysis. *Chaos Solitons Fractals* **2016**, *93*, 14–19. doi:10.1016/j.chaos.2016.08.010. [CrossRef]
52. Prodanov, D. Regularization of derivatives on non-differentiable points. *J. Phys. Conf. Ser.* **2016**, *701*, 012031. [CrossRef]

Article

Non-Commutative Worlds and Classical Constraints

Louis H. Kauffman [1,2]

[1] Department of Mathematics, Statistics and Computer Science, University of Illinois at Chicago, 851 South Morgan Street, Chicago, IL 60607-7045, USA; kauffman@uic.edu
[2] Department of Mechanics and Mathematics, Novosibirsk State University, Novosibirsk 630090, Russia

Received: 30 May 2018; Accepted: 19 June 2018; Published: 21 June 2018

Abstract: This paper reviews results about discrete physics and non-commutative worlds and explores further the structure and consequences of constraints linking classical calculus and discrete calculus formulated via commutators. In particular, we review how the formalism of generalized non-commutative electromagnetism follows from a first order constraint and how, via the Kilmister equation, relationships with general relativity follow from a second order constraint. It is remarkable that a second order constraint, based on interlacing the commutative and non-commutative worlds, leads to an equivalent tensor equation at the pole of geodesic coordinates for general relativity.

Keywords: discrete calculus; iterant; commutator; diffusion constant; Levi-Civita connection; curvature tensor; constraints; Kilmister equation; Bianchi identity

---

1. Introduction

Aspects of gauge theory, Hamiltonian mechanics, relativity and quantum mechanics arise naturally in the mathematics of a non-commutative framework for calculus and differential geometry. In this paper, we give a review of our previous results about discrete physics and non-commutative worlds and an introduction to recent work of the author and Anthony Deakin [1]. In examining the foundations of that work, we find new points of view and clarity of proofs as expressed in the later sections of this paper. A key feature of the present paper is a new and concise derivation of the second constraint in Section 4 and a detailed derivation of the related Kilmister equation in Section 5. In Section 6, we determine the third constraint by similar means. At this time, physics associated with the higher order constraints are not known.

We begin by examining discrete dynamical systems. In our exposition, the simplest discrete system corresponds to the square root of minus one, seen as an oscillation between one and minus one. This way, thinking about i as an *iterant* is explained below. By starting with a discrete time series of positions, one has immediately a non-commutativity of observations, since the measurement of velocity involves the tick of the clock and the measurement of position does not demand the tick of the clock. Commutators that arise from discrete observation suggest a non-commutative calculus, and this calculus leads to a generalization of standard advanced calculus in terms of a non-commutative world. In a non-commutative world, all derivatives are represented by commutators. We review how non-commutative worlds are related to quantum physics and classical physics and review our version of the Feynman-Dyson derivation of the formalism of electromagnetic gauge theory. The rest of the paper then investigates algebraic constraints that bind the commutative and non-commutative worlds. These constraints are demands that time derivatives behave in the non-commutative world analogous to their counterparts in standard advanced calculus. It is one constraint of this type that gives rise to our version of the Feynman-Dyson derivation of electromagnetic formalism. The standard first order constraint requires a quadratic Hamiltonian and so begins a story showing how classical physics arises mathematically from the constraints. The second order constraint turns out, remarkably, to be equivalent to a tensor equation at the pole of canonical coordinates in a relativistic framework. We call this tensor equation the *Kilmister equation* and it is studied in Section 5 and in our paper with Deakin [1].

Section 2 is a self-contained version of the concepts in this paper, starting with the non-commutativity of discrete measurements, the introduction of time-shifting operators and the square root of minus one seen as a discrete oscillation, a clock. We proceed from there and analyze the position of the square root of minus one in relation to discrete systems and quantum mechanics. We end this section by fitting together these observations into the structure of the Heisenberg commutator

$$[p, q] = i\hbar.$$

Section 3 is a review of the context of non-commutative worlds with discussion of the Feynman-Dyson derivation. This section generalizes the concepts in Section 2 and places them in the wider context of non-commutative worlds. The key to this generalization is our method of embedding discrete calculus in the non-commutative context. Section 4 discusses constraints on non-commutative worlds that are imposed by asking for correspondences between forms of classical differentiation and the derivatives represented by commutators in a correpondent non-commutative world. This discussion of constraints parallels work of Tony Deakin [2,3] and is continued in joint work of the author and Deakin [1]. At the level of the second constraint we encounter issues related to general relativity and find that, at the pole of a canonical system of coordinates, the second order constraint is equivalent to the Kilmister equation

$$K_{ab} = g^{ef}(R_{ab;ef} + \frac{2}{3}R_{ae}R_{fb}) = 0,$$

where $a, b, e, f = 1, 2, \ldots, 4$ and R is the curvature tensor corresponding to the metric g_{ab} on spacetime. Section 5 gives a derivation of the Kilmister equation and its relation to the second order constraint, following the original observations of Kilmister [4]. In the present paper, we give a proof that the second order constraint is equivalent to the Kilmister equation. One can regard the Kilmister equation $K_{ab} = 0$ as a higher order replacement for the vacuum Einstein equation $R_{ab} = 0$ (the vanishing of the Ricci tensor). In [1], this approach to modifying general relativity, and some of its consequences are explored in detail.

Section 5 continues the constraints discussion in Section 4, showing how to generalize to higher-order constraints and obtains a commutator formula for the third order constraint. Appendix A is a very condensed review of the relationship of the Bianchi identity in differential geometry and the Einstein equations for general relativity. We then observe that every derivation in a non-commutative world comes equipped with its own Bianchi identity. This observation suggests another way to investigate general relativity in the non-commutative context.

2. Time Series and Discrete Physics

Consider elementary discrete physics in one dimension. Consider a time series of positions

$$x(t) : t = 0, \Delta t, 2\Delta t, 3\Delta t, \ldots$$

We can define the velocity $v(t)$ by the formula

$$v(t) = (x(t + \Delta t) - x(t))/\Delta t = Dx(t),$$

where D denotes this discrete derivative. In order to obtain $v(t)$, we need at least one tick Δt of the discrete clock. We define a time-shift operator to handle the fact that once we have observed $v(t)$, the time has moved up by one tick.

We adjust the discrete derivative. We shall add an operator J that in this context accomplishes the time shift:

$$x(t)J = Jx(t + \Delta t).$$

We then redefine the derivative to include this shift:

$$Dx(t) = J(x(t + \Delta t) - x(t))/\Delta t.$$

This readjustment of the derivative rewrites it so that the temporal properties of successive observations are handled automatically.

Discrete observations do not commute. Let A and B denote quantities that we wish to observe in the discrete system. Let AB denote the result of first observing B and then observing A. The result of this definition is that a successive observation of the form $x(Dx)$ is distinct from an observation of the form $(Dx)x$. In the first case, we first observe the velocity at time t, and then x is measured at $t + \Delta t$. In the second case, we measure x at t and then measure the velocity.

We measure the difference between these two results by taking a commutator

$$[A, B] = AB - BA$$

and we get the following computations where we write $\Delta x = x(t + \Delta t) - x(t)$,

$$x(Dx) = x(t)J(x(t + \Delta t) - x(t))/\Delta t = Jx(t + \Delta t)(x(t + \Delta t) - x(t))/\Delta t,$$

$$(Dx)x = J(x(t + \Delta t) - x(t))x(t)/\Delta t,$$

$$[x, Dx] = x(Dx) - (Dx)x = (J/\Delta t)(x(t + \Delta t) - x(t))^2 = J(\Delta x)^2/\Delta t.$$

This final result is worth recording:

$$[x, Dx] = J(\Delta x)^2/\Delta t.$$

From this result, we see that the commutator of x andDx will be constant if $(\Delta x)^2/\Delta t = k$ is a constant. For a given time-step, this means that

$$(\Delta x)^2 = k\Delta t$$

so that

$$\Delta x = \pm\sqrt{(k\Delta t)}.$$

This is a Brownian process with diffusion constant equal to k.

Thus, we arrive at the result that any discrete process viewed in this framework of discrete observation has the basic commutator

$$[x, Dx] = J(\Delta x)^2/\Delta t,$$

generalizing a Brownian process and containing the factor $(\Delta x)^2/\Delta t$ that corresponds to the classical diffusion constant. It is worth noting that the adjusment that we have made to the discrete derivative makes it into a commutator as follows:

$$Dx(t) = J(x(t + \Delta t) - x(t))/\Delta t = (x(t)J - Jx(t))/\Delta t = [x(t), J]/\Delta t.$$

By replacing discrete derivatives by commutators, we can express discrete physics in many variables in a context of non-commutative algebra. We enter this generalization in the next section of the paper.

A simplest and fundamental instance of these ideas is seen in the structure of $i = \sqrt{-1}$. We view i as an *iterant* [5–11], a discrete elementary dynamical system repeating in time the values $\{\ldots, -1, +1, -1, +1, \ldots\}$. One can think of this system as resulting from the attempt to solve $i^2 = -1$ in the form $i = -1/i$. Then, one iterates the transformation $x \longrightarrow -1/x$ and finds the oscillation from a starting value of $+1$ or -1. In this sense, i is identical in concept to a *primordial time*. Furthermore, the algebraic structure of the complex numbers emerges from two conjugate views of this discrete series as $[-1, +1]$ and $[+1, -1]$. We introduce a temporal shift operator η such that $\eta[-1, +1] = [+1, -1]\eta$ and $\eta^2 = 1$ (sufficient to this purpose). Then, we can define $i = [-1, +1]\eta$, endowing it with one view of the discrete oscillation and the sensitivity to shift the clock when interacting with itself or with another operator. Note that if $e = [-1, +1]$ and we take $[a, b][c, d] = [ab, cd]$ and $-[a, b] = [-a, -b]$, then

$$e^2 = \eta^2 = 1$$

and

$$e\eta + \eta e = 0.$$

Hence, with

$$i = e\eta,$$

we have

$$i^2 = e\eta e\eta = -e^2\eta^2 = -1.$$

Here we see i emerge in the non-commutative context of the Clifford algebra generated by e and η, and we see that, in this way, i becomes inextricably identified with elemental time, and so the physical substitution of it for t (Wick rotation) becomes, in this epistemology, an act of recognition of the nature of time. One does not have an increment of time all alone as in classical t. One has it, a combination of an interval and the elemental dynamic that is time. With this understanding, we can return to the commutator for a discrete process and use $i\Delta t$ for the temporal increment.

We found that discrete observation led to the commutator equation

$$[x, Dx] = J(\Delta x)^2/\Delta t,$$

which we will simplify to

$$[q, p/m] = (\Delta x)^2/\Delta t,$$

taking q for the position x and p/m for velocity, the time derivative of position and ignoring the time shifting operator on the right-hand side of the equation.

Understanding that Δt should be replaced by $i\Delta t$, and that, by comparison with the physics of a process at the Planck scale, one can take

$$(\Delta x)^2/\Delta t = \hbar/m,$$

we have

$$[q, p/m] = (\Delta x)^2/i\Delta t = -i\hbar/m,$$

whence

$$[p, q] = i\hbar,$$

and we have arrived at Heisenberg's fundamental relationship between position and momentum. This mode of arrival is predicated on the recognition that $i\Delta t$ represents an interactive interval of time. In the notion of time, there is an inherent clock and an inherent shift of phase that enables a synchrony, a precise dynamic beneath the apparent dynamic of the observed process. Once this substitution is made, once the imaginary value is placed in the temporal circuit, the patterns of quantum mechanics appear. In this way, quantum mechanics can be seen to emerge from the discrete.

3. Review of Non-Commutative Worlds

Now, we begin a general introduction to non-commutative worlds and to a non-commutative discrete calculus. Our approach begins in an algebraic framework that naturally contains the formalism of the calculus, but not its notions of limits or constructions of spaces with specific locations, points and trajectories. Many patterns of physical law fit well into such an abstract framework. In this viewpoint, one dispenses with continuum spacetime and replaces it by algebraic structure. Behind that structure, space stands ready to be constructed, by discrete derivatives and patterns of steps, or by starting with a discrete pattern in the form of a diagram, a network, a lattice, a knot, or a simplicial complex, and elaborating that structure until the specificity of spatio-temporal locations appear.

Poisson brackets allow one to connect classical notions of location with the non-commutative algebra used herein. Below the level of the Poisson brackets is a treatment of processes and operators as though they were variables in the same context as the variables in the classical calculus. In different degrees, one lets go of the notion of classical variables and yet retains their form, as one makes a descent into the discrete. The discrete world of non-commutative operators is a world linked to our familiar world of continuous and commutative variables. This linkage is traditionally exploited in quantum mechanics to make the transition from the classical to the quantum. One can make the journey in the other direction, from the discrete and non-commutative to the "classical" and commutative, but that journey requires powers of invention and ingenuity that are the subject of this exploration. It is our conviction that the world is basically simple. To find simplicity in the complex requires special attention and care.

In starting from a discrete point of view, one thinks of a sequence of states of the world $S, S', S'', S''', \ldots,$ where S' denotes the state succeeding S in discrete time. It is natural to suppose that there is some measure of difference $DS^{(n)} = S^{(n+1)} - S^{(n)}$, and some way that states S and T might be combined to form a new state ST. We can thus think of world-states as operators in a non-commutative algebra with a temoporal derivative $DS = S' - S$. At this bare level of the formalism, the derivative does not satisfy the Leibniz rule. In fact, it is easy to verify that $D(ST) = D(S)T + S'D(T)$. Remarkably, the Leibniz rule, and hence the formalisms of Newtonian calculus can be restored with the addition of one more operator J. In this instance, J is a temporal shift operator with the property that $SJ = JS'$ for any state S. We then see that, if $\nabla S = JD(S) = J(S' - S)$, then $\nabla(ST) = \nabla(S)T + S\nabla(T)$ for any states S and T. In fact, $\nabla(S) = JS' - JS = SJ - JS = [S, J]$, so that this adjusted derivative is a commutator in the general calculus of states. This, in a nutshell, is our approach to non-commutative worlds. We begin with a very general framework that is a non-numerical calculus of states and operators. It is then fascinating and a topic of research to see how physics and mathematics fit into the frameworks so constructed.

Constructions are performed in a Lie algebra $\mathcal{A}$. One may take $\mathcal{A}$ to be a specific matrix Lie algebra, or abstract Lie algebra. If $\mathcal{A}$ is taken to be an abstract Lie algebra, then it is convenient to use the universal enveloping algebra so that the Lie product can be expressed as a commutator. In making general constructions of operators satisfying certain relations, it is understood that one can always begin with free algebra and make a quotient algebra where the relations are satisfied.

On $\mathcal{A}$, a variant of calculus is built by defining derivations as commutators (or more generally as Lie products). For a fixed N in $\mathcal{A}$, one defines

$$\nabla_N : \mathcal{A} \longrightarrow \mathcal{A}$$

by the formula

$$\nabla_N F = [F, N] = FN - NF.$$

∇_N is a derivation satisfying the Leibniz rule.

$$\nabla_N(FG) = \nabla_N(F)G + F\nabla_N(G).$$

Discrete Derivatives are Replaced by Commutators. There is a lot of motivation for replacing derivatives by commutators. If $f(x)$ denotes (say) a function of a real variable x, and $\tilde{f}(x) = f(x+h)$ for a fixed increment h, define the *discrete derivative* Df by the formula $Df = (\tilde{f} - f)/h$, and find that the Leibniz rule is not satisfied. One has the basic formula for the discrete derivative of a product:

$$D(fg) = D(f)g + \tilde{f}D(g).$$

Correct this deviation from the Leibniz rule by introducing a new non-commutative operator J with the property that

$$fJ = J\tilde{f}.$$

Define a new discrete derivative in an extended non-commutative algebra by the formula

$$\nabla(f) = JD(f).$$

It follows at once that

$$\nabla(fg) = JD(f)g + J\tilde{f}D(g) = JD(f)g + fJD(g) = \nabla(f)g + f\nabla(g).$$

Note that

$$\nabla(f) = (J\tilde{f} - Jf)/h = (fJ - Jf)/h = [f, J/h].$$

In the extended algebra, discrete derivatives are represented by commutators, and satisfy the Leibniz rule. One can regard discrete calculus as a subset of non-commutative calculus based on commutators.

Advanced Calculus and Hamiltonian Mechanics or Quantum Mechanics in a Non-Commutative World. In $\mathcal{A}$, there are as many derivations as there are elements of the algebra, and these derivations behave quite wildly with respect to one another. If one takes the concept of *curvature* as the non-commutation of derivations, then $\mathcal{A}$ is a highly curved world indeed. Within $\mathcal{A}$, one can build a tame world of derivations that mimics the behaviour of flat coordinates in Euclidean space. The description of the structure of $\mathcal{A}$ with respect to these flat coordinates contains many of the equations and patterns of mathematical physics.

The flat coordinates Q^i satisfy the equations below with the P_j chosen to represent differentiation with respect to Q^j:

$$[Q^i, Q^j] = 0,$$

$$[P^i, P^j] = 0,$$

$$[Q^i, P^j] = \delta^{ij}.$$

Here δ^{ij} is the Kronecker delta, equal to 1 when $i = j$ and equal to 0 otherwise. Derivatives are represented by commutators:

$$\partial_i F = \partial F/\partial Q^i = [F, P^i],$$

$$\hat{\partial}_i F = \partial F/\partial P^i = [Q^i, F].$$

Our choice of commutators guarantees that the derivative of a variable with respect to itself is one and that the derivative of a variable with respect to a distinct variable is zero. Furthermore, the commuting of the variables with one another guarantees that mixed partial derivatives are independent of the order of differentiation. This is a flat non-commutative world.

Temporal derivative is represented by commutation with a special (Hamiltonian) element H of the algebra:

$$dF/dt = [F, H].$$

(For quantum mechanics, take $i\hbar dA/dt = [A, H]$.) These non-commutative coordinates are the simplest flat set of coordinates for description of temporal phenomena in a non-commutative world.

Hamilton's Equations are Part of the Mathematical Structure of Non-Commutative Advanced Calculus.

$$dP^i/dt = [P^i, H] = -[H, P^i] = -\partial H/\partial Q^i,$$

$$dQ^i/dt = [Q^i, H] = \partial H/\partial P^i.$$

These are exactly Hamilton's equations of motion. The pattern of Hamilton's equations is built into the system.

The Simplest Time Series Leads to the Diffusion Constant and Heisenberg's Commutator. Consider a time series $\{Q, Q', Q'', \dots\}$ with commuting scalar values. Let

$$\dot{Q} = \nabla Q = JDQ = J(Q' - Q)/\tau,$$

where τ is an elementary time step (If Q denotes a times series value at time t, then Q' denotes the value of the series at time $t + \tau$.). The shift operator J is defined by the equation $QJ = JQ'$ where this refers to any point in the time series so that $Q^{(n)}J = JQ^{(n+1)}$ for any non-negative integer n. Moving J across a variable from left to right corresponds to one tick of the clock. This discrete, non-commutative time derivative satisfies the Leibniz rule.

This derivative ∇ also fits a significant pattern of discrete observation. Consider the act of observing Q at a given time and the act of observing (or obtaining) DQ at a given time. Since Q and Q' are ingredients in computing $(Q' - Q)/\tau$, the numerical value associated with DQ, it is necessary to let the clock tick once.

Thus, if one first observe Q and then obtains DQ, the result is different (for the Q measurement) if one first obtains DQ, and then observes Q. In the second case, one finds the value Q' instead of the value Q, due to the tick of the clock.

1. Let $\dot{Q}Q$ denote the sequence: observe Q, then obtain $\dot{Q}$.
2. Let $Q\dot{Q}$ denote the sequence: obtain $\dot{Q}$, then observe Q.

The commutator $[Q, \dot{Q}]$ expresses the difference between these two orders of discrete measurement. In the simplest case, where the elements of the time series are commuting scalars, one has

$$[Q, \dot{Q}] = Q\dot{Q} - \dot{Q}Q = J(Q' - Q)^2/\tau.$$

Thus, one can interpret the equation

$$[Q, \dot{Q}] = Jk$$

(k a constant scalar) as

$$(Q' - Q)^2/\tau = k.$$

This means that the process is a walk with spatial step

$$\Delta = \pm\sqrt{k\tau},$$

where k is a constant. In other words, one has the equation

$$k = \Delta^2/\tau.$$

This is the diffusion constant for a Brownian walk. A walk with spatial step size Δ and time step τ will satisfy the commutator equation above exactly when the square of the spatial step divided by the time step remains constant. This shows that the diffusion constant of a Brownian process is a structural property of that process, independent of considerations of probability and continuum limits.

Thus, we can write (ignoring the timeshift operator J)

$$[Q, \dot{Q}] = (\Delta Q)^2/\tau.$$

If we work with physics at the Planck scale, then we can take τ as the Planck time and ΔQ as the Planck length. Then,

$$(\Delta Q)^2/\tau = \hbar/m,$$

where m is the Planck mass. However, we shall also Wick rotate the time from τ to $i\tau$ justifying $i\tau$ on the principle (described above) that τ should be multiplied by i to bring time into coincidence with an elemental time that is both a temporal operator (i) and a value (t). With this, we obtain

$$[Q, \dot{Q}] = -i\hbar/m$$

or

$$[m\dot{Q}, Q] = i\hbar,$$

and, taking $P = m\dot{Q}$, we have finally

$$[P, Q] = i\hbar.$$

Heisenberg's commutator for quantum mechanics is seen in the nexus of discrete physics and imaginary time.

Schroedinger's Equation is Discrete. Here is how the Heisenberg form of Schroedinger's equation fits in this context. Let $J = (1 - \frac{i}{\hbar}H\Delta t)$. Then, $\nabla\psi = [\psi, J/\Delta t]$, and we calculate

$$\nabla\psi = \frac{1}{\Delta t}[\psi J - J\psi]$$

$$= \psi[(1 - \frac{i}{\hbar}H\Delta t)/\Delta t] - [(1 - \frac{i}{\hbar}H\Delta t)/\Delta t]\psi = -\frac{i}{\hbar}[\psi, H].$$

Thus,

$$\nabla\psi = -\frac{i}{\hbar}[\psi, H].$$

This is exactly the form of the Heisenberg equation.

Another way to think about this operator $J = (1 - \frac{i}{\hbar}H\Delta t)$ is as an approximation to $e^{-\frac{i}{\hbar}H\Delta t}$. We can then see our discrete model behaving *exactly* in the framework of a calculus using *square zero infinitesimals* [12]. Let us recall the bare bones of this model for calculus. We utilize an algebraic entity denoted here by dt such that $(dt)^2 = 0$ and an extended real number system

$$R^\sharp = \{a + bdt\},$$

where it is understood that a and b are standard real numbers and that $a + bdt = a' + b'dt$ if and only if $a = a'$ and $b = b'$. It is given that $dt > 0$ and $dt < r$ for any positive real number r. We multipy by assuming distributivity and using the nilpotence of dt. Thus,

$$(a + bdt)(e + fdt) = ae + (af + be)dt.$$

The special infinitesimal dt is not invertible, but, for those functions that have a well-defined extension to $R^\sharp$, we can define the derivative by the formula

$$F(t + dt) = F(t) + \dot{F}(t)dt.$$

In the case of the exponential function, we have

$$e^{r+sdt} = e^r(1+sdt)$$

as the *definition* of this extension of the exponential function. The reader should note that this means that

$$e^{sdt} = 1+sdt$$

and that this is exactly the result obtained by substitution into the power series

$$e^x = 1 + x + x^2/2! + x^3/3! + \dots$$

and using the nilpotency of the dt. Thus, we find that $e^{a(t+dt)} = e^{at}(1+adt) = e^{at} + ae^{at}dt$ and therefore the derivative of e^{at} with respect to t is ae^{at}, as expected. In the same vein,

$$e^{idt} = 1+idt$$

and

$$e^{idt} = \cos(dt) + i\sin(dt),$$

from which we conclude that

$$\cos(dt) = 1$$

and

$$\sin(dt) = 0.$$

With this rapid course in infinitesimal calculus, we return to time shifter J.

In the nilpotent infinitesimal calculus, we have

$$J = (1 - \frac{i}{\hbar}Hdt) = e^{-\frac{i}{\hbar}Hdt}$$

and

$$J^{-1} = (1 + \frac{i}{\hbar}Hdt) = e^{+\frac{i}{\hbar}Hdt}.$$

Note that we can formally multiply $(1 - \frac{i}{\hbar}Hdt)(1 + \frac{i}{\hbar}Hdt)$ and obtain 1 since $(dt)^2 = 0$. We continue to think of dt as a discrete increment, even though it is infinitesimal. Our time-shift formula is

$$J\psi(t+dt) = \psi(t)J$$

or, equivalently,

$$\psi(t+dt) = J^{-1}\psi(t)J.$$

With this in mind, we calculate and find:

$$\psi(t+dt) = (1 + \frac{i}{\hbar}Hdt)\psi(t)(1 - \frac{i}{\hbar}Hdt) = (\psi(t) + \frac{i}{\hbar}dtH\psi(t))(1 - \frac{i}{\hbar}dtH)$$

$$= \psi(t) + \frac{i}{\hbar}dt(H\psi(t) - \psi(t)H) = \psi(t) - \frac{i}{\hbar}[\psi, H]dt.$$

Thus,

$$\psi(t+dt) = \psi(t) - \frac{i}{\hbar}[\psi, H]dt$$

from which we conclude that

$$\dot{\psi} = -\frac{i}{\hbar}[\psi, H],$$

arriving again at the Heisenberg version of Schroedinger's equation in the context of nilpotent calculus.

Dynamical Equations Generalize Gauge Theory and Curvature. One can take the general dynamical equation in the form

$$dQ^i/dt = \mathcal{G}_i,$$

where $\{\mathcal{G}_1, \ldots, \mathcal{G}_d\}$ is a collection of elements of $\mathcal{A}$. Write $\mathcal{G}_i$ relative to the flat coordinates via $\mathcal{G}_i = P_i - A_i$. This is a definition of A_i and $\partial F/\partial Q^i = [F, P_i]$. The formalism of gauge theory appears naturally. In particular, if

$$\nabla_i(F) = [F, \mathcal{G}_i],$$

then one has the curvature

$$[\nabla_i, \nabla_j]F = [R_{ij}, F]$$

and

$$R_{ij} = \partial_i A_j - \partial_j A_i + [A_i, A_j].$$

This is the well-known formula for the curvature of a gauge connection. Aspects of geometry arise naturally in this context, including the Levi-Civita connection (which is seen as a consequence of the Jacobi identity in an appropriate non-commutative world).

One can consider the consequences of the commutator $[Q^i, \dot{Q}^j] = g^{ij}$, deriving that

$$\ddot{Q}^r = G_r + F_{rs}\dot{Q}^s + \Gamma_{rst}\dot{Q}^s\dot{Q}^t,$$

where G_r is the analogue of a scalar field, F_{rs} is the analogue of a gauge field and Γ_{rst} is the Levi-Civita connection associated with g^{ij}. This decompositon of the acceleration is uniquely determined by the given framework [13–15].

Non-Commutative Electromagnetism and Gauge Theory. One can use this context to revisit the Feynman-Dyson derivation [16,17] of electromagnetism from commutator equations, showing that most of the derivation is independent of any choice of commutators, but highly dependent upon the choice of definitions of the derivatives involved. Without any assumptions about initial commutator equations, but taking the right (in some sense simplest) definitions of the derivatives one obtains a significant generalization of the result of Feynman-Dyson. We give this derivation in [18] and in [13–15] using diagrammatic algebra to clarify the structure. In this section, we use X to denote the position vector rather than Q, as above, and the partial derivatives $\{\partial_1, \partial_2, \partial_3\}$ are each covariant derivatives represented by commutators with $\dot{X}_1, \dot{X}_2, \dot{X}_2$, respectively.

Theorem 1. *With the appropriate [see below] definitions of the operators, and taking*

$$\nabla^2 = \partial_1^2 + \partial_2^2 + \partial_3^2, \ B = \dot{X} \times \dot{X} \text{ and } E = \partial_t \dot{X}, \text{ one has}$$

1. $\ddot{X} = E + \dot{X} \times B,$
2. $\nabla \bullet B = 0,$
3. $\partial_t B + \nabla \times E = B \times B,$
4. $\partial_t E - \nabla \times B = (\partial_t^2 - \nabla^2)\dot{X}.$

The key to the proof of this Theorem is the definition of the time derivative. This definition is as follows:

$$\partial_t F = \dot{F} - \Sigma_i \dot{X}_i \partial_i(F) = \dot{F} - \Sigma_i \dot{X}_i [F, \dot{X}_i]$$

for all elements or vectors of elements F. The definition creates a distinction between space and time in the non-commutative world. It can be regarded as an articulation of one extra constraint of the first order in the sense that we describe in the next section, Section 4, of this paper.

A calculation reveals that

$$\ddot{X} = \partial_t \dot{X} + \dot{X} \times (\dot{X} \times \dot{X}).$$

This suggests taking $E = \partial_t \dot{X}$ as the electric field, and $B = \dot{X} \times \dot{X}$ as the magnetic field so that the Lorentz force law

$$\ddot{X} = E + \dot{X} \times B$$

is satisfied.

This result can be applied to produce many discrete models of the Theorem. These models show that, just as the commutator $[X, \dot{X}] = Jk$ describes Brownian motion in one dimension, a generalization of electromagnetism describes the interaction of triples of time series in three dimensions.

Taking $\partial_t F = \dot{F} - \Sigma_i \dot{X}_i \partial_i(F) = \dot{F} - \Sigma_i \dot{X}_i [F, \dot{X}_i]$ as a definition of the partial derivative with respect to time is a natural move in this context because there is *no time variable t* in this non-commutative world. A formal move of this kind, matching a pattern from the commutative world to the mathematics of the non-commuative world, is the theme of the next section of this paper. In that section, we consider the well known way to associate an operator to a product of commutative variables by taking a sum over all permutations of products of the operators corresponding to the individual variables. This provides a way to associate operator expressions with expressions in the commuative algebra, and hence to let a classical world correspond or map to a non-commutative world. To bind these worlds more closely, we can ask that the formulas for taking derivatives in the commutative world should have symmetrized operator product correspondences in the non-commutative world. In Sections 4 and 5, we show how the resulting *constraints* are related to having a quadratic Hamiltonian (first order constraint) and to having a version of general relativity [1–3] (second order constraint). Such constraints can be carried to all orders of derivatives, but the algebra of such constraints is, at the present time, in a very primitive state. We discuss some of the complexities of the constraint algebra in Section 6 of this paper.

Remark 1. *While there is a large literature on non-commutative geometry, emanating from the idea of replacing a space by its ring of functions, work discussed herein is not written in that tradition. Non-commutative geometry does occur here, in the sense of geometry occuring in the context of non-commutative algebra. Derivations are represented by commutators. There are relationships between the present work and the traditional non-commutative geometry, but that is a subject for further exploration. In no way is this paper intended to be an introduction to that subject. The present summary is based on [13,14,18–26] and the references cited therein.*

The following references in relation to non-commutative calculus are useful in comparing with the present approach [27–30]. Much of the present work is the fruit of a long series of discussions with Pierre Noyes, Clive Kilmister and Anthony Deakin. The paper [31] also works with minimal coupling for the Feynman-Dyson derivation. The first remark about the minimal coupling occurs in the original paper by Dyson [16], in the context of Poisson brackets. The paper [32] is worth reading as a companion to Dyson. It is the purpose of this summary to indicate how non-commutative calculus can be used in foundations.

4. Constraints—Classical Physics and General Relativity

The program here is to investigate restrictions in a non-commutative world that are imposed by asking for a specific correspondence between classical variables acting in the usual context of continuum calculus, and non-commutative operators corresponding to these classical variables. By asking for the simplest constraints, we find the need for a quadratic Hamiltonian and a remarkable relationship with Einstein's equations for general relativity [2,3]. There is a hierarchy of constraints of which we only analyze the first two levels. An appendix to this paper indicates a direction for exploring the algebra of the higher constraints.

If, for example, we let x and y be classical variables and X and Y the corresponding non-commutative operators, then we ask that x^n correspond to X^n and that y^n correspond to Y^n for positive integers n. We further ask that linear combinations of classical variables correspond to linear combinations of the corresponding operators. These restrictions tell us what happens to products. For example, we have classically that $(x+y)^2 = x^2 + 2xy + y^2$. This, in turn, must correspond to $(X+Y)^2 = X^2 + XY + YX + Y^2$. From this, it follows that $2xy$ corresponds to $XY + YX$. Hence, xy corresponds to

$$\{XY\} = (XY + YX)/2.$$

By a similar calculation, if $x_1, x_2, \ldots, x_n$ are classical variables, then the product $x_1 x_2 \ldots x_n$ correspondst to

$$\{X_1 X_2 \ldots X_n\} = (1/n!)\Sigma_{\sigma \in S_n} X_{\sigma_1} X_{\sigma_2} \ldots X_{\sigma_n},$$

where S_n denotes all permutations of $1, 2, \ldots, n$. Note that we use curly brackets for these symmetrizers and square brackets for commutators as in $[A, B] = AB - BA$.

We can formulate constraints in the non-commutative world by asking for a correspondence between familiar differentiation formulas in continuum calculus and the corresponding formulas in the non-commutative calculus, where all derivatives are expressed via commutators. We will detail how this constraint algebra works in the first few cases. Exploration of these constraints has been pioneered by Anthony Deakin [2,3,33]. The author of this paper and Tony Deakin have written a paper on the consequences of these contraints in the interface among classical and quantum mechanics and relativity [1].

Recall that the temporal derivative in a non-commutative world is represented by commutator with an operator H that can be intrepreted as the Hamiltonian operator in certain contexts.

$$\dot{\Theta} = [\Theta, H].$$

For this discussion, we shall take a collection $Q^1, Q^2, \ldots, Q^n$ of operators to represent spatial coordinates $q^1, q^2, \ldots, q^n$. The Q^i commute with one another, and the derivatives with respect to Q^i are represented by operators P^i so that

$$\partial\Theta/\partial Q^i = \Theta_i = [\Theta, P^i].$$

We also write

$$\partial\Theta/\partial P^i = \Theta^i = [Q^i, \Theta].$$

Note that if Θ had indices of its own, then we would use a comma to separate indices indicating a derivative from the given indices. Thus,

$$\partial F_a/\partial Q^i = [F_a, Q^i] = F_{a,i}.$$

We assume that $[Q^i, P^j] = \delta^{ij}$ and that the P^j commute with one another (so that mixed partial derivatives with respect to the Q^i are independent of order of differentiation).
Note that

$$\dot{Q}^i = [Q^i, H] = H^i.$$

It will be convenient for us to write H^i in place of $\dot{Q}^i$ in the calculations to follow.

The First Constraint. The *first constraint* is the equation

$$\dot{\Theta} = \{\dot{Q}^i \Theta_i\} = \{H^i \Theta_i\}.$$

This equation expresses the symmetrized version of the usual calculus formula $\dot{\theta} = \dot{q}^i \theta_i$. It is worth noting that the first constraint is satisfied by the quadratic Hamiltonian

$$H = \frac{1}{4}(g^{ij}P^iP^j + P^iP^jg^{ij}),$$

where $g^{ij} = g^{ji}$ and the g_{ij} commute with the Q^k. One can show that a quadratic Hamiltonian is necessary for the first order constaint to be satisfied [1,3,13,14]. The fact that the quadratic Hamiltonian is equivalent to the first constraint shows how the constraints bind properties of classical physics (in this case Hamiltonian mechanics) to the non-commutative world.

The Second Constraint. The *second constraint* is the symmetrized analog of the second temporal derivative:

$$\ddot{\Theta} = \{\dot{H}^i\Theta_i\} + \{H^iH^j\Theta_{ij}\}.$$

However, by differentiating the first constraint, we have

$$\ddot{\Theta} = \{\dot{H}^i\Theta_i\} + \{H^i\{H^j\Theta_{ij}\}\}.$$

Thus, the second constraint is equivalent to the equation

$$\{H^i\{H^j\Theta_{ij}\}\} = \{H^iH^j\Theta_{ij}\}.$$

We now reformulate this version of the constraint in the following theorem.

Theorem 2. *The second constraint in the form* $\{H^i\{H^j\Theta_{ij}\}\} = \{H^iH^j\Theta_{ij}\}$ *is equivalent to the equation*

$$[[\Theta_{ij}, H^j], H^i] = 0.$$

Proof. We can shortcut the calculations involved in proving this Theorem by looking at the properties of symbols A, B, C such that $AB = BA$, $ACB = BCA$. Formally these mimic the behaviour of $A = H^i, B = H^j, C = \Theta_{ij}$ in the expressions $H^iH^j\Theta_{ij}$ and $H^i\Theta_{ij}H^j$ since $\Theta_{ij} = \Theta_{ji}$, and the Einstein summation convention is in place. Then,

$$\{A\{BC\}\} = \frac{1}{4}(A(BC + CB) + (BC + CB)A) = \frac{1}{4}(ABC + ACB + BCA + CBA),$$

$$\{ABC\} = \frac{1}{6}(ABC + ACB + BAC + BCA + CAB + CBA).$$

Thus,

$$\begin{aligned}
\{ABC\} - \{A\{BC\}\} &= \frac{1}{12}(-ABC - ACB + 2BAC - BCA + 2CAB - CBA) \\
&= \frac{1}{12}(ABC - 2ACB + CAB) \\
&= \frac{1}{12}(ABC - 2BCA + CBA) \\
&= \frac{1}{12}(A(BC - CB) + (CB - BC)A) \\
&= \frac{1}{12}(A[B,C] - [B,C]A) \\
&= \frac{1}{12}[A,[B,C]].
\end{aligned}$$

Thus, the second constraint is equivalent to the equation

$$[H^i, [H^j, \Theta_{ij}]] = 0.$$

This in turn is equivalent to the equation

$$[[\Theta_{ij}, H^j], H^i] = 0,$$

completing the proof of the Theorem. □

Remark 2. *If we define*

$$\nabla^i(\Theta) = [\Theta, H^i] = [\Theta, \dot{Q}^i],$$

then this is the natural covariant derivative that was described in our discussion of non-commutative electromagnetism in Section 3 of this paper. Thus, the second order constraint is

$$\nabla^i(\nabla^j(\Theta_{ij}) = 0.$$

A Relationship with General Relativity. We choose a non-commutative metric representative g^{ij} in the non-commutative world with an inverse g_{ij} so that $g^{ij} = g^{ji}, g_{ij} = g_{ji}$, and $g^{ik}g_{kj} = \delta^i_j$. We can use the quadratic Hamiltonian $H = \frac{1}{4}(g^{ij}P^iP^j + P^iP^jg^{ij})$ as previously discussed, but we simplify the calculations below by taking $H = \frac{1}{2}(g^{ij}P^iP^j)$. No essential difference ensues in the results. We assume that the g^{ij} commute with the coordinate representatives Q^k so that $[g^{ij}, Q^k] = 0$ for all choices of i, j, k and similarly for the g_{ij}. We take P^i and Q^j as described at the beginning of this section. It is then an easy calculation to verify that

$$[Q^i, \dot{Q}^j] = g^{ij}.$$

More generally, we have the

Lemma 1. $\nabla^i(\Theta) = [\Theta, \dot{Q}^i] = g^{ij}[\Theta, P^j] = g^{ij}\Theta_j$ *for an arbitrary element* Θ *in the non-commutative world algebra that commutes with the* g^{ij}.

Proof.

$$\nabla^i(\Theta) = [\Theta, \dot{Q}^i] = [\Theta, [Q^i, H]] = [\Theta, [Q^i, \frac{1}{2}(g^{ab}P^aP^b)]] = \frac{1}{2}g^{ab}[\Theta, [Q^i, P^aP^b]].$$

Note that

$$[Q^i, P^aP^b] = Q^iP^aP^b - P^aP^bQ^i = Q^iP^aP^b - P^aQ^iP^b + P^aQ^iP^b - P^aP^bQ^i$$

$$= [Q^i, P^a]P^b + P^a[Q^i, P^b] = \delta^{ia}P^b + P^a\delta^{ib}.$$

Therefore,

$$\nabla^i(\Theta) = \frac{1}{2}g^{ab}[\Theta, \delta^{ia}P^b + P^a\delta^{ib}] = \frac{1}{2}g^{ib}[\Theta, P^b] + \frac{1}{2}g^{ai}[\Theta, P^a] = g^{ij}\Theta_j.$$

This completes the proof of the Lemma. □

Remark 3. *By similar algebra it can be verified that* $\dot{\Theta} = \{\dot{Q}^i\Theta_i\}$ *for all* Θ *such that* Θ *and* Θ_k *commute with all* g^{ij}. *Thus such* Θ *satisfy the first constraint.*

As we have seen in this section, the second order constraint is

$$\nabla^i(\nabla^j(\Theta_{ij}) = 0.$$

Using the explicit form of the covariant derivative derived in the previous paragraph, we have

$$\nabla^i(\nabla^j(\Theta_{ij}) = \nabla^i(g^{jk}\Theta_{ijk}) = g^{il}(g^{jk}\Theta_{ijk})_l.$$

With $\Theta = g_{ab}$, the second constraint becomes the equation

$$g^{il}(g^{jk}g_{ab,ijk})_l = 0.$$

We call this equation the *specialized second order constraint*. Kilmister observed in correspondence with Deakin [4] that this last equation is, at the pole of canonical coordinates, equivalent to a fourth order version of Einstein's field equation for vacuum general relativity:

$$K_{ab} = g^{ef}(R_{ab;ef} + \frac{2}{3}R_{ae}R_{fb}) = 0,$$

where $a, b, e, f = 1, 2, \ldots n$ and R is the curvature tensor corresponding to the metric g_{ab}. This equation has been studied by Deakin in [2,3,33] and by Deakin and Kauffman in [1]. It remains to be seen what the full consequences for general relativity are in relation to this formulation, and it remains to be seen what the further consequences of higher order constraints will be. The algebra of the higher order constraints is under investigation at this time.

5. The Kilmister Equation

In this section, we derive the Kilmister equation

$$K_{ab} = g^{ef}(R_{ab;ef} + \frac{2}{3}R_{ae}R_{fb}) = 0,$$

where $a, b, e, f = 1, 2, \ldots, 4$ and R is the curvature tensor corresponding to the metric g_{ab}. The derivation is based on explicating these tensors at the origin (pole) of canonical geodesic coordinates for spacetime with respect to the given metric. See Eddington [34] (p. 79) for a detailed explanation of canonical coordinates. We will show that Kilmister's equation is, at the pole, equivalent to the specialized second order constraint equation

$$g^{ef}(g^{cd}g_{ab,cde})_f = 0$$

as explained in Section 4 of this paper. This is a remarkable coincidence of structure and suggests that the Kilmister equation should be investigated in the context of general relativity and cosmology. Deakin and Kauffman have begun this investigation in [1]. In this section, we give a complete derivation of the Kilmister equation based on the symmetries of the curvature and connection tensors. More work is needed to understand the relationship between this derivation and the structure of the second order constraint as described in the previous section of this paper.

Calculus in this section is classical continuum calculus. We use the standard notation

$$F, a = \partial F / \partial x^a,$$

where x denotes a point in 4-dimensional spacetime with x^4 the temporal coordinate. We use $F; a$ for the corresponding covariant derivative, which will be made explicit in the calculations below.

In order to perform Kilmister's derivation, we need to recall properties of the canonical coordinates and the basic symmetries of the Riemann tensor. For the present section, we will refer to formal properties of the Riemann tensor and Levi-Civita connection as we need them. See the Appendix A for more details or Dirac's book on general relativity [35] for specifics about these tensors.

Eddingtion observes that, in geodesic coordinates for four-dimensional spacetime, we may assume that the components Γ^a_{bc} of the Levi-Civita connection

$$\Gamma^a_{bc} = \frac{g^{ak}}{2}(g_{kb,c} + g_{kc,b} - g_{bc,k})$$

vanish at that pole. Note that, in general, $\Gamma^a_{bc} = \Gamma^a_{cb}$. Eddington further observes that one can assume, without constraining the curvature tensor, that

$$\Gamma^a_{bc,d} + \Gamma^a_{cd,b} + \Gamma^a_{db,c} = 0$$

at the pole. Since the general formula for the Riemann tensor is

$$R^a_{bcd} = \Gamma^a_{bd,c} - \Gamma^a_{bc,d} + \Gamma^k_{bd}\Gamma^a_{kc} + \Gamma^k_{bc}\Gamma^a_{kd}.$$

We know that at the pole

$$R^a_{bcd} = \Gamma^a_{bd,c} - \Gamma^a_{bc,d}.$$

The general symmetries of the Riemann tensor that we use are:

1. $R_{abcd} = R_{cdab} = R_{dcba}$,
2. $R_{abcd} = -R_{bacd} = -R_{abdc}$.

Lemma 2. *At the pole of the canonical coordinates,* $\Gamma^a_{bc,d} = \frac{1}{3}(R^a_{bdc} + R^a_{cdb})$.

Proof. At the pole,

$$\Gamma^a_{bc,d} + \Gamma^a_{cd,b} + \Gamma^a_{db,c} = 0$$

and at the pole

$$R^a_{bcd} = \Gamma^a_{bd,c} - \Gamma^a_{bc,d}.$$

Thus,

$$R^a_{bcd} = \Gamma^a_{bd,c} + \Gamma^a_{cd,b} + \Gamma^a_{db,c} = 2\Gamma^a_{bd,c} + \Gamma^a_{cd,b}.$$

Hence, we have

$$R^a_{bcd} = 2\Gamma^a_{bd,c} + \Gamma^a_{cd,b}$$

and

$$R^a_{cbd} = 2\Gamma^a_{cd,b} + \Gamma^a_{bd,c}.$$

Therefore,

$$2R^a_{cbd} - R^a_{bcd} = 4\Gamma^a_{cd,b} + 2\Gamma^a_{bd,c} - 2\Gamma^a_{bd,c} - \Gamma^a_{cd,b} = 3\Gamma^a_{cd,b}.$$

However, at the pole (and more generally),

$$R^a_{bcd} + R^a_{cdb} + R^a_{dbc} = \Gamma^a_{bd,c} - \Gamma^a_{bc,d} + \Gamma^a_{cb,d} - \Gamma^a_{cd,b} + \Gamma^a_{dc,b} - \Gamma^a_{db,c} = 0.$$

Therefore,

$$\begin{aligned} 3\Gamma^a_{cd,b} &= 2R^a_{cbd} - R^a_{bcd} \\ &= 2R^a_{cbd} + R^a_{cdb} + R^a_{dbc} \\ &= R^a_{cbd} + R^a_{dbc} \end{aligned}$$

since $R^a_{cbd} + R^a_{cdb} = 0$ by anti-symmetry in the indices b and d. Thus, we have shown that

$$\Gamma^a_{bc,d} = \frac{1}{3}(R^a_{bdc} + R^a_{cdb}).$$

This completes the proof of the Lemma. □

Lemma 3. *At the pole of the canonical coordinates,*

$$g_{ab,cd} = \frac{1}{3}(R_{cbad} + R_{cabd}).$$

Proof. It is generally true that

$$g_{ab,c} = g_{pb}\Gamma^p_{ac} + g_{ap}\Gamma^p_{bc}.$$

Thus, at the pole,

$$\begin{aligned} g_{ab,cd} &= g_{pb}\Gamma^p_{ac,d} + g_{ap}\Gamma^p_{bc,d} \\ &= g_{pb}[\frac{1}{3}(R^p_{adc} + R^p_{cda})] + g_{ap}[\frac{1}{3}(R^p_{bdc} + R^p_{cdb})] \\ &= \frac{1}{3}(R_{badc} + R_{bcda} + R_{abdc} + R_{acdb}) \\ &= \frac{1}{3}(R_{cbad} + R_{cabd}) \end{aligned}$$

(using the symmetries of the Riemann tensor). This completes the proof of the Lemma. □

Definition 1. *Recall the definition of the Ricci Tensor:*

$$R_{ab} = g^{ij}R_{iabj} = R^j_{abj}.$$

Note that

$$R_{ab} = g^{cd}R_{cabd} = g^{cd}R_{dbac} = R_{ba},$$

proving the symmetry of the Ricci Tensor.

Remark 4. *Since it is generally true that*

$$g_{ab,c} = g_{pb}\Gamma^p_{ac} + g_{ap}\Gamma^p_{bc},$$

we know that at the pole

$$g_{ab,c} = 0,$$

since the Christoffel symbols vanish at the pole. Note that it follows from this vanishing result that

$$g^{ab}_{,c} = 0$$

at the pole. Higher derivatives may not be zero, as in the above Lemma.

Lemma 4. *At the pole of the canonical coordinates,*

$$g^{cd}g_{ab,cd} = \frac{2}{3}R_{ab}.$$

Proof. By the previous Lemma,

$$g^{cd}g_{ab,cd} = \frac{1}{3}(g^{cd}R_{cbad} + g^{cd}R_{cabd}) = \frac{1}{3}(R_{ba} + R_{ab}) = \frac{2}{3}R_{ab}.$$

This completes the proof of the Lemma. □

Now we are ready to obtain the Kilmister equation. For this, we need to invoke the covariant derivative, $R_{ab;e}$, designated by a semi-colon, not a comma, and the basic formula

$$R_{ab,e} = R_{ab;e} + \Gamma^p_{ae} R_{pb} + \Gamma^p_{be} R_{ap}.$$

From this, it follows that, at the pole,

$$R_{ab,ef} = (R_{ab;e} + \Gamma^p_{ae} R_{pb} + \Gamma^p_{be} R_{ap})_{,f}$$

$$= R_{ab;ef} + \Gamma^p_{ae,f} R_{pb} + \Gamma^p_{be,f} R_{ap}.$$

Note that the other terms in this covariant derivative involve the Christoffel symbols and these vanish at the pole. Thus, we have

$$\begin{aligned} R_{ab,ef} &= R_{ab;ef} + \Gamma^p_{ae,f} R_{pb} + \Gamma^p_{be,f} R_{ap} \\ &= R_{ab;ef} + \frac{1}{3}(R^p_{afe} R_{pb} + R^p_{efa} R_{pb} + R^p_{bfe} R_{ap} + R^p_{efb} R_{ap}). \end{aligned}$$

Hence,

$$\begin{aligned} g^{ef} R_{ab,ef} &= g^{ef} R_{ab;ef} + \frac{1}{3}(g^{ef} R^p_{afe} R_{pb} + g^{ef} R^p_{efa} R_{pb} + g^{ef} R^p_{bfe} R_{ap} + g^{ef} R^p_{efb} R_{ap}) \\ &= g^{ef} R_{ab;ef} + \frac{1}{3}(0 + g^{ef} R^p_{efa} R_{pb} + 0 + g^{ef} R^p_{efb} R_{ap}) \\ &= g^{ef} R_{ab;ef} + \frac{1}{3}(R^p_a R_{pb} + R^p_b R_{ap}) \\ &= g^{ef} R_{ab;ef} + \frac{1}{3}(g^{ef} R_{ae} R_{fb} + g^{ef} R_{eb} R_{af}) \\ &= g^{ef} R_{ab;ef} + \frac{2}{3}(g^{ef} R_{ae} R_{fb}). \end{aligned}$$

Thus, we have shown

Theorem 3. *At the pole of the canonical coordinates,*

$$g^{ef} R_{ab,ef} = K_{ab} = g^{ef}(R_{ab;ef} + \frac{2}{3} R_{ae} R_{fb})$$

and

$$g^{ef} R_{ab,ef} = \frac{3}{2} g^{ef}(g^{cd} g_{ab,cd})_{ef}.$$

Proof. This result follows from the discussion above and the fact at the pole of the canonical coordinates,

$$g^{cd} g_{ab,cd} = \frac{2}{3} R_{ab}.$$

□

Remark 5. *Thus, we have shown that*

$$\frac{2}{3} K_{ab} = g^{ef}(g^{cd} g_{ab,cd})_{ef}$$

$$= g^{ef}(g^{cd}_{,e} g_{ab,cd})_f + g^{ef}(g^{cd} g_{ab,cde})_f = g^{ef}(g^{cd} g_{ab,cde})_f,$$

since $g^{cd}_{,e} = 0$. *Thus, we have proved:*

Theorem 4. *Let*

$$K_{ab} = g^{ef}(R_{ab;ef} + \frac{2}{3}R_{ae}R_{fb}).$$

Then,

$$\frac{2}{3}K_{ab} = g^{ef}(g^{cd}g_{ab,cde})_f.$$

Thus, the Kilmister equation $K_{ab} = 0$ is equivalent to the second order constraint articulated in Section 4 of the present paper.

This completes our description of Clive Kilmister's remarkable derivation of the relationship of the second order constraint with general relativity. All these considerations are motivation for considering the Kilmister tensor equation $K_{ab} = 0$ as a refined version of the vacuum equations for general relativity. In [1], we explore some of the consequences of the Kilmister equation. The exact relationship of the constraint equation and the Kilmister equation remains mysterious. More work needs to be done in this domain and in exploring the relationship of non-commutative worlds and the tensor geometry of classical spacetime.

6. On the Algebra of Constraints

We have the usual advanced calculus formula $\dot{\theta} = \dot{q}^i\theta_i$. We shall define $h^j = \dot{q}^i$ so that we can write $\dot{\theta} = h^i\theta_i$. We can then calculate successive derivatives with $\theta^{(n)}$ denoting the n-th temporal derivative of θ :

$$\theta^{(1)} = h^i\theta_i,$$

$$\theta^{(2)} = h^{i(1)}\theta_i + h^ih^j\theta_{ij},$$

$$\theta^{(3)} = h^{i(2)}\theta_i + 3h^{i(1)}h^j\theta_{ij} + h^ih^jh^k\theta_{ijk}.$$

The equality of mixed partial derivatives in these calculations makes it evident that one can use a formalism that hides all the superscripts and subscripts $(i, j, k, \dots)$. In that simplified formalism, we can write

$$\theta^{(1)} = h\theta,$$

$$\theta^{(2)} = h^{(1)}\theta + h^2\theta,$$

$$\theta^{(3)} = h^{(2)}\theta + 3h^{(1)}h\theta + h^3\theta,$$

$$\theta^{(4)} = h^4\theta + 6h^2\theta h^{(1)} + 3\theta h^{(1)2} + 4h\theta h^{(2)} + \theta h^{(3)}.$$

Each successive row is obtained from the previous row by applying the identity $\theta^{(1)} = h\theta$ in conjunction with the product rule for the derivative.

This procedure can be automated so that one can obtain the formulas for higher order derivatives as far as one desires. These can then be converted into the non-commutative constraint algebra and the consequences examined. Further analysis of this kind will be done in a sequel to this paper.

The interested reader may enjoy seeing how this formalism can be carried out. Below we illustrate a calculation using *Mathematica*^{TM} (Wolfram, Champaign, Illinois, US), where the program already knows how to formally differentiate using the product rule and so only needs to be told that $\theta^{(1)} = h\theta$. This is said in the equation $T'[x] = H[x]T[x]$ where $T[x]$ stands for θ and $H[x]$ stands for h with x a dummy variable for the differentiation. Here $D[T[x], x]$ denotes the derivative of $T[x]$ with respect to x, as does $T'[x]$.

In the calculation below, we have indicated five levels of derivatives. The structure of the coefficients in this recursion is interesting and complex territory. For example, the coefficients of

$H[x]^n T[x] H'[x] = h^n \theta h'$ are the triangular numbers $\{1, 3, 6, 10, 15, 21, \ldots\}$, but the next series are the coefficients of $H[x]^n T[x] H'[x]^2 = h^n \theta h'^2$, and these form the series

$$\{1, 3, 15, 45, 105, 210, 378, 630, 990, 1485, 2145, \ldots\}.$$

This series is eventually constant after four discrete differentiations. This is the next simplest series that occurs in this structure after the triangular numbers. To penetrate the full algebra of constraints, we need to understand the structure of these derivatives and their corresponding non-commutative symmetrizations:

T'[x]:=H[x]T[x]

D[T[x],x]
D[D[T[x],x],x]
D[D[D[T[x],x],x],x]
D[D[D[D[T[x],x],x],x],x]
D[D[D[D[D[T[x],x],x],x],x],x]

$H[x]T[x]$

$H[x]^2T[x] + T[x]H'[x]$

$H[x]^3T[x] + 3H[x]T[x]H'[x] + T[x]H''[x]$

$H[x]^4T[x] + 6H[x]^2T[x]H'[x] + 3T[x]H'[x]^2 + 4H[x]T[x]H''[x] + T[x]H^{(3)}[x]$

$H[x]^5T[x] + 10H[x]^3T[x]H'[x] + 15H[x]T[x]H'[x]^2 + 10H[x]^2T[x]H''[x] + 10T[x]H'[x]H''[x] + 5H[x]T[x]H^{(3)}[x] + T[x]H^{(4)}[x]$

Algebra of Constraints

In this section, we work with the hidden index conventions described before in the paper. In this form, the classical versions of the first two constraint equations are

1. $\dot{\theta} = \theta h$,
2. $\ddot{\theta} = \theta h^2 + \theta \dot{h}$.

In order to obtain the non-commutative versions of these equations, we replace h by H and θ by Θ where the capitalized versions are non-commuting operators. The first and second constraints then become

1. $\{\dot{\Theta}\} = \{\Theta H\} = \frac{1}{2}(\Theta H + H\Theta)$,
2. $\{\ddot{\Theta}\} = \{\Theta H^2\} + \{\Theta \dot{H}\} = \frac{1}{3}(\Theta H^2 + H\Theta H + H^2\Theta) + \frac{1}{2}(\Theta \dot{H} + \dot{H}\Theta)$.

Proposition 1. *The Second Constraint is equivalent to the commutator equation*

$$[[\Theta, H], H] = 0.$$

Proof. We identify

$$\{\dot{\Theta}\}^\bullet = \{\ddot{\Theta}\}$$

and

$$\{\dot{\Theta}\}^\bullet = \{\{\Theta H\}H\} + \{\Theta \dot{H}\}.$$

Thus, we need

$$\{\Theta H^2\} = \{\{\Theta H\}H\}.$$

The explicit formula for $\{\{\Theta H\}H\}$ is

$$\{\{\Theta H\}H\} = \frac{1}{2}(\{\Theta H\}H + H\{\Theta H\}) = \frac{1}{4}(\theta HH + H\Theta H + H\Theta H + HH\Theta).$$

Thus, we require that

$$\frac{1}{3}(\Theta H^2 + H\Theta H + H^2\Theta) = \frac{1}{4}(\theta HH + H\Theta H + H\Theta H + HH\Theta),$$

which is equivalent to

$$\Theta H^2 + H^2\Theta - 2H\Theta H = 0.$$

We then note that

$$[[\Theta, H], H] = (\Theta H - H\Theta)H - H(\Theta H - H\Theta) = \Theta H^2 + H^2\Theta - 2H\Theta H.$$

Thus, the final form of the second constraint is the equation

$$[[\Theta, H], H] = 0.$$

□

The Third Constraint. We now go on to an analysis of the third constraint. The third constraint consists in the two equations

1. $\{\dddot{\Theta}\} = \{\Theta H^3\} + 3\{\Theta H\dot{H}\} + \{\Theta\ddot{H}\}$,
2. $\{\dddot{\Theta}\} = \{\ddot{\Theta}\}^\bullet$, where

$$\{\ddot{\Theta}\}^\bullet = \{\{\Theta H\}H^2\} + 2\{\Theta H\dot{H}\} + \{\{\Theta H\}\dot{H}\} + \{\Theta\ddot{H}\}.$$

Proposition 2. *The Third Constraint is equivalent to the commutator equation*

$$[H^2, [H, \Theta]] = [\dot{H}, [H, \Theta]] - 2[H, [\dot{H}, \Theta]].$$

Proof. We demand that $\{\dddot{\Theta}\} = \{\ddot{\Theta}\}^\bullet$ and this becomes the longer equation

$$\{\Theta H^3\} + 3\{\Theta H\dot{H}\} + \{\Theta\ddot{H}\} = \{\{\Theta H\}H^2\} + 2\{\Theta H\dot{H}\} + \{\{\Theta H\}\dot{H}\} + \{\Theta\ddot{H}\}.$$

This is equivalent to the equation

$$\{\Theta H^3\} + \{\Theta H\dot{H}\} = \{\{\Theta H\}H^2\} + \{\{\Theta H\}\dot{H}\}.$$

This, in turn, is equivalent to

$$\{\Theta H^3\} - \{\{\Theta H\}H^2\} = \{\{\Theta H\}\dot{H}\} - \{\Theta H\dot{H}\}.$$

This is equivalent to

$$(1/4)(H^3\Theta + H^2\Theta H + H\Theta H^2 + \Theta H^3) - (1/6)(H^2(H\Theta + \Theta H) + H(H\Theta + \Theta H)H + (H\Theta + \Theta H)H^2)$$
$$= (1/2)(\dot{H}(1/2)(H\Theta + \Theta H) + (1/2)(H\Theta + \Theta H)\dot{H}) - (1/6)(\dot{H}H\Theta + \dot{H}\Theta H + H\dot{H}\Theta + H\Theta\dot{H}$$
$$+ \Theta H\dot{H} + \Theta\dot{H}H).$$

This is equivalent to

$$3(H^3\Theta + H^2\Theta H + H\Theta H^2 + \Theta H^3) - 2(H^3\Theta + 2H^2\Theta H + 2H\Theta H^2 + \Theta H^3)$$

$$= 3(\dot{H}H\Theta + \dot{H}\Theta H + H\Theta\dot{H} + \Theta H\dot{H}) - 2(\dot{H}H\Theta + \dot{H}\Theta H + H\dot{H}\Theta + H\Theta\dot{H} + \Theta H\dot{H} + \Theta\dot{H}H).$$

This is equivalent to

$$H^3\Theta - H^2\Theta H - H\Theta H^2 + \Theta H^3$$
$$= (\dot{H}H\Theta + \dot{H}\Theta H + H\Theta\dot{H} + \Theta H\dot{H}) - 2(H\dot{H}\Theta + \Theta\dot{H}H).$$

The reader can now easily verify that

$$[H^2, [H, \Theta]] = H^3\Theta - H^2\Theta H - H\Theta H^2 + \Theta H^3$$

and that

$$[\dot{H}, [H, \Theta]] - 2[H, [\dot{H}, \Theta]] = (\dot{H}H\Theta + \dot{H}\Theta H + H\Theta\dot{H} + \Theta H\dot{H}) - 2(H\dot{H}\Theta + \Theta\dot{H}H).$$

Thus, we have proved that the third constraint equations are equivalent to the commutator equation

$$[H^2, [H, \Theta]] = [\dot{H}, [H, \Theta]] - 2[H, [\dot{H}, \Theta]].$$

This completes the proof of the Proposition. □

Each successive constraint involves the explicit formula for the higher derivatives of Θ coupled with the extra constraint that

$$\{\Theta^{(n)}\}^{\bullet} = \{\Theta^{(n+1)}\}.$$

We conjecture that each constraint can be expressed as a commutator equation in terms of Θ, H and the derivatives of H, analogous to the formulas that we have found for the first three constraints. This project will continue with a deeper algebraic study of the constraints and their physical meanings.

Funding: This work was supported by the Laboratory of Topology and Dynamics, Novosibirsk State University (contract No. 14.Y26.31.0025 with the Ministry of Education and Science of the Russian Federation).

Acknowledgments: It gives the author great pleasure to thank A.D. for many conversations and to remember the contributions and joint work with P.N. and C.K. that are integral parts of the present paper.

Conflicts of Interest: The author declares no conflict of interest.

Appendix A. Einstein's Equations and the Bianchi Identity

The purpose of this section is to show how the Bianchi identity (see below for its definition) appears in the context of non-commutative worlds. The Bianchi identity is a crucial mathematical ingredient in general relativity. We shall begin with a quick review of the mathematical structure of general relativity (see for example [36]) and then turn to the context of non-commutative worlds.

The basic tensor in Einstein's theory of general relativity is

$$G^{ab} = R^{ab} - \frac{1}{2}Rg^{ab},$$

where R^{ab} is the Ricci tensor and R the scalar curvature. The Ricci tensor and the scalar curvature are both obtained by contraction from the Riemann curvature tensor R^a_{bcd} with $R_{ab} = R^c_{abc}$, $R^{ab} = g^{ai}g^{bj}R_{ij}$, and $R = g^{ij}R_{ij}$. Because the Einstein tensor G^{ab} has vanishing divergence, it is a prime candidate to be proportional to the energy momentum tensor $T^{\mu\nu}$. The Einstein field equations are

$$R^{\mu\nu} - \frac{1}{2}Rg^{\mu\nu} = \kappa T^{\mu\nu}.$$

The reader may wish to recall that the Riemann tensor is obtained from the commutator of a covariant derivative ∇_k, associated with the Levi-Civita connection $\Gamma^i_{jk} = (\Gamma_k)^i_j$ (built from the space-time metric g_{ij}). One has

$$\lambda_{a:b} = \nabla_b \lambda_a = \partial_b \lambda_a - \Gamma^d_{ab} \lambda_d$$

or

$$\lambda_{:b} = \nabla_b \lambda = \partial_b \lambda - \Gamma_b \lambda$$

for a vector field λ. With

$$R_{ij} = [\nabla_i, \nabla_j] = \partial_j \Gamma_i - \partial_i \Gamma_j + [\Gamma_i, \Gamma_j],$$

one has

$$R^a_{bcd} = (R_{cd})^a_b.$$

Here, R_{cd} is *not* the Ricci tensor. It is the Riemann tensor with two internal indices hidden from sight.

One way to understand the mathematical source of the Einstein tensor, and the vanishing of its divergence, is to see it as a contraction of the Bianchi identity for the Riemann tensor. The Bianchi identity states

$$R^a_{bcd:e} + R^a_{bde:c} + R^a_{bec:d} = 0,$$

where the index after the colon indicates the covariant derivative. Note also that this can be written in the form

$$(R_{cd:e})^a_b + (R_{de:c})^a_b + (R_{ec:d})^a_b = 0.$$

The Bianchi identity is a consequence of local properties of the Levi-Civita connection and consequent symmetries of the Riemann tensor. One relevant symmetry of the Riemann tensor is the equation $R^a_{bcd} = -R^a_{bdc}$.

We will not give a classical derivation of the Bianchi identity here, but it is instructive to see how its contraction leads to the Einstein tensor. To this end, note that we can contract the Bianchi identity to

$$R^a_{bca:e} + R^a_{bae:c} + R^a_{bec:a} = 0,$$

which, in the light of the above definition of the Ricci tensor and the symmetries of the Riemann tensor, is the same as

$$R_{bc:e} - R_{be:c} + R^a_{bec:a} = 0.$$

Contract this tensor equation once more to obtain

$$R_{bc:b} - R_{bb:c} + R^a_{bbc:a} = 0,$$

and raise indices

$$R^b_{c:b} - R_{:c} + R^{ab}_{bc:a} = 0.$$

Further symmetry gives

$$R^{ab}_{bc:a} = R^{ba}_{cb:a} = R^a_{c:a} = R^b_{c:b}.$$

Hence, we have

$$2R^b_{c:b} - R_{:c} = 0,$$

which is equivalent to the equation

$$(R^b_c - \frac{1}{2} R \delta^b_c)_{:b} = G^b_{c:b} = 0.$$

From this, we conclude that $G^{bc}_{:b} = 0$. The Einstein tensor has appeared on the stage with vanishing divergence, courtesy of the Bianchi identity!

Bianchi Identity and Jacobi Identity. Now lets turn to the context of non-commutative worlds. We have infinitely many possible convariant derivatives, all of the form

$$F_{:a} = \nabla_a F = [F, N_a]$$

for some N_a elements in the non-commutative world. Choose any such covariant derivative. Then, as in the introduction to this paper, we have the curvature

$$R_{ij} = [N_i, N_j]$$

that represents the commutator of the covariant derivative with itself in the sense that $[\nabla_i, \nabla_j]F = [[N_i, N_j], F]$. Note that R_{ij} is not a Ricci tensor, but rather the indication of the external structure of the curvature without any particular choice of linear representation (as is given in the classical case as described above). We then have the Jacobi identity

$$[[N_a, N_b], N_c] + [[N_c, N_a], N_b] + [[N_b, N_c], N_a] = 0.$$

Writing the Jacobi identity in terms of curvature and covariant differention, we have

$$R_{ab:c} + R_{ca:b} + R_{bc:a}.$$

Thus, in a non-commutative world, every covariant derivative satisfies its own Bianchi identity. This gives an impetus to study general relativity in non-commutative worlds by looking for covariant derivatives that satisfy the symmetries of the Riemann tensor and link with a metric in an appropriate way. We have only begun this aspect of the investigation. The point of this section has been to show the intimate relationship between the Bianchi idenity and the Jacobi identity that is revealed in the context of non-commutative worlds.

References

1. Deakin, A.M.; Kauffman, L.H. Cosmological theories of the extra terms. *arXiv* **2018**. [CrossRef]
2. Deakin, A.M. Where does Schroedinger's equation really come from? In *Aspects II—Proceedings of ANPA 20*; Bowden, K.G., Ed.; ANPA: Cambridge, UK, 1999.
3. Deakin, A.M. Progress in constraints theory. In *Contexts—Proceedings of ANPA 31*; Ford, A.D., Ed.; ANPA: Cambridge, UK, 2011; pp. 164–201.
4. Kilmister, C. Letter to A. M. Deakin. Unpublished work, 2005.
5. Kauffman, L.H. Iterant algebra. *Entropy* **2017**, *19*, 347. [CrossRef]
6. Kauffman, L.H. Sign and Space. In *Religious Experience and Scientific Paradigms: Proceedings of the IASWR Conference 1982*; Institute of Advanced Study of World Religions: New York, NY, USA, 1985; pp. 118–164.
7. Kauffman, L.H. Self-reference and recursive forms. *J. Soc. Biol. Struct.* **1987**, *10*, 53–72. [CrossRef]
8. Kauffman, L.H. Special relativity and a calculus of distinctions. In Proceedings of the 9th Annual International Meeting of ANPA, Cambridge, UK, 23 September 1987; pp. 290–311.
9. Kauffman, L.H. Imaginary values in mathematical logic. In Proceedings of the Seventeenth International Conference on Multiple Valued Logic, Boston, MA, USA, 26–28 May 1987; IEEE Computer Society Press: Washington, DC, USA; pp. 282–289.
10. Kauffman, L.H. Knot Logic. In *Knots and Applications*; Kauffman, L., Ed.; World Scientific Publishing: Singapore, 1994; pp. 1–110.
11. Kauffman, L.H. Biologic. *Gen. Theor. Phys.* **2002**, *304*, 313–340.
12. Bell, J.L. *A Primer of Infinitesimal Analysis*; Cambridge University Press: Cambridge, UK, 1998.
13. Kauffman, L.H. Glafka-2004: Non-commutative worlds. *Int. J. Theor. Phys.* **2006**, *45*, 1443–1470. [CrossRef]
14. Kauffman, L.H. Non-commutative worlds. *New J. Phys.* **2004**, *6*, 173. [CrossRef]
15. Kauffman, L.H. Differential geometry in non-commutative worlds. In *Quantum Gravity—Mathematical Models and Experimental Bounds*; Fauser, B., Tolksdorf, J., Zeidler, E., Eds.; Birkhauser: Basel, Switzerland, 2007; pp. 61–75.
16. Dyson, F.J. Feynman's proof of the Maxwell Equations. *Am. J. Phys.* **1990**, *58*, 209–211. [CrossRef]
17. Tanimura, S. Relativistic generalization and extension to the non-Abelian gauge theory of Feynman's proof of the Maxwell equations. *Ann. Phys.* **1992**, *220*, 229–247.
18. Kauffman, L.H.; Noyes, H.P. Discrete Physics and the Derivation of Electromagnetism from the formalism of Quantum Mechanics. *Proc. R. Soc. Lond. A* **1996**, *452*, 81–95. [CrossRef]
19. Kauffman, L.H. *Knots and Physics*; World Scientific Publishing: Singapore, 1991.
20. Kauffman, L.H. Time imaginary value, paradox sign and space. *AIP Conf. Proc.* **2002**, *627*, 146.
21. Kauffman, L.H.; Noyes, H.P. Discrete Physics and the Dirac Equation. *Phys. Lett. A* **1996**, *218*, 139–146. [CrossRef]
22. Kauffman, L.H. Quantum electrodynamic birdtracks. *Twist. Newslett.* **1996**, *41*, 15–22.
23. Kauffman, L.H. Noncommutativity and discrete physics. *J. Phys. D* **1998**, *120*, 125–138. [CrossRef]
24. Kauffman, L.H. Space and time in discrete physics. *Int. J. Gen. Syst.* **1998**, *27*, 241–273. [CrossRef]

25. Kauffman, L.H. A non-commutative approach to discrete physics. In Proceedings of the Annual International Meeting of the Alternative Natural Philosophy Association, Cambridge, UK, 3–8 September 1998; pp. 215–238.
26. Kauffman, L.H. Non-commutative calculus and discrete physics. *arXiv* **2003**. [CrossRef]
27. Connes, A. *Non-Commutative Geometry*; Academic Press: Cambridge, MA, USA, 1990.
28. Dimakis, A.; Müller-Hoissen, F. Quantum mechanics on a lattice and q-deformations. *Phys. Lett.* **1992**, *295B*, 242. [CrossRef]
29. Forgy, E.A. Differential Geometry in Computational Electromagnetics. Ph.D. Thesis, University of Illinois Urbana-Champaign, Champaign, IL, USA, 2002.
30. Müller-Hoissen, F. Introduction to non-commutative geometry of commutative algebras and applications in physics. In Proceedings of the 2nd Mexican School on Gravitation and Mathematical Physics, Tlaxcala, Mexico, 1–7 December 1996; Science Network Publication: Kostanz, Germany, 1998.
31. Montesinos, M.; Perez-Lorenzana, A. Minimal coupling and Feynman's proof. *Int. J. Theor. Phys.*, **1999**, *38*, 901.
32. Hughes, R.J. On Feynman's proof of the Maxwell Equations. *Am. J. Phys.* **1992**, *60*, 301–306. [CrossRef]
33. Deakin, A.M. Constraints theory brief. In *Scientific Essays in Honor of H. Pierre Noyes on the Occasion of His 90th Birthday*; Amson, J.C., Kauffman, L.H., Eds.; World Scientific Publishing: Singapore, 2014; pp. 65–76.
34. Eddington, A.S. *The Mathematical Theory of Relativity*; Cambridge University Press: Cambridge, UK, 1965.
35. Dirac, P.A.M. *General Theory of Relativity*; Wiley & Sons: New York, NY, USA, 1975.
36. Foster, J.; Nightingale, J.D. *A Short Course in General Relativit*; Springer: Berlin, Germany, 1995.

Review

The Montevideo Interpretation of Quantum Mechanics: A Short Review

Rodolfo Gambini [1] **and Jorge Pullin** [2,*]

[1] Instituto de Física, Facultad de Ciencias, Iguá 4225, esq. Mataojo, Montevideo 11400, Uruguay; rgambini@fisica.edu.uy
[2] Department of Physics and Astronomy, Louisiana State University, Baton Rouge, LA 70803-4001, USA
* Correspondence: pullin@lsu.edu; Tel.: +1-225-578-0454

Received: 2 May 2018; Accepted: 28 May 2018; Published: 29 May 2018

Abstract: The Montevideo interpretation of quantum mechanics, which consists of supplementing environmental decoherence with fundamental limitations in measurement stemming from gravity, has been described in several publications. However, some of them appeared before the full picture provided by the interpretation was developed. As such, it can be difficult to get a good understanding via the published literature. Here, we summarize it in a self-contained brief presentation including all its principal elements.

Keywords: quantum mechanics; decoherence; interpretations

1. Introduction: The Measurement Problem

Although quantum mechanics is a well-defined theory in terms of providing unambiguous experimental predictions that can be tested, several physicists and philosophers of science find its presentation to be unsatisfactory. At the center of the controversy is the well-known measurement problem. In the quantum theory, states evolve unitarily, unless a measurement takes place. During a measurement, the state suffers a reduction that is not described by a unitary operator. In traditional formulations, this non-unitary evolution is postulated. Such an approach makes the theory complete from a calculational point of view. However, one is left with an odd formulation: a theory that claims our world is quantum in nature, yet its own definition requires referring to a classical world, as measurements are supposed to take place when the system under study interacts with a classical measurement device.

More recently, a more careful inspection of how the interaction with a measurement device takes place has led to a potential solution to the problem. In the decoherence program (for a review and references, see [1,2]), the interaction with a measurement device and, more generally, an environment with a large number of degrees of freedom, leads the quantum system to behave almost as if a reduction had taken place. Essentially, the large number of degrees of freedom of the measurement device and environment "smother" the quantum behavior of the system under study. The evolution of the combined system plus measurement device plus environment is unitary, and everything is ruled by quantum mechanics. However, if one concentrates on the wavefunction of the system under study only, tracing out the environmental degrees of freedom, the evolution appears to be non-unitary and very close to a reduction.

The decoherence program, suitably supplemented by an ontology like the many worlds one, has not convinced everyone (see for instance [3,4]) that it provides a complete solution to the measurement problem. Objections can be summarized in two main points:

1. Since the evolution of the system plus environment plus measuring device is unitary, it could happen that the quantum coherence of the system being studied could be recovered.

Model calculations show that such "revivals" could happen, but they would take a long time for most realistic measuring devices. However, it is therefore clear that the picture that emerges is slightly different from the traditional formulation where one can never dial back a reduction. A possible answer is that for most real experimental situations, one would have to wait longer than the age of the universe. Related to this is the point of when exactly does the measurement take place? Since all quantum states throughout the evolution are unitarily equivalent, what distinguishes the moment when the measurement takes place? Some have put this as: "in this picture nothing ever happens". A possible response is that after a certain amount of time, the state of the system is indistinguishable from the result of a reduction "for all practical purposes" (FAPP) [5]. However, from a conceptual point of view, the formulation of a theory should not rely on practical aspects. One could imagine that future scientists could perhaps find more accurate ways of measuring things and be able to distinguish what today is "FAPP" indistinguishable from a reduction.

A related point is that one can define global observables for the system plus measuring device plus environment [3,6]. The expectation value for one of these observables takes different values if a collapse takes place or not. That could allow in principle to distinguish the FAPP picture of decoherence from a real collapse. From the FAPP perspective, the answer is that these types of observables are very difficult to measure, since this requires measuring the many degrees of freedom of the environment. However, the mere possibility of measuring these observables is not consistent with a realistic description. This point has recently been highlighted by Frauchiger and Renner [7], who show that quantum mechanics is inconsistent with single world interpretations.

2. The "and/or" problem [8]: Even though the interaction with the environment creates a reduced density matrix for the system that has an approximate diagonal form, as all quantum states, the density matrix still represents a superposition of coexisting alternatives. Why is one to interpret it as exclusive alternatives with given probabilities? When is one to transition from an improper to a proper mixture, in d'Espagnat's terminology [3].

   The Montevideo interpretation [9] seeks to address these two criticisms. In the spirit of the decoherence program, it examines more finely what is happening in a measurement and how the theory is being formulated. It also brings into play the role of gravity in physics. It may be surprising that gravity has something to do with the formulation of quantum mechanics as one can imagine many systems where quantum effects are very important, but gravity seems to play no role. However, if one believes in the unity of physics, it should not be surprising that at some level, one needs to include all of physics to make certain situations work. More importantly, gravity brings to bear on physics important limitations on what can be done. Non-gravitational physics allows one to consider in principle arbitrarily large amounts of energy in a confined region, which is clearly not feasible physically if one includes gravity. This in particular places limitations on the accuracy with which we can measure any physical quantity [10,11]. Gravity also imposes limitations on our notions of space and time, which are absolute in non-gravitational physics. In particular, one has to construct measurements of space and time using real physical (and in this context, really quantum) objects, as no externally-defined space-time is pre-existent. This forces subtle changes in how theories are formulated. In particular, unitary theories do not appear to behave entirely unitarily since the notion of unitary evolution is defined with respect to a perfect classical time that cannot be approximated with arbitrary accuracy by a real (quantum) clock [12,13]. Notice that the role of gravity in this approach is different than in Penrose's [14]. Here, the emphasis is on limitations to clocks due to the intrinsically relational nature of time in gravity, whereas in Penrose's differences in time in different places is what is the basis of the mechanism.

These two new elements that the consideration of gravity brings to bear on physics will be key in addressing the two objections to decoherence that we outlined above. Since the evolution of systems is not perfectly unitary, it will not help to revive coherence in quantum systems to wait. Far from seeing coherence restored, it will be progressively further lost. The limitations on measurement will impose fundamental constraints on future physicists in developing means of distinguishing the quantum states produced by decoherence from those produced by a reduction. It will also make it impossible to measure global observables that may tell us if a reduction took place or not. Notice that this is not FAPP: the limitations are fundamental. It is the theories of physics that tell us that the states produced by decoherence are indistinguishable from those produced by a reduction. There is therefore a natural definition of when "something happens". A measurement takes place when the state produced by decoherence is indistinguishable from a reduction according to the laws of physics [15]. No invocation of an external observer is needed. Measurements (more generally events) will be plentiful and happening all the time around the universe as quantum systems interact with the environment irrespective of whether or not an experimenter or measuring device is present. The resulting quantum theory can therefore be formulated on purely quantum terms, without invoking a classical world. It also naturally leads to a new ontology consisting of quantum systems, states and events, all objectively defined, in terms of which to build the world. One could ask: Were systems, states and events not already present in the Copenhagen interpretation? Could we not have used them already to build the world? Not entirely, since the definition of event used there required the existence of a classical external world to begin with. It therefore cannot be logically used to base the world on.

In this short review, we would like to outline some results supporting the above point of view. In the next section, we discuss how to use real clocks to describe physical systems where no external time is available. We will show that the evolution of the states presents a fundamental loss of coherence. Notice that we are not modifying quantum mechanics, just pointing out that we cannot entirely access the underlying usual unitary theory when we describe it in terms of real clocks (and measuring rods for space if one is studying quantum field theories). In the following section, we discuss how fundamental limitations of measurement prevent us from distinguishing a state produced by a reduction and a state produced by decoherence. Obviously, given the complexities of the decoherence process, we cannot show in general that this is the case. We will present a modification of a model of decoherence presented by Zurek [16] to analyze this type of situation to exhibit the point we are making. The next section discusses some philosophical implications of having a realist interpretation of quantum mechanics like the one proposed. We end with a summary.

2. Quantum Mechanics without an External Time

When one considers a system without external time, like when one studies cosmology, or model systems like a set of particles with fixed angular and linear momentum assuming no knowledge of external clocks (see [17] for references), one finds that the Hamiltonian does not generate evolution, but becomes a constraint that can be written generically as $H = 0$. One is left with what is called a "frozen formalism" (see [18,19] and the references therein). The values of the canonical coordinates at a given time $q(t), p(t)$ are not observable, since one does not have access to t. Physical quantities have to have vanishing Poisson brackets with the constraint; they are what is known as "Dirac observables", and the canonical coordinates are not. The resulting picture is very different from usual physics, and it is difficult to extract physical predictions from it since the observables are all constants of the motion, as they have vanishing Poisson brackets with the Hamiltonian. People have proposed several possible solutions to deal with the situation, although no general consensus on a solution exists. We will not summarize all proposals here, in part because we will not need most of them and for reasons of space. We will focus on two proposals that, when combined, we claim provide a satisfactory solution to how to treat systems without external time when combined with each other. For other approaches, the review by Kuchař is very complete [19].

The first proposal we call "evolving Dirac observables". It has appeared in various guises over the years, but it has been emphasized by Rovelli [20]. The idea is to consider Dirac observables that depend on a parameter $O(t)$. These are true Dirac observables, they have vanishing Poisson brackets with the constraint, but their value is not well defined till one specifies the value of a parameter. Notice that t is just a parameter; it does not have to have any connection with "time". The definition requires that when the parameter takes the value of one of the canonical variables, the Dirac observable takes the value of another canonical variable, for example, $Q(t = q_1) = q_2$. This in part justifies why it is a Dirac observable. Neither q_1 nor q_2 can be observed since we do not have an external time, but the value q_2 takes when q_1 takes a given value is a relation that can be defined without referring to an external time, i.e., it is invariant information. As an example, let us consider the relativistic particle in one dimension. We parameterize it, including the energy as one of the canonical variables, p_0. One then has a constraint $\phi = p_0^2 - p^2 - m^2$. One can easily construct two independent Dirac observables: p and $X \equiv q - pq^0/\sqrt{p^2 + m^2}$ and verify that they have vanishing Poisson brackets with the constraint. An evolving constant of the motion could be,

$$Q\left(t, q^a, p_a\right) = X + \frac{p}{\sqrt{p^2 + m^2}} t, \tag{1}$$

and one would have that when the parameter takes the value q^0, the evolving constant $Q\left(t = q^0, q^a, p_a\right) = q$ takes the value of one of the canonical variables. Therefore, one now has an evolution for the system, the one in terms of the parameter t. However, problems arise when one tries to quantize things. There, variables like q_1 become quantum operators, but the parameter remains unquantized. How does one therefore make sense of $t = q_1$ at the quantum level when the left member is a classical quantity and the right a quantum operator (particularly when the quantum operator is not a Dirac observable and therefore not defined on the physical space of states of the theory)?

The second approach was proposed by Page and Wootters [21]. They advocate quantizing systems without time by promoting all canonical variables to quantum operators. Then, one chooses one as a "clock" and asks relational questions between the other canonical variables and the clock. Conditional probabilities are well defined quantum mechanically. Therefore, without invoking a classical external clock, one chooses a physical variable as a clock, and to study the evolution of probabilities, one asks relational questions: what is the expectation value of variable q_2 when variable q_1 (which we chose as clock) takes the value 3:30 p.m.? Again, because relational information does not require the use of external clocks, it has an invariant character, and one can ask physical questions about it. However, trouble arises when one actually tries to compute the conditional probabilities. Quantum probabilities require computing expectation values with quantum states. In these theories, since we argued that the Hamiltonian is a constraint $H = 0$, at a quantum level, one must have $\hat{H}|\Psi\rangle = 0$; only states that are annihilated by the constraint are permissible. However, such a space of states is not invariant under multiplication by one of the canonical, variables, i.e., $\hat{H}q_1|\Psi\rangle \neq 0$. Therefore, one cannot compute the expectation values required to compute the conditional probabilities. One can try to force a calculation pretending that one remains in the space, but then one gets incorrect results. Studies of model systems of a few particles have shown that one does not get the right results for the propagators, for example [19].

Our proposal [13] is to combine the two approaches we have just outlined: one computes conditional probabilities of evolving constants of the motion. Therefore, one chooses an evolving constant of the motion that will be the "clock", $T(t)$, and then, one chooses a variable one wishes to study $O(t)$ and computes,

$$P\left(O \in [O_0 - \Delta_1, O_0 + \Delta_1] | T \in [T_0 - \Delta_2, T_0 + \Delta_2]\right) = \lim_{\tau\to\infty} \frac{\int_{-\tau}^{\tau} dt \mathrm{Tr}\left(P_{O_0}^{\Delta_1}(t) P_{T_0}^{\Delta_2}(t) \rho P_{T_0}^{\Delta_2}(t)\right)}{\int_{-\tau}^{\tau} dt \mathrm{Tr}\left(P_{T_0}^{\Delta_2}(t)\rho\right)}, \tag{2}$$

where we are computing the conditional probability that the variable O takes a value within a range of width $2\Delta_1$ around the value O_0 when the clock variable takes a value within a range of width $2\Delta_2$ around the value T_0 (we are assuming the variables to have continuous spectra, hence the need to ask about ranges of values) on a quantum state described by the density matrix ρ. The quantity $P_{O_0}^{\Delta_1}$ is the projector on the eigenspace associated with the eigenvalue O_0 of the operator $\hat{O}$ and similarly for $P_{T_0}^{\Delta_2}$. Notice that the expression does not require assigning a value to the classical parameter t, since it is integrated over all possible values.

We have shown [13] using a model system of two free particles where we use one of them as the "clock" that this expression, provided one makes judicious assumptions about the clock, indeed reproduces to leading order the correct usual propagator, not having the problems of the Page and Wootters' proposal.

The above expression in terms of conditional probabilities may look unfamiliar. It is better to rewrite it in terms of an effective density matrix. Then, it looks exactly like the ordinary definition of probability in quantum mechanics,

$$P\left(O_0|T_0\right) = \frac{\mathrm{Tr}\left(P_{O_0}^{\Delta_1}\left(0\right)\rho_{\mathrm{eff}}\left(T_0\right)\right)}{\mathrm{Tr}\left(\rho_{\mathrm{eff}}\left(T_0\right)\right)}, \tag{3}$$

where on the left-hand side, we shortened the notation omitting mention of the intervals, but they are still there. The effective density matrix is defined as,

$$\rho_{\mathrm{eff}}\left(T\right) = \int_{-\infty}^{\infty} dt U_s\left(t\right)\rho_s U_s^{\dagger}(t)\mathcal{P}_t\left(T\right), \tag{4}$$

where we have assumed that the density matrix of the total system is a direct product of that of the subsystem we use as clock ρ_{cl} and that of the subsystem under study ρ_s, and a similar assumption holds for their evolution operators U. The probability,

$$\mathcal{P}_t\left(T\right) = \frac{\mathrm{Tr}\left(P_{T_0}^{\Delta_2}(0)U_{\mathrm{cl}}\left(t\right)\rho_{\mathrm{cl}}U_{\mathrm{cl}}^{\dagger}\left(t\right)\right)}{\int_{-\infty}^{\infty} dt \mathrm{Tr}\left(P_{T_0}^{\Delta_2}\left(t\right)\rho_{\mathrm{cl}}\right)}, \tag{5}$$

is an unobservable quantity since it represents the probability that the variable $\hat{T}$ take a given value when the unobservable parameter is t.

The introduction of the effective density matrix clearly illustrates what happens when one describes ordinary quantum mechanics in terms of a clock variable that is a real observable, not a classical parameter. Examining Equation (4), we see in the right-hand side the ordinary density matrix evolving unitarily as a function of the unobservable parameter t. If the probability $\mathcal{P}_t\left(T\right)$ were a Dirac delta, then the effective density matrix would also evolve unitarily. That would mean that the real clock variable is tracking the unobservable parameter t perfectly. However, no physical variable can do that, so there will always be a dispersion, and the probability $\mathcal{P}_t\left(T\right)$ will have non-vanishing support over a range of T. What this is telling us is that the effective density matrix for the system at a time T will correspond to a superposition of density matrices at different values of the unobservable parameter t. The resulting evolution is therefore non-unitary. We see clearly the origin of the non-unitarity: the real clock variable cannot keep track of the unitary evolution of quantum mechanics.

In fact, if we assume that the clock variable tracks the unobservable parameter almost perfectly by writing:

$$\mathcal{P}_t\left(T\right) = \delta(t-T) + b(T)\delta''(t-T) + \ldots, \tag{6}$$

(a term proportional to $\delta(t-T)'$ only adds an unobservable shift), one can show that the evolution implied by (4) is generated by a modified Schrödinger equation,

$$-i\hbar\frac{\partial\rho}{\partial T} = \left[\hat{H},\rho\right] + \sigma(T)\left[\hat{H},\left[\hat{H},\rho\right]\right] + \ldots, \tag{7}$$

where $\sigma(T) = db(T)/dT$ is the rate of spread of the probability $\mathcal{P}_t\left(T\right)$ and $\rho = \rho_{\mathrm{eff}}(T)$.

Therefore, we clearly see that when describing quantum mechanics in terms of a real clock variable associated with a quantum observable rather than with a classical parameter, the system loses unitarity, and it is progressively worse the longer one waits.

The existence of the effect we are discussing is not controversial. In fact, one can make it as large as one wishes simply choosing a bad clock. Bonifacio et al. [22–24] have reinterpreted certain experiments with Rabi oscillations as being described with an inaccurate clock, and indeed, experimentally, one sees the loss of coherence described above. More recently, it has been demonstrated with entangled photons, as well [25].

However, the question still remains: can this effect be made arbitrarily small by a choice of the clock variable? If one takes into account gravity, the answer is negative. Using non-gravitational quantum physics, Salecker and Wigner [10,11] examined the question of how accurate a clock can be. The answer is that the uncertainty in the measurement of time is proportional to the square root of the length of time one desires to measure and inversely proportional to the square root of the mass of the clock. Therefore, to make a clock more accurate, one needs to make it more massive. However, if one takes gravity into account, there clearly is a limitation as to how massive a clock can be: at some point, it turns into a black hole. Several phenomenological models of this were proposed by various authors, and they all agree that the ultimate accuracy of a clock goes as some fractional power of the time to be measured times a fractional power of Planck's time [26–30]. Different arguments lead to slightly different powers, but the result is always that the longer one wishes to measure time, the more inaccurate the clocks become. For instance, in the phenomenological model of Ng and Van Dam [26–30], one has that $\delta T \sim T^{1/3} T_{\text{Planck}}^{2/3}$. Substituting that in the modified Schrödinger equation, its solution can be found in closed form, in an energy eigenbasis,

$$\rho(T)_{nm} = \rho_{nm}(0) \exp\left(-i\omega_{nm} T\right) \exp\left(-\omega_{nm}^2 T_{\text{Planck}}^{4/3} T^{2/3}\right), \tag{8}$$

where ω_{nm} is the Bohr frequency between the two energy eigenstates n and m. We see that the off-diagonal terms of the density matrix die off exponentially. Pure states evolve into mixed states.

3. Completing Decoherence: The Montevideo Interpretation

3.1. Decoherence with Clocks Based on Physical Variables

In this section, we would like to analyze how the use of a physical clock in the description of quantum mechanics we introduced in the last section, combined with other limitations in measurement, will help address the objections to environmental decoherence as a solution to the measurement problem. We start by illustrating the idea of decoherence (and the objections) using a well-known model of environmental decoherence due to Zurek [16], possibly one of the simplest models one can consider that still captures the complexities involved.

3.1.1. Zurek's Model

It consists of a spin one half system representing the microscopic system plus the measuring device, with a two-dimensional Hilbert space $\{|+\rangle, |-\rangle\}$. It interacts with an "environment" given by a bath of many similar two-state "atoms", each with a two-dimensional Hilbert space $\{|+\rangle_k, |-\rangle_k\}$. If there is no interaction with the environment, the two spin states may be taken to have the same energy; we choose it to be zero, and all the atoms also are chosen with zero energy. The interaction Hamiltonian is given by:

$$H_{\text{int}} = \hbar \sum_k \left(g_k \sigma_z \otimes \sigma_z^k \otimes_{j \neq k} I_j \right). \tag{9}$$

σ_z is a Pauli matrix acting on the state of the system. It has eigenvalues $+1$ for the spin eigenvector $|+\rangle$ and -1 for $|-\rangle$. The operators σ_z^k are similar, but acting on the state of the k-th atom. I_j denotes the identity matrix acting on atom j, and g_k is the coupling constant. It has dimensions of frequency and characterizes the coupling energy of one of the spins k with the system. The model can be thought physically as providing a representation of a photon propagating in a polarization analyzer.

Through the interaction, the initial state, which we can take as,

$$|\Psi(0)\rangle = (a|+\rangle + b|-\rangle) \prod_{k=1}^{N} \otimes [\alpha_k|+\rangle_k + \beta_k|-\rangle_k], \tag{10}$$

with a, b, α_k and β_k complex constants, evolved using the Schrödinger equation, becomes,

$$\begin{aligned} |\Psi(t)\rangle &= a|+\rangle \prod_{k=1}^{N} \otimes [\alpha_k \exp(ig_k t)|+\rangle_k + \beta_k \exp(-ig_k t)|-\rangle_k] \\ &\quad + b|-\rangle \prod_{k=1}^{N} \otimes [\alpha_k \exp(-ig_k t)|+\rangle_k + \beta_k \exp(ig_k t)|-\rangle_k]. \end{aligned} \tag{11}$$

From it, one can construct a density matrix for the system plus environment, and tracing out the environmental degrees of freedom, one gets a reduced density matrix for the system,

$$\rho_c(t) = |a|^2|+\rangle\langle+| + |b|^2|-\rangle\langle-| + z(t)ab^*|+\rangle\langle-| + z^*(t)a^*b|-\rangle\langle+|, \tag{12}$$

where:

$$z(t) = \prod_{k=1}^{N} \left[\cos(2g_k t) + i\left(|\alpha_k|^2 - |\beta_k|^2\right)\sin(2g_k t)\right]. \tag{13}$$

The complex valued function of time $z(t)$ determines the values of the off-diagonal elements. If it vanishes, the reduced density matrix could be considered a "proper mixture" representing several outcomes with their corresponding probabilities.

We claim that with the modified evolution we discussed in the previous section, the usual objections to decoherence do not apply. Recall which are the usual objections:

1. The quantum coherence is still there. Although a quantum system interacting with an environment with many degrees of freedom will very likely give the appearance that the initial quantum coherence of the system is lost (the density matrix of the measurement device is almost diagonal), the information about the original superposition could be recovered for instance carrying out a measurement that includes the environment. The fact that such measurements are hard to carry out in practice does not prevent the issue from existing as a conceptual problem.
2. The "and/or problem": Since the density matrix has been obtained by tracing over the environment, it represents an improper, not proper, mixture: looking at Equation (12), there is no way to select (even in some conceptual sense) one of the components of the density matrix versus the others.

Let us discuss now the problem of revivals. In the model, the function $z(t)$ does not die off asymptotically, but is multi-periodic; after a very long time, the off-diagonal terms become large. Whatever definiteness of the values of the preferred quantity we had won by the end of the measurement interaction turns out in the very long run to have been but a temporary victory. This is called the problem of revivals (or "recurrence of coherence", or "recoherence"). This illustrates that the quantum coherence persists, it was just transferred to the environment and could be measured using global observables.

3.1.2. A More Realistic Model and Real Clocks

To analyze the effects of limitations of measurement and the use of real clocks in detail, we will need to consider a more realistic model of spinning particles [31]; the previous model is too simple to capture the effect of the use of real clocks. Although this model is "almost realistic", it has the property that the system, environment and measurement apparatus are all under control, as one would need to measure a global observable, for instance. It consists of a spin S in a cavity with a magnetic field pointing in the z direction. A stream of N "environmental" spins flows sideways into the cavity and eventually exits it, and the interactions last a finite time determined by the time spent in the cavity. The flow of particles that represents the environment is sufficiently diluted such that we can ignore interactions among themselves.

The interaction Hamiltonian for the k-th spin of the environment is,

$$\hat{H}_k = \hat{H}_k^B + \hat{H}_k^{\text{int}}, \tag{14}$$

with,

$$\hat{H}_k^B = \gamma_1 B\hat{S}_z \otimes \hat{I}_k + \gamma_2 B\hat{I} \otimes S_z^k, \tag{15}$$

and:

$$\hat{H}_k^{\text{int}} = f_k\left(\hat{S}_x\hat{S}_x^k + \hat{S}_y\hat{S}_y^k + \hat{S}_z\hat{S}_z^k\right), \tag{16}$$

where f_k are the coupling constants between the spin and each of the particles of the environment, γ_1 and γ_2 are the magnetic moments of the central and environment spins, respectively, and the $\hat{S}$ are spin operators.

For the complete system, one can define an observable considered by d'Espagnat [3]. It has the property that its expectation value is different depending on if the state has suffered a quantum collapse or not. It definition is,

$$\hat{M} \equiv \hat{S}_x \otimes \prod_k^N \hat{S}_x^k. \tag{17}$$

One has that $\langle \hat{M} \rangle_{\text{collapse}} = 0$, whereas,

$$\langle \psi | M | \psi \rangle = ab^* \prod_k^N \left[\alpha_k \beta_k^* + \alpha_k^* \beta_k\right] e^{-2i\Omega_k \tau} + a^* b \prod_k^N \left[\alpha_k \beta_k^* + \alpha_k^* \beta_k\right] e^{2i\Omega_k \tau} \neq 0, \tag{18}$$

with $\Omega_k \equiv \sqrt{4f_k^2 + B^2(\gamma_1 - \gamma_2)^2}$ and τ is the time of flight of the environmental spins through the chamber. One can therefore determine experimentally if a collapse or not took place measuring this observable.

However, if one considers the corrections to the evolution resulting from the use of physical variables as clocks as we discussed in the previous section, one has that,

$$\begin{aligned}\langle \hat{M} \rangle &= ab^* e^{-i2N\Omega T} e^{-4NB^2(\gamma_1-\gamma_2)^2\theta} \textstyle\prod_k^N \left[\alpha_k \beta_k^* e^{-16B^2\gamma_1\gamma_2\theta} + \alpha_k^* \beta_k\right] \\ &\quad + ba^* e^{i2N\Omega T} e^{-4NB^2(\gamma_1-\gamma_2)^2\theta} \textstyle\prod_k^N \left[\alpha_k \beta_k^* + \alpha_k^* \beta_k e^{-16B^2\gamma_1\gamma_2\theta}\right],\end{aligned} \tag{19}$$

where $\Omega \equiv B(\gamma_1 - \gamma_2)$, $\theta \equiv \frac{3}{2} T_{\text{P}}^{4/3} \tau^{2/3}$, τ is the time of flight of the environment spins within the chamber and T is the length of the experiment.

There exists a series of conditions for the experiment to be feasible that imply certain inequalities,

$$(a) \qquad 1 < f\tau = \frac{\mu\gamma_1\gamma_2}{\hbar}\frac{\tau}{d^3}, \tag{20}$$

$$(b) \qquad \Delta x \sim \sqrt{\frac{\hbar T}{m}}, \tag{21}$$

$$(c) \qquad f \ll |B(\gamma_1 - \gamma_2)|, \tag{22}$$

$$(d) \qquad \langle \hat{M} \rangle \sim \exp\left(-6NB^2(\gamma_1 - \gamma_2)^2 T_{\text{Planck}}^{4/3} \tau^{2/3}\right), \tag{23}$$

with f the interaction energy between spins, which was assumed constant through the cell, μ the permeability of the vacuum, d the impact parameter of the spins of the environment, m their mass and Δx the spatial extent of the environment particles.

Condition (*a*) makes the coupling of the spins strong enough for decoherence to occur; (*b*) is to prevent the particles of the environment from dispersing too much and therefore making us unable to find them within the detectors at the end of the experiment; (*c*) is the condition for decoherence to be in the z basis, as was mentioned; (*d*) is an estimation of the the expectation value of the observable when the effect of the real clock is taken into account. For details of the derivation of these conditions, see our previous paper [32].

Therefore, the expectation value is exponentially damped, and it becomes more and more difficult to distinguish it from the vanishing value one has in a collapse situation. A similar analysis allows one to show that revivals are also prevented by the modified evolution. When the multi-periodic functions in the coherences tend to take again the original value after a Poincaré time of recurrence, the exponential decay for sufficiently large systems completely hides the revival under the noise amplitude.

Thus, the difficulties found in testing macroscopic superpositions in a measurement process are enhanced by the corrections resulting from the use of physical clocks.

3.2. Why the Solution Is Not FAPP

Although temporal decoherence involves exponentials and the troublesome terms of decoherence become exponentially small, how does this observation help to solve the problem of outcomes? In what follows, we will provide a criterion for the occurrence of events based on the notion of undecidability.

When one takes into account the way that time enters in generally covariant systems including the quantum fluctuations of the clock, the evolution of the total system (system plus measurement apparatus plus environment) becomes indistinguishable from the collapse. This is also true for revivals and the observation of the coherences of the reduced density matrix of the system plus the measuring device. We call such a situation "undecidability". We are going to show that undecidability is not only for all practical purposes (FAPP), but fundamental.

From the previous discussion, one can gather that as one considers environments with a larger number of degrees of freedom and as longer time measurements are considered, distinguishing between collapse and unitary evolution becomes harder. However, is this enough to be a fundamental claim?

Starting from (19) and using the approximations (20)–(23), one can show [32] that,

$$\langle \hat{M} \rangle \sim \exp\left(-6NB^2(\gamma_1-\gamma_2)^2 T_{\text{Planck}}^{4/3}\tau^{2/3}\right) \equiv e^{-K}. \tag{24}$$

with

$$K \gg \frac{N^5 T_{\text{Planck}}^{4/3}\hbar^{20/3}}{m^4(\gamma_1\gamma_2)^{8/3}\mu^{8/3}}. \tag{25}$$

Is it possible to build a very large ensemble allowing one to distinguish this value from zero?

Brukner and Kofler [33] have recently proposed that from a very general quantum mechanical analysis together with bounds from special and general relativity, there is a fundamental uncertainty in the measurements of angles even if one uses a measuring device of the size of the observable Universe.

$$\Delta\theta \gtrsim \frac{l_P}{R}, \tag{26}$$

where $l_P \equiv \sqrt{\hbar G/c^3} \approx 10^{-35} m$. If we take the radius of the observable universe as a characteristic length, $R \approx 10^{27} m$, we reach a fundamental bound on the measurement of the angle,

$$\Delta\theta \geq 10^{-62}. \tag{27}$$

To distinguish $\langle M \rangle$ from zero, one needs to take into account that the observable will have an error that depends on $\Delta\theta$ (since for instance $\hat{S}_x$ and $\hat{S}_y$ will get mixed). If the error is larger than $\langle M \rangle$, there is no way of distinguishing collapse from a unitary evolution *for fundamental, not practical reasons*. Therefore, the solution is not FAPP.

The expectation value of the observable is [32],

$$\langle \hat{M}^{\Delta\theta} \rangle \gtrsim e^{-K} \pm (\Delta\theta)^{2N} + \langle E(\Delta\theta) \rangle. \tag{28}$$

with:

$$K \gg \frac{N^5 T_{\text{Planck}}^{4/3} \hbar^{20/3}}{m^4 (\gamma_1\gamma_2)^{8/3} \mu^{8/3}}. \tag{29}$$

Therefore, for,

$$N > \left(\frac{2 \ln\left(\frac{R}{\ell_P}\right) \left(m \left(\gamma_1\gamma_2\right)^4\right)^{2/3} \mu^{8/3}}{T^{4/3}\hbar^{20/3}} \right)^{1/4} m \left(\gamma_1\gamma_2\right)^2 \sim 10^7, \tag{30}$$

it becomes undecidable whether collapse has occurred or not. That means that no measurement of any quantity, even in principle, can ascertain whether the evolution equation failed to hold. Notice that the above discussion was restricted to a given experiment. Our present knowledge of quantum gravity and the complexities of the decoherence process in general does not allow us to prove undecidability for an arbitrary experimental setup. Even models slightly more elaborate than the one presented here can be quite challenging to analyze. A different model, involving the interaction of a spin with bosons, has also been analyzed with similar results [34].

This model exhibits the difficulties of trying to obtain generic results concerning decoherence. Notice that expression (30) depends on the magnetic moments of the spins $\gamma_{1,2}$. If they were very large, decoherence would not take place. One would be in the presence of a macroscopic system exhibiting quantum behavior. One does not expect such systems to exist, at least in the terms described in the model, but the model does not rule them out.

3.3. The Problem of Outcomes, Also Known as the Issue of Macro-Objectification

The problem of macro-objectification of properties may be described according to Ghirardi as follows: how, when and under what conditions do definite macroscopic properties emerge (in accordance with our daily experience) for systems that, when all is said and done, we have no good reasons for thinking they are fundamentally different from the micro-systems from which they are composed?

We think that undecidability provides an answer to this problem. We will claim that events occur when a system suffers an interaction with an environment such that one cannot distinguish by any physical means if a unitary evolution or a reduction of the total system, including the environment, took place. This provides a criterion for the production of events, as we had anticipated. In addition, we postulate (we call this the ontological postulate in [15]) that when an event occurs, the system chooses randomly (constrained by their respective probability values) one of the various possible outcomes. Having an objective criterion for the production of events based on undecidability answers the objections raised by [7] since the observer and the "super observer" now have consistent descriptions.

Philosopher Jeremy Butterfield, who has written an assessment of the Montevideo interpretation [35], has observed that up to now, we have only provided precise examples of undecidability for spinning particles. In that sense, he considers that the fundamental loss of coherence due to the use of quantum clocks and to the quantum gravitational effects should be used in the context of a many worlds interpretation because it helps to answer some of the long-held obstructions to the combination of the decoherence program with the many worlds approach.

After a detailed analysis, we do not believe that conclusion is inescapable. Let us assume the worse case scenario: that there are no further quantum gravitational limitations for the measurements of other variables as the ones obtained for the spin by Kofler and Bruckner (even though there have been many proposals to alter uncertainty relations; see the references in [36]). However, given the fact that the distinction between a unitary evolution that includes quantum time measurements or a quantum reduction would require an exponentially growing number of individual measurements in order to have the required statistics for distinguishing a non-vanishing exponentially small mean value from zero, limitations referring to the existence of a finite number of physical resources in a finite observable Universe would be enough to ensure undecidability. Obviously, further investigations are needed, but in a sense, this is the fate of all studies involving decoherence; it is just not possible to develop general proofs given the complexities involved.

4. Some Philosophical Implications

If the fundamental nature of the world is quantum mechanical and we adopt an interpretation that provides an objective criterion for the occurrence of events, we are led to an ontology of objects and events. The interpretation here considered makes reference to primitive concepts like systems, states, events and the properties that characterize them. Although these concepts are not new and are usually considered in quantum mechanics, one can assign them a unambiguous meaning only if one has an interpretation of the theory. For example, events could not be used as the basis of a realistic ontology without a general criterion for the production of events that is independent of measurements and observers.

On the other hand, the concepts of state and system only acquire ontological value when the events also have acquired it, since they are both defined through the production of events. Based on this ontology, objects and events can be considered the building blocks of reality. Objects will be represented in the quantum formalism by systems in certain states and are characterized by their dispositions to produce events under certain conditions. In the new interpretation, events are the actual entities. Concrete reality accessible to our senses is constituted by events localized in space-time. As Whitehead [37] recognized: "the event is the ultimate unit of natural occurrence". Events come with associated properties. Events and properties in quantum theory are represented by mathematical entities called projectors. Quantum mechanics provides probabilities for the occurrence of events and their properties. When an event happens, like in the case of the dot on the photographic plate in the double-slit experiment, typically many properties are actualized. For instance, the dot may be darker on one side than the other, or may have one of many possible shapes.

Take for instance the hydrogen atom. It is a quantum system composed by a proton and an electron. A particular hydrogen atom is a system in a given state; it is an example of what we call an object, and it has a precise disposition to produce events. Russell in The Analysis of Matter [38], asserts that "the enduring thing or object of common sense and the old physics must be interpreted as a world-line, a causally related sequence of events, and that it is events and not substances that we perceive". He thus distinguishes events as basic particulars from objects as derived, constructed particulars. We disagree with this point of view because it ignores the role of the physical states. He adds: "Bits of matter are not among the bricks out of which the world is built. The bricks are events and bits of matter are portions of the structure to which we find it convenient to give separate attention". This is not the picture provided by quantum mechanics. An independent notion of object is required: one can even have event-less objects in quantum mechanics. For instance, when not measured, the hydrogen atom is an object according to the definition above even though it is not producing an event. The resulting ontology is such that objects and events are independent concepts; they are not derived one from the other.

This is only a sketch of philosophical issues raised by the new interpretation. We have a more complete discussion in [39].

5. Summary

We have presented an easy to follow guide to the Montevideo interpretation. Readers interested in an axiomatic formulation should consult our previous paper [15]. All the bibliography can be found in [9].

To summarize, the use of real physical variables to measure time implies a modification to how one writes the equations of quantum mechanics. The resulting picture has a fundamental mechanism for loss of coherence. When environmental decoherence is supplemented with this mechanism and taking into account fundamental limitations in measurement, a picture emerges where there is an objective, observer independent notion for when an event takes place. The resulting interpretation of quantum mechanics, which we call the Montevideo interpretation, is formulated entirely in terms of quantum concepts, without the need to invoke a classical world. We have been able to complete this picture for a simple realistic model of decoherence involving spins. Studies of more elaborate models are needed to further corroborate the picture.

Author Contributions: Both authors contributed equally to the work.

Acknowledgments: This work was supported in part by Grant NSF-PHY-1305000, CCT-LSU, Pedeciba and the Horace Hearne Jr. Institute for Theoretical Physics at LSU.

Conflicts of Interest: The authors declare no conflict of interest.

References

1. Schlosshauer, M.A. *Decoherence and the Quantum-to-Classical Transition*; Springer: Berlin/Heidelberg, Germany, 2007.
2. Joos, E.; Zeh, H.D.; Kiefer, C.; Giulini, D.; Kupsch, J.; Stamatescu, I.O. *Decoherence and the Appearance of a Classical World in Quantum Theory*, 2nd ed.; Springer: Berlin, Germany, 2003. Availabel online: http://www.decoherence.de (accessed on 28 May 2018).
3. D'Espagnat, B. *Veiled Reality*; Addison Wesley: New York, NY, USA, 1995.
4. Girardi, G. *Sneaking a Look at God's Cards, Revised Edition: Unraveling the Mysteries of Quantum Mechanics*; Princeton University Press: Princeton, NJ, USA, 2007.
5. Wallace, D. Decoherence and ontology, or how I learned to stop worrying and love FAPP. In *Many Worlds?: Everett, Quantum Theory, and Reality*; Saunders, S., Barrett, J., Kent, A., Wallace, D., Eds.; Oxford University Press: Oxford, UK, 2010.
6. D'Espagnat, B. Towards a separable "empirical reality"? *Found. Phys.* **1990**, *20*, 1147–1172. [CrossRef]
7. Frauchiger, D.; Renner, R. Single-world interpretations of quantum theory cannot be self consistent. *arXiv* **2016**, arXiv:1604.07422.
8. Bell, J.S. Against measurement. In *Sixty Two Years of Uncertainty*; Miller, A.I., Ed.; Plenum: New York, NY, USA, 1990.
9. The Complete Bibliography on the Montevideo Interpretation. Available online: http://www.montevideointerpretation.org (accessed on 28 May 2018).
10. Wigner, E.P. Relativistic invariance and quantum phenomena. *Rev. Mod. Phys.* **1957**, *29*, 255. [CrossRef]
11. Salecker, H.; Wigner, E.P. Quantum limitations of the measurement of space-time distances. *Phys. Rev.* **1958**, *109*, 571, doi:10.1103/PhysRev.109.571. [CrossRef]
12. Gambini, R.; Porto, R.A.; Pullin, J. Fundamental decoherence from quantum gravity: A pedagogical review. *Gen. Rel. Grav.* **2007**, *39*, 1143–1156. [CrossRef]
13. Gambini, R.; Porto, R.A.; Pullin, J.; Torterolo, S. Conditional probabilities with Dirac observables and the problem of time in quantum gravity. *Phys. Rev. D* **2009**, *79*, 041501, doi:10.1103/PhysRevD.79.041501. [CrossRef]
14. Penrose, R. The road to reality: A Complete Guide to the Laws of the Universe. *Math. Intell.* **2006**, *28*, 59–61.

15. Gambini, R.; García-Pintos, L.P.; Pullin, J. An axiomatic formulation of the Montevideo interpretation of quantum mechanics. *Stud. Hist. Philos. Mod. Phys.* **2011**, *42*, 256–263. [CrossRef]
16. Zurek, W.H. Environment-induced superselection rules. *Phys. Rev. D* **1982**, *26*, 1862, doi:10.1103/PhysRevD.26.1862. [CrossRef]
17. Anderson, E. Triangleland: I. Classical dynamics with exchange of relative angular momentum. *Class. Quant. Grav.* **2009**, *26*, 135020. [CrossRef]
18. Anderson, A. Thawing the frozen formalism: The Difference between observables and what we observe. In *Directions in General Relativity*; Hu, B.L., Jacobson, T.A., Eds.; Cambridge University Press: Cambridge, UK, 2005; Volume 2, pp. 13–27.
19. Kuchař, K.V. Time and interpretations of quantum gravity. *Int. J. Mod. Phys. D* **2011**, *20*, 3–86. [CrossRef]
20. Rovelli, C. Time in quantum gravity: An hypothesis. *Phys. Rev. D* **1991**, *43*, 442, doi:10.1103/PhysRevD.43.442. [CrossRef]
21. Page, D.N.; Wootters, W.K. Evolution without evolution: Dynamics described by stationary observables. *Phys. Rev. D* **1983**, *27*, 2885, doi:10.1103/PhysRevD.27.2885. [CrossRef]
22. Meekhof, D.M.; Monroe, C.; King, B.E.; Itano, W.M.; Wineland, D.J. Generation of Nonclassical Motional States of a Trapped Atom. *Phys. Rev. Lett.* **1996**, *76*, 1796, doi:10.1103/PhysRevLett.76.1796. [CrossRef] [PubMed]
23. Brune, M.; Schmidt-Kaler, F.; Maali, A.; Dreyer, J.; Hagley, E.; Raimond, J.M.; Haroche, S. Quantum Rabi Oscillation: A Direct Test of Field Quantization in a Cavity. *Phys. Rev. Lett.* **1996**, *76*, 1800, doi:10.1103/PhysRevLett.76.1800. [CrossRef] [PubMed]
24. Bonifacio, R.; Olivares, S.; Tombesi, P.; Vitali, D. A model independent approach to non dissipative decoherence. *Phys. Rev. A* **2000**, *61*, 053802, doi:10.1103/PhysRevA.61.053802. [CrossRef]
25. Moreva, E.; Brida, G.; Gramegna, M.; Giovannetti, V.; Maccone, L.; Genovese, M. Time from quantum entanglement: An experimental illustration. *Phys. Rev. A* **2014**, *89*, 052122, doi:10.1103/PhysRevA.89.052122. [CrossRef]
26. Károlhyázy, F.; Frenkel, A.; Lukács, B. *Quantum Concepts in Space and Time*; Penrose, R., Isham, C., Eds.; Oxford University Press: Oxford, UK, 1986.
27. Amelino-Camelia, G. Limits on the measurability of space-time distances in (the semiclassical approximation of) quantum gravity. *Mod. Phys. Lett. A* **1994**, *9*, 3415–3422. [CrossRef]
28. Ng, Y.J.; van Dam, H. Limitation to Quantum Measurements of Space-Time Distances. *Ann. N. Y. Acad. Sci.* **1995**, *755*, 579, doi:10.1111/j.1749-6632.1995.tb38998.x. [CrossRef]
29. Ng, Y.J.; van Dam, H. Limit to space-time measurement. *Mod. Phys. Lett. A* **1994**, *9*, 335.
30. Lloyd, S.; Ng, Y.J. Black hole computers. *Sci. Am.* **2004**, *291*, 52–61. [CrossRef] [PubMed]
31. Gambini, R.; Pintos, L.P.G.; Pullin, J. Undecidability and the problem of outcomes in quantum measurements. *Found. Phys.* **2010**, *40*, 93, doi:10.1007/s10701-009-9376-8. [CrossRef]
32. Gambini, R.; García-Pintos, L.P.; Pullin, J. Undecidability as solution to the problem of measurement: Fundamental criterion for the production of events. *Int. J. Mod. Phys. D* **2011**, *20*, 909–918. [CrossRef]
33. Brukner, C.; Kofler, J. Are there fundamental limits for observing quantum phenomena from within quantum theory? *arXiv* **2010**, arXiv:1009.2654.
34. García-Pintos, L.P. Una Interpretación de la Mecánica Cuántica Basada en Considerar un Tiempo Físico. Master's Thesis, Universidad de la República, Montevideo, Uruguay, 2011. (In Spanish)
35. Butterfield, J. Assessing the Montevideo interpretation of quantum mechanics. *Stud. Hist. Philos. Mod. Phys.* **2015**, *52*, 75–85. [CrossRef]
36. Pikovski, I.; Vanner, M.R.; Aspelmeyer, M.; Kim, M.S.; Brukner, Č. Probing Planck-scale physics with quantum optics. *Nat. Phys.* **2012**, *8*, 393, doi:10.1038/nphys2262. [CrossRef]
37. Whitehead, A.N. *Science and the Modern World*; Free Press: New York, NY, USA, 1997.
38. Russell, B. *The Analysis of Matter*; Spokesman Books: Nottingham, UK, 2007.
39. Gambini, R.; Pullin, J. *A hospitable Universe: Addressing Ethical and Spiritual Concerns in Light of Recent Scientific Discoveries*; Imprint Academic: Upton Pyne, UK, 2018.

MDPI

Article

A Quantum Ruler for Magnetic Deflectometry

Lukas Mairhofer [1], Sandra Eibenberger [2], Armin Shayeghi [1] and Markus Arndt [1,*]

[1] Faculty of Physics, University of Vienna, Boltzmanngasse 5, A-1090 Wien, Austria; lukas.mairhofer@univie.ac.at (L.M.); armin.shayeghi@univie.ac.at (A.S.)

[2] Fritz-Haber-Institut der Max-Planck-Gesellschaft, Faradayweg 4-6, D-14195 Berlin, Germany; eibenberger@fhi-berlin.mpg.de

* Correspondence: markus.arndt@univie.ac.at; Tel.: +43-1-4277-51210

Received: 15 June 2018; Accepted: 6 July 2018; Published: 9 July 2018

Abstract: Matter-wave near-field interference can imprint a nano-scale fringe pattern onto a molecular beam, which allows observing its shifts in the presence of even very small external forces. Here we demonstrate quantum interference of the pre-vitamin 7-dehydrocholesterol and discuss the conceptual challenges of magnetic deflectometry in a near-field interferometer as a tool to explore photochemical processes within molecules whose center of mass is quantum delocalized.

Keywords: molecule interference; matter-waves; metrology; magnetic deflectometry; photochemistry

---

1. Quantum Interference of Organic Molecules

In quantum mechanics we attribute both wave and particle properties to the basic entities of the theory, and following Louis de Broglie [1] we associate an oscillatory phenomenon of wavelength $\lambda_{dB} = h/mv$ to the center-of-mass motion of any particle of mass m and velocity v, even if it has a rich internal structure and exhibits internal excitations. This can be proven in a very intuitive way using matter-wave diffraction [2,3], in analogy to Young's famous double slit experiment. In realizations with complex molecules the de Broglie wavelength is typically of the order of a few picometers while the molecular wave function can be delocalized by more than 10^5 λ_{dB}, and hundreds or thousand times the size of the particle. At the same time a single, complex molecule can be composed of hundreds or even a thousand atoms, and each atom itself is composed of dozens of nuclei and electrons. This physical picture is complemented by acknowledging the presence of hundreds of vibrational modes and excited rotational states. At molecular temperatures around 500 K most of these modes are excited, leading to molecular rotation frequencies around $\Omega \simeq 2\pi \times 10^9$ rad/s and structural or conformational changes on the sub-nanosecond time scale. The molecules can thus be prepared in superpositions of position and momentum even though we can assign classical attributes such as internal temperatures, polarizabilities, dipole moments, magnetic susceptibilities and so forth to them. This philosophical aspect of macromolecular interferometry has very practical applications in metrology for the measurement of electronic, optical, and even magnetic molecular properties. Earlier work has shown that such parameters can be readily measured both in classical beam experiments [4,5] and in Talbot-Lau deflectometry [6–8]. Here, we propose that the de Broglie interference can also be a promising tool for photochemistry. The optically induced change of molecular geometry is often well-understood in solution, but little explored in the gas phase. We are interested in how such atomic rearrangements influence magnetic properties and study this for the example of 7-dehydrocholesterol (7-DHC) where isomerization causes a ring opening and a change of the particle's magnetic susceptibility. We describe successful matter-wave interference with 7-DHC and a thought experiment that exploits the fact that modifications of the magnetic susceptibility will be seen as a relative shift of the de Broglie interference fringe pattern in an external magnetic field.

2. The Quantum Wave Nature of 7-Dehydrocholesterol

Matter-wave physics with complex molecules is most conveniently realized using a near-field interferometer [9], for instance the Kapitza–Dirac–Talbot–Lau setup (KDTLI, see Figure 1), which is the basis here for our discussion [10]. This device is appealing for high-mass quantum experiments [10–12] since it is rugged, compact, and compatible with spatially incoherent particle sources. It has been used for demonstrating the quantum wave nature of organic molecules [13] even with masses beyond 10,000 amu [14] and as a tool for metrology [15,16]. Here we explore its potential for tracking optically induced changes in magnetic susceptibility.

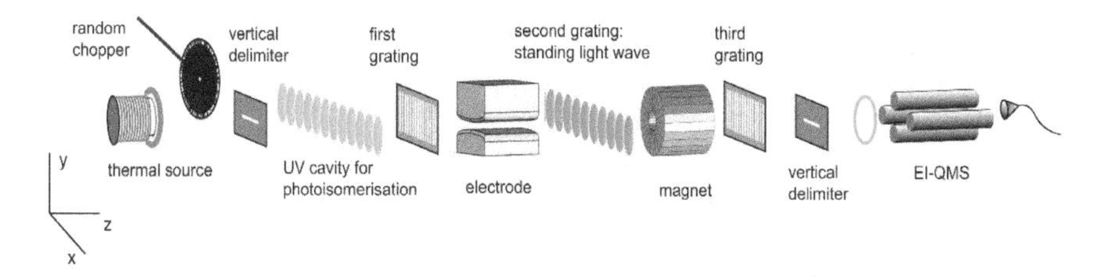

Figure 1. Sketch of Kapitza–Dirac–Talbot–Lau (KDTLI) interferometry, including the proposed extensions for magnetic deflectometry. The molecules evaporate to form a molecular beam in high vacuum. The molecular beam velocity is selected by its free-fall parabola using three horizontal slits. The molecular v-distribution can be recorded by chopping the beam in a pseudorandom sequence and measuring its arrival time at the quadrupole mass detector. The KDTLI comprises two nanofabricated absorptive masks, G1 and G3, and one optical phase grating G2. A tailored magnetic field (Halbach magnet) can exert a homogeneous force onto the molecules and deflect the molecular beam in proportion to the particles' magnetic susceptibility. If the molecules exhibit a permanent magnetic dipole moment, the interference fringes will broaden, and contrast will be reduced.

The 7-DHC beam is prepared by sublimating molecules from a ceramic crucible at a maximal temperature of 460 K. Several delimiters shape a molecular beam of approximately 1 mm width and 200 μm height. They also select a well-defined velocity distribution $f(v_z)$ from an almost thermal initial beam. At the end of the setup, the molecules are ionized by electron impact and the ions are separated and counted by a quadrupole mass spectrometer. A mechanical chopper with a pseudo-random slit sequence imprints a time code onto the molecular beam and allows resolving its time-of-flight and velocity with a selectivity up to 1% [17].

The interferometer consists of three gratings, all with a period of $d = 266$ nm and positioned at equal distance $L = 0.105$ m to one another. The first grating G1 is a periodic slit array in a 160 nm thick membrane of SiN_x. Each slit is nominally 110 nm wide and the confinement of the molecular wavefunction in any slit suffices to expand its coherence function by several orders of magnitude further downstream. The center-of-mass wave function of wavelength λ_{dB} diffracted at each slit of width s thus obtains a spatial coherence $W_c \simeq 2\lambda_{dB}L/s$, which grows with distance to the source, such that the center-of-mass coherence function spreads over the extension of at least two slits when arriving at the second grating. The standing light wave G2 is obtained by retro-reflecting a 532 nm laser beam at a plane mirror. In the antinodes of the grating the light shifts the phase of the transmitted matter-wave mainly by the optical dipole potential, but the full quantum model includes absorption of photons as well [17]. At the center of the Gaussian laser beam, this phase depends on the power P and the vertical beam waist w_x of the laser, as well as on the molecular optical polarizability α (532 nm), and the forward velocity of the molecule v_z. The coherent evolution of the molecules in phase-space leads to the formation of a molecular density pattern of period $d = 266$ nm, which can be sampled by the mechanical mask G3. This pattern forms periodically along the beam line and consecutive patterns are separated by the Talbot length $L_T = d^2/\lambda_{dB}$ [18]. Tracing the number of transmitted molecules as a function of the position of G3, one finds a nearly sinusoidal fringe pattern, as shown in Figure 2a with a visibility $V = (S_{max} - S_{min})/(S_{max} + S_{min})$, where S_{max} and S_{min} are the maximal and minimal count rates.

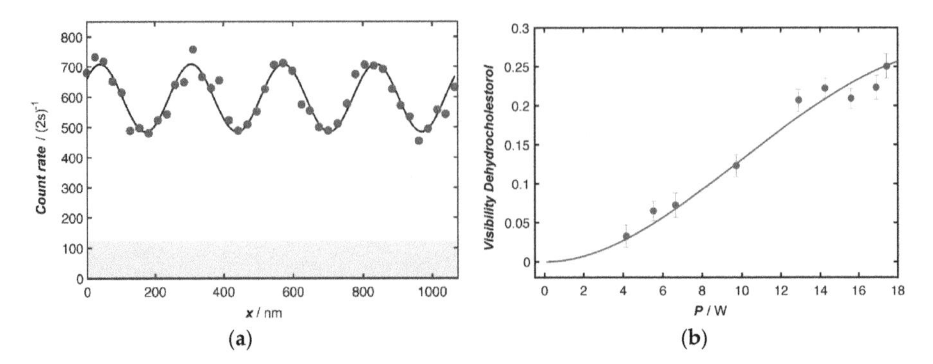

Figure 2. (**a**) Matter-wave interference of 7-dehydrocholestorol with a molecular beam velocity $v_{mean} = 212 \pm 78$ m/s (FWHM). The dots show the molecular count rate at the respective position of the third grating, and the continuous line is a sinusoidal fit to the data exhibiting a fringe contrast of $23.1 \pm 1.5\%$. The grey shaded area indicates the dark counts of the detector; (**b**) The interference contrast varies with the laser power in the diffraction grating G2, following the line shape of the quantum model. We compared the achieved fringe contrast to the theoretical maximum by calibration measurements with the well characterized fullerene C_{60} and found a reduction of 10%, which we attribute to grating misalignment. This is still well compatible with fringe-assisted molecule metrology.

In our experiments, 7-DHC had a mean de Broglie wavelength of $\lambda_{dB} \simeq 4.9$ pm and showed a maximal fringe contrast of about $V = 23\%$. Earlier experiments have shown that understanding such molecular density patterns requires quantum mechanics [3,10–12,19]. We confirm this here, by tracing the interference contrast as a function of the diffracting laser power (Figure 2b). While a fringe pattern could be mistaken as a classical Moiré shadow, the detailed dependence of the fringe visibility $V(P)$ on the diffracting laser power can only be reproduced by quantum theory [18]. The quantum model assumes that the molecular wave function is delocalized over at least two nodes of the standing light wave, that is 200 times the molecular diameter, which has triggered philosophical discussions on the interpretation of quantum mechanics and the reality of the "position" of objects that we would see with 1 nm diameter in surface probe microscopy [20]. However, independent of this important question at the heart of physics, the predicted nanoscale molecular density pattern that arises as a consequence of quantum interference is an experimental fact, as shown in Figure 2a. And it is this nanoruler that we can use to extract even information about intra-molecular properties. Moiré deflectometers have been successfully used in the past to measure small forces on atoms [21] and they are interesting for advanced anti-matter experiments [22]. However, when aiming at higher force sensitivity and using smaller fringe periods such devices automatically become matter-wave interferometers which require quantum physics for a correct description.

3. Photo-Switching

Photoactive molecules are interesting candidates for optically addressable memories, switches in organic electronics, and molecular motors [23]. Diarylethenes [24], fulgides [25], and spiropyrans [26] are common representatives. In solution, they are known to undergo photoisomerization associated with a ring opening or closure. Such photoisomerization is also known for resveratrol [27] and 7-dehydrocholestorol [28]. While most studies have been performed in solution, molecular beam experiments can shed light on the molecular excited state dynamics in a solvent-free environment. Photoisomerization in the gas phase has been demonstrated for spiropyran using electron-diffraction [29]. Here, we want to lay out a new perspective.

In Figure 3 we show 7-dehydrocholesterol (7-DHC) as a prototypical molecule of biological relevance. It plays a vital role in the human metabolism and transforms into vitamin D3 via one photo-induced and one thermal isomerization process. The barrier for the required ring-opening is sufficiently high for the molecule to persist in closed-ring form, even when heated to 500 K.

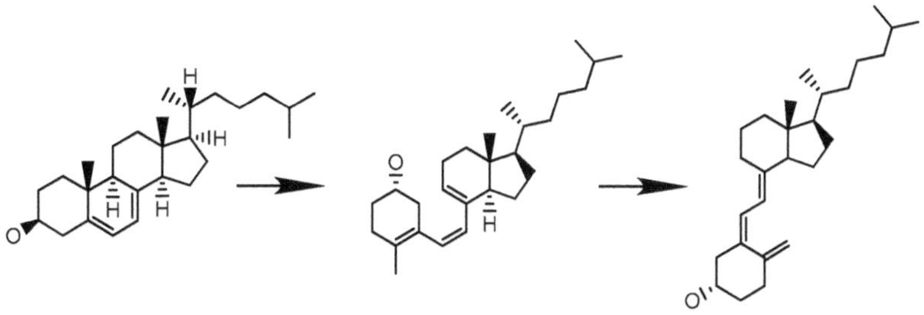

7-Dehydrocholesterol Previtamin D3 Cholecalciferol,Vitamin D3

Figure 3. The photoisomerization (1) from 7-dehydrocholesterol (7-DHC, molecular weight MW = 384 amu) to previtamin D3 is well understood in solution, but little studied in the gas phase. This is also true for the spontaneous isomerization (2) from previtamin D3 to vitamin D3 (cholecalciferol).

When 7-DHC absorbs light in the wavelength range of 260–310 nm it can undergo photoisomerization, as shown in Figure 3 [30]. We assume the absorption cross section in solution $\sigma_{\text{abs}} \simeq 2 \times 10^{-17}$ cm^2 to be also a good approximation for molecules in the gas phase at T = 450 K. Recent experiments with photo-cleavable peptides showed that ultraviolet (UV) absorption cross sections of molecules in this complexity range can be comparable in the gas phase and in solution [31]. When a $v = 100$ m/s fast 7-DHC molecule traverses a gaussian laser beam of power P and waist $w_0 = 0.3$ mm it will absorb on average $n = \frac{2P}{\pi w_0} \frac{\lambda}{hc} \frac{\sigma_{abs}}{v}$ photons. The average $n = 1$ is reached for $\lambda = 266$ nm and $P = 40$ W. Single pass frequency doubling of a green solid state lasers can reliably generate ultraviolet light of $P = 1$ W and a power enhancement of 50–80 is conceivable in low finesse UV cavities, even in high vacuum where UV optics often suffer from outgassing [32]. Also, commercial high-power nanosecond lasers can produce up to 30 W average power at 266 nm and even 200 W at 355 nm, with repetition rates of 100 kHz. This is sufficient to ensure that all molecules interact with the laser beam. Positioned before the first grating, one or even two photoisomerization processes can be completed before the molecules enter the interferometer region. The following considerations focus on the feasibility of detecting such state changes via an interferometric monitoring of a change in molecular magnetism.

4. Magnetic Manifestations of Molecular Photoisomerization in the Gas Phase

Since the days of Stern and Gerlach, when magnetic deflection was used to demonstrate the discreteness of spin orientations [33], beam deflection experiments have become the basis for measuring atomic hyperfine structure [34], the realization of atomic clocks [35], or for studies of cluster magnetism [36,37]. The permanent magnetic moment of radicals has also been used to slow and cool beams of small molecules [38,39]. The magnetic manipulation of complex molecules is much harder to achieve, since their total orbital or spin angular momentum either vanishes or is too small in relation to the molecular mass. Here, we explore, whether the high force and position sensitivity of matter-wave fringes can provide additional information about the magnetic properties of molecules, which can also be a signature for photoisomerization processes.

To understand the different contributions to molecular magnetism, we invoke second order perturbation theory to distinguish the possible responses of a molecule to an external B-field [40,41]. This quantifies the energy shift of a molecule with vanishing total spin as

$$\Delta E_n = \mu_B B \langle n|\Lambda|n\rangle + \frac{e^2}{8m_e} B^2 \left\langle n \left| \sum_k \left(x_k^2 + y_k^2\right) \right| n \right\rangle + \mu_B^2 B^2 \sum_{n' \neq n} \frac{|\langle n|\Lambda|n'\rangle|^2}{E_n - E_{n'}}. \tag{1}$$

Here n designates the electronic quantum number, Λ the quantum number of the projected angular orbital momentum, μ_B Bohr's magneton, and B the modulus of the magnetic flux density. The mass and coordinates of the electrons are m_e, x_k and y_k. The magnetic susceptibility χ_{mag} is the second derivative of the energy shift with respect to the magnetic field strength $\boldsymbol{H}$, with $(\boldsymbol{H}+\boldsymbol{M})\mu_0 = \boldsymbol{B}$, and $\mu_0 = 4\pi \times 10^{-7}\ N/A^2$ the vacuum permeability:

$$\chi_{mag} = \frac{1}{\mu_0 V} \frac{\partial^2 \Delta E_n}{\partial H^2}. \tag{2}$$

The first term in Equation (1) represents the Langevin paramagnetic response for a particle with finite total angular momentum $\boldsymbol{J}$. The magnetic moment $\boldsymbol{\mu}_J$ interacts with the flux density $\boldsymbol{B}$ and experiences an orientation-dependent force $\boldsymbol{F} = -\nabla(\boldsymbol{\mu}_J \boldsymbol{B})$, which will pull an aligned magnetic dipole towards the field maximum and push the anti-aligned particle away. A thermal beam of molecules with random orientations of their figure and rotation axes will therefore be broadened, when exposed to a B-field gradient. In matter-wave interferometry, this broadening will reduce the interference fringe contrast. This resembles the observations for electric dipole moments in electric fields [8,42,43]. In the gas phase first order paramagnetism will always dominate over all other magnetic effects, unless the magnetic dipole moment vanishes. In the following we focus on those molecules, with $\boldsymbol{J} = 0$ in the ground state.

The second term of Equation (1) represents the diamagnetic contribution. A diamagnetic molecule of susceptibility χ_{dia} responds to an external B-field like a particle of polarizability α in an electric field. However, while an electric field induces and aligns a dipole moment such as to attract it to higher fields, according to $\boldsymbol{F}_{el} = \alpha(\boldsymbol{E}\nabla)\boldsymbol{E}$, the induced magnetic moment will be expelled from regions of higher magnetic field strength with a force described by

$$\boldsymbol{F}_{dia} = -\beta(\boldsymbol{B}\nabla)\,\boldsymbol{B} \tag{3}$$

with $= \chi_{mol}^{dia}\, \mu_0^{-1} N_A^{-1}$, χ_{mol}^{dia} the molar diamagnetic susceptibility and N_A the Avogadro number. The experiment will be sensitive to the orientational average of the magnetic polarizability, since the molecules will arrive with an isotropic distribution of initial orientations and rotation axes, and their rotation rate is fast compared to the transit time through the magnet.

The third term of Equation (1) is the second order contribution to paramagnetism, the van Vleck paramagnetism. The van Vleck force is often comparable in magnitude to the diamagnetic component but pointing in the opposite direction.

Finally, in molecules we must also account for nuclear spins of different isotopes: Natural hydrocarbons contain ^{13}C with an abundance of 1.1%. In natural fullerene for instance, 48% of all C_{60} molecules hold at least one nuclear spin and 10% even exactly two. In 7-DHC, still 26% of all molecules hold at least one nuclear spin. Since ^{13}C has a nuclear spin of $\frac{1}{2}$ and a nuclear magnetic moment of $\mu_{C13} = +0.7\mu_N$ the nuclear response will be about two thousand times weaker than that of a single unpaired electron, but nuclear paramagnetism can actually be comparable to electron diamagnetism or van Vleck paramagnetism and must not be ignored for $\boldsymbol{J} = 0$.

We set the scene by estimating the B-field configuration that is required to shift the interference pattern by 1/10 of the full interference fringe; i.e., by $\approx$26 nm. A constant force is achieved in a field of constant $(\boldsymbol{B}\nabla)B_x$. The fringe deflection Δx depends on the molecular mass and velocity, the length L_1

of the magnet and the distance L_2 of its closest edge to G2, as well as on the total interferometer length L through the geometry factor $K = \left(L_1^2/2 - L_1 L_2 + L_1 L\right)$:

$$\Delta x = K \frac{\beta}{mv^2} (\boldsymbol{B}\nabla) B_x \tag{4}$$

We estimate the effect for isotopically pure $^{12}C_{60}$ fullerenes whose magnetic response represents a lower limit to most of the interesting aromatic molecules. The molar magnetic susceptibility of C_{60} has been measured to be $\chi_{C60} = -1.08 \times 10^{-9}$ $m^3 \cdot mol^{-1}$ [44]. This translates into a molecular magnetic polarizability of $\beta_{C60} = 1.4 \times 10^{-27}$ $Am^4 \cdot V^{-1} \cdot s^{-1}$. For $L = 0.2$ m, $L_1 = 0.04$ m, $L_2 = 0.04$ m, $v = 100$ m/s, $m = 720$ amu, and $K = 0.003$, the interference fringe can be shifted by about 25 nm for $(\boldsymbol{B}\nabla)B_x = 70$ $T^2 \cdot m^{-1}$. If a field of that order of magnitude can be prepared, the fringe shift can still be resolved, the interferometer can still be sensitive to χ_{C60}. The case of fullerene C_{60} gives a conservative limit, since the deflection depends on the magnetic polarizability-to-mass ratio β/m. For example, the fully aromatic molecule benzene C_6H_6 exhibits five times greater β/m.

While a full quantum chemical assessment of the magnetic properties of 7-DHC exceeds the scope of this work, we expect the ring opening to induce magnetic susceptibility changes to be at least on the order of the effect estimated here. Since the fringe shift grows linearly with the interferometer length and quadratically with the length of the magnet, future long-baseline interferometers will be ten times more sensitive, at least, and certainly allow measuring even such tiny magnetic susceptibilities.

5. Design of the Required Magnetic Structures

Such a high $(\boldsymbol{B}\nabla)\boldsymbol{B}$ field can be realized using a modified Halbach cylinder, as shown in Figure 4, which we have simulated using the finite element package COMSOL 4.0 multiphysics simulation package (COMSOL AB, Stockholm, Sweden). The arrangement of permanent magnets from neodymium-iron-boron alloy, with a remanent magnetization of 1.3 T and a coercive field strength of 100 kA/m, can guide the field lines inside the cylinder and generate the required field. Figure 4b shows that one can realize a region with $(\boldsymbol{B}\nabla)B_x = 70$ T^2/m that is homogeneous within 2% of its peak value across an area of 1000 × 200 μm²; i.e., across the full molecular beam profile inside a KDTL interferometer.

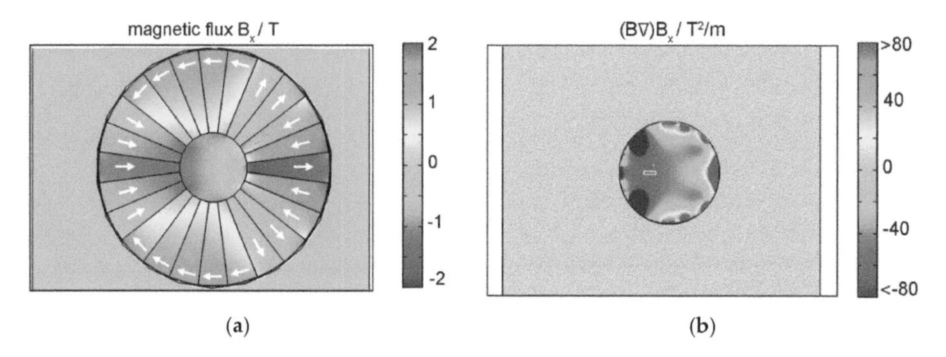

Figure 4. Finite element simulation of the modified Halbach cylinder. (**a**) Magnetic flux: the arrows show the direction of magnetization of the individual segments; (**b**) magnetic force field: the deflection of a molecular beam is proportional to $(\boldsymbol{B}\nabla)B_x$. The diameter of the magnet is 55 mm with an inner bore of 16 mm. The white rectangle indicates the location of the molecular beam, where the force is constant within 2%.

6. Discussion

Our experimental data demonstrate that complex, thermal biomolecules can show quantum interference and be delocalized by a few hundred times their own size. We have also seen that the free-flying molecular nanostructure is a sensitive ruler to measure interference fringe displacements, which can quantify internal molecular properties in the presence of external perturbations. Here we have focused on the role of magnetic fields and showed that even very small magnetic contributions can become accessible in matter-wave assisted deflectometry.

This can open an entire new range of experiments with photo-isomerization groups in spiropyrans, fulgids, and diarylethenes. Spiropyran, for instance, isomerizes to blue merocyanin upon absorption of a UV-photon around 365 nm and the reaction can even be reversed by irradiation with visible light [45]. See Figure 5.

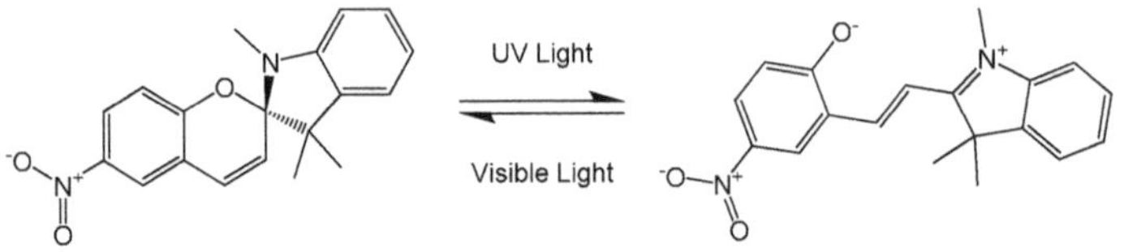

Figure 5. Spiropyran can isomerize to merocyanine upon absorption of a UV photon. This opens one ring which we expect to significantly change the magnetic susceptibility. In contrast to the case of 7-DHC, the process changes the electric dipole moment here by a large factor, from 7 Debye for spiropyran [46] to between 20–50 Debye for merocyanine [26]. Such huge changes will be easily detectable in interferometric electric deflectometry [8].

Merocyanine is zwitterionic with a large electric dipole moment [26] and the isomerization should also be readily detected in interference-assisted electric deflectometry. Thus, a combination of electric and magnetic deflectometry will give insights into the molecular dynamics in the gas phase. Since spiropyrans can thermally isomerize to merocyanine above room temperature [45], optical switching experiments will be best performed with internally cold molecules [47]. The scheme can be generalized to a wide class of molecular systems.

Author Contributions: M.A., L.M. and S.E. conceived and designed the experiments; L.M. and S.E. performed the experiments; L.M. and S.E. and A.S. analyzed the data; L.M. and M.A. wrote the paper with input by all authors.

Funding This research was funded by the European Research Council (ERC) in grant number 320694, the Austrian Science Fund (FWF) in grant numbers, W1210-N25 and P30176. A.S. acknowledges funding by the Austrian Science Fund (FWF) within the Lise-Meitner fellowship M 2364.

Acknowledgments: We thank Philip Rieser for test measurements on the deflection magnet.

Conflicts of Interest: The authors declare no conflict of interest. The founding sponsors had no role in the design of the study; in the collection, analyses, or interpretation of data; in the writing of the manuscript, and in the decision to publish the results.

References

1. De Broglie, L. Waves and Quanta. *Nature* **1923**, *112*, 540. [CrossRef]
2. Juffmann, T.; Milic, A.; Müllneritsch, M.; Asenbaum, P.; Tsukernik, A.; Tüxen, J.; Mayor, M.; Cheshnovsky, O.; Arndt, M. Real-time single-molecule imaging of quantum interference. *Nat. Nanotechnol.* **2012**, *7*, 297–300. [CrossRef] [PubMed]
3. Arndt, M.; Nairz, O.; Voss-Andreae, J.; Keller, C.; van der Zouw, G.; Zeilinger, A. Wave-particle duality of C60 molecules. *Nature* **1999**, *401*, 680–682. [CrossRef] [PubMed]
4. Antoine, R.; Dugourd, P.; Rayane, D.; Benichou, E.; Broyer, M.; Chandezon, F.; Guet, C. Direct measurement of the electric polarizability of isolated C60 molecules. *J. Chem. Phys.* **1999**, *110*, 9771–9772. [CrossRef]

5. Antoine, R.; Compagnon, I.; Rayane, D.; Broyer, M.; Dugourd, P.; Sommerer, N.; Rossignol, M.; Pippen, D.; Hagemeister, F.C.; Jarrold, M.F. Application of Molecular Beam Deflection Time-of-Flight Mass Spectrometry to Peptide Analysis. *Anal. Chem.* **2003**, *75*, 5512–5516. [CrossRef] [PubMed]
6. Berninger, M.; Stefanov, A.; Deachapunya, S.; Arndt, M. Polarizability measurements of a molecule via a near-field matter-wave interferometer. *Phys. Rev. A* **2007**, *76*, 013607. [CrossRef]
7. Gring, M.; Gerlich, S.; Eibenberger, S.; Nimmrichter, S.; Berrada, T.; Arndt, M.; Ulbricht, H.; Hornberger, K.; Müri, M.; Mayor, M.; et al. Influence of conformational molecular dynamics on matter wave interferometry. *Phys. Rev. A* **2010**, *81*, 031604. [CrossRef]
8. Eibenberger, S.; Gerlich, S.; Arndt, M.; Tüxen, J.; Mayor, M. Electric moments in molecule interferometry. *New J. Phys.* **2011**, *13*, 043033. [CrossRef]
9. Hornberger, K.; Gerlich, S.; Haslinger, P.; Nimmrichter, S.; Arndt, M. Colloquium: Quantum interference of clusters and molecules. *Rev. Mod. Phys.* **2012**, *84*, 157–173. [CrossRef]
10. Gerlich, S.; Hackermüller, L.; Hornberger, K.; Stibor, A.; Ulbricht, H.; Gring, M.; Goldfarb, F.; Savas, T.; Müri, M.; Mayor, M.; et al. A Kapitza-Dirac-Talbot-Lau interferometer for highly polarizable molecules. *Nat. Phys.* **2007**, *3*, 711–715. [CrossRef]
11. Brezger, B.; Hackermüller, L.; Uttenthaler, S.; Petschinka, J.; Arndt, M.; Zeilinger, A. Matter-Wave Interferometer for Large Molecules. *Phys. Rev. Lett.* **2002**, *88*, 100404. [CrossRef] [PubMed]
12. Haslinger, P.; Dörre, N.; Geyer, P.; Rodewald, J.; Nimmrichter, S.; Arndt, M. A universal matter-wave interferometer with optical ionization gratings in the time domain. *Nat. Phys.* **2013**, *9*, 144–148. [CrossRef] [PubMed]
13. Gerlich, S.; Eibenberger, S.; Tomandl, M.; Nimmrichter, S.; Hornberger, K.; Fagan, P.; Tüxen, J.; Mayor, M.; Arndt, M. Quantum interference of large organic molecules. *Nat. Commun.* **2011**, *2*, 263. [CrossRef] [PubMed]
14. Eibenberger, S.; Gerlich, S.; Arndt, M.; Mayor, M.; Tüxen, J. Matter-wave interference of particles selected from a molecular library with masses exceeding 10,000 amu. *Phys. Chem. Chem. Phys.* **2013**, *15*, 14696–14700. [CrossRef] [PubMed]
15. Mairhofer, L.; Eibenberger, S.; Cotter, J.P.; Romirer, M.; Shayeghi, A.; Arndt, M. Quantum-Assisted Metrology of Neutral Vitamins in the Gas Phase. *Angew. Chem. Int. Ed.* **2017**, *56*, 10947–10951. [CrossRef] [PubMed]
16. Gerlich, S.; Gring, M.; Ulbricht, H.; Hornberger, K.; Tüxen, J.; Mayor, M.; Arndt, M. Matter-wave metrology as a complementary tool for mass spectrometry. *Angew. Chem. Int. Ed. Engl.* **2008**, *47*, 6195–6198. [CrossRef] [PubMed]
17. Cotter, J.P.; Eibenberger, S.; Mairhofer, L.; Cheng, X.; Asenbaum, P.; Arndt, M.; Walter, K.; Nimmrichter, S.; Hornberger, K. Coherence in the presence of absorption and heating in a molecule interferometer. *Nat. Commun.* **2015**, *6*, 7336. [CrossRef] [PubMed]
18. Hornberger, K.; Gerlich, S.; Ulbricht, H.; Hackermüller, L.; Nimmrichter, S.; Goldt, I.; Boltalina, O.; Arndt, M. Theory and experimental verification of Kapitza-Dirac-Talbot-Lau interferometry. *New J. Phys.* **2009**, *11*, 043032. [CrossRef]
19. Nairz, O.; Brezger, B.; Arndt, M.; Zeilinger, A. Diffraction of Complex Molecules by Structures Made of Light. *Phys. Rev. Lett.* **2001**, *87*, 160401. [CrossRef] [PubMed]
20. Juffmann, T.; Truppe, S.; Geyer, P.; Major, A.G.; Deachapunya, S.; Ulbricht, H.; Arndt, M. Wave and Particle in Molecular Interference Lithography. *Phys. Rev. Lett.* **2009**, *103*. [CrossRef] [PubMed]
21. Oberthaler, M.K.; Bernet, S.; Rasel, E.M.; Schmiedmayer, J.; Zeilinger, A. Inertial Sensing with Classical Atomic Beams. *Phys. Rev. A* **1996**, *54*, 3165–3176. [CrossRef] [PubMed]
22. Aghion, S.; Ahlen, O.; Amsler, C.; Ariga, A.; Ariga, T.; Belov, A.S.; Berggren, K.; Bonomi, G.; Braunig, P.; Bremer, J.; et al. A moire deflectometer for antimatter. *Nat. Commun.* **2014**, *5*, 4538. [CrossRef] [PubMed]
23. Browne, W.R.; Feringa, B.L. Making molecular machines work. *Nat. Nanotechnol.* **2006**, *1*, 25–35. [CrossRef] [PubMed]
24. Irie, M. Diarylethenes for Memories and Switches. *Chem. Rev.* **2000**, *100*, 1685–1716. [CrossRef] [PubMed]
25. Yokoyama, Y. Fulgides for Memories and Switches. *Chem. Rev.* **2000**, *100*, 1717–1740. [CrossRef] [PubMed]
26. Berkovic, G.; Krongauz, V.; Weiss, V. Spiropyrans and Spirooxazines for Memories and Switches. *Chem. Rev.* **2000**, *100*, 1741–1754. [CrossRef] [PubMed]
27. Yang, I.; Kim, E.; Kang, J.; Han, H.; Sul, S.; Park, S.B.; Kim, S.K. Photochemical generation of a new, highly fluorescent compound from non-fluorescent resveratrol. *Chem. Commun.* **2012**, *48*, 3839–3841. [CrossRef] [PubMed]

28. Fuss, W.; Höfer, T.; Hering, P.; Kompa, K.L.; Lochbrunner, S.; Schikarski, T.; Schmid, W.E. Ring Opening in the Dehydrocholesterol–Previtamin D System Studied by Ultrafast Spectroscopy. *J. Phys. Chem.* **1996**, *100*, 921–927. [CrossRef]
29. Gahlmann, A.; Lee, I.R.; Zewail, A.H. Direct structural determination of conformations of photoswitchable molecules by laser desorption-electron diffraction. *Angew. Chem. Int. Ed. Engl.* **2010**, *49*, 6524–6527. [CrossRef] [PubMed]
30. Tang, K.C.; Rury, A.; Orozco, M.B.; Egendorf, J.; Spears, K.G.; Sension, R.J. Ultrafast electrocyclic ring opening of 7-dehydrocholesterol in solution: The influence of solvent on excited state dynamics. *J. Chem. Phys.* **2011**, *134*, 104503. [CrossRef] [PubMed]
31. Debiossac, M.; Schätti, J.; Kriegleder, M.; Geyer, P.; Shayeghi, A.; Mayor, M.; Arndt, M.; Köhler, V. Tailored photocleavable peptides: Fragmentation and neutralization pathways in high vacuum. *Phys. Chem. Chem. Phys.* **2018**, *20*, 11412–11417. [CrossRef] [PubMed]
32. Gangloff, D.; Shi, M.; Wu, T.; Bylinskii, A.; Braverman, B.; Gutierrez, M.; Nichols, R.; Li, J.; Aichholz, K.; Cetina, M.; et al. Preventing and reversing vacuum-induced optical losses in high-finesse tantalum(V) oxide mirror coatings. *Opt. Express* **2015**, *23*, 18014–18028. [CrossRef] [PubMed]
33. Gerlach, W.; Stern, O. Der experimentelle Nachweis des magnetischen Moments des Silberatoms. *Z. Phys.* **1922**, *8*, 110–111. [CrossRef]
34. Rabi, I.I.; Millman, S.; Kusch, P.; Zacharias, J.R. The Molecular Beam Resonance Method for Measuring Nuclear Magnetic Moments. The Magnetic Moments of $_3Li^6$, $_3Li^7$and $_9F^{19}$. *Phys. Rev.* **1939**, *55*, 526–535. [CrossRef]
35. Ramsey, N.F. A Molecular Beam Resonance Method with Separated Oscillating Fields. *Phys. Rev.* **1950**, *78*, 695–699. [CrossRef]
36. Moro, R.; Xu, X.; Yin, S.; de Heer, W.A. Ferroelectricity in free niobium clusters. *Science* **2003**, *300*, 1265–1269. [CrossRef] [PubMed]
37. Rohrmann, U.; Schafer, R. Stern-Gerlach experiments on $Mn@Sn_{12}$: Identification of a paramagnetic superatom and vibrationally induced spin orientation. *Phys. Rev. Lett.* **2013**, *111*, 133401. [CrossRef] [PubMed]
38. Narevicius, E.; Parthey, C.G.; Libson, A.; Riedel, M.F.; Even, U.; Raizen, M.G. Towards magnetic slowing of atoms and molecules. *New J. Phys.* **2007**, *9*. [CrossRef]
39. Akerman, N.; Karpov, M.; David, L.; Lavert-Ofir, E.; Narevicius, J.; Narevicius, E. Simultaneous deceleration of atoms and molecules in a supersonic beam. *New J. Phys.* **2015**, *17*. [CrossRef]
40. Vleck, J.H.V. *The Theory of Electric and Magnetic Susceptibilities*; Oxford University Press: London, UK, 1965.
41. Atkins, P.W.; Friedman, R. *Molecular Quantum Mechanics*, 4nd ed.; Oxford University Press: Oxford, UK, 2005.
42. De Heer, W.A.; Kresin, V.V. Electric and magnetic dipole moments of free nanoclusters. In *Handbook of Nanophysics*; Sattler, K.D., Ed.; CRC Press: Boca Raton, FL, USA, 2011.
43. Heiles, S.; Schäfer, R. *Dielectric Properties of Isolated Clusters Beam Deflection Studies*; Springer: Heidelberg, Germany, 2014.
44. Elser, V.; Haddon, R.C. Icosahedral C60: An aromatic molecule with a vanishingly small ring current magnetic susceptibility. *Nature* **1987**, *325*, 792. [CrossRef]
45. Lin, J.-S.; Chiu, H.-T. Photochromic Behavior of Spiropyran and Fulgide in Thin Films of Blends of PMMA and SBS. *J. Polym. Res.* **2003**, *10*, 105–110. [CrossRef]
46. Weinberger, C.R.; Tucker, G.J. *Multiscale Materials Modeling for Nanomechanics*; Springer: New York, NY, USA, 2016.
47. Geyer, P.; Sezer, U.; Rodewald, J.; Mairhofer, L.; Dörre, N.; Haslinger, P.; Eibenberger, S.; Brand, C.; Arndt, M. Perspectives for quantum interference with biomolecules and biomolecular clusters. *Phys. Scr.* **2016**, *91*. [CrossRef]

Article

When Photons Are Lying about Where They Have Been

Lev Vaidman [1,*] and Izumi Tsutsui [2]

[1] Raymond and Beverly Sackler School of Physics and Astronomy, Tel-Aviv University, Tel-Aviv 69978, Israel
[2] Theory Center, Institute of Particle and Nuclear Studies, High Energy Accelerator Research Organization (KEK), Tsukuba 305-0801, Japan; izumi.tsutsui@kek.jp
* Correspondence: vaidman@post.tau.ac.il; Tel.: +972-5-4590-8806

Received: 1 June 2018; Accepted: 16 July 2018; Published: 19 July 2018

Abstract: The history of photons in a nested Mach–Zehnder interferometer with an inserted Dove prism is analyzed. It is argued that the Dove prism does not change the past of the photon. Alonso and Jordan correctly point out that an experiment by Danan et al. demonstrating the past of the photon in a nested interferometer will show different results when the Dove prism is inserted. The reason, however, is not that the past is changed, but that the experimental demonstration becomes incorrect. The explanation of a signal from the place in which the photon was (almost) not present is given. Bohmian trajectory of the photon is specified.

Keywords: past of the photon; Mach–Zehnder interferometer; Dove prism; photon trajectory

---

1. Introduction

This work describes peculiar behaviour of photons in the modification of the experiment of Danan et al. [1] proposed by Alonso and Jordan (AJ) [2]. In the Danan et al. experiment, photons were asked where exactly they have been inside a nested interferometer tuned in a particular way. The AJ modification makes photons to tell that they have been in a place in which, according to the narrative of the two-state vector formalism (TSVF) [3], they could not have been. Note that this work is only slightly related to the results presented by one of the authors (L.V.) at "Emergent Quantum Mechanics" that have been already published [4,5].

Textbooks of quantum mechanics teach us that we are not supposed to ask where the photons passing through an interferometer were. Wheeler [6] introduced the delayed choice experiment in an attempt to analyze this question. Vaidman [3] suggested a different approach. He proposed a definition according to which a quantum particle was where it left a trace and showed that the past of the particle can be easily seen in the framework of the TSVF [7] as regions of the overlap of the forward and backward evolving quantum states. Vaidman, together with his collaborators, performed an experiment demonstrating a surprising trace of the photons in nested interferometers [1] (see Figure 1). These results became the topic of a very large controversy [8–46].

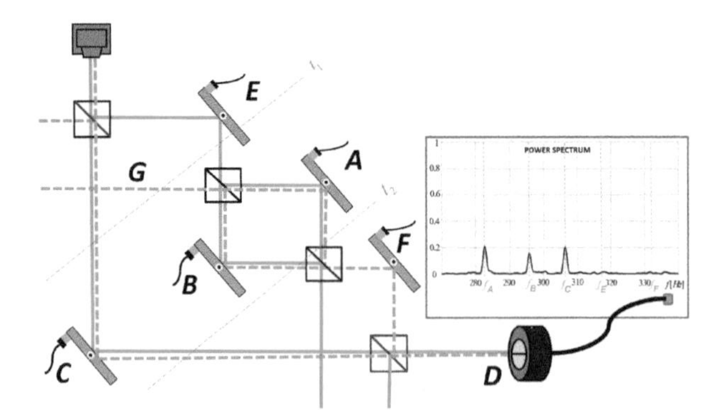

Figure 1. Nested Mach–Zehnder interferometer with inner interferometer tuned to destructive interference towards F. Although our 'common sense' suggests that the only possible path for the photon detected in D is path C, the trace was found also inside the inner interferometer supporting the TSVF proposal according to which the particle was present in the places where forward (red continuous line) and backward (green dashed line) evolving wavefunctions overlap. The latter is demonstrated by the results of the measurement by Danan et al. [1].

2. Alonso and Jordan Modified Interferometer

Here we analyze, in our view, the most interesting objection which was made by Alonso and Jordan [2]. They suggested inserting a Dove prism inside one of the arms of the inner interferometer (see Figure 2). They asked: "Can a Dove prism change the past of a single photon?". Their analysis of this modified experiment was correct. Although the formalism suggested that the past of the photon remains the same as in the original experiment, i.e., the photon was present near mirrors C, A, B but not near mirrors E and F, the experiment should show, in addition to frequencies f_C, f_A, f_B, also the frequency f_E. This is in contradiction with the fact that the photons, according to Vaidman, were not present near mirror E.

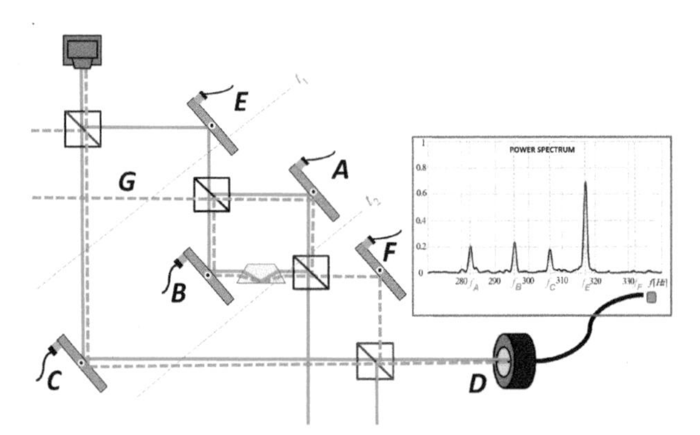

Figure 2. Nested Mach–Zehnder interferometer with a Dove prism inside the inner interferometer as suggested by Alonso and Jordan [2]. The region of the overlap of the forward and the backward evolving states remains the same, but predicted results of an experiment similar to [1] include a signal from mirror E where the photon was not supposed to be.

The experiment of Danan et al. was not a direct measurement of the trace left by the photons. The reason is that such direct measurement is very difficult, as it requires collecting data about the trace conditioned on detection of the photon by a particular detector. In the actual experiment, instead of measuring the trace on the external system (as in a recent experiment [47]), the trace was 'written' on the photons themselves, on the degree of freedom of their transverse motion. Observing this degree of freedom of post-selected particle replaced the coincidence counting in the experimental setup. Although indirect, the experiment [1] was correct. A local trace created at mirrors was read later on the quad-cell detector. We argue that introducing a Dove prism [2] spoils the experiment, making the signal at the quad-cell detector no longer a faithful representation of the trace created at mirror E.

Although the signal in the Danan et al. experiment was appearing as a particular frequency in the output of the quad-cell detector, the frequency was not an actual trace written on each photon. Wiggling with different frequencies was a trick that allowed in a single run to see records made at different mirrors. (It also improved significantly the signal-to-noise ratio, since noise had no preference for the frequencies of the wiggling mirrors.) The physical signal in the Danan et al. experiment (Figure 1) originated from the shift of the beam direction at a mirror. It corresponded to the transversal kick in the momentum δ_{p_x}. This momentum shift translated into a position shift of the beam, which was read in the quad-cell detector. The property which allowed to observe the trace was that the change δ_{p_x} in the transversal momentum had no change when the beam evolved towards port D from all mirrors and through all possible paths.

This is no longer the case when the Dove prism is introduced (Figure 2). For mirrors A and C, it is still true, since the modes do not pass through the Dove prism. For mirror B, there is a difference in that the Dove prism flips the sign of the signal. However, since we measure just the size of the signal, this change is not observable, and the peak at frequency f_B correctly signifies the presence of the photon in B. The only problem occurs with the mirror E. The beam from E reaches the detector through A and through B. The shifts are in opposite directions, so the reading position of the beam on the detector does not tell us what the shift of the transversal momentum in E was. Therefore, we should not rely on the result of the experiment with the setup of the Danan et al. experiment when the Dove prism is present.

Note that a simple modification will restore the results of the Danan et al. experiment even with the presence of the Dove prism. If the wiggling of mirrors is made such that the beam is shifted in the direction perpendicular to the plane of the interferometer, the Dove prism will not cause flipping of the direction of the shift and the peak at f_E will disappear.

3. The Trace Analysis

We have explained that the AJ modification of the Danan et al. experiment is not a legitimate experiment for measuring the presence (according the local trace definition) of the particle near mirror E. Still it is of interest to understand how a strong signal with frequency f_E is obtained in this modification. For this, we need a more detailed analysis of traces in the nested Mach–Zehnder interferometer (MZI) experiments.

We consider, for simplicity, an experiment in which only one particular mirror changes its angle at every run. The shift of the beam direction created at the mirror, characterized by the transversal momentum kick δp_x, leads to the shift of the beam position on the detector. This creates the signal: the difference in the current of the upper and the lower cells of the detector.

Let χ_0 be the original mode of the photons without shifts. The photons in a shifted beam will then be in a superposition of the original mode χ_0 and a mode $\chi_\perp$, orthogonal to χ_0:

$$|\chi'\rangle = \frac{1}{\sqrt{1+\epsilon^2}}\left(|\chi_0\rangle + \epsilon|\chi_\perp\rangle\right). \tag{1}$$

For small signals which appeared in the Danan et al. experiment, the momentum kick is proportional to the relative amplitude ϵ of the orthogonal mode [48]:

$$\delta p_x = 2\epsilon \mathrm{Re}\left[\langle \chi_0 | p_x | \chi_\perp \rangle\right] + \mathcal{O}(\epsilon^2). \tag{2}$$

Note that, for a Gaussian beam (which is a good approximation of the beam in the experiment), higher order contributions do not appear [48].

What is important for our analysis is that χ_0 is symmetric with respect to the center of the beam in the transverse direction, while $\chi_\perp$, which can be approximated as a difference between two slightly shifted Gaussians, is an antisymmetric mode. Indeed, in momentum representation, we have

$$\chi_0 \simeq \mathcal{N}_0 \, e^{-\frac{p_x^2+p_y^2}{2\Delta^2}}, \qquad \chi_\perp \simeq \mathcal{N}_\perp \, p_x e^{-\frac{p_x^2+p_y^2}{2\Delta^2}}, \tag{3}$$

where Δ is the momentum uncertainty of the Gaussian beam, and $\mathcal{N}_0, \mathcal{N}_\perp$ are the normalization constants.

In the Danan et al. experiment (Figure 1), the trace of the photon was read as the shift of the beam on the detector. This shift is proportional to the strength of the trace quantified by the value of the relative amplitude ϵ of the orthogonal component. The original mode χ_0 and the orthogonal mode $\chi_\perp$ evolve towards port D from all mirrors and through all possible paths in an identical manner, so the position shift on the detector faithfully represents a locally created trace.

This is no longer the case when the Dove prism is introduced (Figure 2). For mirror B, there is a difference: mode χ_0 is unaffected by the presence of the prism, while mode $\chi_\perp$ flips the sign. The shift on the detector changes its direction. This change, however, is not observable in the experiment, since the frequency spectrum is sensitive only to the size of the signal. The observable difference appears for mirror E. There are two paths from E to the output port D, one passing through mirror A and another passing through mirror B. The original symmetric mode χ_0 would reach D undisturbed both on path A and on path B, while the orthogonal mode $\chi_\perp$ would reach D undisturbed on path A but with a flipped sign on path B. When combined, there exists a phase difference π between path A and path B, which leads to destructive interference of the original symmetric mode and constructive interference of the orthogonal antisymmetric mode at the output port towards mirror F. As a result, out of the modes of the light reflected by the mirror E, only the mode $\chi_\perp$ reaches D.

If we send the photon only in path A, and do not move mirror A, only mode χ_0 reaches the detector. Adding a small rotation of mirror A will lead to appearance of mode $\chi_\perp$ with relative amplitude ϵ. If, instead, in an undisturbed interferometer, we send the photon only in path E, and nothing will reach the detector. A small rotation of mirror E will lead to appearance of mode $\chi_\perp$ on the detector and only mode $\chi_\perp$. This mode by itself does not lead to a shift of the center of the beam on the detector. In the experiment, the photon is in a superposition of two states, one coming from path C and the other from path E. From path C, we get mode χ_0 with the same amplitude as it comes from path A. It is the interference of mode χ_0 coming through C and mode $\chi_\perp$ coming through F on the surface of the detector that yields the shift of the center of the beam. The resultant shift is larger than the shift created by the same rotation of mirror A because, first, the intensity in E is twice the intensity in A so the amplitude of the mode $\chi_\perp$ created at E is larger than the amplitude of $\chi_\perp$ created at A, and, second, the amplitude is not reduced at the second beam splitter of the inner interfereometer as it happens for the mode created at A, due to the constructive interference of $\chi_\perp$ mode in the inner interferometer with the Dove prism. This explains the larger signal observed at f_E.

4. Do the Photons Have Any Presence in *E*?

Our analysis above shows that the experiment with the Dove prism does not contradict Vaidman's proposal [3] demonstrated in the Danan et al. experiment, and explains using standard quantum mechanics the appearance of the signal at frequency f_E. Thus, it provides a satisfactory reply to Alonso and Jordan. However, it will also be of interest to explain the predicted results of Danan's setup with the Dove prism using Vaidman's approach.

Let us quote the Danan et al. Letter [1]:

> "The photons themselves tell us where they have been. And the story they tell is surprising. The photons do not always follow continuous trajectories. Some of them have been inside the nested interferometer (otherwise, they could not have known the frequencies f_A, f_B), but they never entered and never left the nested interferometer, since otherwise they could not avoid the imprints of frequencies f_E and f_F of mirrors E and F leading photons into and out of the interferometer."

With the Dove prism present, however, we do get frequency f_E. How can it happen if the photons were not in E as we argued here? Let us analyse the situation, in which only mirror E changes its angle by a small amount leading to the superposition (1) of the modes of the photon.

We start by repeating the analysis of the setup without the Dove prism in the framework of the TSVF [7]. After passing the mirror E, at time t_1, the forward evolving state is (see Figure 2)

$$|\Psi\rangle_{t_1} = \sqrt{\frac{2}{3(1+\epsilon^2)}}|E\rangle\left(|\chi_0\rangle + \epsilon|\chi_\perp\rangle\right) + \frac{1}{\sqrt{3}}|C\rangle|\chi_0\rangle, \tag{4}$$

where we split which path and the mode degrees of freedom of the photon. The forward evolving state, at time t_2, in the middle of the interferometer is then

$$|\Psi\rangle_{t_2} = \frac{1}{\sqrt{3(1+\epsilon^2)}}\left(|A\rangle + i|B\rangle\right)\left(|\chi_0\rangle + \epsilon|\chi_\perp\rangle\right) + \frac{1}{\sqrt{3}}|C\rangle|\chi_0\rangle. \tag{5}$$

Since in the experiment we use photon degrees of freedom for the measurement, we do not postselect on a particular state but rather on a space of states corresponding to all modes reaching detector D. Thus, strictly speaking, there is no definite backwards evolving state. However, we can use a standard 'trick' [4], in which we consider a hypothetical additional verification measurement of the mode state after the postselection on the path D. We verify that the state which we calculate will surely be there, and this verification measurement, together with the path post-selection, defines the backward evolving state.

The wave packets from A and B destructively interfere toward F even when mirror E is slightly rotated, so the only mode reaching D is coming from C, which is χ_0. Therefore, the backward evolving state starts from $\langle D|\langle\chi_0|$, which in the middle of the interferometer turns into

$$\langle\Phi|_{t_2} = \frac{1}{\sqrt{3}}\left(\langle A| - i\langle B| + \langle C|\right)\langle\chi_0|. \tag{6}$$

There is here destructive interference of the backward evolving quantum state toward E, so, at time t_1, the backward evolving state is

$$\langle\Phi|_{t_1} = \frac{(\sqrt{2}\langle G| + \langle C|)\langle\chi_0|}{\sqrt{3}}. \tag{7}$$

Thus, the weak value of the projection operator $\mathbf{P}_E = |E\rangle\langle E|$ at E is

$$(\mathbf{P}_E)_w = \frac{\langle\Phi|\mathbf{P}_E|\Psi\rangle_{t_1}}{\langle\Phi|\Psi\rangle_{t_1}} = 0. \tag{8}$$

Therefore, at time t_1 the photons have no presence in E, not even a "small" presence.

With the Dove prism inside, this is no longer the case. Instead of (5), we obtain

$$|\Psi'\rangle_{t_2} = \frac{1}{\sqrt{3(1+\epsilon^2)}}[(|A\rangle + i|B\rangle)|\chi_0\rangle + \epsilon(|A\rangle - i|B\rangle)|\chi_\perp\rangle)] + \frac{1}{\sqrt{3}}|C\rangle|\chi_0\rangle. \tag{9}$$

The wave packets from A and B destructively interfere towards F for mode χ_0, while the mode $\chi_\perp$ interferes constructively towards F. As a result, the backward evolving state (given the proper hypothetical measurement) starts approximately as

$$\frac{1}{\sqrt{1+2\epsilon^2}}\langle D|(\langle\chi_0| + \sqrt{2}\epsilon\langle\chi_\perp|). \tag{10}$$

Evolving it backwards until time t_1, we obtain approximately:

$$\langle\Phi'|_{t_1} = \frac{1}{\sqrt{3(1+2\epsilon^2)}}[(\sqrt{2}\langle G| + \langle C|)\langle\chi_0| + \sqrt{2}\epsilon((\langle C| + \sqrt{2}\langle E|)\langle\chi_\perp|]. \tag{11}$$

The Dove prism does not change the forward evolving state at t_1, so, even with the Dove prism, the state is still given by (4). Calculating now the weak value of projection on E yields

$$(\mathbf{P}_E)_w = \frac{\langle\Phi'|\mathbf{P}_E|\Psi\rangle_{t_1}}{\langle\Phi'|\Psi\rangle_{t_1}} \simeq 2\sqrt{2}\epsilon^2. \tag{12}$$

The photon in the experiment with the Dove prism and the tilted mirror E does have some presence in E. Thus, there is no clear paradox in obtaining the frequency f_E that was present only in E in the framework of the TSVF.

One might wonder why there is no signal at f_F similar to that at f_E in spite of the apparent symmetry of the experiment in the time symmetric TSVF. When the mirror F is tilted instead of mirror E, inserting the Dove prism spoils the destructive interference of the backward evolving wave function towards E similarly to spoiling interference toward F by tilting mirror E. However, more careful analysis shows that the symmetry is not complete. Titling mirror E also changes the effective backward evolving state, while tilting mirror F does not change the forward evolving state. See details in the next section.

5. Quantifying the Presence of Photons

The explanation of the peak at the frequency f_E which we wish to provide is that the photon has a small presence there, but the experimental records imprinted on the pre- and postselected photon reaching the detector are strong, so the size of the peak is similar to that of frequencies f_C, f_B, and f_A, where the photon presence is strong, but the record is weak. However, the second order in ϵ for the presence of the photon in E looks too small for this to be the case. In more detail, for mirrors A, B, and C, the presence of the photon is of order 1 while the strength of the record is of size ϵ. For mirror E, on the other hand, the presence characterized by the weak value of projection operator (12) is apparently only of size ϵ^2. The size of the record of an interaction is characterized by the created relative amplitude of the orthogonal component (see [48]). In our case, the record created at E which reaches the detector D is represented by the orthogonal component $|\chi_\perp\rangle$ and it is the only component reaching the detector, since the symmetric component $|\chi_0\rangle$ is 'filtered out' by the inner interferometer. Thus, we can say that the size of the record created at E which reaches the detector is of order 1. This naive consideration tells us that the peak at f_E should be of order ϵ^2 while other peaks are of order ϵ, in contradiction with predicted results of the experiment which show that the peaks are of the same order.

It is true that the weak effects which depend only on the presence of the photon in E, such as the momentum transferred to the mirror E by the photon, are proportional to $(\mathbf{P}_E)_w$, but the presence of a particle is defined according to *all* local traces it leaves (see Section 6 of [3]). In our case, the weak value of the projection operator $\mathbf{P}_E$ is not the correct parameter to quantify the presence of the particle. It is so when the pre- and post-selection is on spatial degrees of freedom only (see [48]). Here, however, due to the postselection on a subspace, effectively, we are required to consider an associated postselection on a particular mode, along with the well defined preselected mode. Let us define an operator O which connects between the mode $|\chi_0\rangle$ and the mode $|\chi_\perp\rangle$, possessing the eigenvalues ± 1 for the states $|\pm\rangle = (|\chi_0\rangle \pm |\chi_\perp\rangle)/\sqrt{2}$. For the experiment without the Dove prism, the weak value of local variable $O\mathbf{P}_E$ still vanishes, but when the Dove prism is present, we have

$$(O\mathbf{P}_E)_w = \frac{\langle\Phi'|O\mathbf{P}_E|\Psi\rangle_{t_1}}{\langle\Phi'|\Psi\rangle_{t_1}} \simeq 2\sqrt{2}\epsilon. \tag{13}$$

Therefore, the presence in E is found to be of the order ϵ rather than ϵ^2, which is obtained when we naively quantify the presence by $(\mathbf{P}_E)_w$. This explains why we obtain the signal from mirror E of the same order as from other mirrors.

The weak value of local operator of order ϵ explains the signal, but, according to the definition of the full presence of photon in a particular place, we require an order 1 weak value of some local variable. In view of this, we have only 'secondary presence' [10] of the photon in E in the present case.

Now, when mirror E is tilted, we get $(O\mathbf{P}_F)_w \simeq 2\epsilon$, indicating that the presence of the photon is of order ϵ also at mirror F. Nonetheless, we do not get the peak at f_F similar to that at f_E by tilting mirror F as well as E. The reason for this is that the record of the interaction reaching the detector from the tilting mirror F is of order ϵ and not of the order 1 as for the signal from mirror E. Note that, when only mirror F is tilted, we have $(O\mathbf{P}_F)_w = 0$.

We have shown that the results of the interference experiment with a nested interferometer and a Dove prism inside it can be explained in the framework of the recently proposed approach [3]. We get signals from mirrors A, B, and C because the photon presence there is of order 1 and the trace recorded on the photon itself is of order ϵ. A similar signal is obtained from mirror E where the presence of the photon is of the order ϵ, but recorded trace is of order 1.

The signal in E should disappear if the mirror will be wiggled in the perpendicular direction. If only this mirror is wiggled and everything else is not, then there will be exactly zero presence at E. If all mirrors are wiggled as in the experiment [1], then the presence will be of order of ϵ, but the record will also be of order ϵ, so the signal will be too small to observe. It will be of interest to perform a nested Mach–Zehnder interferometer experiment with wiggling mirrors and the Dove prism to demonstrate these effects.

6. Bohmian Trajectory

Before concluding, let us analyse this nested interferometer in the framework of the Bohmian interpretation of quantum mechanics [49]. While Bohr preached to not ask where the particle inside the interferometer was, Wheeler suggested a 'common sense' proposal based on classical intuition. While we have suggested relying on the weak trace that the particle leaves using the TSVF, Bohm has a proposal for a deterministic theory which associates a unique trajectory for every particle. In a particular case of nested interferometer which we consider, with or without Dove prism, the particle detected in D has a well defined trajectory (see Figure 3). Note that it corrects an erroneous trajectory in Figure 2 of [50]. The simplest way to understand why Bohmian trajectory must be as shown is the observation that Bohmian trajectories do not cross [51]. The probability to reach detector D is only $1/9$, while the probability to be in path A is $1/3$. Thus, every Bohmian trajectory which reaches D had to pass through A.

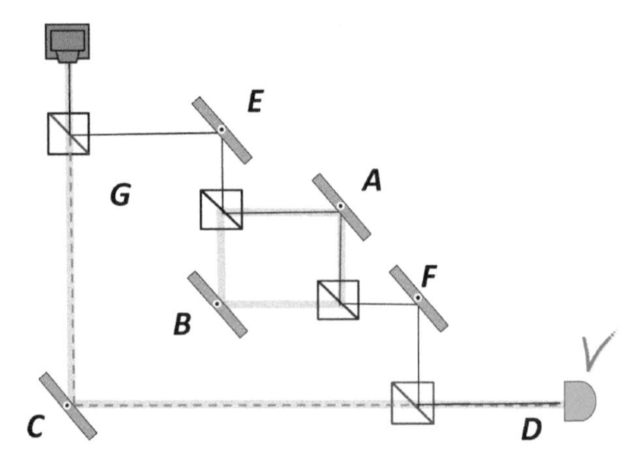

Figure 3. Nested Mach–Zehnder interferometer tuned to destructive interference towards F when a single photon is detected in D. The dashed line represents a common sense proposal by Wheeler, the thick gray line describes the past according to Vaidman's proposal as places where the particle leaves a weak trace, the continuous line represents the Bohmian trajectory.

Bohmian trajectories are entities beyond the standard quantum theory. One of us (L.V.) had the privilege to spend a day of discussions with David Bohm (Charlestone, SC, 1989). I remember telling him what I liked in his theory: a consistent deterministic theory of everything, a candidate for a final theory. However, he completely dismissed this approach. For him, it was nonsense to look for a final theory. He explained to me that his theory is just another step in an infinite search for a better understanding of nature. He was certain that quantum theory is not the last word, and for finding a deeper and more precise theory, quantum theory has to be reformulated. His theory was a counter example to the widespread belief generated by the von Neumann no-go theorem that it would be impossible to extend quantum mechanics consistently by adding hidden variables. Indeed, it opened new horizons for research.

7. Conclusions

Unless a quantum particle is described by a well localized wave packet, the standard quantum theory cannot tell us where the particle was. Vaidman [3] proposed the definition of where a quantum particle was according to the local trace it left: the particle was in a place where the trace is of the order of the trace a single localized particle would leave. In the Danan et al. experiment, photons told us where they have been (according to the trace definition) in a specially tuned nested interferometer. The AJ modification of this experiment, i.e., placing a Dove prism in one of the arms of the inner interferometer did not change significantly the past of the photons, but photons told a different story: they were also near mirror E in spite of the fact that, according to Vaidman's narrative, they were not present there. We conclude that the photons were lying about their presence in E, in the sense that, although the trace they left there was much smaller than the trace that a localized photon would leave, the signal provided by the photons was large as if they had fully been present in E.

How could the photons produce the signal with frequency f_E which was larger than any other signal? In the original and the modified experiments, local traces were not observed. Instead, locally created traces were 'written' on the transversal degree of freedom of the photon itself. In the original experiment, the transversal degree of freedom was not distorted until it reached the detector, so these local traces were faithfully read by the detector. In the modified experiment, the Dove prism influenced the transversal degree of freedom spoiling the faithful readout of local traces by the detector. In fact,

AJ mentioned such an interpretation in [2] as one of the options: "one possible response to this result is that we have improperly read off the past of the photon by letting it suffer further interactions with the environment before reading the weak trace after it was written, so our weak measurement was a bad one for inferring the past of the photon."

Apart from the explanation of the experimental results by the presence of the particle defined through the weak trace, Danan et al. presented a simpler argument of the presence of the photon in A, B and C. The detected photons had to be there because they brought to the detector information which was only there. However, the same should hold for the modified experiment: the particles had to be in E because they brought information about f_E that was present only in E. Sections 4 and 5 explain how it happens in spite of the fact that the trace left by the particles at E was very small. It was small, but not exactly zero, as in the original experiment when only mirror E was wiggling. The Dove prism did change the past of the photons a little.

Introducing a Dove prism not only spoiled faithful transmission of the transverse degree of freedom of the photon to the detector, it also made the inner interferometer extremely sensitive for the misalignment of the input beam. The strength of the signal in the experiment was proportional to the relative amount of the orthogonal component created by local interaction. This component was the asymmetric mode with which the Dove prism passed in full through the inner interferometer, while the reference, the symmetric mode, did not pass at all due to the destructive interference. This explains how a small presence of the photons in E caused a strong signal with frequency f_E.

Note that the Bohmian trajectory did pass through E. However, it also passed through F, although no frequency f_F was observed. It is well known, starting from 'surrealistic trajectories' [51], that we cannot view quantum particles as acting locally in their Bohmian positions (see also [52]).

We have observed that introducing the Dove prism into an inner interferometer of the Danan et al. experiment creates a tiny presence of the photons in E. However, we argue that from this we should not tell that the Dove prism changes the past of a photon in the nested interferometer proposed in [3]. In fact, the origin of the presence of the photons can be found in the disturbance of the mirror E. The weak value of any local operator at E is strictly zero in an ideal interferometer where no mirror is tilted, even if the Dove prism is there.

Author Contributions: The authors worked together on all aspects of the paper.

Funding: This work has been supported in part by the Israel Science Foundation Grant No. 1311/14, the German-Israeli Foundation for Scientific Research and Development Grant No. I-1275-303.14, and the KEK Invitation Program of Foreign Scholars. The publication of this work was supported by the Fetzer Franklin Fund of the John E. Fetzer Memorial Trust.

Conflicts of Interest: The authors declare no conflict of interest.

References

1. Danan, A.; Farfurnik, D.; Bar-Ad, S.; Vaidman, L. Asking Photons Where They Have Been. *Phys. Rev. Lett.* **2013**, *111*, 240402. [CrossRef] [PubMed]
2. Alonso, M.; Jordan, A. Can a Dove prism change the past of a single photon? *Quantum Stud. Math. Found.* **2015**, *2*, 255–261. [CrossRef]
3. Vaidman, L. Past of a quantum particle. *Phys. Rev. A* **2013**, *87*, 052104. [CrossRef]
4. Vaidman, L.; Ben-Israel, A.; Dziewior, A.J.; Knips, L.; Weißl, M.; Meinecke, J.; Schwemmer, C.; Ber, R.; Weinfurter, H. Weak value beyond conditional expectation value of the pointer readings. *Phys. Rev. A* **2017**, *96*, 032114. [CrossRef]
5. Piacentini, F.; Avella, A.; Rebufello, E.; Lussana, R.; Villa, F.; Tosi, A.; Gramegna, M.; Brida, G.; Cohen, E.; Vaidman, L.; et al. Determining the quantum expectation value by measuring a single photon. *Nat. Phys.* **2017**, *13*, 1191–1194. [CrossRef]
6. Wheeler, J.A. The 'Past' and the 'Delayed-Choice Double-Slit Experiment'. In *Mathematical Foundations of Quantum Theory*; Marlow, A.R., Ed.; Academic Press: New York, NY, USA, 1978; pp. 9–48.

7. Aharonov, Y.; Vaidman, L. Properties of a quantum system during the time interval between two measurements. *Phys. Rev. A* **1990**, *41*, 11. [CrossRef] [PubMed]
8. Li, Z.H.; Al-Amri, M.; Zubairy, M.S. Comment on "Past of a Quantum Particle". *Phys. Rev. A* **2013**, *88*, 046102. [CrossRef]
9. Vaidman, L. Reply to the "Comment on 'Past of a Quantum Particle'". *Phys. Rev. A* **2013**, *88*, 046103. [CrossRef]
10. Vaidman, L. Tracing the past of a quantum particle. *Phys. Rev. A* **2014**, *89*, 024102. [CrossRef]
11. Saldanha, P.L. Interpreting a nested Mach–Zehnder interferometer with classical optics. *Phys. Rev. A* **2014**, *89*, 033825. [CrossRef]
12. Li, F.; Hashmi, F.A.; Zhang, J.-X.; Zhu, S.-Y. An Ideal Experiment to Determine the 'Past of a Particle' in the Nested Mach–Zehnder Interferometer. *Chin. Phys. Lett.* **2015**, *32*, 050303. [CrossRef]
13. Ben-Israel, A.; Knips, L.; Dziewior, J.; Meinecke, J.; Danan, A.; Weinfurter, H.; Vaidman, L. An Improved Experiment to Determine the 'Past of a Particle' in the Nested Mach–Zehnder Interferometer. *Chin. Phys. Lett.* **2017**, *34*, 020301. [CrossRef]
14. Potoček, V.; Ferenczi, G. Which-way information in a nested Mach–Zehnder interferometer. *Phys. Rev. A* **2015**, *92*, 023829. [CrossRef]
15. Vaidman, L. Comment on "Which-way information in a nested Mach–Zehnder interferometer". *Phys. Rev. A* **2016**, *93*, 017801. [CrossRef]
16. Salih, H. Commentary: "Asking photons where they have been"—Without telling them what to say. *Front. Phys.* **2015**, *3*, 47. [CrossRef]
17. Vaidman, L.; Danan, A.; Farfurnik, D.; Bar-Ad, S. Response: Commentary: "Asking photons where they have been"—Without telling them what to say. *Front. Phys.* **2015**, *3*, 48.
18. Bartkiewicz, K.; Černoch, A.; Javůrek, D.; Lemr, K.; Soubusta, J.; Svozilík, J. One-state vector formalism for the evolution of a quantum state through nested Mach–Zehnder interferometers. *Phys. Rev. A* **2015**, *91*, 012103. [CrossRef]
19. Vaidman, L. Comment on 'One-state vector formalism for the evolution of a quantum state through nested Mach–Zehnder interferometers'. *Phys. Rev. A* **2016**, *93*, 036103. [CrossRef]
20. Bartkiewicz, K.; Černoch, A.; Javůrek, D.; Lemr, K.; Soubusta, J.; Svozilík, J. Reply to "Comment on 'One-state vector formalism for the evolution of a quantum state through nested Mach–Zehnder interferometers'". *Phys. Rev. A* **2016**, *93*, 036104. [CrossRef]
21. Bartkiewicz, K.; Černoch, A.; Javůrek, D.; Lemr, K.; Soubusta, J.; Svozilík, J. Measuring evolution of a photon in an interferometer with spectrally resolved modes. *Phys. Rev. A* **2016**, *94*, 052106.
22. Hashmi, F.; Li, F.; Zhu, S.-Y.; Zubairy, M. Two-state vector formalism and quantum interference. *J. Phys. A Math. Theor.* **2016**, *49*, 345302. [CrossRef]
23. Vaidman, L. Comment on 'Two-state vector formalism and quantum interference'. *J. Phys. A Math. Theor.* **2018**, *51*, 068002. [CrossRef]
24. Hashmi, F.; Li, F.; Zhu, S.-Y.; Zubairy, M. Reply to the comment on 'Two-state vector formalism and quantum interference'. *J. Phys. A Math. Theor.* **2018**, *51*, 068001. [CrossRef]
25. Wu, Z.-Q.; Cao, H.; Huang, J.-H.; Hu, L.; Xu, X.-X.; Zhang, H.-L.; Zhu, S.-Y. Tracing the trajectory of photons through Fourier spectrum. *Opt. Exp.* **2015**, *23*, 10032–10039. [CrossRef] [PubMed]
26. Griffiths, R. Particle path through a nested Mach–Zehnder interferometer. *Phys. Rev. A* **2016**, *94*, 032115. [CrossRef]
27. Vaidman, L. Comment on "Particle path through a nested Mach–Zehnder interferometer". *Phys. Rev. A* **2017**, *95*, 066101. [CrossRef]
28. Griffiths, R. Reply to "Comment on 'Particle path through a nested Mach–Zehnder interferometer'". *Phys. Rev. A* **2017**, *95*, 066102. [CrossRef]
29. Svensson, B. Non-representative Quantum Mechanical Weak Values. *Found. Phys.* **2015**, *45*, 1645–1656. [CrossRef]
30. Ben-Israel, A.; Vaidman, L. Comment on 'Non-representative Quantum Mechanical Weak Values'. *Found. Phys.* **2017**, *47*, 467–470. [CrossRef]
31. Svensson, B. Response to Comment on 'Non-representative Quantum Mechanical Weak Values'. *Found. Phys.* **2017**, *47*, 1258–1260. [CrossRef]

32. Zho, Z.-Q.; Liu, X.; Kedem, Y.; Cui, J.-M.; Li, Z.-F.; Hua, Y.-L.; Li, C.-F.; Guo, G.-C. Experimental observation of anomalous trajectories of single photons. *Phys. Rev. A* **2017**, *95*, 042121. [CrossRef]
33. Sokolovski, D. Asking photons where they have been in plain language. *Phys. Lett.* **2017**, *A381*, 227–232. [CrossRef]
34. Vaidman, L. A Comment on "Asking photons where they have been in plain language". *arXiv* **2017**, arXiv:1703.03615.
35. Nikolaev, G. Paradox of photons disconnected trajectories being located by means of "weak measurements" in the nested Mach–Zehnder interferometer. *JETP Lett.* **2017**, *105*, 152–157. [CrossRef]
36. Vaidman, L. A comment on "Paradox of photons disconnected trajectories being located by means of 'weak measurements' in the nested Mach–Zehnder interferometer". *JETP Lett.* **2017**, *105*, 473–474. [CrossRef]
37. Nikolaev, G. Response to the comment on "Paradox of photons disconnected trajectories being located by means of 'weak measurements' in the nested Mach–Zehnder interferometer". *JETP Lett.* **2017**, *105*, 475. [CrossRef]
38. Duprey, Q.; Matzkin, A. Null weak values and the past of a quantum particle. *Phys. Rev. A* **2017**, *95*, 032110. [CrossRef]
39. Sokolovski, D. Comment on "Null weak values and the past of a quantum particle". *Phys. Rev. A* **2018**, *97*, 046102. [CrossRef]
40. Duprey, Q.; Matzkin, A. Reply to Comment on "Null weak values and the past of a quantum particle". *Phys. Rev. A* **2018**, *97*, 046103. [CrossRef]
41. Englert, B.; Horia, K.; Dai, J.; Len, Y.; Ng, H. Past of a quantum particle revisited. *Phys. Rev. A* **2017**, *96*, 022126. [CrossRef]
42. Peleg, U.; Vaidman, L. Comment on "Past of a quantum particle revisited". *arXiv* **2018**, arXiv:1805.12171.
43. Bernardo, B.; Canabarro, A.; Azevedo, S. How a single particle simultaneously modifies the physical reality of two distant others: A quantum nonlocality and weak value study. *Sci. Rep.* **2017**, *7*, 39767. [CrossRef] [PubMed]
44. Paneru, D.; Cohen, E. Past of a particle in an entangled state. *Int. J. Quantum Inf.* **2017**, *15*, 1740019. [CrossRef]
45. Aharonov, Y.; Cohen, E.; Landau, A.; Elitzur, A. The Case of the Disappearing (and Re-Appearing) Particle. *Sci. Rep.* **2017**, *7*, 531. [CrossRef] [PubMed]
46. Geppert-Kleinrath, H.; Denkmayr, T.; Sponar, S.; Lemmel, H.; Jenke, T.; Hasegawa, Y. Multifold paths of neutrons in the three-beam interferometer detected by a tiny energy kick. *Phys. Rev. A* **2018**, *97*, 052111. [CrossRef]
47. Hallaji, M.; Feizpour, A.; Dmochowski, G.; Sinclair, J.; Steinberg, A.M. Weak-value amplification of the nonlinear effect of a single photon. *Nat. Phys.* **2017**, *13*, 540–544. [CrossRef]
48. Dziewior, J.; Knips, L.; Farfurnik, D.; Senkalla, K.; Benshalom, N.; Efroni, J.; Meinecke, J.; Bar-Ad, S.; Weinfurter, H.; Vaidman, L. Universality property of local weak interactions and its application for interferometric alignment. *arXiv* **2018**, arXiv:1804.05400.
49. Bohm, D. A Suggested Interpretation of the Quantum Theory in Terms of "Hidden" Variables. I, II. *Phys. Rev.* **1952**, *85*, 166. [CrossRef]
50. Vaidman, L. Surrealistic trajectories. In *Quantum Paths: Festschrift in Honor of Berge Englert on His 60th Birthday*; Ng, H.K., Han, R., Eds.; Worlds Scientific: Hackensack, NJ, USA, 2015; pp. 182–186.
51. Englert, B.G.; Scully, M.O.; Sussmann, G.; Walther, H. Surrealistic Bohm trajectories. *Z. Naturforsch. A* **1992**, *47*, 1175–1186. [CrossRef]
52. Naaman-Marom, G.; Erez, N.; Vaidman, L. Position Measurements in the de Broglie-Bohm Interpretation of Quantum Mechanics. *Ann. Phys.* **2012**, *327*, 2522–2542. [CrossRef]

Article

A Method for Measuring the Weak Value of Spin for Metastable Atoms

Robert Flack *,†, Vincenzo Monachello †, Basil Hiley † and Peter Barker †

Department of Physics and Astronomy, University College, Gower Street, London WC1E 6BT, UK; vincenzo.monachello.14@ucl.ac.uk (V.M.); b.hiley@bbk.ac.uk (B.H.); p.barker@ucl.ac.uk (P.B.)

* Correspondence: r.flack@ucl.ac.uk; Tel.: +44-207-679-3425

† These authors contributed equally to this work.

Received: 27 April 2018; Accepted: 27 July 2018; Published: 30 July 2018

Abstract: A method for measuring the weak value of spin for atoms is proposed using a variant of the original Stern–Gerlach apparatus. A full simulation of an experiment for observing the real part of the weak value using the impulsive approximation has been carried out. Our predictions show a displacement of the beam of helium atoms in the metastable 2^3S_1 state, Δ_w, that is within the resolution of conventional microchannel plate detectors indicating that this type of experiment is feasible. Our analysis also determines the experimental parameters that will give an accurate determination of the weak value of spin. Preliminary experimental results are shown for helium, neon and argon in the 2^3S_1 and $^3P_2$ metastable states, respectively.

Keywords: weak measurement; transition probability amplitude; atomic metastable states

---

1. Introduction

The notion of a weak value introduced by Aharonov, Albert and Vaidman [1,2] has generated wide interest by, not only providing a new possibility of understanding quantum phenomena, but also by generating new experiments to explore deeper aspects of quantum processes. Although Aharonov et al. [1] specifically applied their ideas to spin, Wiseman [3] and Leavens [4] have shown that when applied to the momentum operator, the weak value of the momentum becomes the local momentum used in the Bohm approach [5]. Flack and Hiley [6] have shown that the weak value of the momentum has a close connection with Schwinger's notion of a transition amplitude [7], a notion that Feynman [8] used to introduce the concept of a path integral. Thus, these ideas open up new ways of thinking about and exploring many puzzling questions that lie at the heart of quantum physics.

Already, Kocsis et al. [9] have carried out a two-slit experiment using single photons to measure the weak value of the transverse momentum, which they then used to construct a series of momentum flow lines that they interpreted as 'photon trajectories'. Unfortunately, such an interpretation immediately presents a difficulty in that, whereas particles with non-zero rest mass can be localised in the classical limit producing a classical trajectory [10], photons with zero rest mass have no such limit, calling in to question the meaning of a photon trajectory. In spite of this, Flack and Hiley [11] have shown that the flow lines arise from the new concept of a weak Poynting vector.

In a later paper, Mahler et al. [12] extended the earlier results of Kocsis et al. [9] and demonstrated the existence of non-locality in entangled states in an entirely new way. Unfortunately, in the same paper, they argued that the results can be used to support the Bohm mechanics [5]. However, the Bohm approach is based on the non-relativistic Schrödinger equation and does not apply to the electromagnetic field. A test for the Bohm model in this case requires a generalisation of the Bohm approach to field theory. Indeed, such an extension was first outlined by Bohm [13] himself and later extended by Bohm, Hiley and Kaloyerou [14], Holland [15] and Kaloyerou [16]. It was on this basis that Flack and Hiley [11] showed that by introducing a new notion of the weak value of the Poynting vector, the flow lines could be understood in terms of momentum flow.

To test the original Bohm approach, one must use non-relativistic atoms. This paper is concerned with the development of an experiment to measure such weak values, confining our attention to spin (an attempt to measure weak values of momentum was being carried out by Morley, Edmunds and Barker [17] using argon atoms and will not be discussed further in this paper). As far as we know at the time of writing, the only measurements of weak values of spin have been performed on neutrons [18]. No experiments have used atoms. Not only is this of interest in its own right, but it will enable us to experimentally verify the predictions of the Bohm, Schiller and Tiomno [19,20] model of spin. In this model, the spin vector is well defined in terms of Euler angles, which appear in the expression for the weak value and can therefore be measured. A series of recent results related to this model have been presented by Hiley and Van Reeth [21], who show that the spin does not 'jump' immediately into an eigenstate. Instead, the spin vector rotates, taking a finite, but measurable time to reach the eigenstate, as originally shown by Dewdney et al. [22–24] and Holland [15]. The paper by Hiley and Van Reeth also shows that it is possible to use the weak value to observe this rotation. Hence, it is important to design an experiment to show whether the spin rotates or 'jumps'.

The preliminary outline of this experiment was first presented in a conference [25]. For the benefit of the reader, we have reproduced the two key Figures 1 and 2 from this paper. In order to carry out such an experiment, it must be realised that the displacements needed to detect these effects are extremely small. It is therefore important to understand which parameters are critical in limiting the resolution of the changes expected. This paper focuses the discussion on these requirements. To this end, we report on simulations that explore how our apparatus will function. Here, we concentrate on the strong stage (see Figure 2), and to ensure that the apparatus is functioning correctly, we present experimental results involving Stern–Gerlach displacements of various metastable gas species and our ability to efficiently spin select the atomic beam.

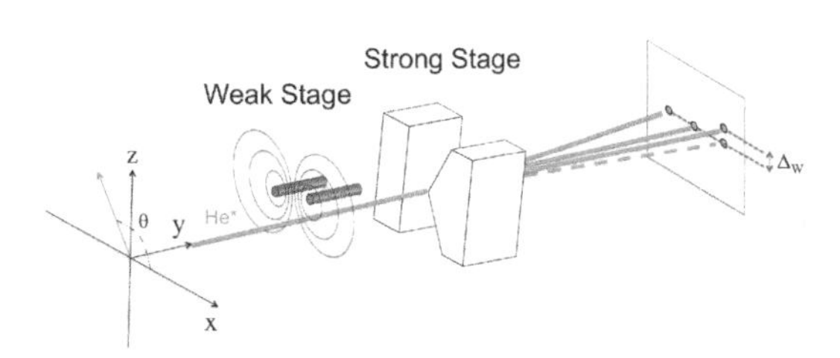

Figure 1. Schematic view of the experimental technique [25]. Helium atoms in the $m_S = +1$ metastable state enter from the left, with spin vector angle θ. The atoms pass through the weak and strong S-G magnets before reaching the detector. The displacement due to the weak interaction is Δ_w, which is a function of the chosen pre-selected spin state. For simplicity, the $m_S = 0$ spin state is not shown.

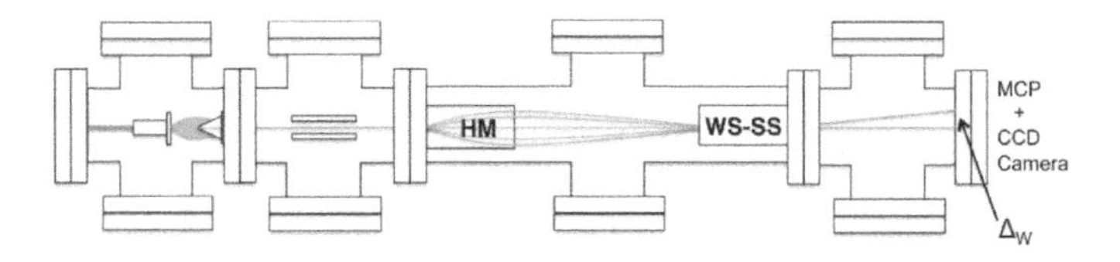

Figure 2. The pulsed helium gas enters from the left. Preparation of the metastable atoms occurs in the first two chambers. In the next chamber, the hexapole magnet (HM) pre-selects the $m_S = +1$ state, which moves onto the weak stage (WS), which is comprised of the magnet, and then on to the strong stage (SS) involving the magnet. Finally, the atoms are detected using a micro-channel plate detector (MCP). This figure is reproduced from [25].

2. Details of the Experimental Apparatus to Determine Weak Values of Spin

2.1. Overview

There are three stages involved in producing the weak value of the spin. Firstly, the atoms are pre-selected in a desired spin state with the spin axis set at a pre-selected angle θ in the x-z plane; see Figure 1. The atoms then propagate through a weak interaction stage, which, in our case, is comprised of two parallel, current-carrying wires producing an S-G -type field gradient that is very small along the z-axis. This stage should not be thought of as constituting a 'measurement'; it simply introduces a unitary Schrödinger interaction, which produces a small phase change in the wave function carrying information about the weak value.

The final stage involves the actual measurement, using a second conventional S-G magnet, with its strong inhomogeneous magnetic field aligned along the x-axis. Note the axes of the weak and strong stage magnets are at right angles to each other. The field of the strong stage magnet must be large enough to clearly separate the spin eigenstates on this axis. It is this separation that enables us to detect the small phase shift, Δ_w, induced by the weak stage, as shown in Figure 1. Since the shift Δ_w is small, we must identify and adjust the relevant experimental parameters to maximise the shift. One of the purposes of this paper is to discuss this optimisation.

2.2. Stern–Gerlach Simulation Using the Impulse Approximation

The simulation is divided into three parts: the initial conditions, the application of the interaction Hamiltonian in the weak stage using the impulsive approximation [26] and, finally, the action of the strong Stern–Gerlach magnet. This approximation neglects the free evolution of the atoms inside the weak magnet, since this produces negligible effects. The analysis follows the scheme outlined in [27], but in our case, we are using spin-one rather than spin-half particles.

2.3. Initial Conditions

Metastable helium atoms in the 2^3S_1 state are initially prepared as a pulsed beam and are described by the normalised Gaussian wave packet at time $t = 0$:

$$\psi(z,0) = \frac{1}{(2\pi\sigma^2)^{\frac{1}{4}}} \exp\left(-\frac{z^2}{4\sigma^2}\right), \tag{1}$$

where σ is the width in position space. The width of the atomic beam is set by passing it through an orifice/skimmer at the entrance of the weak stage. We parametrise the spinor in terms of polar angles θ and ϕ in the following form [28],

$$\xi_{\mathrm{i}}(\theta,\phi,0) = \begin{bmatrix} \frac{1}{2}(1+\sin(\theta))e^{-i\phi} \\ \frac{1}{\sqrt{2}}\cos(\theta) \\ \frac{1}{2}(1-\sin(\theta))e^{i\phi} \end{bmatrix} = \begin{bmatrix} c_+ \\ c_0 \\ c_- \end{bmatrix}. \tag{2}$$

The initial orientation of the spin vector angle θ can be seen in Figure 1, where the azimuthal angle ϕ (not shown) is the corresponding angle in the x-y plane. We set $\phi = 0$ and only consider variations of the angle θ. Therefore, the initial wave function prior to entering the weak stage is:

$$\Psi_{\mathrm{i}}(z,0) = \psi(z,0)\xi_{\mathrm{i}}(\theta). \tag{3}$$

2.4. Theory of the Weak Stage Process

The atoms then traverse the weak stage magnet, where the wave function evolves under the interaction Hamiltonian, weakly coupling the spin to the centre-of-mass wave function. The interaction Hamiltonian is given by:

$$H_I = \mu\left(\hat{\mathbf{s}}.\boldsymbol{B}\right), \tag{4}$$

where μ is the magnetic moment, $\hat{\mathbf{s}}$ is the spin vector and $\boldsymbol{B}$ the magnetic field. If Δt is the time that the atom spends in the weak field, the wave function as it leaves the weak stage is:

$$\Psi_{\mathrm{f}}(z,\Delta t) = \xi_{\mathrm{f}}^{\dagger}\exp\left(-i\frac{\mu\Delta t\frac{\partial B}{\partial z}z\hat{s}_z}{\hbar}\right)\psi(z,0)\xi_{\mathrm{i}}(\theta) \tag{5}$$

where we have used the dominant term in the interaction Hamiltonian $B_z = \frac{\partial B}{\partial z}z$ [29].

2.5. Extracting the Weak Value of Spin

The exponential (phase shift) in Equation (5) can be Taylor expanded:

$$\Psi_{\mathrm{f}}(z,\Delta t) = \langle S_{\mathrm{f}}|\left[1 - i\frac{\mu\Delta t\frac{\partial B}{\partial z}z\hat{s}_z}{\hbar} - \frac{1}{2}\left(\frac{\mu\Delta t\frac{\partial B}{\partial z}z\hat{s}_z}{\hbar}\right)^2 + \ldots\right]|S_{\mathrm{i}}\rangle\psi(z,0), \tag{6}$$

where for convenience, we have written $|S_{\mathrm{i}}\rangle$ for ξ_{i} and $\langle S_{\mathrm{f}}|$ for $\xi_{\mathrm{f}}^{\dagger}$. Hence:

$$\Psi_{\mathrm{f}}(z,\Delta t) = \left[\langle S_{\mathrm{f}}|S_{\mathrm{i}}\rangle - i\frac{\mu\Delta t\frac{\partial B}{\partial z}z}{\hbar}\langle S_{\mathrm{f}}|\hat{s}_z|S_{\mathrm{i}}\rangle - \frac{1}{2}\left(\frac{\mu\Delta t\frac{\partial B}{\partial z}z}{\hbar}\right)^2\langle S_{\mathrm{f}}|\hat{s}_z^2|S_{\mathrm{i}}\rangle + \ldots\right]\psi(z,0). \tag{7}$$

In order to neglect higher order terms in Equation (7), the following inequalities must hold for $n \geq 2$ [27,29],

$$\left|\left(\frac{\mu\Delta t\frac{\partial B}{\partial z}z}{\hbar}\right)^n\langle S_{\mathrm{f}}|\hat{s}_z^n|S_{\mathrm{i}}\rangle\right| << |\langle S_{\mathrm{f}}|S_{\mathrm{i}}\rangle| \tag{8}$$

and:

$$\left|\left(\frac{\mu\Delta t\frac{\partial B}{\partial z}z}{\hbar}\right)^n\langle S_{\mathrm{f}}|\hat{s}_z^n|S_{\mathrm{i}}\rangle\right| << \left|\left(\frac{\mu\Delta t\frac{\partial B}{\partial z}z}{\hbar}\right)\langle S_{\mathrm{f}}|\hat{s}_z|S_{\mathrm{i}}\rangle\right|. \tag{9}$$

In this case, Equation (7) can be expanded to first order:

$$\Psi_{\mathrm{f}}(z,\Delta t) = \left(\langle S_{\mathrm{f}}|S_{\mathrm{i}}\rangle - i\frac{\mu\Delta t\frac{\partial B}{\partial z}z}{\hbar}\langle S_{\mathrm{f}}|\hat{s}_z|S_{\mathrm{i}}\rangle\right)\psi(z,0), \tag{10}$$

and the transition probability amplitude $\langle S_{\mathrm{f}}|S_{\mathrm{i}}\rangle$ can be factored out:

$$\Psi_{\mathrm{f}}(z,\Delta t) = \langle S_{\mathrm{f}}|S_{\mathrm{i}}\rangle\left(1 - i\frac{\mu\Delta t\frac{\partial B}{\partial z}z}{\hbar}\frac{\langle S_{\mathrm{f}}|\hat{s}_z|S_{\mathrm{i}}\rangle}{\langle S_{\mathrm{f}}|S_{\mathrm{i}}\rangle}\right)\psi(z,0). \tag{11}$$

Note that the weak value of the spin, $W = \frac{\langle S_{\mathrm{f}}|\hat{s}_z|S_{\mathrm{i}}\rangle}{\langle S_{\mathrm{f}}|S_{\mathrm{i}}\rangle}$ is in general a complex number with real and imaginary parts. In this case, we are only considering the real part, W_{Re}, which becomes,

$$\Psi_{\mathrm{f}}(z,\Delta t) = \langle S_{\mathrm{f}}|S_{\mathrm{i}}\rangle\left(1 - i\frac{\mu\Delta t\frac{\partial B}{\partial z}z}{\hbar}W_{Re}\right)\psi(z,0). \tag{12}$$

Using the post-selected state, $\xi_f^{\dagger} = [1/2, 1/\sqrt{2}, 1/2]$, the real part of the weak value becomes,

$$W_{Re} = \tan\left(\frac{\theta}{2}\right). \tag{13}$$

In order to cast Equation (12) into an exponential form, the following inequality must be met,

$$L = \left| \frac{\mu \Delta t \frac{\partial B}{\partial z} z}{\hbar} W_{Re} \right| << 1 \tag{14}$$

where $L << 1$ is a limit to be determined [27,29].

As the spread along the z-axis is related experimentally to the width of the atomic beam in question [27], z can be replaced by σ; therefore, the inequality becomes,

$$L = \left| \frac{\mu \Delta t \frac{\partial B}{\partial z} \sigma}{\hbar} \tan\left(\frac{\theta}{2}\right) \right| << 1. \tag{15}$$

The final wave function after the Gaussian wave packet has traversed both the weak and strong magnets is,

$$\Psi_f(z, \Delta t) = \langle S_f | S_i \rangle \exp\left(-i \frac{\mu \Delta t \frac{\partial B}{\partial z} z}{\hbar} \tan\left(\frac{\theta}{2}\right) \right) \psi(z, 0). \tag{16}$$

In this experiment, the real part of the weak value of spin will be measured by setting $\phi = 0$ and varying the angle θ between zero and π.

2.6. Free Evolution of the Gaussian Wave Packet at the Detector

After the strong stage, the problem is treated as the free evolution of a Gaussian wave packet by solving the Pauli equation using well-known methods [26]. The probability density can now be computed, giving the form of the wave function as seen by the detector:

$$|\Psi_D(z,t)|^2 = |\langle S_f| \, S_i \rangle \,|^2 \left[2\pi\sigma^2 \left(1 + \frac{\hbar^2 t^2}{4m^2\sigma^4} \right) \right]^{-\frac{1}{2}} \exp\left[\frac{-\left(z + utW_{Re}\right)^2}{2\sigma^2 \left(1 + \frac{\hbar^2 t^2}{4m^2\sigma^4}\right)} \right], \tag{17}$$

where t is the time of flight from the exit of the strong magnet to the detector. The mean of the post-selected wave function shifts by the value $\Delta_w = (utW_{Re}) = \left(\frac{\mu}{m}\frac{\partial B}{\partial z}\Delta t\right) t \tan\left(\frac{\theta}{2}\right)$, where u is the transverse velocity of the helium atoms. This is in contrast to the standard S-G experiment, where the shift is only ut.

As the pre- and post-selected spin states approach orthogonality, θ tends to π and Δ_w increases, but the transition probability decreases. This reduces the number of post-selected events of interest, leading to the need for longer experimental runs. Again, it is important to understand that this effect only arises when the phase shift acquired at the first stage is sufficiently small; see Equation (15). The centre-of-mass wave function is displaced, but its overall shape is maintained after exiting the weak stage.

2.7. The Limit and Its Validity

In the literature, the real part of the weak value is given as $\tan(\theta/2)$. This functional dependence is for an ideal case when the limit in Equation (15) is equal to, or smaller than, an optimal value, which we will call L_o. For this experiment, it is crucial to know L_o in order to successfully measure the well-known $\tan(\theta/2)$ dependence. If L exceeds L_o, then this will not give the weak value $\tan(\theta/2)$ because higher order terms begin to dominate. In our case, L_o can be determined by analysing two Gaussian wave packets, one describing the first order approximation given by Equation (17) and the other the exact case when no approximation is used, derived from Equation (5).

L_o is calculated by increasing the inhomogeneous magnetic field in the weak stage only, thus increasing the limit shown in Equation (15); all other variables are held constant. Figure 3 illustrates the behaviour of the two Gaussians. For small values of L, the two curves strongly overlap;

the point just before the two wave packets deviate is the optimal limit, L_o. Beyond, L_o the first order approximation continues to move to the left, while the full order approximation slowly reverts to that of a standard S-G measurement. Note: this optimal limit is only valid if $\theta > \pi/2$.

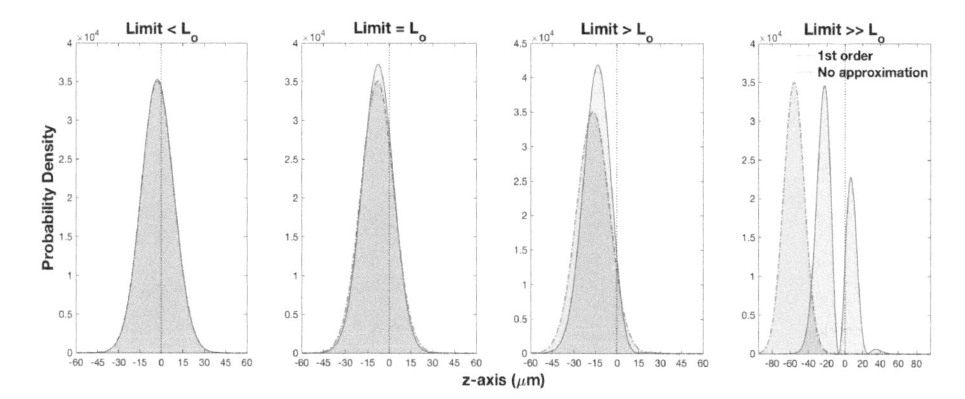

Figure 3. A series of plots showing how the displacement, Δ_w, of the Gaussian wave packet is constrained by various limits. The red curve is the first order approximation, which is dominated by $\tan(\theta/2)$. The blue curve is the exact treatment of the system taking into account all higher order terms. The red and blue curves coincide when the limit $L = L_o = 0.37$; this is the maximum limit for which the first order approximation holds.

By finding this limit, $L_o = 0.37$, experimental parameters can be tailored in order to maximise the atomic beam's displacement due to the weak stage. This is important as certain values of θ produce shifts, which are on the limit of the resolution of our detector. By adjusting experimental parameters in order to meet this limit, displacements for the θ values that would have previously caused an issue can be resolved. As the optimal limit is now fixed, we can rearrange the wave packet deviation Δ_W with respect to this fixed limit.

$$\Delta_w = \frac{\mu \frac{\partial B}{\partial z}(\Delta t)t}{m} \tan\left(\frac{\theta}{2}\right) = \frac{\hbar t}{\sigma m} L_o. \tag{18}$$

This shows that the maximum deviation of the wave packet depends on t and σ. By changing θ and adjusting other experimental parameters so that $L = L_o$, for all values of $\theta > \pi/2$, we will measure the same displacement, a maximal displacement, and from this, the functional dependence $\tan(\theta/2)$ can be observed. This is important if we are measuring θ as outlined in Hiley and Van Reeth [21]. Using parameters from our proposed experiment, of which the most important are the atomic velocity of the beam, 1717 m/s, the free flight distance, 2.4 m, the optimal limit, $L_o = 0.37$, and the width of the beam, $\sigma = 0.5$ μm, our expected displacement, Δ_w, is of the order of 20 μm.

3. Method for the Weak Measurement of Spin for Atomic Systems: Experimental Realisation

3.1. Schematic Lay-Out of the Apparatus

A schematic diagram showing the various stages of the measurement is shown in Figure 2. The first step is to produce a beam of metastable helium in the 2^3S_1 triplet state. Helium gas at high pressure enters the apparatus from the left and is pulsed into the chamber using an electromagnetic valve, producing a pulsed supersonic beam. The atomic beam is excited using an electron-seeded discharge. Here, the atoms collide with a stream of energetic electrons in a 300 V/cm electric field [30]. The excited gas then passes through a 2 mm-diameter skimmer and travels between two electrically-charged plates to remove the unwanted ionised atoms and free electrons.

The next step is to select a single spin state, in our case the $m_S = +1$ state. To do this, we use a hexapole magnet, which focuses this state on to the weak stage magnet (see Figure 2). During this process, the atoms in the $m_S = -1$ state are defocused. The $m_S = 0, 2^1S_0$ singlet state and photons are left untouched, but can be removed from the beam by placing a needle across the centre of the magnet. After the beam exits the hexapole magnet, but before it enters the weak stage, it passes through a 50-μm slit; its rotation about the y-axis, sets the spin vector angle θ. The pre-selected atomic beam is then passed through a final slit, setting the beam width as required in the limit. The beam width at this point of the process is 0.5 μm before entering the weak stage (see Figure 1).

Upon exiting the weak stage, the atomic beam enters the strong stage. Subsequently, the atoms propagate freely onto a detector that consists of two micro-channel plates in a chevron configuration, coupled to a phosphor screen and CCD camera, enabling a resolution of 5 μm using centroiding techniques. The measured deflection, Δ_w, will be proportional to the weak value of the atomic spin.

3.2. Experimental Data Confirming the Correct Functioning of the Last (Post-Selection) Stage

We check that each stage of the experiment is functioning correctly. Having successfully produced and controlled the metastable helium atoms, we test the functioning of the last stage i.e., the final strong S-G measurement. Here, it is important to ensure that the displacement produced by the strong S-G magnet, for each angular momentum eigenstate, is large enough to be easily resolved. To ensure this, we have used a permanent S-G magnet of length 100 mm. The magnet assembly consists of N38-, N40- and N50-grade Nd-Fe-B magnets, arranged in such a way as to produce a constant field gradient, dB/dx, of 100 T/m over a length of 70 mm (see Figure 4). The force, $F_x = -\mu_x \, dB/dx$, experienced by an atom in this field is proportional to the magnetic moment of the atom, $\mu_x = -g_J \mu_B m_J$, where:

$$g_J = \frac{3}{2} + \frac{S(S+1) - L(L+1)}{2J(J+1)} \tag{19}$$

is the Landé g-factor [31].

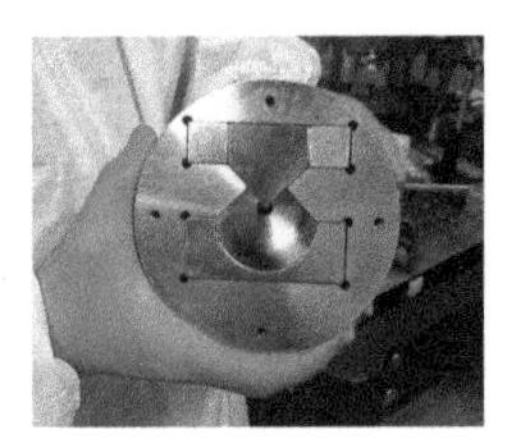

Figure 4. The S-G magnet showing the various grades/shapes of the Nd-Fe-B magnets in the setup in order to achieve a constant field gradient, dB/dx, of 100 T/m.

To carry out this test, we have chosen to the use metastable helium (He*), neon (Ne*) and argon (Ar*). For example, He* has a magnetic moment of $\mu = \pm 2\mu_B$, while other noble gases, such as Ne* and Ar*, have magnetic moments of $\mu = \pm 3\mu_B, \pm\frac{3}{2}\mu_B$ and 0, depending on the atoms', m_J, state. While He* is in a pure spin state, the other two have a combination of spin and orbital angular momentum.

Experimental S-G distributions for He*, Ne* and Ar* have been measured after first travelling through a collimation region consisting of a 100-μm and 10-μm slit separated by 306.5 mm, producing an atomic beam with an angular divergence of 0.36 mrad. The atoms then travel approximately 2 m before hitting the detector.

Figure 5 shows the results, confirming that the spin eigenstates for all the gases are sufficiently resolved, giving a displacement of 7.8 mm, for He*, between the $m_S = \pm 1$ and $m_S = 0$ eigenstates, and 10 and 10.4 mm for Ne* and Ar*, respectively, between the $m_J = \pm 2$ and $m_J = 0$

eigenstates. For all systems, the $m = 0$ state, centred at 0 mm, is unaffected by the magnetic field gradient. The observed separations between the states agree with the theoretical predictions, confirming that the strong stage is working correctly.

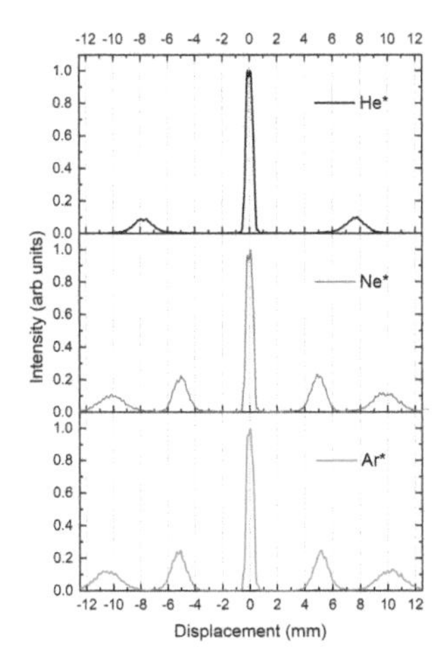

Figure 5. Distribution of three metastable species along the x-axis as they travel through a strong S-G magnet and are detected via an MCP detector. From top to bottom, metastable helium (He*) in the 2^3S_1 triplet state with $m_S = \pm 1, 0$, metastable neon (Ne*) and argon (Ar*) in the $3P_2$ state with $m_J = \pm 2, \pm 1, 0$. The states are clearly delineated, indicating that they would be good candidates for measuring weak values of angular momentum. The central peak contribution is larger for all cases due to the double contribution from the $m = 0$ state and photons.

We have chosen to use metastable helium in the 2^3S_1 state as our preferred atom as this gives several advantages:

1. Its magnetic dipole moment, μ, has a magnitude of two Bohr magnetons $\mu = \pm 2\mu_B$ [30,32], which allows for sufficient displacement between its three spin eigenstates at the detector.
2. It has a lifetime of approximately 8000 s [33], being unable to decay via electric dipole transitions and the Pauli exclusion principle, i.e., its decay is doubly forbidden. This lifetime is clearly large enough for the atoms to pass through all the stages of the apparatus before decaying. Furthermore, this allows scope for increasing the flight distance with no depreciable effects.
3. Metastable helium atoms have an internal energy of 19.6 eV, the highest of any metastable noble gas species. Upon collision with any surface, it will easily ionise, and the emitted electron is observed with higher efficiency at the microchannel plate (MCP) detector.

All of these characteristics combine to enhance the overall signal strength and sensitivity of the experiment.

3.3. The Functioning of the Hexapole Stage

The hexapole magnet contains an array of $M = 12$ segmented nickel-plated N42H-grade permanent magnets, and the array has an ID of 11 mm, an OD of 40 mm and is 60 mm long. The magnetisation direction for each segment is rotated by 120° with respect to the last. The hexapole magnet is shown in Figure 6, with each individual segment located in a 316LN SShousing.

Figure 6. Manufactured hexapole magnet showing the $M = 12$, N42H-grade permanent magnets.

The magnetic field experienced by an atom in a permanent multipole magnet (produced from M segmented pieces) is detailed by Halbach [34] and is shown below:

$$B(r) = B_{rem}\left(\frac{r}{r_1}\right)^{n-1}\frac{n}{n-1}\left[1-\left(\frac{r_1}{r_2}\right)^{n-1}\right]\cos^n\left(\frac{\epsilon\pi}{M}\right)\frac{\sin\left(\frac{n\epsilon\pi}{M}\right)}{\frac{n\pi}{M}}, \tag{20}$$

where $r = \sqrt{x^2+z^2}$ is the atom's radial distance from the magnet's centre. The inner and outer boundaries of the magnet are r_1 and r_2, respectively; B_{rem} is the magnetic remanence of the 12 segmented N42H pieces; and for a hexapole magnet, $n = 3$.

The atomic beam is collimated before entering the hexapole magnet by a 5-mm pin hole at its entrance and the 2-mm skimmer, which was located shortly after the supersonic expansion. The two orifices in this collimation region are separated by 440 mm.

A hexapole magnet utilising these parameters produces a focal point, for He*, which is located approximately 365 mm from the exit of the magnet; see Figure 7. This magnet is also used to reduce the angular divergence of the beam before it passes through our final collimation slit, 1 μm, in order to minimise scattering and maximise flux through the slit region.

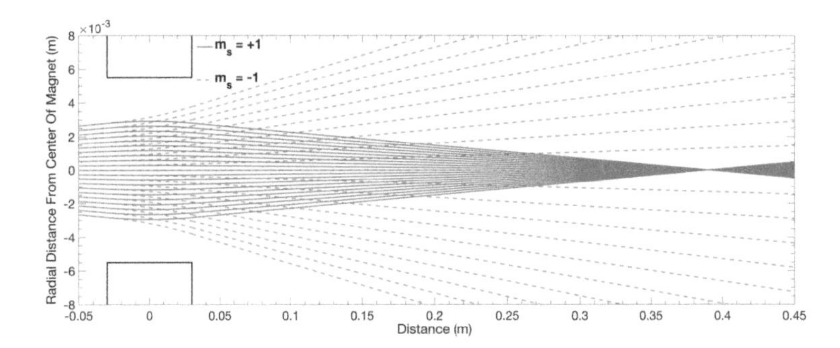

Figure 7. Simulation of a He* beam travelling through the designed hexapole magnet; the dashed red lines signify the $m_s = -1$ defocused state, while the blue solid lines signify the $m_s = +1$ focused state.

Shortly after leaving this hexapole field, the beam then traverses the strong S-G magnet, producing a well-defined separation of the $m_S = +1$ state with complete removal of the $m_S = -1$ state, as seen in Figure 8. As can be seen from this figure, the experiment now produces a highly-efficient spin-selected atomic beam, which is required for part of the pre-selection phase of the experiment.

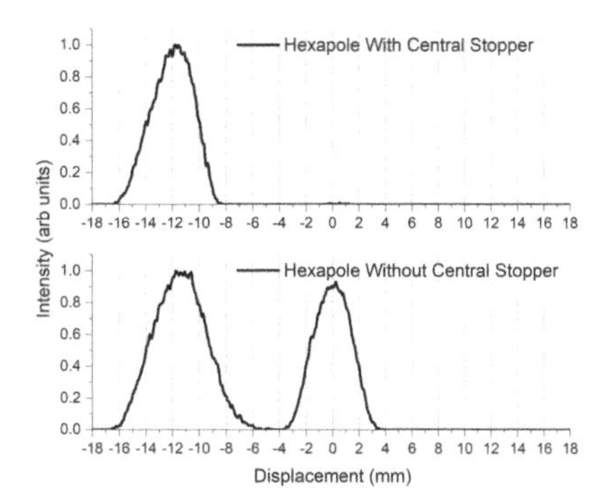

Figure 8. Distribution of the $m_S = +1$ and $m_S = 0$ spin states of the system along the x-axis. When a He* beam travels through a permanent hexapole magnet, the $m_S = -1$ spin state is defocused and lost to the magnet and the vacuum chamber walls. Note: the width of the atom beam is larger here due to the removal of the collimation region before the S-G magnet for test purposes.

4. Conclusions

The experiment described in this paper is designed to measure the real part of the weak value of spin for an atomic system. A full simulation of the process has been carried out giving a prediction of the magnitude of the displacement, Δ_w. A limit, L_o, has been determined defining the range over which the first order approximation holds. Furthermore, we have analysed and optimised the experimental parameters to achieve the largest possible displacement.

We have now been able to sufficiently resolve the spin eigenstates for He* in the x-basis, showing that our post-selection region is working as intended. The ability to excite other noble gas species to metastable levels, and sufficiently resolve their angular momentum eigenstates, allows for flexibility in future experiments. Likewise, part of the pre-selection stage is operational, producing a highly spin-selected He* beam with the ability to remove the $m_S = 0$ and singlet state atoms easily and efficiently from the beam line.

The polarisation mechanics are still to be implemented, allowing us to precisely select the spin vector orientation of the atomic beam, θ. With this, the pre-selection stage is complete. The weak stage S-G magnet has been built and will shortly be introduced into the system. These two extra components complete the main regions of theory and will enable the weak value of spin for He* to be measured.

Using the parameters of our experiment, a shift, Δ_w, of the order of 20 μm is predicted, which is within our experimental resolution. There is also scope to increase Δ_w by cooling the atomic beam, thus reducing the velocity of the atoms and by reducing the width of the beam before the weak stage. These refinements can increase Δ_w to 20–40 μm. Our experiment is designed to vary the angle θ and thereby show its relationship with Δ_w, i.e., $\tan(\theta/2)$. This means that the weak value can be used to measure the angle θ when it is initially unknown. It is this feature that will enable us to track the change of orientation of the spin vector as outlined in Hiley and Van Reeth [21].

Author Contributions: Conceptualization, R.F. and B.H.; Methodology, V.M., R.F. and P.B.

Funding: This research was funded by the Fetzer Franklin Fund of the John E. Fetzer Memorial Trust.

Acknowledgments: The authors would like to thank the Fetzer Franklin Fund of the John E. Fetzer Memorial Trust for their continued generous support.

Conflicts of Interest: The authors declare no conflict of interest.

References

1. Aharonov, Y.; Albert, D.Z.; Vaidman, L. How the Result of a Measurement of a Component of the Spin of a Spin-1/2 Particle Can Turn Out to be 100. *Phys. Rev. Lett.* **1988**, *60*, 1351–1354. [CrossRef] [PubMed]
2. Aharonov, Y.; Vaidman, L. Properties of a quantum system during the time interval between two measurements. *Phys. Rev.* **1990**, *41*, 11–19. [CrossRef]
3. Wiseman, H. Grounding Bohmian mechanics in weak values and Bayesianism. *Phys. Lett. A* **2003**, *311*, 285–291. [CrossRef]
4. Leavens, C.R. Weak Measurements from the point of view of Bohmian Mechanics. *Found. Phys.* **2005**, *35*, 469–491. [CrossRef]
5. Bohm, D.; Hiley, B.J. *The Undivided Universe: An Ontological Interpretation of Quantum Mechanics*; Routledge: London, UK, 1993.
6. Flack, R.; Hiley, B.J. Feynman Paths and Weak Values. *Entropy* **2018**, *20*, 367. [CrossRef]
7. Schwinger, J. The Theory of Quantum Fields III. *Phys. Rev.* **1953**, *91*, 728–740. [CrossRef]
8. Feynman, R.P. Space-Time Approach to Non-Relativistic Quantum Mechanics. *Rev. Mod. Phys.* **1948**, *20*, 367–387. [CrossRef]
9. Kocsis, S.; Braverman, B.; Ravets, S.; Stevens, M.J.; Mirin, R.P.; Shalm, L.K.; Steinberg, A.M. Observing the Average Trajectories of Single Photons in a Two-Slit Interferometer. *Science* **2011**, *332*, 1170–1173. [CrossRef] [PubMed]
10. Hiley, B.J.; Aziz Mufti, A.H. *The Ontological Interpretation of Quantum Field Theory Applied in a Cosmological Context, Fundamental Problems in Quantum Physics*; Ferrero, M., van der Merwe, A., Eds.; Kluwer: Dordrecht, The Netherlands, 1995; pp. 141–156.
11. Flack, R.; Hiley, B.J. Weak Values of Momentum of the Electromagnetic Field: Average Momentum Flow Lines, Not Photon Trajectories. *arXiv* **2016**, arXiv:1611.06510.
12. Mahler, D.H.; Rozema, L.A.; Fisher, K.; Vermeyden, L.; Resch, K.J.; Braverman, B.; Wiseman, H.M.; Steinberg, A.M. Measuring Bohm trajectories of entangled photons. In *Lasers and Electro-Optics (CLEO)*; IEEE: Piscataway, NJ, USA, 2014; pp. 1–2.
13. Bohm, D. A Suggested Interpretation of the Quantum Theory in Terms of Hidden Variables, II. *Phys. Rev.* **1952**, *85*, 180–193. [CrossRef]
14. Bohm, D.; Hiley, B.J.; Kaloyerou, P.N. An Ontological Basis for the Quantum Theory: II-A Causal Interpretation of Quantum Fields. *Phys. Rep.* **1987**, *144*, 349–375. [CrossRef]
15. Holland, P.R. The de Broglie-Bohm theory of motion and quantum field theory. *Phys. Rep.* **1993**, *224*, 95–150. [CrossRef]
16. Kaloyerou, P.N. The Causal Interpretation of the Electromagnetic field. *Phys. Rep.* **1994**, *244*, 287–358. [CrossRef]
17. Morley, J.; Edmunds, P.D.; Barker, P.F. Measuring the weak value of the momentum in a double slit interferometer. *J. Phys. Conf. Ser.* **2016**, *701*, 012030. [CrossRef]
18. Sponar, S.; Denkmayr, T.; Geppert, H.; Lemmel, H.; Matzkin, A.; Tollaksen, J.; Hasegawa, Y. Weak values obtained in matter-wave interferometry. *Phys. Rev. A* **2014**, *92*, 062121. [CrossRef]
19. Bohm, D.; Schiller, R.; Tiomno, J. A Causal Interpretation of the Pauli Equation (A). *Nuovo Cim. Supp.* **1955**, *1*, 48–66. [CrossRef]
20. Bohm, D.; Schiller, R. A Causal Interpretation of the Pauli Equation (B). *Nuovo Cim. Supp.* **1955**, *1*, 67–91. [CrossRef]
21. Hiley, B.J.; van Reeth, P. Quantum Trajectories: Real or Surreal? *Entropy* **2018**, *20*, 353. [CrossRef]
22. Dewdney, C.; Holland, P.R.; Kyprianidis, A. What happens in a spin measurement? *Phys. Lett. A* **1986**, *119*, 259–267. [CrossRef]
23. Dewdney, C.; Holland, P.R.; Kyprianidis, A. A Causal Account of Non-local Einstein-Podolsky-Rosen Spin Correlations. *J. Phys. A Math. Gen.* **1987**, *20*, 4717–4732. [CrossRef]
24. Dewdney, C.; Holland, P.R.; Kyprianidis, A.; Vigier, J.-P. Spin and non-locality in quantum mechanics. *Nature* **1988**, *336*, 536–544. [CrossRef]
25. Monachello, V.; Flack, R. The weak value of spin for atomic systems. *J. Phys. Conf. Ser.* **2016**, *701*, 012028. [CrossRef]
26. Bohm, D. *Quantum Theory*; Prentice Hall: New York, NY, USA, 1951.

27. Duck, I.M.; Stevenson, P.M.; Sudarshan, E.C.G. The sense in which a "weak measurement" of a spin-1/2 particle's spin component yields a value 100. *Phys. Rev. A* **1989**, *40*, 2112–2117. [CrossRef]
28. Ballentine, L.E. *Quantum Mechanics: A Modern Development*; World Scientific Publishing: New York, NY, USA, 1998.
29. Pan, A.K.; Matzkin, A. Weak values in nonideal spin measurements: An exact treatment beyond the asymptotic regime. *Phys. Rev. A* **2012**, *85*, 022122. [CrossRef]
30. Halfmann, T.; Koensgen, J.; Bergmann, K. A source for a high-intensity pulsed beam of metastable helium atoms. *Meas. Sci. Technol.* **2000**, *11*, 1510–1514. [CrossRef]
31. Bleaney, B.I.; Bleaney, B. *Electricity and Magnetism*; Oxford University Press: London, UK, 1965.
32. Baldwin, K. Metastable helium: Atom optics with nano-grenades. *Contemp. Phys.* **2005**, *46*, 105–120. [CrossRef]
33. Hodgman, S.S.; Dall, R.G.; Byron, L.J.; Baldwin, K.G.H.; Buckman, S.J.; Truscott, A.G. Metastable helium: A new determination of the longest atomic excited-state lifetime. *Phys. Rev. Lett.* **2009**, *103*, 053002. [CrossRef] [PubMed]
34. Halbach, K. Design of permanent multipole magnets with oriented rare earth cobalt material. *Nuclear Instrum. Meth.* **1980**, *169*, 1–10. [CrossRef]

MDPI
St. Alban-Anlage 66
4052 Basel
Switzerland
Tel. +41 61 683 77 34
Fax +41 61 302 89 18
www.mdpi.com

Entropy Editorial Office
E-mail: entropy@mdpi.com
www.mdpi.com/journal/entropy

Lightning Source UK Ltd.
Milton Keynes UK
UKHW051051090619
343990UK00003BA/22/P

9 783038 976165